From AD 1 to AD 1000

ASIA	EUROPE	REST OF THE WORLD
8 End of Western Han dynasty		
25 Chinese Eastern Han dynasty		
30 Death of Jesus Christ		
	43 Roman invasion of Britain	50 Rise of kingdom of Axum in north-east Africa
70 Destruction of Temple in Jerusalem	60 Boudicca's revolt	
	79 Destruction of Pompeii	
	117 Highpoint of Roman empire under Trajan	
132 Jewish rebellion against Rome	122-6 Building of Hadrian's Wall	
	143 Building of Antonine Wall	
220-80 Three Kingdoms period in China		
224 Foundation of Sassanian empire		
265 Chinese Western Jin dynasty	290 Roman empire reorganized by Diocletian	300 Rise of Maya civilization in central America
317 Chinese Eastern Jin dynasty	313 Edict of Milan	
320 Foundation of Gupta empire		
	378 Battle of Adrianople	
	405 Completion of Latin Bible	
420 End of Eastern Jin dynasty	410 Visigoths sack Rome	430 Death of St Augustine of Hippo
	450 Start of Anglo-Saxon conquest of Britain	
	453 Death of Attila the Hun	
	476 Fall of last Roman emperor	
	486 Foundation of Frankish kingdom	
	526 Death of Theodoric the Greek	
550 End of Gupta empire	535 Byzantine conquest of Italy	534 Byzantine conquest of North Africa
	563 St Columba founds Iona	
589 Chinese Sui dynasty	590 St Gregory becomes Pope	
607 Unification of Tibet		600 Rise of Ayamará in Bolivia
618 Chinese Tang dynasty		
632 Death of Muhammad, start of Arab expansion		639 Arab conquest of Egypt
642 Fall of Sassanian empire	664 Synod of Whitby	
656 Foundation of Umayyad dynasty		697 Arabs destroy Carthage
	711 Muslim invasion of Spain	
	732 Battle of Tours	
750 Establishment of Abbasid caliphate		
	790 Start of Viking raids on Britain	
	800 Charlemagne crowned emperor	
809 Death of Harun al-Rashid	811 Bulgars defeat Byzantines	
826 Foundation of Novgorod		
849 Start of unification of Burma	843 Treaty of Verdun	850 Collapse of first Maya civilization
858 Fujiwara ascendancy in Japan		
	871 Accession of Alfred the Great	
907 End of Tang dynasty		909 Foundation of Fatimid dynasty
918 Koryo kingdom in Korea		
945 Chinese Liao dynasty		
	955 Battle of Lechfeld	
960 Chinese Northern Song dynasty	962 Otto I crowned emperor	
		969 Fatimids conquer Egypt
985 Rise of Cholas in south India	987 Capetian dynasty in France	985 Viking settlements in Greenland
		1000 Vikings reach North America

OXFORD
ILLUSTRATED ENCYCLOPEDIA

Volume 3
WORLD HISTORY
FROM EARLIEST TIMES TO 1800

OXFORD
ILLUSTRATED ENCYCLOPEDIA

General Editor Harry Judge

Volume 3

WORLD HISTORY
FROM EARLIEST TIMES TO 1800

Volume Editor Harry Judge

OXFORD

OXFORD UNIVERSITY PRESS

NEW YORK MELBOURNE

Oxford University Press, Walton Street, Oxford, OX2 6DP

Oxford New York Toronto
Delhi Bombay Calcutta Madras Karachi
Petaling Jaya Singapore Hong Kong Tokyo
Nairobi Dar es Salaam Cape Town
Melbourne Auckland

and associated companies in
Berlin Ibadan

Oxford is a trademark of Oxford University Press

© *Oxford University Press 1988*
Reprinted 1990, 1991

British Library Cataloguing in Publication Data

Oxford illustrated encyclopedia.
Vol. 3: World history from earliest
times to 1800
1. encyclopedias and dictionaries
I. Judge, Harry
032   AE5
ISBN 0-19-869135-1

Library of Congress Cataloging in Publication Data
Oxford illustrated encyclopedia.
Contents: v. 1. The physical world | volume editor,
Sir Vivian Fuchs—v. 2. The natural world | volume
editor, Malcolm Coe—v. 3. World history, from
earliest times to 1800—
1.   Encyclopedias and dictionaries.
I. Judge, Harry George.   II. Toyne, Anthony.
AE5.094   1985   032   85-4876
ISBN 0-19-869129-7 (v. 1)

Text processed by the Oxford Text System
Printed in Hong Kong

General Preface

The *Oxford Illustrated Encyclopedia* sets out, in its eight independent but related volumes, to provide a clear and authoritative account of all the major fields of human knowledge and endeavour. Each volume deals with one such major field and is devoted to a series of short entries, arranged alphabetically, and furnished with illustrations. One of the delights of the Encyclopedia is that it serves equally those who wish to dip into a particular topic (or settle a doubt or argument) and those who wish to acquire a rounded understanding of a major field of human knowledge.

The first decision taken about the Encyclopedia was that each volume should be dedicated to a clearly defined theme—whether that be the natural world, or human society, or the arts. This means that each volume can be enjoyed on its own, but that when all the volumes are taken together they will provide a clear map of contemporary knowledge. It is not, of course, easy to say where one theme begins and another ends and so each volume foreword (like the one that follows) must therefore attempt to define these boundaries. The work will be completed with a full index to guide the reader to the correct entry in the right volume for the information being sought.

The organization within each volume is kept as clear and simple as possible. The arrangement is alphabetical, and there is no division into separate sections. Such an arrangement rescues the reader from having to hunt through the pages, and makes it unnecessary to provide a separate index for each volume. We wanted, from the beginning, to provide as large a number as possible of short and clearly written entries, since research has shown that readers welcome immediate access to the information they seek rather than being obliged to search through longer entries. The decision, which has been consistently applied, must obviously determine both the number and nature of the entries, and (in particular) the ways in which certain subjects have been divided for ease of reference. Care has been taken to guide the reader through the entries, by providing cross-references to relevant articles elsewhere in the same volume.

This is an illustrated encyclopedia, and the illustrations have been carefully chosen to expand and supplement the written text. They add to the pleasure of those who will wish to browse through these varied volumes as well as to the profit of those who will need to consult them more systematically.

Volumes 1 and 2 of the series deal respectively with the physical world and the natural world. Volumes 3 and 4 are concerned with the massive subject of the history of the world from the evolution of *Homo sapiens* to the present day. Volume 5 introduces the reader to the visual arts, music, and literature, and Volume 6, on technology, both historical and contemporary, provides a guide to human inventiveness. The last two volumes help to place all that has previously been described and illustrated in wider contexts. Volume 7 offers a comprehensive guide to the various ways in which people live throughout the world and so includes many entries on their social, political, and economic organization. The series is completed with Volume 8, which places the life of our shared world in the framework of the universe, and includes entries on the origins of the universe as well as our rapidly advancing understanding and exploration of it.

I am grateful alike to the editors, consultants, contributors, and illustrators whose common aim has been to provide such a stimulating summary of our knowledge of the world.

HARRY JUDGE

CONTRIBUTORS

Dr H. Ainsley

Dr S. R. Airlie

John Bailie

Camilla Boodle

Dr Robin Briggs

Sarah Bunney

Dorothy Castle

Dr Steven Collins

Dr Malcolm Cooper

David Doran

Eda Forbes

John Fox

Dr G. S. P. Freeman-Grenville

Susan Gillingham

Alastair Gray

Dr David M. Jones

Patrick Keeley

Donald Lindsay

The Revd R. Mason

Haydn Middleton

Fr Gareth Moore OP

The Revd J. Morgan-Wynne

John Poulton

Dr A. A. Powell

The Revd William Price

Dr Nigel Ramsay

Martin Roberts

Miss M. Robson

Nick Russel

The Revd Canon G. N. Shaw

Dr Andrew Sheratt

The Revd Peter Southwell

Dr R. F. Stapley

David Stockton

John Stokes

The Revd W. A. Strange

Richard Tames

Dr Norman Tanner

Peter Teed

Dr Paul Thomas

Dr Rosalind Thomas

R. F. Thompson

Peter Tickler

Dr D. H. Trump

Dr P. Woodland

Foreword

Only a very bold person would claim to know what history is. There is, indeed, one important sense in which all eight volumes of this Encyclopedia could be described as history in the broadest sense—everything that has been invented, written, done, created, fought over, argued about, believed. Most of us, however, have a general sense, even if it changes over time, of what counts as 'history'. It is certainly not just about battles and famous people but at its core lies the story of the public life of societies.

This volume, the first of two devoted to history, provides an overview of the most significant developments, events, and individuals in the history of the world from the evolution of *Homo sapiens* up to the end of the 18th century. Its canvas is the world, and not one part of it, so that as well as detailed coverage of British and European history, there is a wealth of material on ancient civilizations, the classical world, America, Africa, the Indian sub-continent, and the Far East. Its entries, in a single alphabetical sequence, cover many aspects of history and not only the dramatic events of politics and war.

The number of potential entries resulting from this policy required two volumes to cover the history of the world, and even then many valuable entries have been squeezed out. We had to decide how best to divide the entries between two volumes and, believing that readers would expect each volume to be self-sufficient (as are the other volumes of the series), we divided the text between the two volumes chronologically rather than alphabetically. After considerable consultation, a somewhat arbitrary date— AD 1800—was chosen to mark the divide. After all, to choose a more precise date (like 1789) would have been to attach undue importance to the history of one part of the world.

Readers will rightly want to know what to expect as a result of this unavoidably arbitrary decision. Major countries have entries on the relevant parts of their histories in both volumes. Where most of the substance for what would otherwise be a double entry falls in the earlier period (the Moguls in India, for example), then its story is brought to its natural conclusion in this volume. Conversely (and it is the only major case), entries on the initial colonization of Australia are found in the next volume, where all the other entries on Australian history are to be found. For a few individuals, such as Napoleon, who did not oblige us by confining their activities to either the 18th or the 19th century, there is an entry in both volumes. The one guiding principle for such decisions has been the convenience of the reader.

There are some fascinating historical questions which cannot be accommodated within an alphabetical encyclopedia made up of short entries. Obvious examples are the history of the family, or of ideas and practices such as democracy. The exclusion of such entries is deliberate. Encyclopedias like this one should concentrate on trying to do what they can do well (as outlined in the General Preface) and on providing accessible, reliable, and clear information on sharply defined topics. Readers who wish for a more discursive account of the history of the world should read any of the general histories of the world.

As this volume is part of a series, certain consequences follow. There are some historical themes which, although they have an undoubted right to be included as of historical importance, find a more natural home in one of the other volumes of this series. The most obvious examples are in the arts or in technology, and it is for this reason that there is no entry here (to take two examples) for William Shakespeare or James Watt.

The unusually wide range of topics covered by this volume has required the editors to consult a large number of experts, many of them from the University of Oxford, and I am most grateful for all the advice that they have provided during the preparation of this volume.

HARRY JUDGE

A User's Guide

This book is designed for easy use but the following notes may be helpful to the reader.

ALPHABETICAL ARRANGEMENT The entries are arranged in a simple letter-by-letter alphabetical order (ignoring the spaces between words) up to the first comma of their headings (thus **Barbados** comes before **Bar Cochba**). When two entry headings are the same up to the first comma, then the entries are placed in alphabetical order according to what follows after the comma (thus **Balliol, Edward** comes before **Balliol, John**). Names beginning with 'St' are placed as though spelt 'Saint', so **St Albans** follows directly after **Safavid**.

ENTRY HEADINGS Entries are usually placed after the key word in the title (the surname, for instance, in a group of names, or the location of a battle, for example, **Tone, Theobald Wolfe**; **Hastings, battle of**). The entry heading appears in the singular unless the plural form is the more common usage. Monarchs and rulers are identified by their names and regnal numbers and are listed in numerical order (and chronologically within any one number), irrespective of country (thus **Henry IV** of England, **Henry IV** of France, **Henry V** of England). When an entry covers a family of several related individuals, important names are highlighted in bold type within an entry.

ALTERNATIVE NAMES Where there are alternative forms of names in common use a cross-reference, indicated by an asterisk (*), directs the reader to the main entry (thus **Cnut** *Canute). In the case of titles, the person appears under the name he or she is most commonly identified by (thus **Burghley, Lord** *Cecil). For the Greek and Roman history entries, Anglicized spellings are used (Rome, not Roma), although on maps showing contemporary events the contemporary spelling has been used. For Chinese names the pinyin system has been used (thus a reader looking up the **Sung** dynasty is directed to **Song** dynasty). Examples of pinyin use that may not be familiar include Beijing (not Peking) and Huang He (not Hwang Ho or Yellow River).

CROSS-REFERENCES An asterisk (*) in front of a word denotes a cross-reference and indicates the entry heading to which attention is being drawn. Cross-references in the text appear only in places where reference is likely to amplify or increase understanding of the entry being read. They are not given automatically in all cases where a separate entry (such as one for a country). Thus it is worth looking for an entry for a name or term in its alphabetical place even if no cross-reference is marked.

ILLUSTRATIONS Pictures and diagrams usually occur on the same page as the entries to which they relate or on a facing page. The picture captions supplement the information given in the text and indicate in bold type the title of the relevant entry. Where a picture shows a work of art of which only one original exists (for example, a painting, sculpture, drawing, or manuscript), the caption concludes, wherever possible, by locating the work of art. Where multiple copies of an original exist (for example, printed books, engravings, or woodcuts), no location has been given, although rare items have sometimes been given a location. There are maps, charts, family trees, and line drawings to illustrate battle campaigns, extents of empires and nations, military and naval history, members of dynasties, and other topics. The time-charts to be found on the endpapers provide easy-to-read information on the major historical events from the evolution of human beings to 1800 world-wide.

WEIGHTS AND MEASURES Both metric measures and their non-metric equivalents are used throughout (thus a measure of distance is given first in kilometres and then in miles). Large measures are generally rounded off, partly for the sake of simplicity and occasionally to reflect differences of opinion as to a precise measurement.

RELATIONSHIP TO OTHER VOLUMES This volume of world history from earliest times to 1800 is Volume 3 of a series, the *Oxford Illustrated Encyclopedia*, which will consist of eight thematic volumes (and an index) in all. It is published with a companion history volume (Volume 4) which covers world history from 1800 to the present day. Each book is self-contained and is designed for use on its own. Readers are advised that under this chronological arrangement the history of countries, for example, is begun in Volume 3 and completed in Volume 4. Further information is contained in the foreword to this volume, which offers a fuller explanation of the book's scope and organization.

The titles in the series are:

1 *The Physical World*
2 *The Natural World*
3 *World History: from earliest times to 1800*
4 *World History: from 1800 to the present day*
5 *The Arts*
6 *The World of Technology*
7 *Human Society*
8 *The Universe*

A

Abbas I (the Great) (1557-1628), Shah of Persia (1588-1628). He ended an inherited war with the *Ottomans by conceding territory (1590) in order to free himself to drive the *Uzbek Turks from north-eastern Persia (1598). By 1618 he had strengthened his army by curbing the Turcoman chiefs who supplied his recruits, and by using foreign advisers, and had reconquered the lands ceded to the Ottomans, but he died before the end of a further war over *Mesopotamia (1623-9).

Abbasid, a Muslim dynasty, ruling most of the central Middle East (750-1258) claiming descent from Abbas, uncle of the Prophet *Muhammad. The Abbasids utilized the Hashimiyya, an extremist group, to capitalize on tribal, sectarian, and ethnic rivalries and overthrow the *Umayyad dynasty, starting in the frontier province of Khurasan. The *caliph al-Mansur established a new capital at Baghdad, which, under *Harun al-Rashid, became a celebrated centre of culture and prosperity. From c.850 central power weakened in the face of local dynasties, as the Aghlabids took power in Ifriqiyya (North Africa) and the Fatimids took over Egypt, while the dynasty itself fell in thrall to the *Shiite Buwayhids. *Seljuk Turks took Baghdad and with it, effectively, the caliphate in 1055, though their rule was soon constrained by Crusader incursions from 1095 and decisively ended by the *Mongol destruction of Baghdad in 1258. A nominal line of Abbasid caliphs continued in Egypt until the *Ottoman conquest of the *Mamelukes in 1517.

Abelard, Peter (or Abailard) (1079-1142), French philosopher and scholar. His teaching of theology in Paris gained him a great reputation and among his students was Héloïse, niece of a canon of Paris, whom he secretly married. When her family discovered this, Abelard was punished with castration and retired to a monastery and Héloïse became a nun. Though the couple were never reunited, their letters show that their deep devotion to each other never faltered. The book *Sic et Non* ('Yes and No'), a collection of apparently conflicting excerpts from scripture and the Church Fathers is among the best known of his works which also include *Ethics or Know Thyself*, an analysis of moral responsibility. His apparent stress upon reason rather than faith led to a clash between Abelard and St Bernard of Clairvaux at Sens in 1140 where Abelard was found guilty of heresy.

Aboukir Bay *Nile, battle of the.

Abraham, first of the patriarchs of *Israel, from whom the Israelites traced their descent. The biblical stories are of varying date and origin, and it is uncertain how much historical fact they contain. According to the book of Genesis (the first book of the Old Testament) Abraham lived in the middle of the 2nd millennium BC at Haran in northern Mesopotamia. He was divinely called to leave his home and family and go to a new land, *Canaan. It is recorded that God made a covenant (or agreement) with him, promising him a multitude of

Abraham's sacrifice of Isaac, a relief on the façade of the Armenian church on the island of Achthamar on Lake Van, AD 915–21. Abraham is on the point of carrying out God's command to offer his only son as a burnt offering. On the left is the ram caught in a thicket which was sacrificed in place of Isaac.

descendants to whom he would give Canaan for ever. Abraham's faith was put to the test by a command to sacrifice his son Isaac. When Abraham showed his readiness to do this, a ram was substituted for the sacrifice and God confirmed his covenant. Through his son Ishmael, he is considered by Muslims an ancestor of the Arabs. He is revered by Jews, Christians, and Muslims.

absentee landlord, a landowner not normally resident on the estate from which he derived income and which was generally managed through an agent. While some landlords cared for the welfare of their tenants, others engaged in such practices as the issue of very short leases, which gave unscrupulous agents opportunities to raise rents frequently and evict anyone unable to pay. Abuses were common in pre-revolutionary France and in Ireland, where successive confiscations had led to Irish estates falling into English hands.

Abu Bakr (c.573-634), first *caliph of Islam (632-4). He was one of the earliest converts to Islam and a close companion of the Prophet *Muhammad, who married his daughter Aisha. When he succeeded to Muhammad's position as temporal leader of the Muslim community, this pious and gentle man was chiefly concerned to reaffirm the allegiance of those Arabian tribes who had withdrawn it at the time of the Prophet's death. These 'wars of apostasy' initiated the *Arab conquests.

Abydos, the name of two ancient cities. Abydos on the south-eastern shore of the *Hellespont stood where those waters are at their narrowest. It was from there that *Xerxes crossed into Europe, via a bridge of boats, in 480 BC. It was also an important base for Antiochus the Great when he crossed into Europe in 197 BC. After his defeat at Magnesia, Abydos probably came under the control of *Pergamum, and subsequently of Rome.

Abydos in Upper Egypt was specially venerated because of its association with the god Osiris. Its remains date from c.3100–500 BC. The earliest pharaohs built funerary monuments there, and later kings such as Seti I and *Ramesses II built temples and sanctuaries there.

Abyssinia   *Ethiopia.

Académie des Sciences, a body of scientists in France receiving pensions from the crown, established (1666) by Jean-Baptiste *Colbert as part of a general campaign to bring intellectual and artistic life under the patronage and control of *Louis XIV. Its most prestigious member in its early years was the Dutch scientist Christiaan Huygens. The Académie was reorganized and given a formal constitution in 1699; in the 18th century it became the leading force in European science, including among its members Lavoisier, Buffon, and Laplace. After a brief disappearance during the *French Revolution it was revived as part of the Institut de France.

Academy, the school established at Athens by *Plato in the 380s BC, probably intended to prepare men to serve the city-state. It was as a philosophical centre that it became celebrated, its students including *Aristotle, *Epicurus, and *Zeno of Citium. Much of its history is obscure, but it survived until its closure by *Justinian in AD 529.

Acadia   *Nova Scotia.

Achaean League, a confederacy of Achaean and other Peloponnesian states in ancient *Greece. Its name derived from the region of Achaea in north-east Greece. In the 4th century BC an alliance was forged which was dissolved in 338 BC. It was refounded in 280 BC, under the leadership of Aratus of Sicyon. It became involved in wars with Macedonia and Sparta, before allying itself with Rome in 198. However, war with Rome in 146 led to defeat and the dissolution of the League.

Achaemenid, the dynasty established by *Cyrus the Great in the 6th century BC and named after his ancestor Achaemenes. Cyrus' predecessors ruled Parsumash, a vassal state of the Median empire, but he overthrew their king Astyages and incorporated the *Medes within his Persian empire, which by his death in 530 BC extended from Asia Minor to the River Indus. His successor Cambyses II (529–521 BC) added Egypt. *Darius I instituted a major reorganization of the administration and finances of the empire, establishing twenty provinces ruled by *satraps. Both he and *Xerxes failed in their attempts to conquer Greece in the early 5th century. By the time *Alexander the Great invaded with his Macedonian army (334 BC) the empire was much weakened. Darius III, defeated at Issus and Gaugamela, was killed by his own men in 330 BC. Achaemenid rule was tolerant of local customs, religions, and forms of government. The construction of a major road system, centred on Susa, facilitated trade and administration. The magnificent remains of *Persepolis provide a glimpse of Achaemenid wealth and power.

Acheh, a sultanate on the northern coast of Sumatra, claimed to be the first Muslim state in south-east Asia. After the fall of *Malacca to the Portuguese in 1511 many Muslim traders moved there. By the late 16th century it had reduced the power of *Johore and controlled much of Sumatra and Malaya, deriving its wealth from pepper and tin. After the Dutch took Malacca in 1641, Acheh consolidated its rule in Sumatra.

Acheulian, a prehistoric culture characterized by hand-axes, named after St Acheul near Amiens, France, but widely distributed across Africa, Europe, the Middle East, and parts of Asia. The handaxes were general-purpose stone tools produced from a cobble or large flake by trimming it to an oval or pear-shaped form. They lacked a handle but were none the less efficient slicing tools. They first appear in Africa around 1.5 million years ago and continued to be made with little modification to the basic shape until 150,000 years ago. Sites with Acheulian tools have provided the earliest certain evidence of the control and use of fire by humans.

acropolis, the *citadel of an ancient Greek city, most notably of Athens. The Athenian citadel was destroyed by the invading Persians in 480 BC, but *Pericles instituted a rebuilding programme. The Parthenon, built 447–432, was a Doric temple containing a gold and ivory statue of Athena. This was followed by the gateway or Propylaea, the temple of Athena Nike (commemorating victory over the Persians), and the Erectheum, which housed the shrines of various cults. Many of the sculptures on the Parthenon were removed by Lord Elgin in 1801–3 and purchased by the British government in 1816. The right to their possession is disputed between Britain and Greece.

The **acropolis** at Athens features a complex of several buildings grouped on an imposing hilltop site. The smaller Erectheum stands to the left of the temple of the Parthenon, built to house a colossal statue of the goddess Athena.

Actium, a promontory on the south headland of the Ambracian gulf, north-west Greece. It was the base of *Mark Antony and Cleopatra in their campaign (31 BC) against Octavian for supremacy in the Roman world. This was the last in the series of civil wars which had begun with Caesar's crossing of the *Rubicon in 49 BC. After blockading Antony's larger fleet, Octavian and his admiral Agrippa scattered it at sea near Actium. Antony and Cleopatra escaped to Egypt and to eventual suicide, leaving their land-forces to surrender. The victory gave Octavian undisputed supremacy in the Roman world and in 27 BC he gained official recognition as Caesar *Augustus. The battle was widely used for some decades in the East as the starting date of a new era.

Adams, John (1735–1826), second American President (1797–1801). He was a young lawyer from Quincy, Massachusetts, who was enlisted for the patriot cause by James *Otis and his cousin Samuel *Adams. His *Dissertation on the Canon and Feudal Law* championed the rights of the individual. He helped to draft the *Declaration of Independence at the second *Continental Congress and the conservative Massachusetts State Constitution (1780), which reflected his fear of popular licence. Service as American representative in France (1778–9), the Netherlands (1780–2), and Britain (1785–8), and as a peace commissioner (1782) led to his election as *Washington's Vice-President (1789–97) and successor as a moderate *Federalist. Rival sympathies in the European war heightened party conflict and precipitated the Alien and Sedition Acts, the Virginia and Kentucky Resolves, and the outrage over the *XYZ Affair. Adams's moral courage prevented *Hamilton's stampede into war against France in 1799. Due to *Jefferson's opposition Adams failed to be re-elected and retired to Quincy.

Adams, Samuel (1722–1803), American patriot, the leader of resistance to Britain in Massachusetts between 1763 and 1776. He founded the *Sons of Liberty in Boston and organized riots, propaganda, and boycotts against tax-raising. He attended the *Continental Congress and signed the *Declaration of Independence. He later served as governor of Massachusetts and drafter of its constitution (1780), but failed to achieve the 'Christian Sparta' of his evangelical dreams. He was only persuaded to support the *Constitution of the USA by the promise of a *Bill of Rights.

Addled Parliament (5 April–7 June 1614), the nickname given to *James I of England's second Parliament. In the absence of effective guidance from crown or councillors, those opposed to the king's policies were able to divert the House of Commons to discussion of grievances, including church reform, impositions (import duties), and court interference at the elections. The king dissolved the Parliament before it had passed any legislation—hence the term 'addled' meaning barren, empty, or muddled—and ruled without one until 1621.

Adrian IV (Nicholas Breakspear) (c.1100–59), Pope (1154–59), the only Englishman to have held the office of pope. He reorganized the church in Norway, and in the bull *Laudabiliter* assisted *Henry II of England to gain control of Ireland. He opposed *Frederick Barbarossa's claims to power, using the interdict (*excommunication) against Frederick's supporters in Rome. The Diet of Besançon in 1157 did not resolve their differences and the dispute was still raging when he died.

Adrianople, battle of (9 August AD 378). The Roman city of Adrianople 480 km. (300 miles) west of Constantinople was the scene of the defeat of the Roman forces by the *Visigoths. Emperor Valens, who had hoped to prevent the Gothic invasion of the Roman empire, was killed.

Aegean civilization   *Mycenaean civilization.

Aegospotami, the site of a naval battle (405 BC) fought in the *Hellespont. It sealed the defeat of Athens in the *Peloponnesian War. For five days the Athenian fleet attempted to draw *Lysander and the Spartan fleet into battle. As the Athenians were disembarking after the fifth attempt, Lysander launched a surprise attack and captured 160 out of 180 *triremes. He executed all the Athenians whom he took prisoner.

Aetolia, a mountainous region in central Greece. In ancient times it was poor, its inhabitants dispersed in small isolated communities. In 426 BC its fierce, skilled troops defeated the invasion force led by the Athenian general *Demosthenes. In the 4th century BC the Aetolian League, a federation of its tribes, was formed. It enjoyed considerable success: in the 3rd century *Delphi fell under its influence, and it also became Rome's first effective ally in Greece. Later it turned against the Romans, but in 189 BC they forced it into submission, breaking its power though allowing it to continue in existence.

Afghanistan, a country in south-central Asia. It was conquered by *Alexander the Great, and after his death became part of the *Bactrian state. A succession of foreign overlords was followed by Arab conquest from the 7th century, the most important Muslim ruler being *Mahmud of Ghazna; in the 11th century the territory was converted to *Islam. It was overrun by *Mongols in 1222, only becoming united under an Afghan leader in 1747, when Ahmad Shah founded the Durrani dynasty at Kandahar.

Africa, the second largest continent. It was the birthplace of the human race, as shown by finds at *Olduvai Gorge and other sites. By the late Stone Age Proto-Berbers inhabited the north, Ethiopians the Nile valley, while *Negroid peoples moved southwards. Pygmies occupied the central forest, and San and Khoikhoi (called Bushmen and Hottentots by white colonists) roamed the south.

By the 4th millennium BC, one of the world's oldest civilizations had developed in *Egypt. In the north *Phoenicians, and then *Carthaginians, organized seaborne empires which fell, with Egypt, to Rome in the last centuries BC. Indigenous kingdoms arose in *Nubia and *Axum. In the 7th century the Arabs seized the north.

In Cameroon c.500 BC a population explosion sent the Bantu eastwards. They slowly occupied most of southern and central Africa, overwhelming the San people. There and in West Africa chieftainships developed, and some empires with sophisticated cultures, especially in Islamic states such as *Mali and in Christian *Ethiopia. Their intricate system of commerce reached from the Medi-

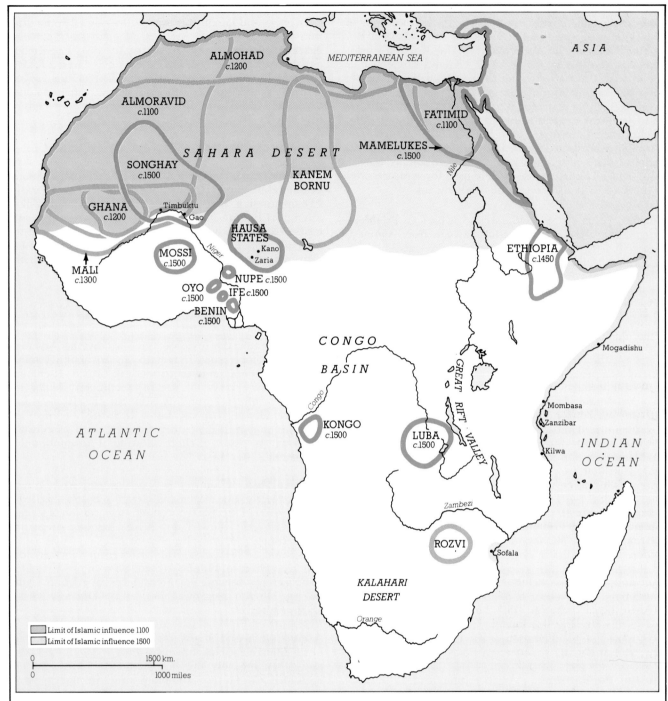

Africa (1100–1500)

The early history of southern and central Africa is much less well known than that of the northern and coastal regions which were visited by merchants. The Saharan trade routes which brought Islam south to western Sudan were also the source of the prosperity and power of a series of vast, though ill-defined, empires that flourished between the 8th and 16th centuries. First Ghana, then Mali, and later Songhay traded their gold and slaves for luxury goods and salt. Merchants also spread Islam to the cosmopolitan cities of the eastern coast which thrived on the Indian Ocean trade.

terranean to Indonesia and China. The Portuguese arrival in the 15th century stimulated trade in the west and centre, but interrupted it in the east. In the 16th century the north fell to the Ottomans, while south of the Sahara Europeans began the *slave trade to the Americas. From the 16th to the 18th century in the present Zaïre and in central Africa Bantu states developed, some of them sizeable empires. During this period the Bantu were pushing their way southwards, but it was not until the 19th century that they began to form recognizable states in the present South Africa. While each tribe or state developed its individual pattern of constitution, some more sophisticated than others, power was generally concentrated in the hands of chieftains and regulated both by tribal conventions and by free public discussion in tribal assemblies.

The battle of **Agincourt**, from the *St Alban's Chronicle* of the 15th century. The English achieved an improbable victory through the expertise of their archers and the immobility of the heavily armoured French knights. (Lambeth Palace Library, London)

Agade *Akkad.

Agincourt, battle of (25 October 1415). The village of Agincourt in northern France was the scene of the defeat of a large French force by an English army led by *Henry V. Henry's force invaded *Normandy in 1415, captured Harfleur, but was intercepted by a large French army after a long march north towards Calais. The English troops, mainly archers and foot soldiers, dug in behind wooden stakes between thickly wooded ground. The next day the French cavalry advanced on a narrow front across muddy ground only to be killed by English archers and infantry. A dozen French notables, including the Constable of France, died, together with perhaps 1,500 knights and 4,500 men-at-arms, including prisoners massacred after French knights made a surprise attack on the English baggage train late in the day. English casualties were light but included the Duke of York and the Earl of Suffolk. The battle was fought on St Crispin's day and there is a famous eve-of-battle speech in Shakespeare's play *Henry V*.

agora, the market-place which stood at the centre of every Greek city, the equivalent of the Roman *forum. In Athens, as well as being a trading centre, it included administrative buildings, temples, meeting places such as the Stoa Poikile where *Zeno of Citium taught, the mint, and, in Roman times, two libraries.

Agricola, Gnaeus Julius (AD 40–93), governor of Roman Britain (78–84). He served with Paulinus against the Iceni queen *Boudicca (61) and commanded the Twentieth Legion in the north-west (70–73). As governor he subjugated the Ordovices of north Wales and extended the frontier north to the rivers Forth and Clyde, defeating the Caledonians in the process. His successes irritated Emperor Domitian who recalled him to Rome in 84. His career, in particular the British governorship, is described in his *Life* by *Tacitus, his son-in-law.

Agricultural Revolution, a process of agricultural change, first occurring in Britain in the 18th century, from which modern systems of farming developed. Some historians stress that agriculture was already undergoing evolutionary change, but that this was speeded up by *enclosure, particularly the parliamentary enclosures of the 18th century. The medieval economy rested on the manorial system and open-field cultivation in *strips which hampered change. The Agricultural Revolution saw this replaced by large-scale farming in consolidated units, the extension of arable farming over heaths and commons, the adoption of intensive livestock husbandry, the conversion of a largely self-subsistent peasantry into a community of agricultural labourers, and considerable attention to the improvement of agricultural techniques like *crop rotation, new crops, for example turnips and potatoes, and improved grasses. Viscount Townshend (1674–1738) and Thomas Coke, Earl of Leicester (1752–1842) were notable for their adoption and promotion of crop rotation; Robert Bakewell (1725–95) was the most famous of the livestock improvers.

agriculture, the process of cultivating land and rearing animals, whose beginnings date from *c*.9000 BC. As the human population grew the traditional methods of hunting animals and gathering wild fruits and plants could not always provide enough food. Thus people began to domesticate animals, for example sheep and goats in the Near East, and cultivate plants from seeds, for example wheat in the Near East, rice in south-east Asia, and maize in the Americas. As settled agricultural communities developed they required more land for cultivation. This was obtained by finding new land (thus spreading areas of human settlement), by using new and more efficient tools (the plough and the wheel were developed in the Near East *c*.4000 BC), and by irrigation. The planning, construction, and control of irrigation systems were major factors in the growth of civilizations in Mesopotamia, Egypt, the Indus valley, and China. Agriculture developed in many ways depending upon climate, soil, and the system of land ownership, as shown, for example by the variety of *field systems. The *Agricultural Revolution of 18th-century Britain introduced new scientific ideas into farming.

A painting of Thomas Coke of Holkham, Earl of Leicester, inspecting some of his Southdown flock, a breed of sheep developed during the **Agricultural Revolution** to raise healthier animals with an increased meat yield.

Agrippa, Marcus Vipsanius (63–12 BC), Roman statesman. He accompanied the young *Augustus to Rome after Julius *Caesar's murder, and later won decisive naval victories over Sextus Pompeius (Naulochus, 36 BC) and *Mark Antony (Actium, 31 BC). Augustus consistently entrusted him with wide military and organizational responsibilities, and marked him out as his heir-apparent. His death in 12 BC was unexpected. A prolific builder, his best-known constructions are the Pantheon at Rome and the Pont du Gard and Maison Carré at Nîmes, southern France.

Ahab (9th century BC), second king of the *Omri dynasty in northern Israel (c.869–c.850 BC). He campaigned in alliance with Syria against Assyria, but was defeated at the battle of Qarqar (853 BC). Three years later he allied with Jehoshaphat, King of Judah, to regain Transjordan from the Syrians. Although temporarily uniting the kingdoms, Ahab was defeated and killed. The stability of his kingdom was undermined by religious disputes. His wife Jezebel was a Phoenician from the city of Tyre. While the marriage brought political advantage, it introduced Phoenician traditions into Hebrew life and religion, and Ahab was publicly denounced by the prophet *Elijah for attempting to unite the Canaanites and Israelites in the worship of Phoenician gods.

Aidan, St (d. 651), Irish missionary. In c.635 while still a monk at the monastery of Iona, he was made a bishop and chosen by King Oswald to act as a missionary in the English kingdom of Northumbria in northern England. With royal support, St Aidan established a monastery on the island of Lindisfarne where Celtic missionaries were trained; they played an instrumental role in promoting Christianity in northern England.

Aix-la-Chapelle, Treaty of (1748), the treaty which concluded the War of the *Austrian Succession. It restored conquered territory to its original owners, with a few exceptions. The terms were drawn up by the British and French and reluctantly accepted by Empress *Maria Theresa of Austria, who had to abandon Silesia to *Frederick II of Prussia. In Italy Don Philip, the younger son of *Philip V of Spain, received Parma. This treaty was a temporary truce in the Anglo-French conflict in India and North America. In North America colonists unwillingly ceded the French fortress of Louisburg, in order to secure the return of Madras to Britain. Prussia's rise to the rank of a great power was strongly resented by Austria. The treaty left many issues of conflict unresolved and war (the *Seven Years War) broke out again eight years later.

Akbar (1542–1605), the third and greatest of the *Moguls, who ruled India from 1555 to 1605. At 13 he inherited a fragile empire, but military conquests brought Rajasthan, Gujarat, Bengal, Kashmir, and the north Deccan under his sway. He introduced a system of civil and military service which ensured loyalty and centralized control, and made modifications to the land revenue system which reduced pressure on peasant cultivators. Although he was illiterate, he had an enquiring mind and extended patronage to artists and scholars of all religions. Underlying his success was a statesmanlike attitude towards the majority Hindu population, sym-bolized by marriage with *Rajput princesses, the sus-

The Mogul emperor **Akbar** receiving the submission of the rebel Bahadur Khan, portrayed in a manuscript dated 1590. Mounted amid warriors and elephants, Akbar confronts the defeated rebel leader beside his beheaded followers. (Victoria and Albert Museum, London)

pension of discriminatory taxes, and the increased employment of Hindus in the imperial service. A man of contradictory and powerful impulses, Akbar created the basis for Mogul control over India until the early 18th century, and also left a distinctive mark on Muslim–Hindu relations in India.

Akhenaten (or Ikhnaton) (d. 1362 BC), Pharaoh of Egypt (1379–62) and husband of *Nefertiti. He was a very early monotheist who prohibited the worship of all gods except Aten, the sun disc. He replaced *Thebes as the capital with his new foundation of Akhetaten, the focal point of which was a roofless temple dedicated to Aten. However, his religious reforms were unpopular, and the influence of the priests of Amon had them overturned at the beginning of *Tutankhamun's reign.

Akkad, an area in central Mesopotamia, named after the city of Agade, which was founded by Sargon I c.2350 BC. Sargon conquered the *Sumerians in southern Mesopotamia after much hard fighting, and in later campaigns penetrated as far as Syria and eastern Asia Minor. He apparently ruled for fifty-six years, though revolts towards the end of his reign indicate the difficulties of holding together such a large empire. Even so, his successors maintained Akkadian supremacy for another hundred years. Thereafter Akkad was simply the name of an area, though Akkadian had become the major spoken language of Mesopotamia; examples of it have been preserved in *cuneiform texts.

Alans, a nomadic people from southern Russia who began to harass the Roman empire's eastern frontier in the 1st century AD. In the 4th century they were forced west by the conquering *Huns, finally reaching Gaul in 406 and Spain in 409. There they merged with the *Vandals and crossed into Africa in 429.

Alaric I (*c.*370-410), King of the *Visigoths. He commanded *Theodosius' Gothic allies and helped put down the Western usurper emperor Eugenius. On Theodosius' death the Eastern and Western Roman empires were formally divided. Alaric revolted against the rule of *Constantinople and moved with his people in search of homelands. He invaded Italy in 401. Twice defeated by Stilicho, the Roman general, he entered a treaty of alliance with him. After the execution of Stilicho by Emperor Honorius, Alaric repudiated the pact and ravaged Italy, laying siege to Rome three times. The city fell in 410. That same year he planned invasions of Sicily and Africa, but his fleet was destroyed by storms. He died at Cosenza and was buried with treasures looted, reputedly from Rome, in the bed of the River Busentus.

Alba, Fernando Alvarez de Toledo, Duke of (*c.*1507-82), Spanish statesman and general. He rose to prominence in the armies of Emperor *Charles V. A stickler for discipline and a master of logistics, he contributed significantly to the defeat of the German Protestants at the battle of Mühlberg (1547). *Philip II sent him as governor-general to deal with unrest in the Netherlands in 1567, but his notorious 'Council of Blood' executed or banished over a thousand men, and was responsible for sparking off the *Dutch Revolts. He was recalled to Spain at his own request in 1573, and in 1580 Philip gave him command of the forces which conquered Portugal.

Albany Congress (1754). At Albany, a town in New York state, the British Board of Trade convened a congress of seven colonies to concert defence against the French and pacify the Iroquois. It was at Albany that Benjamin *Franklin presented his 'Plan of Union' for a Grand Council elected by the thirteen colonies to control defence and Indian relations, but this first step towards American unity was rejected by the individual colonies.

Albemarle, George, 1st Duke of *Monck.

Alberoni, Giulio (1664-1752), Italian cardinal and statesman. In 1713 he arranged the marriage of the Duke of Parma's niece Elizabeth *Farnese with *Philip V. He became effective ruler of Spain in 1715 and strengthened royal power in Spain at the expense of the nobles. His chief aims were to strengthen Spain, nullify the Peace of *Utrecht, and crush *Habsburg power in Italy. He was doubtful about the wisdom of declaring war on Austria in July 1717 and it proved to be a disastrous decision, mainly because of British and French intervention against Spain, and resulted in his dismissal by Philip in 1719. He retired to Rome.

Albigensians, followers of a form of the *Cathar heresy; they took their name from the town of Albi in Languedoc in southern France. There and in northern Italy the sect acquired immense popularity. The movement was condemned at the Council of Toulouse in 1119 and by the Third and Fourth *Lateran councils in 1179 and 1215, which opposed it not only as heretical but because it threatened the family and the state. St *Bernard and St *Dominic were its vigorous opponents. Between 1209 and 1228 the wars known as the Albigensian Crusade were mounted, led principally by Simon de *Montfort. By 1229 the heretics were largely crushed and the Treaty of Meaux delivered most of their territory to France.

Albion, an ancient name, probably pre-Celtic, for the island now comprising England, Wales, and Scotland. Roman writers linked it with the Latin *albus* ('white') and the chalk cliffs and downlands of southern England. The Latin 'Britannia' soon replaced the term.

Albuquerque, Alfonso de (1453-1515), governor of the Portuguese empire in Asia (1508-15). He captured key ports on the sea route to India and created a Portuguese monopoly of trade. After establishing a fortified factory at Cochin on the west coast of India (1505), he disrupted Arab trade with India by fortifications at key points on the East African coast and Persian Gulf. In 1510 he conquered *Goa, which then expanded into a trading centre. By seizing *Malacca on the Malay peninsula (1511), he laid the basis for Portuguese monopoly of the spice trade of the south-east Asian archipelago, but died at sea, in disgrace, before the full results of his empire-building initiatives were appreciated.

Alcázar (Arabic *al-kasr*, 'the palace'), a type of fortress in Spain, built by the Christians during their 14th- and

The **Alcázar** of Seville, built in 1364–6 for King Pedro the Cruel by Mudejar architects. The influence of Muslim art on Christian Spain can be clearly seen.

15th-century wars against the *Moors. It was usually rectangular with great corner towers, and contained an open space or patio, surrounded by chapels, hospitals, and salons. The most renowned is the Alcázar of Seville, built by King Pedro the Cruel (1334-69). The most splendid Muslim fortress-palace in Spain is the Alhambra ('the red'), built by the Moorish monarchs of Granada, chiefly between 1238 and 1358.

Alcazarquivir (locally called al-Kasr al-Kabir), a city in northern *Morocco, famous for the battle of the Three Kings (1578). From the 15th century it suffered attacks from Portuguese coastal ports, and became an advanced post of the *mujahidin*, 'warriors for the faith', in retaliation. In 1578 Sebastian I of Portugal attacked with 20,000 men, but the Moroccans, mustering 50,000, defeated them utterly. Sebastian himself perished, with 8,000 men, and in 1580 Portugal fell into the hands of Spain.

Alcibiades (c. 450-404 BC), an Athenian leader in the *Peloponnesian War. His plans to defeat Sparta on land faltered at the battle of Mantinea (418). He advocated and was appointed one of the commanders of the ill-fated expedition to Sicily (415-413), but was charged with the desecration of sacred statues and recalled to Athens. He fled to Sparta, where he gave advice that was seriously damaging to his own city. Nevertheless, after breaking with the Athenian oligarchs in 411, he later won the confidence of the officers of the Athenian fleet based at Samos, leading it with conspicuous success. A defeat in his absence led to his downfall. He was murdered in Phrygia, Asia Minor.

Alcuin (c.735-804), English scholar and theologian. He was educated in the cathedral school at York, later becoming its head. In 782 was employed by Emperor *Charlemagne as head of his palace school at Aachen where his pupils included many of the outstanding figures in the 'Carolingian Renaissance'. Alcuin played a central role in fostering this cultural revival. In 796 he became abbot of St Martin at Tours where he continued his work until his death.

alderman (Old English, *ealdorman*, 'elderman'), a title dating from the Anglo-Saxon period when ealdormen, nobles by birth, exercised considerable powers. They were initially appointed by the crown to administer the shire system (particularly the shire moot or assembly and *fyrd). By the 10th century their influence extended beyond the shire, and in the early 11th century their title evolved into 'earl'. Under the Norman kings the senior shire official was the sheriff, and the title alderman later came to apply to those who held municipal office.

Alexander I (c.1078-1124), King of Scotland (1107-24). He succeeded his brother Edgar although the regions of Strathclyde, Lothian, and Cumbria were ruled with Anglo-Norman support by his younger brother (later *David I). Educated in England, Alexander encouraged the feudalization of his country while still retaining its independence of England. After crushing a Celtic revolt (c.1115) he was styled 'the Fierce' although he was a pious man, founding Augustinian houses at Scone (1115) and Inchcolm. He refused the Bishop of St Andrews (1120, 1124) leave to acknowledge the ecclesiastical authority of York or Canterbury.

Alexander II (1198-1249), King of Scotland (1214-49). He succeeded William the Lion. After supporting the English barons in the first *Barons' War against King John he had to suppress revolts in Moray (1221), Argyll (1222), Caithness (1222), and Galloway (1224). His campaigns against England and the Norse in the Western Isles (1249) were motivated by territorial ambitions.

Alexander III (the Great) (356-323 BC), King of Macedonia (336-323). He succeeded his father *Philip II in 336 BC, and conquered the whole of the *Achaemenid Persian empire in the course of his short reign. He inherited a highly professional army and commanded it with tactical brilliance: major victories were won at Granicus, Issus, Gaugamela, and Hydaspes, and his capture of the island-city of Tyre was a masterpiece of siege warfare. His conquest of Persia, however, was not merely the expedition of revenge which Greeks had talked about since the *Greek-Persian wars. Rather than overthrow the Persian empire, he determined to rule it in co-operation with the Persian nobles, some of whom he appointed as his governors. He also drafted many non-Greeks into his army and adopted much of Persian court ceremonial. Such policies were resented by his fellow Macedonians who were reluctant to share power with the conquered barbarians.

His death at the age of 32, whether from fever or, less likely, poison, was untimely, but nevertheless his short career changed the course of history. His conquests spread the Greek language and culture (*Hellenistic civilization) over much of the known world. In areas as far east as Afghanistan and northern India the remains of Greek-type cities have been uncovered. After his death his empire was bitterly contested by his generals, and the resultant 'successor' kingdoms—*Antigonid Macedonia, *Seleucid Asia, and Ptolemaic *Egypt—were often at war with each other until each in turn was conquered by Rome.

Alexander III (1241-86), King of Scotland (1249-86). He defeated Haakon of Norway at the battle of Largs (1263) and received the Hebrides by the Treaty of Perth. Despite close ties with England (his father-in-law was *Henry III) Alexander resisted English claims to the Scottish kingdom. The early death of his children left the succession to his granddaughter *Margaret of Norway.

Alexander Nevsky (c.1220-63), Russian soldier, Grand Duke of Vladimir (1252-63). Born in Vladimir, son of the Grand Duke Jaroslav II of Novgorod, he acquired his second name after his defeat of the Swedish army on the banks of the River Neva in 1240. Wars against the Germans and Lithuanians culminated in a battle with the *Teutonic Knights on the frozen Lake Peipus which he won decisively. After his death he was canonized as a saint of the Russian Orthodox Church.

Alexandria, the name of six cities founded or refounded by *Alexander the Great throughout the conquered Persian empire. In their architecture and their institutions they were very much Greek cities, and contributed greatly to the spread of *Hellenistic civilization.

The most successful foundation was that established in Egypt in 331 BC. It was ideally situated—at the western edge of the Nile delta—to become the main port of Egypt, and under Ptolemy I it replaced Memphis as the

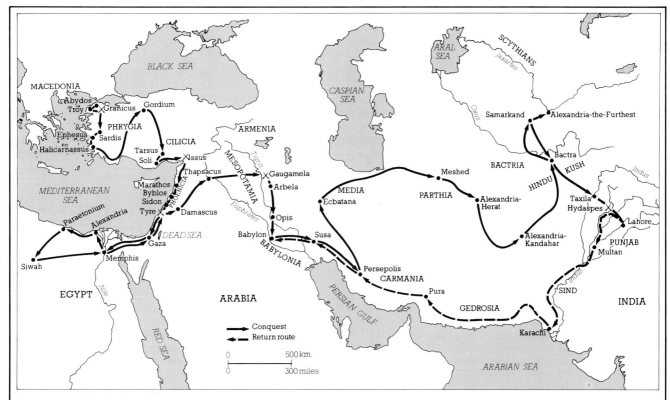

Alexander III (the Great)

Alexander's victorious progress through Asia Minor against the Persians culminated in the defeat of King Darius at Issus (333 BC). Advancing through Phoenicia, where he met resistance from Tyre and Gaza, he reached Egypt in 332, and was welcomed as a liberator. After founding Alexandria he turned eastwards and won a decisive battle at Gaugamela (331). Babylonia offered no serious resistance and he advanced through Bactria to the Jaxartes, founding Alexandria-the-Furthest as a defence against nomadic Scythians. His defeat of the Indian ruler Porus on the Hydaspes (326) took him into new territory, but his weary soldiers persuaded him to turn back. On the return journey he sailed down the Hydaspes to the Indus delta, as he was anxious both to conquer Sind and to see 'Ocean', which the Greeks believed encircled the world. After a gruelling desert march he reached Susa in 324. Alexander died at Babylon in 323, in his 33rd year.

capital city of that country. It prospered throughout the *Ptolemaic dynasty, and continued to do so after Egypt became a Roman province. A substantial Jewish population settled in one quarter of the city: at one time the city was the capital of the Jewish world. Violent anti-Jewish riots were frequent. Enhanced by its library and museum, Alexandria prospered culturally as well as economically. Arabs captured the city in AD 641 and destroyed the library, said to contain some 700,000 items. In the 14th century the canal to the Nile silted up, and at the end of the 15th the Cape of Good Hope was discovered as a new European route to the East.

Alfonso V (the Magnanimous) (1396–1458), King of *Aragon (1416–58), and of *Naples (1443–58). He pursued a foreign policy committed to territorial expansion, particularly in Italy. Joanna II, Queen of *Naples, adopted him as her heir and on her death he transferred his court to Naples in 1443, which he developed as a centre of *Renaissance culture. His patronage earned the admiration of contemporary *humanists.

Alfonso X (the Wise) (1221–84), King of *Castile and León (1252–84). His reign was a contrast between the failure of his political ambitions and his scholarly success as a law-giver. He spent fruitless years trying to become Holy Roman Emperor and failed to complete his father's

Crusade against the *Moors in southern Spain. His indecision caused his son, Sancho IV, to rebel and isolate him in Seville. Of real importance was his *Siete Partidas*

Alfonso X at his court in Toledo, a centre of learning and a haven for Arab and Jewish as well as Christian scholars. From the richly illuminated 13th-century *Cantigas of Alfonso X*, a collection of over 400 poems in which the king recounted the miracles attributed to the mother of Jesus Christ. (Library of Escorial, Madrid)

Widely believed to have belonged to **Alfred** the Great, this late 9th-century ornament was found near the Isle of Athelney in Somerset in 1693. It bears the inscription 'Aelfred mec heht gewyrcan' ('Alfred ordered me to be made') and the portrait, in cloisonné enamel, of a man holding two sceptres, may well be of Alfred himself. (Ashmolean Museum, Oxford)

(1256), a collection of constitutional, civil, and criminal law, the first such work to be written in Spanish.

Alfred (the Great) (849–99), King of *Wessex (871–99). He became king in 871 at the time of Danish invasions. Earlier (865), the other English kingdoms of *East Anglia, *Mercia, and *Northumbria had been overwhelmed by the Danes who had even begun (871) to challenge the defences of Wessex. Three fierce Danish attacks (871–8) followed, the most serious being in 877 when, under Guthrum, they drove Alfred into hiding in the marshes of Athelney. Alfred's counter-offensive produced a decisive victory at Edington (878) and the Treaty of Wedmore (Chippenham) by which Guthrum agreed to withdraw to East Anglia and to become a Christian. In 886 Alfred, having captured London, defined with Guthrum the boundaries between Wessex and the *Danelaw. Alfred improved the defences of England by reorganizing the *fyrd, developing a navy, and improving the system of fortified towns.

Alfred also encouraged learning. He translated the Latin texts of *Boethius, *Bede, Orosius, and *Gregory the Great, and may have suggested the writing of the *Anglo-Saxon Chronicle. He and other members of his family established fortified towns (burghs), many of which are still county towns. He was also renowned for the laws which he made. The probably apocryphal story of Alfred burning the cakes is told in the 12th century *Chronicle of St Neot's.*

Algeria, a North African country whose boundaries were established after French conquest in the 1830s. The native people were Berbers, but the coast was colonized by the Phoenicians in the 9th century BC. In the 2nd century BC the Romans incorporated the whole region into the province of Africa. In the 7th century AD the Romanized Berbers resisted the Arab invasion fiercely. Once conquered they were converted to Islam, and became members of the extreme Kharijite sect. From the 11th century they were repeatedly ravaged by the Banu Hilal and other Arabs, and ruled by a series of dynasties until the Ottoman conquest.

Algonquin, an eastern Canadian group of Algonquian-speaking Indians inhabiting the Ottawa valley and adjacent regions in the 17th century, and southern Ontario in the 18th. They fought the *Iroquois, especially the *Mohawk, from 1570 in alliance with the Montagnais, to the east, and from 1603 as allies of the French in a conflict over the European fur trade that lasted into the 18th century. They were essentially a hunting-fishing people, though some grew corn, beans, and squash in cleared areas of the bush. They lived in small groups in dwellings—pointed tipis or dome-shaped wigwams—made from birchbark laid over poles, and are associated with the light birchbark canoe, and with snowshoes, sleds, and toboggans, which they used for winter travel. A small number (c.2000) survive today as hunters' guides and trappers.

Allen, Ethan (1738–89), American patriot. He migrated from Connecticut to the 'New Hampshire Grants', later *Vermont, in 1769. With his younger brothers, Ira and Levi, he organized the *Green Mountain Boys to combat New York's claims to the region. In 1775, together with Benedict *Arnold, he helped overrun *Ticonderoga, but was captured while leading an expedition against Canada. Released in 1778, he continued to campaign for Vermont independence, even considering making it into a British province.

Almohad (from the Arabic al-Muwahhiddun, 'unitarian'), a Berber dynasty that originated in the Atlas Mountains of North Africa c.1121. The founder, Ibn Tumart, claimed to be Mahdi ('the divinely guided one'), whose coming was foretold by *Muhammad, and he preached an extreme puritanical form of Islam. His successor, Abd al-Mumin, seized all North Africa and then southern Spain from the *Almoravids in 1145. The Almohad empire was a great Islamic and Mediterranean power. A centrally directed administration with a professional civil service collected taxes and maintained a large fleet and army. Besides fine architecture the empire also produced influential international scholars, like the Spanish-Arabian philosopher *Averroës (Ibn Rushd). War in Spain proved disastrous, and the defeat at Las Navas de Tolosa (1212) ended the Almohad regime. The dynasty survived in Marrakesh until 1269.

Almoravid (from the Arabic al-Murabitun, 'member of a religious group'), a Berber dynasty that originated in North Africa (1061–1145), and in Spain from 1086. It originated among the Lamtuna Tuareg, whose extreme Islamic faith compelled even men to wear veils. The founder, Abu Bakr, built Marrakesh in 1070, and his cousin, Yusuf ibn Tashfin, conquered all of north-west Africa, and then Spain, to which he was invited in 1086 by al-Mutamid of Seville. His Berber army defeated Alfonso VI of León at Zallaqah, near Badajoz, in 1086. In 1088 he returned to Spain and began his new campaign by taking Granada. He took Seville and Córdoba in 1091, Badajoz in 1094, Valencia in 1102, and Saragossa in 1110, but he failed to overcome Alfonso VI, or to take Toledo. Yusuf and his followers were primarily fighting men and their time in Spain was one of confusion and cultural stagnation. It was a time of suffering and persecution for many Christians, Jews, and even liberal Muslims—divines as respected as al-Ghazzali had their works proscribed. The end of the dynasty came in 1145, with a further Berber invasion under the *Almohads, who succeeded to the conquests of the Almoravids.

almshouse, a sanctuary for the reception and succour of the poor. Almshouses were originally those sections of medieval monasteries where alms (food and money) were distributed. Most medieval foundations were made by clergymen, like Bishop Henry of Blois, who set up the Hospital of St Cross in Winchester, England, *c.*1135. The term was also used to describe privately financed dwellings, usually for the support of the old and infirm. Wealthy merchants, as individuals or in corporations, became especially active in endowing almshouses, as a way of showing their charitable intentions. From the 16th century the charitable relief supplied by almshouses was supplemented by a series of *Poor Laws.

Alva, Duke of *Alba.

Amboina (or Ambon), an island in the *Moluccas, a major centre for clove production. In 1521 the Portuguese established themselves there but were driven out by the Dutch in 1605. An English attempt to set up a trading station for spices ended in 1623 when all their men were tried and executed for conspiring against the Dutch in the so-called 'Amboina massacre'. Dutch missions were active among the Ambonese, who became the most loyal of the Netherlands' overseas subjects

Ambrose, St (*c.*340–97), Bishop of Milan. He was acclaimed bishop in 374 when he was Roman governor of the region and awaiting baptism. A renowned preacher and writer, he attacked *Arianism, the pagan revival, and the Jews. Three Roman emperors came under his influence: *Theodosius was even compelled to perform penance for a massacre. He saw church and state as complementary spiritual and temporal institutions facing shared threats from heretics and barbarians.

American Philosophical Society, the first and still the most illustrious American scholarly association, founded in Philadelphia in 1743 by Benjamin *Franklin. The inventor and astronomer David Rittenhouse (1791) and Thomas *Jefferson (1797) succeeded him as presidents. The Society has a major collection of scientific literature and sponsors academic research.

American Revolution *Independence, American War of.

Amerindians, the indigenous peoples of North and South America. They are usually classified as a major branch of the *Mongoloid peoples but are sometimes described as a distinct racial group. With the Inuits and Aleuts (who are unquestionably Mongoloids), they were the inhabitants of the New World at the time of the first European exploration in the late 15th century. Their forebears came from north-eastern Asia, most probably taking advantage of low sea levels during the last Ice Age to cross the Bering Strait on land. The earliest certain evidence suggests that people were in America by 15,000 years ago but an earlier date seems increasingly likely. Recent controversial archaeological finds in Mexico, Chile, Brazil, and elsewhere suggest a human presence as early as 30,000 or more years ago. There could have been several separate colonizations; the

Inuits and Aleuts are the descendants of the most recent one, within the past 10,000 years. The first colonizers brought little with them other than simple stone tools and perhaps domesticated dogs for hunting. As hunters and gatherers they spread quickly south. There was plentiful game to hunt using fine stone projectile points such as those of the *Clovis tradition.

The cultural development of Amerindians provides an interesting comparison with the Old World. Agriculture, which started developing 7000 or more years ago, was based on maize, squash, and beans, with manioc being grown in tropical forest regions. With no suitable animals to domesticate, apart from the llama and the guinea-pig, and no draught animals to pull the plough, the development of more mixed farming was gradual. In the Andes, an advanced metallurgical technology developed from 1000 BC. Complex societies developed in many areas, which grew into sophisticated civilizations, for example, the *Aztecs and *Incas, but most collapsed after the arrival of the *conquistadores and other European explorers in the 16th century.

Amherst, Jeffrey, Baron (1717–97), British general. He commanded the combined operation which captured *Louisburg in 1758. On his appointment as commander-in-chief in America, he applied widespread pressure on the French. His own army advanced northward up the Hudson Valley, taking *Ticonderoga and Crown Point in 1759 and Montreal in 1760, thus ending French control of Canada. He was then made governor of Virginia, but failed to contain *Pontiac's Indian Rebellion in 1763 and was recalled. He refused to fight against Americans in 1775, but advised on strategy.

amphora, a two-handled ancient Greek or Roman pottery jar with a pointed or knobbed bottom to facilitate its transport, used for the storage of wine and olive oil.

Roman **amphorae** or two-handled storage jars from Carthage, showing the variety of shapes and sizes that were generally used.

The amphora was often stamped with a mark—a sign that it conformed to local regulations on size. This mark might include the potter's name, the name of the relevant annual magistrate, and the country of origin. Often located by marine archaeologists in wrecks, they provide valuable evidence of trade links.

Anabaptism, a Christian religious doctrine which centred on the baptism of believers, and held that people baptized as infants must be rebaptized as adults. Anabaptists or 'Re-baptists' formed part of the radical wing of the 16th-century *Reformation. The sects originated mainly in Zürich in the 1520s, with the aim of restoring the spirit and institutions of the early Church. Their belief that the Church was not an earthly institution and that true earthly institutions were by their nature hopelessly corrupt, led them to a repudiation of the very basis of the authority of the civil power. Thus, though they were often law-abiding, they reserved the right in conscience to disobey the law, and for this they were feared and persecuted. They managed to establish centres in Saxony, Austria, Moravia, Poland, the Lower Rhine, and the Netherlands, but made almost no headway in the French-speaking world. In the 17th century the *Mennonites preserved some of the best of the Anabaptist traditions, which made a significant contribution to the religious history of modern Europe and America.

Anasazi, a North American Indian culture centred in the 'four corners' region of modern Utah, Colorado, Arizona, and New Mexico, USA. It began c.500 BC when a variant, called San Jose (c.500–100 BC), of the hunter-gatherer *Desert cultures took the first steps towards agriculture and village life. Their abundant basketry has given the name *'Basket-makers' to their early stages. By AD 450 they were making pottery, and by 700–900 great kivas (round ceremonial chambers) were being built. In the 13th to 15th centuries droughts, crop failures, and the influx of Athapascan tribes (*Navajo and *Apache) led to the abandonment of many of their settlements and the building of *cliff-dwellings, or pueblos, for defence. They were visited by Francisco Vasquez de Coronado's expedition of 1540–2, by which time they had begun to resettle some of their old territory. Initial relations were friendly, but a shortage of food led to resistance, ended by Coronado's mass execution of Indians. Spanish missions came in the 17th century, and attempts were made to expel the missionaries in the 1680s, followed by severe Spanish reprisals from 1692. They became known as the Pueblo Indians.

ancien régime, a term used to describe the political and administrative systems in France in the 17th and 18th centuries under the Bourbon kings, before the *French Revolution; it is also applied more widely to much of the rest of Europe. The monarch had (in theory) unlimited authority, including the right to imprison individuals without trial. There was no representative assembly. Privilege, above all, was the hallmark of the *ancien régime*: the nobility were privileged before the law, in matters of taxation, and in the holding of high offices. This was particularly resented by the increasingly prosperous middle class and in fact limited the power of the monarchy. The clergy were equally privileged and the Roman Catholic Church had extensive land-holdings in France. The peasants were over-taxed and though

French society of the **ancien régime**, as depicted in a contemporary cartoon. It illustrates representatives of the church, the military, the peasantry, and the legal bureaucracy. The privileges enjoyed by church and nobility formed the basis of the *ancien régime* and against these two groups, the middle class struggled for a political voice to match its growing economic power.

they were not *serfs, their landlords were able to exercise many rights over them.

The *ancien régime* was inefficient: reforms in law, taxation, and local government were long overdue, and it was government bankruptcy which was to be one of the causes of the Revolution. In spite of restraints on the freedom of the press by both church and state, criticism was vigorous and widespread. The regime acted as a straitjacket on a society that was evolving rapidly and it was destroyed by the French Revolution.

Andalusia, a large region in southern Spain bordered by Portugal in the west and the Atlantic and Mediterranean to the south. As the Roman province of Baetica it was the birthplace of the emperors *Trajan and *Hadrian. During the dissolution of the Roman empire it fell to the *Vandals. In 711 it was captured by the Muslims and it remained in their hands for many centuries. Internal strife between rival emirs allowed the recovery of much of the territory by Ferdinand II of Castile, but Granada continued to resist until 1492 when it too fell, to Ferdinand and Isabella of Spain. Following the Spanish conquest of South America it enjoyed a period of prosperity and many of the *conquistadores were recruited from the region. In 1609 the Moriscos, Christian converts from the Muslim faith, were expelled, to the economic detriment of the area. In 1704 Gibraltar was lost to the British.

Andrewes, Lancelot (1555–1626), English prelate, successively Bishop of Chichester (1605), Ely (1609), and Winchester (1619). A celebrated scholar and famous preacher, he was prominent at the courts of *Elizabeth I and *James I. He was a key figure at the *Hampton Court Conference (1603–4), and was closely involved in producing the Authorized Version of the English Bible (1611). He played an important part in developing the theology of the *Anglican Church.

Angevin, the dynasty of the counts of Anjou in France which began with Fulk I (the 'Red'), under the Carolingian emperors of the 9th century. Their badge, a sprig of the broom plant *Genista*, gave rise later to the name of Plantagenet. Geoffrey of Anjou married *Matilda, the daughter of Henry I of England, in 1128, and their son, as *Henry II of England, was the first of an English royal dynasty. The power of the Angevins under Henry was formidable, overshadowing the *Capetian kings of France. Anjou remained in English hands until 1203 when Philip Augustus wrested it from John. Louis gave the Angevin title to his brother Charles who, as King of Naples and the Two Sicilies, established the second Angevin dynasty. In 1328 *Philip IV inherited it together with Maine from his mother and thus it passed directly to the French crown.

Angkor, in Kampuchea (formerly Cambodia), the site of several capitals of the *Khmer empire. It is renowned for the temples which the Khmers built between the 9th and 12th centuries for their god-kings to live in after death. At Angkor Thom was the grandiose Bayon (temple) of Jayavarman VII (1181–c.1220); on pinnacle after pinnacle the king's features live on in the faces of the Buddha. Under this Buddhist king, Angkor Thom reached its zenith. The city, with its 13 km. (8 miles) of moated walls and position on the shores of the vast inland lake of Tonlé Sap, lay at the heart of an elaborate irrigation system which was partially laid out and controlled by the Khmer kings. For centuries the city and the great temples, with their bas-reliefs recording sacred myths and the daily lives and bloody battles of the Khmers, were lost to the jungle. After their rediscovery in the 19th century they were much restored.

Angles, a Germanic tribe closely linked to the *Jutes and *Saxons, thought to have originated in Schleswig-Holstein or Denmark. In the 5th century they settled in eastern Britain in *East Anglia and *Northumbria. Because of their presence, the land of the *Anglo-Saxons later became known as 'Englaland' and thereby England.

Anglican Church, the Church of England, which was established during the 16th century Protestant *Reformation. Although Henry VIII broke with the Roman Catholic Church and *Edward VI made moves to establish Protestant doctrines and practices, the formulation of Anglican principles dates from the reign of *Elizabeth I. The second Book of Common Prayer of Edward VI's reign was revised with modifications (1559) and its use enforced by an Act of *Uniformity. In 1563 the *Thirty-Nine Articles were issued by Convocation (the highest assembly of the Church) and finally adopted by the Church of England (1571) as a statement of its beliefs and practices. The aim was to set up a comprehensive, national, episcopal church with the monarch as supreme governor. Those who refused to attend church services were fined. The *Puritans were dissatisfied with the Elizabethan religious settlement but the queen opposed all their attempts to modify her Anglican Church.

The 'Catholicization' of the Church in the 1630s under Archbishop *Laud exacerbated Puritan antipathy to the bishops, and religion was a crucial factor in the outbreak of the *English Civil War. Although Anglicanism was banned during the Commonwealth and Protectorate, it

The approach to the great temple at **Angkor Wat**. Built in the 11th century by the Khmer ruler Suryavarman II, the enormous structure forms one of the world's greatest religious buildings. Rising in three concentric enclosures, its complex of richly decorated courts, terraces, staircases, porticoes, and towers represents the outer world surrounding the shrine as the focal point of the universe.

returned with vigour at the Restoration (1660). The Clarendon Code and *Test Acts created a breach between establishment Anglicanism and *Nonconformists, and James II's pro-Catholic policies played a significant part in provoking the *Glorious Revolution. The *Toleration Act (1689) secured limited toleration for Nonconformists, although clergymen refusing to swear the oath of allegiance to William III were deprived of their office. (Catholics were not emancipated until 1829.)

The 18th century witnessed disputes between High Anglicans, who maintained Laud's conservatism, and Low Anglicans, or Latitudinarians who were less concerned with forms of worship. Opposition to the evangelism of John *Wesley led to the establishment of an independent *Methodist Church in 1791.

Anglo-Dutch wars, three maritime wars, 1652–4, 1665–7, and 1672–4, fought between the United Provinces and Britain on grounds of commercial and naval rivalry. The Dutch navy was commanded by able admirals but the prevailing westerly winds gave the English sailors a significant advantage.

The first war began when the Dutch carrying-trade was undermined by the English *Navigation Acts of 1651, and the Dutch refused to salute the English flag in the English Channel. *Tromp defeated *Blake off Dungeness in December 1652, but convoying Dutch merchant ships through the Channel proved difficult and de *Witt settled for reasonable peace terms from Cromwell in 1654. The Dutch recognized English sovereignty in the English Channel, gave compensation for the massacre at *Amboina, and promised not to assist the exiled *Charles II. An encounter off the African coast began the second war, followed by the fall of New Amsterdam (renamed New York) to the English, who also defeated the Dutch off Lowestoft in June 1665. However in 1666 Charles II was in financial difficulties, Cornelius Tromp and *Ruyter won the Four Days War, and Ruyter made his celebrated raid on the English dockyards at Chatham. Peace was made at Breda in 1667. The Navigation Acts were modified in favour of the Dutch and territories gained during the war were retained, the Dutch keeping Surinam and the British, Delaware and New England. In 1672 Charles II, dependent on French subsidies, supported *Louis XIV against the Dutch. The Dutch admirals had the advantage and the Treaty of Westminster signed in 1674 renewed the terms of Breda.

Anglo-Saxon Chronicle, a collection of seven manuscripts written in Anglo-Saxon (Old English) which together provide a history of England from the beginning of the conversion to Christianity up to 1154. The major text (known as the *Parker Chronicle*) appears to have been written by one clerk until 891. Most of the copies end in

The **Anglo-Dutch wars** produced notable naval engagements including the Four Days War (11–14 June 1666), depicted here in a contemporary painting by Storck. This was a fierce and, at times, confused encounter between the British, under Albemarle and Prince Rupert, and the Dutch fleet, led by admirals Ruyter and Tromp. The British suffered the heavier losses, although both sides were content to return to their respective harbours at the battle's conclusion. (National Maritime Museum, London)

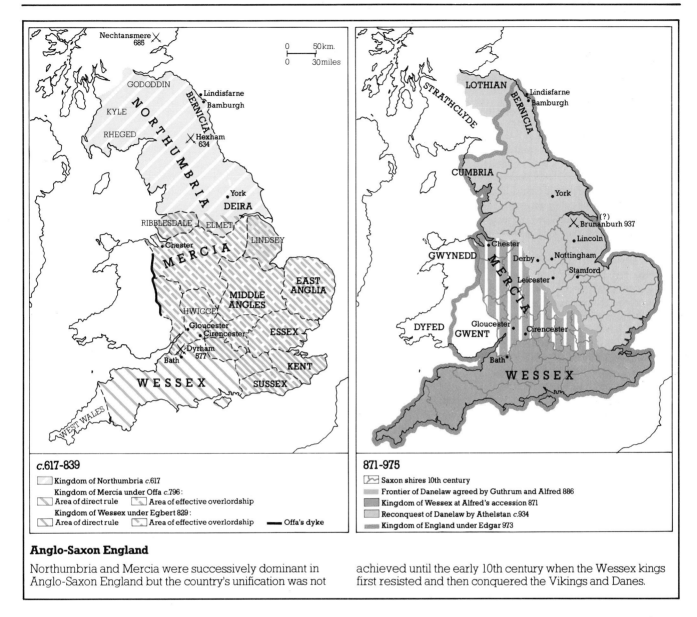

c.617–839

▫ Kingdom of Northumbria *c.*617
Kingdom of Mercia under Offa *c.*796:
▨ Area of direct rule   ◩ Area of effective overlordship
Kingdom of Wessex under Egbert 829:
▨ Area of direct rule   ◩ Area of effective overlordship   ▬ Offa's dyke

871–975

▨ Saxon shires 10th century
░ Frontier of Danelaw agreed by Guthrum and Alfred 886
▓ Kingdom of Wessex at Alfred's accession 871
░ Reconquest of Danelaw by Athelstan *c.*934
▥ Kingdom of England under Edgar 973

Anglo-Saxon England

Northumbria and Mercia were successively dominant in Anglo-Saxon England but the country's unification was not achieved until the early 10th century when the Wessex kings first resisted and then conquered the Vikings and Danes.

the 11th century; after 1079 only the *Peterborough Chronicle* continued, breaking off abruptly with an unfinished entry for 1154. The *Chronicle* probably originated as notes inserted in the tables used by the Christian Church when calculating the date of Easter.

Anglo-Saxons, the name given to the *Angles, *Saxons, and *Jutes whose invasions of Britain (*c.*450–*c.*600) began with the departure of the Roman legions. They were probably joined by *Frisians, *Swabians, and settlers from southern Sweden.

The monk *Bede in his *Historia Ecclesiastica Gentis Anglorum* ('Ecclesiastical History of the English People') explained the origins of the independent kingdoms which emerged. The East Angles (East Anglia), Middle Angles (East Midlands), Mercians (Midlands), and those who lived north of the Humber were descended from the Angles. The Saxons established the kingdoms of the East Saxons (Essex), the South Saxons (Sussex), and the West Saxons (Wessex). The Jutes settled in Kent, on the Isle of Wight, and in Hampshire. In the 7th century Northumbria under Edwin, *Oswald, and *Oswy claimed authority (bretwaldaship) over all Anglo-Saxons.

By the 8th century this hegemony had passed to Mercia under Ethelbald and *Offa and finally, by the 9th century, to Wessex, whose kings were able to resist the Viking invasions. Under *Alfred, *Edward the Elder, *Athelstan, and *Edgar, Wessex established an undisputed claim to the overlordship of England. The renewal of Scandinavian raids led to a Danish king, *Canute (Cnut), becoming King of England in 1016. Although the West Saxon line was restored temporarily with *Edward the Confessor in 1042, it finally ended with the Norman Conquest.

Until the conversion to Christianity of Ethelbert of Kent by St *Augustine (597) the Anglo-Saxons had been pagan. Thereafter their conversion proceeded rapidly. Church art and architecture contain some of the finest expressions of Anglo-Saxon culture and the monasteries, especially that at Winchester, produced beautiful illuminated manuscripts, noted particularly for their decorated initial letters.

Anjou, historically a province of western France, whose principal city was Angers. The Celtic inhabitants, the Andecavi, came under Roman domination during Julius

*Caesar's conquest of Gaul but seceded from the Roman empire in the 5th century. In the 9th and 10th centuries it suffered depredations, first at the hands of the Vikings and then from the Normans.

In the 12th century the title of count passed to *Henry II of England and remained in English hands until 1203 when it reverted to France. In 1366 it became a duchy of the kingdom of Naples but in 1480 was restored to France under Louis XI. It ceased to exist as a province in 1790, as part of the French Revolutionary changes.

annals (Latin *annus*, 'year'), the yearly records kept by the priests in Rome from the earliest times. They noted ceremonies, state enactments, and the holders of office. The high priest (Pontifex Maximus) was responsible for maintaining the records in his official residence. The accumulated material (mainly dating from after 300 BC) was published in eighty books known as the *Annales Maximi c.*123 BC. *Cato the Elder, *Livy, and *Tacitus among other writers drew heavily on these records. The name came to be applied generally to the writing of history in strict chronological order.

Annam (Chinese, 'Pacified South'), the central region of Vietnam. It formerly covered Tonkin and northern Annam—those parts of Vietnam conquered by the Chinese in 111 BC. The original home of the Vietnamese was centred on the Red River and its delta in northern Vietnam. In 939 the Annamese drove out the Chinese and established an independent kingdom, though it continued to pay tribute to China until the 19th century. By 1471 it had conquered the *Champa. There followed a period of unrest and after a rebellion in 1558, the kingdom was divided in two. The Trinh ruled from Hanoi over an area that became known as Tonkin, and the Nguyen ruled southern Annam from Hué. The Nguyen expanded their rule over *Cochin China in the 18th century and in 1802 Nguyen Anh, an Annamese general, reunited both parts of Annam with French assistance.

Anne (1665–1714), Queen of England, Scotland, and Ireland (1702–7) and of Great Britain and Ireland (1707–14), the last *Stuart sovereign. She was the younger daughter of the Roman Catholic *James II and his first wife Anne Hyde, but was brought up as a Protestant and in 1683 married the Protestant Prince George of Denmark (d. 1708). During the *Glorious Revolution she abandoned her father's cause and from 1689 gave support to her brother-in-law *William III. The death of her last surviving child in 1700 led to the Act of *Settlement by which, after Anne's death, Parliament bestowed the succession to the throne on her Hanoverian (and Protestant) relations. Her reign was dominated on the one hand by the War of the *Spanish Succession (known in the American colonies as *Queen Anne's War), and on the other hand by the struggle between the Whigs and Tories to control the queen's government. At first the Whigs dominated through the influence on the queen of Sarah Churchill, Duchess of *Marlborough. From 1707 she was ousted by a new royal favourite, Abigail Masham, and in 1710 the Tories came to power, retaining it until Anne's sudden and fatal illness.

Anne was not a politically gifted woman and was easily influenced by the opinions of others and by her own prejudices. She was the last English monarch to veto an Act of Parliament and caused controversy when she

A portrait of Queen **Anne** of England by Sir Godfrey Kneller, court painter to William III, who retained the position under Anne. This official study, done early in her reign, shows Anne in state robes. The queen's head was used as a model for her coinage and medals. (Royal Collection)

created a dozen peers to give the Tories a working majority in the House of Lords in 1712. Her reign was notable for its many literary figures, including Alexander Pope and Jonathan Swift, which helped to justify its title of the 'Augustan Age', implying comparison with the greatest age of the *Roman empire.

Anne of Austria (1601–66), wife of Louis XIII of France whom she married in 1615. She was the daughter of Philip III of Spain. Her friend Madame de Chevreuse was involved in plots against *Richelieu, and she was accused of encouraging the advances of the Duke of *Buckingham. When her 4-year-old son succeeded to the throne as *Louis XIV in 1643 she was declared regent and gave her full support to *Mazarin during the *Fronde. She influenced her son until her death, though her regency ended in 1651.

Anne of Cleves (1515–57), German princess, Queen consort of *Henry VIII. She was chosen by Thomas *Cromwell to marry Henry VIII of England, and thereby seal a diplomatic alliance with the German Protestants. After a proxy courtship she became the king's fourth wife in January 1540. Six months later Henry had the marriage annulled, alleging that the marriage was unconsummated. She stayed in England, unmarried, on a generous pension for the rest of her life.

Anselm, St (1033–1109), philosopher, theologian, and Archbishop of Canterbury. He was a Benedictine monk at the monastery of Bec, in Normandy, where *Lanfranc was prior, and he succeeded him as prior in 1063. Over the next thirty years Anselm established Bec as a major centre of scholarship and himself as 'the Father of Scholasticism'. He argued that faith was a necessary prerequisite for the understanding of the scriptures and was not itself dependent upon that understanding—although reason could be used constructively within the context of faith. Christian truths, he believed, were capable of rational exposition.

Anselm also succeeded Lanfranc as Archbishop of Canterbury, in 1093. His relationship with both *William II and *Henry I was uneasy, particularly over the election of bishops (*investiture) without interference from the crown. Twice Anselm went into voluntary exile over this issue (1097–1100, 1103–6). Finally he accepted a compromise under which the king surrendered his claim to invest bishops but retained the right to their homage for lands held by virtue of their office.

Anson, George, Baron Anson (1697–1762), English admiral, remembered for his circumnavigation of the world, 1740–4. Due to shipwreck and scurvy among his crew he returned with only one of his original six ships though with almost £500,000 worth of Spanish treasure. In 1747, off Cape Finisterre, he captured six French enemy warships during the War of the *Austrian Succession. Later, at the Board of Admiralty, he created the corps of marines, and by his reforms and effective planning, played a major part in securing Britain's naval successes in the *Seven Years War.

anticlericalism, an attitude of hostility toward the Christian clergy's involvement in political and secular affairs. It has existed since the earliest days of institutionalized Christianity. In medieval Europe it was unorganized and apolitical. Pluralism (the simultaneous holding of several church offices) and absenteeism (the non-residence of clergy) were prime targets of criticism, along with immoral conduct, the abuse of *indulgences, and the excessive powers of church courts. Anticlerical sentiment helped to pave the way for the Reformation, as it coalesced with the calls for doctrinal reform from *Wyclif and *Luther. Modern political anticlericalism dates from the *Enlightenment.

Antigonus I (the 'One-eyed') (c. 382–301 BC), an officer in the army of *Alexander the Great. After the latter's death (323), and that of the Macedonian regent, Antipater (319), he attempted, in the subsequent power struggle, to re-establish Alexander's empire under his own sole leadership, declaring himself king (306). His considerable success induced his rivals—Ptolemy, Seleucus, Cassander, and Lysimachus—to combine, defeat, and kill him at the 'battle of the kings' at Ipsus.

Antioch, founded 300 BC by Seleucus I, one of the most prosperous cities of antiquity. It stood strategically on the trade routes linking the East with the Mediterranean, and was the capital city of the *Seleucids. In 64 BC it was annexed by Pompey for Rome and remained a flourishing commercial and intellectual centre, though it had declined by the time of its capture by the Arabs in AD 637–8. The Crusaders captured it in 1098 and it remained an important Christian city until it fell in 1268 to the *Mamelukes of Egypt. It came under *Ottoman rule from 1516.

antipope, a person who claims or exercises the office of pope (*papacy) in opposition to the true pope of the time. There have been about 35 antipopes in the history of the Catholic Church, the last being Felix V (1439–49). There have been two main causes. First, a disputed election, in which there was disagreement among the electors or other interested parties as to which person was elected pope. Secondly, the desire of various *Holy Roman Emperors to have a more pliable person as pope, and their setting up of antipopes for this purpose. In some cases, especially during the *Great Schism of 1378–1417, it is very difficult to say which person was the true pope and which was the antipope.

Antonines, the name of a Roman imperial dynasty beginning with Titus Aurelius Antoninus (AD 86–161). He succeeded *Hadrian in 137 and was entitled 'Pius' (Latin, 'the Devout') by the *Roman Senate. His reign was peaceful, by virtue of his respect for the traditional role of the Senate. The administration was streamlined and centralized and the weaker points of Rome's vast imperial frontiers were secured. The remains of a column and temple to his memory still exist in Rome. His nephew and son-in-law *Marcus Aurelius was named his adopted son and heir. Aurelius' son, Commodus, was technically the last of the dynasty; but Lucius Septimius *Severus adopted himself into the line. Severus' son 'Caracalla' and great-nephew Elagabalus continued to use the name and the title 'Pius'.

Antonine Wall, the Roman empire's northernmost fixed frontier in Britain, built c.AD 143 in what is now southern Scotland. Constructed by detachments of all three legions of the Roman army then stationed in Britain, it was a turf barricade standing some 3 m. (10 ft.) high. A ditch 12 m. (40 ft.) wide and over 3 m. (12 ft.) deep along its northern front was lined with sharp stakes. Nineteen forts marked its 60-km. (37-mile) length, along with advanced outposts and signal towers. As a line of defence it was evacuated in about AD 155, almost certainly for good, although there may have been a brief reoccupation some twenty years or so later.

Antony, Mark   *Mark Antony.

Anyang, a city in Henan province, China. It is an important *Longshan site, and also the site of the most famous (c.1300 BC) *Shang capital, which has been extensively excavated. Chinese characters scratched on to *oracle bones found there, some stored in 'archive' pits, show that writing had already advanced well beyond the pictographic stage and that bamboo-slip books, writing brushes, and ink were in use. Excavated pit tombs provide evidence of the slaughter of human beings and of horses at royal funerals. Bronze vessels in these tombs are of sophisticated design and show great skill in the technique of bronze casting; distinctive shapes, in particular a tripod with udder-like hollow legs, link this mature Shang metallurgy with the shapes of earlier Longshan pottery.

Apache, a North American Indian tribe. In prehistoric times they were nomads using the dog-travois (sledge) on the central and southern Great Plains, gradually moving southwards into the semi-deserts during the 9th to 15th centuries. They and the *Navaho raided towns of the *Anasazi as early as *c*.1275 and were partly responsible for their downfall. Spanish explorers found them well established in Arizona, New Mexico, Texas, and northern Mexico in the late 16th century and regular contact with Spanish settlements was developed by the early 17th century. As they, and numerous other tribes on the eastern edges of the plains, acquired horses, competition for buffalo hunting became fierce and the *Comanche eventually drove them off the Great Plains into the deserts by the mid-18th century.

Appian Way, ancient Rome's earliest major military road, named after Appius Claudius Caecus, who authorized its construction in 312 BC. It was described by the Roman poet Statius as the 'Queen of Roads'. The first stage, from Rome to Capua, was constructed during the *Samnite wars; it was later extended to Taranto and Brindisi. It was paved with cobblestones, and marked by milestones. *Trajan added a spur-road to Bari, the Appia Traiana, in AD 109. The first few miles from Rome still preserve many of the ancient tombs which lined the road and some of the original stone paving.

Aquinas, Thomas, St (*c*.1225–74), philosopher and theologian. He was born to a noble Italian family and despite opposition became a *Dominican friar, studying under Albertus Magnus of Cologne, one of the best teachers of the day. He then taught in Paris and was for a time attached to the papal court. He was the greatest theologian of the medieval church and a powerful force in the movement known as *scholasticism. His main arguments are set out in his *Summa Theologica* which covers the whole range of theology. He deduced the existence of God from what could be seen in the nature of the world. He made the work of Aristotle acceptable in Christian Western Europe; his own metaphysics, his account of the human mind, and his moral philosophy were a development of Aristotle's. He was also indebted to Arab philosophers. His followers are known as Thomists and his work remains the respected basis for much Catholic philosophy and theology up to the present day.

Aquitaine, a province in south-western France, originally the Roman Aquitania. Between the 3rd and 7th centuries it suffered from German and Gascon invasions. The attempts of the region to preserve some political independence were frustrated by the *Carolingians, who made it part of their empire in the 8th century. It remained semi-independent, however, and, after the collapse of Carolingian power, emerged as a duchy in the 10th century under the counts of Poitiers. It passed briefly to France when Duchess *Eleanor married Louis VII, but they divorced in 1152 and when her new husband became *Henry II of England it came to the English crown. Nevertheless Henry had to do homage for it as a vassal of the King of France. Effective control of the territory fluctuated between France and England until the middle of the 15th century when all but the port of Calais had been recovered by France.

Arab conquests, wars which, in the century after the death of *Muhammad in 632, created an empire stretching from Spain to the Indus valley. Beginning as a *jihad (holy war) against the apostasy of the Arabian tribes that had renounced *Islam they acquired a momentum of their own as the Arabs, inspired by the

A 15th-century stained glass window of the theologian St Thomas **Aquinas** by Domenico Ghirlandaio in the church of Santa Maria Novella, Florence.

prospect of vast booty and the belief that death in battle would gain them instant admission to paradise, confronted the waning power of *Byzantium and *Persia.

In Syria and Egypt the conquerors allowed both Christians and Jews to keep their faiths as *dhimmi* (protected peoples) upon payment of a discriminatory tax. Local resistance in Persia and North Africa made the conquests there slower. After the first civil war (656–61) the Arab capital was moved from Medina to Damascus by the *Umayyads, and under the *Abbasids to the new city of Baghdad where, with the en-

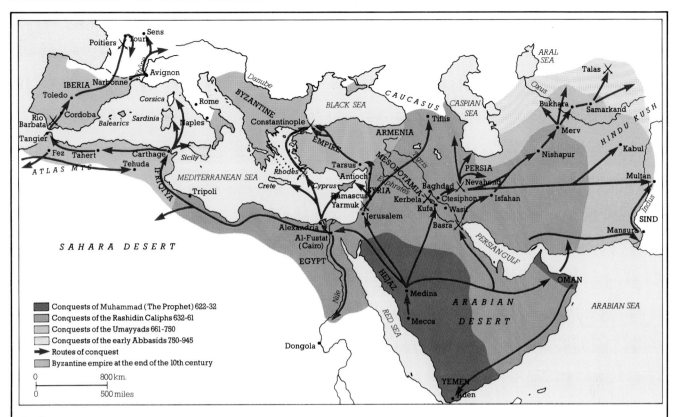

Arab conquests: the spread of Islam between the 7th and 10th centuries

The Arab conquests, fuelled by religious zeal and Arab expansionism, were facilitated by the relative weakness of the Byzantine and Persian empires. The main era of conquest lasted about a century, being checked in Europe by the Frankish victory at Poitiers (732). Byzantium held out against siege, and progress in the east was halted in the Indus valley. The Arab empire was never a unified state. Local dynasties established in the west were virtually independent by the 10th century, and the authority of the caliph declined after the death of the Harun al-Rashid (809).

Legend:
- Conquests of Muhammad (The Prophet) 622-32
- Conquests of the Rashidin Caliphs 632-61
- Conquests of the Umayyads 661-750
- Conquests of the early Abbasids 750-945
- Routes of conquest
- Byzantine empire at the end of the 10th century

0   800 km.
0   500 miles

couragement of the caliphs *Harun al-Rashid and al-Mamun, Islamic culture flowered. The political unity of this empire was short-lived—rival *caliphates appeared in North Africa and Spain in the 9th and 10th centuries —but cultural coherence was maintained by the universality of the Arabic language and Islamic law (*shariah*), and by the traffic of traders, scholars, and pilgrims which these made possible.

Aragon, formerly a kingdom, now a province in northern Spain. It became part of the Roman empire under Augustus, was taken by the Visigoths in the 5th century and by the Moors in the 8th century. In 1137 it was united with Catalonia by the marriage of the monarchs, and under James I acquired the Balearic Islands. During the 14th century Sardinia, Naples, and Sicily were added. In 1469 *Ferdinand II married *Isabella of Castile, uniting the kingdoms of Aragon and Castile.

Arakan, the western coastal strip of Burma, extending along the Bay of Bengal. It was formerly an independent kingdom, but in about 1600 its ruler, with help from Portuguese mercenaries, invaded Burma and a Portuguese adventurer established himself as 'king' on the Irrawaddy delta. In 1784 Bodawpaya, King of Burma, conquered Arakan, bringing the Burmese frontier up to Chittagong. This threat to British India was followed by the First Anglo-Burmese War (1824-6) and Arakan's cession to the British.

Arapaho, a North American Indian tribe who inhabited the hills bordering the northern Great Plains, along with the *Blackfoot and Gros Ventre. In the late 17th century, they changed from being agriculturalists and seasonal buffalo hunters to nomads on horseback following the great herds. Alongside the Kiowa and *Comanche they helped drive the *Apache off the Great Plains by the mid-18th century, and also fought the *Crow, *Pawnee, and Cheyenne as they moved on to the plains.

Arcadia, a mountainous region in the Peloponnese, Greece. It had no great political unity or strength in ancient times. Many of its communities were small and scattered, and its two leading cities, Tegea and Mantinea, were often at odds with each other. By the 480s BC it was a major supplier of mercenaries, but rose to political prominence only when Megalopolis was founded in 370-362 by the Theban general Epaminondas. This city led resistance to *Sparta, was well disposed towards *Macedonia, and in 235 joined the *Achaean League. Arcadia was identified in classical times and during the Renaissance (for example, in Sir Philip Sidney's *Arcadia*) as an earthly paradise.

archaeology, the study of the past of mankind, especially in the prehistoric period, and usually by excavation. Archaeological research includes four stages. The most obvious is recovery of material by excavation, chance find, surface survey, and observation from the

Archaeology

Archaeology is the study of the history and way of life of past cultures through the systematic and scientific excavation and analysis of remains in the ground. These remains may come from buildings, territorial boundaries, domestic, craft or industrial activities, or may merely be discarded debris. Location, excavation, interpretation, and restoration of archaeological remains involves many disciplines.

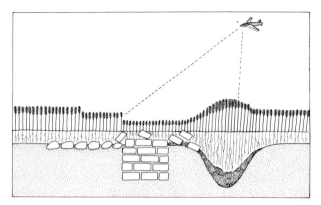

Archaeological artefacts in the ground may be caught by the plough and brought to the surface, or, if the ground is unploughed, features may survive as mounds or hollows, under a grass mantle. In this example, the crop will grow differently over the hidden cobbles, wall, and ditch. Under the right conditions, this pattern may be seen from the air, indicating the location of a site.

The process of excavating a site involves taking off the topsoil and cleaning, recording, drawing, and photographing the different sediments and features, together with any artefacts, such as pot sherds, bones, and metal objects that are found. Sediments may be sampled for microanalysis of their composition, and artefacts will be taken to laboratories for analysis and conservation.

The excavated data, including written records, plans, photographs, and artefacts are subsequently brought together to form an idea of what the site might have looked like in the past. In this example, the evidence of the excavations suggests that there was a wall and gateway, with cobbled roads, and ditches for rubbish.

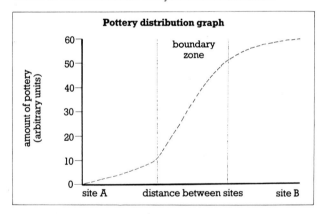

The data gathered at one site may be analysed within a wider cultural context. Here, the amount of a particular type of pottery found at two neighbouring contemporary sites varies dramatically, and increases rapidly in the area between the sites. This intermediate area is a cultural or trading boundary zone separating the sites.

air. Digging remains crucial because it alone can recover the precise context of finds, without which they lose much of their significance. It can take a wide variety of forms depending on the nature of the site—an isolated grave, a long-occupied cave, a wreck on the sea-bed, a standing building, a modern construction site, and many more. Then, finds have to be turned into evidence by analysis. Their form, composition, date, and associations all have information to impart. Typology (study of changes in forms) can link finds from different sites. A whole battery of scientific and mathematical aids can be brought to bear at this stage. Thirdly, the results have to be built into a coherent story to give an account of what happened when. Finally, and often the most difficult task, reasons must be sought for the processes of cultural change.

archer, a soldier armed with bow and arrows. Archers have practised their deadly skill since prehistory in most parts of the world, for example, the Romans employed *Scythian archers on horseback. In the Middle Ages the cumbrous but powerful crossbow was widely used in continental Europe, despite being forbidden against all save infidels by the Lateran Council of 1139. In England the potential of the longbow was discovered in the time of Edward I, but it was in Edward III's reign that full use was first made of it; nearly 2 m. (6 ft.) long, and made of yew, oak, or maple, it enabled accurate firing of arrows at a range of up to about 320 m. (350 yards), and it gave England such victories as *Crécy in 1346 and *Poitiers in 1356. Archery became the English national sport; Roger Ascham, tutor to the future

*Elizabeth I, published *Toxophilus*, a treatise on archery (1545). The musketeer superseded the archer in Europe from the later 16th century, but in 19th-century North America the Indians proved how devastating the mounted archer could be, even against men armed with rifles.

Areopagus, a council which met on the hill of that name in ancient *Athens. Drawn in the beginning from the richest class, the Eupatridae, it was originally an advisory body to the kings, but by the 7th century BC virtually ruled Athens. Its influence was still considerable in the early 5th century. Ephialtes' removal of its 'guardianship of the laws' in 462-1 marked the beginning of the radical *Athenian democracy. It continued to judge some criminal and religious cases, but power thereafter lay with the popular assembly and the lawcourts.

Argentina, a South American country, long considered a frontier outpost of the Spanish empire in America. Argentine colonists dedicated themselves to stock raising on the fertile pampas and agriculture in the areas of Salta, Jujuy, and Córdoba. Not until 1776 was the region given viceregal status. With its capital in Buenos Aires, the viceroyalty of La Plata comprised, in addition to Argentina, Uruguay, Paraguay, and Bolivia.

Argos, an important Peloponnesian city-state in ancient Greece. It reached a peak of power in the 7th century BC, possibly under King Pheidon. After his death it declined in influence, its position of supremacy in the Peloponnese being usurped by *Sparta. Argos remained neutral throughout the *Greek-Persian wars, and over the next century made unsuccessful attempts to reassert itself in the Peloponnese. It supported *Philip II of Macedonia and finally joined the *Achaean League.

Arianism, the teaching of Arius (AD 250-336), a Libyan priest living in Alexandria, who preached a Christian heresy. He declared *Jesus Christ was not divine, simply an exceptional human being. His teachings reached a wide audience, from the imperial household down to humble citizens. In 325 the Council of *Nicaea excommunicated and banished him. After *Constantine's death the Roman empire was divided on the issue, and another condemnation was issued at Constantinople in 381. Germanic invaders of the empire generally adopted Arianism as it was simpler than orthodox Christianity. It spread throughout western Europe and persisted in places until the 8th century.

Aristides (5th century BC), Athenian general and statesman. He fought at the battle of *Marathon, was exiled in 482, but was recalled in time to fight at *Salamis. He led the Athenian contingent at *Plataea. He assessed the levels of tribute to be paid by members of the *Delian League and enjoyed a leading role in the early development of the *Athenian empire. He was famed for being scrupulously fair—hence his nickname 'the Just'.

Aristotle (384-322 BC), one of the most celebrated Greek philosophers. At the age of 17 he joined Plato's *Academy, where he stayed until shortly after Plato's death in 348-7. He was later (343-2) appointed tutor to *Alexander the Great. In 335 he returned to Athens,

A Roman copy of a Greek statue of the philosopher **Aristotle**. Aristotle so impressed his tutor Plato when studying at his Academy that Plato called him 'the mind of the school'. Aristotle himself later opened a school in the Lyceum at Athens. (Galleria Spada, Rome)

where he established a school and a collection of manuscripts which was the model for later libraries. He organized research projects, the fruit of one being a comparative study of 158 Greek constitutions. Following the death of Alexander in 323, he was charged with impiety and left Athens, dying soon afterwards in Chalcis.

His output was enormous: dialogues which exist only in fragments; collections of historical information; the extant *Constitution of the Athenians* (though the authorship of this is now doubted); and scientific and philosophical works which are mostly extant, such as the *Nicomachaean Ethics*, the *Politics*, and the *Metaphysics*, some of which reveal the influence of Plato. His work was characterized by a love of order, which was manifested in his careful classification of the different areas of science. It was rediscovered by Arab scholars, and, translated into Latin, shaped the development of medieval thought in the arts and science. St Thomas *Aquinas reconciled the Aristotelian doctrines with those of Christian theology and they remained a key part of higher education in Europe from the 13th to the 17th centuries.

Army

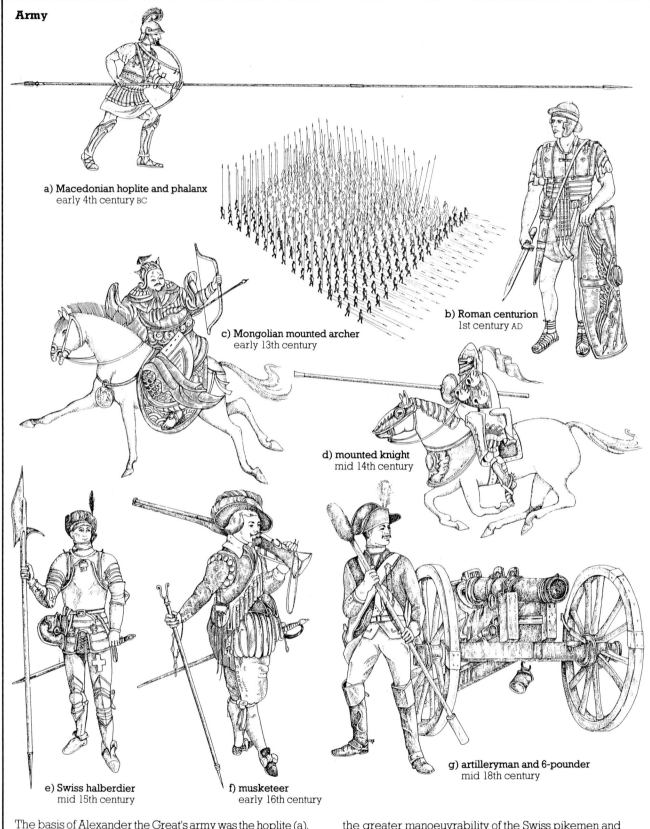

a) **Macedonian hoplite and phalanx**
early 4th century BC

b) **Roman centurion**
1st century AD

c) **Mongolian mounted archer**
early 13th century

d) **mounted knight**
mid 14th century

e) **Swiss halberdier**
mid 15th century

f) **musketeer**
early 16th century

g) **artilleryman and 6-pounder**
mid 18th century

The basis of Alexander the Great's army was the hoplite (a), organized into phalanxes and supported by other lightly armed troops and cavalry. The Roman foot-soldiers (b) were also backed by cavalry. Mongol archers (c) used three types of arrow, suitable for different ranges. Chain mail (d) evolved to full plate armour by the 15th century, but the greater manoeuvrability of the Swiss pikemen and halberdiers (e) eventually led to the demise of the armoured knights. The invention of gunpowder contributed to the replacement of the archers and the foot-soldiers by musketeers (f), and during the reign of Frederick the Great quite sophisticated artillery (g) was being used.

Armagnac, a region in Gascony in south-west France. It was disputed by the English and French kings from the 12th century and gained independence from frequent changes of allegiance. Under English suzerainty by the Treaty of Calais in 1360 it continued to change hands in the *Hundred Years War. The early 15th century saw a civil war fought between the Armagnacs against the Burgundians. The Armagnacs, led by their count, were supported by the nobility and the southern regions, while the Burgundians derived support in the north. In 1607 the whole region passed to the French crown.

Armenians, a people of Indo-European origin who entered eastern Anatolia in the 8th and 7th centuries BC. They were incorporated within various empires, from that of the *Achaemenids on, finally breaking free of the *Seleucids after the battle of Magnesia in 189 BC. Tigranes the Great (c.140–55 BC), who allied himself to *Mithridates, briefly carved out a sizeable empire, but in 66 BC he was defeated by *Pompey and became a client king of Rome. Armenia was subsequently a battle-ground for *Parthia and Rome, and was divided between the *Sassanids and *Byzantium in AD 387. It had been converted to Christianity and its church opposed the *Nestorians, but following the Council of Chalcedon in 451 it broke with Western Christianity, and in c.506 the Gregorian Church was established. The Arabs conquered the country c.653, but in 885 an indigenous dynasty, the Bagatids, gained control. This lasted until the Seljuk Turks invaded, though many Armenians fled to *Cilicia. Most of Armenia itself was ruled by the Ottomans from 1516.

Arminius (c.18 BC–AD 19), leader of the Germanic resistance to Roman colonization. Son of a noble family, he served as an officer in the Roman auxiliary forces and became a Roman citizen. But he turned against Rome, and in AD 9 annihilated Quinctilius Varus and his three legions, thereby wrecking *Augustus' German policy; and again in 16 he thwarted the attempt of *Tiberius' nephew Germanicus to renew the conquest. However, he failed to unite the fragmented Germanic tribes; in 19 his own aspirations to kingship encountered popular opposition, and he was murdered. He was a superb tactician and a master of the surprise attack. *Tacitus hailed him as 'the Liberator of Germany'.

Arminius, Jacobus (or Jakob Harmensen) (1560–1609), Dutch theologian, the founder of the theological movement known as Arminianism. He studied at Utrecht, Leiden, Basle, and Geneva before being ordained in 1588. The last six years of his life were spent as professor of theology at Leiden University, where he became involved in theological controversies with a colleague, Franciscus Gomarus. These centred on his rejection of the strict *Calvinist doctrine of absolute predestination or election and his arrival at a more liberal view of the relation between salvation and free will. Arminianism subsequently gave rise to the Dutch Remonstrant movement and in England it influenced Archbishop *Laud.

army, an organized force of men armed for fighting on land. Armies came into existence with the earliest states, and underpinned the great empires of antiquity: Egypt, Babylon, and Assyria. The essential components of armies in early history were infantry, with some chariots, and cavalry. In ancient Greece the tendency towards greater professionalism reached its climax with the Macedonian army of *Alexander the Great. From this time on, the development of siege techniques was an important part of military practice. The generals of Carthage, especially *Hannibal, hired mercenaries to great effect in their forces, but it was the armies of Rome, gradually evolving into fully professional standing forces, which dominated Europe from the 2nd century BC to the 5th century AD. Less organized but swiftly moving armies then came to the fore in the *Dark Ages, from those of *Attila the Hun to the *Mongols. In Europe in the Middle Ages the limitations of the heavily armoured mounted knight were finally exposed by Swiss infantry armed with pikes or halberds and English infantry armed with longbows. The use of mercenaries (*condottiere) became commonplace.

The major advances of the 15th and 16th centuries were the invention of gunpowder and the development of cannon. Organization, discipline, and further advances in weaponry led to the creation of highly efficient armies, most notably those of *Frederick the Great of Prussia. The first great conscript army was raised by *Napoleon.

Arnold, Benedict (1741–1801), American soldier and traitor. He was a hero of the early stages of the War of *Independence, serving with conspicuous valour at *Ticonderoga, the invasion of Canada, and *Saratoga. After 1778, possibly persuaded by his loyalist wife, he began plotting with *Clinton to deliver West Point to the British. When his courier, Major André, was captured, he fled to the British, for whom he fought thereafter. He died, neglected, in England.

Artaxerxes II (c.436–358 BC), King of Persia (404–358), the son of Darius II. He crushed the rebellion of his younger brother *Cyrus at Cunaxa in 401. By the peace of Antalcidas, made with the Spartans in 386, he recovered the Greek cities of Asia Minor, but he was unsuccessful in his attempts to repossess Egypt, and he put down the *satraps' revolt of 366–358 only with difficulty. His son, Artaxerxes III, killed his brothers and crushed two rebellious satraps in order to establish his power. In 343 he finally forced Egypt back into the empire, but his reign was one of terror and he was murdered by his minister Bagoas in 338.

Arthur, a legendary figure, supposedly King of the Britons, for whom there is probably some historical basis. Writers have portrayed Arthur as a powerful medieval king attracting to his court at Camelot an élite corps of knights (the 'Round Table') bound by the ideals of *chivalry and a semi-mystical form of Christianity. This romance has little connection with any historical Arthur. The writer *Nennius (9th century) claimed that Arthur commanded a mixed Roman–British force against the Saxons, whose raids on Britain increased with the departure of the Roman legions (c.450). According to Nennius, Arthur inflicted a major defeat on the *Saxons at Mount Badon (c.518), but was mortally wounded in a later battle at Camlan (c.537). *Gildas, writing before 547, mentioned Badon but did not connect Arthur with the victory, and neither Mount Badon nor Camlan has been identified. The great number of places in Britain associated with Arthur indicates that belief in his heroic deeds was widespread from the 6th century onwards.

The legends of King **Arthur** developed in western Europe from the 6th century onwards. By the 12th century Geoffrey of Monmouth's *Historia Regum Britanniae* related many of the famous elements of the story. This Italian manuscript illustration, 1370–80, shows the young Galahad's introduction to the knights of the Round Table. (Bibliothèque Nationale, Paris)

Aryans, ancient settlers in the Indian subcontinent, speakers of an Indo-European language. The Aryans, a people of uncertain racial origin, invaded India from Persia, gradually conquering the resident Dravidian and Munda peoples in the period 2000–1200 BC. The following millennium witnessed gradual eastwards expansion and progressive absorption of the indigenous population, as well as the evolution of the Vedic caste system, but the emerging hierarchical agrarian society remained split between warring petty states. In the late 4th century BC northern India was united for the first time by the *Mauryan dynasty, founded by Chandragupta, which ruled over an area stretching from Herat to the Ganges delta from its capital at Pataliputra, establishing an extensive administrative system and a regular army before being overthrown in the 2nd century BC by the Sunga dynasty. The upsurge of *Buddhism led in the 3rd century BC to the establishment of the short-lived empire of *Asoka which at the pinnacle of its strength included most of the subcontinent, but the west remained the stronghold of the original Aryan culture, and served as the home for the emerging *Hindu religion.

Asante (Ashanti), West African chiefdom. It emerged, under the Asantehene Osei Tutu in the 1670s, as a powerful kingdom, the Asante Confederacy, ruled by the Asantehene from Kumasi (now in Ghana). The wealth of the confederacy was based on the control of trade, particularly of cola nuts, and of gold mines, and by selling slaves for European goods to the European trading stations established along the Gold Coast of West Africa.

Ascham, Roger (1515–68), English scholar, an influential exponent of the *New Learning. From 1540 he was reader in Greek at St John's College, Cambridge and in 1545 he published a popular treatise on archery, *Toxophilus*, which earned him a royal pension. He was tutor to the future *Elizabeth I for two years and subsequently engaged in diplomatic missions for *Mary I and Elizabeth I. *The Scholemaster* (published in 1570) advocated an enlightened approach to education including the teaching of the English language as well as the classics.

Ashikaga (also called Muromachi, from the district in Kyoto where, after 1392, the shoguns lived), the *shogunate in Japan from 1339 to 1573. In 1333 Ashikaga Takanju (1305–58) overthrew the *Hojo, who had acted as regents for the *Kamakura shoguns. Soon after he drove the emperor Go-Daigo from the capital of Kyoto. In the ensuing dynastic dispute he installed Koyo as emperor, and in return the emperor appointed him shogun in succession to the Kamakura shogunate, which had become ineffectual. He moved the shogunate from Kamakura to Kyoto. The Ashikaga shoguns never exercised great power as the shogunate witnessed much fighting between rival *daimyo and their *samurai armies. The increasing disorder of the Ashikaga shogunate ended in 1573 when *Oda Nobunaga and his army drove the shogun from Kyoto.

Ashley Cooper, Anthony *Shaftesbury.

Ashurbanipal, ruler of the *Assyrian empire (669–626 BC). His reign was the apogee of Assyrian power, although it collapsed soon after his death. Early in his reign, Egypt

An **Asante** ceremonial helmet, indicative of the wealth of this West African chiefdom. Mounted with a pair of golden horns, the helmet is made of antelope skin and is decorated with golden jaws and trophy heads. (Museum of Mankind, London)

rebelled successfully and broke away, but when one of his half-brothers attempted the same thing in Babylon several years later, he crushed him. Although he campaigned vigorously in other parts of his empire too, he was no mere warrior-king. A man of culture, and a patron of the arts, he assembled a large and wide-ranging library in Nineveh.

asiento de negros, a contract made between Britain and Spain in 1713 for the sale of slaves to the Spanish American colonies. In the Peace of *Utrecht (1713) Spain granted Britain a monopoly of the supply of slaves to the Spanish American colonies of 144,000 slaves at 4,800 a year for thirty years, with other privileges. They were the origin of the speculation which resulted in the *South Sea Bubble. They led to endless disputes, and to the War of *Jenkins's Ear between England and Spain in 1739. The treaty was ended by agreement in 1750.

Askia Muhammad I (d. 1528), Emperor of *Songhay (1493–1528) in West Africa. Originally named Muhammad Turé, he was *Sonni Ali's best general. He usurped Songhay from Ali's son in 1493, thus founding a new dynasty, and took the title Askia. He was a convert to Islam, but tolerant towards pagans, and made the pilgrimage to *Mecca, meeting many notable men, especially the great Muslim teacher al-Maghili. He had close political and commercial relationships with Morocco and Egypt, organized an efficient administration, and an army and a navy on the River Niger, and made *Timbuktu the capital of the Songhay empire and an important intellectual and religious centre.

Asoka, the last great *Mauryan Emperor of India (ruled c.265–c.238 BC). He is remembered chiefly for his patronage of Buddhism and his high ethical standards as a ruler. Adoption of the Buddhist *dharma* (teaching on religious truth) led him to an overriding concern for the spiritual and material welfare of his subjects, and to toleration of other religions.

He inherited an empire which already extended over the entire subcontinent except the extreme south. After one successful campaign he renounced warfare because of the suffering it generated. Although his empire disintegrated soon after his death, he is regarded as one of the greatest of the subcontinent's early rulers. Knowledge of his empire derives mainly from inscriptions carved on rocks and pillars in far-flung parts of the subcontinent and indicate that he was an efficient ruler who strengthened and humanized a remarkable administrative system, incorporating a standing army of 700,000, a widespread secret service, and a large bureaucracy. Asoka's lasting influence on India is embodied in the country's adoption of his Sarnath lion capital as one of its national emblems.

assassin (from the Arabic *hashishiyun*, 'smoker of hashish'), a member of a secret sect of the *Ismaili branch of Shiite Islam. It was founded by Hasan ibn al-Sabbah in 1078 to support the claim of Nizar to the *Fatimid caliphate, and established a headquarters at Alamut in north-west Persia. The assassins wielded influence through suicide squads of political murderers, confident of earning a place in paradise if they died while obeying orders. The *Mongol Hulagu took Alamut in 1256, executing the grand master of the order. Their last Syrian strongholds fell to the *Mameluke Baybars in 1273. (The

The Sarnath lion capital executed towards the end of **Asoka**'s reign. The four lions standing back to back on the pedestal once supported a stone wheel. On the sides of the pedestal are carvings of a horse, a bull, an elephant, and a lion. The column was erected to mark the spot where the Buddha publicly preached his doctrine.

widely scattered Nizari branch of the Ismailis, who revere the Aga Khan, are their spiritual descendants.)

assizes, a procedure introduced into English law in the later 12th century by *Henry II. The Assize of Clarendon (1166), which dealt with criminal trials and the Assize of Arms (1181), which reorganized local defence and police measures, were enactments made at sessions of the king's council. The assizes of novel disseisin and mort d'ancestor (both relating to tenancy), and the Grand Assize (to determine titles to disputed lands) were introduced by sessions of Henry II's council (1166, 1176, and the late 1170s); these procedures remained important throughout the Middle Ages.

Travelling justices were established in the 13th century; these justices came to be called justices of assize, and their sessions were called assizes. A system of such judicial sessions was regularized (1293–1328) and judicial circuits were established that remained in force until a new system of Crown Courts was set up in 1971.

Assyria, an area of northern Mesopotamia centred on the city of Ashur (or Assur), which became a Semitic state. The first Assyrian empire was established early in the 2nd millennium BC, and many documents discovered in Anatolia attest to vigorous commercial intercourse with that part of the world. Shamshi-Adad I (ruled c. 1813-1781 BC) brought Mesopotamia under his control, but after his death his empire collapsed. It was attacked by *Hammurabi of *Babylon and then fell to the Mitanni, a people from the west. Assyria re-emerged as a political power under Ashur-uballit I (ruled c.1362-1327) and his successors. The Mitanni were conquered, northern Mesopotamia was secured, and under Tukulti-Ninurta I (ruled 1242-1206) Babylon was captured. Following this king's death Assyrian fortunes declined until Tiglath-Pileser I (ruled c.1114-1076) revived them, although the pressure of Aramaean nomads migrating from the east posed a threat to the stability of the Assyrian civilization. The years 911 to 824 saw Assyrian expansion, with the empire extending to the Mediterranean coast. Iron was extensively traded and was the main source of wealth. The peak of Assyrian power and civilization began with the reign of Tiglath-Pileser III (ruled 744-727), who reconquered Babylon but allowed it to retain limited autonomy. This policy did not ensure peace, and Babylon was destroyed in 689 by *Sennacherib who made Nineveh his capital. His son Esarhaddon even conquered Egypt, which he ruled through native princes, but the Egyptians rebelled against his successor *Ashurbanipal, who suffered other revolts which further weakened his empire. Finally Nabopolassar, a Chaldean, took control of Babylon in 625 BC, and under attack from him and the Medes Assyrian power collapsed. The

A relief depicting Ashurbanipal's siege of Hamanu, from the North Palace at Nineveh, the city which became the capital of **Assyria** after the destruction of Babylon in 689. (British Museum, London)

Assyrians were famed as ruthless soldiers, armed with iron weapons, and were sophisticated not only in military technique, but also in administration, art, and architecture.

Atahualpa (d.1533), the last ruler of the *Inca empire, son of Huayna Capac. Ruling from 1525 in *Quito, he defeated Huáscar, his half-brother and co-ruler in *Cuzco, whom he killed after the battle of Huancavelica in 1530. In 1532 he marched against *Pizarro and remnants of the Huáscar faction, who had allied themselves to the Spaniards; at Cajamarca he was drawn into an ambush, captured, and held for ransom. He ordered a room to be filled with gold and silver objects while another army secretly marched to free him, but was murdered when Pizarro learned of it. Shortly thereafter Pizarro captured Cuzco and within a few years Spain ruled the lands of the Incas.

Athanasius, St (c.295-373), Bishop of Alexandria. He was a vehement opponent of *Arianism and refused *Constantine's request to restore the excommunicated Arius and was subsequently himself exiled five times by Constantine and later emperors. His theological position was confirmed at the first Council of Constantinople in 381 when Arianism was finally prohibited and the Nicene Creed, a statement of Christian belief, was approved.

Athelstan (895-939), King of the English (926-39). As King of Wessex from 925 he established Wessex's supremacy throughout England, Wales, and southern Scotland, and defended it in an overwhelming defeat of a combined force of Scots and Danes at an unidentified place called Brunanburh (937) as recorded in the *Anglo-Saxon Chronicle. Athelstan's fame was not due solely to his military exploits. He provided sound government, reformed the coinage, granted charters to towns, and, in a century of legal reform, issued six series of laws.

Athenian democracy, a form of popular government established in Athens by Cleisthenes in the last decade of the 6th century BC. At first, the *Areopagus retained considerable influence, during the *Greek-Persian wars, and it was only after Ephialtes stripped it of its powers in 462 that a more radical democracy came into being. Ephialtes died suddenly in mysterious circumstances but *Pericles pushed through further reforms, establishing in particular the important principle of pay for jury service.

The principal organ of democracy was the popular assembly (*ekklesia*), which was open to all Athenian male citizens aged over 18. All members had the right to speak, and it was the assembly which decided all legislative and policy matters. The council of 500 (*boule*), elected by lot for a year from Athenian male citizens over the age of 30, was an executive body which prepared business for the assembly and then saw that its decisions were carried out. Pericles dominated the democracy until his death in 429, but none of the 'demagogues' who followed him achieved the same level of influence. Radical democracy had its dangers, however: established laws were sometimes overridden by the assembly, and skilled orators could easily manipulate listeners' emotions.

Athenian empire, the cities and islands mainly in the Aegean area that paid tribute to Athens in the 5th century BC. It developed out of the *Delian League as

Athens, by virtue of its great naval superiority, imposed its will on its allies. A significant step was the transference of the League's treasury from Delos to Athens probably in 454 BC, since this ensured for Athens absolute control of the tribute. Inscriptions and literary sources reveal the means by which Athens controlled its subjects: the installation of garrisons; the establishment of clenruchies (colonies) of Athenian citizens in important or rebellious areas; the encouragement of local democracies; the referral of important judicial cases to Athens; the imposition of Athenian weights and measures throughout the empire, and officials to keep an eye on subject cities.

As long as it had a strong navy, Athens could crush revolts and enforce its will throughout the Aegean, but the empire died with Athens' final defeat in the *Peloponnesian War. Nevertheless it did establish the Second Athenian Confederacy in 377 BC, trying to avoid the mistakes of the 5th century.

Athens, the capital of modern Greece, historically an ancient Greek city-state. It was formed as a result of the unification of a number of small villages of *Attica. It was first under the rule of hereditary kings, and monarchy was followed by a long-lived aristocracy, first successfully challenged by *Solon in 594 BC. Tyranny was established by *Pisistratus, temporarily in 561 and more permanently in 546, until his son Hippias was driven out in 510. Within a few years Cleisthenes had put the *Athenian democracy on to a firm footing.

In 490 and 480-479 the city-state enjoyed success in the *Greek–Persian wars. Subsequently its rulers transformed the *Delian League into the *Athenian empire. The city supported brilliant artistic activity, attracting artists from throughout the Mediterranean. However, it was defeated by Sparta in the *Peloponnesian War, losing by 404 the empire, almost all its fleet, and the city walls. It recovered remarkably in the 4th century and led the resistance to *Philip II of Macedonia. The city was a centre of philosophy, science, and the arts, centred on the *Academy.

Athens was prey to the successors of Alexander the Great, losing its independence in 262, though regaining it in 228. After supporting *Mithridates, King of Pontus, (120-63 BC) against Rome, it was successfully besieged by his antagonist, *Sulla, and sacked (87-86). From then on its importance was as a university town which attracted many young men, particularly Romans. This apart, the city underwent a prolonged period of historical obscurity and economic decline. It was captured by the Turks in 1456, and suffered during the Venetian siege of 1687.

Atlantis, according to a myth told by *Plato, a large island lying in the Atlantic Ocean, west of the mountains in north-west Africa, where Atlas supposedly supported the heavens. It was said to have disappeared beneath the sea, which has led some to equate it with the *Minoan civilization of Crete, which they believe was destroyed by the eruption of the volcano of Thera (modern Santorini) c.1500 BC. Classicists generally give no credence to the Platonic myth of Atlantis, though archaeologists continue to search for evidence to support the legend.

attainder, the extinction of civil rights and powers when judgement of death or outlawry was recorded against a person convicted of treason or felony. It was the severest English common law penalty, for an attainted person lost all his goods and lands to the crown. Procedure by Act of Attainder became common in the Wars of the *Roses, when because it was reversible it could be used as a powerful threat. Of the 397 people condemned by process in Parliament between 1453 and 1509, over 250 ultimately had their attainders reversed. Acts of Attainder came to be disapproved of because an opportunity for defence was not necessarily given; they became rare in the 18th century and ceased after 1798.

Attica, a region comprising the south-eastern portion of central Greece. In ancient times it contained a number of small villages and towns and these gradually were united politically into the city-state of *Athens, the process being completed by the 7th century BC. Major land-owning families continued to live outside the city, though in the time of the *Peloponnesian War (431-404 BC) the countryside was abandoned temporarily to the depredations of the invading Spartans. Attica was rich in natural resources, notably clay for a thriving pottery industry, marble, lead, and the silver which financed the Athenian navy.

Attila (c.406-53), King of the *Huns. He became king in 434 jointly with his brother Bleda, whom he murdered in 445. He forcibly united the Hun tribes into a vast horde based in Hungary; they raided from the Rhine to the Caspian Sea, ravaging the divided *Roman empire, and exacting tribute as the price of peace treaties. The historian Priscus has left a vivid description of the squat and wily Mongoloid conqueror and his court. Invading Gaul in 451, the 'Scourge of God' met his only defeat on the *Catalaunian Fields near Châlons at the hands of combined Roman, Frankish, and Visigothic forces. He then turned on Italy but spared Rome, perhaps in response to pleas or ransom from Pope Leo I. His sudden death was followed by the collapse of his dominion.

Augsburg, a city in Bavaria, from 1276 one of the great free imperial cities of Germany. It was an important member of the *Swabian League and was famous in the 16th century for its merchant princes, the *Fuggers and Welsers. It was often chosen by *Charles V as the venue for negotiations which might lead to religious unity. In 1530 the **Confession of Augsburg** was drawn up by *Melanchthon; in moderate language it gave a statement of essential *Lutheran doctrines, such as justification by faith. It was not accepted by the Roman Catholics but remains the chief standard of faith in the Lutheran churches. In 1548 the **Interim of Augsburg**, a Catholic statement with modest concessions to the Lutherans, was not accepted in Protestant areas. In 1555 the **Peace of Augsburg** was concluded by the new emperor Ferdinand and the *Electors: Catholicism and Lutheranism (but not *Calvinism) were recognized and each prince could impose the faith of his choice on his territories (the principle of *cuius regio, eius religio*, the ruler may dictate the religion); in free cities such as Augsburg both faiths could be practised; all ecclesiastical land already obtained by Protestants could be retained, but in future ecclesiastical princes who were converted had to give up the Church lands in their possession. This principle was often not observed by Protestant princes and was applied sporadically.

In 1686 the defensive **League of Augsburg** was formed by Emperor Leopold and some German princes to resist *Louis XIV's advance into the Rhineland; it was joined by the Holy Roman Emperor, the Dutch, Spain, and Sweden. After the French invaded the *Palatinate and *William III became King of England a new Grand Alliance was formed; the ensuing *Nine Years War is also known as the War of the League of Augsburg or the War of the Grand Alliance.

augur, an official diviner or soothsayer in ancient Rome, usually of noble birth. The augur's task was to watch for indications of the attitude of the gods towards proposed activities of the state or its officers. They played a key role in choosing or 'inaugurating' successive non-hereditary kings of early Rome. The name was thought to be linked with birds (Latin, *aves*), since they scrutinized the activities of birds, besides other animals, as well as accidents, and dreams, particularly on the eve of military expeditions and at the moment of important births. All political assemblies of the Roman people were preceded by the taking of 'auspices'. Ancient Roman society set much store by omens: after an augur warned Caesar against appearing publicly on the 'Ides of March' (15 March), a wren was seen being torn to pieces by other birds near the spot where Caesar was about to die.

Augustine of Canterbury, St (d.*c.*605), the first Archbishop of Canterbury. He was chosen (596) by Pope *Gregory the Great to convert the English to Christianity. With forty monks Augustine came first to Kent (597) and converted King Ethelbert, whose wife was already a Christian. Consecrated archbishop (597), Augustine organized the church into twelve dioceses (598) but failed at a meeting with the Celtic bishops in 603 to resolve the differences between the Roman and Celtic churches, although these differences were resolved at the Synod of *Whitby (664). Augustine's work was instrumental in the re-establishment of Christianity in England.

Augustine of Hippo, St (AD 354-430), Christian bishop, one of the outstanding theologians of the early Christian church. Born in North Africa of a pagan father and a Christian mother, he was early attracted to

In Rome the duties of an **augur** included the examination of animal entrails in order to determine the likelihood of a favourable outcome to any major proceeding. Such an examination (of a bull) is depicted in this Roman relief. (Musée du Louvre, Paris)

*Manichaeism, which he later rejected. He taught rhetoric in Rome and Milan, where he was influenced by Bishop *Ambrose. Augustine lived a monastic life for some time, then becoming a priest and (after 395) Bishop of Hippo in North Africa. His episcopate there was marked by controversy with followers of heretical Christian sects (Manichees, Donatists, and Pelagians) and of pagan philosophies. He was an upholder of order in a time of political strife caused by the disintegration of the *Roman empire and he died as the *Vandals reached Hippo.

De Civitate Dei ('The City of God'), a vindication of the Church against paganism, is perhaps his most important work, apart from his *Confessions* which contains a striking account of his early life and conversion. His writings reached a wide audience after his death and were known in Anglo-Saxon England in the reign of King *Alfred. His theology influenced much subsequent Christian theology; the 'rule of St Augustine' (*Augustinian) was based on his writings.

Augustinian, a member of one of the religious orders following the rule or code of conduct laid down by St *Augustine of Hippo. Augustine's rule proved practical and adaptable to changing conditions over many centuries and was used by both St *Dominic and St *Francis as a model upon which to construct constitutions for their own orders. Essentially it required men to live a communal life apart from the world but allowed for involvement in missionary work and care of the sick. The rule was endorsed and promoted by the Fourth *Lateran Council of 1215.

Augustus (Gaius Julius Caesar Octavianus) (63 BC–AD 14), first Roman emperor (27 BC–AD 14). Born as Octavian into the *Caesar family, he was Julius Caesar's nephew and was adopted by Caesar in his will and entered the power struggle after Caesar's death in 44 BC. Despite *Mark Antony's opposition, he gained the consulship and soon after joined Antony and Lepidus in the 'Second Triumvirate', an alliance of three dictators. Their republican opponents, *Brutus and Cassius, were defeated (battle of *Philippi) and Antony and Octavian ruled the empire between them, Antony the eastern part, and Octavian the western territories. The battle of *Actium in 31 BC gave Octavian victory, and, after Antony's suicide, total supremacy. In 28 BC at the request of the *Senate, he took over responsibility for Rome's military provinces and their garrisons. His pre-eminence was reflected in the title of 'Augustus' bestowed upon him, which later became synonymous with 'emperor'. His control of the army was the basis of his power and his generals were victorious in Asia, Spain, Pannonia, Dalmatia, and Gaul. The only reversal was defeat by the Germans under *Arminius.

He enjoyed formidable power, but although an absolute ruler in all but name, he was careful to preserve the institutions of republican government. His achievement was the creation of the Principate, a system of stable and effectively monarchic government, after half a century of strife. An efficient administration was formed out of the old governing class. The 'Augustan Age' brought security and prosperity to the Roman empire. His court circle provided patronage for writers such as Virgil, Horace, and Livy, and he adorned Rome and other cities with beautiful buildings.

Augustus II (the Strong) (1670-1733), King of Poland (1696-1733). He was Elector of Saxony from 1694 and succeeded John III (John *Sobieski) as King of Poland in 1696. He joined Russia and Denmark against *Charles XII of Sweden without Polish support but was defeated. Charles had him banished and Stanislaus Leszczynski elected king in his place. Augustus recovered his position after Charles's defeat at *Poltava (1709) and for the rest of his reign brought some economic prosperity to Saxony and Poland, although renewed war with Sweden lasted until 1718. A ruler of considerable extravagance, supposed to be the most dissolute monarch in Europe, he was a patron of the arts and gave special support to the Dresden and Meissen china factories.

Aurangzeb (1618-1707), known also as Alamgir ('world-holder'), Mogul Emperor of India (1659-1707). He was the last great ruler in his line, though the signs of subsequent decline first became apparent during his reign. After eliminating his brothers, he seized the throne from his father, and then pushed the boundaries of Mogul India to their fullest extent by defeating the Muslim rulers of Bijapur and Golconda. However, his obsession with expansion into the *Deccan severely overextended his resources. Desiring to reassert Muslim orthodoxy, he reversed the conciliatory gestures, notably in taxation, which had won Hindu support, particularly since *Akbar's reign. The result was intensified opposition to Mogul rule among newly assertive groups such as the *Marathas and the *Sikhs. His achievements and failures remain controversial.

Aurelian (Lucius Domitius Aurelianus) (AD 215-75), Roman emperor (270-5). After a career as a cavalry commander, he was adopted as co-emperor by Claudius II, who died that year. Under Aurelian's rule the rivers Rhine and Danube were temporarily regained as the empire's natural frontiers against the barbarian Alemanni and *Goths. Internal threats from the Gallic usurper Tetricus (defeated at Châlons in 274) and the kingdom of Palmyra in Syria were suppressed. Aurelian was murdered while campaigning against the Persians.

Aurignacian *Upper Palaeolithic.

Australoids, the indigenous (aboriginal) peoples of Australia and Tasmania. The ancestors of modern Australoids could have been among the first seafarers, as their settlement of Australia, probably 50,000-40,000 years ago, involved a sea crossing of 50-100 km. (30-60 miles). By 20,000 years ago the population ranged from New Guinea in the north to the tip of western Australia. The earliest known fossil evidence of human occupation comes from Lake Mungo in western New South Wales where remains of anatomically modern *Homo sapiens* dating between 30,000 and 28,000 years ago have been found. Archaeological sites, with stone tools and hearths dating from 40,000 years ago, are also known in Australia.

There are several explanations as to who were the original colonists. One suggestion is that there were waves of migration by different groups, who subsequently interbred to produce the modern Australoid physique. Another possibility is that there was only one source—Indonesia. For thousands of years the Australian Aborigines were isolated from outside contact. Their tools were relatively simple, though they had multiple uses, but their sacred rituals and social organization were very sophisticated. This richness of Aboriginal culture in the past is only now becoming appreciated. When Europeans first reached Australia about 200 years ago, there were at least 300,000 Aborigines in the continent.

australopithecines, early members of the human line of evolution. The first australopithecine fossil was discovered at Taung in southern Africa in 1924 and named *Australopithecus africanus* ('southern ape of Africa'). Since then, australopithecine fossils have been found in southern and eastern Africa but the relationships between the different forms is still far from clear.

Current opinion divides them into two, or perhaps three main groups that date from over 5 million to nearly 1 million years ago. The oldest (5-3 million years ago) and most ape-like is *Australopithecus afarensis*, now known from eastern African sites including *Hadar and *Laetoli. This species is often linked closely to *Australopithecus africanus* (3-2 million years ago), best represented at *Sterkfontein and Makapansgat in southern Africa. Some authorities consider both these species are human ancestors; others rule out *A. africanus*, some even discount both from being human ancestors. These lightly built australopithecines are described as 'gracile' to distinguish them from the more heavily built, 'robust' forms—*Australopithecus robustus* (2-1.5 million years ago) from southern Africa (for example from Swartkrans), and the even heavier *Australopithecus boisei* (2.5-1.2 million years ago) from eastern Africa. These robust forms are widely regarded as off the direct human lineage. Australopithecines were clearly capable of walking upright but their brains were still ape-like. It is uncertain if they made tools.

Austrian Succession, War of the (1740-48). A complicated European conflict in which the key issue was the right of *Maria Theresa of Austria to succeed to the lands of her father, Emperor Charles VI, and that of her husband Francis of Lorraine to the imperial title. Francis's claims (in spite of the *Pragmatic Sanction) were disputed by Charles Albert, Elector of Bavaria, supported by Frederick II of Prussia and Louis XV of France. Additionally Philip V of Spain and Maria Theresa were in dispute over who should have control of Italy, and Britain was challenging France and Spain's domination of the Mediterranean (War of *Jenkins's Ear), and fighting for control of India and America (*King George's War).

After the death of Charles VI in 1740 war was precipitated by Frederick II of Prussia, who seized Silesia. The war began badly for Austria: the French seized Prague, a Spanish army landed in north Italy, Charles Albert was elected Holy Roman Emperor, and Silesia was ceded by treaty to Frederick II in 1742. Britain now supported Austria by organizing the so-called Pragmatic Army (Britain, Austria, Hanover, and Hesse) and under the personal command of George II it defeated the French at *Dettingen in 1743. Savoy joined Austria and Britain (Treaty of Worms, September 1743) and the tide of war began to turn in Austria's favour. In 1744-5 Frederick II re-entered the war, determined to retain Silesia. Meanwhile Charles Albert died and Francis was elected Holy Roman Emperor in exchange for the return of the lands of Bavaria to the Elector's heir. Frederick II won a series of victories against Austria, and the Treaty of Dresden (1745) confirmed his possession of Silesia.

A painting by van Blarenberghe of the battle of Fontenoy (1745) in the War of the **Austrian Succession**. Fontenoy was a bloody battle with heavy losses on both sides. The painting shows the massed infantry formations which were a feature of 18th-century engagements. (Musée de Versailles)

The struggle between France and Britain intensified. The French supported the Jacobite invasion of Britain (the *Forty-Five) and in India the French captured the British town of Madras (1746). The British won major victories at sea: off Cape Finisterre, Spain and Belle-Ile, France in 1747.

By 1748 all participants were ready for peace, which was concluded at *Aix-la-Chapelle. The war had been a long and costly effort by Maria Theresa to keep her Habsburg inheritance intact and in this she largely succeeded. But Austria was weakened and Prussia, which held Silesia, consolidated its position as a significant European power.

Avars, Central Asian nomads whose irruption into eastern Europe in 568 led to the creation of a large empire centred on the Danube valley. They drove the southern Slavs into the Balkans and the Germanic Lombards into northern Italy, thereby depriving *Byzantium of its Latin-speaking subjects and confirming its Greek character. Their empire was destroyed by Charlemagne in 791, and the resulting rebellions of their subject peoples led to their complete extermination.

Averroës (Ibn Rushd) (1126–98), Islamic philosopher. He was born in Córdoba in Muslim Spain. He had a comprehensive knowledge of the religious and philosophical works available in his day, and his commentaries on Aristotle greatly influenced the Christian *scholastics.

He came to believe that religion and philosophy both embody truths but are none the less separate and may appear contradictory. His work had relatively little effect on later Muslim thought but persisted in Christian circles in spite of its condemnation by the Bishop of Paris in 1270. It remained influential in Italy until the *Renaissance.

Axum (or Aksum), a town in northern Ethiopia. It was the capital of a trading kingdom between the 3rd and 1st centuries BC, selling ivory, gums, and spices at the port of Adulis, chiefly to Egyptian merchants. In the 4th century AD King Ezana minted its first gold coinage, and signed a treaty with Byzantium. The country was then slowly converted to Christianity. To this time belong the huge obelisk tombs for which Axum is famous. In the 6th century Axum controlled western Arabia for a short time, but lost it to the rising power of Persia. By 1000 the kingdom had collapsed.

Ayuthia *Siam.

Azores, nine islands in the Atlantic Ocean belonging to Portugal. They were discovered in the 14th century. They were uninhabited, and the first colonists came from Portugal and Flanders and the Portuguese crown took control after 1494. In 1581 the battle of Salga between the Spanish and Portuguese fleets took place off Terceira Island, and in 1591 Sir Richard *Grenville, in the *Revenge*, fought the entire Spanish fleet in a battle which immortalized his name.

Aztec, the Central American Indian people anciently known as the *México* (whence Mexico) or *Tenochea* (whence *Tenochtitlán, their capital). Like their *Toltec predecessors, they originated somewhere in northern

Mexico. During the early 13th century they migrated into the valley of Mexico and eventually settled on an island in Lake Texcoco (later mostly drained by the Spaniards), after the instructions of their gods to settle where they witnessed an eagle sitting on a cactus devouring a snake. From this base they formed an alliance with two other cities—Texcoco and Tlacopán—against Atzcapotzalco, defeated it, and proceeded through the 15th century to conquer the other cities of the valley, then to carve out an empire stretching from coast to coast, north to the deserts, and south to the *Maya kingdoms of *Yucatán, by the early 16th century.

Three things characterized their culture: their religion which demanded large-scale human sacrifices particularly to their god of war, Huitzilopochtli; their efficient use of *chinampas* (artificial garden islands built in regular grids out into the lake with canals dividing them) to feed their vast population; and their widespread trading network and system of tribute administration.

However, many of their subjects had only recently been subdued, and, eager to lose their tributary status, became allies of *Cortés in 1519. The Aztec ruler *Montezuma considered that the Spaniards were descendants of the god-king *Quetzalcóatl, and did not appreciate the danger his kingdom faced. He received Cortés, who subsequently held him as a hostage. In 1520 there was an Aztec revolt and Montezuma was killed. His successor, Cuauhtémoc, the last Aztec ruler, resisted the invaders, but in 1521, Cortés captured Tenochtitlán and subdued the empire.

The Aztecs

The Aztec empire was controlled from Tenochtitlán, to which conquered tribes delivered large amounts of tribute. Engineering, astronomy, agriculture, and the arts were all highly developed by the Aztecs, despite their lack of either the wheel or a written language. Religion was central to their culture, which was particularly influenced by the belief that the god Huitzilopochtli required constant nourishment by human blood. War was endemic as a means of providing a supply of prisoners for this rite.

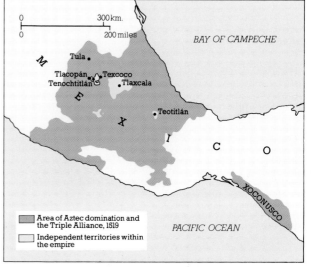

Quetzalcóatl
the 'plumed serpent god', and a legendary god-king

Tonacatecutli
'Lord of our flesh', one of many Aztec gods

Having settled at Tenochtitlán as mercenaries of the dominant Tepanecs, the Aztecs joined with Texcoco and Tlacopán and made war on their former patrons, subjugating central Mexico.

Chalchithuitlicue
goddess of storms and fresh water

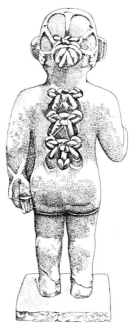

Xipe Totec
'the flayed god', a god of regeneration and spring growth, the subject of a grisly cult in which sacrificial victims were flayed and their skins worn by priests. In this statuette of the god, the gaping mouth (above) represents the stretching of the flayed skin, with the skin of the victim's body sewn up neatly at the back (left).

B

Babeuf, François Noël (1760–97), French Revolutionary who called himself 'Caius-*Gracchus, tribune of the people'. A domestic servant before the French Revolution, he moved to Paris in 1794. There he started to publish the *Journal de la liberté de la presse*, in which he argued that the Revolution should go further than establishing political equality. He formed a small group (the Equals) of discontented artisans and soldiers and campaigned for the equal ownership of property by all. This idea thrived in the turmoil following *Robespierre's execution but secret agents learnt of his plans for an armed rising on 11 May 1796. He was captured and was executed and his followers were executed or deported. His conspiracy influenced later revolutionary socialists, particularly in the 1830s.

Babington Plot (1586), a conspiracy to co-ordinate a Spanish invasion of England with a rising of English Catholics, to assassinate *Elizabeth I, and to replace her

A miniature painting of **Babur**, c.1590 by Bishan Das, with the faces by Nanha. It shows the Mogul emperor making the Garden of Fidelity at Kabul, which he seized in 1504 at the start of his rise to power. A lover of nature, Babur constructed gardens wherever he went. (Victoria and Albert Museum, London)

on the throne with *Mary, Queen of Scots. Sir Anthony Babington (1561–86) was the go-between in the secret preparations. *Walsingham monitored Babington's correspondence with the captive Queen Mary until he had enough evidence of her treasonable intentions to have her tried and executed in 1587, Babington having been executed after torture at Tyburn.

Babur (1483–1530), the first *Mogul Emperor of India (1526–30). He was born in Ferghana, Central Asia, in a princely family of mixed Mongol and Turkish blood. Failure to recover his father's lands caused him to turn reluctantly south-east, for India seemed to present the last hope for his ambitions. Defeat of Ibrahim Lodi, the Afghan ruler of Delhi, at the battle of *Panipat in 1526 initiated 200 years of strong Mogul rule in India. Having conquered much of northern India, Babur ruled by force, lacking any civil administration. In addition to his military genius, he possessed a love of learning and wrote his own memoirs.

Babylon, an ancient city in Mesopotamia on the River Euphrates c.90 km. (55 miles) south of modern Baghdad. It was twice the centre of a major empire. The first was established by the Amorite Sumuabum in 1894 BC and augmented by his successors, above all *Hammurabi. Following a Hittite attack c.1595, it fell under the rule of the Kassites (another Indo-European people) c.1570 for over 400 years (although the *Assyrians captured it briefly in 1234), before being looted by the Elamites (also from western Persia) in 1158. They were in turn defeated by Nebuchadnezzar I (ruled 1124–1103) who established a dynasty that lasted into the early 10th century. Thereafter Babylon came much more under Assyrian influence, and *Sennacherib, despairing of controlling the local tribesmen, destroyed the city in 689. His successor Esarhaddon had it rebuilt.

In 625 BC the Chaldeans, led by Nabopolassar, took control of Babylon, and under his son *Nebuchadnezzar II the second great Babylonian empire was extended as far as Palestine and Syria. Babylon itself was fortified with extensive new walls and was the scene of much building activity. The Chaldean dynasty continued until *Cyrus the Great captured the city in 539 BC. It flourished under the *Achaemenids but never again became an independent power in its own right.

The city of Nebuchadnezzar II's time was the largest of the ancient world and included the Hanging Gardens, (*Seven Wonders of the World), and the temple of Marduk with its ziggurat (pyramidal tower), popularly identified with the Tower of Babel described in the Bible. It derived much wealth from its position on the north–south and east–west trade routes. There are considerable and impressive remains still extant.

Bacon, Francis, 1st Baron Verulam, Viscount St Albans (1561–1626), English statesman and philosopher. He became a barrister in 1582 and entered Parliament two years later. In the 1590s he prospered as a member of the *Essex faction at court, and published his first edition of witty, aphoristic *Essays* (1597). Under James I he had to compete with *Coke for high legal office, but when Coke began to oppose the crown, Bacon was taken up by *Villiers and rose to be Lord Chancellor (1618). In 1621 he was impeached by Parliament for accepting bribes and his political career was ruined. In retirement he devoted himself to literary and philosophical work. His emphasis on the observation and classification of the natural world, in *The Advancement of Learning* (1605), *Novum Organum* (1620), and *New Atlantis* (1627), contributed significantly to the European *Scientific Revolution.

Bacon's Rebellion (1676), an uprising in *Virginia, North America, led by an English immigrant, Nathaniel Bacon. Dissident county leaders and landless ex-servants followed his opposition to Sir William *Berkeley. Though he was initially successful, Bacon died soon after the passage of reforms in the Virginian Assembly. Underlying the rebellion were problems caused by depressed tobacco prices and lack of colonial autonomy.

Bactria, an area famed in antiquity for its fertility and approximating to northern Afghanistan and its environs. It was a *satrapy (province) under the *Achaemenids, and put up fierce resistance to *Alexander the Great, who conquered it in 329 BC. In the mid-3rd century BC Diodotus I established a kingdom independent of the *Seleucids, and although his successor Euthydemus I (ruled 235–200 BC) was compelled to acknowledge Antiochus the Great as his overlord, Menander (ruled 155–130) brought under his control an area comprising all of modern Afghanistan and parts of Pakistan and southern Russia. In *c.*130 BC the northern part was conquered by the Yueh-Chih (a nomadic tribe from western China) and *c.*100 the southern part by the Sakas (a Scythian tribe). In the late 1st century BC part of the Chinese tribe, known as the Kushan dynasty, became dominant and made Bactria a centre of the growing faith of *Buddhism. It subsequently came under the Sassanians, the White Huns, and in the late 7th century, Arabs.

Bahamas, a group of islands off Florida. They were discovered by Columbus in 1492, his first landfall in the New World being on an island he named San Salvador (generally considered to be Guanahani), then inhabited by Arawak Indians. In 1629 the Bahamas were included in the British Carolina colonies, but actual settlement did not begin until 1648, with settlers from *Bermuda. Possession was disputed by Spain, but was acknowledged in 1670 under the Treaty of Madrid. The islands were sugar-producing and featured prominently in the *slave trade and in piracy, as well as being of major strategic importance during the 18th and 19th centuries.

Bahmani, a dynasty of sultans of the *Deccan plateau in central India (1347–1518). The dynasty was founded by Ala-ud-din Bahman Shah, who in 1347 rebelled against his Delhi suzerain. His successors expanded over the west-central Deccan, reaching a peak in the late 15th century under Mahmud Gawan, who successfully held encroaching Hindu and Muslim powers at bay. During the early 16th century the Hindu empire of *Vijayanagar to the south expanded at the Bahmanis' expense, and between 1490 and 1518 the sultanate gradually dissolved into five successor Muslim states, Bijapur, Ahmadnagar, Golconda, Berar, and Bidar.

bailiff, the estate manager of the lord of the *manor in England from the 11th century. In each medieval village there was usually one salaried bailiff (or else a serjeant or reeve, of lesser status). He would make sure that his

A reconstruction of the Ishtar Gate at **Babylon**, built in the reign of Nebuchadnezzar II. Standing 10 m. (30 ft.) high, with towers twice that height, it was one of eight fortified gates in the walls, decorated with bulls, lions, and dragons and symbols of the Babylonian god, Marduk. (Staatliche Museen zu Berlin)

village reached its annual production target for the lord, for he might be sued if it did not, while he gained the surplus if it was exceeded. The word 'bailiff' gradually shifted its meaning, and in the later Middle Ages, when lords more commonly let out their manors to farmers, the bailiff was one of the lesser officials of the sheriff. The bailiff had always had a legal role in the village, prosecuting on its behalf or acting as the legal representative of individual villagers, but from the 14th century he was primarily the sheriff's representative, authorized to make arrests, summon people to court, and seize property. Farmers and urban landlords also employed him as a rent-collector, knowing that his legal skills could be drawn on in cases of non-payment.

Balboa, Vasco Núñez de (*c.*1475–1517), Spanish conquistador, the discoverer of the Pacific Ocean. He first arrived in the New World in 1501 and, after an unsuccessful period as a plantation owner in Hispaniola, he joined and soon became leader of an expedition that founded the town of Darien in the Gulf of Uraba in 1511. There he heard of an ocean beyond the mountains and of the gold of Peru, and in 1513 led an expedition of 190 Spaniards (including Francisco *Pizarro) and 1,000 Indians across the mountains to the ocean. He named it the 'Great South Sea' and took formal possession of it in the name of *Ferdinand of Spain. The king made him admiral of the South Sea and governor of Panama and Coyba, and sent out a governor to take charge of

Darien itself. This new governor, jealous of Balboa and encouraged by Pizarro, had him seized, charged with treason, and executed.

Baldwin I (c.1058–1118), King of Jerusalem (1100–18). On the death of his brother, *Godfrey of Bouillon, he was crowned first King of Jerusalem. He foiled the ambitions of the Patriarch Daimbert and ensured that Jerusalem would become a secular kingdom with himself as its first monarch. His control of the Levantine ports secured vital sea communications with Europe, and by asserting his suzerainty over other Crusader principalities he consolidated the primacy of the Latin Kingdom of Jerusalem.

Bali, an island lying immediately east of Java. Controlled by Hindu states, it resisted *Islam, which spread through Java in the 16th century. Refugee princes from *Majapahit in the early 16th century and later from Belambangan in east Java, strengthened Hindu influence in the island, although the caste system was never fully adopted. The Balinese were skilled agriculturists, who built remarkable rice terraces. They were divided into nine warring states and also engaged in piracy, slave trading, and shipwrecking, which brought them into conflict with the Dutch.

Balkans, a region in south-eastern Europe, now comprising Albania, Greece, Bulgaria, European Turkey, Yugoslavia, and Romania. The Balkans have been inhabited since c.200,000 BC and there is archaeological evidence of the *Aurignacian and *Gravettian cultures. By 7,000 *Neolithic culture had developed, including a fine decorated pottery. The area was then settled by semi-nomadic farmers from the Russian steppes (c.3,500 BC) and then by central European *Urnfield peoples. It was part of several empires, being dominated by the Persians, the Greeks, the Romans, and, in the early Middle Ages, by the Byzantines. Although *Serbs, *Bulgarians, and *Magyars were able to carve brief empires in the area, by the late 14th century it yielded to the westward expansion of the *Ottomans. They reached the Dardanelles in 1354, *Macedonia by the 1370s, and, after the battle of *Kossovo in 1389, marched on to *Serbia. The balance of power changed after the last siege of Vienna in 1683, and the Turks were driven back by a revived *Habsburg empire, and by Russia, which championed the Balkan peoples, many of whom were either *Slavs or *Orthodox Christians.

Ball, John (d. 1381), English rebel. He was a priest who combined the religious teachings of John *Wyclif with an egalitarian social message. He was imprisoned for heresy, but released in June 1381 by the Kentish rebels in the *Peasants' Revolt. Outside London he preached to them and incited them to attack anyone opposed to his ideal of social equality. A month later he was captured, tried, and hanged as a traitor.

Balliol, Edward (d.c.1364), King of Scotland (1332–6), son of John Balliol. In 1332 he landed in Fife to reclaim the throne his father had given up. He defeated the Scots at Dupplin and was crowned at Scone. Within three months he was forced to flee but returned with the help of *Edward III of England after his victory at Halidon Hill. In 1341 Balliol was again expelled from

Scotland and in 1356 he resigned his claim to the Scottish throne.

Balliol, John (c.1240–1314), King of Scotland (1292–6). He was descended through his maternal grandmother from *David I of Scotland, and in 1291–2 his claim to the crown was upheld in a trial between him and Robert Bruce, Lord of Annandale. The trial was arranged by *Edward I of England, and less than a month after his coronation (30 November 1292) Balliol grudgingly did homage to Edward as his superior. In 1295 he attempted to ally with France, which resulted in an English invasion of Scotland. Balliol was forced to give up his kingdom to Edward and was taken as a captive to England, before retiring to his estates in France.

Bancroft, Edward (1744–1821), American traitor. Born in Massachusetts, he studied medicine in London. He met Benjamin *Franklin, and during the War of Independence was a friend of Arthur Lee and Silas Dean when they were negotiating an alliance with the French and became secretary to the American peace delegation. Bancroft was actually an agent for England at an annual fee of £500 and a life pension and it was almost a century before his activities as a British agent were discovered.

An 18th-century engraving of the **Bank of England**. In 1734, forty years after its foundation, the bank moved from its original site, the Grocers' Hall in London, to its present building in Threadneedle Street. Apart from the governors and directors, the bank's initial staff included seventeen clerks and two doorkeepers.

S. Wale delin.
The Bank
J. Green sc. Oxon

bandkeramik, the German term for a kind of pottery (decorated with incised ribbon-like ornament) which was characteristic of the first *Neolithic settlers in central Europe and which gives its name to their culture. These peoples spread up from the Balkans c.5000 BC. They occupied small plots of land on fertile soil near rivers, where they built wooden longhouses for themselves and their livestock, and cultivated cereals which they introduced to this area. They formed the basis of later Neolithic populations.

Bank of England, the British central bank, popularly known as the 'Old Lady of Threadneedle Street', where its London office stands. It was founded in 1694 as an undertaking by 1268 shareholders to lend £1,200,000 to the government of William III to finance his wars against France. In return it received 8 per cent interest and the right to issue notes against the security of the loan. These privileges were confirmed in 1708 when its capital was doubled, and it was given a monopoly of joint-stock banking which lasted until 1826, thus preventing rival banks from having large numbers of shareholders. However, small family concerns multiplied and there were over 700 banks in London by 1800. The Bank played an important role in the *South Sea Bubble crash of 1720, this time increasing its capital in order to take over some of the South Sea Company's obligations.

The foundation of the Bank of England, the institution of the National Debt, a debt secured against the national income (1694), and the setting up of the Stock Exchange (1773) were part of a financial revolution in England. The rewards from financial speculation led investors to look away from land, the traditional source of wealth, towards the City of London which became a centre of commercial activity and prosperity.

Bannockburn, battle of (24 June 1314). At Bannockburn, about 4.5 km. (2 miles) from Stirling in Scotland, a major battle was fought between *Edward II of England and *Robert the Bruce. Edward's large invading army, perhaps 20,000 strong, was outmanoevred and forced into the Bannock burn (or river) and adjacent marshes; it was a disastrous defeat for the English, and Edward was lucky to be able to flee to safety.

Bantam (or Bantĕn), a port in Java, commanding the Sunda Straits. It was a village when *Malacca fell to the Portuguese in 1511 and it attracted traders unfriendly to the Portuguese. After its conversion to *Islam in the early 16th century, it became a centre for Islamic ideas. At its height, Bantam's control extended over the major pepper-growing areas of southern Sumatra. In 1604 the English built a trading post and the subsequent Dutch conquest of Malacca (1641) and *Macassar (1667–8) increased the importance of Bantam. Its sultans stirred up anti-Dutch movements among Sumatran pepper-growers. The power of the Bantamese sultans was already waning, however, due to internal rivalries, which the Dutch exploited. In 1684 the Dutch expelled the British. After a rebellion in 1752 Bantam became a vassal of the *Dutch East India Company. Its sultanate was finally suppressed in 1832.

Baptist Church, a Christian Protestant sect practising baptism of adults by immersion in water. In the 16th century various religious groups in Europe (*Anabaptism), established the ritual of adult baptism. Baptists date their beginnings from the English church established in Amsterdam in 1609 by John Smyth (1554–1612) and the church in London under Thomas Helwys (1612). They were 'General' or *Arminian Baptists, as opposed to 'Particular' or Calvinist Baptists, who evolved between 1633 and 1638. After the *Restoration they moved closer to the *Presbyterians and Independents and were recognized as dissenters from the Protestant faith. America's first Baptist church was probably the one established at Providence, Rhode Island, with the help of Roger Williams (1639). From 1740, under the influence of the *Great Awakening, the movement made considerable headway, especially in the southern states.

Barbarossa (Turkish, Khayr ad-Din Pasha) (c.1483–1546), a famous *corsair, and later grand admiral of the *Ottoman fleet. He and his brother Aruj first came to fame for their success against Christian vessels in the eastern Mediterranean. In 1516 Algiers appealed to Aruj to save it from Spain, and made him sultan. Aruj was killed fighting in 1518, and his brother Khayr ad-Din diplomatically ceded Algiers and its territory to the Ottoman sultan. He served as viceroy until 1533, when he was made grand admiral. In 1534 he took Tunis, but *Charles V expelled him in 1535. After a number of minor engagements he retired in 1544.

Barbados, the most easterly island of the West Indies. It may have been discovered by the conquistador Rodrigo de Bastidas in 1501, though the island that he called Isla Verde (because of its luxuriant vegetation) could have been Grenada rather than Barbados. *Carib and Arawak Indians had both once inhabited the island but it was uninhabited when the first British colony was established in 1627. Black slaves were imported from Africa and a substantial cane sugar industry established, to remain the principal product of the island throughout the 17th, 18th, and 19th centuries, even after slavery was abolished in the British empire in 1834. It was also a strategic port for the British navy.

Bar Cochba (Simeon bar Kosiba, 'Simon son of the

A painting by Nigari of a white-bearded **Barbarossa** after his retirement from naval service to the Ottoman sultan. (Topkapi Palace Museum, Istanbul)

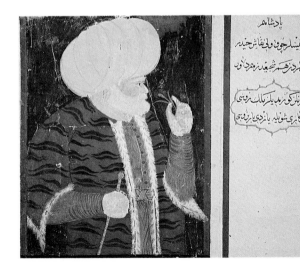

star') (d. 135), leader of a Jewish rebellion against Roman rule in 132. He failed to gain general support and was killed with the fall of his stronghold, Betar. His life and military exploits have become legendary.

Barebones Parliament, the assembly summoned by Oliver *Cromwell in July 1653, after he had dissolved the *Rump Parliament. It consisted of 140 members chosen partly by the army leaders and partly by congregations of 'godly men'. Known initially as the Parliament of Saints, it was later nicknamed after 'Praise-God' Barbon, or Barebones (c.1596–1679), one of its excessively pious leaders. Its attacks on the Court of Chancery and on the Church of England alarmed both Cromwell and its more moderate members. The dissolution of this Parliament was followed by the Instrument of Government and the proclamation of *Cromwell as Lord Protector.

Barnet, battle of (14 April 1471). A battle in the Wars of the *Roses fought between the Lancastrian forces, led by Richard Neville, Earl of *Warwick ('the Kingmaker') and the Yorkist troops of *Edward IV. Both sides suffered heavy losses, but Warwick was slain and Edward's recovery of his throne was made almost certain.

baron, a member of the lowest rank of the English peerage. Barons were originally the military tenants-in-chief of the crown. The title was introduced in England with the Norman Conquest and signified the vassal of a lord. Its limitation to those who held land directly from the king in return for military service occurred early. *Magna Carta (1215) made the distinction between the lesser baronage, summoned to the Great Council (Parliament) by general writ, and the greater baronage, called by personal writ, which was regarded as having conferred a hereditary peerage on the recipient. In 1387 Richard II made John de Beauchamp Baron of Kidderminster by letter patent (a letter from the sovereign) which gave this rank to Beauchamp and to his heirs. This new practice was regularly followed from 1446, establishing the principle that the peerage could be created either by writ of summons or by letters patent, the latter being more customary.

Barons' wars (1215–17 and 1264–7), the civil wars fought in England between King *John and the barons. In June 1215 at Runnymede, King John, faced by the concerted opposition of the barons and church, conceded *Magna Carta. He failed to honour his promise and thereby provoked the barons to offer the crown to Louis, Dauphin of France, who landed in Kent in May 1216. John's death (October 1216) and the reissue of Magna Carta by the regent of his son *Henry III prevented a major civil war. With his defeat at Lincoln and the capture of his supply ships off Sandwich, Louis accepted the Treaty of Kingston-upon-Thames in September 1217.

Baronial opposition to the incompetent Henry III led to his accepting a programme of reform, the Provisions of Oxford (1258). Henry's renunciation of those reforms led to civil war in 1264, the baronial forces being led by Simon de *Montfort. The king's capture at the battle of Lewes (May 1264) began a brief period of baronial control when de Montfort sought to broaden his support by extending parliamentary franchise to the shires and towns (1265). After his defeat and death at Evesham

(August 1265), the struggle was continued unsuccessfully until 1267 by his supporters.

barrow (or tumulus), an earthen mound raised over a grave (if of stone, 'cairn' is the usual term). Grave-mounds of this type were characteristic throughout Europe and parts of central and southern Asia during the *Neolithic and *Bronze Ages, and in places much later. They occurred less frequently in other parts of the world.

Mounds could be raised over either inhumation or cremation burials, or sometimes over elaborate mortuary structures of stone or wood, houses of the dead. The best known examples include the Neolithic long barrows of Atlantic Europe, many enclosing spectacular chambers of *megaliths; the round barrows of the 2nd millennium across most of the Continent; the ship burials of the *Vikings and related peoples in the north; the richly furnished Scythic tombs on the Altai Mountains of central Asia; and the burial mounds of the Indians of the *Mississippi valley in the USA.

basket-makers, a group of North American Indians in Colorado and neighbouring areas who developed out of the *Desert cultures in the last centuries BC. They adopted farming from *Mexico. The baskets they used instead of pottery have been preserved in the dry climatic conditions of their territory. In about AD 900, improving agricultural techniques and increasing population led to the *Anasazi culture of the Pueblos.

Bastille, a fortress in eastern Paris that was the French state prison in the 17th and 18th centuries, holding prisoners held under *lettre de cachet* or royal orders. In 1789 it held only seven prisoners but it was none the less regarded as a symbol of royal despotism and was believed to contain arms and ammunition. On 14 July it was attacked by crowds, including mutineers from the Gardes Françaises. The governor, De Launay, surrendered, but he and his men were killed and the fortress demolished. The event is celebrated as the beginning of the French Revolution and Bastille Day (14 July) is a French national holiday.

Batavia   *Dutch East Indies.

The fall of the **Bastille** on 14 July 1789 from a contemporary engraving. The storming of the fortress, its bombardment by mutinous artillerymen, and its firing and capture by the mob, served to rally the French Revolutionary forces. (Musée Carnavalet, Paris)

Bavarian Succession, War of the (1778-9), a war between Austria and Prussia resulting from *Joseph II's ambition to add Bavaria to the *Habsburg dominions. The childless Maximilian Joseph of Bavaria (d. 1777) had designated Karl Theodore, Elector Palatine, as his heir. He was a weak man with illegitimate sons to provide for, and in January 1778 he agreed to sell a third of Bavaria to Joseph. *Frederick II headed the opposition to this and the 'Potato War' (so called because the Prussian troops occupied their time by picking potatoes) took place in Bohemia: no battles were fought. Russia advanced into Poland, France offered mediation, and the Peace of Feschen was signed, under which Austria gained only the very small area of Innviertel.

Baxter, Richard (1615-91), English Puritan minister. He was ordained as an Anglican clergyman, but rejected belief in episcopacy and became a *Nonconformist. In 1645 he became chaplain to a *Roundhead regiment. He published the first of some 150 pamphlets in 1649 and in 1650 *The Saints' Everlasting Rest*, an important devotional work. During the Commonwealth period, his appeals for tolerance did not succeed. At the Restoration he became a royal chaplain but refused a bishopric. The 1662 Act of *Uniformity forced his resignation, and in about 1673 he took out a licence as a Nonconformist minister. In 1685 he was imprisoned and fined by Judge *Jeffreys for 'libelling the Church'.

Bayard, Pierre Terrail, seigneur de (c.1473-1524), French knight. He was celebrated by contemporary chroniclers as the outstanding knight of his generation '*sans peur et sans reproche*' ('without fear and without reproach'). He accompanied Charles VIII and Louis XII of France in their invasions of Italy. He was renowned for a heroic defence of Mézières against Emperor *Charles V in 1521, which saved France from *Habsburg conquest. He died in battle.

Bayezid I (1347-1403), *Ottoman sultan (1389-1402), known as Yildirim ('Thunderbolt'). He succeeded his father *Murad I and absorbed rival Turkish principalities in western Asia Minor, took Trnovo in Bulgaria (1393), and Thessaloniki in Greece (1394), blockaded Constantinople (1394-1401), and defeated a Christian army at Nicopolis in 1396. His thrust into eastern Asia Minor, however, brought him to disaster at Ankara in 1402, and he died a captive of his conqueror, *Tamerlane.

Bayezid II (c.1447-1512), Ottoman sultan (1481-1512). He wrested the throne from his brother Jem on the death of their father *Mehmed, fought inconclusively with the *Mamelukes (1485-91), gained Greek and Adriatic territories from Venice, and was faced with the emerging power of *Ismail Safavi. He abdicated a month before his death in favour of his youngest son, Selim.

Bayle, Pierre (1647-1706), French philosopher and critic. He was professor of philosophy and history at Rotterdam University and an influential writer on scientific subjects who wished to encourage a sceptical attitude of mind. He argued that religion and morality were independent of one another, and championed the cause of universal religious toleration. His most famous work was a historical and analytical dictionary, the *Dictionnaire historique et critique* (1696).

A handled beaker, c.1800–1600 BC, typical of the **Beaker cultures**, incised with lattice-work decoration, cordons, and running chevrons. It stands 15 cm. (6 in.) high and was found near Abingdon in Oxfordshire. (British Museum, London)

Beaker cultures, the term for the people in many parts of western Europe at the end of the *Neolithic period (c.2600-2200 BC) who made and used a particular type of decorated pottery drinking-vessel. It was shaped like an inverted bell, with or without handles, and ornamented with zones of stamped impressions. The style of 'Beaker' pottery seems to have developed in the Lower Rhine area, though it absorbed motifs from other areas with which it was in contact by sea and river routes. These pots were valuable to their owners, and are often found as grave-goods in male burials, along with weapons such as a copper dagger or the remains of archery equipment. Their wide distribution, from the western Mediterranean to northern Germany, led earlier investigators to postulate a 'Beaker Folk' spreading up from Portugal, or perhaps from central Europe. They are now seen more simply as part of a general trend to ostentatious display of personal wealth, introduced at that time from central Europe. These included copper metallurgy and horses for riding. Such personal wealth was buried with the individual in a new form of single grave under its own mound, a custom which superseded older funerary practices based on communal burial in *megalithic monuments.

Beaton, David (or Bethune) (1494-1546), Scottish churchman. He worked for the preservation of the Catholic religion, leading the anti-Protestant faction at court, and favouring the 'Auld Alliance' with France. Created cardinal in 1538, he succeeded his uncle as Archbishop of St Andrews in 1539 and was made Chancellor in 1543. His harsh persecution of Protestant preachers culminated in the execution of George *Wishart, and Beaton was assassinated by Protestant nobles.

Beaufort, an English family descending from three illegitimate sons of John of *Gaunt (fourth son of Edward III) and Katherine Swynford. The children were legitimated in 1407 but with the exclusion of any claim to the crown. Their father and their half-brother *Henry IV made them powerful and wealthy: **Thomas** (d. 1427) became Duke of Exeter, **John** (c.1371-1410) was made

Lord High Admiral and Earl of Somerset, and **Henry** (d. 1447) was Bishop of Winchester and later a cardinal. As a court politician he led the so-called constitutional party against *Humphrey, Duke of Gloucester. The *Yorkists had no love for the Beauforts, and by 1471 all three of the Earl of Somerset's grandsons had been killed in battle or executed. The male line thus ended, but their niece Margaret Beaufort (1443-1509), daughter of John, Duke of Somerset, who married Edmund *Tudor, enjoyed a life of charity and patronage of learning after her son became king as *Henry VII.

Becket, Thomas à, St (1118-70), Archbishop of Canterbury (1162-70). Born in London, and educated in Paris and Bologna, Becket became archdeacon of Canterbury (1154) shortly before *Henry II made him Chancellor of England (1155) and he served the king as statesman and diplomat. However, when Becket became archbishop (1162) he became a determined defender of the rights of the church. He came into conflict with the king in the councils of Westminster, Clarendon, and Northampton (1163-4), particularly over Henry's claim to try in the lay courts clergy who had already been convicted in an ecclesiastical court. Refusing to endorse the Constitutions of Clarendon (1164), Becket went into exile in France.

A miniature from a 13th-century Book of Psalms depicting the murder of Thomas à **Becket**. Becket's shrine, a celebrated place of pilgrimage for almost 400 years, was totally destroyed in 1538 during the Reformation. On the orders of Henry VIII the archbishop's bones were removed and burned. (British Library, London)

On his return (1170), apparently reconciled to Henry, Becket suspended those bishops who had accepted these Constitutions. Henry's rage over this action was misinterpreted by four knights who assumed he would approve their murder of Becket in Canterbury Cathedral. Acclaimed a martyr, and canonized (1173), his shrine became a centre of Christian *pilgrimage.

Bede (673-735), English scholar and historian. He was celebrated as the author of the most famous work on Anglo-Saxon England, the *Historia Ecclesiastica Gentis Anglorum* ('Ecclesiastical History of the English People') (731). From the age of 7, he was educated at the Northumbrian monasteries of Wearmouth and Jarrow, where he remained as a monk for the rest of his life. He wrote (in Latin) lives of five abbots of his monastery and numerous biblical works, hymns, verse, books on astronomy, letters, and a martyrology of 114 saints. He popularized a new chronology using *anno domini* (AD) in dating. Although he probably did not travel further than York or Lindisfarne, Bede was the best-known Englishman of his time, and a synod at Aachen (836) awarded him the title 'Venerable'.

Bedford, John, Duke of (1389-1435), Regent of France (1422-35). He was the third son of *Henry IV. While his brother *Henry V was in France, Bedford was appointed Guardian of England on several occasions between 1415 and 1421, and on Henry's death in 1422, he was made governor of Normandy and Regent of France. He succeeded in retaining England's French territories, despite the campaign of *Joan of Arc and insufficient funds; by his first marriage, to Anne of Burgundy, daughter of Duke John the Fearless, he cemented England's crucial Burgundian alliance.

Belgae, German and Celtic tribes who inhabited the Rhine estuary, the Low Countries, and north-east France. Subdued by Julius *Caesar in the *Gallic wars their name was given to 'Gallia Belgica', one of the three main administrative units of Gaul established by *Augustus. During the last century BC a number of Belgae settled in southern Britain. They led resistance against the invasion of Britain in 43 AD, most notably the Catuvellauni under *Caratacus.

Belgium, a country in north-west Europe on the North Sea. It takes it name from the *Belgae, one of the peoples of ancient Gaul, but by the 5th century immigrations from the north had resulted in a large settled German population. After several centuries under the Franks the region split into independent duchies and, especially in *Flanders, free merchant cities. In the 15th century all of what is now Belgium became part of the duchy of *Burgundy, but the Low Countries (which included Belgium) in 1477 passed by marriage to the Habsburg empire of *Maximilian I, and were later absorbed into the Spanish empire. Belgium was occupied by France in 1795 during the French Revolutionary wars.

Belisarius (AD 505-65), Roman general under *Justinian. He was instrumental in halting the collapse of the Roman empire, if only temporarily. In 530 he defeated the Persians in the east, although they quickly reasserted themselves in Syria. Six years later he conquered *Vandal North Africa, capturing its king. In 535-40 he took back

Bede, a late 12th-century illumination from his *Life of St Cuthbert*. Bede is known as the 'Father of English History' and some of his works were translated into Old English from Latin by King Alfred. (British Library, London)

Italy from the *Ostrogoths, advancing as far north as Ravenna, taking their king prisoner, and followed it with a second Italian campaign a few years later. He took Rome in 549 but was dismissed and even charged with conspiracy by a jealous Justinian, though reinstated in 564.

Belshazzar (6th century BC), son of Nabonidus, the last of the Chaldean kings of *Babylon. When his father embarked on a prolonged campaign in Arabia c.556 BC, he remained behind as ruler (probably with the title of king) in Babylon. The city was captured by *Cyrus the Great in 539 BC and he was killed.

Benares *Varanasi.

Benbow, John (1653-1702), English admiral. He was prominent in sea battles against the French for control of the English Channel during the early 1690s. He served in the West Indies for most of the period 1698-1702, where in his last engagement his daring plans for pursuing the retreating French were defied by his own captains. He died of his wounds in Jamaica, leaving a reputation for vigour, toughness, and bravery.

Benedict of Nursia, St (c.480-c.550), founder of the *Benedictine order of monks. He built his monastery at Monte Cassino in central Italy in c.525. Monasticism

then was lacking order and regulation, and he introduced his 'rule', which was a definition of the qualities and actions required of a monk. These were principally humility, prayer, obedience, silence, and solitude. St Benedict of Aniane (c.750-821) systematized the Benedictine rule and his *capitulare monasticum* received official approval in 817 as the basis for the reform of French monastic houses.

Benedictine, the name for a monk or nun of an order following the rule of St *Benedict. From the original Benedictine foundations at Subiaco and Monte Cassino in Italy the number of monastic houses in Europe grew to many thousands. The order reached its peak of prestige and influence in the 10th and 11th centuries, with the abbey of *Cluny in Burgundy its most prestigious foundation. The basic concept of Benedictine monasticism was that it should encourage a way of life separated from the world, within which monks could achieve a life devoted to prayer.

benefit of clergy, the privilege whereby a cleric, on being accused of a crime, could claim to be exempted from trial by a secular court, and to be subject only to the church courts, which usually dealt with him more leniently. It was a system open to abuse, especially when clerics were numerous and difficult to identify with certainty, as was the case in the Middle Ages. Indeed the mere ability to read was often accepted as proof of clerical status. In England it was a principal issue in the controversy between Archbishop Thomas à *Becket and *Henry II and the privilege was largely conceded by the crown in the aftermath of Becket's murder in 1170; later its application was limited by various Acts of Parliament and it was finally abolished in 1827.

Bengal, the delta region of the Ganges and Brahmaputra rivers. In early times the region was incorporated in some of the great *Buddhist and *Hindu empires of northern India, but a sense of separate identity was fostered by periods of independent rule. From the 13th century a series of Muslim invasions annexed Bengal to the *Delhi sultanate. In succeeding centuries local governors maintained considerable autonomy until *Mogul conquest in 1576 brought subordination to Delhi. On Mogul decline in the 18th century the Bengali nawabs (governors) enjoyed a shortlived era of renewed autonomy. This was ended by the *East India Company's ambitions. Its merchants had been trading in Calcutta since 1690, but in 1757 Robert *Clive took advantage of the *Black Hole of Calcutta incident to force a trial of strength. After his victory at *Plassey the Company gradually ousted the local nawabs and their Mogul suzerains from the economic and judicial control of the province. By the 1790s the British controlled Bengal.

Benin, West African kingdom based on Benin City, now in southern Nigeria, probably founded in the 13th century. It developed by trading in ivory, pepper, cloth, metals, and, from the 15th century, slaves. The kingdom achieved its greatest power under Oba Equare, who ruled from about 1440 to 1481. With his powerful army he conquered Yoruba lands to the west and Lower Niger to the east. He initiated administrative reforms, established a sophisticated bureaucracy, and ensured that the Portuguese, who arrived on the coast in 1472, did

A bronze relief of musicians, from the African kingdom of **Benin**. Most early Benin bronze sculptures were in the form of hollow-cast heads, often representing dead kings, which were placed on altars for use in ancestral rites. In later periods, figurative plaques were added to buildings as decoration. The first Benin casts were probably made in the 13th century, the process having been imported from neighbouring Ife. (Museum of Mankind, London)

not establish control over Benin. The kingdom expanded further in the 16th century but by the 18th century its power waned with the growing strength of *Oyo and other Yoruba states. Its iron work and bronze and ivory sculptures rank with the finest art of Africa. The republic of *Dahomey subsequently took the name Benin.

Bennington, battle of (16 August 1777). A battle of the American War of *Independence in which 1,600 *Green Mountain Boys under General John Stark overwhelmed 1,200 German mercenaries of *Burgoyne's army. Encouraged by this success, the Americans forced Burgoyne to surrender after the defeat at *Saratoga.

Berbers, the people who have occupied northern and north-western Africa since prehistoric times. *Herodotus recorded that they were found in various tribes. They do not seem commonly to have formed kingdoms, although they co-operated on occasions, for example against Roman rule. Their extreme independence and austerity were exemplified by the *Donatist *circumcelliones* (violent bands of marauders) of the 4th and 5th centuries, by the Kharijite sect of early Islam, and by the cults of marabouts, Islamic holy men of ascetic devotion and organizers of fraternities. In this way they both resisted the *Arab conquest and transformed Islam to suit their own tastes. They supported the *Umayyads in Spain, and the *Fatimids in Morocco, and then set up several dynasties of their own, of whom the *Almohads and the *Almoravids were the most important.

Berkeley, George (1685-1753), Irish bishop and philosopher who devoted much of his life to the encouragement and development of missionary work. He went to America in 1728 to found a missionary college, but had to return to Ireland because of shortage of funds. His philosophy was based on the ideas of empiricists such as John *Locke, but he asserted that, since phenomena are defined only by their perceived effects, what is not perceived does not exist. His principal writings include *An Essay Towards a New Theory of Vision* (1709), and *Principles of Human Knowledge*.

Berkeley, Sir William (1606-77), governor of Virginia (1642-52, 1660-76). A royalist aristocrat, he was removed from the governorship during the *Commonwealth. At the Restoration, Charles II renewed his commission and he won colonist popularity by his attempts to protect tobacco prices and gain a charter, but his regime grew corrupt and oligarchical, triggering *Bacon's Rebellion in 1676. His cruelty against the rebels led to his recall.

Bermuda, a cluster of some 150 islands in the west Atlantic, discovered by a Spaniard, Juan de Bermundez. First settled by the Virginia Company, they have the oldest parliament in the New World, dating to 1620. The islands produced tobacco, using Indian and African slave labour. From the 18th century Bermuda served as a naval base.

Bernard of Clairvaux, St (1090-1153), theologian and reformer, one of the most influential figures of the Middle Ages. He was born into a noble family in Burgundy but rejected this privileged existence to adopt a life of religious study and austerity. He became abbot of Clairvaux, a *Cistercian monastery which he founded and which became a model for reformed monastic houses. The Cistercian order grew rapidly under his influence. In the disputed papal election of 1130 he supported Pope Innocent II and throughout his life he attacked heresy. His preaching in support of Innocent II brought him a European-wide reputation, seen in the support he attracted for the Second *Crusade (1140). A fervent mystical religious thinker he clashed with the more intellectual approach of *Abelard, and helped to bring about his downfall.

Berwick, Treaty of. Three treaties were named after Berwick, a town in Northumberland, sited on the border between England and Scotland. The first (3 October 1357) arranged for the release from captivity of David II of Scotland in return for a large ransom to be paid to Edward III of England, but this debt was never fully discharged. The second (27 February 1560) committed the English to send the Scottish Protestants military aid to help overthrow the Roman Catholic regent Mary of Guise. The third (18 June 1639) ended the first *Bishops' War between Charles I and Scottish Covenanters, although it did not fully resolve the conflict and was regarded as unsatisfactory by both parties.

Bhagavadgita (Sanskrit, 'Song of the Lord'), part of the sixth book of the Sanskrit epic poem, the Mahabharata, and the most famous religious text of *Hinduism.

Composed between the 2nd century BC and the 2nd century AD, it took the form of a dialogue between the Pandava prince, Arjuna, and his charioteer, Krishna, an incarnation of the god Vishnu. On the eve of a great battle Arjuna hesitated, horrified at the prospect of killing his kinsmen, and Krishna reminded him of his duty to fight. He then expounded on the nature of God and the paths to eternal union with Brahman. His emphasis on loving devotion as the highest path, and one open to all, was the earliest exposition of devotional worship (*bhakti*) of a supreme god.

Bible, the, the collection of books consisting of the Old Testament (the scriptures of *Judaism), the New Testament, and the Apocrypha. It forms the scriptures of the Christian Church. The Old Testament contains thirty-nine books. The first five books ('the Law', the Torah, or Pentateuch) describe the origins of the Jewish people. 'The Prophets' give a history of the settlement in *Canaan, the period of the kingdom of *Israel, and prophetic commentaries. 'The Writings' consist of the remainder of the books including the Psalms, Job, and Daniel. The final content of the Hebrew Old Testament was probably agreed *c.*AD 100. The New Testament consists of twenty-seven books. The four Gospels (meaning 'good news'), attributed to Matthew, Mark, Luke, and John, record the life, death, and resurrection of *Jesus Christ. The Acts of the Apostles traces the development of the early Christian Church and the Epistles (or Letters), notably those of St *Paul, contain advice on worship, conduct, and organization for the first Christian communities. The Book of Revelation gives a prophetic description of the end of the world. The final content of the New Testament was probably agreed in 382 in Rome. The Apocrypha contains twelve books that were included in the Septuagint, a Greek translation of the Old Testament of the 3rd and 2nd centuries BC that was used by the early Christian Church. These books do not appear in the Hebrew Old Testament and are not accepted by all Christian churches.

The Bible was originally written in Hebrew, Aramaic, and Greek. The first translation of the whole book was the Vulgate (AD 405) of St *Jerome. The first translation into English was undertaken by John *Wyclif and his followers (1382–8). The development of *printing stimulated the production of vernacular editions. Martin *Luther translated the New Testament into German in 1522 and William *Tyndale into English in 1525–6. William Coverdale's edition of the Bible, drawing heavily on Tyndale's work, was first published in 1535 and revised as the Great Bible in 1539. The Authorized or King James Version (1611), named after *James I who agreed to a new translation at the *Hampton Court Conference, was produced by about fifty scholars and remained for centuries the Bible of every English-speaking country. There are now translations of all or part of the Bible in over 1760 languages.

Bihar, a region in India comprising the middle Ganges plains and the Chota Nagpur plateau in north-eastern India. The region had its 'golden age' during the evolution of early Indian civilization. Among its ancient kingdoms was *Magadha, where both Gautama *Buddha and the *Jain seer, Mahavira, preached. Its capital, Pataliputra (now Patna), was adopted by several notable empire builders, including the *Mauryas and the *Guptas.

About 1200 it came under Muslim influence and remained subservient to the *Delhi sultans until becoming a province of the *Mogul empire in the 16th century. In 1765 British victories resulted in its amalgamation with Bengal and the introduction of indigo plantations.

Bijapur, a city and former state on the Deccan plateau, south-western India. It was the capital of a Muslim kingdom, founded by the Yadava dynasty in the 12th century. It fell under the control of the *Bahmani Muslims in the 14th century. Its era of independent splendour was from 1489 to 1686 when the Adil Shahi sultans made it their capital and were responsible for Islamic architecture of outstanding quality. In 1686 the Mogul emperor *Aurangzeb defeated Bijapur, but was unable to exert firm control and the region soon fell under *Maratha sway, from which it passed into East India Company hands in the early 19th century.

Bill of Rights (1689), a declaration of the conditions upon which *William and Mary were to become joint sovereigns of England, Scotland, and Ireland which became an Act of Parliament. Its major important provisions were that the king could not levy taxes without the consent of Parliament, that he no longer had the power to suspend or dispense with the laws, and that there was to be no peacetime standing army without Parliament's consent. These terms dealt with issues that had been raised by the actions of *James II and were seen as a guarantee of Englishmen's liberties, helping to justify the name *Glorious Revolution for the events of 1688–9. American patriots often referred to the Bill of Rights when claiming, in the dispute with Britain in the late 18th century, that their liberties had been undermined.

Bill of Rights (1791), the first ten amendments to the *Constitution of the USA. The constitutional arrangements of 1787 were assumed to guarantee human and civil rights, but omission of specific rights led to criticism. To prevent this issue jeopardizing ratification, a Bill of Rights was adopted in 1791. Based on features of the English *Bill of Rights (1689) and common law principles, it guaranteed freedom of speech, press, worship, assembly, and petition (the first amendment). Americans had the right to speedy and fair trial, reasonable bail, and to bear arms. They could not be forced to incriminate themselves (the fifth amendment) or suffer unwarranted search and seizure or cruel and unusual punishments.

Bishops' wars (1639–40), two brief conflicts over *Charles I's attempt to impose Anglicanism on the Scots, and important as a factor leading to the outbreak of the *English Civil War. Since 1625 the king had been trying to take back former church lands from Scottish noblemen, provoking great bitterness. In 1637, a modified version of the English Prayer Book was introduced in Scotland. This spurred the *Covenanters into abolishing the episcopacy. The first war (May–June 1639) was a bloodless fiasco. Charles had refused to call a Parliament to vote funds and, acknowledging that his new recruits were no match for the Covenanters, he made peace at Berwick. For the second war (August–September 1640), refused supplies by the English 'Short Parliament', he obtained money from the Irish Parliament, but his army was routed by the Covenanters at Newburn, near Newcastle upon Tyne. With the Scots occupying Northumberland and Durham, Charles was forced to make peace at Ripon, and to call the *Long Parliament.

Black Death (1347–50), a later name for the most virulent epidemic of bubonic and pneumonic plague ever recorded. It reached Europe from the *Tartar armies, fresh from campaigning in the Crimea, who besieged the port of Caffa (1347). Rats carrying infected fleas swarmed aboard trading vessels, thus transmitting the plague to southern Europe. By 1348 it reached France, Spain, and England; a year later Germany, Russia, and Scandinavia. Numbers of dead cannot be exact but up to 25,000,000 may have died in Europe; perhaps one-third of the population in England.

Effects were profound and lasting. Shortage of manpower put those peasants who survived in a strong position. The demand for labour led to the substitution of wages for labour services and peasants' agitation for further improvements led to agrarian revolts. The church, too, was adversely affected, with inadequately trained clerics being ordained to replace dead parish priests. Mass hysteria caused by fear and helplessness was reflected in art and literature. Outbreaks of plague continued in Europe until the 17th century (*Great Plague).

Blackfoot, a North American Indian tribe who inhabited southern Alberta and north-western Montana, originally a hunter-gatherer people similar to the *Cree. The Shoshone of Idaho became nomadic once they acquired horses and firearms in the 18th century, and raided the horseless Blackfoot. The Blackfoot retaliated by stealing horses from the Shoshone and trading for firearms with the Cree. Thereafter they rapidly adopted nomadism on the northern Great Plains.

Black Hole of Calcutta, a prison room at Fort William, Calcutta, India, so called after the alleged suffocation there in 1756 of some English prisoners. They had been incarcerated by the nawab, Siraj ud-Daula, in retaliation for extending the fort against previous agreements. The incident has an important place in British imperial mythology, for British accounts grossly exaggerated both the smallness of the room and the number of prisoners, thus suggesting an act of barbarism on the nawab's part.

Black Prince *Edward the Black Prince.

Blake, Robert (1599–1657), English admiral. He was a member of the *Long Parliament and fought for the *Roundheads during the *English Civil War. He achieved successes against the Royalists (1649–51), the Dutch (1652–4), and Spain (1656–7). His involvement in the preparation of the *Fighting Instructions* and *Articles of War* was crucial to the developing professionalism of the English navy, as was his association with the building of large, heavily armed vessels.

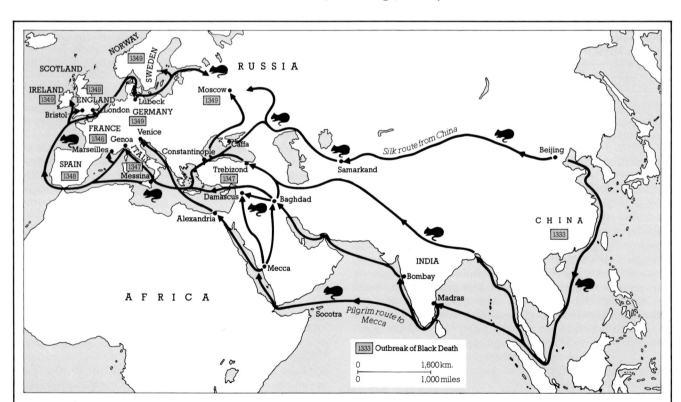

Black Death

The Black Death was the most devastating of many outbreaks of plague. Although its origins are uncertain, it is believed to have come from the Far East and to have been carried westward to Europe by merchants, pilgrims, and other travellers. It spread especially fast along sea trade routes, transmitted by the fleas of rats on board ship. The particular virulence of this epidemic may have been due to the presence of the more deadly pneumonic variety of plague, the only form that can be directly transmitted from one human to another (by sneezing, for example). It is estimated that as much as one-third of the population of Europe and the Near East died as the result of this outbreak in less than 20 years.

Blenheim, battle of (13 August 1704). A Bavarian village on the north bank of the River Danube gave its name to a major battle of the War of the *Spanish Succession. In 1704 *Louis XIV's armies were advancing towards Vienna and the French leader, Marshal Tallard, occupied the village while his ally, the Elector of Bavaria, held Lutzingen. John Churchill, Duke of *Marlborough, captain-general of the allied armies, had made a brilliant march down the Rhine and joined forces with the Austrian commander, Prince *Eugène. On 13 August Marlborough overwhelmed the French while Prince Eugène defeated the Elector. After heavy losses (12,000 Allies; c.30,000 French) on both sides Tallard surrendered: Vienna was saved and Bavaria conquered.

Blenheim Palace (a large *country house) was built at Woodstock in Oxfordshire as a gift to the Duke of Marlborough from the grateful British nation.

Blood, Thomas (c.1618–80), Irish colonel and adventurer. He lost his estates at the *Restoration in 1660 and hoped to persuade the authorities to return them by his attack on Dublin castle in 1663. His most famous exploit was the theft of the English crown jewels from the Tower of London in 1671. *Charles II, who examined Blood personally after his arrest, was so impressed with his audacity that he was pardoned and his estates restored.

Bloody Assizes, a series of trials held in 1685 to punish those who took part in *Monmouth's Rebellion, conducted by Judge *Jeffreys, the Lord Chief Justice, in those centres of western England most affected by the rebellion. Of 1,400 prisoners brought before him, 300 were hanged and 800 more were sold as slaves in the colonies, some of the profits from this enslavement going to courtiers, and even to the queen and her ladies. The severity of the sentences helped to mobilize support for *William of Orange in the West Country in 1688.

Boadicea *Boudicca.

Bodin, Jean (1530–96), French political philosopher and economist. In 1576 he published his great work on limited monarchy, *Les Six Livres de la République* ('Six Books of the Commonwealth'). Its argument that sovereignty arose from human needs rather than divine institution influenced the later English philosopher Thomas *Hobbes. Unlike other contemporary *Protestant writings, it held that citizens were never justified in rebelling against their ruler. Bodin was also a pioneer of the study of money and its effect on prices.

Boeotia, a region in central Greece whose cities in classical times included *Thebes and *Plataea. It contained much good agricultural land, suitable for the growing of corn and rearing of horses. A Boeotian League was established c.447 BC and lasted until 387, the city of Thebes controlling four of the eleven federal areas. In 424 Boeotian forces won an important victory over the invading Athenians at Delium, but Boeotia's strength ultimately depended on that of Thebes.

Boethius, Aricius Mantius Severinus (c.475–525), one of the most influential authors of the Middle Ages. He was employed at the court of *Theodoric the Great in Rome but was accused of treason, imprisoned, and executed. In prison he wrote the *De Consolatione Philosophiae*

Prince Eugène of Savoy at the battle of **Blenheim**. The cavalry of Marlborough and Prince Eugène pushed back the French–Bavarian forces towards the river to prevent the collapse of the Austrian defence of Vienna. (Plâs-Newydd, Anglesey)

('The Consolations of Philosophy') which owed much to Neoplatonic thought and discussed the value of philosophy to those Christians who suffer in a troubled world. The work was translated into English by *Alfred and later by Chaucer, who incorporated some of it in the *Canterbury Tales*. His translation of Aristotle's work on logic into Latin gave medieval scholars access to Greek philosophical thought. He also wrote at length on theology, mathematics, logic, art, and music.

Bohemia, a region in central Europe, now part of Czechoslovakia. Bohemia was established as a duchy by the Premyslid dynasty in the 9th century, but as a result of the rising power of the Ottonians was forced to accept the suzerainty of the German *Holy Roman Emperors in the following century. Combined with the neighbouring region of Moravia, Bohemia remained under the Premyslids until 1306, becoming a kingdom in 1198, and at the height of its power, under Ottakar II, also controlling the duchies of Austria. In the later Middle Ages, Bohemia was ruled by a number of families, most notably the German Luxemburgs and the Polish *Jagiellons and played a central role in the turbulent politics of the empire and papacy. In the early 15th century the martyrdom of the Prague religious reformer John *Huss (1415) solidified the identification of religious reform with an emerging popular nationalism, and the Hussite wars of 1420–33 marked the departure of Bohemia from the German orbit and its assumption of a more overtly Slavic identity. In 1526 the kingdom was inherited by the imperial *Habsburg dynasty, but in the 17th century it was once again at the centre of politico-religious upheaval within the empire, Protestantism and resurgent nationalism helping precipitate a revolt against imperial power which led to the *Thirty Years War (1618–48). Bohemia remained part of the Habsburg domain until the destruction of the empire in 1918.

Bohemond I (*c*.1056–1111), *Norman Prince of Antioch, the eldest son of Robert *Guiscard. He fought for Guiscard against the Byzantine emperor, Alexius *Comnenus; after his father's death he joined the First *Crusade and played a prominent part in the capture of the Syrian city of Antioch. He established himself as prince in Antioch but was captured by the *Turks and imprisoned for two years. In 1107 he led an expedition against the *Byzantine empire and was defeated by Alexius, making peace at the Treaty of Devol (1108).

Boleyn, Anne (1507–36), second wife of *Henry VIII and mother of *Elizabeth I. By 1527 the king was conducting a secret affair with Anne and contemplating divorce from his wife, Catherine of Aragon, who had failed to provide him with a male heir. His liaison with Anne continued and in January 1533 they were secretly married. Archbishop *Cranmer annulled Henry's first marriage in May 1533, and Anne was crowned in June, three months before the birth of Elizabeth. Anne and her family were supporters of the *Protestant religion and she helped to promote *Reformation doctrines at court. Anne, too, failed to produce a son, and in May 1536 she was tried on dubious charges of adultery. She was beheaded in the Tower of London.

Bolingbroke, Henry of   *Henry IV.

Bolingbroke, Henry St John, 1st Viscount (1678–1751), English politician. He entered Parliament as a Tory in 1701, became Secretary of State following the Tory triumph of 1710, and was responsible for negotiating the Peace of *Utrecht in 1713, though he angered Britain's allies by abandoning the military effort before peace had been finally agreed. In the power struggle among the Tories shortly before the death of Queen *Anne he was victorious over *Harley. Dismissed by George I in 1714, and impeached by the Whig Parliament of 1715, he fled to France, where he joined James Edward Stuart, but soon became disillusioned with the *Pretender's cause. In 1723 he was pardoned by George I

and allowed back into England. Though he was refused permission to return to Parliament, he remained politically active by contributing from 1725 to *The Craftsman*, a journal which attacked *Walpole and political patronage. His *Idea of a Patriot King* was written in 1738 to flatter Prince *Frederick Louis, but his political influence ceased after Walpole's fall in 1742.

Bolivia, a land-locked country of central South America. It became an important Ayamará Indian state between 600 and 1000 AD but was conquered by the growing *Inca state *c*.1200. Some Ayamará continued to resist, however, and were not completely subdued until the late 15th century. Spanish conquest followed six years after *Pizarro's landing in Peru in 1532, and in 1539 the capital at Charcas (modern Sucre) was founded. The discovery of silver deposits in the Potosí mountains in 1545 led to the establishment of the Audiencia (a high court with a political role) of Charcas, under the viceroyalty of Peru. Revolutionary movements against Spain occurred here earlier than anywhere else in South America: at La Paz in 1661, Cochabamba in 1730, and Charcas, Cochabamba, La Paz, and Oruro in 1776–80, but all failed.

Boniface VIII (1235–1303), Pope (1294–1303). He was a papal diplomat and lawyer who travelled widely. He succeeded Pope Celestine V. He quarrelled disastrously with *Philip IV of France when he asserted papal authority to challenge Philip's right to tax the clergy. In response Philip had him siezed in 1303. The shock hastened the pope's death and contributed towards the transfer of the papacy from Italy to Avignon in France.

Bonnie Prince Charlie   *Pretender, *Stuart.

Boone, Daniel (*c*.1735–1820), American pioneer. Born in Pennsylvania, he made trips into unexplored areas of Kentucky from 1767 onwards, organizing settlements and defending them against hostile Indians. In 1775 he opened the *Wilderness trail, which became the main route from Virginia to Kentucky. He later moved further west to Missouri, being granted land there by the Spanish in 1799. He remained there after Missouri became part of the United States in the Louisiana Purchase (1803).

Borgia, Cesare, Duke of Romagna (*c*.1475–1507), the illegitimate son of Pope Alexander VI and his principal mistress, Vanozza dei Caltaneis. His career provides a classic example of how 15th-century popes used their families to recover power and prestige lost during the *Great Schism (1378–1417). He was made archbishop and cardinal but his lack of religious vocation quickly became notorious. He went as papal legate to France where he married the daughter of the King of Navarre. His support of his father's ambitions was seen during the French invasions of Italy (after 1494) and in his attempted reconquest of papal lands in central Italy. His hopes of carving out his own kingdom there ended with the death of Alexander (1503). Imprisoned by Pope *Julius II, he escaped to Navarre and was later killed fighting in Castile. Ruthless and unscrupulous, he served as a model for *Machiavelli's book *The Prince*.

Borgia, Lucrezia, Duchess of Ferrara (1480–1519), the beautiful illegitimate daughter of Pope Alexander

Boniface VIII presiding over the Sacred College of Cardinals, a detail of a miniature from the *Decretals of Boniface VIII*, a 14th-century Italian manuscript. (British Library, London)

A portrait possibly of Lucrezia **Borgia**, known as the *Young Courtesan*, by the early 16th-century Italian painter Bartolomeo Veneto. Lucrezia was associated by reputation with her brother Cesare's crimes. Her third husband, Alfonso d'Este, was heir of the Duke of Ferrara and her court became a centre for artists, poets, and scholars. (Städelsches Kunstinstitut, Frankfurt)

VI, sister of Cesare Borgia. To further the ambitions of her father and brother she was married three times—to Giovanni Sforza of Pesaro (1493), a marriage annulled to enable that with Alfonso of Aragon (1500). Her third husband was Alfonso d'Este (1501) who became Duke of Ferrara (1505), ruling over a brilliant court. Henceforth she devoted herself to the patronage of art and literature, to works of charity, and the care of her children.

Boris Godunov (*c.*1551–1605), Tsar (Emperor) of Russia (1598–1605). He began his career of court service under *Ivan the Terrible, became virtual ruler of Muscovy during the reign of his imbecile son Fyodor (1584–98), and engineered his own elevation to the Tsardom. He conducted a successful war against Sweden (1590–5), promoted foreign trade, and dealt ruthlessly with those *boyar families which opposed him. In 1604 boyar animosity combined with popular dissatisfaction ushered in the 'Time of Troubles'—a confused eight-year dynastic and political crisis.

Borneo, an island east of the Malay peninsula in south-east Asia. Its name is derived from the sultanate of Brunei (of which Borneo is a European corruption), which reached the height of its influence in the 16th century just as Europeans were beginning to appear in south-east Asia. The name was extended to cover the whole island. Its Negrito inhabitants were dispersed by Malays who began arriving *c.*2000 BC. Living in river valleys separated by mountains and jungle, these early immigrants developed as isolated communities, some in longhouses, some in *kampongs* (villages) built over water, some nomadic. Apart from Indians and Chinese seeking kingfisher feathers, birds' nests for soup, and jungle produce, there was little outside contact until the 16th century when Portuguese and Spaniards began arriving. Brunei fended off Spanish attacks from the Philippines. In the 17th and 18th centuries the Dutch and British gained footholds, and after the fall of *Johore in the late 17th century many settlers came from the Malay peninsula. From the 16th century Chinese began establishing pepper plantations and mining communities.

Bornu *Kanem-Bornu.

borough, a town in England, enjoying particular privileges. The boroughs evolved from the Anglo-Saxon *burhs* and from the 12th century benefited from royal and noble grants of *charters. Their representatives attended Parliament regularly from the 14th century, having first been summoned in 1265 when Simon de *Montfort called two representatives from each city and borough. The Scottish equivalent of the English borough was the burgh, of which there were three types. The Burgh of Barony was located within a barony and controlled by magistrates; the Burgh of Regality was similar, with exclusive legal jurisdiction over its land; the Royal Burgh had received its charter from the crown. Freeholders in a borough were known as burgesses.

Boscawen, Edward (1711–61), British admiral, known as 'Old Dreadnought'. He served in the West Indies during the War of *Jenkins's Ear and the War of the *Austrian Succession, and was in charge of naval operations at the siege of *Louisburg, Nova Scotia, in

The **Boston** massacre, an engraving by Paul Revere (1735–1818). Printed and sold in Boston, it shows the first casualties of the American War of Independence. A minor incident in itself, the Boston massacre added fuel to the revolutionary fervour of the American colonies. (Metropolitan Museum of Art, New York)

1758, where his success opened the way for the conquest of Canada. His most famous exploit was the destruction of the French Mediterranean fleet off the Portuguese coast at Lagos in 1759, which helped to establish British naval supremacy in the *Seven Years War.

Bosnia and Hercegovina, a region in eastern Europe now in Yugoslavia. First inhabited by the *Illyrians, the region became part of the Roman province of Illyricum. *Slavs settled in the 7th century and, in 1137, it came under Hungarian rule. The Ottomans invaded in 1386 and after much resistance made it a province in 1463. They governed through Bosnian nobles, many of whom became Muslim, though much of the population

became rebellious as Ottoman power declined. During the early 18th century Austrian forces began to push the Turks back.

Boston, Massachusetts, USA, a city founded on a peninsula at the mouth of the Charles River by Puritan emigrants in 1630. As the colony capital, it housed some 10,000 people by 1700 and was the leading trading centre of New England. Bostonians took the lead in resisting British attempts at taxation with the *Stamp Act Riots (1765). In 1770 troops threatened by a mob opened fire, killing five people in the Boston massacre. When tea ships from England in 1773 threatened other tea importers and clandestine revenue-raising under the *Townshend

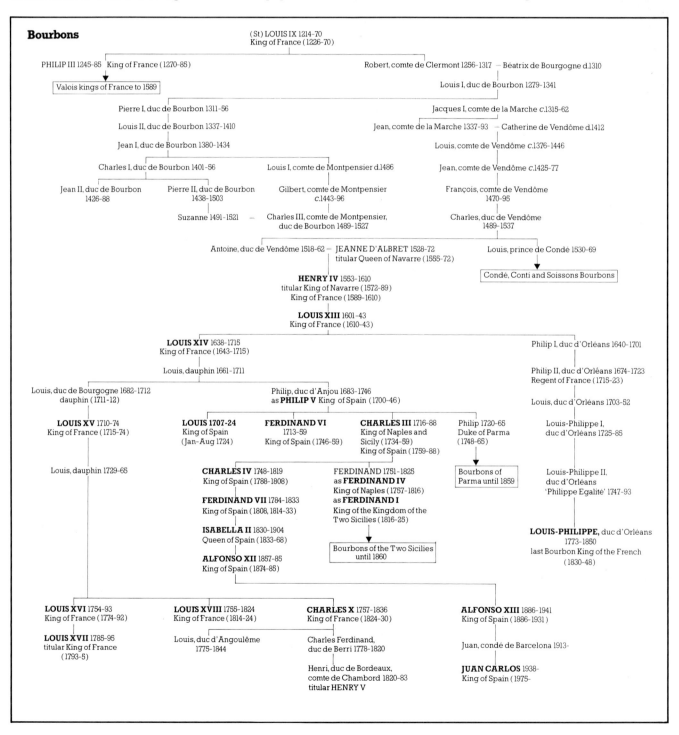

Acts, the *Sons of Liberty threw the cargo overboard in what was known as the Boston Tea Party. The city, having been evacuated by the British, was entered by *Washington in 1776, and was a *Federalist stronghold in the early republic.

Bosworth Field, battle of (22 August 1485). A battle fought close to the English town of Bosworth in Leicestershire, its outcome was to establish *Henry VII and the Tudor dynasty on the English throne. Just over a fortnight after Henry had landed on the Welsh coast, he and his army of Welsh followers were met in battle by *Richard III's larger army. The issue was uncertain when Lord Stanley arrived and with his followers went over to Henry's side; Henry was victorious, Richard was killed, and at the end of the day Stanley placed Richard's crown on Henry's head.

Bothwell, James Hepburn, 4th Earl of (1536–78), Scottish Protestant nobleman, the third husband of *Mary, Queen of Scots. He was a supporter and adviser of Mary, while she was married to *Darnley. In 1567 he was acquitted of Darnley's murder but then his swift divorce, promotion to the dukedom of Orkney and Shetland, and marriage to Mary caused the Scottish lords to rise against him. He fled from Scotland after the battle of Carberry Hill (June 1567), when Mary's forces were defeated. He turned to piracy, but was captured in Norway, and died in a Danish prison.

Boudicca (or Boadicea) (d. AD 60), Queen of the *Iceni in East Anglia. She succeeded her husband Prasutagus, who had left his kingdom jointly to his daughters and the Roman emperor, on his death in AD 60. The Romans displayed greed and brutality in taking over the kingdom, and her tribe was reduced to the status of an occupied people. She took advantage of the absence of the main Roman forces on a campaign in Wales and led the Iceni and Trinovantes in revolt. Roman settlements at Camulodunum (Colchester), Verulamium (St Albans), and Londinium (London) were burnt. Defeated by the governor Suetonius Paulinus, she committed suicide by taking poison.

Bourbon, a great European ruling dynasty, founded when **Robert of Clermont** (1256–1317), the sixth son of Louis IX of France, married the heiress to the lordship of Bourbon. The first duke was their son, **Louis I** (1279–1341). In 1503 the title passed to the Montpensier branch of the family, but in 1527 headship of the house of Bourbon passed to the line of Marche-Vendôme. **Antoine de Bourbon** (1518–62), duc de Vendôme, became King Consort of *Navarre, while his brother Louis (1530–69) was made Prince of *Condé. On the death of the last *Valois king in 1589, Antoine's son became King of France as *Henry IV (ruled 1589–1610). His heirs ruled France uninterruptedly until 1792: Louis XIII (ruled 1610–43), Louis XIV (ruled 1643–1715), Louis XV (ruled 1715–74) and Louis XVI (ruled 1774–92). The latter was overthrown during the *French Revolution, and Louis XVII (titular king 1793–5) died without reigning; Louis XVI's brothers Louis XVIII (ruled 1814–24) and Charles X (ruled 1824–30) both ruled after the Bourbon restoration. Louis-Philippe (ruled 1830–48), the last Bourbon King of France, was a member of the cadet *Orléans branch of the family.

A **Bow Street Runner**, described as 'Mr Townsend, Police-officer, Bow Street'. A force of men responsible for law and order in the 18th century, they were nicknamed 'Robin Redbreasts' after their distinctive red waistcoats.

In 1700 Louis XIV's second grandson became Philip V (ruled 1700–46) of Spain, thus setting in train the War of the *Spanish Succession. His successors have held the Spanish throne ever since (excepting the republican period, 1931–75).

bowl cultures   *Western Neolithic.

Bow Street Runners, the first organized police force, based at Bow Street Magistrates' Court in London, recruited by the magistrate (and novelist) Henry Fielding from the 1740s to augment the forces at his disposal. Their functions included serving writs and acting as detectives. They gained a reputation for efficiency and were much feared and respected by criminals. The formation of the London Metropolitan Police Force in 1829 brought their separate existence to an end.

boyar, a member of the highest non-princely class of medieval Russian society. In the 10th to 12th century the boyars formed the senior levels of the princes' retinues. They received large grants of land, and exercised considerable independent power during the period of decentralization after the 13th-century Mongol conquest;

Artillery bombardments in the battle of the **Boyne**, depicted by Jan Wyck in 1693. The battle proved a decisive victory for William III's army which included Dutch, Danish, and Huguenot battalions. (National Gallery of Ireland, Dublin)

but as the grand princes of Muscovy consolidated their own power, they managed to curb boyar independence.

From the 15th to the 17th century Muscovite boyars formed a closed aristocratic class drawn from about 200 families. They retained a stake in princely affairs through their membership of the boyar *duma* or council. *Ivan the Terrible (ruled 1547-84) reduced their power significantly by relying on favourites and locally elected officials. Their social and political importance continued to decline throughout the 17th century, and *Peter the Great eventually abolished the rank and title.

Boyne, battle of the (1 July 1690), a major defeat for the Stuart cause which confirmed *William III's control over Ireland. It took place near Drogheda, where the recently deposed *James II and his Irish and French forces were greatly outnumbered by the Protestant army led by William III. When William attacked across the River Boyne James's troops broke and fled. He returned to exile in France, and William's position as King of England, Scotland, and Ireland was immeasurably strengthened. The victory is still commemorated annually by the Orange Order, a political society founded in 1795 to support Protestantism in Ireland.

Braddock, Edward (1695-1755), English general, who commanded British forces in America against the French and Indians in 1755. In his advance on Fort Duquesne (Pittsburg) he allowed himself to be ambushed crossing the Monongahela River. He died in the battle of the Wilderness and *Washington took command of the retreat of the defeated army.

Bradford, William (1590-1657), *Pilgrim leader. He was born in Yorkshire and escaped to Holland with the Scrooby separatists. After the *Mayflower* reached *Plymouth, he was elected governor and guided the colony until his death. He pacified the Indians, achieved financial independence from the London merchants, and wrote his *History of Plimmoth Plantation*.

Braganza, the ruling dynasty of Portugal (1640-1910). Alfonso, an illegitimate son of John I of Portugal, was made first Duke of Braganza (1442). His descendants became the wealthiest nobles in the kingdom, and, by marriage into the royal family, had a claim to the Portuguese throne before the Spaniards took control of the country in 1580. When the Portuguese threw off Spanish rule in 1640, the 8th Duke of Braganza ascended the throne as John IV. The title of Duke of Braganza was thenceforth borne by the heir to the throne.

Brandenburg, a German state, the nucleus of the kingdom of *Prussia. German conquest of its Slavic population began in the early 12th century. The margravate (established c.1157) took its name from the town of Brandenburg, west of Berlin. In 1356 the *margrave's status as an imperial *Elector was confirmed by the Golden Bull of *Charles V.

Strong central government began with the advent of the *Hohenzollern dynasty in 1415. Brandenburg accepted the Lutheran *Reformation after 1540, and in the early 17th century it acquired further territories in western Germany and also Prussia (1618). After an initial period of neutrality during the *Thirty Years War, *Frederick William (the 'Great Elector') (1620-88) entered the fighting and secured excellent terms at the Treaty of *Westphalia (1648). He subsequently achieved full sovereignty in Prussia (1660) and turned Brandenburg-Prussia into a centralized European power with a highly effective army and bureaucracy. He used the opportunity offered him by Louis XIV's persecution of the *Huguenots to develop his country's industry and trade. In 1701 Elector Frederick III (1657–1713) was granted the title of King in Prussia, and from that time Brandenburg was a province of the Prussian kingdom.

Brandywine, battle of (11 September 1777). An engagement in the American War of *Independence when British forces were attacking Philadelphia. Washington was outmanoeuvred by Sir William Howe and was forced to retreat with heavy losses. The British occupied Philadelphia on 27 September, but this victory was offset by *Saratoga less than a month later.

Brazil, the largest country in South America, and the only one originally established as a Portuguese colony, having been awarded to the Portuguese crown by the Treaty of *Tordesillas (1494). Settlement of Brazil began in 1532 with the foundation of São Vicente by Martim Afonso de *Sousa. During the first half of the 16th century twelve captaincies were established. No centralized government was established until 1549 when Thomé de Sousa was named governor-general and a capital was established at Salvador (Bahia). The north-eastern coast was lost to the Dutch briefly in the 17th century but was regained.

Breakspear, Nicholas *Adrian IV.

Breda, a Dutch city in North Brabant, historically an important frontier town close to the Belgian border. The **Compromise of Breda** in 1566 was a league formed by Protestant and Catholic nobles and burghers to fight against *Philip II's policies in the Netherlands. The most dramatic event in its history was its surrender to the Spanish commander Spinola in 1625, the subject of a famous picture by Velázquez; it was retaken by the Dutch in 1636. It finally became part of the Netherlands in 1648. The **Declaration of Breda** was made by *Charles II in 1660 just before his Restoration, promising an amnesty, religious toleration, and payment of arrears to the army. The **Treaty of Breda** (1667) ended the second *Anglo-Dutch War.

Breitenfeld, battles of, two battles during the *Thirty Year War, which take their name from a village near Leipzig (now in East Germany). The first was fought on 17 September 1631, between Count Johannes *Tilly's Catholic forces and the Protestant army of *Gustavus Adolphus of Sweden. Despite an early advantage, Tilly's traditional infantry squares were overwhelmed by the Swedes' flexible linear tactics. Gustavus's victory was the first major Protestant success of the war, and it announced the arrival of Sweden as a power on the European stage. The second battle, on 2 November 1642, ended in another Swedish victory.

Brendan, St (the Navigator) (c.484–577), founder and abbot of the monastery of Clonfert in Galway, Ireland (c.560). According to tradition he visited holy sites in Ireland and western Scotland, including that of St Columba on the island of Iona. His travels were fictionalized in a remarkable 11th century work, *Brendan's Voyage*, which makes use of tales from Irish mythology. It relates some astonishing adventures when St Brendan, with a group of monks, sailed in a leather boat to a

The Surrender of Breda, 1635, a painting by Diego Velázquez. It illustrates the handing over of the keys of **Breda** fortress in 1625 by the Dutch general Justin of Nassau to the commander of the opposing Spanish army, Ambrosio Spinola. (Museo del Prado, Madrid)

'Land of Promise' in the Atlantic which has been identified with a number of places, including the Canary Islands, and even Newfoundland.

Brétigny, Treaty of (1360), concluded between Edward III of England, and John II of France following John's defeat and capture at *Poitiers. It released John on payment of a ransom of three million crowns, brought the *Hundred Years War temporarily to a halt, and saw the English renounce claims to Anjou and Normandy while retaining Gascony and Guyenne. It was never fully implemented, and Anglo-French hostilities broke out again in 1369.

Brian Boru (c.926–1014), the last High King of Ireland (1011). He had previously made himself ruler of Munster and Limerick in southern Ireland. In doing so Brian, ruler of the Dal Cais dynasty of Munster, overcame the influence of the powerful Uú Néill dynasty which had dominated Ireland for three centuries. In 1012 the men of Leinster, and supported by the Norse settlers of Dublin, rose in revolt. The battle of Clontarf brought victory to Brian's forces, though he was killed in the fighting.

Bridgewater, Francis Egerton, 3rd Duke of (1736–1803), British landowner, pioneer of canal construction. His estates included coal-mines and to move his coal cheaply, he financed the cutting of the Bridgewater canal from Worsley to Manchester. James Brindley (1716–72) was the gifted engineer he employed for this project, completed in 1772, which not only caused a dramatic reduction in the price of coal there, but introduced new methods of canal construction and inaugurated the great era of English canal-building in the 1770s and 1780s.

Brigantes ('mountain folk'), the Celtic inhabitants of northern Britain between the Humber and the Tyne. After the Roman invasion in AD 43, Emperor Claudius formed an alliance with their queen Cartimandua. Roman troops helped suppress at least three revolts against her; she also handed over the refugee *Caratacus. During the *Roman civil wars 68–9 she was expelled by her anti-Roman husband. Petilius Cerealis was made governor (legatus) of Britain by Vespasian, advanced north *c.*AD 71–4, and established Eboracum (York) as a permanent legionary fortress for the Ninth Legion in this former tribal territory.

Brissot de Warville, Jacques Pierre (1754–93), French journalist, social reformer, and politician. He abandoned the legal profession in 1776 for a career in journalism. He was the founder of the French anti-slavery movement, and wrote pamphlets and newspapers in England, Switzerland, and America before founding an extremist newspaper, *Le Patriote Français* in 1789. He was a prominent member of the *Jacobin Club and was elected to the Legislative Assembly, where he pressed for war against the central European countries. His influence and that of his supporters (the Brissotins, later called *Girondins) declined after the military failures of 1792, and he died on the guillotine.

Britain, Great, the countries of England, Wales, and Scotland, and small adjacent islands, linked together as a political and administrative unit. *Wales was incorporated into England in the reign of Henry VIII. In 1604 James I was proclaimed 'King of Great Britain', but although his accession to the English throne (1603) had joined the two crowns of *England and *Scotland the countries were not formally united. In the aftermath of the English Civil War, Oliver Cromwell effected a temporary union between England and Scotland, but it did not survive the *Restoration. The countries were joined by the Act of *Union (1707) which left unchanged the Scottish judicial system and the Presbyterian church.

Brittany, an isolated region in north-west France which has always enjoyed considerable autonomy, retaining a distinct local language and culture. After the Roman conquest in 56 BC it became part of the province of Armorica. Celtic missionaries brought Christianity by the 8th century and it developed a regional church distinguished by local saints and customs. It was able to withstand assault by *Merovingians and *Carolingians, achieved unity in the 9th century, and passed to the dukes of Brittany in the 10th century.

A civil war between English and French candidates for the succession to the French crown took place during the *Hundred Years War, and Brittany passed to the French crown (1488; formally incorporated 1532) through the marriages of Breton heiress Anne to Charles VIII and Louis XII of France. Retaining an independent tradition, it was a centre of opposition to the *French Revolutionaries in the 1790s.

Broken Hill, a mine at Kabwe in central Zambia where a fossilized human skull and other bones representing three or four individuals were found in 1921. Bones of extinct animals and stone tools of late *Acheulian type were found with the remains. The skeletal material, once called Rhodesian Man, but now usually referred to as Kabwe or Broken Hill Man, is important because it represents a population in Africa 400,000–200,000 years ago that is transitional between late *Homo erectus* and early or 'archaic' *Homo sapiens*. This skull, and one from Bodo in Ethiopia, may have been ancestral to two lineages—anatomically modern humans and European *neanderthals—or just to modern humans.

Bronze Age, the prehistoric period when bronze, an alloy of copper and tin, was the principal material used for cutting tools and weapons. Appreciation of the advantages of alloying came slowly, and various mixes were tried before the optimum 10 per cent tin was arrived at. The transition from the *Copper Age is therefore difficult to fix, as is that to the *Iron Age which followed. It is now accepted that the technological advance to bronze was made on several separate occasions between 3500 and 3000 BC in the Near East, the Balkans, and south-east Asia, and not until the 15th century AD among the Aztecs of Mexico. Knowledge of the new alloy spread slowly, mainly because of the scarcity of tin, so the Bronze Age tends to have widely different dates in different parts of the world. Indeed sub-Saharan Africa and Australasia, nearly all of America, and much of Asia never experienced a Bronze Age at all.

Although much more metal came into circulation in Bronze Age cultures, the high cost of tin led to two significant results. International trade increased greatly in order to secure supplies, and greater emphasis on social stratification is noticeable practically everywhere following the introduction of bronze, as those able to produce or obtain it strengthened their power over those without it.

A late **Bronze Age** incense burner stand (12th century BC) found at Episkopi, Cyprus. The stand consists of a ring supported on four legs. Each of the four side panels depicts a man wearing the long robes favoured by the Phoenicians and carrying a different object—in this case a copper ingot in the shape of an oxhide. Cyprus was a major source of copper for Phoenician traders. (British Museum, London)

Browne, Robert (*c*.1550–1633), English Protestant Nonconformist, founder of a religious sect, the 'Brownists'. His followers were the first to separate from the Anglican Church after the Reformation. His treatise *Reformation without Tarrying for Any* (1582) called for immediate separatism and doctrinal reform. Mental instability undermined his leadership, and by 1591 he was reconciled to the Anglican Church. He is seen by English and American *Congregationalists as the founder of their principles of church government.

Bruce, Robert   *Robert I (the Bruce).

Brunhilda (d.534–613), Visigothic queen of the *Merovingian kingdom of Austrasia. After her husband's assassination she tried to rule in the name of her son Childebert II but, faced with internal revolts and the opposition of the King of Neustria, she fled to Burgundy. In old age she claimed Burgundy and Austrasia in the name of her great-grandson, but Chlothar of Neustria defeated her. She is alleged to have been executed by being dragged to death by wild horses.

Bruno, Giordano (*c*.1548–1600), Italian philosopher, who influenced later philosophers, notably Leibniz (1646–1716) and *Spinoza. He entered the *Dominican order in 1562 and was ordained priest in 1572. He was charged with heresy in 1576, and entered upon a peripatetic career of teaching and writing. He accepted the main astronomical teachings of Copernicus, criticized *Aristotle, and published many books, the ideas of which about the nature of truth and of the universe angered orthodox Roman Catholics. Denounced to the Inquisition in 1592 he was burned to death for heresy.

Bruno of Cologne, St (*c*.1032–1101), founder of the *Carthusian order of monks at La Grande Chartreuse in France, *c*.1084, where he was the first prior. He advocated the separation of his monks from the secular world and adherence to a very strict regime of self-denial. Respect for the order grew rapidly and he was summoned to Rome around 1090 to advise Pope Urban II about the state of the church. In England Carthusian houses were called charterhouses; they were forcibly closed during the dissolution of the monasteries in the 16th century.

Brutus, Marcus Junius (*c*.85–42 BC), Roman soldier, one of the assassins of Julius *Caesar. He was the nephew of *Cato and a conservative republican Roman. He took *Pompey's side against Caesar in the *Roman civil wars. Pardoned by Caesar after *Pharsalus, he became governor of Cisalpine *Gaul, and then urban praetor in 44 through Caesar's favour. Together with Cassius he plotted Caesar's death. It was Brutus' idealism which confined the conspirators' action to the single act of killing Caesar: they thereby lost the political initiative to the consul Antony, whom they had spared, and were compelled to flee, afterwards forming a fleet and army in Greece against *Mark Antony and *Octavian. Defeat at *Philippi in 42 was followed by his suicide.

buccaneer, a pirate or privateer who preyed on Spanish shipping and settlements in the Caribbean and South America in the 17th century. Mainly of British, French, and Dutch stock, buccaneers made their headquarters first on Tortuga Island off Haiti and then on Jamaica.

In wartime they formed a mercenary navy for Spain's enemies, fighting with reckless bravery. Their triumphs included the sackings of Porto Bello, Panama, Chagres, New Segovia, and Maracaibo. Henry *Morgan was their most famous commander. After 1680 they penetrated to the Pacific coast of South America. Their power and prosperity rapidly declined in the early 18th century.

Buckingham, George Villiers, 1st Duke of (1529–1628), English statesman, favourite of James I and Charles I. In 1615 James appointed Villiers, a young man of no distinction but attractive to the king, to the office of Gentleman of the Bedchamber. Thereafter he amassed fortune and power for himself and his followers by distributing offices and favours. His personal extravagance, promotion of Archbishop *Laud, and political incompetence combined to tarnish the reputation of the court. He accompanied Charles, Prince of Wales, to Madrid in 1623 in the hopes of arranging a marriage for him with the Spanish Infanta, an expedition which served only to fuel hostile rumours of Charles's conversion

A portrait of George Villiers, 1st Duke of **Buckingham**, *c*.1616, attributed to William Larkin. As a favourite of the Stuart kings, Buckingham virtually ruled England during the period of his ascendancy until his murder in 1628. (National Portrait Gallery, London)

to the Catholic faith. After Charles's accession in 1625 Buckingham remained the king's policy-maker, ignored Parliament's hostility towards war, and insisted on a costly campaign against Spain which ended with a disastrous expedition to Cadiz in 1625. Parliament attempted to impeach him, charging him with corruption and financial mismanagement, but the king dissolved Parliament and Buckingham pursued campaigns against both France and Spain, personally leading an unsuccessful expedition to the relief of the *Huguenots at La Rochelle. In 1628 he was murdered by a soldier aggrieved by the mismanagement of the war.

Buddha (Siddhartha Gautama) (*c*.563–483 BC). He was born a prince near Kapilavastu close to the modern India–Nepal border in the Himalayan foothills. Tradition says that despite his luxurious and happy upbringing, when he saw sickness, old age, and death, and a serene holy man, Siddhartha decided to renounce the world at the age of 29, to find the answer to the riddle of existence. Leaving his wife and child, he became a wandering *ascetic, but found no answer to his quest in extreme asceticism. He then chose a 'middle way' between this self-torture and the self-indulgence of the world, and at last achieved enlightenment under the Bo-Tree. Henceforth he was called the Buddha, 'the Enlightened One'. For forty-five years he travelled in north India, preaching and gathering a large number of monks, nuns and lay-followers. On his death he is believed to have passed into nirvana, complete liberation.

The Buddha's teaching is set in the context of the Indian belief in a cycle of rebirth, and the aim to find a

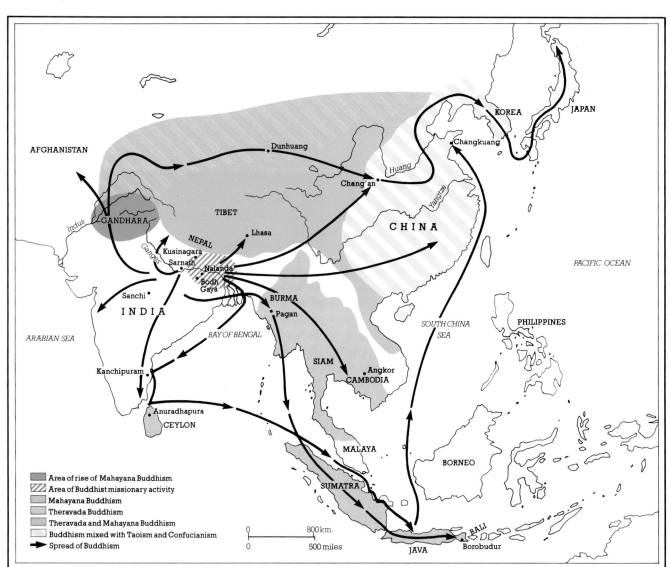

Buddhism: its expansion to *c*.AD 1000

The expansion of Buddhism beyond India can be traced to the reign of Asoka (273–232 BC), a convert to the religion. He sent out missions, notably to Anuradhapura, capital of Ceylon (Sri Lanka), where Buddhism was established by a relative, perhaps a son. The adoption of Buddhism in other countries was gradual. By the 1st century AD it was spreading throughout central Asia along the Silk Route, and was known in China, although it did not become widely popular there for another four centuries. From China it reached Korea in the 4th century, and Japan in the 6th century, challenging Shinto for a time. It was established in Sumatra and Java in the 5th century and had also spread to Siam (Thailand), Cambodia, and Tibet by the 7th century. Meanwhile Buddhism was in decline in India, where Hinduism was becoming the dominant religion.

complete liberation from it. Although he shared many ideas with other Indian religions, such as *Jainism and *Hinduism, his central teaching of the non-existence of any real soul or self (anatman) set him apart from them. His first sermon outlines the 'Four Noble Truths': the omnipresence of suffering; its cause, desire; the elimination of suffering by the elimination of desire; and the Eightfold Path leading to this end. The Eightfold Path comprises Right Understanding, Intention, Speech, Conduct, Livelihood, Effort, Mindfulness, and Meditation. Anyone can reach nirvana, by following this Path of Morality, Meditation and Wisdom. The Buddha is widely represented in art and sculpture, often in the seated, cross-legged posture of meditation.

Buddhism, the religion derived from the teachings of the *Buddha. After his death in 483 BC, Buddhism spread widely in India, receiving considerable impetus from the support of the emperor *Asoka in the 3rd century BC. Offering a way to salvation that did not depend on caste or the ritualism and sacrifices of the Brahmin priesthood of Hinduism, and strengthened by a large, disciplined monastic order, the *sangha*, it made a very great impact; but by the end of the 1st millennium AD it had lost ground to a resurgent Hinduism centring on devotional worship, and the subsequent Muslim invasions virtually extinguished it in India. Meanwhile however, monks had taken the faith all over Asia, to central and northern areas now called Afghanistan, Tibet, Mongolia, China, Japan, Korea and Vietnam; and in south and south-east Asia to Sri Lanka, Burma, Thailand, Cambodia, and Laos.

During the centuries following the Buddha's death, many different schools emerged, usually grouped into two 'vehicles': the Mahayana, 'the great vehicle', dominant in North Asia, and what it called 'the lesser vehicle', Hinayana, of which the only form now remaining is the Theravada, 'the way of the elders', of south and south-east Asia. There are many religious and philosophical differences between the various schools and sects of the Mahayana, which include the Pure Land School and the *Zen sect, but they all differ from the Theravada in giving a greater status to the Buddha, who is sometimes seen as an eternal and transcendent being rather than a man, and in making it the aim of all Buddhists to become not an Arhat, an individual saint, but a Bodhisattva, a being who works for the salvation of all.

Bugis, Muslim mercenaries and traders of south-east Asia. They were enterprising seamen and traders living in villages in Sulawesi (Celebes). When *Macassar fell to the Dutch (1667) they lost their livelihood. Thereafter they sought employment as mercenaries and engaged in piracy in Borneo, Java, Sumatra, and Malaya. They fought for and against the Dutch. They suffered a reverse when their leader Raja Haji was killed while assaulting *Malacca (1784), but went on to found states, like Selangor and Riao, on the Malay peninsula. Their prahus, boats with a triangular sail and a canoe-like outrigger, continued to trade throughout the archipelago.

Bulgaria, a country in south-east Europe, on the Black Sea. It was settled by central Asian tribesmen in the 5th century, colonized by the Romans, and then invaded by *Slav Bulgars. They killed Emperor Nicephorus in AD 811, and captured *Adrianople in 813. Christianity was introduced in the 9th century. Greek and *Magyar threats were repulsed but rebellion, and incursions by Greeks, Russians, and *Serbs resulted in the kingdom being divided into three in the 11th century. It was annexed by the *Ottoman empire and ruled by the Turks from 1396 to 1876.

Bunker Hill, battle of (17 June 1775), a battle in the American War of *Independence ending in a British

An engraving of the battle of **Bunker Hill**, the first pitched battle of the American War of Independence. The high casualties suffered by the regular British troops as they occupied Bunker and Breed's hills overlooking Boston, led the American colonists to claim a morale-boosting triumph for themselves.

victory. Thomas *Gage, the British commander besieged in Boston, sent 2,400 troops (redcoats) to take the heights occupied by 1,600 Americans under William Prescott. Only after three bloody uphill assaults, costing 1,000 British against 400 American casualties, were they successful. The impressive defence undermined the myth of redcoat invincibility and encouraged colonial unification.

Burghley, Lord   *Cecil.

Burgoyne, John (1722–92), British general, who fought against the colonial army in the American War of *Independence. As commander-in-chief of the northern army, 'Gentleman Johnny' led the attack southward from Montreal towards the Hudson Valley in 1776 and 1777. He was unable to adapt to American conditions and was defeated at *Saratoga in 1777; this led to his recall and France's alliance with America.

Burgundy, a former duchy in south-central France. The Burgundii, a Germanic tribe, settled there in the 5th century. It was under Merovingian control and then absorbed into the *Carolingian empire. During the reign of strong Holy Roman Emperors most of it was under imperial control but in the late Middle Ages it was ruled by a series of strong dukes. *Philip the Bold acquired Flanders and John the Fearless the Netherlands. Geographically the separation of territories made government difficult and *Charles the Bold tried, but failed, to unite the northern and southern parts by annexing Lorraine. He was killed in 1477, leaving no son to succeed, and Louis XI of France claimed the duchy. The final subjection to France occurred when Louis XIV seized Franche-Comté.

During its history the duchy had achieved great power and influence, its court in the 15th century the most splendid in Europe. Certainly some of its dukes were more powerful than many kings of France and when they allied themselves with the English, as they did during the *Hundred Years War, they posed a real threat to the security of the French monarch. The court of the dukes of Burgundy was renowned for its artistic patronage; the name 'School of Burgundy' is applied to a group of Flemish panel painters and miniaturists working for them between 1390 and 1420.

Burke, Edmund (1729–97), British statesman and political theorist. He was the son of an Irish lawyer, and went to London in 1750. In 1765 he became private secretary to the Prime Minister, the Marquis of *Rockingham, soon afterwards entering Parliament, where he quickly gained a reputation as a skilful debater. On the American question he argued that Britain ought to abandon the abstract right to tax the Americans, on the ground of expediency. He became a friend of Charles James *Fox in 1774, and in defence of liberty he attacked political corruption and injustice while demanding parliamentary reform. He took a leading role in the unsuccessful prosecution (1788–95) of Warren *Hastings, former governor-general of India.

Burke split with Fox over the *French Revolution. While Fox could see little to criticize in the events in France, Burke saw liberty only in law and order. In his *Reflections on the Revolution in France* (1790) he denounced events in France as mob rule, and by supporting reform through evolution, not revolution, he laid down the basis of modern British conservatism. Burke's views on the French Revolution were more widely accepted than those of Fox's supporters, and at his retirement in 1794 he was granted a large pension by George III.

Burma, a country in south-east Asia. There was a *Mon kingdom, Prome, there in the 5th century AD. After the arrival of the Burmans in the 9th century there was much hostility between them and the indigenous peoples. Following a period of Mon ascendancy the Burmans of *Pagan unified the country for a time (c.849–1287). From the Mons an Indian script and Theravada *Buddhism spread to Pagan and thence throughout Burma. During the 16th century the country was re-united under the *Toungoo, but wars against Thai kingdoms and *Laos exhausted it. The last dynasty, the Konbaung, founded by Alaungpaya in 1757, was constantly engaged in wars against *Siam which led to the fall of the Siamese state of Ayuthia in 1767. In 1770 Burma repelled a Chinese invasion. The conquest of *Arakan brought the Burmese border to the boundary of British India.

Burnet, Gilbert (1643–1715), Scottish churchman and historian. He sought advancement in England from 1674, but moved to the Continent on the accession of James II. He became adviser to *William of Orange, accompanying him to England in 1688, and was rewarded with the bishopric of Salisbury in 1689. His greatest work was *The History of My Own Times*, published after his death, a valuable source of information on contemporary events, but coloured by Burnet's strong Whig bias.

Bute, John Stuart, 3rd Earl of (1713–92), Scottish courtier and statesman. He joined the household of *Frederick, Prince of Wales, in 1747, and after Frederick's death exercised much influence over his widow Augusta and her son who became *George III in 1760. The king appointed him Secretary of State in 1761, and in 1762 he succeeded the Elder *Pitt as Prime Minister. Bute was widely disliked and was lampooned by radicals such as *Wilkes. He resigned after forcing through the unpopular Treaty of *Paris and a contentious additional excise duty on cider. His influence with the king soon waned, and he retired from public life.

Butler, Joseph (1692–1752), English bishop and moral philosopher. He wrote *The Analogy of Religion* (1736), a defence of revealed religion against the deists and rationalists, whose attacks on religion were an important aspect of the Age of *Reason. Butler's own reasoning was rationalist in method, and he was much admired for coming as close to proving the existence of God by argument as would ever be possible. He is said to have declined the post of Archbishop of Canterbury in 1747, but became Bishop of Durham in 1750.

Buxar, battle of (22 October 1764). The town of Buxar in north-east India was the site of a decisive battle, which confirmed the *East India Company's control of Bengal and Bihar. Facing the Company were the combined forces of the *Mogul emperor (Shah Alam), the governor of Oudh (Shuja ad-Daula), and the dispossessed governor of Bengal (Mir Qasim). The Company's victory achieved recognition of its predominance in the region, demonstrated by the transfer of the *diwani* (revenue collecting powers) to the Company's agents in 1765.

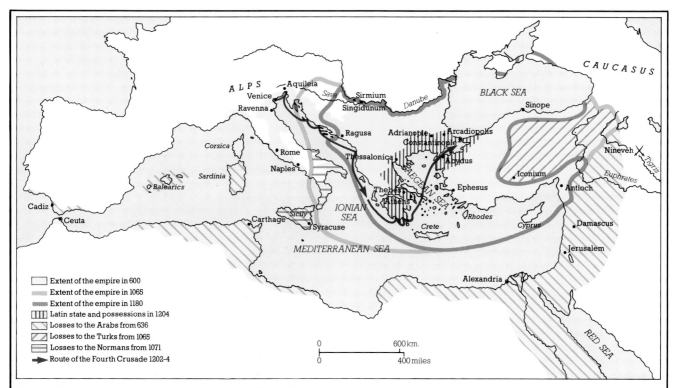

Byzantine empire: its greatness and decline 628–1204

The Byzantine empire underwent a revival during the 6th century under Justinian's rule, but in the early 7th century much of the empire was overrun by the Persians. Heraclius restored the frontiers, crushing the Persians at Nineveh (628), but the empire was soon threatened by Muslim Arabs, who won Syria, Palestine, Egypt, Africa, and Sicily before being checked in 717 at Constantinople. The empire experienced a period of prosperity until the 11th century, when an equally formidable Muslim power, the Seljuk Turks, conquered the Asian provinces, and the Normans, having captured Byzantine territory in southern Italy, attacked Greece and the Balkans. The Normans were repelled in 1085, but the empire was now in an inevitable gradual decline. Help from the Christian west against Islam was double-edged: Constantinople was seized during the Fourth Crusade, the emperor was killed, and a Latin state was established which lasted until 1621.

Byng, George, Viscount Torrington (1663–1733), English admiral. He received promotion for his loyalty to William of Orange, and gained a great reputation for his successes in the War of the *Spanish Succession. His most famous battle was in 1718 at Cape Passaro, when he sank a Spanish fleet which was attempting to take Sicily.

His son **John** (1704–57) owed his rapid and somewhat undeserved promotion to his father's influence. He was sent with an inadequate force in 1756 to save Minorca, then under siege by the French, and to protect Gibraltar, but returned to England having failed to do either. He was court-martialled for negligence and sentenced to death, providing a useful scapegoat for the government's mismanagement. His execution prompted Voltaire's famous remark that in England 'they like to shoot an admiral from time to time to encourage the others'.

Byzantine empire, the eastern half of the Roman empire. Emperor *Constantine (306–34) had reunited the two halves, divided by Diocletian (284–305), and had refounded the Greek city of Byzantium as his eastern capital, calling it *Constantinople (330). At his death in 395 Emperor *Theodosius divided the empire between his sons. After the fall of Rome to the *Ostrogoths (476) Constantinople was the capital of the empire and was famous for its art, architecture, and wealth. While barbarian invaders overran the Western empire, the Byzantine emperors always hoped to defeat them and reunite the empire. Emperor *Justinian reconquered North Africa and part of Italy, making Ravenna the western capital, but his success was shortlived.

After *Muhammad's death (632) Muslim Arab forces swept through Persia and the Middle East, across North Africa, and into Spain. By 750 only the Balkans and Asia Minor remained unconquered. From the 9th century *Charlemagne's Frankish empire dominated the West. In the 8th and 9th centuries religious disunity, notably the *Iconoclastic controversy, weakened the empire. Theological and political differences between Rome and Constantinople led to the *East–West Schism between Latin and Orthodox Christianity. (1054). The vigorous emperor Alexius *Comnenus (1081–1118) defeated barbarian attacks from the north and appealed to the Franks for help against the *Seljuk Turks. In the 12th century, some reconquests were made in Asia Minor and the period was one of achievement in literature and art, only brought to an end by the Frankish sack of Constantinople in 1204. The failure to achieve any united Christian opposition to the Turks and the growing independence of the Balkan princedoms weakened the empire. Ottoman incursions in the 14th and 15th centuries culminated in the capture of Constantinople in 1453 and the end of the empire.

Byzantium   *Constantinople.

C

cabal, a group or association of political intriguers. In England the term was generally pejoratively applied to the inner circle of the more important ministers, those who would be fully informed of all government secrets. Thus in the 17th century it was a precursor of the English *cabinet, but in modern times the term is applied to any political group which pursues its aims by underhand methods. From 1667 to 1673 the word (though of older origin) was used somewhat misleadingly of Charles II's ministers, who did not really form a united group but the initials of whose names—Clifford, Ashley, Buckingham, Arlington, and Lauderdale—happened to spell CABAL.

Cabeza de Vaca, Alvar Núñez (*c.*1490–*c.*1557), Spanish soldier. He pursued a military career, serving in Europe before joining Pánfilo de *Narváez in an expedition to Florida in 1527. When it failed, he and three other survivors spent ten years trekking 6,000 miles through the south of North America and back to New Spain. He hoped to command another expedition, but delays in returning to Spain lost him the opportunity. Instead he was made governor of Rio de La Plata, and led two 1,000-mile expeditions through the jungles and up the Rio Paraguay in 1541 and 1542. Arrested in 1543 by jealous colleagues, whom he had prohibited from looting and enslaving the local Indians, he was returned to Spain in chains. His sentence of eight years' exile in Africa was annulled, however, and a royal pension enabled him to write his *Commentarios* on his South American treks.

cabinet, the group of ministers responsible for implementing government policy. The kings of England always had advisers, but it was not until the *Restoration in 1660 that a cabinet (or cabinet council) developed, consisting of the major office-bearers, and the king's most trusted members of the Privy Council, meeting as a committee in a private room (the cabinet, whence its name) and taking decisions without consulting the full Privy Council. In the time of Queen *Anne it became the main machinery of executive government and the Privy Council became formal. From about 1717 the monarch *George I ceased to attend, and from that time the cabinet met independently. *George III became obliged, through insanity and age, to leave more and more to his ministers, but it was not until after the Reform Act of 1832 that the royal power was dissolved and cabinets came to depend, for their existence and policies, upon the support of the majority in the House of Commons.

Cabot, John (*c.*1450–98), Venetian navigator who settled in Bristol in 1484, hoping to find sponsors for a voyage in search of a route to the Orient across the Atlantic. Before he set out news came that *Columbus had already sailed and in 1496 *Henry VII granted him permission for a voyage of general exploration. He discovered Newfoundland (which he believed to be off the coast of China) and took possession of it in the king's

The English prime minister, Sir Robert Walpole, addressing his **cabinet** colleagues. This early 18th-century cartoon by Joseph Goupy emphasizes Walpole's domination of the governments he led.

name. He set off on a second expedition in 1498, but his fleet was never heard of again. His son **Sebastian** (*c.*1485–1557) led a Spanish expedition (1526) to La Plata in South America, but was turned back by hostile Indians on the Paraguay River. He returned to Bristol and after 1548 several expeditions were sent out under his auspices in search of a north-east passage (the route to the Orient to the north of Europe and Asia).

Cade, Jack (d. 1450), English rebel leader. In June 1450 he led a band of rebels from Kent in a protest against the financial oppression and general incompetence of *Henry VI's government. They occupied London for three days and put to death both the treasurer of England and the sheriff of Kent. Their rising differed from the *Peasants' Revolt of 1381 in having considerable support from the landed classes and clearer aims, but like Wat *Tyler, Cade was captured and killed.

Cadwalader (or Cadwallon) (d.633), King of Gwynedd, north Wales. His hatred of the Anglo-Saxon kingdom of *Northumbria intensified when his attempts at invasion (629) failed and he was forced to flee to Ireland. Although a Christian, he next allied with the heathen King *Penda of Mercia. Their victory at Hatfield Chase (632) over Edwin of Northumbria was followed by the devastation of Northumbria. Thereafter Northumbrian fortunes recovered and Cadwalader was killed in battle by Edwin's nephew Oswald at Heavenfield, near Hexham.

Caesar, Gaius Julius (100–44 BC), Roman general and dictator. Born into a *patrician family, he became Pontifex Maximus (High Priest) in 63 BC as part of a deal with *Pompey and *Crassus, the so-called 'First Triumvirate'; as consul in 59 he obtained the provinces of Illyricum and Cisalpine and Transalpine *Gaul. A superb general, able to inspire loyalty in his soldiers, he subjugated Gaul, crossed the River Rhine, and made two expeditions to Britain. He refused to surrender command until he had secured a second consulship for 48 BC, which would render him immune from prosecution by his enemies, by now including Pompey. When the Senate delivered an ultimatum in January 49, he crossed the *Rubicon, took Rome, and defeated Pompey at *Pharsalus in 48. He demonstrated clemency by permitting those who wished to do so to return to Italy. After campaigns in Asia Minor, Egypt, Africa, and Spain he returned to Rome in 45.

He governed Rome as dictator, finally as 'perpetual' dictator. His wide-ranging programme of reform, which included the institution of the Julian *Calendar, reveals his breadth of vision, but he flaunted his ascendancy and ignored republican traditions. It was alleged that he wanted to be king, although this was anathema to the Romans. A conspiracy was formed, led by *Brutus and Cassius, and he was assassinated on the Ides (15th) of March 44. He was later deified and a temple was dedicated to his worship in the Forum.

Caesars, a branch of the aristocratic Roman Julia clan, the name of which passed from its most famous member Julius *Caesar to become an imperial title. Julius Caesar had no legitimate sons, and his young son Caesarion (by

A basalt bust of Gaius Julius **Caesar**, the Roman statesman best remembered for his outstanding military career and conquest of Gaul. (Museo Baracco, Rome)

Cleopatra) was not recognized in Roman law. Octavian *Augustus took the name as his adoptive son. It was used by the Julio-Claudian dynasty until the line died out with *Nero in 68. All succeeding Roman emperors adopted it, conferring the title on their designated heirs so that it came signify a 'prince'. The name and title was used in the Eastern empire as 'Kaisaros'. From this were later derived the imperial Russian and German titles Tsar and Kaiser.

Caledonia, the Roman name for Scotland north of the *Antonine Wall, approximating to the Scottish Highlands. In AD 83 the governor of Britain, Agricola, invaded the territory of the Caledonii, ancestors of the later *Picts and defeated them, though they remained a constant threat to the Roman frontier in Britain, necessitating a further campaign by *Severus.

calendar, a method of counting days and other divisions of time and relating them to the phases of the moon and sun and the changing seasons of the year. In ancient Egypt a particular year was identified by reference to a king or official. In Rome the years were numbered from the foundation of the city (AUC, *ab urbe condita*), traditionally 753 BC, or, more usually, calculated by reference to the names of the consuls in office and emperors' reigns. The birth of *Jesus Christ was adopted in the 6th century as a fixed base for dates but an error of several years was made in the calculation. Muslims use *Muhammad's flight from Mecca (622) as a starting point.

Rome's earliest calendar was a lunar one of ten months, but later a twelve-month division totalling 355 days was adopted. The pontiffs inserted an extra month from time to time to keep the years roughly in line with the solar year. Orginally March was the first month (hence 'September' means 'seventh month') but had been superseded by January by 153 BC. The months still bear their Roman names. July (originally Quinctilis) was renamed after Julius *Caesar in his lifetime and August (originally Sextilis) after Augustus. Each month had three named days, the Kalends (1st), Nones (5th or 7th), and Ides (13th or 15th). In 45 BC Julius Caesar introduced the solar calendar still in use. The months were given their present number of days and the leap year every fourth year was devised to cope with the extra six hours by which the solar year exceeds 365 days. But this overcompensated by eleven minutes, fourteen seconds, which had amounted to ten full days by 1582. Pope Gregory XIII decreed their omission for that year. Catholic countries adopted the modification but it was only in 1752 that Britain came into line by striking out eleven days. The Russians persisted with the old calendar until 1918. The calendar has been modified so that for century years only those divisible by 400 are leap years.

Caligula (Gaius Julius Caesar Germanicus) (AD 12–41), Roman Emperor (37–41). His nickname derived from the miniature army-boots (*caligae*) which he wore as a child when his father Germanicus was commander-in-chief on the Rhine. He was the great-grandson of both *Augustus and *Mark Antony, and the great-nephew and successor of *Tiberius. His reign was brief, bloody, and authoritarian, scarred by mental instability, personal excesses, and delusions of divinity. He was murdered in his palace together with his (fourth) wife and only child.

caliphate, formerly the central ruling office of Islam. The first caliph (Arabic, *khalifa*, 'deputy of God' or, 'successor of his Prophet') after the Prophet Muhammad's death in 632 was his father-in-law *Abu Bakr, and he was followed by *Umar, *Uthman, and Ali: these four are called the Rashidun (rightly guided) caliphs. When Ali died in 661 *Shiite Muslims recognized his successors, the imams, as rightful possessors of the Prophet's authority, the rest of Islam accepting the *Umayyad dynasty. They were overthrown in 750 by the *Abbasids, but within two centuries they were virtually puppet rulers under Turkish control. Meanwhile an Umayyad refugee had established an independent emirate in Spain in 756 which survived for 250 years, and in North Africa a Shiite caliphate arose under the *Fatimids, the imams of the Ismailis (909–1171). After the Mongols sacked Baghdad in 1258 the caliphate, now only a name, passed to the *Mameluke rulers of Egypt, and from the *Ottoman conquest of Egypt in 1517 the title was assumed by the Turkish sultans, until its abolition in 1924.

Calvin, John (1509–64), French theologian, the leading figure in the second generation of *Protestant reformers. He was the son of a clerk and was educated at Paris, Orleans, and Bourges. About 1533 he became a convert to the reformed faith. His *Institutes of the Christian Religion* (1536) was a lucid exposition of Reformed theology, minimizing the freedom of the human will. This was followed by his *Ecclesiastical Ordinances* (1541), in which he set forth a form of church government which subsequently became a model for *Presbyterians.

A portrait of the French theologian John **Calvin** by J. Faber. According to Calvin the virtues of thrift, hard work, and sobriety were essential if the reign of God on earth was to come to pass.

From 1536 to 1564, with a three-year interval in Strasburg from 1538, he devoted himself to imposing his version of liturgy, church organization, doctrine, and moral behaviour upon the Swiss city of Geneva. It became a haven and inspiration for many European Protestants, and a base for world-wide missionary activity. Calvin's creed was particularly influential in France, the Netherlands, and Scotland. It also formed the bedrock of the *Puritan movement in England and North America. The Calvinist doctrines of predestination (God's foreordaining of what will come to pass) and legitimate resistance to 'ungodly authority' gave encouragement to Protestants who found themselves suffering at the hands of unsympathetic lay rulers.

Cambrai, League of (1508), an alliance of the *papacy, the *Holy Roman Empire, France, and Spain against *Venice. In 1529 the 'Ladies' Peace (the Peace of Cambrai) temporarily halted the Habsburg–Valois wars.

Camden, battle of (16 August 1780). A battle of the American War of *Independence in which some 2,000 American militiamen under Horatio *Gates were defeated when attacked by *Cornwallis's army 193 km. (120 miles) north-north-west of Charleston, South Carolina. Gates was replaced by Nathanael *Greene as commander of the Southern Army, which revenged itself at the battle of King's Mountain in October.

Camisards, French Protestants, who in 1702 defied *Louis XIV in the Cévennes, a mountainous region of southern France with a strong tradition of independence. The loss of their leaders in 1704 was followed by a period of savage persecution, but the rebels were bought off rather than defeated, and the authorities subsequently preferred to leave the area largely alone. Their name may come from the 'camise' or shirt they wore over their clothes.

Camperdown, battle of (11 October 1797). A naval battle fought off the coast of Holland in which the British fleet destroyed the Dutch fleet. The Dutch tried to lure the British commander on to the shoals, but he accepted the risk, chased them, and captured nine ships. This victory, and the defeat of the Spanish fleet in February at *Cape St Vincent, ended *Napoleon's hopes of invading England and enabled *Pitt the Younger to negotiate the formation of another coalition.

Campion, Edmund, St (1540–81), English *Jesuit scholar and Catholic martyr. He was ordained in the Church of England in 1568. In 1571 he left England for Douai in the Low Countries, where he joined the Roman Catholic Church; in Rome, two years later, he became a Jesuit. In 1580 he participated in the first secret Jesuit mission to England. Although he claimed that he came only to teach and minister to the Catholic community, he was arrested, tortured, tried, and executed for treason.

Canaan, an ancient name for *Palestine, the 'Promised Land' of the Israelites, promised by their God to *Abraham and his descendants. The Canaanites inhabited the area by about 2000 BC and from 1500 BC were periodically subject to the Egyptians and Hittites. The arrival of the Israelites in the 13th century following their

*Exodus from Egypt and their gradual conquest of the area confined the Canaanites to the coastal strip of *Phoenicia. The religion of the Canaanites, including the worship of local deities called Baals, sacred prostitution, child sacrifice, and frenzied prophecy, all incurred the condemnation of the Hebrew prophets, particularly in the 9th century, the period of *Elijah and Elisha.

Canada, a country that occupies the northern half of North America. Originally inhabited by *North American Indians and by Inuit in the far north, in the 10th century *Vikings established a settlement at *L'Anse aux Meadows. John *Cabot landed in Labrador, Newfoundland, or Cape Breton Island, in 1497 and in 1534 Jacques *Cartier claimed the land for France. The first French settlement was begun by *fur traders in *Acadia in 1604. In 1608 Samuel de *Champlain founded *Quebec on the St Lawrence River. Governor *Frontenac defended Quebec against Sir William Phips (1691) and led a successful campaign against the hostile *Iroquois (1696). Explorers followed the routes of the Great Lakes and the Mississippi Valley—*La Salle reached the mouth of the Mississippi in 1682—and the name Canada came to be used interchangeably with that of *New France, which referred to all French possessions in North America. Conflict between Britain and France was mirrored in Canada in the *French and Indian wars. By the Peace of *Utrecht (1713) France gave up most of Acadia, Newfoundland, and Hudson Bay. The remainder of New France was conquered by Britain and ceded in 1763. During or immediately after the American War of *Independence some 40,000 *United Empire Loyalists arrived in Nova Scotia (formerly Acadia) and present-day Ontario.

Cannae, battle of (216 BC). The village of Cannae in southern Italy was the site of one of the classic victories in military history. The Carthaginian general, *Hannibal, his infantry considerably outnumbered, but stronger in cavalry, stationed his troops in a shallow crescent formation. The densely-packed Roman legionaries, under the consuls Aemilius Paullus and Terentius Varro, charged Hannibal's centre, forced it back, but failed to break it. As it slowly and deliberately gave ground, and the Romans pushed deeper, Hannibal effected his brilliant double-encirclement: his cavalry, having defeated the opposing right and left wings, closed the trap and assaulted the Romans from flanks and rear. Out of some 50,000 men the Romans lost 35,000 killed or captured, Hannibal only 5,700. Rome's hold on Italy was imperilled, and many of its allies in central and southern Italy defected to Hannibal.

canonist, a compiler of or commentator on canon law, the law of the Western Christian Church. Church law evolved to deal with matters of discipline, organization, and administration, as well as of general morality and liturgy. Guidance was received from the scriptures, from the influence of Roman law, from church councils and from the writings of St *Paul and of the Church Fathers. Important figures in the development of a body of law enforced by the *papacy included Pope Gregory IX whose *Decretals* of 1234 represented the first official collection of papal law. Significant adjustments were made within the Roman Catholic Church at the Council of *Trent, in response to the Protestant *Reformation, and at the Vatican Council.

Canossa, a castle in the Apennines, in Italy, where, in the winter of 1077, the German emperor *Henry IV waited for three days until Pope *Gregory VII granted absolution and removed a ban of *excommunication from him. Henry had been at odds with the papacy over ultimate control within the Holy Roman Empire. His penance greatly strengthened his hand against the German princes who threatened him, for they had been allies of the pope and when Henry was absolved the princes withdrew their support for Gregory.

Canton   *Guangzhou.

Canute (or Cnut) (c.994-1035), King of England, Denmark, Norway, and Sweden, one of the most powerful

King **Canute** and his concubine Aelgifu of Northampton placing a cross on the altar of Hyde Abbey. This full-page drawing has been reproduced from the *Liber Vitae*, the abbey register, which was completed in c.1020. (British Library, London)

rulers in Europe. He accompanied his father, *Sweyn, in the invasion of England (1013) and was chosen King of Denmark on Sweyn's death in the following year. After a long struggle with *Ethelred II and his successor *Edmund II of Wessex which ended with Edmund's murder (1016), he became King of England (1017), marrying Emma, the widow of Ethelred. His reign was marked by legal and military reforms and by internal peace, Canute wisely using both Englishmen and Danes as advisers. The legendary incident in which Canute commanded the tide to recede was related in Henry of Huntingdon's *Historia Anglorum.*

Cao Cao (or Ts'ao Ts'ao) (155–220), Chinese general. One of China's greatest soldiers, he unified much of northern China after the collapse of the *Han dynasty. His conquests enabled his son to found the Wei kingdom, one of the *Three Kingdoms. His campaigns and adventures are recorded in one of the classics of Chinese literature, *The Romance of the Three Kingdoms.*

Cape St Vincent, battle of (14 February 1797). A naval battle off the south-west coast of Portugal in which *Nelson and *Jervis defeated a combined French and Spanish fleet of twenty-seven ships. The British were outnumbered almost two to one, but the disorder of the Spanish fleet cancelled out its advantage in numbers. After this victory the British fleet was able to continue its blockade of Cadiz and to re-enter the Mediterranean in pursuit of Napoleon in Egypt.

Capetian (987–1328), the dynasty of French kings who succeeded the *Carolingians. Not until the reign of Louis VI (1108–1137), did the dynasty manage to establish firm control over its own territories around Paris and begin the slow process of gaining real power in France. Philip Augustus (1180–1223) seized Normandy and recovered many other areas which had been occupied by, or were under the influence of, the English crown. This effectively doubled the size of the country. Paris became the true centre of government. By the end of the reign of Philip IV (1285–1314) France had achieved a great degree of stability and acquired many of the legal and governmental systems which were to survive up to the time of the French Revolution. On the death of Charles IV in 1328 the throne passed to the *Valois House but they, as also the later *Bourbons, could claim indirect descent from Hugh Capet (ruled 987–996), the first of the line.

Cappadocia, the central area of ancient Asia Minor. Its *satrap Ariarathes resisted *Alexander the Great's Macedonians until he was killed in 322 BC. After 301 his descendants re-established control and Ariarathes IV fought against the Romans at Magnesia in 190, though he and his successors thereafter aligned themselves with Rome. Cappadocia suffered badly at the hands of the neighbouring Armenians during the Mithridatic War, though the Roman general *Pompey aided recovery by granting loans for town building. The area was annexed to the Roman empire in AD 17.

Caratacus (or Caractacus or Caradog) (d. AD 54), King of the Catuvellauni, a tribe of southern Britain. He was the son of *Cunobelinus. After his father's death c.AD 40 and the Roman capture of Camulodunum he resisted the Romans, but was defeated and fled to the *Brigantes, who handed him to *Claudius, in 51. He and his family were kept as respected hostages in Rome.

caravanserai (Persian, *karwansaray*, 'caravan place'), an inn along Asia's caravan routes. The shelters, often under municipal supervision, provided a place for resting, making repairs, and preparing for the next stage of a journey. They were centres of news, information, and often espionage. At terminals like Damascus they were 'mansions', accommodating several hundred camels or mules, with storerooms at ground level and sleeping quarters above. On desert routes, at points where there was a little brackish water, they offered the barest facilities. Generally located outside the town walls, they were constructed round a courtyard, sometimes arcaded, usually with a well at the centre.

caravel, a small light ship, used especially by the Spanish and Portuguese in the 15th–17th centuries. Caravels were common as trading ships in the Mediterranean and were used for the early voyages of discovery. Christopher *Columbus's flagship of 1492, the *Santa Maria*, was a 29 m. (95 ft.) long caravel, and Bartholomew Diaz rounded the Cape of Good Hope in a similar vessel in 1488. For long ocean voyages they were developed into three-masted ships.

Caribbean, a sea and its islands on the Atlantic side of Central America. Archaeologists have established that by c.5000 BC the Ciboney, hunter-gatherer-fisher people from South America, had crossed the seas to *Hispaniola and *Cuba, and eventually occupied the other Antilles. From c.1000 BC to c. AD 200 they developed agriculture and pottery, and through the period c. AD 200–c.1000 the Ciboney were mostly replaced by migrating Arawak Indians from north-eastern South America. The influence of Mesoamerica (Mexico and northern Central America) is seen in some of their religious practices, in ceremonial centres of platform-mounds round plazas, and in the growing of maize as a secondary crop. This 'Taino' Arawak culture persisted in the Greater Antilles, but was

A guard keeps watch by the door of an open **caravanserai**, an illustration from an Italian manuscript. Inside the inn travellers enjoy warmth and shelter, their weapons hung up on the wall and their horses safely tethered. (Museo Correr, Venice)

eliminated by the warlike *Caribs, also from northern South America, in the Lesser Antilles from c.1000 AD.

Columbus and others found and settled the islands in the late 15th–16th centuries. By the 18th century the indigenous peoples were nearly extinct. Black slaves were imported in huge numbers from Africa as sugar, tobacco, and coffee *plantations dominated the economies, and Spain, France, Britain, and Holland fought for possession of various islands. Through official military, privateering, and piratical methods, many islands changed hands several times during the 17th and 18th centuries.

Caribs, people of South American origin, who migrated to the islands of the Lesser Antilles from c. AD 1000. There they replaced the agricultural Arawak culture, killing off most of the men and taking their women. Many Arawak men were eaten in ritual cannibalism. The Caribs were the first 'Indians' discovered by Columbus.

Carmelite, a monk or nun who is a follower of the Order of Our Lady of Mount Carmel. Carmelites obey the strict monastic 'rule' of St Albert of Jerusalem. They originated in *Palestine c.1154 but came to western Europe when Palestine was conquered by the Muslims. Their order was approved by Pope Honorius III in 1226. The rule required strict isolation from the secular world but was later relaxed to allow a degree of involvement with the care of the laity. In 1452 the Carmelite Sisters was formed. In 1594 a reformed group of the order was established, the Discalced Carmelites, but it remained essentially similar in organization and objectives. The order has produced celebrated mystics including St *John of the Cross and St *Teresa of Ávila.

Carnac, a town in Brittany, France, near a major centre of ritual activity between the 5th and 3rd millennia BC. A peninsula is marked off by rows of *megaliths, presumably as some sort of sanctuary. There are numerous megalithic tombs in the area and nearby, at Locmariaquer, is the largest known *menhir, originally standing 20 m. (65 ft.) high.

Carnot, Lazare Nicolas Marguerite (1753–1823), French general and military tactician who was drawn into political life after the *French Revolution. He was a republican who voted for the execution of the king. His reorganization of the fighting methods and administration of the army was mainly responsible for the successes of the Revolutionary armies. He was a member of the *Committee of Public Safety and then of the *Directory, and was in charge of the war department. After being forced to flee from France in 1797, he returned to become Minister of War in 1800 and to continue with his administrative reforms. His resignation was reluctantly accepted in 1801 and he spent most of his retirement writing books on military fortifications. His heavy detached fortress wall (Carnot's Wall) and principle of active defence influenced modern fortifications. He interrupted his retirement to serve Napoleon in 1814 and 1815.

Caro, Joseph (1488–1575), Jewish scholar and mystic, who wrote the last great codification of Jewish law. His family was among the Jews expelled from Spain in 1492. Caro finally settled in Safed in Palestine c.1537, where he produced *House of Joseph* (1559); his popular con-

Avenues of megaliths at **Carnac** in Brittany, c.2500 BC. The standing stones start as large boulders, gradually decreasing in size to stones about 1 m. (3 ft.) high at the end of each row.

densation of this work, *Prepared Table* (1565), was initially opposed by some rabbinical authorities, but has since become the authoritative standard of *Judaism. In 1646 his *Preacher of Righteousness*, a diary of his mystical experiences, was published.

Caroline of Ansbach (1683–1737), German princess, Queen consort of *George II, whom she married in 1705. She was a cultivated woman, possessed of much common sense and considerable political skill, and used her influence in support of *Walpole. She was a popular queen, and during the king's absences in Hanover, she was four times appointed 'Guardian of the Realm'. She was responsible for enclosing 121 ha. (300 acres) of London's Hyde Park to form Kensington Gardens.

Carolingian empire, the collection of territories in Western Europe ruled by the family of *Charlemagne (AD 768–814) from whom the dynasty took its name. Charlemagne's ancestors, Frankish aristocrats, fought their way to supreme power under the *Merovingian kings, the last of whom was deposed by Charlemagne's father *Pepin III in 751. Under Charlemagne, the empire covered modern-day France, part of Spain, Germany to the River Elbe, and much of Italy. Charlemagne was crowned Emperor of the West by Pope Leo III in 800 and made his court a centre of learning (the 'Carolingian Renaissance'). After the division of the empire by the Treaty of *Verdun in 843, civil war among the Carolingians, *Viking raids, and the ambitions of rival families subjected the empire to intolerable strains. Nevertheless, Carolingians reigned in Germany till 911 and in France till 987 and they left behind a prestige which later kings of the Middle Ages sought to emulate.

Carthage, a city on the coast of North Africa (in modern Tunisia). It was founded by colonists from *Tyre (though later than the traditional date of 814 BC) and developed into one of the leading trading cities of the western Mediterranean, with footholds in Spain, Sardinia, and Sicily. Between the 5th and 3rd centuries BC it engaged in frequent hostilities with the Greeks of Sicily. It was on that island that Carthage first clashed with Rome, and the three *Punic Wars ended with the razing of the city in 146 BC.

It was refounded as a Roman colony by *Caesar and

A Roman mosaic showing the Earth goddess Tannit, the most venerated in **Carthage**. The mosaic dates from around the 4th century AD.

*Augustus, and again achieved great prosperity. As a centre of Christianity it opposed the *Donatists. The Vandal *Genseric occupied it in 439 and established it as his capital, but in 533-4 *Belisarius overthrew the Vandals and from then until its capture and destruction by the Arabs in 697 it remained part of the Byzantine empire.

Carthusian, a member of the monastic order founded by St *Bruno of Cologne at Chartreuse in France in 1084. Their 'rule' is extremely severe, requiring solitude, abstinence from meat, regular fasting, and silence except for a few hours each week. Nuns affiliated to the order may eat together and, uniquely within the Roman Catholic Church, are allowed to become deaconesses. Lay brothers and sisters tend their needs and provide the minimum necessary contact with the outside world. They are governed by the general chapter, consisting of the priors of all houses together with the community of La Grande Chartreuse itself.

Cartier, Jacques (1491-1557), French navigator, the discoverer of the St Lawrence River. Like other explorers of his period, he was seeking a north-west passage to China when in 1534 he reached Newfoundland and discovered the mouth of the great river. On his next voyage two years later he sailed up the river as far as the Île d'Orléans, and continued in longboats to the Indian village which became the site of Montreal. On his third voyage (1541-2) he wintered on the river but made no new geographical discoveries. Further exploration of Canada was undertaken by *Champlain.

Carver, John (c.1576-c.1621), *Pilgrim leader. He had been deacon of the separatist church in Leiden and led the migration in the *Mayflower* in 1620. Elected first governor of Plymouth Plantation, he died shortly afterwards and was succeeded by William *Bradford.

Casanova de Seingalt, Giovanni Giacomo (1725-98), Venetian adventurer. He led a wandering existence as a gambler and spy. He had considerable personal charm and was notorious for the number of his seductions. He is celebrated for his escape (1756) from the state prison in Venice. His voluminous memoirs, written in French and published posthumously, give a lively account of his adventures and amours all over Europe as well as an entertaining picture of 18th-century society.

Casimir III (the Great) (1310-70), King of Poland (1333-70). He consolidated the achievements of his predecessor, Wladyslaw I, reorganizing the country's administration, codifying the law, and acquiring territory through diplomacy. Links with Lithuania, Hesse, Silesia, Brandenburg, and the Holy Roman Empire were forged through marriage. He successfully fought against *Russia, the *Teutonic Knights, and the *Bohemians.

Castile, a former kingdom in northern Spain, the name deriving from the castle-building activities there of Garcia of León in the early 10th century. The period which followed was a confused one in which alliances with the Spanish *Moors alternated with expansion at their expense, especially during the reigns of Alfonso VI and Alfonso VII. León was gained, lost, and finally reunited with the kingdom under Ferdinand III in the 13th century. Under *Alfonso X, the cultural life of the country developed but a long period of weak rule and internal turmoil followed. In 1469 *Isabella of Castile, having married *Ferdinand II of Aragon, inherited the crown and thus the two greatest kingdoms of Spain were united.

castle, a fortified building for the defence of a town or district, doubling as the private residence of a *baron in the Middle Ages. Although also called 'castles', Celtic hill-forts, Roman camps, and Saxon burhs were designed to provide refuge for whole populations; archaeological evidence suggests that in England fortified private residences date from the 9th century. The 'motte and bailey' design of the 11th century comprised a palisaded 'motte' (a steep-sided earthen mound) and a 'bailey' (an enclosure or courtyard) separated from the motte by a ditch. Both were surrounded by a second ditch. Initially timber-built, and often prefabricated for rapid assembly, many were later rebuilt in stone. Design modifications in the 12th century included stone tower keeps (as at Rochester, c.1130, and Castle Hedingham c.1140) to replace the motte. The keep (the rounded form was called a shell keep) combined strong defence with domestic quarters. The need to extend these quarters meant that the courtyard had to be protected by a line of towers joined by 'curtain' walls. In the 12th century the concentric castle (one ring of defences enclosing another) was developed from the model of the castles built by the Crusaders, who themselves had copied the Saracens. Two of the greatest castles of the late 12th century were Château Gaillard in Normandy and the Krak des Chevaliers in Syria. At the end of the 13th century, *Edward I of England, following a policy of subduing north Wales, built a series of castles at Caernarvon, Conway, Harlech, and Beaumaris. Design improvements saw the further development of rounded towers, which were more difficult to undermine, machicolations, which enabled objects to be dropped or poured on the besiegers,

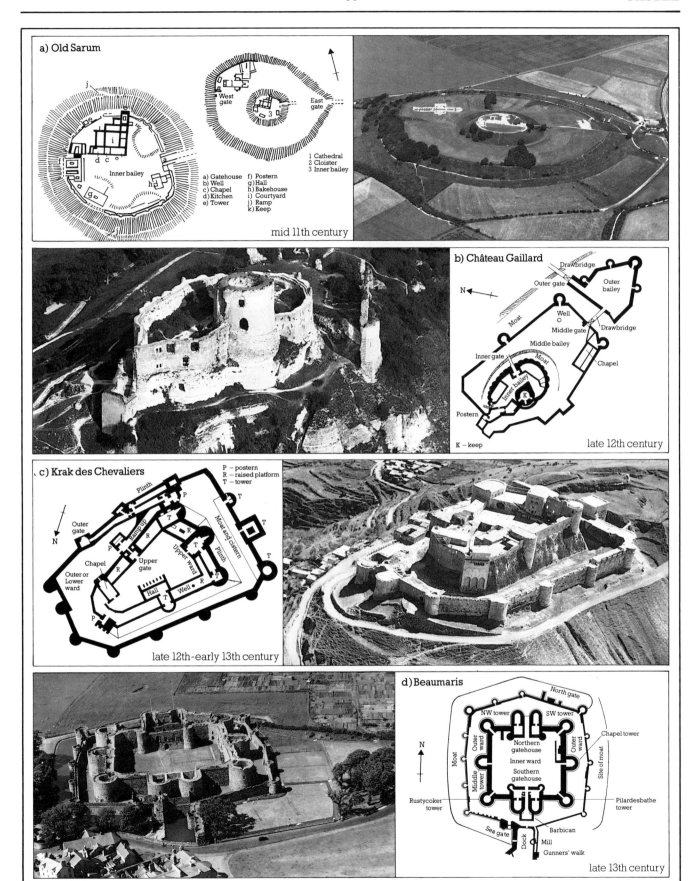

a) Old Sarum

West gate
East gate

Inner bailey

1 Cathedral
2 Cloister
3 Inner bailey

a) Gatehouse f) Postern
b) Well g) Hall
c) Chapel h) Bakehouse
d) Kitchen i) Courtyard
e) Tower j) Ramp
k) Keep

mid 11th century

b) Château Gaillard

Drawbridge
Outer gate
Outer bailey
N
Moat
Well
Middle gate
Drawbridge
Middle bailey
Inner gate
Moat
Chapel
Inner bailey
K
Postern

K – keep
late 12th century

c) Krak des Chevaliers

P – postern
R – raised platform
T – tower

Plinth
Outer gate
Ramp up
Moat and cistern
N
Chapel
Upper gate
Upper ward
Outer or Lower ward
Plinth
Hall
Well

late 12th–early 13th century

d) Beaumaris

North gate
NW tower
SW tower
N
Chapel tower
Outer ward
Northern gatehouse
Moat
Inner ward
Middle tower
Southern gatehouse
Site of moat
Rustycoker tower
Pilardesbathe tower
Sea gate
Barbican
Dock
Mill
Gunners' walk

late 13th century

Castle

(a) a prehistoric fortress, converted into a 'motte and bailey' castle by the Normans. (b) a modified 'bailey' castle, an early European example of the concentric form. (c) the best preserved of the Crusader castles, with a concentric scheme as a result of successive rebuildings. (d) a concentric fortress, designed as a symmetrical, integrated unit but not completed.

The interior of a **catacomb** in Rome. These subterranean burial galleries were narrow tunnels with side recesses for tombs, and were dug at different levels.

massive gatehouses, and refinements to the battlements, or crenellations, along the walls. Castles were built throughout Europe during the Middle Ages but although they continued to be built and improved in England in the 14th and 15th centuries, a more settled society encouraged the building of unfortified *manor houses. The invention of gunpowder had made castles obsolete for defensive purposes by the middle of the 16th century.

catacomb, an underground burial gallery of early Christian Rome. Catacombs were named after the best known example, St Sebastian in the Hollow (*ad Cata-cumbas*). Forty such subterranean chambers are known in Rome, tunnelled through soft rock outside the ancient city boundaries. The anniversaries of *martyrs were celebrated at the graves. Looted by barbarians and subject to collapse, they were virtually forgotten until their accidental rediscovery in the 16th century. Similar ones are also found as far apart as Salzburg and Malta.

Catalaunian Fields (or Plains), the site of a major battle in AD 451, reputedly near Châlons-sur-Marne in France, but placed by some nearer Troyes. The Roman general Aetius with a combined force of Romans, Goths, and Burgundians defeated *Attila the Hun, forcing his retreat from Gaul. He was expelled from Italy the following year.

Çatal Hüyük, a large Neolithic settlement (about 13 ha., 32 acres), near Konya in south-central Turkey, which dates from *c*.6500 BC. Its small houses, built of mud bricks, were so close together that all access had to be by way of the flat roofs. A high proportion of rooms are believed to have been shrines, on the basis of numerous bulls' horns mounted in benches. There were also wall decorations, votive offerings, and richly furnished burials beneath the floors. Only a small part of the site has been excavated.

Catalonia, a semi-autonomous region in north-east Spain. Once a Roman colony, it was overrun by the *Visigoths in the 5th century and became independent under the counts of Barcelona from 874. Linked by

marriage alliance to *Aragon in 1137, and to *Castile in 1479, the Catalans rebelled in 1462–72 and 1640–52. In the second outbreak they appealed to the French for help. The French invaded from 1689 to 1697. During the War of the *Spanish Succession, support for the Austrian candidate, and defeat by the *Bourbons led to loss of privileges. Catalan, a Romance language similar to Provençal, is the traditional language of Catalonia, Andorra, and the Balearic Islands.

Cathar (Greek *katharos*, 'pure'), a member of a medieval sect seeking to achieve a life of great purity. Cathars believed in a 'dualist' heresy. Their basic belief was that if God, being wholly good, had alone created the world it would have been impossible for evil to exist within it, and that another, diabolical, creative force must have taken part. They held that the material world and all within it were irredeemably evil. The human body and its appetites were despised. Marriage was rejected and suicide by starvation was admired. A pure life was impossible to all but a very few called the 'perfect', and the rest—known simply as 'believers'—could live as they wished. Salvation was assured if they took a form of confirmation known as the 'consolamentum' before death. This ceremony was to be delayed as long as possible to reduce the chance of the recipient's sinning further before he or she died. The heresy originated in Bulgaria and appeared in western Europe in the 1140s. In southern France this Christian heresy was called *Albigensianism.

Catherine of Aragon (1485–1536), Spanish princess, the first wife of *Henry VIII of England to whom she was married in 1509. She bore him a daughter, the future *Mary I, but no male heir survived; the importance of the Spanish alliance diminished, and by 1527 Henry, infatuated with Anne *Boleyn, sought a papal annulment, claiming that Catherine's marriage in 1501 to his elder brother Arthur rendered his own marriage invalid. The pope was uncooperative, and *Cranmer annulled the king's marriage in 1533. Thereafter Catherine lived in seclusion in England.

Catherine I (*c*.1684–1727), Empress of Russia (1725–7). She was a Lithuanian servant girl who was first the mistress and then the second wife of *Peter I. On his death she was proclaimed ruler with the support of her husband's favourite, Menshikov, and the guards regiments. Menshikov became the effective head of government, working through the newly established Privy Council, but fell from power on Catherine's death. Her daughter *Elizabeth became empress (1741–62).

Catherine II (the Great) (1729–96), Empress of Russia (1762–96). She was a German princess, Sophia of Anhalt-Zerbst, who in 1745 married the future emperor Peter III, a cruel and feeble-minded young man. In 1762, six months after his accession, he was murdered, and she was proclaimed empress with the support of the guards regiments of St Petersburg, and she ruled for forty-four years. She was an intelligent and ambitious woman who corresponded with *Voltaire and considered herself a disciple of the *Enlightenment. However, her much-heralded Legislative Commission (1765–74) achieved nothing; and her charter of 1785 established the nobility as a privileged class and caused serfdom to be extended and made harsher. The revolt of *Pugachev (1773–4) was difficult to suppress. Her claim to greatness rests mainly on her foreign policy: with the help of *Potemkin and Suvorov she obtained most of Poland in the partitions of 1772, 1793, and 1795, gained Azov in the first Turkish War, annexed the Crimea, and by 1792 the whole northern shore of the Black Sea.

Catherine of Braganza (1638–1705), Portuguese princess and Queen consort of *Charles II, whom she married in 1662, bringing Tangier and Bombay as part of her dowry. The marriage was childless and she had to tolerate the king's infidelities. As a Roman Catholic, she was unpopular and there were attempts to implicate her in the *Popish Plot. In 1692, as a widow, she returned to Portugal, where she died.

Catholic League   *Holy League.

Catholic emancipation, the granting of full political and civil liberties to Roman Catholics. Although religious toleration was practised in Britain from the late 17th century, the *Test Acts limited holders of public office to communicant Anglicans and placed additional disabilities on members of other churches. Until 1745 the *Jacobite threat seemed to justify continued discrimination against Roman Catholics, and fears of Catholic emancipation led to the *Gordon riots in 1780. The chief agitation for reform came from *Ireland, where successive minor concessions were made from 1778, and from 1812 the issue was regularly debated in Parliament. In 1828 Daniel O'Connell was elected to Parliament, leading to the passing of the Catholic Relief Act (1829), which repealed most of the discriminatory laws.

Cato, Marcus Porcius (the Elder) (*c*.234–149 BC), 'the Censor'. He was a politically conservative and very influential Roman statesman. As consul in 195 he suppressed revolt in former Carthaginian Spain with severity; as censor in 184 he was equally severe against private extravagance. *Delenda est Carthago* ('Carthage must be destroyed') became his slogan, although he did not live to see the event in 146. He prosecuted *Scipio for corruption. His book on agriculture (*De agri cultura*) survives but of his history of Rome, the *Origines*, there remain only a few fragments.

Cato, Marcus Porcius (the Younger) (95–46 BC), the great-grandson of Cato the Elder. He was known posthumously as 'Uticensis' after the place of his death. A conservative republican, he long opposed *Pompey, but finally sided with him against Julius *Caesar. He committed suicide at Utica in northern Africa after Caesar's victory at Thapsus rather than seek Caesar's pardon. Less noteworthy than his great-grandfather, he nevertheless became proverbial as an exemplar of republican and traditional Roman values.

Caucasoids, fair-skinned 'European' people, named after the Caucasus Mountains between the Black and Caspian seas. They occupy Europe, Africa as far south as the Sahara, the Middle East, and the Indian subcontinent; in the past five centuries they have spread world-wide. In parts of Central Asia they were replaced in historic times by *Mongoloids. There was always admixture with, and incomplete differentiation from, neighbouring races, making a coherent story of the origin

Catherine II portrayed during the first decade of her long reign as Empress of Russia, by the Danish court painter Vigilius Erichsen. (Private collection)

and dispersal of Caucasoids difficult. Modern-looking people (**Homo sapiens sapiens*) had appeared in the Middle East by 50,000 years ago and in Europe by 35,000 years ago. It is now widely believed that these people came from Africa. They presumably interbred with and replaced the *Neanderthal groups then occupying these regions. The *Cro-Magnon race of people, the forebears of modern Europeans, resulted.

Cavalier Parliament (or the Long Parliament of the Restoration) (1661–79), the first in *Charles II's reign to be elected by royal writ. Strongly Royalist and Anglican in composition, it contained 100 members from the *Long Parliament of Charles I. Its long duration enabled the Commons to claim a large part in affairs, despite being in session for only sixty months of the eighteen years. Its early years were marked by harsh laws against Roman Catholics and Protestant Dissenters. As its membership changed it became increasingly critical of royal policy.

Cavaliers (French *chevalier*, 'horseman'), the name of the Royalist party before, during, and after the English *Civil War. Opponents used the word from about 1641 as a term of abuse: later it acquired a romantic aura in contrast to the image of puritanical *Roundheads. The party, made up of all social classes, but dominated by the country gentry and landowners, was defined by loyalty to the crown and the Anglican Church. The Restoration brought the Royalists back to power— the Parliament of 1661–79 is called the *'Cavalier Parliament'.

cave-dwellers, the name for the people who first used caves as shelters. This became widespread during the Middle and Upper *Palaeolithic periods, when humans penetrated for the first time into the northern tundra environments in front of the ice-sheets of the last glaciation. Since the remains of open-air campsites, such as wind-breaks or tents, are generally less well preserved and less likely to be discovered than bones and tools incorporated in cave sediments, early investigators imagined that *Stone Age people lived entirely in caves. This has now been refuted by the excavation of huts and tent-foundations preserved under wind-blown sediments in the Ukraine, central Europe, and France.

Caxton, William (c.1422-91), English printer. He spent most of his life as a mercer (cloth merchant), but in 1476 he set up the first printing press in England, in the precincts of Westminster Abbey. He was successful in producing a series of books in English, some from his own translations, which appealed to the taste of the court and gentry.

Cecil, William, 1st Baron Burghley (1520-98), English statesman. He trained as a lawyer and held office under *Henry VIII, *Edward VI, and finally as *Elizabeth I's Secretary of State from 1558. Politically adept, he formulated the queen's policy at home and abroad and was rewarded by the offices of Master of the Courts of Wards and Liveries (1561) and Lord Treasurer (1572). He was created Lord Burghley in 1571.

For forty years he ensured the stability of the Elizabethan regime. He promoted the *Anglican Church by commissioning Bishop *Jewel to write his *Apologia Ecclesiae Anglicanae*. More Protestant in sympathy than the queen, he persuaded her to aid the French Huguenots (1567) and the Dutch Calvinists (1585). He exercised control of appointments to the universities of Oxford and Cambridge and was responsible for ordering the execution of *Mary, Queen of Scots, whose existence he perceived as a threat to the state. He encouraged new industries, particularly glass-making, and introduced financial reforms. He

The 8th-century Ardagh chalice, one of the finest surviving artefacts of the **Celtic Church**, was discovered in an Irish potato field. Made from gold and silver, it is decorated with typical Celtic motifs. (National Museum of Ireland, Dublin)

profited handsomely from his career and used his wealth to build the mansions of Burghley House, Lincolnshire, and Theobalds in Hertfordshire.

Cecil, Robert, 1st Earl of Salisbury and 1st Viscount Cranborne (1563-1612), English statesman. The son of William Cecil, Lord Burghley, he succeeded his father as *Elizabeth I's chief minister in 1598. He was responsible for ensuring the succession of *James I in 1603. He was created Viscount Cranborne (1604) and Earl of Salisbury (1605). He was made Lord Treasurer in 1608 and was faced with crown debts of nearly a million pounds. He increased the king's income by introducing additional *customs duties (impositions) and attempted to improve the administrations of crown lands and revenues. In 1610 he proposed the 'Great Contract' by which Parliament would vote revenues annually to the king but it refused to ratify the scheme and he had to continue raising money by unpopular means, principally a forced loan (1611), and the sale of titles. After his death in 1612 expenditure not only continued to exceed income but increased under James I and his adviser *Buckingham.

Celtic Church, a description principally applied to the Christian Church in Ireland, whose inhabitants were *Celts, from the 5th to the 10th centuries, but also applied to the Christian Church in other parts of the British Isles where Celts dwelt during this period. It is probable that Ireland had early contacts with Christianity through Roman Britain, but the widespread conversion of the country to Christianity appears to have occurred in the 5th century, notably under St *Patrick (c.390-460). For the next three centuries Ireland was the most important centre of Christianity in north-western Europe. In the evangelization of Scotland, Irish missionaries played a prominent role, notably St *Columba (c.521-97), who established many monasteries including one on the island of Iona. Celtic missionaries from Ireland also played a prominent role in the re-conversion of England in the 7th century. Others, such as St *Columbanus (c. 543-615), established monasteries in Gaul and north Italy. Celtic Christianity had its own strong characteristics. Its life-style was evangelical and ascetic, as shown by surviving Celtic 'Penitential Codes'; it was noted for its missionary zeal; and its organization was monastic rather than diocesan or parochial, with even bishops being subject to the abbots of monasteries. There flourished a rich culture, most famously the 'Book of Kells', the 8th-century illuminated manuscript of the gospels. Gradually the Celtic Church lost many of its distinctive features, as Ireland was absorbed into the mainstream of Western Christendom: the decisive moment in England was the Synod of *Whitby in 664, when the Roman date for Easter was preferred to the Irish; in Ireland and Scotland the decline was later.

Celts (often also called 'Gauls'), a group of peoples identifiable by common cultural and linguistic features. Their earliest archaeological traces are found in the Upper Danube region (13th century BC), from where branches spread to Galatia in Asia Minor, Gaul (modern France), North Italy, Galicia and Celtiberia in Spain, and the British Isles. Their main expansion seems to have occurred after 800 BC. They sacked Rome in 390, and Delphi about a century later, when they also reached

Asia Minor. Their artefacts are conventionally divided into *Urnfield (earlier Danubian phase), *Hallstatt, and (after c.500 BC) three successive *La Tène periods. They were gifted craftsmen and fierce fighters; but they lacked political cohesion, and by the end of the 1st millennium BC were increasingly 'squeezed' between the expanding power of Rome and migratory Germanic tribes, and settled in the remote areas of Europe (Brittany, Wales, and Ireland), where their dialects have survived.

Central America, the land mass, today comprising Panama, Costa Rica, Nicaragua, El Salvador, Honduras, Guatemala, and Belize (British Honduras), which connects North and South America. It was populated by diverse aboriginal groups at the time of the first European contact in the early 16th century, and its colonization was the product of Spanish expansion from the Caribbean settlements of *Hispaniola and *Cuba. From 1535 to 1810, except for Panama, Central America was part of the viceroyalty of *New Spain and subject to the jurisdiction of the viceroy in Mexico City.

centurion, a professional middle-ranking officer of the Roman army. The title means 'leader of a hundred'. A century was a military unit, based on early citizen lists drawn up for military service. In earlier days most centurions rose from the ranks; but in the later republic

The predominantly Indian culture of the **Champa** is evident in this sandstone bas-relief. The figure probably represents both the god Siva and a deified king. (Musée Guimet, Paris)

and under the emperors (Principate) some men enlisted directly as centurions. The rigorous discipline, leadership and experience of the centurions made them a vital factor in the success of the professional army.

ceorl, a free peasant farmer of Anglo-Saxon England. In status ceorls were above the *serfs but below the *thanes (noblemen), with a *wergild of usually 200 shillings. They were liable to military service in the *fyrd and to taxation. Although they could own land, they were often forced by economic pressures and by reasons of security to place themselves in the control of the richer landowners. After the Norman Conquest their status diminished rapidly and the term 'churl' came to mean an ill-bred serf.

Ceylon   *Sri Lanka.

Chaeronea, the northernmost city of *Boeotia, central Greece, the scene of two important battles. In 338 BC *Philip II of Macedonia crushed the Thebans, Athenians, and their allies there, and so brought mainland Greece under his control. An enormous stone lion, commemorating the site of the fighting, can still be seen.

In 86 BC two armies of *Mithridates, King of Pontus combined there against the Roman forces of *Sulla, but were defeated despite a considerable numerical superiority. A further Roman victory at Orchomenus ensured the ejection of the Pontic forces from Greece.

Chalcedon, Council of (451). The city of Chalcedon, in Greece, was the scene of the fourth ecumenical council of the Christian Church. This rejected the view expressed by a meeting—convened without papal approval—at Ephesus in 449 which declared *Jesus Christ to have a single nature, asserting instead that Christ's nature was both human and divine.

Chalcis, in ancient times, the leading city of the island of Euboea, off the eastern coast of central Greece. It was a trading city, famous for its metal goods, and was important in transmitting Greek civilization to Sicily and southern Italy. In 506 BC it was forced to cede some of its territory to Athenian settlers. In 446 an attempt to secede from the Athenian empire was crushed. In 338 *Philip II of Macedonia installed a garrison there— one of the 'fetters' of Greece—but the city thrived subsequently until it was partly destroyed by the Romans after having sided with the *Achaean League against them in 146.

Chaldea, an area at the head of the Persian Gulf. It was attacked by the Assyrian Shalmaneser II c.850 BC. In 721–710 a Chaldean king managed to wrest Babylon from the Assyrians, but was later ejected. It was not until 626 that Nabopolassar established the great Chaldean dynasty of *Babylon, of which the most famous representative was *Nebuchadnezzar II. After it had been overthrown by Cyrus the Great in 539 BC the term 'Chaldean' became equivalent to 'Babylonian'.

Champa, the kingdom of the Chams, a Malay people, said to have been founded in the 2nd century AD in Vietnam. It was frequently at war with the *Khmers to its west, and succumbed in the 15th century to *Annam to the north. Its people are commemorated in *Angkor:

in a bas-relief of a battle against the Khmers they are distinguished by their flat hats, each with a flower in it.

Champagne, a province of north-eastern France adjoining Lorraine. International trade *fairs were held there in the Middle Ages. In 1284 the marriage of Jeanne, daughter of Henry III, the last count, to *Philip the Fair led to union with France. The discovery of the method of making its celebrated sparkling wine, champagne, is attributed to a Benedictine monk, Dom Perignon (1668–1715).

Champlain, Samuel de (1567–1635), French explorer of *Canada. After leading a Spanish expedition to the West Indies in 1599, he made eleven voyages in French service to Canada, following *Cartier's discovery of the St Lawrence River. He explored the New England coast and later founded a settlement at what is now Quebec, which he used as a base for exploring the Canadian interior, discovering the lake to which he gave his own name. He believed that the Great Lakes must lead to a way through to the Pacific, but was deflected from further exploration by his role as governor of the French colony. In 1629 Quebec was captured by the English and he remained for three years captive in England. On the return of Canada to France in 1633 he returned to Quebec and died there.

chancery (from the Latin *cancella*, 'screen', hence a screened-off place, or office), the writing-office attached to the court of a ruler—emperor, pope, or king. Since it supplied the writ necessary for a lawsuit to be heard by the king's judges, it came to be a law court itself, presided over by its head, the Chancellor. From the late 14th century in England its legal business grew rapidly; by the 16th century, it was notorious for delays and it was reformed in the 19th century.

Chandragupta Maurya *Mauryan empire.

Changamire (*fl. c.*1500), East African ruler, the name taken by Changa, son of *Mwene Mutapa Matope, by adding the Arabic title *amir* (commander) to his given name. On his father's death he killed Nyahuma, the lawful successor. His own son fought Chikuyo, Nyahuma's son, until 1502. He began the dismemberment of the *Rozvi empire. His kingdom was known also as Butwa, and lasted until the early 19th century, when the Nguni destroyed it. His successors built a number of stone monuments and added to the Great *Zimbabwe.

chariot, a fast, two-wheeled, horse-drawn vehicle. They were originally designed for use in war, and developed from the battle-wagons used by the *Sumerians *c.*2500 BC, which had four wheels, were drawn by onagers (wild asses), and served as mobile fighting platforms. The use of horses, and light two-wheeled vehicles adapted to them, was introduced to the Near East from the region between the Black Sea and the Caspian *c.*2000 BC. (Horses, which were only the size of ponies, were rarely used for riding.)

Chariots were the prized possessions of potentates, as much for prestige as for warfare, from Greece and Egypt as far as China throughout the 2nd millennium BC, but declined in importance in the 1st millennium when heavier horses made cavalry possible. By the time Julius

The coronation of **Charlemagne**, from a 12th-century French manuscript. It shows Charlemagne kneeling before Pope Leo III during the ceremony which took place in Rome on Christmas Day 800. (Musée Goya, Castres)

*Caesar went to Britain in 55 BC, their use in war had ceased on the Continent, and he was interested to find them used by the Britons to give warriors mobility on the battlefield. In Rome, they continued to be used as sporting vehicles, in chariot races.

Charlemagne (*c.*742–814), Frankish king and Emperor of the West (*Holy Roman Empire) (800–14). He was the son of *Pepin the Short and grandson of *Charles Martel. As King of the Franks, reigning at first jointly with his brother Carloman, who died in 771, he set about the formidable task of imposing his rule. The Franks had long suffered from weak government and continuous invasions from the barbarian north and east and the Muslim south. His long campaign began in 772, directed initially at the pagan Saxons, then against the Avars to the east. Bavaria and Lombardy came under his control and he was then able to strengthen and support the papacy by restoring papal lands in Italy. On Christmas Day 800 he was crowned as Emperor of the West by Pope Leo III.

The palace school at his capital city, Aachen, became the most important centre of learning in western Christendom, and there the emperor brought many great scholars and teachers including *Alcuin. He founded schools in cathedrals and monasteries throughout the Empire, initiating a revival in scholarship whose effects were profound and lasting, and a prime contribution to what is called the 'Carolingian Renaissance'. Charlemagne enjoyed a heroic posthumous reputation throughout the Middle Ages, witnessed in the poem the *Chanson de Roland*.

Charles I (1600–49), King of Great Britain and Ireland (1625–49). He was the second son of *James I and Anne of Denmark. He was neglected by his father in favour of his favourite, *Buckingham, who also dominated Charles in the opening years of his reign. A disastrous foreign policy, Charles's illegal levying of *tunnage and poundage, and the mildness of his policy towards Roman Catholics (*recusants) culminated in the forcing through by a hostile Parliament of the *Petition of Right (1628). From 1629 he ruled without a Parliament.

Charles was a man of strong religious conviction: he was also stubborn and politically naïve. During the 'Eleven Year Tyranny' (1629–40) he relied increasingly on *Laud, *Strafford, and his French Catholic queen, *Henrietta Maria; their influence, and the king's use of unconstitutional measures, deepened the widespread antagonism to the court, especially after the *Ship Money crisis (1637). The fiasco of the *Bishops' Wars drove him to recall Parliament in 1640. The *Long Parliament forced him to sacrifice Laud and Strafford, who were impeached and executed. He had to accept severe limitations of his powers, but an open breach came in January 1642, when he tried to arrest *five members of the House of Commons, a blunder which united the Lords and Commons against the king and made the *English Civil War inevitable.

The royal standard was raised at Nottingham in August. Charles was soundly beaten at *Marston Moor (1644) and *Naseby (1645) and in 1646 surrendered to the Scots near Newark, was handed over to Parliament the following year, and subsequently captured by the

The execution of **Charles I**, King of England, outside the Banqueting Hall in Whitehall, a painting by Weesop. The insets depict the king (top left), his walk from St James's to the scaffold (below), Thomas Fairfax (top right), and citizens gathering relics (below). (Private Collection, on loan to National Gallery of Scotland, Edinburgh)

Parliamentary army. After escaping to Carisbrooke Castle, Isle of Wight, he signed the 'Engagement' with the Scots (1647) that enabled him to renew the war with their help, but with little success. He was recaptured in 1648, tried, and publicly executed in London.

Charles I of Anjou (1226–85), King of Naples and Sicily (1266–85), son of Louis VIII of France. He acquired *Provence by marriage in 1246. Pope Urban IV was under severe threat from the *Hohenstaufens and gave him the kingdom of Sicily in order to curtail their power. He defeated and killed *Manfred at Benevento, effectively ending Hohenstaufen influence, but then went on to take Naples as well as most of northern Italy, himself becoming a real threat to papal interests. His ambitions were ended by the uprising known as the *Sicilian Vespers in which he was assassinated and the French expelled.

Charles II (the Bald) (823–77), King of the West Franks (843–77), Emperor of Germany (875–7). He was the son of Emperor Louis the Pious. After the death of their father he and his brother, Louis the German, made war on their eldest brother Lothair, who had inherited the title of King of the West Franks. By the Treaty of *Verdun in 843 Charles gained that kingdom. He and Louis divided Lothair's central kingdom between them in 870 by the Treaty of Mersen, and Charles gained the imperial title in 875. The internal conflicts of his reign were further complicated by *Viking incursions. He was a noted patron of scholarship and the arts.

Charles II (1630–85), King of England, Scotland, and Ireland (1660–85), the son of *Charles I. In the first phase of the *English Civil War he was at the battle of Edgehill (1642), and then took refuge in the west of England (1645–6) until he escaped to France. After his father's execution he was crowned in Scotland (1651),

having signed the *Solemn League and Covenant, which he later repudiated. *Cromwell had already defeated the Scottish army at Dunbar, and Charles's advance into England was halted at the battle of Worcester. He was on the run for six weeks, before escaping to the Continent. The efforts of General *Monck were largely responsible for his *Restoration to the English throne in May 1660.

Charles shrewdly adopted conciliatory policies, offering indemnity to all but the regicides (those responsible for the death of his father) and attempting to wean England from religious prejudice. His marriage in 1662 to *Catherine of Braganza increased anti-Catholic feeling, however; the union was childless, and the heir apparent, the king's brother James, Duke of York, was also a Roman Catholic. His foreign policy was originally directed against France in the Triple Alliance (1668) with Sweden and the United Provinces. He then signed the Treaty of Dover (1670) and agreed to support Louis XIV against the Dutch in return for financial and territorial gains; a secret clause committed him to announcing his conversion to Catholicism. He undermined the *Clarendon Code by Declarations of *Indulgence, provoking the *Popish Plot (1678) and the *Rye House Plot (1683), as well as the exclusion crisis of 1679–81 led by his chief opponent in Parliament, the Earl of *Shaftesbury. After March 1684 he did not call Parliament though this was illegal under the terms of the Triennial Act. In the country he was a popular monarch and, although he is commonly remembered for his mistresses and his horse-racing, he was also a notable patron of the arts and sciences and a sponsor of the *Royal Society.

Charles III (the Fat) (832–88), King of the Franks (884–7), Emperor of Germany (882–8). He was the youngest son of Louis the German. He inherited Swabia and acquired both east and west Frankish kingdoms by 884 after his older brothers died. He was unsuccessful in repelling *Saracen invaders and was obliged to buy a respite from attacks by the *Vikings and so was deposed in 887. His death marked the end of the Carolingian monopoly of kingship over the Franks.

Charles III (1716–88), King of Spain (1759–88) and of Naples and Sicily (1734–59). His enlightened policies

A woodcut of **Charles V**, the last Holy Roman Emperor to be crowned by the pope, accompanied by Pope Clement VIII, riding to the imperial coronation at Bologna in 1530. (Graphische Sammlung, Albertina, Vienna)

met with opposition in Spain. He tried to improve agriculture and industry, reformed the judicial system, and reduced the *Inquisition's powers. His foreign policy was dominated by alliance with France (the Family Compact, 1761). He lost Florida in 1763 but regained it with Minorca in 1783. In 1779 he began a three-year siege of Gibraltar but failed to retake it from Britain.

Charles IV (1316–78), King of Bohemia (1346–78), *Holy Roman Emperor (1347–78). He acquired authority over Austria and Hungary in 1364, and received the imperial crown from the pope, in Rome, in 1355. The Golden Bull (1356) issued in his reign formed the imperial constitution, regulating the duties of the seven *Electors. He was an intellectual, interested in the development of the German language, and founded the University of Prague in 1348.

Charles V (the Wise) (1337–80), King of France (1364–80). He earned his nickname from his intellectual pursuits which included book-collecting and artistic patronage, his religious piety, and his cautious adoption of delaying and 'scorched-earth' tactics in fighting the English during the *Hundred Years War. Assuming responsibility as Regent of France in 1356 when his father, John II was captured at *Poitiers, he quelled revolt in Paris and from the *Jacquerie and, aided by the Constable of France, Bertrand du *Guesclin, was able to recover most of France from the invading English forces.

Charles V (1500–58), Holy Roman Emperor (1519–56) and (as Charles I) King of Spain (1516–56). The son of *Philip the Handsome and Joanna of Spain, and grandson of *Emperor Maximilian I, Charles came to the throne of Spain in 1516 and united it with that of the empire when he inherited the latter in 1519. Tied down by such wide responsibilities, and hampered by the fact that his authority in his separate territories was established on different bases, Charles was never able to give proper attention to national and international problems. His achievements were none the less considerable. In Spain he survived an early revolt and laid the foundations of the strong government which underpinned Spanish greatness in the century after his death, while in Italy he overcame papal resistance to the establishment of Spanish hegemony. While his long war with *Francis I of France was not decisive, it did weaken France to the extent that it was unable to challenge Spain again before the outbreak of the *Thirty Years War. He blunted the *Ottoman offensive against Christian Europe, and maintained his authority under difficult circumstances in the Netherlands. His greatest failure was in Germany, where he was unable either to check the spread of Protestantism or curb the independence of the local princes. Charles handed Naples (1554), the Netherlands (1555), and Spain (1556) over to his son *Philip, and the imperial crown (1556) to his brother Ferdinand, and retired to a monastery in Spain.

Charles VII (1403–61), King of France (1422–61). During his youth France was badly ruled by his father Charles the Mad and much territory was lost. Internal quarrels and war with England dominated his reign. He was not crowned until 1429, and then only thanks to *Joan of Arc. He established greater control over the church in the Pragmatic Sanction of Bourges of 1438, which upheld the right of the French church to administer its property and nominate clergy to benefices, independently of the papacy. He brought the *Hundred Years War to an end, having recovered most of his land and established his authority.

Charles XII (1682–1718), King of Sweden (1697–1718). The story of his reign is reflected in the progress of the *Northern War. He was attacked by a coalition of enemies and won a series of victories; then in 1707 he invaded Russia and was defeated at *Poltava. He took refuge in Turkish territory was imprisoned and escaped and was finally killed while on another military campaign. His wars left Sweden financially drained and no longer one of the great powers of Europe.

Charles Martel (c.688–741) (French, *martel*, 'hammer'), Frankish leader. He was the son of Pepin II, 'mayor of the palace' under *Merovingian rule. He gained control of the Austrasian province and defeated the Neustrian mayor. Burgundy and Aquitaine were also acquired. His greatest achievement, and the one which made him a traditional French hero, was his defeat of the Muslim forces between Poitiers and Tours in 732, which signalled the end of their northward expansion.

Charles the Bold (1433–77), Duke of *Burgundy (1467–77). He was the greatest of the dukes of Burgundy, and almost succeeded in creating a kingdom independent of France. He tried to persuade the *Holy Roman Emperor to grant him the title of king in 1473. He supported the League of the Public Weal against the French king, Louis XI, and, after 1467, concentrated with successful results on expansion into the Rhineland and Alsace. After 1475, war with the Swiss and defeat in battle culminated in his own death in battle. His realm was absorbed by the French and by *Maximilian I.

charter (Latin, *carta*, 'written document'), a legal document from a ruler or government, conferring rights or laying down a constitution. Charters in England date from the 7th century, when they were used to confirm grants of land, usually recorded in Latin. Borough charters granting towns specific privileges, which could include self-government and freedom from certain fiscal burdens, were regularly awarded by English kings between 1066 and 1216 (over 300 were issued). *Magna Carta (1215) was a charter which sought to regularize the feudal contract between the crown and its *barons.

The commercial and colonial expansion of England from the 16th century led to the use of charters to authorize the trading ventures of companies (*chartered company) and to form the first constitutions of the English colonies in America. Such colonial charters were in the form of a grant to a company (Virginia Company 1606), or gave recognition to the self-governing status of existing colonies (as with Connecticut in 1662). The importance of these charters was recognized by the Americans during the War of *Independence.

chartered company, a form of trading company which developed from the European medieval trading guilds, which was prominent in the late 16th and 17th centuries. The discovery by explorers of India and America stimulated individual merchants into forming groups, safeguarded by royal charter in order to monopolize trade.

Governments awarded exclusive trading rights in a particular area to a few rich merchants. Such companies were easy to control and, with their specially granted diplomatic, legislative, and military authority, they acted as virtual representatives of the crown. Since the companies were so restrictive, they could arouse considerable domestic opposition.

The *Dutch East India Company (founded 1602) probably had the best record of profit of all the 'joint-stock' ventures. These were companies in which members held shares entitling them to a proportion of the profits. They differed significantly from the earlier 'regulated' companies, which were associations of individuals who traded alone with their own stock and employees, subject to the company's regulations.

Many European chartered companies were costly failures. In the 17th century the French monarchy set up some thirty—including its own *French East India Company (1664)—most of which were unprofitable. Individual French traders were more successful in Haiti, Martinique, and Guadeloupe. The heyday of the chartered company was before 1800.

Chassey culture *Western Neolithic.

Chatham, 1st Earl of *Pitt.

Chavín culture, a civilization which flourished in Peru 1000–200 BC. The culture was based on the ceremonial centre of Chavín de Huantar, high in the Andes 280 km. (175 miles) north of Lima. It united an area 800 km. (500 miles) along the Peruvian coast in a common culture, and its influence spread almost as far again. The unifying force was probably religious rather than political, the most characteristic feature being figures, presumably gods, with jaguar fangs projecting from their lips. Notable advances included improved maize, the back-strap loom, and metallurgy. As its religious authority waned, regional groups appeared, that dominated Peru for the next thousand years.

Cheng Ho *Zheng He.

A Peruvian bowl or mortar in the shape of an animal, probably used as a sacrificial vessel, from the middle **Chavín** period, 1000–700 BC.(Dallas Museum of Fine Arts)

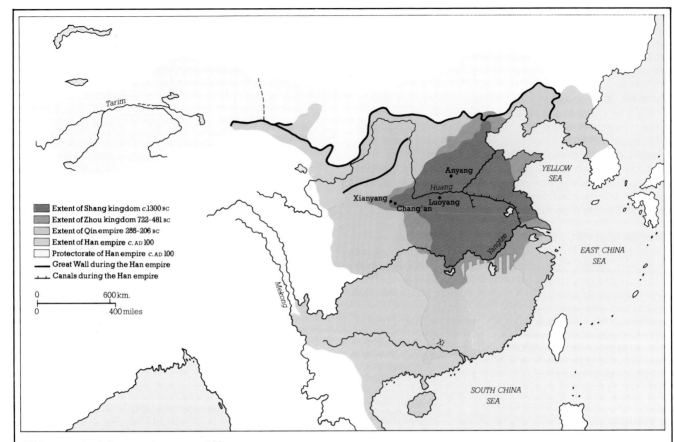

Key:
- Extent of Shang kingdom c.1300 BC
- Extent of Zhou kingdom 722–481 BC
- Extent of Qin empire 255–206 BC
- Extent of Han empire c.AD 100
- Protectorate of Han empire c.AD 100
- Great Wall during the Han empire
- Canals during the Han empire

0 600 km.
0 400 miles

China: territorial expansion to c.AD 100

Early Chinese civilization was based on river valleys, spreading from the Huang and, later, the Yangtze. Over the centuries it experienced successive periods of expansion under strong dynasties interspersed with periods of disorder. Neither the Shang nor the Zhou (Chou) controlled a truly unified state, but their cultural influence extended from Manchuria to the Yangtze valley. Under the Qin (Ch'in) unity was established in a Chinese state extending as far south as Guangdong (Kwangtung). Expansion continued under the Han, who eventually ruled an area as large as the Roman empire, including Korea and Annam and reaching into Kansu in the west in the first century BC.

Dynasties of China

Xia (Hsia) c.21st century–c.16th century BC	Chinese claim as first dynasty. Founder said to be Yu, master of flood control and irrigation
Shang c.16th century–c.11th century	The first verifiable dynasty, noted for its bronze vessels and weapons. First evidence of a written language
Zhou (Chou) c.11th century–256	Despite feudal wars, growth of cities from c.8th century BC, with emerging merchant class. Use of iron implements and metal coins. Flowering in literature and philosophy: Confucius, Mencius, and Taoism
Qin (Ch'in) 221–206	First to bring unity to China with central control, country-wide administration and major public works. State walls joined to form Great Wall. Standard weights and measures and writing introduced
Han 202 BC–AD 220	Continued Qin expansion with conquest of Korea and western territory. Developments in art and technology (including invention of paper). Buddhism introduced. Beginning of Chinese civil service
The Three Kingdoms 220–280	The start of three centuries of periodic warfare and disruption, with rival Han generals striving for supremacy
Western Jin (Western Chin) 265–316	China briefly united
Northern and Southern dynasties 317–589	Succession of petty dynasties. Buddhism's influence strong
Sui 581–618	Re-united China after preceding 'dark ages'. Important canal building projects and extension of Great Wall
Tang (T'ang) 618–907	Distinguished in many achievements, the Tang for a period ruled the world's largest empire. Growth in the arts and science; printing and gunpowder invented. International trade developing through Canton: much traffic along the Silk Route
The Five Dynasties 907–960	A disruptive period, ruled by brief military dictatorships
Song (Sung) 960–1279	A dynasty obliged to rule a divided China, first from the north, then from the south, after northern invasions by the Jin. Developments in painting and poetry; the compass invented. Marco Polo carried news to the west of China's remarkable level of civilization
Liao 945–1125	A dynasty of Tartar nomads, ruling parts of north China and Manchuria
Jin (Chin) 1126–1234	Juchen nomads ruling much of the north until overthrown by the Mongols. The Juchen were gradually absorbed into Chinese culture
Yuan (Mongols) 1279–1368	Under Kublai Khan, the Yuan dynasty conquered China. Administration improved through province system and post-road networks. International trade expanded. Marco Polo in Kublai Khan's service
Ming 1368–1644	A native-born Chinese dynasty that drove out the Mongols, the Ming saw orderly government and prosperity. Voyages of exploration to Persian Gulf and Africa. First Jesuits arrived in China
Qing (Ch'ing) (Manchus) 1644–1912	A dynasty of northern invaders who mainly adopted Chinese ways. Final century of its reign marked by great impact of western powers through 'treaty ports', leading to declaration of republic in 1912

Cherokee, a North American Indian tribe which traditionally inhabited a region stretching across western Virginia and the Carolinas, eastern Kentucky and Tennessee, and northern Georgia and Alabama. Their prehistoric ancestors built ancient Etowah (Georgia), an important ceremonial centre of the eastern *Mississippi cultures, visited by Hernando *De Soto in his explorations of 1540-2. Smallpox and other European-introduced diseases had greatly reduced their population by the 17th century, when French and English traders made contact. Conflict with white settlers moving westwards led to several wars, but the Cherokee adapted quickly to white life-styles after the American War of *Independence.

Chesterfield, Earl of *Stanhope.

Chickasaw, a North American Indian tribe which inhabited the region of modern northern Alabama and Mississippi, and southern Tennessee, and were descendants of the late prehistoric *Mississippi cultures. In 1739 they were attacked by a French campaign from Montreal and Fort Michilimackinac (Michigan), aided by several Great Lakes tribes; and as allies of the British based in Charlestown they fought in the 18th century with pro-French tribes such as the Illinois in the north and the Choctaw in the south.

Chien-lung *Qianlong.

Children's Crusade (1212), a pathetic episode in the *Crusades, growing out of simple faith and fanatical zeal for the recapture of *Palestine from the Saracens. Some 50,000 children, mainly from France and Germany, are said to have taken part in the expedition which probably included poor adults. It was doomed from the start. Those who did manage to embark from the ports of France and Italy were dispatched to Muslim slave markets. Very few ever returned to their homes. The legend of the Pied Piper of Hamelin telling of the loss of 130 children who followed a mysterious flautist 'to Calvary' may derive from this episode. Robert Browning based his poem (1842) on the legend.

Chile, a South American country between the Pacific and the Andes. At the time of the first Spanish contact in 1536 the dominant Indian group, the Araucanians, were theoretically subject to the *Inca empire, but in practice they retained considerable independence within the Inca realm. Though they resisted Spanish encroachments, the Araucanians were gradually pushed south of the Bío Bío River where they were more or less kept under control. Spanish colonization began with the foundation of Santiago in 1541. The colony grew moderately but did not prosper for the next two centuries as it was overshadowed by wealthier Peru. Politically Chile became part of the viceroyalty of Peru.

Chimú, the most powerful state of the north coast of Peru between c.AD 1000 and 1476, when it was conquered by the *Incas. Its capital was Chan Chan (near modern Trujillo), a vast city with ten large rectangular enclosures measuring 400 by 200 m. (1300 by 650 ft.). These were built in sequence as the ruler died and was buried, and comprised a royal compound complete with residences, administrative rooms, gardens, kitchens, storerooms, and the royal tomb. Associated artefacts included a distinctive mould-made black pottery and some of the finest gold, silver, and bronze-work known from the New World.

Chin dynasty *Jin.

Ch'in dynasty *Qin.

China, a country in East Asia. It has a recorded history beginning nearly 4,000 years ago, with the *Shang who settled in the Huang He (Yellow River) valley. Under the Eastern *Zhou, from the 6th century BC, *Confucius and *Mencius formulated ideas that became the framework of Chinese society. *Taoism appeared during the 3rd century BC. Gradually Chinese culture spread out from the Huang He valley. A form of writing with characters representing meanings rather than sounds— and required by *Shi Huangdi, the first ruler of a unified China, to be written in a uniform style—bound together people divided by geography and different spoken dialects. From the *Qin the concept of a unified empire prevailed, surviving periods of fragmentation and rule by non-Chinese dynasties such as the *Yuan. Under strong dynasties such as the *Han and the *Tang China's power extended far west into *Turkistan and south into *Annam. On its neighbours, particularly *Korea and Annam, it exercised a powerful influence. Barbarian invaders and dynasties usually adopted Chinese cultural traditions.

The ideas of *Buddhism began to reach China from the 1st century AD and were gradually changed and assimilated into Chinese culture. The Chinese people, showing remarkable inventiveness, were ahead of the West in technology until about the end of the *Song dynasty. However, after the *Mongol conquest the country drew in on itself. Learning, in high esteem from early times, became rooted in the stereotyped study of the Confucian classics, for success in examinations based on the classics was for centuries the means to promotion in the civil service. In time, study of the classics had a deadening intellectual influence.

Throughout history, China, the 'Middle Kingdom', as it is called by the Chinese, regarded itself as superior to all others—a view shared by philosophers of the *Enlightenment. Western countries attempted to establish trading links with the *Qing dynasty but with little success. As the power of the Qing dynasty weakened towards the end of the 18th century, Western pressure for change built up, leading to direct European involvement in China in the 19th century.

Ch'ing dynasty *Qing.

chivalry, the code of behaviour practised in the Middle Ages, especially in the 12th and 13th centuries, by the mounted soldier or *knight. The chivalric ethic represented the fusion of Christian and military concepts of conduct. A knight was to be brave, loyal to his lord, and the protector of women. The songs of the *troubadours celebrated these virtues.

It was a system of apprenticeship: as boys, knights' sons became pages in the castles of other knights; from the age of 14 they learnt horsemanship and military skills, and were themselves knighted at the age of 21. The *Crusades saw the apogee of the chivalric ideal, as new Christian orders of knights (*Knights Templars, *Knights Hospitallers), waged war in *Palestine against the *Muslims. During times of peace, the *tournament was the

The age of **chivalry** in medieval Europe inspired an outburst of literature extolling the virtues of courtly love. This illustration is from a 15th-century manuscript of the celebrated *Roman de la Rose*, the allegorical romance on the 'Art of Love' by Guillaume de Loris, written *c*.1240. (British Library, London)

setting for displays of military and equestrian skill. The 15th century saw a decline in the real value of chivalry, and though new orders, such as the Order of the Golden Fleece (Burgundy) were created, tournaments survived merely as ritualized ceremonies.

Choctaw, a North American Indian tribe who inhabited modern Mississippi south of the *Chickasaw, and whose ancestors formed the southern extent of the *Mississippi cultures. Hernando *de Soto explored their territory in 1540-2, and they competed with the *Natchez and Chickasaw for trade with early Spanish settlers, and later with the French at New Orleans in the 17th and 18th centuries. Conflict with American colonists increased after the American War of *Independence.

Choiseul, Étienne François, duc de (1719-85), French statesman, Secretary of State for Foreign Affairs (1758-70). He concluded the Family Compact of 1761 with *Charles III of Spain and, considering the weakness of the French position, was a successful negotiator at the Treaty of *Paris in 1763. He then tried to reform the army and navy, but was dismissed in December 1770 when he tried to persuade *Louis XV to support Spain against Britain over the *Falkland Islands. Lorraine and Corsica were both annexed during his period in office.

Chola, a Tamil Hindu dynasty dominant in south India from the 9th to the 13th century. Their origins are uncertain, but they were influential from at least the 3rd century AD, becoming an imperial power on the overthrow of their *Pallava neighbours in the late 9th century. Victory over the *Pandyas followed, and then expansion into the Deccan, Orissa, and Sri Lanka. Their peak was during the reigns of Rajaraja I (985-1014) and Rajendra I (1014-44), when Chola armies reached the Ganges and the Malay archipelago. The dynasty remained the paramount power in south India until the mid-13th century when *Hoysala and Pandya incursions and the rise of *Vijayanagar eventually destroyed its claims.

Chou dynasty   *Zhou.

Christ   *Jesus Christ.

Christian I (1426-81), King of Denmark and Norway (1448-81), and Sweden (1457-64), founder of the Oldenburg dynasty. Elected to power by the Danish Rigstad, and confirming his status by marriage to his predecessor's widow, he gained the Swedish throne after the war of 1451-7, but lost control to the Swedish nobility later. He also gained Schleswig and Holstein, and was at war with England (1469-74). Strongly Catholic, he founded the Catholic University of Copenhagen in 1479.

Christianity, the religion based on the belief that *Jesus Christ was the Son of God and on his teaching. At first it was the faith of a group of Palestinian *Jews who believed that Jesus was the Messiah or Christ (from the Hebrew and Greek respectively for 'anointed one') who would bring freedom to the Jews. The teachings of Jesus began to spread, particularly through the missionary travels of the former pharisee, *Paul, who visited Asia Minor, Greece, and Rome. Paul's message that faith in Jesus was open to everyone brought Christianity to Gentiles (non-Jews) who were not willing to accept the ritual obligations of Judaism and enabled Christianity to spread rapidly. Initially Christians experienced intermittent harassment by the Roman authorities though there was no clear legal basis for this until the reign of Emperor Decius, who began systematic persecution of the Christians in AD 250. By the 3rd century, Christianity was widespread throughout the Roman empire; in 313 *Constantine ended persecution and in 380 Theodosius recognized it as the official religion of the empire. By this time Christianity had also reached Armenia, Egypt, Persia, and probably southern India.

Around 200, the Church leaders began to collect together the most authoritative Christian writings into the New Testament of the *Bible, the final selection being agreed by 382; and in 325 at the Council of *Nicaea a statement of Christian belief was agreed. As the Church grew, however, there were disputes between Christians on matters of doctrine and later over church organization. A division, originally cultural and linguistic, grew between the Eastern Church based at Constantinople and the Western Church at Rome, culminating in the *East-West Schism of 1054 and sealed by the sacking of Constantinople by the *Crusaders in 1204. In the West the unity of the Church, focused on the *papacy in Rome, was challenged by the Protestant *Reformation in the 16th century and the emergence of autonomous reformed churches. This period also saw renewed missionary activity, particularly by Catholic religious orders, as European countries colonized other parts of the world. By 1800 the political influence of the Church was waning

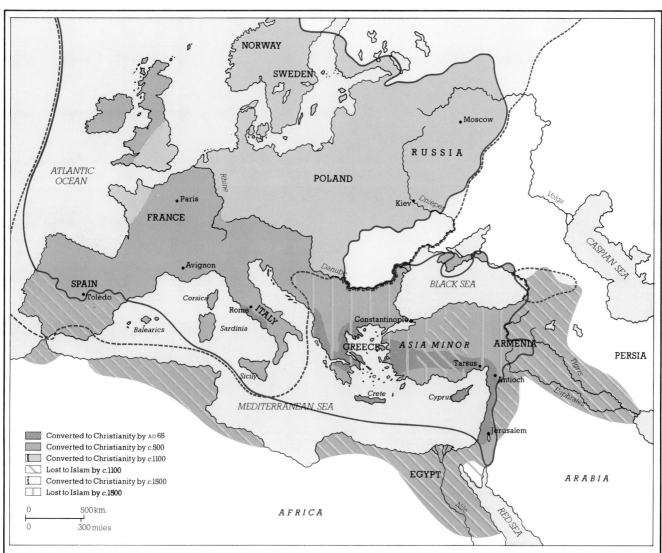

Christianity: its expansion to *c.* 1500

By AD 65, the Christian doctrines had been spread by the teachings of St Paul and other missionaries to a number of independent, often isolated 'churches', of which Palestine was predominant. By the 4th century, after a period of persecution by the Romans, Christianity had become the official religion of the Roman empire, but by the end of the 5th century it was already under threat, with the collapse of the empire and the Germanic invasions. A powerful rival appeared in the east with the Arab-led rise of Islam in the 7th century. By 1100 Islam was predominant in North Africa, much of the Middle East, Spain, and the Balkans. In the west, Celtic missionaries spread Christianity to England in the 7th century, and through northern Europe and Scandinavia during the 9th and 10th centuries. By 1500 Spain had been regained, but Islam had consolidated its hold on the Middle East and, represented by the Ottoman empire, was encroaching further on eastern Europe.

but its moral teachings continued to guide the behaviour of individuals.

Christina (1626–89), Queen of Sweden (1632–54). She was the daughter and successor of *Gustavus Adolphus. During her minority, the kingdom was governed mainly by Chancellor Axel *Oxenstierna. When she assumed power in 1644, she showed herself to be clever, restless, and headstrong. She attracted many foreign artists and scholars (including *Descartes) to her court, but after a serious constitutional crisis in 1650, she made plans to abdicate in favour of her cousin, Charles X. This was partly because of the pressure of social unrest within Sweden, and partly because of her secret conversion to the proscribed Roman Catholic faith. In 1654 she abdicated, and spent most of her remaining years in Rome. Her time was occupied in patronizing the arts and in intriguing for the crowns of Naples and Poland.

Chuang-tzu   *Zhuangzi.

Church of England   *Anglican Church.

Church, Benjamin (1639–1718), American soldier, a Rhode Island militia captain in *King Philip's War (1675–6). He cornered Philip in the Great Swamp near Kingston, destroying the remnant of his force. In 1705 he joined a New England expedition against the French in Nova Scotia. His *Entertaining Passages Relating to King Philip's War* appeared in 1716.

His grandson, **Benjamin** (1734–76) was a leading Boston doctor and patriot who in 1775 betrayed the

The abdication of Queen **Christina** of Sweden, a public ceremony in the royal palace at Uppsala in 1654. This engraving by Willem Swidde was published in Germany in 1697. All the notables of the realm, including Chancellor Oxenstierna, are present to witness the abdication. (Uppsala Universitetsbibliotek)

American cause to the British. Paroled from life imprisonment, he died en route to the West Indies.

churl *ceorl.

Cicero, Marcus Tullius (106–43 BC), Roman orator, statesman, and writer. He first made his name as a lawyer in civil and criminal trials. His brilliant prosecution of Verres, the corrupt Roman governor of Sicily, in 70, established his reputation and the nobility came to see in him a strong candidate for the consulship of 63. As consul he outmanoeuvred Catiline and his fellow conspirators, who were plotting to take over Rome. He spoke out against the First Triumvirate, and was exiled on the charge of executing the Cataline conspirators without trial. The exertions of his friends secured his recall in 57 amid popular acclaim. He returned to Rome just before the eruption of civil war, which he did his best to avert.

His allegiance lay with Pompey and the senatorial cause but he became disillusioned by Pompey's leadership and after his defeat at *Pharsalus, returned to Italy. Caesar admired him greatly and valued his political support but he rejected Caesar and immersed himself in his philosophical writings. After Caesar's assassination it fell to Cicero to rally the Senate. He denounced *Mark Antony in the *Philippicae* ('Philippic Orations') and hoped

that he could revive the republic. In 43 Antony, Octavian, and Lepidus ordered him to be put to death and he was captured and executed.

Although Cicero rated his political role most highly his lasting claim to greatness rests on his writings. He was the finest Roman orator and his speeches, correspondence, and treatises have been a major influence on Western thought and literature up to the present day.

Cid Campeador *El Cid.

Cilicia, a country in south-eastern Turkey, geographically divided into two distinct parts. The western area was mountainous, while the eastern consisted of a fertile plain with Tarsus as its main city. It came under the control of the *Hittites, the *Assyrians, the *Achaemenids, and *Alexander the Great and was fought over by the *Seleucids and *Ptolemies. In the 2nd century BC it became a haven for pirates, who were finally crushed by the Roman general Pompey in 67 BC, and by the end of the century it was part of the *Roman empire. It was occupied by migrating Armenians in 1080. In 1375 it was conquered by the Mamelukes of Egypt and in 1515 by the *Ottoman Turks.

Cincinnatus, Lucius Quinctius (c.519–438 BC), Roman republican hero famous for his devotion to the republic in times of crisis. Appointed dictator in 458 when a Roman army was trapped in battle by the Aequi tribe, he won a crushing victory and rescued the beleaguered troops. After this success, he resigned his command and returned to farm his small estate. *Cato the Elder and other later republicans regarded him as the ideal statesman, representing the old Roman values of rustic frugality, duty to fatherland, courage, and lack of personal ambition; consequently much of what was told about him by the historian *Livy was strongly infected by romantic invention, as in the famous picture of his being 'called from the plough' in 458 to take the supreme command. The 18th century also idealized him as the citizen-soldier *par excellence* and Cincinatti (USA) was so named to honour the contribution of volunteer officers in the War of *Independence.

Cinque Ports, a confederation of coastal towns in south-east England. They provided the crown with ships and men to patrol the Channel and to convey its armies to the Continent, from the 11th to 16th centuries. The original five (French, *cinque*) ports were Hastings, Romney, Hythe, Sandwich, and Dover—known collectively as the 'head' ports. They were joined by thirty-two other ports, known as 'limbs'. By the 14th century Winchelsea and Rye were also recognized as head ports. The ports of the confederation received privileges including exemption from taxes, the right to return members to Parliament (retained until the 19th century), and the honour of attending on the monarch at the coronation. Burgesses of the ports were in 1205 granted the title of 'barons'. A Warden of the Cinque Ports was created in 1268 as an extension of the powers of the Constable of Dover Castle. A royal *charter incorporating these privileges was granted in 1278. The Cinque Ports declined with the setting up of a permanent navy and the Warden's title became honorary.

circus, an entertainment of ancient Rome, which took its name from the long racing 'circuit' of the arena. *Chariot races became the chief attraction of the programme. Four rival teams represented the elements, wearing Green, Red, Blue, and White, and each drove four horses representing the seasons. A race lasted seven laps, and each day had twenty-four races. Assassination and riots occurred, though 'bread and circuses' were supposed to keep the Roman mob quiescent. Ninety days a year came to be devoted to circuses. The Circus Maximus hippodrome held 350,000, one-third of Rome's population.

Cistercian, a member of a monastic order founded by St Robert of Molesme in 1098 at Cîteaux in France. Cistercians followed a strict interpretation of the 'rule' of St *Benedict; their constitution was laid down in the *Carta Caritatis* ('The Charter of Charity'). St *Bernard founded a daughter house at Clairvaux which rapidly gained a great reputation. The Cistercians followed a life of strict austerity and during the great spread of monastic houses in the 11th and 12th centuries they led the movement to bring formerly unproductive land (marsh and moor) into use for agriculture, pioneering many new techniques including the employment of water power, and became very wealthy as sheep farmers and wool traders. The monks are now divided into two observances, the strict observance (following the original rule), known as Trappists, and the common observance, which allows certain relaxations.

A statue of the Roman statesman **Cicero**. His surviving writings include his *Letters*, a record of his personal and political activities. (Museo Nazionale, Naples)

citadel, a key feature of a Greek city, being the stronghold around which large communities originally developed. When a city expanded, and a protective encircling wall was built to protect the citizens' houses, the citadel lessened in importance, though it often became a religious centre and housed the public treasury. The *acropolis of Athens is the most famous example.

civilization, a stage in a society's development characterized by central economic, administrative, and religious facilities. In ancient *Mesopotamia the development of irrigation and farming techniques resulted in advancement from subsistence farming for all towards specialization of labour in an ordered society. Builders, craftsmen, and priests all performed their particular tasks while farmers provided food for the community. Towns and cities developed, under the rule of a king or priest, who controlled the economy; written records were kept, and ceremonial buildings for religious worship erected. By 3000 BC town life was also developing in ancient *Egypt and it emerged independently in the Indus valley, China, Mexico, and Peru. From the towns of these civilizations economic and social organization spread to the surrounding lands.

Civil War, English   *English Civil War.

civil wars, French   *French Wars of Religion.

civil wars, Roman   *Roman civil wars.

clan, a group of families with a common ancestor. Its members share a common surname and show fierce loyalty to their chief. Clans have always been politically significant in Scottish history and clan support was vital to the Scottish king, who often played upon clan rivalries to maintain his power. These rivalries, especially between the Highland and Lowland clans, intensified at the time of the Reformation. The Highland clans retained their Roman Catholic faith, they fought for the Royalists during the English Civil War, and their reluctance to accept *William of Orange led to the *Glencoe massacre in 1692. They also took the lead in the *Jacobite rebellions of 1715 and 1745, after which an attempt was made by the British government to break up the clans by banning the wearing of the kilt and by undermining the system of communal clan ownership of land.

In Ireland a clan-based social system prevailed until, after the 16th-century rebellions against English rule, their influence was progressively destroyed by military suppression and a policy of wholesale land confiscation.

Clarence, George Plantagenet, Duke of (1449-78), one of *Edward IV's younger brothers. He intrigued with the Burgundians and fell out with both Edward and his other brother, Richard, Duke of Gloucester (*Richard III); he was found guilty of high treason and is supposed to have been drowned in a butt of malmsey wine.

Clarence, Lionel, 1st Duke of (1338-68), the second surviving son of *Edward III and Philippa of Hainaut, known as Lionel of Antwerp from his birth in Antwerp. From about 1341 it was arranged that he would marry the Anglo-Irish heiress Elizabeth de Burgh (1332-63); he was created Earl of Ulster and in 1361 was sent to Ireland as governor, to reassert English rule there. In 1362 he was created Duke of Clarence, the title being derived from his wife's inheritance of the lordship of Clare in Suffolk. After her death another rich marriage was arranged for him, to Violante, the only daughter of Galeazzo Visconti, Lord of Pavia; he died only a few months after this wedding.

The title of Clarence was revived in 1412 for **Thomas of Lancaster** (1389-1421), second son of *Henry IV and Mary de Bohun; it lapsed again after his death.

Clarendon, Constitutions of, a document presented by *Henry II of England to a council convened at Clarendon, near Salisbury, in 1166. The king sought to define certain relationships between the state and the church according to established usage. Churchmen, in particular Thomas à *Becket, saw it as state interference. The most controversial issue, *benefit of clergy, concerned Henry's claim to try in his law courts clerics who had already been convicted in the ecclesiastical courts. After Becket's murder in 1170 Henry conceded the benefit of clergy, but not other points at issue.

Clarendon, Edward Hyde, 1st Earl of (1609-74), English statesman and historian. He began his political career in the Short and Long Parliaments as an opponent of royal authority, but in 1641 he refused to support the Grand Remonstrance, changed sides, and became a trusted adviser of Charles I, and later of Charles II, with whom he shared exile. At the *Restoration Charles II made him Lord Chancellor; he helped to carry out the king's conciliatory policies, and his influence reached its peak when his daughter Anne married the heir apparent, James, Duke of York. Clarendon had little sympathy with the so-called Clarendon Code (1661-5), a series of laws aimed at Roman Catholics and dissenters, but he enforced them against the king's wishes. He was popularly blamed for the naval disasters of the second *Anglo-Dutch war. He fell from power in 1667 and fled to France to avoid impeachment. His *History of the Rebellion* (published 1702-4) is a masterly account of the English Civil War, written from a royalist standpoint but with a considerable degree of objectivity.

Claudius (Tiberius Claudius Drusus Nero Germanicus) (10 BC–AD 54), Roman emperor (41-54). He was the nephew of Tiberius and the uncle of Gaius *Caligula. An intelligent man of poor physique, he devoted himself to scholarship before sharing consular office with Caligula. After his nephew's murder the *Praetorians proclaimed his accession. He set about repairing the damage of the previous reign, taking an interest in the army, the Senate, and the administration. He added Britain and Mauretania (in north Africa) to the empire. He took a formal personal part in the invasion of Britain and added 'Britannicus' to his son's names to indicate the Roman possession of Britain. In later years his court was dominated by freedmen. His third wife Messalina, the mother of his children Britannicus and Octavia, was notorious for her infidelities and was eventually put to death. His

Cleopatra VII with Caesarion, the son she claimed was fathered by Julius Caesar. This sandstone relief from the wall of the Temple of Hathor at Denderah in Egypt portrays Caesarion in the attire of a pharaoh; Cleopatra stands behind him offering incense.

fourth wife Agrippina, mother of *Nero, is said to have poisoned him with mushrooms to hasten her son's succession. He was declared a god although *Seneca in his satire *Apocolocyntosis* mocked the deification ceremony.

Cleopatra VII (69–30 BC), the last of the *Ptolemies. She became co-ruler of Egypt with Ptolemy XIII in 51, but was driven out in 48. She was restored by Julius *Caesar, and in 47 bore a son whom she said was his. In 46 they both accompanied him to Rome. After his assassination in 44 she returned to Egypt, and in 41 met *Mark Antony at Tarsus. He spent the following winter with her in Alexandria, and she in due course gave birth to twins. In 37 Antony acknowledged these children and restored the territories of Cyrene and elsewhere to her; she pledged Egypt's support to him. In 34 they formally announced the division of *Alexander the Great's former empire between Cleopatra and her children. In 32 Octavian declared war on her, and in the following year the battle of *Actium resulted in the collapse of her fortunes. In 30 she committed suicide and Egypt passed into Roman hands.

cliff-dwellers, a generalized term for the *Anasazi peoples living at Mesa Verde, Colorado, and similar sites such as Montezuma Castle, Arizona, or sites in the Canyon de Chelly, Arizona, especially after *c*. AD 1150. Such sites were partly defensive against internal Pueblo warfare and against *Apache and *Navajo raids. Perhaps the most famous site is Cliff Palace, Mesa Verde, comprising several tiers of adobe brick structures under an overhanging cliff face—stout terrace walls, square and round apartment towers with over 200 rooms, and 23 large round *kiva* ceremonial chambers.

Clinton, George (1739–1812), American patriot leader. He controlled the popular anti-British faction in New York City from 1768. After attending the second *Continental Congress (1775–6) he helped draft the state's constitution and served as governor for fifteen years. An opponent of *Hamilton, he joined *Jefferson in founding the Democratic-Republican party, and was Vice-President (1805–12) to both Jefferson and Madison.

Clinton, Sir Henry (*c*.1738–95), British general. He was promoted to commander-in-chief in America after fighting at *Bunker Hill and the capture of New York. Although victorious at Monmouth (1778) and Charleston (1780), he was hampered by problems of supply and by the jealousy of *Cornwallis. His failure to prevent Franco-American concentration of troops at *Yorktown or to reinforce Cornwallis contributed to Britain's defeat and his own resignation. In 1794 he was appointed governor of Gibraltar, where he died.

Clive of Plassey, Robert, Baron (1725–74), British general and first British governor of *Bengal (1757–60 and 1765–7). Sent to south India at 18 as an *East India Company clerk, he demonstrated such military prowess against the French (notably in the siege of Arcot, 1751), that he soon rose to the position of governor of Madras. He was responsible for securing south-east India as a sphere of British influence. In 1756 his decision to transfer Company forces to Bengal was the first step in securing control of the region. He defeated the Bengal nawab, Siraj ud-Daula, at *Plassey in 1757, and assumed the

The adobe brick house of **cliff-dwellers** at Mesa Verde, Colorado. The grandeur of Anasazi architecture grew out of the prehistoric basket-maker culture.

governorship of Bengal. After consolidation of the Company's hold, during which he amassed a personal fortune, he returned home to a peerage.

He was recalled to Bengal in 1765 to extricate the Company from its growing economic difficulties. He attempted to reform the Company's exploitative government, but paid the price for his own earlier manipulations when Parliament censured him after his final return home. Not unusually dishonest for his day, he was the victim of changing, more critical attitudes to overseas ventures, and of jealousy of his *nabob life-style. Parliament reversed its verdict, but Clive, always melancholic, committed suicide in 1774.

Clodius, Publius Claudius Pulcher (*c*.92–52 BC), Roman statesman. Though of *patrician birth, in 59 *Caesar, *Pompey, and *Crassus assisted his adoption into a plebeian family; thus he became eligible to stand for election as a tribune of the *plebs, which office he held in 58. Active in Caesar's and Crassus' support, but also personally ambitious, he enacted several populist laws, including one for free corn distributions to the urban poor and another to exile *Cicero. He was a brilliant organizer and director of para-military groups, and for several years Rome became a battle-ground of rival political gangs, until early in 52 he was killed in an ambush. The resultant chaos subsequently led to Pompey's appointment as sole consul to break the power of the gangs.

Clovis (*c*. AD 466–511), the founder of the Frankish kingdom. He ruled a tribe of Salian *Franks at Tournai in what is now Belgium and defeated the last Roman governor of *Gaul at Soissons in 486, bringing the area between the rivers Loire and Seine under his control. In 496 he acquired the upper Rhineland by defeating the tribe of the Alemanni. He conquered the *Visigoths near Poitiers, extending his lands south as far as the Pyrenees. Small independent kingdoms in northern France were absorbed into his domain. His conversion to Christianity gained him the support of the Catholic Church, and thus strengthened his dynasty, the *Merovingians.

Clovis culture, a prehistoric culture in North America, characterized by lance-shaped stone points, 7–12 cm. (3–5 in.) long, fluted near the base. The tools are often found in association with bones of large mammals, such as bison and extinct mammoth, and are assumed to have been used as spear heads. Named after a town in western New Mexico, they are found at sites throughout the mid-west and south-west, USA from a period between 12,000 and 10,000 years ago. At one time, Clovis hunters were regarded as typifying the first *Amerindians, but there is increasing evidence that people were in the Americas long before, perhaps by 30,000 years ago.

Cluny, Order of, a reformed Benedictine monastic order, whose mother house was the abbey of Cluny in France, founded by Duke William of Aquitaine in 909. Under Abbot Odilo (994–1048) Cluny became the head of a system of dependent 'daughter houses' throughout western Europe. In the 11th and 12th centuries Cluny became a spiritual and cultural centre of vast influence; four of its members became popes and it inspired the zealous reforming innovations of Pope *Gregory VII from 1073. The monastery church at Cluny was a model for much ecclesiastical building in Europe. By the 13th century the period of greatest achievement was over, as the monastic ideal suffered from too close an involvement with the secular world and too great a share of its wealth and worldly power.

Cnut *Canute.

Cochin *Kerala.

Cochin China, the southern region of Vietnam, centred on the Mekong delta. It was so called to distinguish it from Cochin in India. The home of people akin to the

Shah Allam II handing **Clive of Plassey** sovereign rights in Bengal in 1765, a painting by Benjamin West. (India Office Library, London)

*Khmers, it was not fully absorbed by the Vietnamese until the 18th century. It was the base from which Gia-Long, as emperor of *Annam, unified Vietnam in 1802.

Coercive Acts (1774), legislation passed by the British Parliament as a punishment for the *Boston Tea Party. They closed the port of Boston pending compensation, amended Massachusetts's charter, allowed trials to be transferred to other colonies, and troops to be quartered at Boston's expense. Though Lord *North's aim was to isolate Massachusetts, the 'Intolerable Acts', as they became known, stiffened American resistance and precipitated the *Continental Congress.

coffee-house, a public place of refreshment where the main beverage was coffee. Coffee had enjoyed varying popularity following its introduction into Europe in the 16th century, but with the opening of the first coffee-house in London in 1652 its future was assured. Most of the cities of Europe had coffee-houses by the late 17th century, and the institution spread into the American colonies in 1689. They became centres where business was transacted, newspapers were read, and literary and political opinions were exchanged. In London Lloyd's coffee-house was the centre for marine insurance, and in New York the Merchants coffee-house, opened in 1737, became of major political importance in the years leading up to the American War of *Independence. The heyday of these establishments was over by the mid-18th century, especially in England, where their social role was taken over by gentlemen's clubs.

Coke, Sir Edward (1552–1634), English lawyer and politician. He rose to the position of Lord Chief Justice (1613), prosecuting such defendants as *Essex (1601) and the *Gunpowder Plot conspirators (1606). In 1616 James I dismissed him, since, at first a supporter of the royal prerogative, Coke had become a defender of the common law against church and crown: as a Member of Parliament he led opposition to James I and *Charles I. He was largely responsible for drafting the *Petition of Right (1628) and wrote commentaries on medieval and contemporary English law.

Colbert, Jean Baptiste (1619–83), a leading minister of France from the mid-17th century. The son of a merchant who became an important financier in Paris, he rose to be one of the chief ministers of *Louis XIV, having shown his ability in building up *Mazarin's private fortune and in profiting from the fall of *Fouquet. He was loyal, dedicated, and hard-working, and when he became Controller-General of Finance in 1665 he halved the expense of tax collection and greatly increased the revenue. Putting *mercantilist theories into practice, he stimulated industry, improved communications, and established trading companies. He was constantly thwarted by the king's costly wars, but his aim was to make France great through the prosperity of the people. He did not remedy the basic weakness of the French fiscal system, however, and tended to burden industry with bureaucratic details, and his tariff policy was aggressive. He encouraged the construction of the Canal du Midi, linking Toulouse to the Mediterranean, and as Secretary of State he restored the French navy. In the artistic sphere Colbert supervised the reorganization of the Gobelins

tapestry factory and the re-establishment of the Royal Academy of Painting and Sculpture. He was a collector himself and it was largely owing to his practical support for the king's ambitions that France replaced Italy as the artistic capital of Europe.

Coligny, Gaspard de, seigneur de Châtillon (1519–72), French nobleman of the House of Montmorency, appointed Admiral of France in 1552. He was captured by the Spaniards in 1557, and during his incarceration in prison he became a committed *Calvinist. His high personal standing subsequently conferred respectability on the *Huguenot cause in the first phase of the *French Wars of Religion. On *Condé's death he was elected commander-in-chief (1569), and then helped to engineer the favourable Peace of St Germain (1570). But his ascendancy over the youthful Charles IX alienated him from Catherine de *Medici, who almost certainly acquiesced in a plan to assassinate him. The *St Bartholomew's Day Massacre seems to have broken out spontaneously in the wake of his assassination.

Colombia, a country in the extreme north-west of the South American continent. It was occupied by the Chibcha Indians before the Spanish conquest. The first permanent European settlements were made on the Caribbean coast, Santa Marta being founded in 1525 and Cartagena eight years later. Colonization of the interior was led by Gonzalo Jiménez de Quesada, who defeated the Chibchas and founded the city of Bogotá in 1538. The region was initially part of the viceroyalty of Peru, but a different political status came with the establishment of the viceroyalty of New Granada in the first half of the 18th century. The viceroy sitting in Bogotá was given jurisdiction over Venezuela, Ecuador, and Panama as well as Colombia.

The lively interior of a London **coffee-house** by an unknown artist, a print of 1668. (British Museum, London)

Colosseum (literally 'Colossal Building'), the name dating from the 8th century, given to the Flavian Amphitheatre in Rome. It was built by the emperors *Vespasian and Titus in front of *Nero's 'Golden Palace'. Prisoners brought to Rome after the suppression of the *Jewish Revolt laboured on it, completing it in AD 80. With a seating capacity of 50,000 on three levels, it was shaded by adjustable awnings and flooded for mock sea-battles. A wooden floor covered in sand was the stage for *gladiators and wild beasts. In turn a fortress and a home for squatters, its preservation was ordered by the pope in the 11th century.

Columba, St (c.521-97), abbot and missionary. He was born in Donegal, Ireland, of the family of the Irish High Kings. In 563 he and twelve monks founded the monastery of Iona, off the west coast of Scotland. For the next thirty years Columba continued the conversion of the heathen Picts. His consecration of King Aidan at the coronation (574) was the first royal consecration in Britain. Through the work of Columba and his successors Iona became the centre of *Celtic Christianity in northern Britain.

Columbanus, St (c.543-615), abbot and missionary, from his youth a monk at Bangor, Ireland. In about 590 he left for France and founded monasteries at Annegray and Luxeuil. His support for the *Celtic practice of Christianity, and especially for the Irish dating of Easter, upset Pope Gregory I and he was ordered back to Ireland (610). He promptly crossed the Alps to Lombardy in Italy and established an abbey at Bobbio (614). However, his austere monasticism lost its appeal before the more practical provisions of St *Benedict.

Columbus, Christopher (1451-1506), Genoese explorer, celebrated as the discoverer of America. His great interest was in what he called his 'Enterprise to the Indies', the search for a westward route to the Orient. For over a decade he tried to get financial support for his 'Enterprise', and at last in 1492 persuaded *Ferdinand and Isabella of Spain to sponsor an expedition. He set out in the *Santa Maria*, with two other small ships, expecting to reach Japan, and when he came on the islands of the Caribbean he named them the West Indies, and the native Arawak people Indians. On Cuba (which he thought was China) tobacco was discovered. For his second voyage a year later he was provided with seventeen ships and expected to trade for gold and establish colonies. He surveyed much of the Caribbean archipelago during the next three years. He organized a third voyage in 1498, and discovered Trinidad and the mouth of the Orinoco River, but the colony he had left on *Hispaniola was seething with rebellion. Ferdinand and Isabella sent a new governor to control it and paid off Columbus by allowing him to fit out a fourth voyage (1502-4) at their expense. He explored much of the coast of Central America until his worm-eaten little ships were unfit for the voyage home. He chartered another vessel and reached Spain ill and discredited, and died forgotten.

Comanche, a North American Indian tribe of the Great Plains. They used the dog-travois (sledge) to follow migrant game herds, but gained greater mobility after the introduction of horses in the 17th century from Spanish settlements in the south-west. Throughout the 18th century warfare was endemic against other tribes.

A portrait, said to be of **Columbus** (artist unknown). Little is known of the navigator's early life before the New World voyages which brought him immortality but neither reward nor honour from his backers. (Civico Museo Storico, Como)

The Comanche obtained firearms from the French, and continued raiding against the Utes and Spanish until 1786, when the Spanish began to give them guns, ammunition, and supplies to make them dependent and cause them to lose their skill as *archers.

comitia, assemblies of Roman citizens meeting for elections or legislation. The oldest assembly, the 'Curiata', consisted of representatives of religious groups based on kinship: it survived later as a body which sanctioned adoptions and ratified wills. The creation of the 'Centuriata' as the assembly of the people in arms was attributed to King Servius Tullius. It originally elected magistrates and legislated. The citizens were organized in 'centuries' according to census rating and military equipment and function, and voting power was weighted in favour of wealth and age. It remained the electoral assembly for consuls and praetors.

The 'Tributa' was the meeting of the people in the thirty-five tribes, based on domicile, in which votes were equal, irrespective of property, but it was the tribal and not the individual vote which counted. Like the 'Centuriata' it was convened by consuls or praetors and became the main legislative body and elected most of the lower magistrates. It was perhaps modelled on the 'Concilium Plebis', under the presidency of the tribunes of the *plebs. This was established early in the conflict between patricians and plebeians. It consisted of plebeians only and was convened by the tribunes. Bills carried here were 'plebiscita'; but after 287 BC 'plebiscita' were accorded the same form as laws of the whole Roman people, and were generally also called 'leges'.

Commines, Philippe de (or Commynes) (*c*.1447-1511), French historian. Born in *Flanders, the son of a noble *Burgundian commander, he was raised at the Burgundian court, joined Louis XI of France in 1472, and was later disgraced for plotting against Charles VIII in 1486. Restored to favour, he joined Charles's invasion of Italy in 1494. He wrote his *Memoires*, six books on Louis and two on Charles during 1489-98.

Committee of Public Safety, an emergency body set up in France in April 1793. It was the first effective executive government of the Revolutionary period and governed France during the most critical year of the Revolution. Its nine members (later twelve) were chiefly drawn from the *Jacobins and it contained some of the ablest men in France, dominated at first by *Danton and then by *Robespierre. It successfully defeated France's external enemies but was largely responsible for the Reign of *Terror, and its ruthless methods, at a time of growing economic distress, led to growing opposition. In March 1794 an attempt to overthrow it, led by *Hébert, was quashed, but four months later the reaction which overthrew Robespierre marked the end of the Committee's power. It was restricted to foreign affairs until its influence was finally ended in October 1795.

common (or common land), usually woodland or rough pasture for the villagers' animals in medieval England. By the Statute of Merton (1236) the lord of the manor or other owner of a village was allowed to enclose waste land for his own use only if he left adequate pasture for the villagers. *Enclosure of common land started in the 12th century, and increased dramatically in the second half of the 18th century, often arousing opposition and claims of theft.

In colonial America many village communities had large areas of common land, partly for defensive purposes as well as for pasturage. These areas sometimes survived to be used for recreation, the best known being Boston Common, today a public park but bought by the town for pasturage in 1634.

Commons, House of, the lower chamber of the British *Parliament. It began as an element of the Parliaments summoned by the king in the later 13th century: both knights of the shire and burgesses of *boroughs were summoned to Simon de *Montfort's Parliament in 1265. It took 350 years for the Commons to become supreme in the tripartite division of power between it, the House of *Lords, and the monarchy. In the 14th century both Houses gained constitutional rights in relation to the monarchy; many of the struggles between *Richard II and his opponents were waged through the Commons—notably in the Merciless Parliament of 1388. The *Lancastrian monarchs were obliged to summon parliaments frequently, for grants of taxes, and *Henry VIII enhanced the importance of Parliament by his use of it during the English *Reformation.

In the early 17th century, when differences between the monarchy and Parliament first surfaced, the Commons took the lead in, for instance, the *Petition of Right (1628), winning *Charles I's acceptance of the principle of no taxation without parliamentary assent. The *Long Parliament (1640-60) abolished the House of Lords and set up the *Commonwealth, and it was the Commons that was instrumental in inviting *Charles II to take up

the throne, just as it promoted the *Bill of Rights (1689) and Act of *Settlement (1701) that defined the relations between Commons, Lords, and the monarchy.

Commonwealth, the republican government of England between the execution of Charles I in 1649 and the restoration of Charles II in 1660. The *Rump Parliament claimed to 'have the supreme power in this nation', and ruled through a nominated forty-man Council of State. In 1650 an 'Engagement' to be faithful to the Commonwealth was imposed on all adult males. While Oliver *Cromwell was eliminating Royalist resistance in Ireland and then Scotland (1649-51), the Rump disappointed expectations of radical reform. Unpopular taxes had to be raised to finance the army's expeditions. Furthermore, the Navigation Acts sparked off the much-resented *Anglo-Dutch War of 1652-4.

Cromwell expelled the Rump in April 1653. He hoped to reach a political and religious settlement through the *Barebones Parliament (July-December 1653), but in December he accepted the necessity of taking the headship of state himself. The period of Cromwellian rule is usually known as the *Protectorate.

commune, a medieval western European town which had acquired specific privileges by purchase or force. The privileges might include a charter of liberties, freedom to elect councils, responsibility for regulating local order, justice, and trade, and powers to raise taxes and tolls. The burghers initially swore an oath binding themselves together. There were regional differences: in northern and central Italy where there was a strong tradition of municipal independence, a commercial boom in the Middle Ages enhanced the wealth of the towns and increased their power, some eventually becoming independent city-republics (*Venice, *Florence). The communes of Flanders, the German cities of the Holy Roman Empire, and of Spain all achieved a measure of independence as the high costs of warfare forced their rulers to surrender direct control in return for financial benefits. The communes often pursued their own diplomatic policies as political alliances shifted. They flourished where central government was weak and became bastions of local power, and after the Reformation, of religious loyalties. The growth of strong national monarchies reduced them in the 16th and 17th centuries.

Comnena, Anna (*c*.1083-1148), historian, daughter of the Byzantine emperor Alexius *Comnenus. She nurtured ambitions that her husband would usurp her brother as emperor but when she was widowed in 1137, she retired to a monastery. She wrote the *Alexiad*, a history in fifteen books which was largely an account of the First *Crusade and panegyric of her father's life.

Comnenus, Alexius (1048-1118), *Byzantine emperor (1081-1118). In the mid-11th century Byzantine politics were dominated by a military aristocracy and court officials; Alexius Comnenus was an army general who forged an alliance between his military supporters and a number of court officials and so won the throne for himself. He succeeded in checking the challenge from the *Normans under *Guiscard in the Mediterranean but was continually harassed by the threat of barbarian invasions. In 1095 he approached Pope Urban II for help in recruiting mercenaries, a call which led to the

First *Crusade which the pope hoped would save the empire from the Seljuk *Turks. The Crusade was to Alexius's advantage and he was able to leave his son John to inherit the Byzantine empire on his death.

Concord *Lexington and Concord, battle of.

Conciliar Movement (1409–49), a church movement centred on the three general (or ecumenical) councils of Pisa (1409), *Constance (1414–18), and Basle (1431–49). Its original purpose was to heal the papal schism caused by there being two, and later three, popes at the

A stone rubbing of **Confucius**, from an original painting attributed to the Chinese artist Wu Daoze, from the early 8th century. (Museum of Fine Arts, Boston)

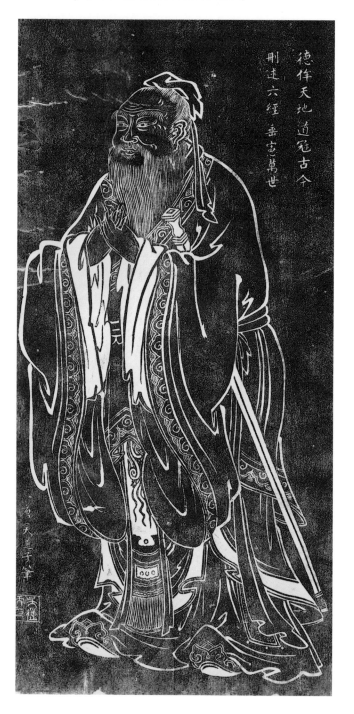

德侔天地道冠古今
删述六经垂宪万世

same time (*antipope). The movement was successful, deposing or accepting the resignation of the popes concerned. It declared the superiority of a general council of the church over the papacy, formulated in the decree *Haec Sancta* (sometimes called *Sacrosancta*) of 1415, and tried to make general councils a regular feature of the Western Church. It also dealt with various heresies, the council of Constance burning John *Huss and condemning John *Wyclif in 1415, and it initiated some reforms. The movement, in so far as it challenged papal authority, was eventually defeated by the papacy, but its long-term influence upon Christian churches was considerable.

Condé, the name of a junior branch of the French royal House of *Bourbon. The name was first borne by Louis I de Bourbon (1530–69), prince de Condé, a military leader of the *Huguenots during the first phase of the *French Wars of Religion. A bitter enemy of the *Guise faction, he was killed at the battle of Jarnac. Henry I de Bourbon (1552–88) took over his father's leadership of the Huguenots. He briefly renounced his faith at the time of the *St Bartholomew's Day Massacre (1572), but subsequently embarrassed his cousin, the future *Henry IV, with his Protestant fanaticism.

Henry II de Bourbon (1588–1646) was brought up as a Catholic; he plotted during the regency of Marie de Medici, and distinguished himself only by fathering Louis II de Bourbon, his successor, known as the Great Condé. The latter married a niece of Cardinal *Richelieu, and excelled as a military commander in the last phase of the *Thirty Years War. During the first *Fronde he sided with the court party; disagreements with *Mazarin led to his arrest and imprisonment (1650), and on the failure of his insurrection against the government (1651–2), he fled and took service in the Spanish armies in the Netherlands. When he was allowed to return to France in 1660, he conquered Franche-Comté for *Louis XIV (1668), and held high command in the war against the *United Provinces (1672); but Louis never really forgave him for his part in the Fronde, and his treasonable defection to the Spaniards.

Condorcet, Antoine Nicolas, marquis de (1743–94), French philosopher and politician. He was the only prominent French *philosophe* to play any real part in the events of the Revolution. As a *Girondin and a friend of *Sieyès and *Brissot he was elected to the National Convention. In October 1793 he was condemned by the Revolutionary Tribunal and eventually poisoned himself to avoid the guillotine. His best-known work *Esquisse d'un tableau historique des progrès de l'esprit humain* ('Sketch for a Historical Picture of the Progress of the Human Mind') was published in 1795.

condottiere (Italian *condotta*, 'contract'), a term for the leader of a medieval mercenary band of soldiers. Mercenaries flourished in the climate of economic prosperity and inter-municipal warfare of 14th- and 15th-century Italy. The earliest such mercenaries were recruited from the unemployed mercenary 'free companies' of the 1360s and included Catalans, the Germans and Hungarians of the so-called Grand Company, and the English Sir John Hawkwood, leader of the White Company in the 14th century. The system was refined in the 15th century by the *Sforzas, although the

CONQVISTA DE MEXICO POR CORTES. N.7

The Spanish soldiers or **conquistadores** under the leadership of Cortés, storming the Aztec capital of Tenochtitlán in 1521. The Spanish, with horses and firearms, defeated the more numerous Aztec forces and the battle proved the climax of their conquest of Mexico. This 17th-century Spanish painting is one of a series of eight by an unknown artist. (Private Collection on loan to British Embassy, Mexico City)

condottiere were always motivated by self-interest and changing of sides and loyalties was frequent. The system died out as a result of the Habsburg–Valois wars of the 16th century, which led to changes in the financing and organization of armies.

Confucius (Kongzi, K'ung Fu Tzu) (c.551–479 BC), the most celebrated Chinese philosopher. Amid the instability of Eastern *Zhou he sought to restore the 'golden age' which he believed existed under the early *Zhou. His teachings are contained within the *Analects* ('Sayings'), probably compiled by his followers after his death. So little is known of the world of men, he said, that speculation concerning spiritual matters is idle. The ideal society, he claimed, could be re-established by harmonization of the 'Five Relationships': between ruler and minister, husband and wife, father and son, elder and younger brother, friend and friend. The inferior must obey, the superior must be righteous, indeed benevolent. From *Han times his ideas, at least in theory, were the basis of conduct throughout China. The core of education was the Confucian classics, with which, except for the *Analects*, his connection is tenuous, and their lengthy commentaries. He became a minister in the state of Lu (in modern Shandong), but when his advice was not taken, he resigned. He then spent years visiting the courts

of the Chinese states, seeking a ruler to practise his precepts and appoint him to office, but had little success.

Congregationalist, a supporter of a form of church organization in which each local church is independent. The system derives from the belief that *Jesus Christ is the sole head of his church, and it is held to represent the original form of the church's organization. Known at different times in England as Separatists or Independents, they can be traced back to the 16th century followers of Robert *Browne, who broke with the *Anglican Church. Driven underground by persecution, they resurfaced in 17th century Holland and America. They were among the *Pilgrim Fathers who sailed to the New World in 1620. Meanwhile in England, after figuring prominently in the *New Model Army, they enjoyed freedom of worship under Oliver *Cromwell. This was abolished by the 1662 Act of Uniformity, then restored by the 1689 Toleration Act. In America they were allowed freedom of worship. Keen educationists, they played a major part in founding the universities of Harvard (1636) and Yale (1701).

Connaught (or Connacht), a province in the mid-west of Ireland. It was one of the five ancient Irish kingdoms, ruled from the 5th century AD by the O'Connors, kings of *Tara. With the 12th-century Anglo-Norman conquest of Ireland the area passed through the hands of various Norman nobles, reverting to the English crown in 1461. In the Tudor period the province was divided into shires, and the Composition of Connaught (1585) confirmed the possession of the land to the local gentry, ensuring their loyalty to Elizabeth I during *Tyrone's rising. Connaught suffered from the extortionary policies of *Charles I.

Connecticut, USA, a New England colony and state. Chartered in 1662, it combined the settlements round Hartford and New Haven. Its first governor was John *Winthrop junior. It was never subject to royal control. In the early national period, it was a leading stronghold of the *Federalist Party.

conquistadores, Spanish soldiers and adventurers in the 16th century. The term stems in part from the *Reconquista* of Spain from Moorish rule, culminating in the capture of *Granada by *Ferdinand and Isabella in 1492 after some 500 years of slow reconquest. The most famous conquistadores were Hérnan *Cortés, the conqueror of Aztec Mexico, and Francisco *Pizarro, the conqueror of Inca Peru; but there were many others. Their discoveries and conquests included the Caribbean, Latin America, southern and south-western USA, and the Philippines. Many would-be conquistadores explored immense areas but conquered nothing and founded no permanent settlements. As proper colonial administrations were established their activity diminished, and they never again found such rich empires as those of the Aztecs or Incas. Not all were soldiers: some were Christian missionaries, often appalled by the heartless Spanish exploitation of the native races. Most notable was Bartolomé de *Las Casas, who became an outspoken critic of the Spanish treatment of the Indians.

Constance, Council of (1414–18). An ecclesiastical council held at Constance, in Germany was called to deal with reform and heresy within the Christian Church. It resolved the *Great Schism, decreed the regular calling of councils, and presided over the trial and burning of John *Huss. It failed, however, to produce effective reform of outstanding abuses in clerical finance and conduct, or to curb papal independence.

Constantine I (the Great) (Flavius Valerius Aurelius Constantinus) (*c*. AD 274–337), Roman emperor (324–37). On the death of his father Constantius I in 306 at Eboracum (York) the army proclaimed him emperor. After a period of political complications, with several emperors competing for power, Constantine and Licinius

A 13th-century wall painting of **Constantine I** shows the emperor on foot leading the mounted Pope Sylvester I into Rome. (Chiesa Santi Quattro Incoronati, Rome)

divided the empire between them, East and West. War was fought between the two rulers (314) and Constantine defeated and killed Licinius (323) and he became sole emperor, founding a new second capital at *Byzantium which he named *Constantinople.

He adopted Christian symbols for his battle standards in 312 prompted by a 'vision' of the sign of the cross in the rays of the sun. In the following year he proclaimed tolerance and recognition of Christianity in the 'Edict' of Milan. Although his own beliefs are uncertain he supported orthodox Christianity in an attempt to maintain the unity of the vast *Roman empire. Sunday was declared a holiday in 326. He and his mother Helena took great interest in the Christian sites of *Rome and *Palestine. Basilicas were built on the site of the stable-cave in Bethlehem, where Jesus Christ was supposed to have been born, his alleged tomb in Jerusalem, and St *Peter's grave on the Vatican hill in Rome, and at Constantinople (St Sophia). The Eastern Church lists him as a saint.

Constantinople, a city in Turkey, formerly Byzantium, founded in 657 BC as a Greek colony. Early in the 4th century AD *Constantine chose the site as the capital for the Eastern Empire in preference to *Diocletian's nearby Nicomedia. It was designed as a new Rome, straddling seven hills and divided into fourteen districts and was renamed Constantinople in 330. The second capital of a single Roman empire ruled by two emperors, in 395 it became the sole capital of the East as Italy and Rome came under barbarian threat.

A city of monuments, churches, but no pagan temples, it was decorated with material from 'old Rome'. The walls of this capital of the *Byzantine empire withstood siege by Goths, Persians, and Arabs but was looted after a horrifying attack by Western Crusaders in 1204. It finally fell to the Ottoman Turks in 1453 and became the capital of the *Ottoman empire until the 19th century. Today it is the Turkish city of Istanbul.

Constitution of the USA, the fundamental written instrument of American government. It replaced the Articles of Confederation (1781–7), a league of sovereign states, with an effective central, national, federal government. Three months' secret debate among the *Founding Fathers at the Federal Constitutional Convention in Philadelphia in 1787 produced a series of modifications to *Madison's original Virginia Plan. The Great Compromise, between large and small states, gave equal representation in the Senate but by population in the House of Representatives. North and South finally agreed to slaves being counted as three-fifths of a person in representation and taxation, continuation of the slave trade until at least 1808, and no taxes on exports. Conflict between state and central power was reconciled by enumerating areas of federal concern. The principle of popular sovereignty with direct biennial election of congressmen was balanced by indirect election of senators and presidents for renewable six- and four-year terms. Within the federal system of executive (President, Vice-President, cabinet, and civil service), legislature (Senate and House of Representatives), and judiciary (Supreme and other federal courts), a series of checks and balances sought to share and divide power between these three components of government. Thus, for instance, the Supreme Court, appointed by the President with Senatorial approval, may declare actions of the executive or legislature unconstitutional, but may not initiate suits or legislation. The 1787 draft required ratification by nine state conventions. Major opposition in Virginia, Massachusetts, and New York came from the anti-Federalist, states' rights advocates like Patrick *Henry or George *Clinton. The constitution was defended in the *Federalist Papers* by *Madison, *Hamilton, and *Jay and came into operation in 1789. Major shortcomings included failure to foresee political parties, initial absence of a *Bill of Rights, complexity of electoral arrangements, and frustration of executive initiative. The loose definition of congressional powers and the persuasive influence of federal grants tipped the balance from state to central supremacy. This flexible guide for government has been amended twenty-six times, most notably to abolish slavery, to add a Bill of Rights, and to give black and female enfranchisement.

Continental Congress (1774, 1775–89), the assembly which first met in Philadelphia to concert a colonial response to the 'Intolerable' *Coercive Acts. At its first session, the radicals, led by delegates from Massachusetts, Virginia, and South Carolina, outmanoeuvred the moderates from New York and Pennsylvania and adopted the Suffolk County (Massachusetts) Resolves, rejecting the Acts as 'the attempts of a wicked administration to enslave America'. The second Congress, convened in the wake of *Lexington and Concord, created a Continental Army under *Washington and, as a result of British intransigence and radical pressure, moved gradually towards the *Declaration of Independence (1776). The Congress undertook the central direction of the War of *Independence, and, under the Articles of Confederation (1781), the government of the USA. Its delegates, however, were little more than ambassadors from the thirteen sovereign states, lacking financial and disciplinary powers. It was superseded by the *Constitution in 1789.

Cook, James (1728–79), British naval captain and explorer. He entered the Royal Navy as an able seaman but was quickly promoted to master in recognition of his skills as a navigator. He chartered the St Lawrence Channel and the coasts of Nova Scotia and Newfoundland during his service in North America (1759–67). He then commanded an expedition in HMS *Endeavour* (1768–71) to Tahiti and continued to chart the coasts of New Zealand and eastern Australia, returning home by way of New Guinea and Batavia.

On his second voyage (1772–5), in the *Resolution*, he was the first navigator to cross the Antarctic Circle and explored vast areas of the Pacific, revisiting New Zealand and Tahiti, and charting Easter Island, the Marquesas and Society Islands, the Friendly Isles (Tonga), and the New Hebrides, and discovering New Caledonia and Norfolk Island. He returned home via Cape Horn, visiting South Georgia and discovering the South Sandwich Islands. He sailed again in 1776 and, after a year in the South Pacific, set off in search of a north-west passage from the Pacific to the Atlantic. He discovered the Hawaiian Islands and charted part of the Alaska coast. On the return voyage he was killed in a skirmish with islanders in Hawaii. His *Journals* give a detailed account of his three voyages to the Pacific.

Cooper, Anthony Ashley   *Shaftesbury.

Copper Age, the stage of technological development between the introduction of copper and the manufacture of bronze (an alloy of copper and tin) in the *Bronze Age. As copper was initially very scarce, the impact of metallurgy was often slight, and it was used only for ornaments and rare daggers or flat axes.

Copper appeared at very different dates in various parts of the world. In some cases there was no separate stage before the adoption of true bronze or even iron: in others, Andean South America for example, it was in use for very much longer. It was often accompanied by gold and occasionally by silver. The beginning of the period is difficult to define, as the occasional trinket of copper is found sometimes many centuries before metal began to displace stone for tools or weapons. The use of cold-hammered native copper is held not to constitute a Copper Age; examples are found among the Eskimos of North America.

Copt, a member of the Coptic Orthodox Church of Egypt and Ethiopia. The word comes from *Aiguptioi*, (Greek, 'Egyptians'). According to tradition they were converted by St Mark. From the 2nd to the 5th century the Catechetical School of *Alexandria was the most important intellectual institution in the Christian Church. Monasticism originated in the Coptic Church in the 3rd century and it spread along North Africa to Rome, and through Palestine to Syria and Asia Minor. The Copts suffered greatly in Diocletian's persecution: their Calendar dates from 284, the Year of the Martyrs. They broke away from the rest of the Church in 451, claiming that *Jesus Christ has a single nature (part divine, part human) only. After the Muslim conquest of Egypt in 641 many Christians converted to Islam. The Coptic Church retained its position in *Ethiopia.

Corday d'Armont, Charlotte (1768–93), French noblewoman, the murderess of *Marat. After a lonely childhood in Normandy she began to attend the meetings of the *Girondins, where she heard of Marat as a tyrant and conceived the idea of assassinating him. She arrived in Paris in 1793 and on 13 July murdered Marat in his bath. A plea of insanity was overruled and she was sentenced to death on the guillotine.

Corinth, a port on the Isthmus of Corinth, southern Greece. Located at the crossroads of much land and sea trade (it was easier and safer for ships to cross or unload at the Isthmus than to risk the trip round the Peloponnese), it was strategically important. Its trade enabled it, by the mid-6th century BC, to become the most prosperous city in Greece, and a prolific exporter of high-quality pottery. Corinthians founded colonies in north-west Greece and in Sicily, thus transmitting Greek influence throughout the Mediterranean. In c.657 BC Cypselus brought Corinth under a tyranny, which form of government lasted into the early 6th century. Its power was not checked until the rise of the *Athenian empire, and Corinth pressed Sparta to embark on the *Peloponnesian War. In the 4th century, however, it fought against Sparta, but was neutral in the face of *Philip II of Macedonia.

It joined the *Achaean League in 243 and was sacked and destroyed by Rome in 146. Rebuilt in 44 BC as a Roman colony, an earthquake destroyed the ancient city in AD 521.

A contemporary watercolour of the landing of British forces under General **Cornwallis** in the Jerseys in 1776 as part of the New Jersey campaigns. The painting is variously attributed to Thomas Davies or to Lord Rawdon, an officer who served under Cornwallis. (Emmet Collection, New York Public Library)

Cornwallis, Charles, 1st Marquis (1738–1805), British general, who fought in the War of *Independence at Long Island and *Brandywine. He took command of the southern campaign in 1780, defeating the Americans at *Camden and Guildford Court House, but by his relentless pursuit into the interior he lost contact with *Clinton and exhausted his troops. His choice of *Yorktown as a base proved disastrous, and he was forced to surrender (1781). Later reinstated, he served as governor-general of India (1786–93, 1805) where he defeated *Tipu Sultan and his Cornwallis Code reformed land tenure. He was also viceroy of Ireland (1798–1801) and negotiator of the Treaty of Amiens (1802).

Coromandel, a name used for the eastern coast of India between the Cauvery and Krishna deltas, where the English and French competed for supremacy in the 18th century. The name may have derived from the great medieval *Chola dynasty which ruled this region ('*chola-mandala*' means 'Chola kingdom').

corsair, a privateer of the Barbary Coast of North Africa, and especially Algiers. Piracy existed here in Roman times, but, after the *Moorish expulsion from Spain, individuals (with government connivance) began attacks on Christian shipping. The early 17th century was the peak of their activity. In Algiers alone 80,000 Christian captives were held as slaves. Britain and France frequently attacked the corsairs and from 1800 to 1815 the USA made war on Tripoli and Algiers. Privateering ceased with the French occupation of Algiers in 1830.

Corsica, the fourth largest island in the Mediterranean, owned by France since 1768. It is about 100 km. (60 miles) south-west of the Tuscan coast of Italy. Greeks, Etruscans, and Carthaginians all made settlements here before the Roman conquest of 259 BC. It became a Roman province, and was used as a place of political banishment. *Vandal conquest in *c.*AD 469 was followed by periods of Byzantine, Gothic, Lombard, Frankish, and Moorish domination. In 1077 the papacy assigned control of the island to the Bishop of Pisa. After a long period of conflict, Pisa's hegemony passed to Geneva in 1284. The Genoese subsequently withstood challenges from Aragon (1297–1453) and France (1553–9), and remained in control throughout the 17th century despite considerable popular discontent. There was an unsuccessful mass revolt in 1729, but in 1755 Pasquale de Paoli established an effectively independent Corsican state. In 1768 the French purchased all rights to the island from Genoa, and in 1769 defeated Paoli's troops. Corsica then became a French province, where on 15 August 1769 *Napoleon was born.

Cortés, Hernan, Marqués del Valle de Oaxaca (1485–1547), *conquistador and conqueror of *Mexico. He was born into a noble Spanish family and at 19 sailed for Hispaniola. In 1511 he joined Diego Velásquez's expedition to conquer Cuba, became a man of means as a result, and began to equip himself with help from Velásquez to lead an expedition to colonize the Mexican mainland. He sailed abruptly, after quarrelling with Velásquez, landing at Vera Cruz on 21 April 1519, and soon met envoys from the *Aztecs.

His arrival coincided with the predicted return of the god-king of Aztec mythology, *Quetzalcóatl, and he was welcomed by the emperor *Montezuma in *Tenochtitlán, the Aztec capital. Suspicions grew among the Aztecs, however, and Cortés seized Montezuma as a hostage, forcing him to parley with his people. Meanwhile Velásquez sent a force under Pánfilo de *Narváez to retrieve Cortés, but he won them over to his side at Vera Cruz in 1520. He then returned to Tenochtitlán, where fighting had broken out in his absence. Montezuma was mortally wounded and the new emperor, Cuauhtémoc, led the Aztecs in driving the Spanish from the city. Cortés returned the following year and the final siege lasted ninety-three days, the city falling on 13 August 1521. The rest of the empire was quickly subdued, but an expedition to Honduras in 1524 was unsuccessful. Cortés was forgiven for his 'rebellion' by Charles I of Spain, and was awarded estates, but spent most of the rest of his life fighting in Mexico and Spain.

Cossack (from the Turkish, adventurer or guerrilla), a people in south Russia. They were descended from refugees from religious persecution in *Poland and *Muscovy, and from peasants fleeing the taxes and obligations of the feudal system. Settling in mainly autonomous tribal groups around the rivers Don and Dnieper, they played an important role in the history of the Ukraine. A frontier life-style encouraged military prowess and horsemanship, males aged 16–60 years being obliged to bear arms. They were democratic, directly electing their leaders or *hetmen*. Their relations with Russia included military service and military alliance, especially against the Turks, but there were rebellions against Russia under the leaderships of Stenka Razin (1667–9), Iran Mazeppa (1709), and *Pugachev (1773–4).

Costa Rica, a Central American country discovered by *Columbus during his fourth voyage to the New World in 1502. Permanent settlement did not occur until 1564 when Juan Vásquez de Coronado, with settlers from Nicaragua, founded Cartago on the Meseta Central. The small Indian population fell victim to disease, leaving the ethnic make-up of the area mostly European.

A contemporary 18th-century engraving of a **Cossack** (right), wielding a lance and sabre in a skirmish with Prussian dragoons. The Cossacks supplied the Russian army with its most effective and feared light infantry.

Counter-Reformation, a revival in the *Roman Catholic Church between the mid-16th and mid-17th centuries. It had its origins in reform movements which were independent of the Protestant *Reformation, but it increasingly became identified with, and took its name from, efforts to 'counter' the Protestant Reformation. There were three main ecclesiastical aspects. First a reformed papacy, with a succession of popes who had a notably more spiritual outlook than their immediate predecessors, and a number of reforms in the church's central government initiated by them. Secondly, the foundation of new religious orders, notably the Oratorians and in 1540 the Society of Jesus (*Jesuits), and the reform of older orders, notably the Capuchin reform of the *Franciscans. Thirdly, the Council of *Trent (1545–63), which defined and clarified Catholic doctrine on most points in dispute with Protestants and instituted important moral and disciplinary reforms within the Catholic Church, including the provision of a better education for the clergy through theological colleges called seminaries. Prominent Counter-Reformation figures were Ignatius of *Loyola (c.1491–1556), founder of the Society of Jesus, Charles Borromeo, Archbishop of Milan 1560–84, Pope Pius V (1566–72), and the Spanish Carmelite mystics *Teresa of Avila (1515–82) and *John of the Cross (1542–91). All this led to a flowering of Catholic spirituality at the popular level, but also to an increasingly anti-Protestant mentality. The movement became political through its links with Catholic rulers, notably *Philip II of Spain, who sought to re-establish Roman Catholicism by force. The stalemate between Catholics and Protestants was effectively recognized by the Treaty of *Westphalia in 1648, which brought to an end the Thirty Years War and in a sense concluded the Counter-Reformation period.

country house, a large house standing in its own park or estate. Country houses in the grand manner were built in England by the wealthy from the 16th century onwards. The stability and prosperity of the *Tudor period meant that these great houses were built for comfort rather than for defence (unlike *castles), and to entertain Elizabeth I when she toured the country on a royal *progress. Longleat House (Wiltshire) and Hardwick Hall (Derbyshire) date from the 16th century, and Hatfield House (Hertfordshire) is a fine Jacobean house of the early 17th century. The great age of country-house building was the 18th century: noble families recovered their estates after the *Restoration, increased their income from agricultural land by *enclosures, and the development of mineral resources, and built magnificent residences to display their collections of paintings, furniture, and silver (*Grand Tour). Blenheim Palace, one of the most splendid, was a gift from the British nation to the Duke of Marlborough after his victory at the battle of *Blenheim.

Courtrai, battle of (11 July 1302), sometimes known as the 'battle of the Spurs'. Philip IV of France had attempted to overrun *Flanders but Flemish troops fought the French at Courtrai. Flemish burghers defeated the French nobility, and, in celebration of victory, hung their spurs from the church in Bruges. The battle of Courtrai was one of the most significant defeats suffered by France in the 14th century. Charles VI of France avenged this insult by sacking Courtrai in 1382.

Covenanter, originally a Scot who opposed the ecclesiastical innovations of *Charles I of England. Drawn from all parts of Scotland and all sections of society, Covenanters subscribed to the National Covenant of 1638. This was a revised version of a previous covenant (1581), which had been signed by James VI of Scotland. They swore to resist 'episcopal' (the church governed by bishops) religious changes, and, in the event of such changes, they set up a full *Presbyterian system and defended it in the *Bishops' wars. They hoped to impose their system on England in 1643, by drawing up the *Solemn League and Covenant with the *Long Parliament. Disappointed in this, they turned in 1650 to *Charles II, who signed the Covenant, but then abjured it at his *Restoration (1660), condemning it as an unlawful oath and ordering all copies of the document to be burned. In Scotland, after the episcopacy was re-established in 1661, Covenanters were severely treated.

Coverdale, Miles (1488–1568), English scholar and translator of the *Bible into English. A priest from 1514, he came under the influence of William *Tyndale, and held increasingly Protestant views. His first translation of the Bible was printed in Zürich (1535), and he is claimed as the author of the 'Great Bible' (1539), commissioned by Thomas *Cromwell, which was placed by order in all parish churches. In 1551, he was made Bishop of Exeter, but fled abroad during the reign of Mary I. He possibly contributed to the Geneva Bible of 1560, and under Elizabeth I was one of the leaders of the *Puritans.

Cranmer, Thomas (1489–1556), English cleric, a founding father of the English Protestant church. He served *Henry VIII on diplomatic missions before becoming Archbishop of Canterbury in 1532. He annulled Henry's marriages to Catherine of Aragon, Anne Boleyn, and Anne of Cleves. During *Edward VI's reign, he was chiefly responsible for liturgical reform including the First and Second English Prayer Books (1549 and 1552) and the Forty-Two Articles (1553). He supported Lady Jane *Grey's succession in 1553; after Queen Mary's accession he was tried for high treason, then for heresy, and finally burnt at the stake in Oxford.

Crassus, Marcus Licinius (c.115–53 BC), Roman general and member of the 'First Triumvirate' with *Pompey and Julius *Caesar. He had fled from *Marius as a young man and joined *Sulla, whose troops he commanded at the battle of the Colline Gate (November 82). Sulla rewarded him and he further increased his fortune by property speculation to become one of the richest men in Rome. After defeating *Spartacus in 71 he and Pompey were elected consuls. Though hostile to each other during the sixties, in 59 they formed a junta or 'triumvirate' with Caesar to dominate Roman politics. Crassus was given command against the Parthians as governor of Syria for a five-year term in 55. Greed and ambition drove him to sack *Jerusalem and then to attack *Parthia. Defeated at Carrhae, he was captured and executed by the *patricians.

Crécy, battle of (26 August 1346). The village of Crécy in northern France was the site for the defeat of the French under Philip VI by the *archers of the English king, Edward III. Edward's raiding army, anxious to avoid pitched battle, was trapped by a numerically superior French force. The English bowmen dug pits to impede advancing cavalry, while the knights dismounted and formed three supporting divisions, their right commanded by Edward's son and heir, the *Black Prince. Genoese crossbowmen in French pay, handicapped by wet bowstrings and the slowness and short range of their weapons, were swiftly dispersed. The French knights charged forward only to have their horses shot from under them by longbow arrows. Wounded and trampled on, some were then killed by English pikes and knives and others held for ransom. Over 1,500 of the French died, including the cream of the nobility, as against 40 English dead. Edward was able to march north and besiege Calais. This was a decisive English victory at the outset of the *Hundred Years War.

Cree, the largest and most widespread group of Algonquian-speaking Indians in Canada. Prehistorically they were hunters in the northern forest and tundra areas of central and mid-western Canada, living in small bands or hunting groups in conical or dome-shaped lodges, travelling by foot or canoe in summer and by snow shoes and toboggan in winter. Following contact with the *Hudson's Bay Company some Cree became trappers, expanding their territory. With the introduction of firearms the Plains Cree became horse-mounted hunters of buffalo, virtually destroying these herds in the 1880s.

Creek (or Muskogee), a tribe of North American Indians who inhabited Alabama and Georgia, and were the prehistoric eastern extension of the *Mississippi cultures. The Spanish explorer Hernando *de Soto visited the region in 1540–2, and found them to have a powerful and highly organized political and social system. They lived in large fortified towns with central ceremonial mounds, linked in a loose confederacy, which bitterly fought European attempts at conquest but were defeated in the late 18th and early 19th centuries.

Crete, the largest of the Greek islands, sited in the southern part of the Aegean Sea. In the pre-historic period it was home to the *Minoan civilization. Its continued importance was ensured by its position on the sea-routes to Egypt, the Levant, and Cyprus. It played a significant role in the development of archaic Greek art, and in early written law-codes, was well known as a home of mercenaries, and maintained an aristocratic society. It was neutral during the *Greek–Persian wars.

A league of Cretan cities was established in the 3rd century BC, and this accepted the protection of Philip V of Macedonia, who encouraged the pirates for which the

The burning of the Anglican reformer Thomas **Cranmer** at Oxford in 1556, a woodcut from John Foxe's *Book of Martyrs* (1570), a history of English Protestant martyrs from the 14th to the 16th century.

island was then infamous. When Cretan pirates later threw their lot in with *Mithridates, an enemy of Rome, they provoked Roman retaliation and in 68-7 the island was subdued and *Knossos destroyed. The island subsequently became a Roman province. After Roman and Byzantine rule, it underwent a period of Arab control before being recaptured by the Byzantines in 960-1. In 1210 it was taken over by the Venetians, who were not ejected until 1669 by the Turks. The extant fortifications by the harbour of Heraklion are an impressive reminder of this period.

Crimea, a peninsula on the northern shore of the Black Sea. It was colonized by the Greeks in the 6th century BC, and became a Roman protectorate in the 1st century AD. It was overrun by Ostrogoths, Huns, and others, and was partly under Byzantine control from the 6th to the 12th century. The 13th century saw it pillaged by Mongols, yet enjoying trade relations with *Kievan Russia and Genoa. In the late 15th century a short-lived Tartar khanate was absorbed by the *Ottomans, until Russia annexed it in 1783.

Croatia, a region of Yugoslavia. Once the Roman province of Illyricum, it suffered successive barbarian invasions, with the Slavs becoming the majority population. Conquered by *Charlemagne, the first Croatian state was formed with its own knezes or princes when the Carolingian empire collapsed. With papal support Kneze Tomislav became the first king. Struggles between *Hungary, *Venice, and the *Byzantine empire resulted in rule by the Hungarian crown until 1301, when the House of Anjou took control. From 1381 there was a long period of civil war. The battle of *Mohács in 1526 brought most of the country under *Ottoman rule with the remainder governed by the *Habsburgs.

Croesus, King of *Lydia (c.560-546 BC). He expanded his domains to include all the Greek cities on the coast of Asia Minor, and the stories of his wealth indicate the extent of his power. However, he was unable to withstand *Cyrus the Great, and after his defeat Lydia entered the Persian empire of the Achaemenids.

Cro-Magnons, early modern people (*Homo sapiens sapiens*) found in Europe until about 10,000 years ago. They were generally more heavily built than humans today but otherwise had the same anatomical characteristics. They appeared around 35,000 years ago and are named after a rock shelter in the Dordogne, France, where four adult skeletons, an infant's skeleton, and other remains were found in 1868. With the skeletons were *Upper Palaeolithic flint tools of Aurignacian type and signs of decorative art in the form of pierced sea shells. The ancestry of Cro-Magnons is unclear. It was once believed these people were the direct descendants of *Neanderthals, but it is now thought that they evolved elsewhere, probably in Africa, and replaced the Neanderthals within a few thousand years of reaching Europe.

Cromwell, Oliver (1599-1658), English statesman and general, Lord Protector of the Commonwealth of England (1653-8). He was born of Huntingdon gentry stock, indirectly descended from Thomas *Cromwell. An opposition member of the Long Parliament, he rose to prominence during the *English Civil War. He helped

to form the Eastern Association and secured East Anglia's support for the Roundheads. Until 1645, he displayed great skill in cavalry organization and tactics in the Army of the Eastern Association. After helping to engineer the *Self-Denying Ordinance, he became Fairfax's second-in-command in the *New Model Army. He played a decisive role (1646-8). He supported the Army in its quarrel with the Parliamentary Presbyterians, and the Army Grandees against the *Levellers. During the war's second phase, he won the battle of Preston, approved of *Pride's Purge, and signed Charles I's death warrant.

Appointed a member of the *Commonwealth's Council of State, he repressed resistance to the regime in Ireland, Scotland, then England. He became the focus of general dissatisfaction with the policies of the Rump Parliament. In April 1653 he expelled it, becoming Lord Protector some months later, and spent the rest of his life vainly attempting to give constitutional permanence to his military regime. A sincere Puritan, he always claimed to be directed by the will of God, and believed in religious toleration. Although in effect dictator, he refused the crown when it was urged on him by Parliament in 1657.

His Protectorate, based on the Instrument of Government, and upheld by the rule of major-generals in the counties, was unpopular. After his death his son Richard *Cromwell failed to keep the senior military commanders under control and factions emerged. Eighteen months after Cromwell's death one section of the army under General *Monck called for free elections and it was voted to recall *Charles II from exile.

Cromwell, Richard (1626-1712), son of Oliver *Cromwell whom he succeeded as Lord Protector of the Commonwealth of England (1658-9). He was more interested in country life than in politics and, incapable of reconciling the military and civilian factions in Parliament, he retired after a few months. At the *Restoration he fled to the Continent, returning c.1689 to spend the rest of his life quietly in Hampshire.

Cromwell, Thomas, Earl of Essex (1485-1540), English statesman. His beginnings are obscure: in early life he appears to have travelled abroad, acquiring skills in commerce, law, languages, and mercenary warfare. After 1520, he obtained the patronage of Thomas *Wolsey, and by 1529, when his master was discredited, was attracting *Henry VIII's attention. By 1531 he belonged to the inner ring of the royal council, and his successive appointments included Chancellor of the Exchequer (1533), Principal Secretary of State (1534), Vicar-General (1535), and Lord Privy Seal (1536). From 1533 to 1540 he was the king's chief minister. His managerial genius accomplished the divorce from Catherine of Aragon, the break with the pope and the Roman Catholic Church, the Dissolution of the *Monasteries, and what amounted to an administrative revolution. His Protestant sympathies led to the choice of *Anne of Cleves as the king's fourth wife, resulting in the loss of royal favour and arousing hostility both inside the court and out. The Catholic *Howard faction contrived his execution for treason just after he had been made Earl of Essex.

crop rotation, the practice of growing different crops in different years on the same land (so as to prevent the soil's goodness from being exhausted), which was widespread in Europe from the time of the *Roman

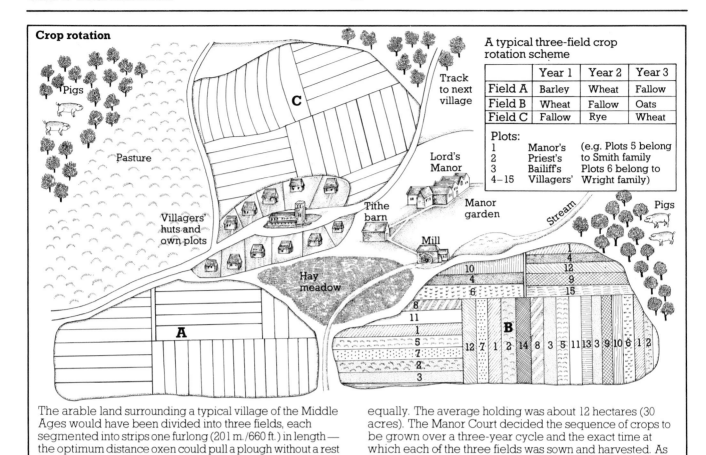

Crop rotation

Pigs

Pasture

Track to next village

A typical three-field crop rotation scheme

	Year 1	Year 2	Year 3
Field A	Barley	Wheat	Fallow
Field B	Wheat	Fallow	Oats
Field C	Fallow	Rye	Wheat

Plots:
1	Manor's	(e.g. Plots 5 belong
2	Priest's	to Smith family
3	Bailiff's	Plots 6 belong to
4–15	Villagers'	Wright family)

C

Lord's Manor

Manor garden

Villagers' huts and own plots

Tithe barn

Mill

Stream

Pigs

Hay meadow

A

B

The arable land surrounding a typical village of the Middle Ages would have been divided into three fields, each segmented into strips one furlong (201 m./660 ft.) in length — the optimum distance oxen could pull a plough without a rest — and one rod (5.03 m./16½ ft.) in width — the length of the ploughman's stick. Each family was allocated random strips in each field, so that the good and bad land was shared equally. The average holding was about 12 hectares (30 acres). The Manor Court decided the sequence of crops to be grown over a three-year cycle and the exact time at which each of the three fields was sown and harvested. As the table shows, cereal crops formed the staple diet, being used for bread and beer.

empire. Two-field rotation was first practised: one half of a farmer's land was planted in the spring or autumn of each year, while the other half was left fallow. In the course of the Middle Ages this gradually gave way to the three-field system where a three-year cycle was followed on each of three fields, with an autumn-sown crop such as rye or winter wheat, a spring-sown crop such as oats or beans, and a year of lying fallow. Two out of three fields were thus in cultivation every year. The three-field system succeeded only in countries with mild climates, such as England. With the *Agricultural Revolution and the acceleration of *enclosures in the 18th century, more scientific methods were applied to crop rotation. A four-course rotation was adopted based on turnips, clover, barley, and wheat. The introduction of root crops improved the soil and hence the quality of harvest and livestock, and ended the practice of leaving fields lying fallow.

Crow and Hidatsa (or Absaroke), North American Indians who inhabited Montana and northern Wyoming. In prehistoric times they lived in permanent villages and practised a well-balanced agricultural economy with seasonal buffalo hunts. When they acquired horses in the 18th century, the Crow abandoned their villages for a nomadic life of full-time buffalo-hunting, trading meat for some of the crops of the farmer Hidatsa.

crucifixion, a form of capital punishment used by various ancient peoples including the Persians, Car-

thaginians, and Romans for criminals, usually applicable only to slaves and other persons with no civil rights. The victim, nailed or roped to a crossbar, was hoisted on to an upright to form a 'T' or cross. Six thousand rebels with *Spartacus were crucified in 71 BC, as was *Jesus Christ (c. AD 30). Romans regarded the cross with horror. Only after Constantine abolished this form of penalty did Christians adopt the cross as a symbol.

Crusades, a series of expeditions (11th–14th century) to secure Christian rule over the Muslim-controlled holy places of *Palestine. (The term is by extension used to describe any religious war or even moral or political movement). The wealthy powerful orders of *Knights Hospitallers and *Knights Templar were created by the Crusades. The First Crusade was called by Pope Urban II, and was provoked by the rise to power of the *Seljuk Turks, which interfered with traditional *pilgrimage to Palestine. The pope promised spiritual benefits to warriors willing to fight under Christian banners. The Crusaders captured *Jerusalem in 1099 and massacred its inhabitants, establishing a kingdom there under *Godfrey of Bouillon. The Second Crusade (1147–9) succeeded only in souring relations between the Crusader kingdoms, the Byzantines, and friendly Muslim rulers. The Third Crusade (1189–92), prompted by *Saladin's capture of Jerusalem, recaptured Acre but achieved little more. The Fourth (1202–4) was diverted by Venetian interests to Constantinople, which was sacked, making the gulf between Eastern and Western churches unbridgeable, though some Crusaders benefited from the division of Byzantine territories known as the Latin empire of the East (1204–61). This briefly replaced the Greek empire at Constantinople until *Michael VIII retook the city. Later expeditions concentrated on North Africa, but to little purpose. The fall of Acre in 1291 ended the Crusader presence in the Levant. All, except the peaceful Sixth Crusade (1228–9), were marred by greed and brutality: Jews and Christians in Europe were slaughtered by rabble armies on their way to the Holy Land. The papacy was incapable of controlling the immense forces at its disposal. However, the Crusades attracted such leaders as *Richard I and *Louis IX, greatly affected European *chivalry, and for centuries, its literature. While deepening the hostility between Christianity and Islam, they also stimulated economic and cultural contacts of lasting benefit to European civilization.

Ctesiphon, an ancient city on the River Tigris, originally established as a military outpost of Parthia. The kings of the Arsacid dynasty of Parthia used it as their winter headquarters, and after nearby Seleucia was destroyed in AD 165, it became the leading city of Babylonia. When Artaxerxes, son of Papek, established the *Sassanian empire in 211–12, Ctesiphon became the capital. It was captured by the Arabs in 636.

Cuba, a large island in the Caribbean. It was first settled by migrating hunter-gatherer-fisher people, the Ciboney from South America, by c.3000 BC. Migrations of agriculturist, pottery-making Arawak Indians from northern South America began to displace them in eastern Cuba after c.1000 BC, but the Ciboney remained in the west. Cuba was discovered by Columbus in 1492 but it was not realized that it was an island until it was circumnavigated in 1508. Spanish settlement began in 1511 when Diego Velásquez founded Havana and several other towns. The Arawak became virtually extinct by the end of the century from exploitation and European-

introduced diseases. Black slaves were imported for the plantations (especially sugar and tobacco) from 1526. Britain seized the island in 1762–3 but immediately exchanged it with the Spanish for Florida.

Culloden, battle of (16 April 1746). On a bleak moor in Scotland to the east of Inverness the *Jacobite forces of Charles Edward Stuart were routed during a sleet storm by the English and German troops led by the Duke of *Cumberland, a victory of trained professionals over enthusiastic amateurs. The battle was followed by ruthless slaughter of the Jacobite wounded and prisoners, with survivors hunted down and killed, earning Cumberland the nickname 'Butcher'. Culloden ended the *Forty-Five Rebellion and virtually destroyed the Jacobite cause.

Culpeper's Rebellion (1677), a demonstration of local antagonism to the syndicate of proprietors who administered the new colony of North Carolina. It was brought to a head by attempts to enforce the *Navigation Acts on tobacco and to collect land taxes, and by the example of *Bacon's Rebellion in Virginia in 1676. A 'parliament' of eighteen proclaimed one of the ringleaders, John Culpeper, governor and he ruled until replaced by a proprietorial nominee in 1679. The factionalism and insubordination of North Carolina continued until 1714, with further rebellions in 1689 and 1711.

Cumberland, William Augustus, Duke of (1721–65), British army commander, second son of *George II. He achieved military success at an early age, was made captain-general of the British army in 1745, and destroyed the Jacobites at *Culloden in 1746. He was less successful against the French in the *Seven Years War, losing the battle of Hastenbeck in 1757 and agreeing by the

English infantry at the battle of **Culloden** advancing on the Highlanders, an 18th-century painting by David Morier. (Royal Collection)

Convention of Klosterseven to withdraw his troops from Hanover and Germany. His political influence with George II resulted in the collapse of the *Pitt-dominated Devonshire ministry in 1757, but his failures on the Continent, the death of his father, and the mistrust of *George III's mother caused his influence to wane after 1760. He was an important focus for the Whig politicians who had been dismissed by George III, but he died soon after their leader, *Rockingham, came into office.

cuneiform, a script, or rather a family of scripts, developed in the Middle East as a result of using split reeds for writing on soft clay. Incised free-hand signs were turned into groups of impressed triangles (cuneiform means wedge-shaped) by the Sumerians c.2500 BC. Thereafter it was adapted for other languages, including Akkadian and Assyrian. All these were elaborate scripts with signs serving many different purposes; practice tablets and glossaries show that it required long training to write properly. The forms were rigidly maintained, even when inscriptions were carved on stone. About 1500 BC in Persia, alphabets of cuneiform signs were invented, eventually to be replaced by derivatives of the Phoenician alphabet.

Cunobelinus, King of the Belgic Catuvellauni in Britain (c.AD5–40). He advanced from his capital at Verulamium (St Albans) to take Camulodunum (Colchester) and control of much of south-east Britain. His growing power was seen by Rome as a threat and prompted *Caligula to contemplate an invasion of Britain. His sons *Caratacus and Togodumnus pursued a policy which prompted Roman invasion in 43. Medieval tradition created the 'Cymbeline' of Shakespeare.

Cursus Honorum, the name given to the ladder of (annual) offices which would-be Roman politicians had to climb. After a prescribed period of military service (though this requirement lapsed in the very late republic), or the tenure of certain minor magistracies, the first major rung was the quaestorship, which before *Sulla effectively, and after Sulla statutorily, gave membership of the *Senate. Thereafter came praetorship and consulship (though not all achieved these offices), and finally the quinquennial office of censor, the crown of a republican politician's career. Other magistracies, the aedileship and the tribunate of the plebs, might be held between quaestorship and praetorship, but were not obligatory. In the middle and late republic, specific minimum ages and intervals between offices were established by statute. Quaestors, praetors, and consuls were often employed after their year of office at Rome as 'pro-magistrates' to administer the provinces of the Roman empire.

customs and excise, duties charged on goods (both home-produced and imported) to raise revenue for governments. In England customs date from the reign of *Edward I, when they were raised on wool and leather. *Tunnage and poundage was introduced under *Edward II. Impositions, additional duties levied by the monarchy without Parliament's consent, became a source of controversy in the early 17th century. In 1606 Bate's case arose when John Bate, a merchant, challenged *James I's right to levy a duty on imported currants. The case was decided in the crown's favour, and *Charles I continued to raise money from this source and from

monopolies, despite opposition; monopolies were abolished by the *Long Parliament in 1643.

Excise was first introduced in 1643 to finance the parliamentary armies in the *English Civil War and was a tax on alcoholic beverages, mainly beer and ale. At the *Restoration Charles II was granted excise duties for life by Parliament. However, the tax always remained unpopular, as *Walpole discovered when he attempted to extend it in 1733. In 1799, *Pitt the Younger introduced income tax to help finance the war effort, and gradually direct tax on income came to represent the main source of revenue for governments rather than indirect taxes on commodities.

Cuzco, a city in Peru. It was established as the *Inca capital c. AD 1200 but from 1438 was largely rebuilt by the emperor Pachacuti. There was an inner city, including the *Huacapata* or Holy Place, palaces, and administrative buildings, the *Sunturhuasi* tower in the square, and the *Coricancha* or Sun Temple. Around this were regularly planned city wards representing all the provinces of the empire. In 1535, after *Pizarro's defeat of the Incas, it was replaced by Lima as the viceregal capital of Peru. Its mountain location was impractical to the Spaniards, who needed a large port city to bring European trade to their colonial possessions. It remained an important provincial governorship, however, and was still sacred to the Incas, who set up a successor state at Vilcabamba in the hills to the north. They tried to recapture it in 1538 and were not finally defeated until 1572.

Cymbeline *Cunobelinus.

Cynic, in ancient times a member of a sect of philosophers popularly thought to have been established by *Diogenes, though his mentor, Antisthenes of Athens, should perhaps be accorded the title of founder. Since the Cynics were never a formal school, with no fully defined philosophy, considerable differences emerged amongst *Diogenes' disciples, who adopted only those ideas which appealed to them. Crates of Thebes was his most faithful follower: he demonstrated how in troubled times happiness was possible for the man who gave up material possessions,

The Inca capital of **Cuzco** was planned and built on impressive lines. Its walls and buildings were constructed of massive blocks of stone which were trimmed precisely to fit without the need for mortar, a considerable achievement in a culture with neither iron nor the wheel.

kept his needs to an absolute minimum, and maintained his independence.

The Cynic philosophy flourished through the 3rd century BC, and the beggar-philosopher, knapsack on his back and stick in hand, became a familiar sight in Greece. A steady decline thereafter was reversed by a temporary revival in the 1st century AD, though the Cynics' readiness to criticize the conduct of the emperors led to many expulsions from Rome. The last recorded beggar-philosopher lived at the end of the 5th century.

Cyprus, a large island in the eastern Mediterranean with Turkey to the north and Syria to the east. A Mycenaean colony in the 14th century BC, it was ruled successively by the Assyrian, Persian, Roman, and Byzantine empires. *Richard I of England conquered it in 1191 and sold it to the French Crusader Guy de Lusignan under whom it became a feudal monarchy. An important base for the *Crusades, it eventually came under the control of Italian trading states, until in 1571 it fell to the *Ottoman empire.

Cyril, St (c.827-69), Greek missionary. Educated in the Byzantine court, he and his brother, St Methodius (c.825-85), were sent to convert the Slavic tribes of Central Europe. Together they began, and Methodius completed, a translation of the liturgy and the Bible into Slavic, adopting what became known as the Cyrillic alphabet, still used in Russia. The introduction of the vernacular in the liturgy aroused opposition, especially from rival Latin missionaries in Bulgaria. This competitive conversion increased the tension between Rome and Constantinople.

Cyrus II (the Great) (d.530 BC), the founder of the *Achaemenid Persian empire after he had overthrown Astyages, King of Media, and taken possession of his capital Ecbatana in c.500. In 546 he defeated *Croesus to take control of Asia Minor and Nabonidus (the last of the Chaldean kings) to add Babylonia, Assyria, Syria, and Palestine to his domains. Further conquests to the north and east created a vast empire. His policies towards his subjects were enlightened and tolerant: the Medians had access to important administrative posts; the Jews were freed from their Babylonian *Exile and allowed to start rebuilding the Temple (their main religious centre) at Jerusalem; and generally he refrained from interfering with native customs and religions. He was probably killed in battle, and was buried at Pasargadae, where his tomb can still be seen.

Cyrus the Younger (d. 401 BC), a Persian prince, son of Darius II, who was given command of Asia Minor in 408 BC. He allied himself with *Lysander, and supported the Spartans financially, enabling them to defeat the Athenians in the *Peloponnesian War. When his elder brother, *Artaxerxes II, ascended the throne, he narrowly avoided execution. He then raised an army which included many Greek mercenaries, among them *Xenophon, and marched east to depose his brother, but was defeated and killed at Cunaxa.

D

Dacian wars, campaigns fought by successive Roman emperors over territory corresponding roughly to modern Romania and part of Hungary. The Dacians threatened the lands south of the River Danube which Rome regarded as a natural frontier. Under Emperor Domitian peace was agreed and considerable financial aid given to the Dacians. Then Emperor *Trajan stopped payments, crossed the Lower Danube, and fought two campaigns AD 101 and 105-6 that were commemorated on Trajan's column in Rome, which is still standing today. Dacia became a Roman province, until Emperor Aurelian abandoned it to the Goths in 270.

da Gama, Vasco (c.1469-1524), Portuguese navigator and conquistador, the first European to discover a sea route to India. In 1497 he was chosen by the King of Portugal to follow up the discovery made by Bartholomew *Diaz of a great ocean east of the Cape of Good Hope. He rounded the Cape and sailed up the east coast of Africa and across the Indian Ocean to the Malabar coast; he returned home with a rich cargo of spices. He was given command of a punitive expedition to India in 1502-3, Muslim traders having attacked a Portuguese settlement at Calicut; he bombarded the town before sailing on to Cochin for another cargo of spices. In 1524 he was recalled from retirement to restore Portuguese authority in the east, but fell ill shortly after his arrival

A relief on Trajan's column, Rome, a pictorial chronicle of the emperor's two campaigns in the **Dacian wars**, AD 101 and 105-6. This section shows a council of war in a Dacian fortress. The column is 30 m. (100 ft.) high and Trajan's ashes once lay beneath it.

A miniature portrait of the Portuguese navigator, Vasco **da Gama** from *The Opening of the World*. His voyages of discovery extended Portugal's power and led to the establishment of its overseas empire. (British Library, London)

at Goa and died at Cochin a few weeks later. His exploits were celebrated in the great epic poem, the *Lusiads* (1572), by Luis de Camoëns.

Dahomey, a kingdom in West Africa. In the 16th century the kingdom of Allada, with which the Portuguese had commercial relations, was founded. Two further kingdoms of Abomey and Adjatché (now Porto Novo), were founded *c.*1625. These were united by conquest by Ouegbadja of Abomey between 1645 and 1685, and renamed Dahomey. The kingdom had a special notoriety with travellers from Europe for its 'customs': the 'grand customs' on the death of a king, and the biennial 'minor customs', at both of which captured slaves were sacrificed in numbers to provide the deceased king with attendants in the spirit world. Women soldiers were first trained by King Agadja (1708–32). French trading forts were established in the 18th and 19th centuries, but the rulers of Dahomey succeeded in limiting their influence and restricting the slave trade.

daimyo (Japanese, 'great names'), Japan's feudal lords of the 14th to 19th centuries. They expanded their *samurai armies during the confusion of the *Ashikaga period, and territorial disputes between daimyo threatened Japan's unity. A re-allocation of fiefs under *Hideyoshi had reduced their power by 1591. The *Tokugawa controlled much of their activity, requiring them to spend alternate years at the Tokugawa base of Edo (Tokyo) and to leave their families there as hostages when they returned to their estates. Under the Tokugawa they could only marry or repair their castles with the shogun's permission.

Dakota Indians, the largest division of a North American Indian group of related tribes, commonly known as the Sioux, who originally inhabited the woodlands of Minnesota and the eastern Dakotas on the fringe of the northern Great Plains. As French and English *fur trade increased, so did intertribal warfare, exterminating some tribes and driving others, including the Dakota, on to the plains. They raided the tribes of the Missouri River to the south-east, and also acted as middlemen, exchanging European goods, especially firearms, for corn, tobacco, and other produce. Traditional enemies and trade rivals were the *Cree and Ojibwa to the north and east. Before they acquired horses, buffalo hunting had been ecologically balanced; seasonal migrations were aided by the travois (sledge), pulled by dogs but later adapted for horses, and the tipi (Dakota for 'they dwell'). Over-hunting with horses began to deplete the herds, further exacerbated by whites moving on to tribal lands and systematically devastating the herds.

Dampier, William (1652–1715), English explorer and adventurer. After a career as a buccaneer, he returned to England in 1691. The success of his book *A New Voyage round the World* (1697) led to his being sent in command of HMS *Roebuck* on a voyage to Australia, where he made a survey of much of the west coast. His harsh treatment of his men resulted in a mutiny and on his return to England he was court-martialled. In 1703 he joined a privateering expedition to the Pacific, and again his crew mutinied. The voyage is remembered as the occasion on which Alexander Selkirk, whose story inspired Daniel Defoe's *Robinson Crusoe*, was marooned at his own request on the island of Juan Fernandez off the coast of Chile. On a later privateering voyage to the Pacific, Selkirk was rescued after four years of solitary life.

Danby, Thomas Osborne, 1st Earl of (1631–1712), English statesman. He entered Parliament in 1665 as a supporter of the restored *Charles II. He received rapid promotion, becoming Secretary of the Navy in 1671 and Lord Treasurer in 1673. His reluctant negotiations with *Louis XIV of France to supply Charles II with money led in 1678 to accusations by Parliament of corruption and he was imprisoned until 1684. In 1688 he signed the invitation to *William of Orange to come to England, regained royal favour, and became Duke of Leeds in 1694, but following further accusations of corruption he retired from public life after 1695.

Dandolo, Enrico (*c.*1108–1205), member of a *Venetian family important in the Middle Ages, and *doge of *Venice. He established military and naval power by personally directing the Fourth *Crusade to attack Dalmatia and sack *Constantinople. Under him, Venice was victorious against Pisa, secured important treaties with Armenia and the *Holy Roman Empire, and reformed its laws.

Danegeld, the tribute paid in silver by *Ethelred II of England to buy peace from the invading Danes. It was raised by a tax levied on land. The first payment (991) was 10,000 pounds in weight of silver (1 pound equals 0.54 kg.); later payments were greater—16,000 pounds (994), 24,000 pounds (1002), 36,000 pounds (1007), and a massive 158,000 pounds (1012). Later (1012-51) it was levied to maintain a navy and the royal bodyguard (housecarls), when it was known as 'heregeld'; when raised by the *Norman kings the levy was used for general as well as military purposes.

Danelaw, the name given to the northern and eastern parts of Anglo-Saxon England settled by the Danes in the late 9th and 10th centuries and where Danish, rather than English, laws and customs applied. The earliest attempt to describe a boundary between these areas was made by King *Alfred and the Danish leader Guthrum (c.886). The Danelaw was identified as being north and east of a line from the Thames estuary (but not including London), through Hertfordshire and Bedfordshire, then along the River Ouse and the Roman Watling Street to Chester. It was granted legal autonomy by King *Edgar.

Danton, Georges Jacques (1759-94), French Revolutionary leader. In 1790 he founded the militant Cordelier Club and took part in *Jacobin debates, where he petitioned for the king's trial and the creation of a republic. Though he was a favourite of the *sans-culottes, he was forced to leave France briefly in 1791 but returned to become Minister of Justice. His calmness and authority during the crises of 1792, including his famous call for 'more daring', underlined his growing importance to the Revolution. Nevertheless, his criticism of the massacre of prisoners led to his resignation. He voted for the king's execution in January 1793, was appointed a member of the *Committee of Public Safety in April, and for three months effectively led the government. His attempts to negotiate and compromise with France's enemies failed and he was not re-elected. His disapproval of the repression of the *Terror and his growing moderation soon brought him into conflict with *Robespierre and led to his execution in April 1794.

Darius I (the Great) (d. 486 BC), ruler of Achaemenid *Persia from 521 to his death. The first years of his reign were overshadowed by uprisings in various parts of the empire, but after quelling them he embarked on a major task of reorganization. *Satraps were established to rule the various provinces, though he allowed considerable local independence. He was tolerant towards the worship of local gods, overhauled the finances of the empire, and built a renowned road system and a palace at *Persepolis.

He conducted various campaigns to secure his frontiers, but was unable to subdue the *Scythians. The revolt of the Ionian Greeks (499-494) was suppressed but attempts to extract revenge on mainland Greece (*Greek–Persian wars) for assisting the rebels met with disaster: storms scattered the fleet in 492, and the army was defeated in battle at *Marathon in 490.

Dark Ages, a term used to describe Europe in the 5th and 6th centuries, referring to the scarcity of historical material from this era. Following the collapse of the Roman empire, many Germanic tribes crossed through Italy, Germany, France, Spain, and North Africa often attacking and destroying towns. Rome was sacked on three successive occasions. Many tribes formed their

A relief from the Treasury at Persepolis, the ancient Persian city which **Darius I** planned as the ceremonial capital of his empire. Behind the seated figure of Darius stands his son Xerxes who succeeded him as King of Persia in 486 BC. (Archaeological Museum, Tehran)

own kingdoms (Vandals in Spain and North Africa; Ostrogoths and Lombards in northern Italy; the *Franks in France and western Germany; the *Anglo-Saxons in England). The Visigoths successfully fought with Rome against the Huns of *Attila in 451. The Ostrogoth *Theodoric ruled in Italy (493-526) as the representative of the *Byzantine empire, retaining Rome's administrative system.

The period of the Dark Ages saw cultural and economic decline though in the past this has been exaggerated. The period saw the foundation of Christian monasteries which kept scholarship alive. The 7th and 8th centuries saw relative stability and the encouragement of learning at the courts of *Charlemagne and *Alfred the Great.

Darnley, Henry Stuart, Lord (1545-67), Anglo-Scottish aristocrat, second husband of *Mary, Queen of Scots. After their union in 1565, Mary produced a son, the future James VI of Scotland and *James I of England. Mary's reliance on her secretary David *Rizzio (who may have been her lover) led Darnley to murder him. Darnley was subsequently murdered in a conspiracy involving the Earl of *Bothwell.

dating, the assignment of dates to artefacts. For the period of recorded history dating can be provided by associating an artefact with recorded monuments, inscriptions, or other written records. In the prehistoric period the assignment of an absolute date is more difficult. Relative dating in an archaeological excavation can be established by stratigraphic succession—for instance, where a layer with one distinctive kind of artefact overlies another with different kinds—or it can be inferred from a sequence of gradual changes in the artefact. Stylistic comparisons with more securely dated objects can some-

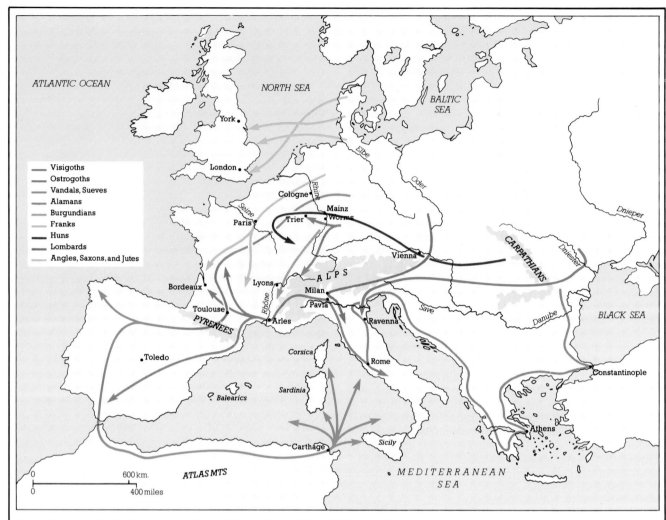

Dark Ages: major barbarian invasions of Europe in the 5th and 6th centuries

The invasions of the Roman empire in the 5th century by the mainly Germanic peoples whom the Romans called barbarians were largely provoked by westward pressure from the Huns, rather than by hostility to Roman civilization itself. The barbarians can be divided into two main groups. The western group, including the Franks and Anglo-Saxons, already had a settled agricultural society. They advanced relatively slowly and made successful, long-lasting settlements, founding the English kingdom and the Frankish empire. The eastern group, including the Goths, Lombards, and Vandals, as well as the Huns themselves, were essentially nomadic. The Vandals, in particular, made spectacular but unconsolidated treks and conquests, leaving no lasting influence on the lands they briefly ruled in Spain and Africa.

times be made in order to establish a rough guide to age. However, more accurate dating requires the use of scientific tests, usually using radioactive substances with a known decay rate. Measurement of radioactivity can then be used to establish a measure of age. The most common, and thus the most useful radioactive isotope in living matter is carbon-14 (radio-carbon), and this is the method most widely used where suitable samples are available. It can provide dates for objects up to 40,000 years old. All these methods are best used in combination, so that some degree of cross-checking is possible.

Dauphiné, a former province in south-east France. It was conquered by the Romans, Burgundians, and Franks. Once a fief of the *Holy Roman Empire, it passed to the kingdom of Arles, and, in 1029 to the counts of D'Albon who, from 1133, took the title of Dauphin of Vienne. By 1282, it had acquired its regional name and it was sold to the future *Charles V of France in 1346.

Thereafter heirs to the French throne assumed the title of dauphin. It acquired a *Parlement in 1453, but was annexed to the crown in 1457 and lost its local privileges, especially during the French *Wars of Religion.

David I (c.1084–1153), King of Scotland (1124–53) in succession to his brother *Alexander I. He had been brought up at the English court of *William II and *Henry I, whom his sister Matilda later married (1100). His own marriage (1113) gave him the earldom of Huntingdon and thus a right to intervene in English politics. David's English upbringing prompted him to introduce into Scotland Norman institutions of government and feudal tenure and to sponsor new monasteries and bishoprics—measures which led to Scottish resentment and rebellion. His claim to the northern counties of England, which was initially checked by a crushing English victory (battle of the Standard, 1138), was eventually recognized by *Henry II.

fita eft fup ripam flumine tyeve. in loco qui c pali fubiectione libam ee conceffit. Ita fcilicet ut

David I, King of Scotland (left) shown with his grandson Malcolm IV, who succeeded him as king, in a decoration from the medieval charter of Kelso Abbey. (Floors Castle, Kelso)

David II (1324–71), King of Scotland (1329–71), the only son of *Robert I (the Bruce). At the age of 4 he was married to Joan (1321–62), daughter of *Edward II of England, and at the age of 5 he inherited the throne of Scotland. In 1334 he was forced off his throne by Edward *Balliol, and fled to France; he returned to Scotland in 1341, invaded England, and was captured at the battle of *Neville's Cross (1346). His subsequent years of captivity Anglicized him, so that when ultimately freed (1357) he found himself at odds with his people and disliked his heir-presumptive Robert the Steward.

David (d. *c*.961 BC), second King of Israel (*c*.1000–*c*.961 BC). In the biblical account he appears initially as a harp player at *Saul's court and as the slayer of the Philistine Goliath. As a military commander David became a friend of Saul's son, Jonathan, and married his daughter, provoking the jealousy of Saul, who exiled him. After Saul's death, he ruled the tribe of Judah while Saul's son Ishbosheth ruled the rest of Israel. On Ishbosheth's death, David was chosen as the king of all Israel and his reign marks a change in the fortunes of the Jews from a confederation of tribes to a settled nation. He moved the capital from Hebron to *Jerusalem, which had no previous tribal loyalties and made it the religious centre of the Israelites by bringing the Ark of the Covenant (their most sacred object) with him. He expanded the territories over which he ruled and brought prosperity to Israel. His later years were troubled by rebellions led by his sons and family rivalries at court. He is traditionally regarded as the author of the Psalms, but only a fraction are thought to be his work.

David, St (d. *c*.601), patron saint of Wales. According to legend he was born in Dyfed, the son of St Non and of Sant, a local prince. He is credited with founding twelve monasteries in Wales, the most important being at Mynyw (St David's) where he resided when he became Bishop of Wales. He took a prominent part in the synod of Llandewi-Brevi (*c*.560) to suppress the *Pelagian heresy, and supposedly persuaded the 'Synod of Victory' at Caerleon (*c*.569) to adopt the teachings of St *Augustine of Hippo instead.

Dead Sea Scrolls, the popular name for a collection of Hebrew and Aramaic manuscripts, the first of which were found in 1947 by shepherds in a cave near the north-western shore of the Dead Sea. They belonged to the library of the Jewish (perhaps Essene) community at nearby Qumran, and were probably hidden shortly before the Roman destruction of AD 68. The scrolls include fragments of nearly every book of the Hebrew *Bible; of special interest are: by far the oldest known manuscript of the book of Isaiah; a commentary on the book of Habakkuk; a manual of teachings and rules of discipline for the community; the Temple Scroll, which lays down in detail how the ideal temple of Jerusalem should be built; and there are also many other biblical, sectarian, and apocryphal writings of varying importance and in varying states of preservation.

Until the discovery of the scrolls, the earliest surviving Hebrew biblical manuscripts dated from the 9th century AD. They are therefore immensely important for scholars of the Old Testament and confirm the care with which the biblical texts were preserved and copied by the Jews.

King **David** playing the harp. An illumination from a medieval French manuscript.

Deccan (Sanskrit *dakshin*, 'south'), a triangular plateau covering most of peninsular India south of the Narbada River. Few empire-builders achieved full control over it, although some, notably *Asoka (*c*.250 BC), *Aurangzeb (late 17th century), and the British (19th century) claimed suzerainty. Between these eras of fragile, externally imposed unity, local dynasties struggled for hegemony, some like the *Cholas (10th to 13th century AD) successfully dominating parts of the region. While Hindu power was maintained by the *Vijayanagara empire in the south (*c*.1347–*c*.1565), Muslim dynasties established the *Bahmani kingdom in the north (*c*.1347). This later split into five sultanates which in turn fell to *Mogul expansion in the 17th century. Hindu strength was reasserted in the 18th century by various *Maratha clans, but after British victories in the early 19th century, most of the remaining kingdoms, Muslim and Hindu, were absorbed into the *East India Company's Madras and Bombay presidencies. Survivors, such as Hyderabad, entered into treaty relationships with the British.

Declaration of Independence, the foundation document of the United States of America, which proclaimed American separation from Britain and was adopted by the *Continental Congress on 4 July 1776. Its principal

The **Declaration of Independence** was signed a month after it had been adopted on 4 July 1776. John Trumbull's painting of the ceremony, begun in 1786 and completed in 1794, shows Thomas Jefferson, the tall figure in the centre, holding the declaration he had drafted with forty-eight of the fifty-six signatories in attendance. Thirty-six were drawn from life straight on to the canvas. The first signature was that of John Hancock, President of the Congress. (Yale University Art Gallery)

author was Thomas *Jefferson, who based its arguments on John *Locke's ideas of contractual government. Its celebrated preamble declared that all men are created equal and have inalienable rights to life, liberty, and the pursuit of happiness. There followed a detailed list of acts of tyranny committed by George III, his ministers, and Parliament against the American people, similar in tone to those in the English *Bill of Rights (1689). The original document had fifty-six signatories whose names were initially kept secret for fear of British reprisals in the event of American defeat.

Declarations of Indulgence, four proclamations issued by *Charles II and *James II of England in an attempt to achieve religious toleration. Charles II issued Declarations in 1662 and 1672, stating that the penal laws against Roman Catholics and Protestant dissenters were to be suspended, but protests by Parliament caused both attempts to be abandoned. James II issued similar Declarations in 1687 and 1688, the latter leading to the trial of the Seven Bishops. James II insisted that the Declaration should be read in all churches; a Tory High Churchman, Archbishop Sancroft and six bishops who refused to do so were tried on a charge of seditious libel and were acquitted. The verdict was a popular one and widespread protest and defiance followed during the months leading up to the *Glorious Revolution of 1688.

Delaware, a state of the USA situated on the east coast between New Jersey and Maryland. Though discovered by Henry Hudson, its name derives from Lord de la Warr, governor of Virginia in 1610. It was first settled as New Sweden in 1638, but came under English control in 1664. As the three 'Lower Counties' of Pennsylvania,

it enjoyed virtual autonomy under the *Penn family but in 1776 it achieved independent statehood. It was the first state to ratify the US constitution in 1787.

Delaware Indians, a North American Indian tribe who lived in the Delaware River valley of eastern Pennsylvania and New Jersey. Contacts with Spanish and Portuguese ships preceded Henry Hudson's arrival in 1609. Dutch settlements at Albany, New York (1614), Burlington, New Jersey (1624), and Fort Amsterdam (Manhattan, 1626) traded European goods for beaver pelts, but open warfare increased towards the middle of the century as more colonists arrived. Peace was officially concluded in 1645, but incidents continued to occur through the 18th century. The tribe was dispersed in the 18th and 19th centuries by the English and Americans.

Delhi, the capital of India, situated on the banks of the River Jumna in the north Indian plains. According to legend the city has changed site seven times, and there is archaeological evidence for at least this number of previous strongholds. The earliest evidence belongs to the 6th century BC. In the 1st century BC a 'Raja Dhilu' gave his name to the site. Little is known until the 12th century when Muslim invaders wrested it from the Hindu king *Prithviraj, but the Tomar Rajputs occupied the site in the 8th century and made it their capital. During the Delhi sultanate era (12th–16th century) it was the capital of a succession of Muslim dynasties (*Mameluke, *Khalji, *Tughluq, *Sayyid, and *Lodi). The first Muslim capital is marked by the Qutb Minar, a carved tower, built by the founder of the Mameluke dynasty, Sultan Qutb ud-Din Aibak. Natural and strategic factors caused movements to new sites, but always within a few miles of the original settlement. In the Mogul era Emperor *Shah Jahan built the Red Fort Palace and mosque which still dominate walled 'Old Delhi'. Mogul decline led to renewed invasions, notably the sack of Delhi by the Persian king, *Nadir Shah (1739). It was captured by the British in 1803.

Delian League, a voluntary alliance formed by the Greek city-states in 478–47 BC to seek revenge for losses suffered during the *Greek–Persian wars. All members paid tribute in the form of ships or money, the latter being stored on the sacred island of Delos, the League's nominal base. At first, under the leadership of Athens, the League actively sought to drive Persian garrisons out of Europe and to liberate the Greek cities of Asia Minor. At Eurymedon in c.466 a Persian fleet and army were crushed. However, Athens, which had by far the largest navy, had begun to dictate matters in a manner not foreseen by its allies. Thus in c.472 Carystus had been compelled to join the League; and in 465–462 the revolt of Thasos was crushed, its mining and other commercial interests on the mainland taken over by Athens. *Pericles encouraged the conversion of the alliance into the beginnings of the *Athenian empire.

Delphi, a site on Mount Parnassus in central Greece. It was the seat of the most respected *oracle of ancient times. Individuals and city-states consulted Pythia, the priestess of Apollo, and her answers were interpreted by a prophet. Delphi endured largely unscathed, from attacks by Persians (480 BC) and Gauls (279), though the Roman emperor *Nero removed 500 statues. The

A contemporary Persian painting of the capture of **Delhi** by Nadir Shah in 1739. The Persian king rides in triumph as his army wins a victory over Delhi's Mogul defenders. (Chelhel Sotun Palace, Isfahan)

oracle survived until closed by *Theodosius in AD 390. Remains of a number of buildings can be seen today, including a temple to Athena, national treasuries, and the *stadion* or running track used for the Pythian Games, a sprinting competition and festival.

demesne, in the Middle Ages, the lands retained by a lord under his direct control. The medieval lord, whether a king or *vassal, needed land to provide food and all other necessities for himself and his own household. Demesnes were the site of his residences which could be manors, palaces, or castles, and possibly all three. Lords with widespread territories would have demesne lands in several areas, especially where there was military threat. The day-to-day running of such estates was carried out by the lord's personal servants. Lords who failed to keep sufficient land in their own hands found themselves in great difficulty when times were troubled.

Demosthenes (384–322 BC), the greatest orator produced by ancient *Athens. In his speeches on public policy he consistently urged the need to resist the encroachments of *Philip II of Macedonia, and he twice served on embassies to that king. He fought at *Chaeronea in 338, and was a leading figure at Athens until the death of *Alexander the Great. After Athens had been defeated by the Macedonians in 322 BC, he committed suicide to avoid execution.

Denmark, a country in northern Europe, which since the 14th century has included the Faeroe Islands and Greenland. There was active Danish participation in the *Viking explorations and conquests after c.800. King *Canute ruled over a great 11th-century empire comprising Denmark, England, Norway, southern Sweden, and parts of Finland. His reign was notable for the spread of Christianity, initially introduced in the 9th century.

After a period of internal disunity, Denmark re-emerged as the leading Scandinavian nation in the 13th century. Civil warfare and constitutional troubles continued, however, until Christopher II (1320-32) made major concessions to the nobles and clergy at the expense of royal authority. His son, Waldemar IV (1340-75), re-established royal power, and his daughter, Margaret I (1387-1412), succeeded in creating the Pan-Scandinavian Union of *Kalmar (1397-1523). In 1448 the House of Oldenburg became the ruling dynasty. The 16th-century Protestant Reformation brought a national Lutheran church, and Christian IV (1588-1648) intervened in the *Thirty Years War as a champion of Protestantism. A sequence of 17th-century wars with Sweden resulted in Denmark's eclipse as the leading Baltic power. *Enlightenment ideas reached Denmark in the late 18th century, leading to major land reforms in favour of the peasants.

Dermot McMurrough (Diarmuid MacMurragh) (c.1110-71), King of Leinster (1126-71) in Ireland. In 1166, after feuding with his neighbours, he was defeated and banished by the Irish High King. He sought support from *Henry II of England and obtained the aid of Richard de Clare, Earl of Pembroke ('Strongbow'), offering him his daughter in marriage and the succession of Leinster. Dermot regained his kingdom in 1170 and after his death Leinster became an English fief and Henry II began to establish English dominance in Ireland.

Derry *Londonderry.

Descartes, René (1596-1650), French mathematician, scientist, and philosopher. He received a Jesuit education before taking his degree in law (1616). From 1617 to 1619 he followed a military career in the armies of the Netherlands and Bavaria. After a period of travelling, he spent some time in Paris and finally settled in the Netherlands (1628), having begun his *Rules for the Direction of the Mind* (published in 1701). He then produced the works which bought him contemporary and posthumous fame: *Discours de la méthode* ('Discourse on Method', 1637), *Meditationes de Prima Philosophia* ('Meditations on First Philosophy', 1641), and *Principia Philosophiae* ('Principles of Philosophy', 1644). His views exposed him to persecution by the theologians, and he accepted Queen *Christina's invitation to take refuge in Sweden, where he died.

By his philosophical method all the sciences, being interconnected, must be studied together, and by a single process designed to distinguish what is certain from what is probable. He based philosophical reasoning on the principles and methods of mathematics, thereby refusing to make any initial metaphysical assumptions. His quest for a basic certitude started from his famous phrase, '*Cogito ergo sum*'('I think, therefore I am'). He advanced mathematics by his development of analytical geometry, and optics by his discovery of the law of refraction. His influence has been profound and can be traced in the works of rationalists, empiricists, materialists, and even of philosophers who rejected his doctrines but benefited from his general intellectual rigour.

Desert cultures, a description applied to early post-glacial groups of hunter-gatherers in the south-western United States and Mexico from c.8000 BC. They lived mostly on vegetables, and digging sticks and grinders are common among the archaeological finds. Spears were used for hunting. The spread of maize cultivation gave rise to the *basket makers. Agriculture started in *Mexico soon after 3500 BC, reaching some areas of the south-western USA around the beginning of the Christian era, but in others the Desert cultures continued into the 19th century.

Desmoulins, Camille (1760-94), French journalist and Revolutionary. He became an advocate in the Paris *Parlement* in 1785, and four years later, after the dismissal of *Necker, he summoned the crowd outside the Palais Royal 'to arms'. Two days later, on 14 July, the mob stormed the *Bastille. Soon afterwards he began to publish his famous journal *Les Révolutions de France*, attacking the *ancien régime*. He married Lucile Duplessis in 1790 and began a close association with *Danton. He voted for the execution of *Louis XVI and campaigned against the *Girondins and *Brissot. His support of Danton's policies of clemency angered *Robespierre and led to his arrest and execution on 5 April 1794. A week later his wife followed him to the guillotine.

de Soto, Hernando (c.1500-42), Spanish conquistador and explorer. He was made governor of Cuba by Emperor *Charles V, with the right to conquer the mainland of America. He landed on the Florida coast in 1539 and reached North Carolina before crossing the Appalachian Mountains and returning through Tennessee and Alabama. In 1541 he led another expedition, crossing the Mississippi (which they were probably the first white men to see) and going up the Arkansas River into Oklahoma. They were seeking gold, silver, and other treasure, but returned disappointed. De Soto died when they reached the banks of the Mississippi.

Despenser, Hugh le, Earl of Winchester (the Elder) (1262-1326). He was loyal to *Edward I and to *Edward II. In 1321 he and his son, Hugh le Despenser (the Younger) were attacked in Parliament for their allegedly evil counselling of the king and were disinherited and exiled from the realm. These sentences were annulled in the following year, but when Edward was put to flight in 1326 both the Despensers were captured and hanged.

Dettingen, battle of (27 June 1743), an important victory for the British over the French in the War of the *Austrian Succession. *George II at the head of 40,000 British, Hanoverian, and Austrian troops marched from the Austrian Netherlands to the banks of the River Main. He was attacked by a larger French army under Noailles, but forced them back across the Main and finally across the River Rhine. George II was the last reigning British sovereign to take command on the battlefield.

Devolution, War of, an attempt by *Louis XIV of France to seize the Spanish Netherlands. In 1665, on the death of his father-in-law, Philip IV of Spain, he invoked dubious laws based on local customs by which a child of a first wife (as was his queen Maria Theresa) inherited titles and territory, rather than the son of a second wife. A campaign under *Turenne alarmed Europe, a defensive Triple Alliance was formed by the United Provinces, England, and Sweden to check the French advance, and Louis made peace. He restored most of his conquests, hoping to obtain part of the Spanish empire peacefully on the death of Charles II.

Diane de Poitiers, duchesse de Valentinois (1499–1566), mistress of *Henry II of France. She came to court during the reign of Francis I (1515–47) and Prince Henry, twenty years her junior, fell passionately in love with her. On his accession she became queen in all but name displacing Henry's wife, Catherine de Medici. A beautiful and cultured woman, she was friend and patron of poets and artists. She played little part in politics, contenting herself with augmenting her income and providing for her family. On Henry's death (1559) Catherine forced her to surrender the crown jewels and banished her to Chaumont.

diaspora (from the Greek, 'dispersion'), the collective term for Jewish communities outside Israel itself. The process began with the Assyrian and Babylonian expulsions of 721 and 597 BC, was continued by voluntary migration, and accelerated by the Roman destruction of the Temple in Jerusalem in AD 70. By the 1st century AD there were Jewish communities from the Levant to Italy and notably in Babylon and Egypt. The diaspora Jews of the Graeco-Roman world were mostly Greek-speaking but remained loyal to their faith, visited Jerusalem, and regarded Israel as their homeland. The existence of these diaspora communities was also an important factor in the spread of Christianity.

By the Middle Ages Spain and France had become major centres of the overseas Jewish community. Anti-Semitism, a product of ethnic and theological prejudices, and economic grievances, emerged. Eastern Europe next became the diaspora's centre of gravity until the pogroms of the 1880s drove many westwards, via Germany and Britain, to the USA. Some Jews interpret diaspora as exile, others as a positive aspect of Judaism's spiritual destiny.

Diaz de Novaes, Bartholomew (*c*.1455–1500), Portuguese explorer and the discoverer of the Cape of Good Hope. On a voyage of exploration down the west coast of Africa his ships were swept southwards for thirteen days by strong gales. His landfall was near the southernmost tip of Africa, which he named Cabo Tormentoso ('Cape of Storms'). On a later voyage, during which Brazil was discovered, he perished off the Cape of Storms.

Diderot, Denis (1713–84), French novelist, dramatist, critic, and philosopher. He was critical of the *ancien régime*, and in 1749 he was imprisoned for a few months by royal order. With d'Alembert he edited the great *Encyclopédie*, in which he expressed his passionate belief in science and his scorn of superstition. His *Salons* (accounts of exhibitions of contemporary art) inaugurated the genre of art criticism in France. He corresponded with *Catherine the Great, who supported him by buying his library from him, leaving it with him, and paying him to act as librarian. He was important for his success in popularizing scientific knowledge and philosophical doctrines.

Diet, a meeting of the representatives of the German states of the *Holy Roman Empire with their emperor. Important meetings took place at Augsburg (1500), Constance (1507), and Frankfurt (1518) between the princes and *Maximilian I, attempting to reform the empire. The most famous was that of *Worms (1521), dealing with the confrontation of Martin *Luther and *Charles V. As the princes used Lutheranism to challenge Charles V's authority, further diets discussed religion. At Speier, a solution offering religious tolerance was submitted in 1526, and a strict Catholic alternative in 1529. The two sides further defined their terms at *Augsburg in 1530, and a last attempt at conciliation was made at Regensberg in 1541. Further meetings at Augsburg in 1547–8 and 1555 brought religious settlement. Another important meeting, at Regensburg, in 1732, saw the princes accept the *Pragmatic Sanction.

Digby, Sir Kenelm (1603–65), English diplomat and writer. He was the son of Sir Everard Digby, executed as a traitor for his involvement in the *Gunpowder Plot. He led a successful privateering expedition in the Mediterranean and Adriatic (1628). Thereafter he supported *Laud and from 1636 worked to achieve toleration for Catholics in England. He was imprisoned by Parliament, then banished, but in the 1650s undertook diplomatic missions in Europe for *Cromwell. An amateur scientist, he was one of the first members of the *Royal Society.

Digger, the name for a radical Puritan, a member of a group which flourished briefly in 1649–50, when England's political future was uncertain. Led by Gerrard *Winstanley, the Diggers began seizing common land and sharing it out. They called themselves the True Levellers, but were opposed and denounced by the

A portrait of the French novelist and philosopher **Diderot** at his writing desk, 1767, by the painter Louis Michel van Loo. (Musée du Louvre, Paris)

*Levellers, who disliked their communistic attitude towards property. Although they themselves rejected the use of force, their settlements in Surrey were dispersed by the authorities in March 1650.

Diocletian (Gaius Aurelius Valerius Diocletianus) (AD 245–316), Roman emperor (284–305). Born in humble circumstances in Dalmatia, he joined the Roman army, rose to become commander of the imperial household troops, and was acclaimed emperor by the army at Nicomedia in 284. His accession restored order after a long and chaotic period of short-lived emperors. He campaigned vigorously and successfully to protect the empire's collapsing frontiers and to repress internal rebellions, most notably in Britain and Egypt. He effected far-reaching and lasting military, financial, and administrative reforms. He established the so-called 'Tetrarchy': himself the 'Augustus', or emperor, in the East with Galerius as his 'Caesar', or deputy, he appointed Maximian as 'Augustus' of the Western Empire with Constantius Chlorus as his 'Caesar'. From 303 he ordered the systematic persecution of the Christians. He abdicated in 305, and apart from a brief re-emergence in 308 lived until his death in 316 in a splendid palace at Salonae (Split, Yugoslavia), much of which survives. He provided the solid base from which Constantius and *Constantine were to continue to restore the empire's fortunes and for most of the 4th century his system of appointing more than one emperor held firm.

Diogenes (c.400–325 BC), Greek philosopher, popularly credited with being the founder of the *Cynics. He came to Athens, where he lived as a pauper and flouted conventional behaviour. He believed that an individual needed only to satisfy his natural needs in the simplest manner possible to be happy. His apparent shamelessness led to his nickname 'the dog' (Greek kuōn, hence 'Cynic'). He is said to have lived in a barrel. His debt to Antisthenes of Athens, himself a devoted disciple of *Socrates, was considerable, and many of the details of his particular philosophy are uncertain.

Directory, French (1795–9), the government of France in the difficult years between the *Jacobin dictatorship and the Consulate. It was composed of two legislative houses, a Council of Five Hundred and a Council of Ancients, and an executive (elected by the councils) of five Directors. It was dominated by moderates and sought to stabilize the country by overcoming the economic and financial problems at home and ending the war abroad. In 1796 it introduced measures to combat inflation and the monetary crisis, but popular distress increased and opposition grew as the Jacobins reassembled. A conspiracy, led by François *Babeuf, was successfully crushed but it persuaded the Directory to seek support from the royalists. In the elections the next year, supported by *Napoleon, it decided to resort to force.

This second Directory implemented an authoritarian domestic policy ('Directorial Terror'), which for a time established relative stability as financial and fiscal reforms met with some success. By 1798, however, economic difficulties in agriculture and industry led to renewed opposition which, after the defeats abroad in 1799, became a crisis. The Directors, fearing a foreign invasion and a Jacobin coup, turned to Napoleon who took this opportunity to seize power.

disciple, pupil or learner, used specifically to describe an original follower of *Jesus Christ. In the Jewish society of Jesus' time many religious teachers attracted disciples who came to be taught their master's intepretation of scriptures. Jesus' followers differed from such groups in several respects. For example, he actively sought out disciples and found many of them among people judged socially or morally as outcasts. The Apostles were the twelve chief disciples: *Peter (the leader), Andrew, James, John, Philip, Bartholomew, Thomas, Matthew, James (the Less), Thaddeus, Simon, and Judas Iscariot. After the suicide of Judas, who betrayed Jesus, his place was taken by Matthias. *Paul and his original companion Barnabas are also considered as Apostles.

Dissolution of the Monasteries *Monasteries, Dissolution of the.

Divine Right of Kings, a European doctrine which taught that monarchy was a divinely ordained institution, that hereditary right could not be abolished, that kings were answerable only to God, and that it was therefore sinful for their subjects to resist them actively. It evolved during the Middle Ages, in part as a reaction to papal intrusions into secular affairs. The extension of the principle, to justify absolute rule and illegal taxation, aroused controversy. *James I of England upheld the doctrine in his speeches and writings and his son *Charles I was executed for refusing to accept parliamentary control of his policies. After the *Glorious Revolution the doctrine was far less influential, yielding to anti-absolutist arguments like those of John *Locke. In late 17th-century France *Louis XIV's monarchy was based on the principle of Divine Right.

doge, the title of the holder of the highest civil office in Venice, Genoa, and Amalfi from the 7th century until the 18th century. The office originated in Venice; in 1032 hereditary succession was formally banned and election was made increasingly complicated to prevent domination by particular factions, although the Participazio and Candiano families provided most candidates in the 9th and 10th centuries, and the Tiepolo and Dandolo in the 13th and 14th. The system ended with the Napoleonic conquest of 1797. The Genoese introduced a similar system after 1339. Democratic until 1515, it became an aristocratic office thereafter and also succumbed to *Napoleon. The first doge's palace in Venice was built in 814 and destroyed in 976. Both in Venice and in Genoa, mercantile riches financed lavish residences for civic leaders; the 14th century palace in Venice was decorated by the painters Tintoretto and Titian.

dolmen *megalith.

Domesday Book, a survey of property in England conducted in 1086. Conceived by *William I, but probably to some extent based on pre-Conquest administrative records, it was the most comprehensive assessment of property and land ever undertaken in medieval Europe. The English shires were visited by royal commissioners and the survey yielded evidence relating to the identity of landholders, their status, the size of their holding, its use, its tax liability, and the number of animals maintained. The information for each shire was then condensed and reorganized into feudal groupings.

The final version comprised two volumes—Little Domesday (Norfolk, Suffolk, and Essex) and Great Domesday (the rest of England except for the four northern shires, London, and Winchester). Its purpose was to maximize the revenues from the land tax and it caused resentment and even riots. It was given its name on account of its definitive nature; today its volumes are housed in the Public Record Office, London.

Dominic, St (1170–1221), founder of the *Dominican order of friars. He was born in Spain, of noble family, but as a young man adopted an austere life, becoming a priest and canon of Osma Cathedral. In 1203 he was a missionary to the *Albigensian heretics of southern France, working also with the Crusaders who were trying to suppress them by force. In 1215 he attended the Fourth *Lateran Council. In that year he founded his own order known as the Dominicans or 'Black Friars'.

Dominican, a member of the Order of Friars Preachers. They were founded by St *Dominic and the order received papal approval in 1216. They are governed by their master-general and wear a white tunic with a black mantle which has given them the name of 'Black Friars'. They were a mendicant or begging order, devoted to teaching and preaching within the world and not confined within the walls of a monastery. Special emphasis on study was based on a teaching system involving houses of study, the 'Studia Generalia'. The philosopher and theologian St Thomas *Aquinas was a member of the order, as were four popes. The order gained great popularity during the 13th century; the popes used the Dominicans for preaching crusades and for the *Inquisition. With the rise of new orders during the *Counter-Reformation their influence was reduced.

Donatism, the beliefs of a group which broke away from the Catholic Church in North Africa in the early 4th century AD. The Donatists were named after the Numidian Donatus, whom they set up as a rival bishop to the bishop of Carthage. Although Donatus was exiled by the Byzantine emperor Constans in 347, the Donatists continued to thrive, especially among the poor inhabitants of rural Numidia and Mauretania. In 411 a meeting of bishops declared against them, and they were persecuted until the *Vandals invaded in 429. Nevertheless, there was a revival of Donatism in the following century.

Doria, Andrea (1466–1560), *doge of Venice (1528–60), an outstanding soldier and admiral. He fought for *Francis I and *Charles V, expelled the French from *Genoa in 1528, and took power himself, creating the aristocratic republic. His descendants contributed six doges and numerous officials to the state.

Dorians, invaders from the north who entered ancient Greece c.1125–1025 BC. They settled first in the Peloponnese then also in the islands of Crete, Melos, and Thera, and the southern coast of Asia Minor. They spoke a dialect of the Greek language and seem to have come from *Epirus and south-west *Macedonia. They destroyed the *Mycenaean civilization, thereby ushering in the Greek 'Dark Age', an obscure period about which little is known. In Sparta and Crete they suppressed their subjects as *helots, but elsewhere a gradual fusion of conquerors and conquered took place.

St **Dominic** burning heretical Albigensian texts. The Dominicans abhorred the Albigensians' beliefs and, with the help of the Crusaders, had largely succeeded in suppressing the movement by force by 1229. (Museo del Prado, Madrid)

Draconian laws, the first written code of laws drawn up at *Athens, believed to have been introduced in 621 or 620 BC by a statesman named Draco. Although their details are obscure, they apparently covered a number of offences. The modern adjective 'Draconian' (excessively harsh) reflects the fact that penalties laid down in the code were extremely severe: pilfering received the same punishment as murder—death. A 4th-century BC politician quipped that Draco wrote his laws not in ink, but in blood.

dragoon, a mounted infantry soldier, named in 16th-century France after the short musket called the 'dragon'. Originally trained to fight on foot, dragoons were organized in infantry companies, not cavalry squadrons, but were progressively trained to cavalry standard. Thus by the early 18th century they were known as medium cavalry in the Prussian army, and light cavalry in the British army. Their versatility on horse or foot made them ideal for the maintenance of public order or for dealing with guerrilla warfare. The 'dragonnades' against the French Huguenots in the 1680s were an example of their brutal effectiveness.

Drake, Sir Francis (c.1543–96), English admiral and explorer, the first Englishman to circumnavigate the world. He sailed on a slave-trading voyage with his cousin, John *Hawkins, in 1567, and spent the next few years in privateering raids on the *Spanish Main. In 1577 he was engaged by a syndicate headed by *Elizabeth I to undertake a voyage of circumnavigation of which the chief object was undoubtedly plunder. In his 100-ton ship the *Golden Hind* he passed through the Magellan Straits, raided the Spanish settlements on the west coast of South America, and crossed the Pacific from California to the Moluccas, returning to Plymouth a rich man.

In 1585 he commanded an expedition which was the first act of open war against Spain, sacking San Domingo, Cartagena, and St Augustine in Florida, and then rescuing *Raleigh's Virginia colonists. He returned to England to hear news of the preparations for the *Spanish Armada, some ships of which he proceeded to destroy at Cadiz in the operation known as 'singeing the King of Spain's beard'. When the Armada sailed in 1588 Drake was appointed vice-admiral of the English fleet at Plymouth under Lord Howard of Effingham. The story of his finishing a game of bowls before going into battle is improbable. In command of the *Revenge*, he took a leading part in the defeat of the Armada and its pursuit into the North Sea. In 1595 he and Hawkins again sailed to the Caribbean, where they both died.

Druid, a member of the ruling caste of the Gallic *Celts. Knowledge of the Druids is derived chiefly from the hostile accounts of them in the Roman authors Julius *Caesar and *Tacitus. Caesar reports that they exercised judicial and priestly functions, worshipped in groves (clearings in forests), and cut mistletoe from the oak tree (sacred to them) with a golden sickle. The religion was stamped out by the Romans, lest it should become a force for resistance to Roman rule. Suetonius Paulinus destroyed the Druid centre at Mona (Anglesey, north Wales) in AD 61, after which there is no further mention of them in England and Wales. The supposed association of the Druids with Stonehenge is rejected by scholars but the modern Druidical order seeks to make use of the site to conduct its annual solstice ceremonies.

du Barry, Marie Jeanne, comtesse (1743–93), favourite of *Louis XV of France. She was a great beauty who in 1769 became the king's mistress and influenced him until his death in 1774. *Choiseul criticized her and she may have helped to bring about his dismissal. During the *French Revolution she was arrested by the Revolutionary Tribunal and guillotined.

Dudley, John   *Northumberland, Duke of.

Dudley, Robert   *Leicester, 1st Earl of.

Dunbar, battle of (3 September 1650). Near the port of Dunbar in Scotland, Oliver *Cromwell's force of 14,000 men won a victory over 27,000 Scots, and enormous numbers were taken prisoner together with all the Scottish guns. Cromwell's victory destroyed the *Stuart cause in Scotland for a decade.

Duncan I (c.1010–40), King of Scotland (1034–40). He was ruler of Strathclyde which was added to the Scottish kingdom inherited from his grandfather Malcolm II. His accession was unpopular with the northern tribes and twice he was defeated by the Earl of Orkney before being killed in battle by the Earl of Moray, *Macbeth.

Duncan II (c.1060–94), King of Scotland (1094). He gained the throne through the support of William II of England, who provided the army with which Duncan defeated his uncle and rival Donald Bane. However, Duncan's English alliance was resented and he was murdered at his uncle's instigation.

Duns Scotus, John (c.1265–1308), scholastic philosopher and theologian. Born in Scotland (hence his name 'Scotus'), he became a Franciscan friar and then a priest and studied at Oxford and Paris. He later returned to both universities as a teacher before finally moving (1307) to Cologne University. In his writings and lectures he stressed the distinction between faith and reason, arguing the limitations of reason and that the will is superior to the intellect. His ideas were different from those of St Thomas *Aquinas, who saw a possible harmony between faith and reason. His ideas were unpopular with 16th-century reformers who regarded his supporters as 'Dunsmen' or 'dunces'.

Dunstan, St (c.909–88), Archbishop of Canterbury (959–88). He was a reformer of organized monasticism in England, and a counsellor to its kings. *Edmund I of Wessex made Dunstan abbot of the Benedictine house of Glastonbury, Somerset. Although exiled (956–7) by King Edwy, Dunstan was recalled by Edwy's successor *Edgar, and made successively Bishop of Worcester and London. With Edgar's support Dunstan founded or re-founded many abbeys and helped draw up a code of monastic observance, the *Regularis Concordis*, at the Synod of Winchester (c.970). His formulation of the ceremony for Edgar's coronation became the basis of all subsequent coronations in England. Dunstan was also a skilful musician, draughtsman, and metalworker.

Dupleix, Joseph-François, marquis de (1697–1763), governor-general of the *French East India Company in India (1742–54). He demonstrated able leadership of France's Indian interests, but was outmanoeuvred by Robert *Clive during wars in south India. In 1754 he was recalled to Paris, where he died in ignominy, aware that Clive's subsequent victories in Bengal had finally destroyed his own dream of extending French influence in India. Historians differ in considering him gifted but unlucky, or rashly overambitious.

Dutch East India Company, a *chartered company established (1602) under the aegis of Prince Maurice of Nassau to co-ordinate the activities of companies competing for trade in the East Indies and to act as an arm of the Dutch state in its struggle against Spain. It was involved in attacks on the Portuguese (then part of the *Spanish empire), and warfare with native rulers, and created a virtual monopoly in trade in fine spices (for example cloves, nutmeg, and mace) grown under its supervision in the *Moluccas and the Banda Islands. In 1619 it made Batavia its headquarters. It ousted the Portuguese from Ceylon, set up trading posts in India, Persia, and Nagasaki, and made the Cape of Good Hope a base for Dutch ships *en route* to and from the East.

Its chief activities were in south-east Asia. One after

another it brought to heel *Amboina and the Moluccas, *Malacca, *Macassar, and *Bantam. The subjection of the kingdom of Mataram in *Java proved more difficult, bringing military, financial, and administrative problems.

Throughout the 18th century the Company operated at a loss. High mortality and inadequate salaries led to corruption among its officials. Nor did it succeed in making its monopoly in the Indies complete. In 1799 it was liquidated, its debts, possessions, and responsibilities being taken over by the Dutch state.

Dutch East Indies, an area formerly under the rule of the *Dutch East India Company, now Indonesia, in south-east Asia. The archipelago has been subject to Indian and Chinese cultural influences. Hinduism, Buddhism, and later Islam reached the islands from India and were grafted on to local cultures. Many Chinese settled in the islands as traders and planters, maintaining regular contact with China. Empires in the Indies based on sea-power were, except for *Srivijaya, relatively ephemeral, for example *Macassar and Ternate in the *Moluccas. The land-based empires, in *Java for example, were inclined to last longer. Islam spread throughout the Indies after 1300 (Hinduism survived only in *Bali), and provided a weapon against the Portuguese and Dutch, who arrived in force in the 16th and 17th centuries. The Portuguese established footholds in the Moluccas but their presence proved to be short-lived. The Dutch, initially a sea-based trading power with many scattered bases, were drawn into land-based responsibilities, especially in Java, where the Dutch East India Company became ever more deeply involved in the politics of the island from the time of the establishment in Batavia (now Jakarta) in 1619.

Dutch empire, the overseas territories of the *United Provinces of the Netherlands. Dutch wealth rested on the fishing and shipping industries, assisted by Holland's position on the chief European trade routes. Amsterdam became the principal warehouse and trading centre for all Europe. Grain and naval stores from the Baltic, much of Spanish, Portuguese, and English trade, and the bulk of French exports were transported in efficiently designed vessels. Modern banking methods developed from Amsterdam's exchange bank (1609). Overseas trade with Asia, America, and Africa grew steadily even during war.

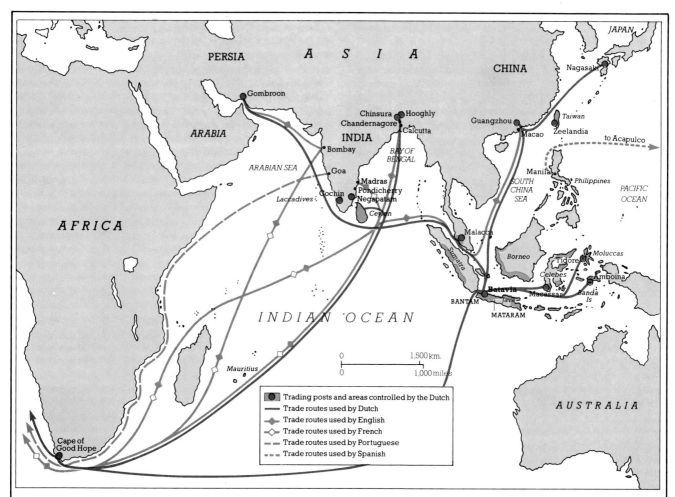

Dutch East India Company: 1602–1799

The Dutch East India Company had two important strategic advantages in the cut-throat competition for trade and constant naval skirmishing among European powers in south-east Asia in the 17th century. Its base at Batavia was ideally placed to defend the company's interests in the Spice Islands, and the acquisition of the Cape of Good Hope in South Africa in 1652 provided a vital staging post for Dutch captains, who followed a direct route to Batavia, taking advantage of the prevailing westerlies. Dutch dominance in the area was largely secured at the expense of the Portuguese.

Spain and Portugal's attempt to exclude the Dutch from the 'New World' prompted them to found the *Dutch East India Company (1602). From headquarters in Java they came to control the Indian Ocean and the spice trade and traded extensively in the China seas. The West India Company (1621) became primarily concerned with the African slave trade. In America a settlement, to become New York, was founded on the Hudson River (1609). Growing rivalry with Britain led to loss of maritime supremacy and of all Dutch colonies except in south-east Asia.

Dutch Revolts (1567–1648), the struggle by the Netherlands for independence from Spain. The Low Countries formed part of the Spanish empire but the tactlessness of the Council of Regency for Philip II alienated the local nobles, who were excluded from government. High taxation, unemployment, and Calvinist fears of Catholic persecution aroused dangerous opposition which the Duke of *Alba came to crush (1567) with a reign of terror and punitive taxation. Open revolt led by *William the Silent followed. He avoided pitched battles with the superior Spanish forces and exploited his local knowledge, saving besieged cities like Leiden (1573–4) by opening the dykes and flooding the countryside. The sack of Antwerp (1576) led to a temporary union of the whole Netherlands in the Pacification of Ghent. Calvinist excesses soon caused the southern provinces to form the Union of Arras (1579) and make peace with Spain. The northern provinces formed the Union of Utrecht and the war became a religious struggle for independence. William held out with foreign aid until assassinated (1584), when the leadership passed to Maurice of Nassau and the politician *Oldenbarneveldt. The United Provinces were saved by Spain's commitment to wars against France, England, and Turkey. A truce (1609) was followed by recognition of full independence at the Peace of *Westphalia (1648).

earl, British nobleman ranking between marquis and viscount, ranked third in the peerage. From Alfred's time *aldermen had charge of shires, but during the 10th century they became more important, with overall control of several shires. King *Canute's dependence on his Scandinavian jarls (earls) gave them territorial power in England over the regions of Northumbria, East Anglia, and Wessex (where Godwin was earl). They presided over the shire court, commanded its *fyrd, and retained one-third of the profits of justice (replaced later by King John with a fixed sum). Under the Norman kings in the 11th century shire administration passed to the sheriffs but the hereditary title of earl survived.

East Anglia, the region of eastern England occupied by the counties of Norfolk and Suffolk. The kingdom of the East Angles was formed from the end of the 5th century by Angles, Saxons, Frisians, and Swabians from the German Baltic coast. Initial settlements were made in Norfolk, but the settlements in Suffolk, especially around Ipswich, were to provide the nucleus of the kingdom, with the royal seat at Rendlesham on the River Deben. Protection against the Britons to the west was provided by the Middle Angles settled in the waterlands of the Fens and by the construction of earthworks (Devil's Dyke, Fleam Dyke, Heydon Dyke).

Little is known about its kings but Raedwald (d. c.625), was briefly a Christian and was styled Bretwalda, or overlord, of the south, by *Bede. His is probably the rich barrow-burial at *Sutton Hoo. Sigeberht, his son, founded the bishopric at Dunwich and a monastery at Burgh Castle. Although never again attaining overlordship, the kingdom continued until overrun by Danes in the later 9th century. It became part of the *Danelaw and eventually (921) one of the four English earldoms.

Easter Island, an island in the South Pacific some 3,800 km. (2,400 miles) west of Chile. It was colonized

The stone heads on **Easter Island** in the Pacific Ocean are carved out of rock cut from the crater of the Rana Raraku volcano and are ranged in terraces. The largest weigh over 50 tonnes. Their origin and purpose remain unclear.

by Polynesians from the nearest occupied islands 2,400 km (1,800 miles) to the west. At the time of its discovery by the Dutch navigator Jakob Roggeveen on Easter Day 1722, the Polynesians were no longer engaged on voyages of exploration, and had developed a complex system of chiefdoms. This involved the carving and erection of extraordinary stone heads, some of which were over 9 m. (30 ft.) high.

East India Company, English, a *chartered company of London merchants which in two and a half centuries transformed trading privileges in Asia into a territorial empire centred on India. Chartered in 1600, the Company soon lost the Spice Islands (*Moluccas) to the Dutch, but by 1700 had secured important trading ports in India, notably Madras, Bombay, and Calcutta. In the mid-18th century Anglo-French hostility in Europe was reflected in a struggle for supremacy with the *French East India Company. The English commander *Clive outmanoeuvred the French governor *Dupleix in south India, then intervened in the rich north-eastern province of *Bengal. Victory over the Bengal ruler in 1757 initiated a century of expansion, the East India Company emerging as the greatest European trader in India, though with strong French competition.

East–West Schism, the schism between the Eastern (or *Orthodox) Church and the Western (or Roman) Church, which became definitive in the year 1054. Tension between the two churches dated back at least to the division of the *Roman empire into an Eastern and a Western part, and the transferral of the capital city from Rome to *Constantinople in the 4th century. An increasingly different mental outlook between the two churches resulted from the occupation of the West by formerly barbarian invaders, while the East remained the heirs of the classical world. This was exacerbated when the popes turned for support to the *Holy Roman Empire in the West rather than to the *Byzantine empire in the East, especially from the time of *Charlemagne onwards. There were also doctrinal disputes, and arguments over the nature of papal authority. Matters came to a head in 1054 when the two churches, through their official representatives, excommunicated and anathematized (formally denounced) each other. The breach was deepened in 1204 when the Fourth *Crusade was diverted to Constantinople and sacked the city, and a Latin (Western) Empire was established there for some time. There were various attempts to heal the schism, notably at the ecclesiastical councils of Lyons II (1274) and Florence (1439), but the reunions proved fleeting. These attempts were effectively brought to an end when the *Ottoman Turks captured Constantinople in 1453 and occupied almost all of the former Byzantine empire for many centuries. It is only in recent years that the dialogue between the two churches to heal the schism has been effectively re-opened.

Ecuador, a country on the north-west coast of South America. In prehistory it formed the southern part of an area between the cultures of the Peruvian Andes and Mesoamerica (Mexico and northern Central America). Its outstanding feature was the early development of metal-working, from c.500 BC. Independent kingdoms developed from c. AD 500, with two cultural regions—a coastal one, adapted to the open sea, and one adapted

East India Company: a view of Fort William from the Hooghly River, painted by Lambert and Scott in 1731. Fort William was built in 1696 to protect the East India Company trading post founded on the river six years earlier. This settlement later grew into the flourishing city of Calcutta. Company trading vessels, seen firing a salute, were as well armed as most warships of the day. (India Office Library, London)

to the interior environment. The Incas conquered the central valley in the 15th century, and their communications network included a road from *Cuzco to *Quito, which they set up as their regional capital. Pizarro united the region to his Peruvian conquests in 1535 and installed his brother, Gonzalo, as governor. Internal dissensions led to a take-over by the Spanish crown and the establishment of Quito as an Audiencia (a high court with a political role) under the viceroy of Peru.

Edgar (c.944–75), King of Northumbria and Mercia (957–75) and of England (959–75). After the political failure of his brother Eadwig, he was chosen king of England north of the Thames (957). The southern part also became his on Edwy's death (959). Edgar's reign saw freedom from Danish raids (due in part to his building of an effective navy), hence his title of 'Edgar the Peaceful'. His authority was acknowledged (973) by the other kings in England, Scotland, and Wales. Edgar supported Saints Dunstan, Ethelwold, and Oswald in their reform of English monasteries. He was succeeded in turn by two of his sons, Edward the Martyr and Ethelred II.

Edgar the Aetheling (from Anglo-Saxon *aetheline*, 'prince') (c.1050–1130), the grandson of Edmund Ironside. His father's death (1057) in exile left Edgar as the heir to *Edward the Confessor in 1066, but because of his youth he was rejected in favour of *Harold II. On Harold's death at the battle of *Hastings the Witan finally chose Edgar as king but, despite his involvement in a rebellion against *William I (1069), he was unable to organize any further resistance and became a member of William's court (1074). He was captured by *Henry I at Tinchebrai (1106) for supporting Henry's brother Robert Curthose, but was later released.

Edgehill, battle of (23 October 1642), the first battle of the *English Civil War. Charles I's Royalists, marching south from Shrewsbury, with the eventual aim of re-capturing London, clashed with the Parliamentarians under the 3rd Earl of *Essex, at Edgehill, near Warwick. Prince *Rupert and his Royalist cavalry gained an early advantage, but the brunt of the fighting was borne by both infantries. Darkness, and the exhaustion of the troops ended the struggle, with no clear victor and heavy losses on both sides.

Edmund I (921–46), King of England (939–46). He succeeded his brother Athelstan as king. In his brief reign he defended Athelstan's territorial gains and reconquered (944) Danish-occupied Mercia. He sought to stabilize his north-west border against Danish raids from Ireland by presenting (945) the region of Strathclyde to his ally the King of Scotland, Malcolm I. Edmund appointed *Dunstan as abbot of Glastonbury and supported his reform of organized monasticism. He was murdered by a convicted robber in a banquet brawl.

Edmund II (Ironside) (c.993–1016), King of Wessex. He was the son of *Ethelred II, who had temporarily (1013–14) lost his throne to Sweyn I of Denmark. Edmund's succession was challenged by Sweyn's son *Canute. Over a period of eight months (1016) they fought six major battles. Defeated at Ashingdon in Essex (October 1016), Edmund agreed to rule southern England and Canute the remainder, either survivor succeeding to the whole country. Edmund died in the following month and Canute became overall ruler.

Edward I (1239–1307), King of England (1272–1307), in succession to his father *Henry III. He was married to Eleanor of Castile (1254), then to Margaret of France (1299). Edward's reputation as a successful ruler rests on his military and legal skills (for which he was called 'the English *Justinian'). His military achievements, which were motivated by a determination to extend royal power, included the defeat of Simon de *Montfort (1265), the conquest of Wales (1277–82), the suppression of rebellions in Wales (1294–5) and Scotland (1296–1305), and the defence of his lands in *Gascony against the French crown (1294–9). His legal reforms covered such matters as feudal administration (Statute of Westminster, 1275), crown lands (Quo Warranto, 1278), and law and order (Statute of Winchester, 1285), and he summoned the *Model Parliament in 1295. He died during a vain expedition to subdue the Scots and was succeeded by *Edward II.

Edward II (1284–1327), the first English Prince of Wales (from 1301) and King of England (1307–27), fourth (but eldest surviving) son of *Edward I and Eleanor of Castile. He was notorious in his own lifetime for his inordinate affection for Piers *Gaveston and for his unhappy marriage with *Isabella, daughter of Philip IV of France. Gaveston dominated Edward by 1304 and helped alienate him from his barons; the barons, led by Edward's cousin, Thomas of Lancaster, hemmed Edward in by a set of Ordinances (1310) and had Gaveston killed in 1312. The king's prestige fell further when he was defeated by *Robert the Bruce at *Bannockburn in 1314, and although he attempted to reassert his royal authority and annulled the Ordinances (1322), his own wife and her lover, Roger *Mortimer, imprisoned him in 1326 and finally had him murdered.

Edward III (1312–77), King of England, Ireland, and France (1327–77). He succeeded his father, *Edward II, though the throne was at first his in little more than name, power remaining in the hands of his mother, *Isabella, and her lover Roger *Mortimer; but in 1330 Edward had Mortimer arrested and began his personal rule. He secured the Scottish frontier with relative ease by the victory of *Halidon Hill (1333); but his initial French strategy was less successful, for although he prudently bought up the allegiance of France's neighbours the cost proved excessive. By 1341 Edward was virtually bankrupt. In 1346 he sought to justify his claim to the French throne by the more direct means of leading a vast army to France, and victories at *Crécy (1346) and in Brittany made him effectively king in France. The English by now were gaining a taste for foreign warfare and booty, and the truce of 1354 was ended by a fresh invasion of France two years later, crowned by the epic victory that Edward's son, *Edward the Black Prince, won at *Poitiers.

The rest of Edward's long reign was less successful—there were failures in France between 1369 and 1375, and after the death of his wife, Philippa of Hainaut in 1369, his health and mind began to deteriorate; he fell under the influence of his mistress, Alice Perrers. Edward left his successor, *Richard II, with a legacy of social discontent in England as well as the possession of vast tracts of France.

Edward IV (1442–83), King of England (1461–70, 1471–83). He was the eldest son of Richard, Duke of *York and so had a clear hereditary right to the throne by descent from Edward III. He gained the throne at the age of 19 with the help of his cousin Richard Neville, Earl of *Warwick, while the Lancastrians hesitated after their victory of *St Albans, and he then defeated them at *Towton. His marriage to Elizabeth Woodville and alliance with Burgundy alienated Warwick, who in October 1470 invaded England from France and secured *Henry VI's nominal restoration, but Edward won back the throne by victories at *Barnet and *Tewkesbury (1471). Thereafter he was a strong ruler and promoter of English commerce, but his dissolute lifestyle probably caused his early death, which left England with a 12-year-old king, *Edward V.

Edward V (1470–83), King of England (1483), the eldest son of *Edward IV and Elizabeth Woodville. His short reign, which began on 9 April 1483, was little more than a power struggle between the Woodvilles and his paternal uncle Richard, Duke of Gloucester. In June, Richard assumed royal dignity as *Richard III, and it was at some time between then and August of the same year that Edward and his brother were murdered in the Tower of London, probably at the instigation of Richard.

Edward VI (1537–53), King of England (1547–53). He was the son of *Henry VIII and Jane Seymour. During his minority effective power was exercised by Edward *Seymour, Duke of Somerset until 1549, and subsequently by John Dudley, Duke of *Northumberland. He favoured the Protestant religion, endorsing Archbishop *Cranmer's English Prayer Books (1549 and 1552). Contemporaries noted his studious, unemotional nature, and a callous streak reminiscent of his father. Always a sickly child, he died of tuberculosis aged 16.

Edward the Black Prince (1330–76), Prince of Wales (1343–76), the eldest son of *Edward III. He was an outstanding example of the chivalric ideal, a military leader who helped restore national pride to the English by a series of victories in the *Hundred Years War. He commanded part of his father's army at *Crécy (1346), and in 1356 won the battle of *Poitiers, capturing John II. In 1367 he restored King Pedro to the throne of Castile, but the campaign in Spain ruined his health. By his love match to Joan, the 'Fair Maid of Kent', he left one son, the future *Richard II.

Edward the Confessor, St (*c*.1003–66), King of England (1042–66). He succeeded King *Canute's Danish heirs as king, temporarily re-establishing the West Saxon monarchy, although he favoured the Normans, among whom he had been brought up. In 1045 he married the daughter of Earl Godwin of Wessex and six years later put down a rebellion by the earl. It will never be known for certain whether, as the Normans claimed, he promised the crown to Duke William before his death; the succession after his death of Harold, Earl Godwin's son, caused the Normans to take England by conquest. Edward was the founder of Westminster Abbey (1045), and had a great reputation for piety. He was canonized in 1161.

Edward the Elder (d. 924), King of Wessex (899–924). He was the eldest son of King *Alfred, whom he

Surrounded by his nobles and bishops, **Edward the Confessor** is anointed during his coronation at Winchester. This detail is from a 13th-century English manuscript. (Cambridge University Library)

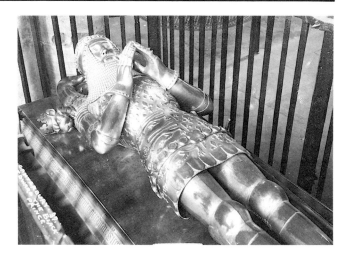

The tomb of **Edward the Black Prince** in Canterbury Cathedral. He had a contemporary reputation as an outstanding military leader, and was called the Black Prince in the 16th century for unknown reasons.

succeeded in 899. He continued his father's policy of repossessing the *Danelaw. A system of fortified towns was developed. A series of victories (909–18) secured the Midlands and the important towns of Derby, Leicester, Lincoln, Stamford, and Nottingham and convinced the Danes of the need to recognize English rule south of the Humber. Edward's authority was also acknowledged in southern Scotland.

Edward the Martyr, St (*c*.963–79), King of England (975–8). He succeeded his father Edgar as king, but his

c u serunse cuc me met   weuenc barmis englar   e unn dist fu id au fir marie

accession was disputed by his stepbrother *Ethelred II, and while visiting him and his stepmother Alfrida at Corfe Castle in Dorset, Edward was murdered. Miracles were reported at his tomb at Shaftesbury and Ethelred had to pronounce the date of Edward's death (18 March) a solemn festival. He was canonized and became the focus of a considerable medieval cult.

Edwards, Jonathan (1703–58), American Puritan divine, the scion of two distinguished ministerial families in the Connecticut valley. He was the leading Puritan intellectual of his generation and one of the pioneers of

Ancient Egypt

Period and Dynasty		Dates	Principal kings	
Archaic	1 & 2	c.3100–2700	Menes	Unification of Upper & Lower Egypt by Menes, the first pharaoh. Memphis founded. Royal Monuments at Saqqara and Abydos
Old Kingdom	3	c.2700–2600	Zoser	Zoser's Step Pyramid, the first monumental building in stone, constructed
	4	c.2600–2500	Snofru, Cheops (Khufu), Chephren (Khafre), Mycerinus (Menkaure)	The great age of pyramid building, including that of Khufu at Giza, the largest ever constructed (147 m./481 ft. high)
	5–6	c.2500–2200	Teti, Pepi I, Pepi II	Growth of power of provincial governors (nomarchs)
First Intermediate Period	7–10	c.2200–2050		Climatic deterioration leading to low inundations and famine. Collapse of central authority. Brief period of anarchy, followed by intermittent civil war between Heracleopolis (Dynasties 9 and 10) and Thebes (Dynasty 11). Thebes triumphant. Country reunited c.2050 by Mentuhotep II
Middle Kingdom	11	c.2050–2000	Mentuhotep II	
	12	c.2000–1750	Amenemhat I–III, Sesostris I–III	Golden age of art and craftsmanship. Curbing of nomarchs' power by Sesostris III. Conquest of Lower Nubia completed
Second Intermediate Period	13–17	c.1750–1550	Hyksos kings; Seqenenre, Kamose	Northern Egypt taken over by Asiatics, later called Hyksos (Dynasties 15 and 16). Thebans remain independent (Dynasty 17), eventually expel Hyksos and reunite the country
New Kingdom	18	c.1550–1300	Amenhotep I–III, Tuthmosis I–IV, Queen Hatshepsut, Akhenaten, Tutankhamun, Horemheb	Great expansion of Egyptian power (Tuthmosis III). Religious 'heresy' of Akhenaten. Egyptian civilization reaches its zenith under Amenhotep III
	19	c.1300–1200	Seti I, Ramesses II, Merenptah	Probable connection between the Exodus and Ramesses II. Ramesses fights the Hittites at Kadesh. Peace treaty later concluded between the two powers. Royal residence in the East Delta
	20	c.1200–1050	Ramesses III	Invasion by Sea Peoples thwarted by Ramesses III. Economic difficulties towards end of dynasty. First recorded strike in history. Tomb robbery trials
Third Intermediate Period	21–25	c.1050–650	Sheshonq I, Taharqa	High-priestly 'dynasty' at Thebes during Dynasty 21. Dynasty 22 kings of Libyan origin. Sheshonq I invades Palestine c.925. Nubians invade and take over Egypt c.750 (Dynasty 25). Assyrians invade c.655 and sack Thebes
Late Period	26–31	c.650–332	Psammeticus I, Nectanebo II	Cultural renaissance during Dynasty 26. Two periods of Persian domination (Dynasties 27 and 31). Last native pharaoh, Nectanebo II. Alexander the Great reaches Egypt, 332, and defeats the Persians

the *Great Awakening and of resistance to *Arminianism. His pastoral style and theological dogmatism led to his dismissal in 1750, and he died in 1758 shortly after becoming the president of Princeton University.

Egbert (or Ecgbert) (d. 839), King of Wessex (802–39). He was elected king after living in exile (789–802) at the court of *Charlemagne, having been expelled from England by *Offa. By 829 he had extended his authority over the other southern English kingdoms although Mercia regained its independence in 830. Egbert styled himself 'King of the English', but his authority was tested by the onset of the Danish raids towards the end of his reign. His victory at Hengist Down in Cornwall (837) provided only a brief respite from attack.

Egmont, Lamoral, Count of, Prince of Gavre (1522–68), Flemish statesman and soldier. He was made statholder (governor) of Flanders and Artois in 1559. Although he was a member of *Philip II of Spain's regency council, he opposed his sovereign's policy of imposing Catholicism and Spanish government on the Netherlands, and helped to oust Cardinal Granvelle from his pre-eminent position in the government of the Netherlands (1564). In 1565 he withdrew from the council with *Horn, but during the first phase of the *Dutch Revolts he vacillated, and refused to join *William the Silent in armed resistance. He was seized by *Alba in 1567, and beheaded on a charge of treason.

Egypt, a country in north-eastern Africa, the site of one of the first civilizations, together with Mesopotamia, of the Old World. Agriculture and metallurgy were both introduced from western Asia, and the great fertility of the Nile floodplain allowed the growth of a highly distinctive cultural tradition. Two kingdoms, one in the Delta (Lower Egypt) and one centred upstream round Thebes (Upper Egypt), were in existence during the 4th millennium BC. These were unified by the conquest of Lower Egypt some time shortly before 3000 BC, initiating the Protodynastic period. The shift of the capital to Memphis, near the head of the Delta, in the Old Kingdom (2700–2200 BC) perhaps indicates the importance of sea-borne trade with the Levant. The major pyramids were constructed here on the desert edge overlooking the river. A period of fragmentation (the first of two 'intermediate' periods) separated the Old from the Middle Kingdom (c.2050–1750), when some expansion into Palestine took place and the Nubian frontier was fortified. After a period of domination by foreign rulers (the *'Hyksos'), the New Kingdom (1550–1050) was a period of imperial expansion when Egypt fought the Asiatic powers for control of Palestine. It was punctuated by the Amarna Period when *Akhenaten founded a new capital and religion. Egypt suffered from attacks of marauding *Sea Peoples in the 12th century BC, but maintained continuity of tradition into the Late Period (c.650–332). However, its independence came to an end with its successive incorporation into Assyrian, Persian, and Hellenistic empires. When the Romans took it, Egypt was virtually self-governing. It was a granary for Rome, and its capital, Alexandria, became the world's chief commercial centre, when, c. AD 106, the sea route to India was opened.

Until 451 Alexandria was the intellectual centre of the Christian Church (*Copt). When the Arab armies

reached Egypt in 639, they had little difficulty in taking the country. Under Arab rule taxes were lighter, administration remained in local hands, and there was little pressure for conversion to Islam. The new capital of Misr, now Old Cairo, was the military base for the Arab conquest of North Africa. In the 9th century the *caliphate gradually weakened, and Ibn Tulun, a Turk, made it independent for a time. In 969 the *Fatimids seized the country, and built a new capital named al-Qahira, Cairo. Local administration continued with little change, and the country's prosperity is reflected in Fatimid art and architecture. In 1171 there followed the Fatimid dynasty of Saladin, and then the *Mamelukes, foreign slave rulers under whom Egypt had the most prosperous period in her history (1250–1517). Then, with the rest of North Africa and the Middle East, Egypt fell to *Ottoman Turkey, although Mamelukes still maintained much local power. In 1798 Napoleon invaded Egypt in an attempt to restrict British trade with the east, but was driven out by the Turkish and British armies in 1801.

El Cid (Campeador) (*c*.1040–99) (Arabic *al-Said*, 'the lord' and Spanish *campeador*, 'champion'), Spanish hero. He was Rodrigo Díaz de Bivar, a Castilian nobleman, who was exiled after the war between the brothers Sancho II of Castile and Alfonso VI of León, becoming a mercenary captain fighting mainly for the *Moors. He captured Valencia on his own behalf but was expelled in 1099, dying shortly afterwards. Many of the legends concerning him bear little relation to historical facts.

El Dorado (Spanish, 'gilded man'), the promised land of legend which may have some foundation in fact. When Spanish *conquistadores defeated the Muisca Indians of central Colombia in the 1530s they heard tales of *el indio dorado* from captives. According to the tales, when a new ruler was appointed at Lake Guatavita he was taken after a period of seclusion to the lake, stripped, covered with mud, then gold dust, and set on a raft laden with golden objects. Pushed out into the middle, he and companion chiefs offered the gifts to the waters. 'He' has since become a mythical person, city, or even kingdom.

Eleanor of Aquitaine, wife of Henry II and mother of Richard the Lionheart. In death, mother and son lie in effigy in the abbey church of Fontevrault, France.

Eleanor of Aquitaine (*c*.1122–1204), a duchess in her own right, who became Queen of France, and subsequently Queen of England. She married *Louis VII of France and accompanied him on the Second *Crusade (1147), but the marriage was annulled and in 1152 she married Henry, the Duke of Normandy and Count of Anjou. When he became the *Angevin king, *Henry II of England, the lands they claimed stretched from Scotland to the Mediterranean. Her ten children included Richard and John, future kings of England whose accession she acted zealously to ensure.

Eleanor of Castile (*c*.1244–90), Queen of England (1272–90). She was the daughter of Ferdinand III of Castile in Spain and married *Edward I of England in 1254. Eleanor bore thirteen children and accompanied her husband on Crusade (1270–3). After her death at Hadby in Nottinghamshire, Edward organized elaborate funeral rites. Her body was embalmed and taken to Westminster Abbey. At each of the ten overnight stopping places Edward ordered a stone cross to be erected to her memory, the 'Eleanor crosses'.

Elector, a prince of the *Holy Roman Empire who had the right to elect the emperor. Although the monarchy was elective by the 12th century, it was not until the contested election of 1257 that the number of Electors was fixed at seven. They were: the Count Palatine of the Rhine (Imperial Steward), the Margrave of Brandenburg (Chamberlain), the Duke of Saxony (Marshal), the King of Bohemia (Imperial Cupbearer), and the Archbishops of Mainz, Trier, and Cologne (Chancellors). Additional Electorates were later created for Bavaria (1623–1778), Hanover (1708), and Hesse-Kassel (1803). The Electors exercised considerable power at disputed successions by reason of their independence, though the imperial crown gradually became, in practice, hereditary in the *Habsburg family. The office of Elector disappeared when Napoleon abolished the empire in 1806.

Elijah (9th century BC), Hebrew prophet at the time of King *Ahab. His mission, as told in the Old Testament of the Bible, was to strengthen the worship of the God of the Israelites, to oppose the worship of all other gods, and to promote moral uprightness and social justice. He rebuked Ahab for his devotion to the fertility god Baal, worshipped by his wife Jezebel. He charged his successor, **Elisha**, with the destruction of the *Omri dynasty. Elisha became involved in court affairs, inspiring revolutions in Syria and Israel and, by anointing Jehu as King of Israel, instigated the downfall of the Omri.

Elizabeth Petrovna (1709–62), Empress of Russia (1741–62). She was the unmarried daughter of *Peter the Great, a beautiful and extravagant woman who seized the throne from the infant Ivan VI. She was more interested in social life and the arts than in affairs of state and government was conducted mainly by her ministers. The court became more westernized and the economy flourished. The nobility added to their privileges by increasing their power over the *serfs. Foreign affairs were managed by Count Bestuzhev until 1758. Russia increased its hold on Poland in the War of the *Polish Succession and the *Seven Years War. On her death Peter III immediately changed sides, thus making *Frederick the Great's ultimate victory possible.

Elizabeth I (1533–1603), Queen of England and Ireland (1558–1603). She was the only child of *Henry VIII and Anne *Boleyn. After her mother's downfall she was temporarily illegitimized, then imprisoned under her half-sister *Mary I for suspected implication in *Wyatt's Rebellion. She was well educated in the humanities, and, succeeding to the throne on Mary's death, she proved an industrious and intelligent monarch. The regime she established with the indispensable aid of William *Cecil enjoyed a considerable degree of popular support: the nation achieved stability and prosperity under her rule and enjoyed a 'golden age' of achievement in art, music, and literature.

She did not please everyone. Her *Anglican Church settlement (1559–63) offended Catholics and Puritans alike by its very moderation. Her refusal to marry and ensure the succession irritated certain Members of Parliament, as did her financial demands and her lengthy procrastination over the execution of *Mary, Queen of Scots. Abroad, meanwhile, her covert aid to the Dutch and French Protestants and her sponsorship of privateering against the Spanish helped to incite *Philip II of Spain to open warfare against England. Yet there was no religious warfare under her rule; *James I succeeded peacefully when she died; and the navy not only resisted the attempted invasion by the *Spanish Armada (1588), but maintained England's advantage in the expensive Anglo-Spanish war which continued until 1604. The problems presented by Puritans, Catholics, and Parliamentary opposition remained to be faced by James I, but by force of character and clever temporizing Elizabeth dominated a remarkably talented array of Englishmen.

El Salvador, the smallest Central American country. It was conquered by Pedro de Alvarado, a lieutenant of Hernan *Cortés. The region formed part of the viceroyalty of *New Spain, but was subject to the jurisdiction of the captain-general sitting in Guatemala City.

One of the less formal portraits of **Elizabeth I**, Queen of England, dancing with her favourite, the courtier Robert Dudley, Earl of Leicester, c.1580, artist unknown. (Penshurst Place, Kent)

emirate, a general term for a *Muslim territory ruled by an emir (Arabic *amir*, 'lord' or 'prince'), often uniting civil and military authority. Depending on the strength of the *caliphate, an emir might be either a diligent subordinate, subject to supervision and removal, as under the early *Abbasids, or a virtually independent princeling, able to defy his nominal master. In the latter case recognition of overlordship was signified merely by symbolic acts, such as acceptance of caliphal confirmation of their office, acknowledgement of his title in Friday congregational prayers, and use of his name and title on coinage. Some dispensed even with these formalities. The title *amir al-muminin*—'commander of the faithful'—was taken by *Umar and borne by all subsequent caliphs, and some monarchs claiming independent authority, such as the kings of Morocco. The term *amir* could also be applied to a specific office such as commander-in-chief of the armies (*amir al-umara*) or leader of the pilgrimage (*amir al-hajj*). It was also applied as a courtesy title to descendants of *Muhammad and is the origin, via medieval Italian, of the English title 'admiral'.

enclosure, an area of land formed as the result of enclosing (with fences, ditches, and hedges) what had usually been *common land so as to make it private property. Enclosures gradually transformed English farming from the medieval system of communally controlled open fields farmed in strips by the villagers, into a system of individually owned fields whose cropping and stocking were their owner's choice. Enclosures were created for different reasons at different times, and the reaction to the process depended on whether it was affecting valuable arable land or common wastes. In Tudor times enclosure was popularly seen as the conversion of the peasants' tilled land to grass on which a landowner's sheep would graze: the sheep were eating men, it was said, because the villagers were losing both their employment and their tillage. Enclosures became a national issue, but although they were denounced by the church (especially by Cardinal *Wolsey and Thomas *More) and were penalized by statutes and royal proclamations, and even provoked Kett's Rebellion (1549), their financial advantages were so strong that they continued to be carried out.

In the second half of the 18th century enclosure by private Act of Parliament increased dramatically, and the General Enclosure Act of 1801 standardized the procedure. Enclosures were less unpopular in the 18th century, as they enabled farmers to introduce improvements in crops and breeding without reference to their neighbours'.

Encyclopedists, the name given to the '*philosophes*' and others who contributed to and otherwise supported the *Encyclopédie*, published in France in thirty-five volumes between 1751 and 1780, one of the great literary achievements of the 18th century. It was a complete review of the arts and sciences of the day, explaining the new physics and cosmology and proclaiming a new philosophy of humanism. It was edited by *Diderot and d'Alembert and articles were contributed by *Voltaire, *Montesquieu, *Rousseau, Buffon, and baron d'Holbach. The strict censorship laws in France prevented direct attacks on church and state but these twin institutions were treated in the *Encyclopédie* with irony and disdain. A decree of 1752 suppressed the first volumes and in 1759

it was placed on the Index (of books forbidden to Roman Catholics), but it continued to circulate. The critical attitudes fostered by the *Encyclopédie* are believed to have contributed to the *French Revolution.

Enghien, Louis Antoine Henri de Bourbon-Condé, duc d' (1772–1804), French émigré, the only son of Henri, prince de Condé. He left France early in the Revolution. In 1792 he joined a force of exiled royalists (known as the *armée des émigrés* or *l'armée des princes*) and commanded them from 1796 until 1799. This was an army of princes who had left France after the Revolution, based at Worms, in Germany. Between 1793 and 1801 it was funded variously by Austria, Russia, and England, and fought against French Republican armies throughout Europe. This force was dissolved after the Peace of Lunéville in 1801 and he retired to Baden. Three years later he was wrongly accused by Napoleon of being involved in a plot to invade France, and he was kidnapped and shot. With his death, the House of Condé ended.

England, the largest political division within the United Kingdom. There were settlements in England from at least palaeolithic times, and considerable remains exist of neolithic and Bronze Age cultures. These were followed by the *Celts whose civilization spread over the whole country. The Romans under Julius *Caesar raided the south of Britain in 55 and 54 BC, but full-scale invasion did not take place until a century later; it was then ruled as a Roman province until the withdrawal of the last Roman garrison in the 5th century. In the 3rd to the 7th century *Angles, *Saxons, and *Jutes raided and settled, establishing independent kingdoms and when that of Wessex became dominant in the 9th century England emerged as a distinct political entity. From 1066 England under its Norman and *Plantagenet kings was closely linked to France.

The neighbouring principality of *Wales was conquered during the Middle Ages and politically incorporated in the 16th century. During the period of *Tudor rule England emerged as a Protestant state with a strong monarchy and as a naval power. In the 17th century the upheavals of the *English Civil War and the period of republican government under *Cromwell gave way to the *Restoration of Charles II and the invitation to *William III. Scotland (ruled from England since 1603) was united with England in 1707 (Act of *Union), and it was then that *Great Britain was created.

English Civil War (1642–9), the armed struggle between the supporters of the king (*Cavaliers) and Parliamentarians (*Roundheads), which erupted in 1642 and continued, with an interruption, until 1648. It arose from constitutional, religious, and economic differences between *Charles I and the Members of the *Long Parliament. Of these the most decisive factor was religion since the attempts of *Laud to impose liturgical uniformity had alienated substantial numbers of clergy, gentry, and craftsmen. All sections of society were affected, though many in the localities desired peace not war, and sometimes families were divided by conflicting allegiances.

The king's primary objective in 1642 was the capture of London, a Parliamentary stronghold. After an indecisive engagement at *Edgehill, he eventually had to take refuge in Oxford, which became his wartime capital. His plan in 1643 to bring together Cavalier armies from Oxford, Newcastle, and the south-west, followed by a march on London, was not realized. Meanwhile the balance was tipping toward the Roundheads, for by the *Solemn League and Covenant they secured Scottish assistance, of value in 1644 at *Marston Moor. Charles's attempt to march on London (1644) was frustrated at the battle of Newbury. With the formation of the *New Model Army, the Roundheads were able to inflict a crushing defeat on the Cavaliers at *Naseby (1645). Charles, having rejected terms previously offered him at the Uxbridge negotiations, eventually surrendered to the Scots near Newark (1646) after Oxford had fallen.

Charles's subsequent attempts to profit from divisions between the Parliamentary factions prevented a settlement from being reached in 1647. His escape to the Isle of Wight and 'Engagement' with the Scots sparked off the second phase of the war (1648). This consisted of unsuccessful Cavalier risings in Wales, Essex, and Kent, and a Scottish invasion which came to grief at *Preston. *Pride's Purge of Parliament then cleared the way for the trial and execution of the king and the establishment of the English *Commonwealth.

Enlightenment, the term used to describe the philosophical, scientific, and rational attitudes, the freedom from superstition, and the belief in religious tolerance of much of 18th century Europe, the 'Age of *Reason'. Its ideas were variously pursued in Germany, France, England, and Scotland. The lodges of the *Freemasons helped to disseminate the new ideas throughout Europe. In Germany the *Aufklarung* ('Enlightenment') which extended from the middle of the 17th century to the beginning of the 19th century was a literary and philosophical movement confined to the universities. The writings of Liebniz and *Kant argued for rationalism and religious tolerance in place of superstition and repression.

In France the Enlightenment was associated with the *philosophes*, the literary men, scientists, and thinkers who were united in their belief in the supremacy of reason and their desire to see practical change to combat

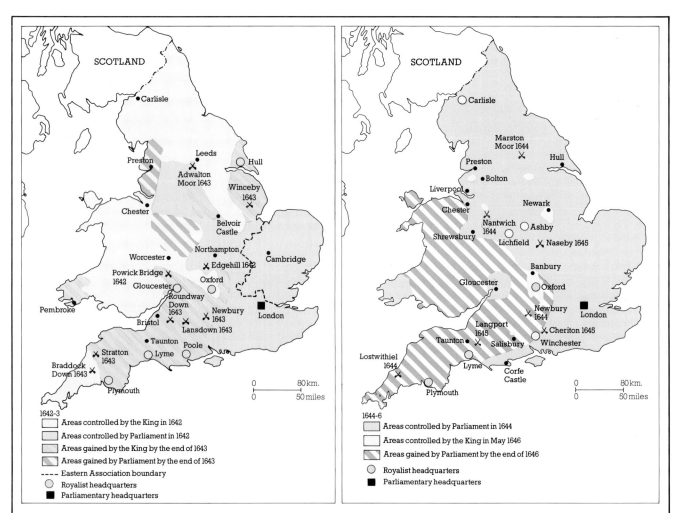

English Civil War (1642–9)

London and south-east England were of particular strategic importance in the English Civil War. Charles I was forced to abandon the capital before the outbreak of fighting in 1642 and his failure to regain it made the Royalist defeat inevitable. The Royalists won a succession of minor victories in the south-west in 1642–3, but by the end of 1644 the Parliamentary forces were in the ascendant, having gained most of the Royalist strongholds in the north. By 1646 the military conflict was essentially at an end. An important factor in Parliament's success was its control of the navy, which denied the King the seaborne reinforcements and foreign subsidies made vital by Parliamentary control of most government revenue.

inequality and injustice. The movement against established beliefs and institutions gained momentum throughout the 18th century under *Voltaire, *Rousseau, *Turgot, *Condorcet, and others. Many were imprisoned for their views but through the *Encyclopédie, their attacks on the government, the church, and the judiciary provided the intellectual basis for the *French Revolution.

The English Enlightenment owed its origin both to the political theories of *Locke, and to the French example. *Paine, an admirer of the French, advocated American independence, and many writers and poets transmitted Enlightenment ideas. In Scotland an intellectual movement flourished in Edinburgh between 1750 and 1800; Its outstanding philosophers were *Hume and Adam *Smith; important scientific advances were made in chemistry, geology, and medicine. The *Encyclopaedia Britannica* began in 1768–71 as a dictionary of the arts and sciences issued by a 'Society of gentlemen in Scotland'.

Ephesus, an ancient city on the west coast of Asia Minor. It was founded by the Ionians, and maintained its independence against the kingdom of Lydia until *Croesus captured it in the middle of the 6th century BC. He assisted in the building of a famous temple there to Artemis, one of the *Seven Wonders of the World, which was destroyed by fire in 356. The city was successively a member of the *Delian League and the *Athenian empire and fell under the overlordship of *Alexander the Great, the Seleucids, and Pergamum, before becoming Roman territory in 133 BC. By this time it had outstripped its rival *Miletus to become one of the most prosperous cities of the eastern Mediterranean. Excavations have revealed much of the city, including shops, streets, temples, and a magnificent theatre. In AD 431 the Council of Ephesus, summoned by Theodosius II, confirmed the Nicean Creed and rejected the doctrine of *Nestorius, who was excommunicated.

Epictetus (c. AD 50–135), Phrygian *Stoic philosopher. He was expelled from Rome c.90 AD when Emperor Domitian proscribed all philosophers. Although he wrote nothing himself, the lecture notes taken by the historian Arrian, survived him. He was contemporary with the rise of Christianity and his Stoicism, combined with a strong belief in one God, has many similarities with the teaching of *Jesus Christ. He insisted that trust in God was the only answer to the mysteries of pain, loss, and death. *Marcus Aurelius admired him greatly.

Epicurus (341–270 BC), the founder of the philosophical school of Epicureans. He was educated at the *Academy in Athens and established his own school there (307–306). His followers (including women and slaves) lived very modestly, but his desire for privacy and his hedonistic doctrine led to many accusations of a selfish pursuit of pleasure from rival philosophers. Epicureans believed that pleasure was the only worthwhile aim in life, but not that life should be an endless search for new pleasures. Rather pleasure was a state of being, with natural and necessary desires being satisfied. Epicureans sought freedom from disturbances, and chose to avoid the stresses associated with involvement in politics and public life, and any deep emotional attachments.

Epidaurus, an ancient Greek city-state in the north-east Peloponnese. It enjoyed close ties with nearby *Argos and was famous for its oracular sanctuary to Asclepius, the god of healing (early 4th century BC). This comprised a large temple housing a gold and ivory statue of the god, and other religious and secular buildings. Its Greek theatre, marvellously well preserved and with excellent acoustics, is still used.

Epirus, a coastal region in north-western Greece. It was famous in antiquity for its oracle at Dodona. The highpoint of its early history was the reign of the mercurial *Pyrrhus who considerably expanded and strengthened its territory. It fell foul of Rome after giving support to Macedonia, and in 167 BC 150,000 of its inhabitants were taken into slavery. Following the sacking of Constantinople by the knights of the Fourth Crusade in 1204, the despotate of Epirus was established by Byzantine Greeks, but in 1337 it returned to the re-established *Byzantine empire. In 1430 it fell to the Ottoman Turks.

Erasmus, Desiderius (c.1467–1536), Dutch *humanist scholar. He was the first major European figure whose fame and influence were based on the printed word. He began life in poverty-stricken obscurity, the illegitimate son of a priest, entered the Augustinian order, but later left his monastery, and travelled extensively in Europe. In 1516 he published his own edition of the Greek New Testament, followed by a Latin translation. This was to have enormous significance to European disciples of the *New Learning. Editions of early Christian authors followed, while his reputation was also enhanced by works like his *Adages* (1500) and *The Praise of Folly* (1511), a witty satire on monasticism and the Church, dedicated to his close friend, Thomas *More. Until 1521

A portrait of the Dutch humanist **Erasmus** by Hans Holbein the Younger, 1523. It depicts Erasmus, the most famous scholar in Europe in his day, in his study. (Musée du Louvre, Paris)

he moved throughout Europe, lecturing, debating, and writing letters to rulers and eminent men. He wished for peaceful, rational reform of the Church, and though he sympathized with Luther initially, he ultimately repudiated the Protestant Reformation. He retired to Basle, disillusioned by the sharpening religious conflict. In 1559 all his works were placed on the *Inquisition's Papal Index of prohibited books.

Erastus, Thomas (Thomas Lieber or Liebler) (1524-83), Swiss physician and Protestant theologian. He was appointed professor of medicine at Heidelberg University (1558), where he became closely associated with the introduction of reformed Protestantism into the Palatinate. He was a follower of *Zwingli, and he tried unsuccessfully to prevent the imposition of a Calvinist system of church government in Heidelberg in 1570. He was excommunicated for two years, and wrote his *Explication of the Gravest Question* at Basle (posthumously published in 1589). Later the term 'Erastian' was applied to those who wished to subordinate the interests and institutions of religion to the state, though Erastus himself never held such an extreme view.

Eric the Red, 10th-century Norwegian explorer. He left Iceland c.984 in search of new land to the west. He discovered *Greenland and returned the following year with a party of colonists from Iceland to found the settlement of Brattahlid. His son **Leif Ericsson** sailed westward from Greenland c.1001 and discovered land, which was probably Baffin Island, Labrador, and Newfoundland or New England; he named the latter *Vinland. He and his crew were the first Europeans to set foot on the American continent.

Ermine Street, a Roman road in Britain which led northwards from Londinium (London) to Eboracum (York) by way of Lindum Colonia (Lincoln). Partly built c. AD 60-70 for the advance north of the *Fosse Way, it followed some of the line of an earlier trackway. Its name came from the later Anglo-Saxons. A 'street' denoted a paved Roman road and it was linked with 'the people of Earna' (Earningas), a Germanic settlement leader.

Essex, Robert Devereux, 2nd Earl of (1567-1601), English courtier, favourite of *Elizabeth I. He distinguished himself as a soldier during the *Dutch Revolt (1586), but earned the queen's displeasure by participating in the disastrous Lisbon expedition (1589) and by marrying Sir Philip Sidney's widow (1590). The love-hate relationship between queen and courtier continued throughout the 1590s. He commanded an English contingent during the *French Wars of Religion (1591-2) and shared in the capture of Cadiz (1596). Gradually, his rivalry with the *Cecil faction grew. In 1599 Elizabeth sent him as Lord Lieutenant of Ireland to put down *Tyrone's rebellion. He failed ignominiously and was stripped of his offices. His subsequent attempt to raise the London people in an anti-Cecil coup (1601) led to his trial and execution for treason.

Essex, Robert Devereux, 3rd Earl of (1591-1646), English soldier, commander of the *Roundheads. Although he served *Charles I in 1625 he opposed him at the outbreak of the *English Civil War and in 1642 was appointed commander of the Roundhead forces, leading

This manuscript illustration from 18th-century **Ethiopia** depicts an Amharan ruler presiding over executions. (British Library, London)

them at the battle of Edgehill. After a number of Roundhead defeats, the *New Model Army was organized in 1645 and Essex resigned his command.

Estates-General *States-General.

Etaples, Treaty of (9 November 1492), a truce concluded between Charles VIII of France and Henry VII of England. The latter had revived claims from the *Hundred Years War and raised an army, but little fighting took place following an invasion and Henry was bought off in return for a sum of 745,000 gold crowns paid in annual instalments. Charles was left free to proceed with his planned invasion of Italy.

Ethelred I (d. 871), King of Wessex (866-71). His rule coincided with unremitting Danish raids that assumed the scale of an invasion. Assisted by his younger brother, *Alfred, Ethelred had some success against those Danes advancing into neighbouring Mercia (868), and into Wessex itself (870). However, three major battles, including a defeat near the Danish base at Reading and a notable victory at Ashdown, and numerous skirmishes, failed to give any advantage to Ethelred who died of wounds received in the battle at Merton.

Ethelred II (the Unready) (c.968-1016), King of England (978-1016). He succeeded his stepbrother *Edward the Martyr, who had been murdered on instructions from Ethelred's mother Alfrida. This inauspicious beginning to the reign was compounded by further blunders which earned Ethelred the title 'unready', meaning 'devoid of counsel'. Encouraged by his misfortunes the Danes renewed their invasions. Ethelred bought them off on five occasions (991, 994, 1002, 1007, 1012) with *Danegeld. His attempt (1002) to massacre all the Danes in his kingdom was answered (1013) by the invasion of the

King of Denmark, *Sweyn I, who ruled England until his death (1014) when Ethelred was restored.

Ethiopia, a country, formerly called Abyssinia, in north-eastern Africa. By the 2nd century AD the kingdom of *Axum had a brisk trade with Egypt, Syria, Arabia, and India in gold, ivory, and incense, and minted a gold currency. In the 4th century the court became Christian. Axum collapsed c.1000, and, after a time of confusion, the *Zagwe dynasty emerged. In 1270 it was replaced by the Solomonic dynasty claiming lineal descent from *Solomon and the Queen of Sheba, bringing the Amharas from the mountains of central Ethiopia to prominence. For Europe in the Middle Ages this was possibly the legendary kingdom of *Prester John. In the 16th century the Muslims of the lowlands attacked the Christian highlands, but were repulsed in 1542 with Portuguese artillery. When Jesuit missionaries came to Ethiopia Emperor Susenyos was converted to Roman Catholicism (1626). His son Fasilidas (1632–67), having forced him to abdicate, made Gondar the capital. Surrounded by Islam, and torn by warring factions, the empire foundered. The only unifying force was the Ethiopian *Coptic Church, and the empire was not reunited until 1855, when Emperor Tewodros II was crowned.

Etruscans, the inhabitants of ancient Etruria (approximating to modern Tuscany, Italy), west of the Apennines and the River Tiber. Twelve independent cities including Vulci, Clusium, and Cortona were formed into a league and came to dominate central Italy in the 7th and 6th centuries BC. Tradition held that they came from Asia Minor in the 10th century BC, though it is now believed that they were native to Italy before that and only culturally influenced by the Greek colonies of south Italy. In the 6th century BC they were driven out of southern central Italy by the Greeks, Latins, and *Samnites. In the following century their navy was defeated off Cumae. Traditionally, in 510 BC the last Etruscan king of Rome, *Tarquin, was expelled. In the 4th century they were driven out of Elba and Corsica, defeated by the Gauls in 390, and finally allied themselves

The **Etruscan** cemetery site of Cerveteri, although looted by tomb robbers, revealed a wealth of pottery and funerary furnishings. Inside the tomb, modelled on an Etruscan domestic interior, the pillars are decorated with carvings of household goods and offerings.

with Rome after defeat in 283. From this time they came under Rome's control and began to lose their unique cultural identity.

Their art is known from many tombs discovered north of Rome, and reveals an aristocratic society in which women enjoyed an emancipated style of life. The technical skill of their bronze and metal work and terracotta statuary is impressive. Their language has so far proved beyond translation; it was still spoken and written in the 1st century AD but no literature survives.

Eugène of Savoy (1663–1736), Prince of the House of Savoy. He was born in Paris; his mother, Olympe Mancini, was a niece of *Mazarin. When Vienna was besieged by the Turks in 1683 he entered the Austrian army and became one of the country's greatest generals. In 1697 he was given command of the Danube army and won a decisive victory over the Turks at Zente. In the War of the *Spanish Succession he was president of the Council of War, co-operated successfully with *Marlborough at *Blenheim and *Oudenarde, and won control of north Italy at the battle of Turin in 1706. In 1716–17 he led another successful campaign against the Turks and recovered Belgrade.

eunuch, a castrated human male. Eunuchs were used as guardians of harems in ancient China and in the Persian empire of the Achaemenids and also at the courts of the Byzantine emperors and the Ottoman sultans. They became the friends and advisers of the rulers of these powers, as they did of Roman emperors. Castration was also imposed as a form of punishment (*Abelard suffered in this way); was practised voluntarily by some Christian sects (the most notable Christian eunuch being the theologian Origen); and was used to produce male adult sopranos in Italy—castrati—until Pope Leo XIII banned the practice in 1878.

Europe, a continent of the northern hemisphere, which throughout history has exerted an influence disproportionate to its size. Its most important ancient civilizations developed in the Mediterranean region. Greek civilization reached its zenith between c.500 and c.300 BC, to be succeeded by that of *Rome. *Christianity became the official religion of the Roman empire in the late 4th century, shortly before the empire's western section succumbed to Germanic invaders. The eastern section lived on as the *Byzantine empire, centred on Constantinople, which eventually fell to the *Ottoman Turks in 1453.

During the *Middle Ages a politically fragmented Europe underwent varying degrees of invasion and colonization from *Moors, *Vikings,*Magyars, and others. The attempt of the powerful *Franks to re-establish the Western Roman empire soon failed, but the year 962 marked the foundation of what later became the *Holy Roman Empire. The *Roman Catholic Church became the unifying force throughout the continent; but in the wake of the *Renaissance, the 16th century bought about a religious schism (the *Reformation) in western Christendom and ushered in an era of national and international politico-religious warfare.

Post-medieval Europe was characterized by the rise of strong individual nation-states such as Spain, France, England, the Netherlands, and eventually Russia. Their influence on the rest of the world was the result of their acquisition of vast empires outside Europe. Imperial expansion continued through the age of European revolutions, of which the *French Revolution was the most momentous.

Evesham, battle of (4 August 1265). A crucial engagement in the second *Barons War (1264–5) when Prince Edward defeated Simon de Montfort and rescued

his father Henry III. Simon's headless corpse was buried in the abbey at Evesham which subsequently became a place of pilgrimage.

Evolution, human, the stages of development whereby humans diverged from ape-like ancestors and took on their present form. The process took at least 5 million years (*hominids, *australopithecines, *Homo habilis, *Homo

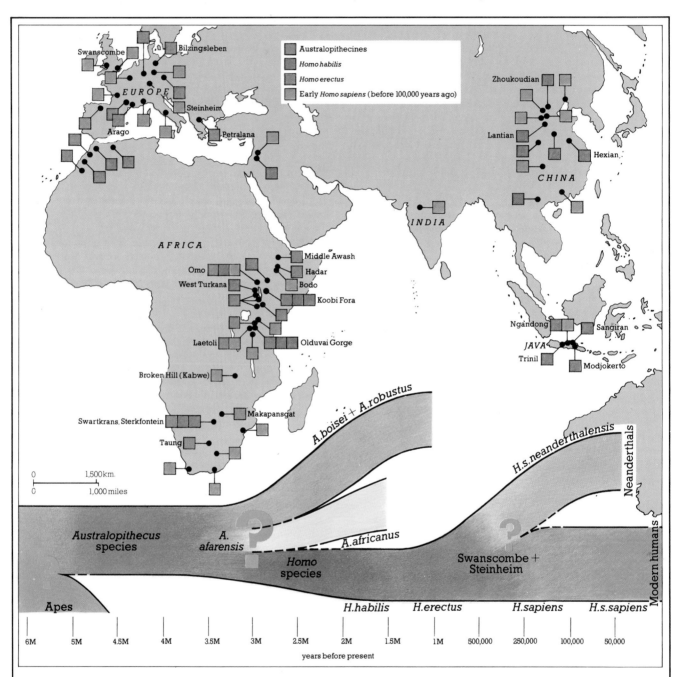

Human evolution

The major finds of early human (hominid) remains date from around 5 million to 100,000 years ago. The identity of some specimens of *Homo erectus*, especially those from Europe and North Africa, is controversial, some authorities preferring to call them early or 'archaic' representatives of our own species *Homo sapiens*. It now seems increasingly likely that our roots lie in Africa. *Homo erectus* had an upright posture in motion and developed the use of fire and tools. It was probably the

first human species to travel to new continents, but the routes taken by this and later species and the timing of the migrations remain unclear. Uncertainty also surrounds the relationship between the various forms of *Australopithecus* and between early and modern forms of *Homo sapiens*. The majority of Neanderthal finds have come from Europe and western Asia, but the origins and extinction of Neanderthals continue to be an enigma.

erectus, *Homo sapiens*, and *Neanderthals). Many details remain uncertain, particularly of the relationship between the australopithecines and the *Homo* lineage, and the position of such remains as *Broken Hill, and the later Neanderthals. However, the general outline is becoming clearer with every new discovery and as *dating methods become more refined.

Exchequer, a former English government department dealing with finance. The Normans created two departments dealing with finance. One was the Treasury, which received and paid out money on behalf of the monarch, the other was the Exchequer which was itself divided into two parts, lower and upper. The lower Exchequer was an office for receiving money and was connected to the Treasury; the upper Exchequer was a court of law dealing with cases related to revenue, and was merged with the High Court of Justice in 1880.

excise   *customs and excise.

Exclusion crisis, the attempt to exclude James, Duke of York, later *James II, from succeeding to the English throne. After the unmasking of the *Popish Plot the Whigs tried in three successive parliamentary sessions to force through a bill to alter the succession but all three attempts (1679, 1680, 1681) failed. The Whig opposition eventually triumphed at the *Glorious Revolution.

excommunication, the exclusion of an individual from membership and especially the sacraments of the Christian Church. The process was first used against individuals holding unorthodox or heretical religious beliefs, but it was later employed as a disciplinary and political weapon against rulers who opposed the church and especially the papacy; Pope Adrian IV was one of the first to use it in this way. It could include releasing subjects from their duty to obey their lord which could seriously threaten a weak king. King John of England was punished in this way, as was the Holy Roman Emperor Henry IV who finally submitted to Pope Gregory VII at *Canossa. Its effectiveness as a weapon depended on the recipient's willingness to be frightened by it, which is why it was frequently employed in the medieval period when the majority of the populace was greatly concerned with its spiritual welfare. The interdict was a less severe punishment that was also used against the laity.

Exile, the captivity of the Jews in *Babylon (the 'Babylonian Captivity'). In 597 BC the Babylonians captured *Jerusalem and took King Jehoiachin and many leaders of the Judaean community, including the prophet *Ezekiel, into exile in Babylon. Following further revolt, they again attacked Jerusalem and, after a three-year siege captured and destroyed it in 586 BC. Many of those taken to Babylon were settled in communities, with the result that distinctive Jewish teaching, religion, and life could continue. In 539 BC Babylon fell to Persia and one year later *Cyrus the Great gave permission for Jews who wished to do so to return home. The number returning was probably small and the return protracted over a long time.

Exodus, the departure of the Israelites under *Moses from their captivity in Egypt *c*.1300 BC, recorded in the

The episode in **Exodus** of the parting of the Reed Sea to allow passage to the fleeing Israelites, from a 5th-century mosaic depicting the life of Moses in the church of Santa Maria Maggiore, Rome.

Old Testament book of Exodus. According to the biblical account, the Israelites were pursued by the pharaoh's army, but were saved by a tidal wave that swept across a region known as the Reed Sea (probably near one of the lakes now joined by the Suez Canal). The Israelites then spent over forty years wandering in the wilderness of Sinai, during which time they received through *Moses the Ten Commandments which established their relationship with their God and between one another. After the death of Moses, Joshua became their leader, and his capture of Jericho led to the occcupation of *Canaan. The variety of sources make it impossible to regard this narrative as a straightforward historical account, but it is central to Jewish history as evidence of God's favour to his chosen people and is commemorated annually in the Passover feast.

exploration, the investigation of undiscovered territories. For the first explorers, travel was easier by sea than by land: *Phoenician traders frequently sailed to Galicia (in Spain) and Brittany, and perhaps even to Cornwall.
    In the early Middle Ages the *Vikings sailed as far west as Greenland and to North America (*Eric the Red) and curiosity about the 'marvels of the east' led the Italian *Marco Polo overland to China (1271-95). The Chinese Ming emperors supported the seven voyages of discovery of *Zheng He (1405-33). Under the patronage of Prince *Henry the Navigator, the Portuguese in the 15th century sailed to the Indian Ocean, and in 1498 Vasco *da Gama crossed the South Atlantic; in 1516 the Portuguese reached China. Probably with the aim of reaching the East, Christopher *Columbus crossed the North Atlantic (1492) and Ferdinand *Magellan found the strait which enabled him to reach the Chinese coast (1521). The North American landmass was such a deterrent that searches were long made for a navigable

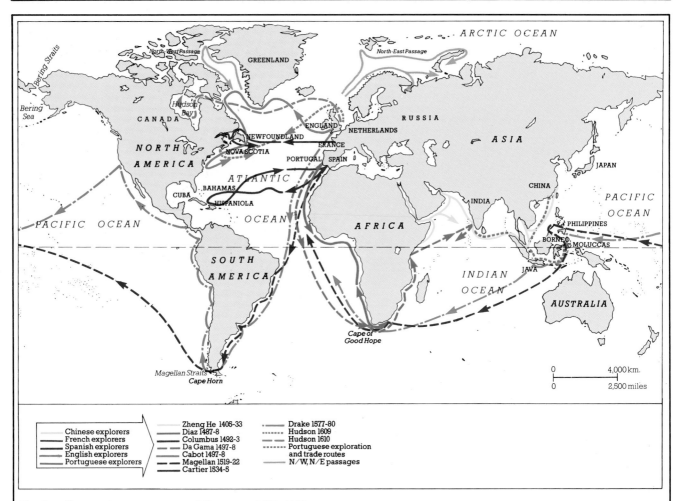

Chinese explorers — Zheng He 1405-33 — Drake 1577-80
French explorers — Diaz 1487-8 — Hudson 1609
Spanish explorers — Columbus 1492-3 — Hudson 1610
English explorers — Da Gama 1497-8 — Portuguese exploration and trade routes
Portuguese explorers — Cabot 1497-8 — N/W, N/E passages
Magellan 1519-22
Cartier 1534-5

Exploration: major sea voyages of discovery 1405–1616

In the 15th century Europeans' knowledge of the world was restricted to their own continent, part of Asia, and the northern fringes of Africa. The Chinese admiral Zheng He made seven voyages of discovery across the Indian Ocean between 1405 and 1433, visiting East Africa before the Portuguese, but Chinese exploration ceased on his death. Driven by the need to find trade routes to the east, European navigators embarked on a series of exploratory voyages which led to the discovery that the oceans of the southern hemisphere were connected. Southerly routes to the east were found relatively quickly but they proved too long to be practical and much effort was spent in the vain attempt to locate a north-west or north-east passage.

passage to the north of it: the search for a north-west passage led *Cabot to what was probably Hudson Bay (1509) and *Cartier along the St Lawrence River (1534-41). North America's interior began to be explored in the 17th century, but it was not until c.1730 that the Rocky Mountains were discovered, and the continent was not crossed until 1793, when Mackenzie traversed Canada. The United States was first crossed by Lewis and Clark, in 1803. Scientific research helped Captain James *Cook secure backing for his voyages to New Zealand and eastern Australia (1769-77).

Ezekiel (6th century BC), Hebrew prophet. He was a priest who was taken into *Exile in Babylon in 597 BC. He denounced religious apostasy and idolatry and prophesized the forthcoming destruction of Jerusalem and the Jewish nation. As soon as Jerusalem fell, in 586 BC, he began to prophesy that God would restore the nation and that the Israelites would begin life back in their land in a new relationship with their God.

Ezra (5th or 4th century BC), Jewish priest instrumental in reforming Judaism after the *Exile in Babylon. Although the Biblical record is unclear, it is thought that he arrived in Jerusalem in 397 BC with authority from the Persian king Artaxerxes II. He reformed the system of worship at the Jerusalem Temple and established a written code of laws and the priestly leadership of Judaism. His work was facilitated by the political successes of *Nehemiah.

F

Fabius, Quintus, Maximus Verrucosus Cunctator ('the Delayer'), (d. 203 BC), Roman general and consul five times between 233 and 209. He was appointed dictator in 221 and again for a second time in 217 after the battle of Trasimene, during the Second *Punic War. Appreciating that the Carthaginian forces were superior to his own, he declined to engage in pitched battles. His unspectacular tactics of slow harrassment against Hannibal's army in Italy at first won little popular support, and the nickname Cunctator was intended as an insult. After the defeat at Cannae (216) the feeling against his strategy waned and the insult became a title of approval. He opposed *Scipio's aggressive war against Carthage on the African mainland. ('Fabian' Socialists are those who seek change by slow democratic means.)

fair (Latin *feriae*, 'holiday'), originally a short-term market, set up at regular intervals. Fairs were held at religious festivals in China (12th century), Greece (including the Olympic Games), and in Rome. The fair of St Denis, near Paris, was probably the first in Western Europe (629). With Europe's economic recovery in the 10th century fairs proliferated—in Italy (Pisa, Venice, Genoa), Flanders (Bruges, Ypres), Germany (Cologne, Leipzig), Russia (Nijni Novgorod), England (Boston, Stourbridge, St Bartholomew's in London). The most important, however, were in Champagne and Brie in France where they were held seasonally at Lagny-sur-Marne (January), Bar-sur-Aube (Lent), St Quirface in Provins (May), Troyes (June and October), and St Ayoul in Provins (September). Merchants from the Middle East and Africa attended these fairs which lasted up to six weeks. Although international trading at fairs generally declined from the 14th century as merchants left long-distance trading to their agents, many continued

St Bartholomew's **fair**, in Smithfield, London, an aquatint of 1728–30, designed to be made into a fan. Booths and sideshows feature a rope dancer; Faux, a conjuror; actors Lee and Harper presenting *Judith and Holophernes*; and an early ferris wheel. (Museum of London)

at such towns as Milan, Frankfurt-on-Main, Brussels, and Paris. Specialist fairs in England were held at Horncastle in Lincolnshire (horses), and at Ipswich in Suffolk (sheep).

Fairfax, Thomas, 3rd Baron Fairfax of Cameron (1612–71), English Parliamentary general. He was largely responsible for the defeat of the Royalists in the *English Civil War. A Puritan, he rose from commander of the Yorkshire cavalry (1642) to commander-in-chief of the *New Model Army (1645). He proved to be heroically brave, and popular with his men. His victories included the battles of Marston Moor (1644) and *Naseby (1645). He disassociated himself from the decision to execute *Charles I, and resigned his command in 1650, rather than lead the campaign against Royalist resistance into Scotland.

Falkirk, battles of. There have been two battles fought at Falkirk, a town 16 km. (10 miles) from Stirling, in Scotland. The first (22 July 1298) resulted in victory for *Edward I of England over Sir William *Wallace, leader of the Scottish resistance to English sovereignty. The second (17 January 1746) was a victory for the Jacobite army of Prince Charles Edward Stuart over the royalist forces, in the *Forty-Five Rebellion.

Falkland Islands (Spanish, 'Islas Malvinas'), a group of islands in the South Atlantic. They have experienced a complicated diplomatic history since their discovery in the late 16th century, having been claimed at various times by the Spanish, the British, the Argentines, and the French. They were first occupied in 1764 when French settlers began grazing sheep there. Within a year the French were dislodged by the British who claimed the islands on the basis of their discovery in 1592 by Captain John Davis and an expedition a century later under the command of Captain John Strong. When the French were driven out they sold their rights to Spain, and conflict continued over their possession.

farmers-general, a group of some forty to sixty financiers in 18th century France. They bought from the crown the right of collecting indirect taxes on wine, tobacco, and salt, a practice known as 'farming' taxes. Employing inspectors to collect the money, they retained the difference between what they paid the crown for this right and what they actually extorted. The salt tax

(*gabelle*) was especially harsh. The system was abolished in the French Revolution.

Farnese, an Italian family which ruled the duchy of Parma from 1545 to 1731. Originating in the 11th century, its first outstanding member was **Alessandro** (1468–1549), who became Pope Paul III in 1534 and created the duchy of Parma and Piacenza. His grandson **Alessandro** (1520–89) was named a cardinal at the age of 14, and remained a powerful figure at the papal court for fifty years; he was a noted patron of the arts.

His nephew **Alessandro** (1545–92), Duke of Parma from 1586, was the family's most distinguished scion. After serving against the Ottomans at the battle of *Lepanto (1571), he succeeded Don *John of Austria as governor-general of the Netherlands and commander-in-chief of the Spanish forces which were dealing with the *Dutch Revolts (1578). By subtle diplomacy he detached the southern provinces from the revolt (1579). Then he conducted a sequence of superbly planned military campaigns further north, including the capture of Antwerp (1585). In 1588 *Philip II diverted him from his campaigns in the north, ordering him to liaise with the *Spanish Armada. In 1590 he was diverted again, this time to intervene in the *French Wars of Religion, where he managed to relieve Paris (1590) and Rouen (1592), but was wounded and died.

fasces, bundles of rods bound with thongs which were the sign of regal or magisterial authority both within and outside Rome. After the expulsion of the *Etruscan kings, consuls had twelve fasces (a dictator twenty-four),

A Roman relief depicting lictors, or attendants, bearing **fasces**, the symbol of authority. The bundle of rods which formed the fasces were made from elm or birch. Tied with a red strap, they were held over the left shoulder. One lictor is shown with the axe head that would originally have been carried inside the rods. (Museo Nazionale Concordiese, Portoguaro)

praetors six, lesser magistrates fewer. Originally axes were included in the bundle; but from the early republic the axe was removed in Rome, in deference to the People's ultimate power in capital cases.

Fatimid, the dynasty which reigned in Morocco from 909 until 969, and in Egypt until 1171. They claimed descent from *Muhammad's daughter Fatima as proof of their right to be caliphs. The founder, Ubaidallah, claimed to be Mahdi, the divinely guided one, and preached an extreme form of *Shiism. He and his successors steadily conquered north-west Africa and finally Egypt, Syria, and western Arabia in 969, abandoning Kairouan for a new capital of al-Qahira (Cairo). The Fatimids were commercially successful, and had generally good relations with their neighbours, but they suffered decline, partly because of the rise of the *Seljuks and the *Crusaders, and partly because from the mid-12th century they no longer had rulers of any ability.

Fawkes, Guy   *Gunpowder Plot.

Federalist Party, American political party. The first political party to emerge after the *Constitution became operative (1789), it took its name from the *Federalist Papers*, a collection of essays written by *Madison, *Hamilton, and *Jay to influence the ratification of the Constitution by New York. The party of George *Washington and John *Adams, it had support in New England and the north-east generally, both from commercial interests and wealthier landowners. It stood for strong central government and the firm enforcement of domestic laws, was pro-British in foreign affairs, and identified itself with the economic policies of Hamilton. The party's role, which would benefit 'the wise, the good, and the rich', was exemplified in the military campaign in 1793 against the refusal of the *Whisky Rebels to pay excise duty. The emergence of new political issues, disagreements over commercial and foreign policy, and the narrowness of its popular appeal gradually undermined the Party, although it continued to elect members to Congress until it finally disappeared in 1825.

Ferdinand II (1578–1637), Holy Roman Emperor (1619–37), King of Bohemia (1617–27) and Hungary (1618–26). He was educated by the Jesuits and developed into a determined spokesman for *Counter-Reformation Catholicism. Before his election to the imperial throne, he used authoritarian measures against the Protestants of Inner Austria, with some success, but in 1619 the largely Protestant Bohemian Diet deposed him in favour of *Frederick V (the Winter King). This crisis was one of the opening moves in the catastrophic *Thirty Years War. The first ten years of the conflict did not go badly for Ferdinand. He reached his high point when he issued the Edict of Restitution (1629), which ordered the return of all Roman Catholic property seized since 1552. Subsequently he was seen as a threat to German liberty and opposed by both Catholic and Protestant princes. The interventions of Sweden and France finally turned the tide of the war against him, and he was forced to abandon his more extreme Catholic absolutist ambitions.

Ferdinand V (the Catholic) (1452–1516), King of Castile and León (1474–1516), King of Aragon as

Ferdinand II (1479–1516), King of Sicily (1468–1516), and King of Naples (1502–16). He was the son of John II of Aragon. In 1469 he married Princess *Isabella of Castile, a significant step towards Spanish unification. He succeeded to the throne of Aragon in 1479, and in the same year helped Isabella to win the war of succession in Castile (1474–9). They began to rule jointly in both kingdoms in 1481, and in 1492 annexed the conquered territory of Granada to Castile. On Isabella's death in 1504, he was recognized as Regent of Castile for his daughter *Joanna the Mad. He subsequently married Germaine de Foix (1506), and incorporated Navarre into Castile (1515), thus becoming personal monarch of all Spain from the Pyrenees to Gibraltar.

A ruthlessly realistic politician, he was especially successful in the conduct of foreign policy. He surrounded France with a network of allies and acquired Naples. At home he modernized Spain's governmental institutions, vested in himself the grand masterships of the wealthy military orders, and won important ecclesiastical concessions from the papacy, including the Bull of 1478 authorizing the *Spanish Inquisition, a powerful council to combat heresy, to be controlled by the crown.

Fertile Crescent, the relatively well-watered area extending from the head of the Persian Gulf via the rivers Tigris and Euphrates westward into northern Syria and then southwards through the *Levant to the lower Nile valley. Western scholars have traditionally regarded it as the cradle of civilization, where empires, such as those of

*Babylon, *Assyria, and *Egypt created monumental buildings, pyramids, and ziggurats, developed sophisticated craft-skills in pottery, weaving, and metal working, and perfected writing-systems for administrative and ritual purposes. Modern archaeological excavations and scientific techniques such as carbon-dating confirm the existence of settled agricultural communities as far back as 8000 BC. The significance of the region for western thinking is confirmed by its associations with Judaism, the achievements of *Hellenistic culture in late antiquity, and the emergence of Christianity. The *Arab conquests claimed the entire region permanently for Islam and it became therefore the major focus for *Crusader enterprise before passing under *Ottoman control.

Feudal system, a medieval European political and economic system based on the holding of lands on condition of homage or military service and labour. Feudalism probably originated in the Frankish kingdom in the 8th century and spread into northern Italy, Spain, and Germany. It was introduced by the *Normans into England, Ireland, Scotland, southern Italy, and Sicily. The nobility held lands from the crown and provided troops for the king in times of war. The *knight was the tenant of the noble and a class of unfree peasants (*villein) lived on the land under the jurisdiction of their lord (*manorial system). Bishops and abbots were invested by secular lords with their livings in return for services and the church received produce and labour from the peasantry. It became a varied and complex system: lords

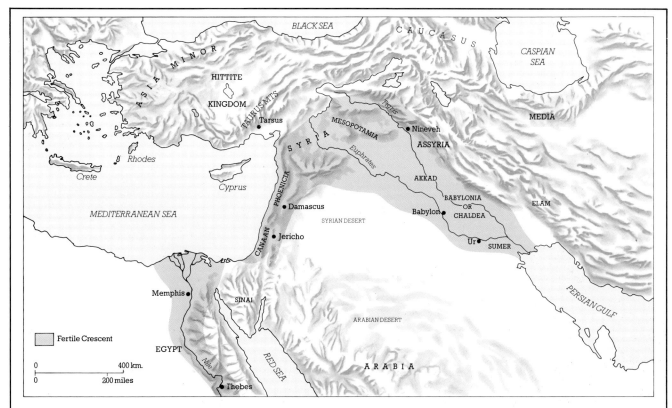

Fertile Crescent

Bounded by mountains in the north and desert in the south, by the Mediterranean in the west and the Persian Gulf in the east, the Fertile Crescent is interrupted by desert in Mesopotamia, Syria, and Sinai. Archaeological research suggests that the development of primitive irrigation, made necessary by the dry summers, probably preceded the rise of the ancient civilizations of Babylon, Assyria, and Egypt.

built up their own military forces and power to the point where they became semi-independent of the king; from the 12th century payments (*scutage) could be substituted for military duties. The system broke down in the 12th and 13th centuries as towns (*commune) and individuals achieved independence from their lords, though *serfdom survived in some countries for much longer.

fief, the land held under the *feudal system by a *vassal from his lord. Fiefs ranged in size from vast duchies down to the area of land needed to support a single knight, called a knight's 'fee'. Large or small, they provided the agricultural produce which was the source of all wealth. During the early Middle Ages areas which had been forest or barren land came under cultivation and were incorporated into the system.

Field of the Cloth of Gold, the name given to the site near Calais, where *Henry VIII of England met *Francis I of France in June 1520, in an attempt to forge a diplomatic alliance. Henry's retinue was made up of more than 5,000 people. He wrestled, danced, jousted, and tilted with Francis for almost a fortnight, amid scenes of great festivity and pageantry. But the two sovereigns retained their initial mutual suspicions, and within days of leaving Francis, Henry met the French king's arch-rival, Emperor Charles V, at Gravelines.

field systems, the visible traces on the landscape of man's present or former agricultural activities. Simple agriculture by digging stick or hoe will leave virtually no trace, and only after prolonged use of the plough, which was available from at least the 4th millennium BC in south-west Asia and Europe, or by deliberate

The meeting at the **Field of the Cloth of Gold** of the kings of England and France in June 1520 was a magnificent formal occasion of state, as this painting by an unknown artist shows. (Royal Collection)

construction will such fields remain visible. Examples are as widely separated as the 'Celtic' fields and lynchets in north-west Europe, where the combined effects of ploughing and gravity on sloping ground transfers soil from the upper to the lower edge of each field, to ancient cultivation plots, levelled and drained, in the forests of Amazonia or Central America. Deliberate terracing leaves a much clearer mark, and is more likely to be maintained in working order since it represents a massive investment of human effort. Andean slopes in South America, vineyards in central and southern Europe, and rice paddies in south-east Asia are major examples still in use. When terraces are abandoned, traces will be largely obliterated over time by different farming practices or natural erosion.

Fifteen, the (1715), a *Jacobite rebellion aimed at removing the Hanoverian *George I from the British throne. Queen *Anne's sudden death in August 1714 had caught the Jacobites by surprise. Their lack of preparedness and their inability to win the English Tories to their cause delayed the rebellion until September 1715. A simultaneous rising in Scotland and England was planned, with Thomas Forster to lead the English northern rebels and the indecisive Earl of Mar to command the Scots. Forster was compelled to surrender his small force at Preston and Mar's inconclusive battle at *Sheriffmuir virtually ended the rebellion.

Many Englishmen may have sympathized with the Jacobite cause, but few were willing to fight for it. Foreign help had been promised, but *Louis XIV's death, storms in the English Channel, and the vigour of the English government doomed the rebellion to failure. James Edward Stuart, the Old *Pretender, arrived in Scotland too late to revive it. By the standards of the day the British government was comparatively lenient towards the rebels: 26 officers suffered the death penalty, and about 700 of the rank and file were sent to the West Indies to serve seven years as indentured servants. Forster

THE PICTVRE OF Y MOST FAMOVS CITY OF LONDON AS IT APPEARED IN Y NIGHT IN THE HEIGHT OF ITS RVINOVS CONDITION BY FIRE

THE   RIVER   THAMES   PARTE OF SOVTHWARK

The **Fire of London** of 1666. A contemporary engraving shows St Paul's Cathedral in flames, and around it many of the old city churches. London Bridge offered the fleeing residents an escape southwards, while many more took to the River Thames in boats and barges. (Society of Antiquaries of London)

escaped from Newgate prison and fled to the Continent, and Mar also died in exile.

Fifth-Monarchy Men, an extreme Puritan sect in England in the mid-17th century. They believed that the rule of Jesus Christ and his saints was imminent, and that it would be the fifth monarchy to rule the world, succeeding those of Assyria, Persia, Greece, and Rome. They hoped that, through the *Barebones Parliament, the rule of the saints would become a reality, but *Cromwell's establishment of the Protectorate turned them against him. Their agitation became a nuisance, their leaders were arrested, and their abortive rebellions in 1657 and 1661 were suppressed.

Finland, a Baltic country, sometimes considered part of *Scandinavia. Occupied between AD 100 and 800 by Finno-Ugrian tribes who drove the original Lapp population into its northernmost regions, Finland was conquered and converted to Christianity by Eric IX of Sweden in the late 1150s, and throughout the Middle Ages found itself at the centre of Swedish–Russian rivalry in the Baltic area. In 1556 *Gustavus Vasa made Finland into a separate duchy for his second son John, and following the latter's succession to the Swedish throne as John III in 1568 it was elevated to a grand duchy. Although still dominated by Sweden, Finland was allowed its own Diet and granted a degree of autonomy. In 1809 Finland was ceded to Russia.

Fire of London, a major fire which devastated London in September 1666. The fire began in a baker's shop and, fanned by an east wind, raged for four days, destroying 87 churches, including St Paul's, and more than 13,000 houses mostly build from wood. It was stopped by blowing up buildings in its path. There are eyewitness accounts in the diaries of *Pepys and Evelyn. Plans for a modern city with wide streets and squares were abandoned, but Sir Christopher Wren rebuilt St Paul's

and a number of other churches and public buildings, and designed the Monument (1677) which commemorates the fire. The fire destroyed the slums where the *Great Plague still lingered, and gave an enormous boost to the growth of fire insurance.

five members of Parliament: Charles I attempted to arrest five members of the English Parliament (4 January 1642). Mounting opposition to the king, culminating in the *Grand Remonstrance (December 1641), had been led by *Pym, supported by *Hampden, Holles, Hesilrige, and Strode. Charles rashly decided to arrest and impeach them. He entered the House of Commons only to find that the members had fled into the City of London; they returned to the House of Commons a week later. This attempted use of force by the king hardened Parliamentary opposition against him and was a factor leading to the outbreak of the *English Civil War. No other monarch has ever set foot in the House of Commons; its independence from interference is fundamental to its existence.

Five Nations   *Iroquois.

flagellant, a religious fanatic who scourged himself in public processions, often to the accompaniment of psalms. Such penitential activities took place in ancient *Sparta and *Rome and subsequently throughout the Christian world from about the 4th century. Flagellants appeared periodically in medieval Europe, usually during times of disorder or natural disaster. Major demonstrations came in Perugia in 1260 during political unrest; in 1349 as hysterical reaction to the *Black Death; and in Germany in 1414. Linked, especially in 1349, with anti-Semitism, these demonstrations were critical of the church establishment, and were condemned by the papacy.

Flanders, historically a region in the south-west of the Low Countries, now part of Belgium. It was taken from Celtic tribes by the *Franks in the 5th century. In the medieval period it became extremely prosperous, especially through its cloth trade with England, and it allied itself with the English during the *Hundred Years War. By 1384 it was a French province, later belonging to *Burgundy and then coming under *Habsburg control. In 1506 it passed to the Spanish crown. After a long

Dyers of medieval **Flanders** at work making the richly coloured cloth for which the area was famous. This detail from a 15th-century manuscript shows the dye being heated in a large wooden vat, ready for prepared textiles to be steeped in the liquid. (British Library, London)

period of revolt Philip IV of Spain was obliged to cede the north-western portion to the *United Provinces, and *Louis XIV of France acquired the southern part by the treaty of Aix-la-Chapelle in 1668.

Fleury, André Hercule de (1653-1743), cardinal and French statesman. He was tutor to *Louis XV of France and in 1726 became his chief minister in all but name. His policies were similar to those of *Walpole: financial retrenchment, encouragement of trade, and peace instead of war. However, he could not prevent France's involvement in the wars of the *Polish and *Austrian Succession, though he limited its intervention in the former. He began the negotiations which led to France peacefully annexing Lorraine.

Flodden Field, battle of (9 September 1513). An important battle between the English and the Scots took place on the Scottish border and resulted in a major English victory that gave *Henry VII security in the north for many years. A large army led by *James IV was defeated by a somewhat smaller force of about 20,000 English soldiers under the command of Thomas Howard, Earl of Surrey. Long Scottish spears were no match for English bills and longbows, and the Scots lost perhaps as many as 10,000 dead, including James IV himself.

Florence, a former city-state in northern Italy. It was founded as a colony for Roman veterans in the 1st century BC, and after the barbarian invasions of the 5th and 6th centuries passed to the *Carolingians in the 8th

century. A bishopric, and, by the 11th century, a *commune, it profited from economic boom in the 12th century to become a great banking city, as well as a centre of the cloth trade. In the 13th century it witnessed the rivalry between the papal and imperial factions (*Guelphs and Ghibellines). It was ruled by the Medici family from 1421 to 1737, and enjoyed its 'golden age' (1400-1520) as a centre of the arts during the *Renaissance. It then passed to the *Habsburgs. There was a brief republican interlude under *Savonarola and the *Medici established a republic in 1527.

Florida, the southernmost state of the US east coast. It was discovered in 1513 by *Ponce de León and first settled by the Spanish as St Augustine in 1565. Under Spanish control until 1763, Florida embraced much of the Gulf and southern Atlantic coastal regions. It reverted to Spain after British rule during 1763-83, and had troubled relations with the United States until its sale in 1819 under the Adams-Onis treaty.

Fludd, Robert (1574-1637), English physician and mystic. He belonged to the school of medieval mystics which claimed to possess the key to universal science. He is probably best remembered for his defence of the *Rosicrucians, especially from the suspicions of theologians, in his *Brief Apology for the Fraternity of the Rosy Cross* (1616). Many contemporaries believed he was a magician, and attacked him for his occult beliefs.

Fontenoy, battle of (11 May 1745). A village in Hainaut in south-west Belgium was the scene of a French victory in the War of the *Austrian Succession. *Saxe, the French commander, had overrun the Austrian Netherlands. The British general, the Duke of *Cumberland, was advancing to the relief of Tournai with a British, Austrian, and Dutch force; he found a gap in the French line of fortifications but was driven back.

Formigny, battle of (15 April 1450), a battle in the *Hundred Years War. English forces were intercepted by French troops on their way to reinforce the garrison at Caen. Despite successes won by their archers, the English were overcome when French reinforcements arrived. This French victory led to the fall of Caen two months later, and the English loss of *Normandy soon after.

Fort Stanwix, a colonial American military stronghold. Named after General John Stanwix, the fort was an important defence point and trading centre located between the Upper Mohawk River and Wood Creek. It fell into disrepair after 1763, but was rebuilt at the beginning of the War of Independence. Fort Stanwix is chiefly remembered as the site of the signing of two treaties with Indian tribes in 1768 and 1784, in the second of which the *Iroquois ceded their territory in Pennsylvania to the US government.

Forty-Five, the (1745), a *Jacobite rebellion in England and Scotland. Its aim was the removal of the Hanoverian *George II from the throne and his replacement by James Edward Stuart, the Old *Pretender. Jacobite hopes centred on the facts that Britain was heavily engaged in the War of the *Austrian Succession, and that the Hanoverians had never been popular. The

Pretender sent his 25-year-old son Charles Edward (Bonnie Prince Charlie, the Young Pretender) to represent him. Most of Scotland was soon overrun and the Jacobite victory at *Prestonpans was followed by the invasion of England. But the English armies of General Wade and the Duke of *Cumberland were closing in and, without any significant numbers of English recruits, Charles was advised by his commanders to return to Scotland. The Jacobites turned back at Derby when barely 160 km. (100 miles) from London, where panic at their advance had caused a run on the Bank of England. The decision to retreat meant that the rebellion was doomed. The last Jacobite army was routed at the battle of *Culloden, which ended any serious Jacobite challenge to the Hanoverian succession.

forum, an open public space in a town or city of the Roman empire. From the 6th century BC the Roman forum was a place for civic meetings and religious and military ceremonial. The Curia (Senate House) and *comitia were situated there, together with markets, libraries, and courts. War trophies were put on display, the most famous being the ram-beaks ('rostra') of Carthaginian galleys taken in the First *Punic War, which decorated the public platform or 'rostra' outside the Senate House. Other forums were built in Rome by early emperors including *Augustus, *Vespasian, and *Trajan. The model was followed in virtually every town of the Roman empire.

The **forum** in Rome, the centre of Roman public life, as seen from Capitoline Hill. The plan of the forum was determined by Julius Caesar and Augustus in the later 1st century BC but considerable alterations were made until as late as the 4th century AD.

Fosse Way, a Roman road which crossed Britain from Isca (Exeter) to Lindum Colonia (Lincoln) by way of Aquae Sulis (Bath) and Ratae (Leicester). It marked the first and southernmost frontier in Britain after the Roman invasion of AD 43 and was probably laid as a military road along the temporary frontier. Northwards from Lincoln it became *Ermine Street. The name comes from the 'fossa' or drainage ditch which ran alongside the road.

Fouché, Joseph, duc d'Otranto (c.1759–1820), French statesman. He was a leading member of the *Jacobin Club in Nantes in 1790. He supported their violent doctrines, demanded the execution of the king, and was used to crush revolts in the west. He helped initiate the atheistical movement which led him into conflict with *Robespierre and to his ejection from the Jacobin Club in 1794. During the next five years his skill and energy enabled him to play a successful part in the coups that overthrew Robespierre and the *Directory. As Minister of Police (1799–1802), and of the Interior under *Napoleon, he was one of the most powerful men in France until his resignation in 1815.

Founding Fathers, the nickname of the fifty-five delegates to the Constitutional Convention of 1787 that drafted the *Constitution of the USA. They included outstanding public officials, of whom the most respected were George *Washington and Benjamin *Franklin, while the leaders were James *Madison and George Mason of Virginia, Governor Morris and James Wilson of Pennsylvania, and Roger Sherman and Elbridge Gerry of Massachusetts. Of the fifty-five delegates, over half were lawyers, while planters and merchants, together with a few physicians and college professors, made

up the rest. Washington was elected president of the Convention and William Jackson secretary. Jackson's notes were meagre, but a report of the debates was given in Madison's journal (and in notes made by other delegates), though, as the Convention was sworn to secrecy, Madison's notes were not published until 1840.

foundling hospital, an institution for the care and upbringing of children who had been abandoned by their parents. The earliest known one was established in Milan in the 8th century. A number were founded in various European countries, including the famous Paris foundling hospital, incorporated in 1670. In 1739 Thomas Coram (1668–1751) set up his foundling hospital in London, which was at first reserved for illegitimate children.

Fouquet, Nicolas, vicomte de Melun et de Vaux, marquis de Belle-Isle (1615–80), French statesman until his dismissal by *Louis XIV. He was Superintendant of Finances in 1653 and a munificent patron of the arts. On *Mazarin's death in 1661 he hoped to become first minister and invited *Louis XIV to a lavish entertainment at his palatial residence at Vaux-le-Vicomte. Colbert persuaded the king that Fouquet's ambitions were a danger to his authority, and Louis XIV, determined not to be controlled by a powerful minister, ordered his arrest. His trial lasted for three years and the eventual sentence was banishment, but Louis had him imprisoned for life.

Fox, Charles James (1749–1806), British statesman. He entered Parliament at the age of 20. At first he supported Lord *North, but soon became a bitter critic of North's American policy and favoured American independence. After North's resignation and *Rockingham's death Fox refused to join the *Shelburne ministry and collaborated with North to bring it down in 1783. The Fox–North coalition lasted only a few months, until Fox's India Bill was defeated in the House of Lords, and *George III took the opportunity to dismiss a coalition he detested. Fox spent most of the rest of his career in opposition to the new Prime Minister, the Younger *Pitt. His friendship with the Prince of Wales, his own blatant dissipations, and his apparently unprincipled efforts to achieve power meant that he never won wide political support, except when his welcome of the French Revolution was briefly in accord with the national mood; but *Burke's attack on the Revolution split the Foxite Whigs and pushed Fox into political isolation. His rivalry with Pitt led him into irresponsible opposition to some of Pitt's earlier reforms, but they shared a dislike of the *slave trade. When Pitt died in 1806, Fox became a leading member of the new ministry, and through his dying efforts Parliament agreed to an anti-slavery Bill.

Fox, George (1624–91), English founder of the *Quakers. The son of a Puritan weaver, he left home in 1643 to lead an itinerant life, arguing with religious radicals, then preaching. By 1655 he had attracted thousands of converts. With the help of his wife, Alice Fell (married 1669), he proved to be a tireless organizer. He undertook missionary journeys throughout Britain, and ventured as far as the West Indies and North America (1671–2) and Holland (1677 and 1684). He was repeatedly imprisoned for his beliefs; his *Journal* was published posthumously in 1694.

Foxe, John (1516–87), English clergyman. He is best known as the author of *Acts and Monuments* (Latin edition, Strasburg 1554), a history of the persecution of Christians, and especially of Protestant martyrs from *Wyclif's time to his own. The English edition of 1563, *Acts and Monuments of matters happening in the Church*, complete with sensational woodcut illustrations, was popular with both the government and the clerical establishment. 'Foxe's Book of Martyrs', as it was known, celebrated the struggle and triumph of reformed doctrine in the English nation; it was banned by Archbishop *Laud.

France, a country in western Europe with coastlines on the Atlantic Ocean and Mediterranean Sea. Prehistoric remains, cave paintings, and megalithic monuments attest to a long history of human settlement. The area known as *Gaul to the Romans was conquered by the armies of Julius *Caesar, and its native inhabitants thoroughly Romanized by centuries of occupation. After 330 it was invaded by *Goths, *Franks, and *Burgundians, and then ruled by Clovis (465–511), a *Merovingian king. It became part of the empire of *Charlemagne and, after repeated assaults from *Vikings and *Saracens, a *Capetian dynasty emerged in 987. Fierce competition with the rival rulers of *Brittany, *Burgundy, and, after 1066, with the Norman and Plantagenet kings of England ensued, culminating in the *Hundred Years War. France did not emerge as a permanently unified state until the ejection of the English and the Burgundians at the end of the Middle Ages. Under the *Valois and *Bourbon dynasties France rose to contest European hegemony in the 16th to 18th centuries, notably in the wars of *Louis XIV. In the 18th century, weak government, expensive wars, and colonial rivalry with England wrecked the monarchy's finances, and mounting popular anger culminated in the *French Revolution. The rule of *Napoleon followed.

Franche-Comté, a French province bordering the Swiss frontier of which the chief town is Besancon. It was within the boundary of the Holy Roman Empire and, while the duchy of *Burgundy was seized by Louis XI in 1477, the Franche-Comté passed eventually to the Spanish *Habsburgs. In the War of *Devolution (1667–8) *Louis XIV laid claim to it; it was overrun by *Condé, but returned to Spain at the Treaty of *Aix-la-Chapelle. In 1674 it was reoccupied and since the Treaty of *Nijmegen (1679) it has been part of France.

Francis I (1494–1547), King of France (1515–47). He was in many respects an archetypal Renaissance prince, able, quick-witted, and licentious, and a patron of art and learning, but he developed into a cruel persecutor of Protestants, and devoted the best part of his reign to an inconclusive struggle with the *Habsburgs. He began by recovering the duchy of Milan (1515), but failed in his bid to be elected Holy Roman Emperor (1519). In 1520 he tried to secure the support of *Henry VIII of England against the successful candidate, *Charles V, at the *Field of the Cloth of Gold, and in 1521 he embarked on the first of four wars against the emperor, which ended with his capture at the battle of *Pavia (1525). He gained his release from captivity in Spain by renouncing his claims in Italy, but hostilities were resumed and continued intermittently, with the recovery of Milan as a main object, until the Peace of Crespy in 1544. His foreign wars were ruinously expensive, and the prodigality of his court foreshadowed that of *Louis XIV. The palace of Fontainebleau was rebuilt during his reign, and the artists Leonardo da Vinci, Benvenuto Cellini, and Andrea del Sarto worked at his court.

Francis of Assisi, St (1182–1226), founder of the Franciscan order of friars. He was born to a wealthy merchant family in northern Italy and initially followed his father's trade. After a period of imprisonment following involvement in a border dispute in 1202 he abandoned this in favour of the religious life. He adopted extreme poverty but remained 'within the world', working and preaching especially to the poor and sick. His example gained for him a large following and Pope *Innocent III approved the *Franciscan order in 1209–10. He was ordained deacon, which position he retained, his humility preventing his acceptance of full priesthood. His teaching reflected a profound love of the natural world and respect for even the lowliest of its creatures. He rejected personal possessions, wearing only the simplest clothing and ruling that his followers should do likewise. He made missionary journeys to southern Europe and visited the sultan al-Kamil in Egypt in an effort to secure peace during the Fifth *Crusade. A number of miracles and visions were attributed to him, and he was canonized in 1228.

Francis Xavier, St (1506–52), missionary priest, known as the 'Apostle of the Indies'. Born into a Spanish noble family, he was the first *Jesuit missionary in the East. He worked among lepers in *Goa, in India, where he baptized thousands of pearl divers, and in the *Moluccas and *Malacca. In 1549 he went to Japan where he remained for two years and made some 2,000 converts. Convinced that the Japanese would readily become Christians if the Chinese were converted, he set out for Guangzhou (Canton) but died on an island near Hong

Kong while awaiting an opportunity to enter China. His voluminous correspondence survives.

Franciscan, a friar of the order founded by St *Francis of Assisi. Their 'rule' which advocated strict poverty, was approved by Pope Innocent III in 1209–10 and confirmed in 1223. They quickly gained popular support through their work with the sick and the poor and others in the secular world. With the spread of the order two factions developed, the 'spirituals', who insisted on a literal interpretation of the rule and the 'moderates', whose views triumphed and allowed the order corporate ownership of property. Reform followed in the 16th century; the Capuchin friars were a reformed group approved by Pope Clement VII in 1528. The Second Order are the nuns known as 'Poor Clares', and the Third Order live and work among the laity.

Franklin, Benjamin (1706–90), American statesman, journalist, and inventor. He was born in Boston of humble parents and trained as a printer. In 1723 he ran away to Philadelphia. After a visit to London, he became publisher of the *Pennsylvania Gazette* (1730–66) and obtained the lucrative contract for government printing. His *Poor Richard's Almanack* (1732–57), full of advice on success in business, won him intercolonial fame. Influenced by Cotton *Mather, Daniel Defoe, and *The Spectator*, he was a proponent of civic improvements and self-help projects such as the Junto (1727), the Library Company (1731), the Fire Company (1737), the *American Philosophical Society (1743), the University of Pennsylvania (1751), and the Philadelphia Hospital (1752). He also pursued scientific inquiries in the spirit of the *Enlightenment, seeking practical improvements like bifocal lenses. His famous electrical experiments with a kite resulted in the development of the lightning conductor. Appointed Deputy Postmaster-General for the Colonies in 1753, he improved intercolonial communications and developed a federal outlook, shown at the *Albany Congress in 1754.

Drawn into politics by opposition to the *Penn family's privileges and to the pacifism of the ruling Quakers, he was an agent in London for long periods between 1757

An illustration of **St Francis of Assisi** from the *Lutterell Psalter*, c.1340. Dressed in a simple habit, he is surrounded by birds, a lion, and oak leaves, all of which symbolize his love of the natural world. (British Library, London)

and 1775, first seeking a new charter and then lobbying against revenue-raising measures. He attended the *Continental Congress before appointment as ambassador to France (1776–85). Posing as a simple citizen at Louis XVI's court, his popularity helped secure the French alliance (1778) and he was able to arrange favourable terms for America at the Peace of *Paris (1783). In France he continued writing his *Autobiography* which, though tinged with self-congratulation, reveals his pragmatic philosophy of personal improvement and natural inquiry. After attending the Constitutional Convention, he died in 1790, the embodiment of the rags-to-riches dream and the public-spirited entrepreneur.

franklin, English freeholder in the 13th and 14th centuries. Franklins paid a rent for their land and did not owe military service to their lord. Their status was above that of the free peasant but below the gentry or nobility. After the *Black Death (1349) the status of many franklins improved in the favourable economic climate. The newly rich and socially ambitious franklin was satirized by Chaucer in the Prologue to his *Canterbury Tales*.

frankpledge (or 'peace-pledge'), a system for preserving law and order in English communities from the 10th century to the 14th, when it was superseded by the appointment of Justices of the Peace. Communities were grouped into associations of ten men (a tithing) under a headman (chief pledge or tithingman) and held responsible for the good behaviour of members. Twice a year, in the 'view of frankpledge', sheriffs examined its effectiveness. Frankpledge was not applied to the aristocracy, to certain freeholders, or to vagrants, nor was it found in northern England where the alternative system of Serjeants of the Peace existed.

Franks, a group of Germanic tribes who dominated Europe after the collapse of the Western *Roman empire. The name was adopted in the 3rd century AD, possibly from their word for a javelin (*franca*). They consisted of Salians from what is now Belgium and Ripuarians from the Lower Rhine. They settled in Gaul by the mid-4th century and ruled it by the following century when the Salians under *Clovis defeated the Romans at Soissons. Gaul became 'Francia', ruled from the old capital Lutetia Parisiorum (Paris) of the Parisii Gauls. The *Merovingian succession continued until 751. Power then passed from the kings to their palace mayors. In 751 Pepin, son of Charles Martel, became the first *Carolingian ruler of the Franks.

Frederick I (Barbarossa) (*c.*1123–90), King of Germany and Italy (1152–90), Holy Roman Emperor (1155–90). He was the son of Frederick II, Duke of Swabia. He put down the rebellion of Arnold of Brescia in northern Italy and in 1155 was rewarded by Pope Adrian IV with the imperial crown. From that point onwards, however, he was at odds with the papacy as he tried to strengthen his empire in Italy. He was opposed by the Lombard League (an alliance of north Italian cities) and Pope Alexander III, whom he was eventually obliged to recognize in 1177. He came to terms with the Lombards at the Peace of Constance in 1183. In Germany he had to assert his power over his rival *Henry the Lion. This achieved, he left the empire in the hands of his son, and

went on *Crusade. He won two victories over the Muslims but was drowned crossing a river in Asia Minor.

Frederick I (1657–1713), Elector of Brandenburg from 1688, King of Prussia (1701–13). He lacked the ability of his father, *Frederick William, the Great Elector, and dissipated funds in display and extravagance. In 1700 he supported the Holy Roman Emperor Leopold I in the War of the *Spanish Succession and with his approval was able to proclaim himself king, taking his title from his territory of East *Prussia. In 1713 he acquired Upper Gelders. With his second wife Sophia Charlotte he developed Berlin and established the Academy of Science and the University of Halle.

Frederick II (1194–1250), Holy Roman Emperor (1220–50). The grandson of *Frederick Barbarossa. Frederick II was known as *Stupor Mundi* ('Wonder of the World') because of the breadth of his power and of his administrative, military, and intellectual abilities. He was crowned King of the Germans in 1215 and Holy Roman Emperor in 1220, but his reign was dominated by a long and ultimately unsuccessful struggle for power with the papacy. Twice excommunicated by Pope Gregory IX, and opposed in Italy by the Lombard League, Frederick devolved a great deal of imperial power within Germany on the lay and clerical princes in an effort to maintain their support, and concentrated on building up a power base in Sicily, a process completed by the Constitution

A contemporary portrait of **Frederick II** (the Great) by J. G. Glume. The founder of Prussian power in the 18th century, Frederick was a fervent admirer of all things French. (Schloss Charlottenburg, Berlin)

of Melfi in 1231. He successfully defeated the Lombard League at Cortenuova in 1237 and humiliated Gregory IX prior to the latter's death in 1241, but failed in his efforts to conciliate Innocent IV who appealed to Germany to revolt at the Synod of Lyons in 1245. Frederick's position was crumbling in the face of revolt, papal propaganda, and military defeat when he died suddenly in 1250, leaving an impossible situation for his less talented heirs to solve.

Frederick II (the Great) (1712–86), King of Prussia (1740–86). He was the son of *Frederick William I and Sophia Dorothea, daughter of *George I of Great Britain. Able, cultured, a hard-working administrator, and a brilliant soldier, he was an example of an 18th-century enlightened despot. He believed that a ruler should exercise absolute power, but exercise it for the good of his subjects. He established full religious toleration, abolished torture, and freed the serfs on his own estates. He continued the state's tradition of efficient administration, fair though heavy taxation, and above all devotion to the army and its needs.

His foreign policy was unscrupulous. In 1740 he attacked *Maria Theresa and gained Silesia. He fought to keep it in the War of the *Austrian Succession and the *Seven Years War, and greatly increased his prestige by his victories over the French at Rossbach (1757) and the Austrians at Leuthen (1757). He obtained the valuable territory of West *Prussia in the first partition of Poland in 1772 and opposed *Joseph II in the war of the *Bavarian Succession. He also built the rococo palace of Sans Souci at Potsdam, played and composed flute music, and corresponded with *Voltaire.

Frederick III (1415–93), King of Germany (1440–93), *Holy Roman Emperor (1452–93). He inherited the *Habsburg domains as Archduke of Austria in 1424. A weak and rather ineffectual figure, he failed to assert family interest in Hungary, and was troubled by rival claimants to his own lands, and by Turkish attacks, which became more threatening after the fall of *Constantinople in 1453. On good terms with the papacy, he was the last Holy Roman Emperor to be crowned by the pope at Rome. He earned unpopularity by his efforts to suppress John *Huss's followers in Bohemia and Hungary. By arranging the marriage of his son *Maximilian I to Mary, daughter of *Charles the Bold, he greatly extended the dynastic power of the Habsburgs.

Frederick V (the Winter King) (1596–1632), Elector Palatine (1610–20) and King of Bohemia (1619–20). In 1613 he married Elizabeth, daughter of *James I of England. He then assumed the leadership of the German Protestant Union, and accepted the Bohemian crown when it was offered following the deposition of *Ferdinand II in November 1619. Thenceforth his fortunes merged with the course of the *Thirty Years War, a struggle in which he took little personal part after his defeat at the battle of the White Mountain (November 1620). He withdrew to The Hague, and forfeited the Palatinate. *George I of Great Britain (1660–1727) was his grandson.

Frederick Louis (1707–51), British prince, the eldest son of *George II, with whom he quarrelled bitterly. As Prince of Wales his home in London, Leicester House, became the meeting place of the opposition leaders who

The coronation of **Frederick III** by Pope Nicholas V in 1452, from a contemporary painting attributed to Simon Marmion. (Germanishes National-Museum, Nuremburg)

helped to bring down Sir Robert *Walpole in 1742. His premature death disappointed the hopes of those politicians who had supported him in the expectation of preferment upon his succession. The throne passed to his eldest son, who reigned as *George III.

Frederick William (the Great Elector) (1620–88), Elector of Brandenburg (1640–8), sometimes called the greatest of the *Hohenzollerns. He succeeded to an inheritance impoverished by the *Thirty Years War but was able, and a good diplomat, and rapidly improved his position and made gains, including East Pomerania, at the Treaty of *Westphalia. By 1688 his standing army of 30,000 troops had become the efficient basis of the state, communications were improved, waste lands cultivated, and his territory of East Prussia freed from Polish overlordship. His victory of Fehrbellin (1675) over the Swedes greatly increased his prestige.

Frederick William I (1688–1740), King of Prussia (1713–40). He was the son of *Frederick I and was known as 'the royal drill-sergeant': he was a strict Calvinist, hardworking, violent tempered, and notorious for his ill-treatment of his son, *Frederick II. He left a model administration, a large revenue, and an efficient and well-disciplined army. He acquired Stettin in 1720.

Frederick William II (1744–97), King of Prussia (1787–97). He was the nephew of *Frederick II and a man of little ability, though a patron of the arts. He fought in the early campaigns against the French Revolutionary armies but became more concerned with Poland gaining land, including Warsaw, in the partitions of 1793 and 1795.

Freemason, a member of an international fraternity for mutual help and fellowship, called 'Free and Accepted Masons', having an elaborate ritual and system of secret signs. The original 'free masons' were probably emancipated skilled itinerant stonemasons who (in and after the 14th century) found work wherever important buildings were being erected, all of whom recognized their fellow craftsmen by secret signs. The 'accepted masons' were honorary members (originally supposed to be eminent for architectural or antiquarian learning) who began to be admitted early in the 17th century. The distinction of being an 'accepted mason' became a fashionable object of ambition, and before the end of the 17th century the purpose of the fraternities seems to have been chiefly social and convivial. In 1717 four of these societies or 'lodges' in London united to form a 'grand lodge', with a new constitution and ritual, and a new objective of mutual help and fellowship among members. The London 'grand lodge' became the parent of other 'lodges' in Britain and around the world, and there are now bodies of Freemasons in many countries of the world.

French and Indian wars (1689–1763), Anglo-French conflicts in North America, part of the international rivalry between the two nations. They consisted of *King William's War (1689–97), *Queen Anne's War (1703–13), *King George's War (1744–8), and the French and Indian War (1755–63), the American part of the *Seven Years War. As a result of an alliance with Prussia, *Pitt was able to devote more British resources to America. In 1755 the British commander, General *Braddock, led forces into Ohio but was defeated at Fort Duquesne. Other forces defeated the French at the battle of Lake George, but no advantage was taken. The French claimed a number of victories in 1756 and 1757 against the British, already weakened by friction between the new commander, Lord John Loudoun, and the states of Massachusetts and Virginia. In 1759, however, *Wolfe defeated the French at the battle of the *Plains of Abraham. Quebec surrendered shortly thereafter, followed by Montreal one year later, all Canada then passing into British hands. By the Treaty of *Paris (1763) Britain gained Canada and Louisiana east of the Mississippi.

French East India Company, a commercial organization, founded in 1664 to compete with *Dutch and English *East India companies. Until the 1740s it was less successful than its rivals, but led by an ambitious governor, *Dupleix, the Company then made a bid to challenge English influence in India, notably by alliances with local rulers in south India. Although a number of trading ports, including *Pondicherry and Chandernagore, remained in French control until 1949, the Company itself collapsed during the French Revolutionary period.

French Revolution (1789), the political upheaval which ended with the overthrow of the Bourbon monarchy in France and marked a watershed in European history. Various groups in French society opposed the *ancien régime with its privileged Establishment and discredited monarchy. Its leaders were influenced by the American Revolution of the 1770s and had much popular support in the 1780s and 1790s. Social and economic unrest combined with urgent financial problems persuaded Louis XVI to summon the *States-General in 1789, an act which helped to set the Revolution in motion. From the States-General emerged the National Assembly and a new Constitution which abolished the *ancien régime*, nationalized the church's lands, and divided the country into departments to be ruled by elected assemblies. Fear of royal retaliation led to popular unrest, the storming of the *Bastille, and the capturing of the king by the National Guard. The National Assembly tried to create a monarchical system in which the king would share power with an elected assembly, but after the king's unsuccessful flight to *Varennes and the mobilization of exiled royalists, the Revolutionaries faced increasing military threats from Austria and Prussia which led to war abroad and more radical policies at home. In 1792 the monarchy was abolished, a republic established, and the execution of the king was followed by a Reign of *Terror (September 1793–July 1794). The Revolution failed to produce a stable form of republican government as several different factions (*Girondins, *Jacobins, Cordeliers, *Robespierre) fought for power. After several different forms of administration had been tried, the last, the Directory, was overthrown by *Napoleon in 1799.

French Wars of Religion, a series of nine religious and political conflicts in France, which took place intermittently between 1562 and 1598. They revolved round the great noble families fighting for control of the expiring *Valois dynasty, supported on one side by the Protestant *Huguenots and on the other by Catholic extremists. The wars were complicated and prolonged by interventions by Spain, Savoy, and Rome on the Catholic side and by England, the Netherlands, and the German princes on the Protestant side. After the turning-point of the *St Bartholomew's Day Massacre (1572), a third party of moderate Catholic 'Politiques' emerged under the Montmorency family. However, its advocacy of mutual religious toleration was undermined in 1576 by the formation of the Catholic extremist *Holy League, which opposed *Henry III's tolerant settlement of the fifth war. The Guise-led League grew more militant after the *Bourbon Huguenot leader Henry of Navarre became heir to the French throne in 1584. The resulting War of the *Three Henrys (1585–9), ended with the assassination of Henry III. Henry of Navarre fought on, overcame the League, and drove its Spanish allies out of the country. He adopted Catholicism (1593), and as *Henry IV was able to establish religious toleration in France with the Edict of *Nantes (1598). At the Peace of Vervins (1598) he reached a settlement with Spain. Then he applied himself to providing the firm monarchical rule which had been so damagingly lacking since the death of Henry II in 1559.

friar (from the French *frère*, 'brother'), a member of a religious order for men. Friars, together with monks, are known as 'regulars' because they follow a written 'rule' for the conduct of their lives, taking vows of poverty, celibacy, and obedience. Unlike monks, who shut themselves off from the secular world, their duty is to work within the world, preaching and healing. They support themselves largely on the gifts they receive from the laity and are therefore sometimes called mendicants. During the 13th century friars achieved immense popularity because of their exemplary lives and the practical help which they offered to the poor and the sick. They helped to counter confusions arising from *heresy and worked to suppress heretical movements.

The four main orders are the *Augustinians or Austin friars, *Franciscans or Grey Friars, *Dominicans or Black Friars, and *Carmelites or White Friars, the last three being named for the colour of their dress or habit. These orders have survived up to the present day.

Frisians, a Germanic seafaring people. In Roman times they occupied northern Holland and north-west Germany. Apart from records of their revolts against Rome between 12 BC and AD 69, little is known about them. Some of them served in the Roman army in Britain in the 3rd century. By the 4th century they were under Saxon domination. The *Franks attempted to convert them to Christianity by force, though this was less successful than the missionary efforts of St Wilfred from England. They were part of the *Carolingian empire, and in the 16th century became part of the Habsburg empire of *Charles V. In 1579 they reluctantly joined the Union of *Utrecht against *Philip II. They continued their independent role in the new state, electing their own Statholder (or President) until 1747.

Froissart, Jean (c.1337–c.1410), Flemish poet and historian. His four *Chroniques* provide a detailed, often eye-witness account of European events from 1325 to 1400. His first book copied the work of an earlier chronicler, Jean le Bel (c.1290–c.1370). The others, drawn from extensive travels, especially to the English court, provide an account of events during the *Hundred Years War. His viewpoint and sympathies were those of the aristocracy and he celebrated the chivalrous ideals and activities of the nobility. His writings were very popular in western Europe in the 15th century.

Fronde (French, *fronde*, 'boys' game with slings'), the name first used by Cardinal de *Retz to describe street fighting in Paris; the word was applied particularly to two revolts against the absolutism of the crown in France between 1648 and 1652 during the minority of *Louis XIV. The First Fronde began as a protest by the *Parlement* of Paris supported by the Paris mob against war taxation. Disaffected nobles joined in and intrigued with France's

Street fighting in the Second **Fronde** during July 1652. The nobles' attempt to resist the king by arms ended in failure later in the year. (Bibliothèque Nationale, Paris)

enemy, Spain. Peace was restored in March 1649. The Second Fronde began in 1651 with *Mazarin's arrest of the arrogant and overbearing *Condé. Throughout France nobles indulged in irresponsible and confused fighting in which certain great ladies, including Condé's sister, Madame de Longueville, and the king's cousin, La Grande Mademoiselle, played a conspiratorial role. Mazarin fled from France, but Condé and the mutinous nobles who supported him soon lost popularity. Mazarin was able to return, giving the command of the army to *Turenne, who had rejoined the royalist party and quickly recovered Paris for the king. The Fronde ended in Paris in October 1652. It was the last attempt of the nobility to resist the king by arms, and resulted in the strengthening of the monarchy. The legacy of the Frondes was Louis XIV's dislike of Paris and the building of the palace of *Versailles at a distance from the city.

Frontenac, Louis de Buade, comte de (1622–98), governor of New France (1672–82, 1689–98). He fought with distinction in the *Thirty Years War. During his first service in Canada, he strengthened the defences of the colony of New France and encouraged exploration, but clashed with Bishop Laval of Quebec and was recalled. Returning to Canada in 1689, he repulsed the New Englanders' attack on Quebec in 1690 and re-imposed French authority against the Iroquois in 1696.

Fugger, a south German family of bankers and merchants, the creditors of many rulers in the Middle Ages. The family first achieved prominence under Jacob I (1410–69), head of the Augsburg weavers' guild. Ulrich (1441–1510) supplied cloth, and then lent money to the *Habsburgs. **Jacob II** (1459–1525), headed the family from 1510 and lent enormous sums to *Maximilian I and *Charles V, financing Charles's candidacy as *Holy Roman Emperor in 1519. Resources came from silver and mercury mines in Germany and later from the Spanish empire in South America. The Fuggers were deeply involved in the finances of the *papacy, and in the sale of *indulgences under *Tetzel. The family was ennobled and, in 1546 attained a peak of prosperity. Thereafter, Habsburg bankruptcies and the ravages of the *Thirty Years War damaged their position. They safeguarded their remaining wealth by shrewd management of their extensive estates.

Fujiwara, a noble Japanese family whose members enjoyed the patronage of the Japanese emperors. In 858 Fujiwara Yoshifusa became the first non-imperial regent for a child emperor. Soon it was customary for every emperor to have a Fujiwara regent. For about three centuries they dominated the court of Kyoto, securing their power by marrying their daughters into the imperial family.

The Fujiwara period (late 9th to late 12th century) saw great artistic and literary development. A court lady, Lady Murasaki, wrote Japan's most famous classic novel *The Tale of Genji* (c.1001–15). But the central administration Prince *Shotoku had initiated in the 7th century was breaking down into a near feudal society. Tax-free estates proliferated and warrior families came to prominence. The establishment of the *Kamakura shogunate marked the end of Fujiwara ascendancy, though they acted as important court officials until the Meiji period in the 19th century.

Fulani, a people of West Africa. Their origin is disputed, but most probably they were pastoralists forced southwards by the gradual desiccation of the Sahara. They expanded eastwards from Senegal to Cameroon from about the 12th century, and spread their belief in Islam. By intermarriage with pre-existing populations, they have produced a number of related groups, of which the most important initiated the empire of Macina (14th century), which reached its apogee under Seku Hamadu (1810–44); of Futa Jallon (1694); of the Hausa emirates established by Uthman dan Fodio in the early 19th century; and of the chiefdoms of Adamawa and Liptako.

Funj, a Sudanese kingdom founded, together with the city of Sennar, by Amara Dunkas *c*.1504–5. Successive rulers expanded it over the Gezira and southern Kordofan, and fought two wars with Ethiopia. In the 17th century the monarchy was kept in power by a great army of slaves, and thereafter declined. When the Egyptians invaded it in 1821, the last king (whose state had been weakened by disunity) offered no resistance.

fur trade, the trading in animal furs, which were used as clothing from the earliest times. *Marco Polo celebrated the rich fur potential of Central Asia, and the Russian penetration of Siberia was pioneered by fur traders. North American settlement was likewise spurred by the profitability of the fur trade, especially beaver pelts popular in hat making. The *Pilgrims, New Englanders, New Yorkers, and French Canadians were all active in the trade. Wars were fought for control of rich fur-bearing areas round the Great Lakes. In the 19th century the trade moved west, with 'mountain men' in the Rockies delivering their pelts at rendezvous in the spring to agents of the *Hudson's Bay or North-West companies or of John Jacob Astor, whose American Fur Company made him the first American millionaire.

fusilier, originally a soldier armed with a 'fusil' (an improved flintlock musket) in the 17th-century French army. The *fusiliers du roi* were an élite force and the first to be armed with bayonets. Later the name was given to various infantry regiments, including machine-gunners.

fyrd, the military force of freemen available to the Anglo-Saxon kings. Military service was one of the three duties (the *trinoda necessitas*) required since the 7th century of all freemen—the other two duties being the maintenance of forts and the upkeep of roads and bridges. It was unusual for the fyrd to serve outside the shire in which it was raised. Modified by *Alfred the Great, it continued to be called even under the Norman kings. Henry II reorganized it in his Assize of Arms (1181) and Edward I in the Statute of Winchester (1285). It eventually became the *militia.

Gage, Thomas (1721–87), British general, appointed British commander in America in 1763, after service in Flanders, at *Culloden, and in the *French and Indian War. His responsibilities shifted from frontier defence to quelling unrest in such towns as New York and Boston. He was appointed governor of Massachusetts in 1774 to enforce the *Coercive Acts, but he bungled *Lexington and Concord, and resigned after *Bunker Hill.

Gaiseric *Genseric.

Galatia, the area in central Asia Minor occupied by the Gauls who crossed the Hellespont from Europe in 278 BC. In 230 they were decisively defeated by Attalus I of Pergamum; Rome later made use of them against Pergamum, and they stayed loyal to Rome during the wars with *Mithridates. The Roman province of Galatia, established in 25 BC, comprised a much wider area.

Galicia, the name of two separate areas. One is a region in north-western Spain. In the 6th century it was a kingdom of the *Goths. It became a centre of resistance against the *Moors in the 8th century and in the 13th century passed to Castile.

The other region extends west of the Ukraine and east of Hungary, Romania, and Czechoslovakia. In the Middle Ages it was an independent kingdom based on Kiev. It extended as far west as Cracow in modern Poland. Under Polish rule from 1349 it passed to Austria in the 18th century.

galleon, a sailing warship of the late 16th century. Galleons eventually replaced the less manoeuvrable carracks as the principal type of European trading ship. About 1550 naval architects began to reduce the top-heaviness of existing larger warships, and, by narrowing the beam to give a galley-like look, they created the galleon. From the designs of Sir John *Hawkins, English galleons were the first to develop the characteristic beaked prow and modestly sized forecastle. These features were later incorporated in the great Spanish and Portuguese galleons which were used for overseas trade.

galley, a single-decked war ship equipped with oars and sail and propelled mainly by oars in battle. The Phoenicians seem to have invented the bireme (with two banks of oars) *c*.700 BC. Ancient *Athens used the *trireme (three banks) as its principal warship—one of the greatest trireme battles being that at *Salamis in 480. In Hellenistic times the quinquereme was popular—it was considerably larger with five men at each oar. The Viking *longship was a small but durable type of galley, simpler than the larger, more complicated designs of the galleys employed by the major Mediterranean powers such as Byzantium and Venice. *Lepanto (1571) was the last great naval battle involving large numbers of galleys.

Gallicanism, in French Roman Catholicism, the tradition of resistance to papal authority; its opposite was

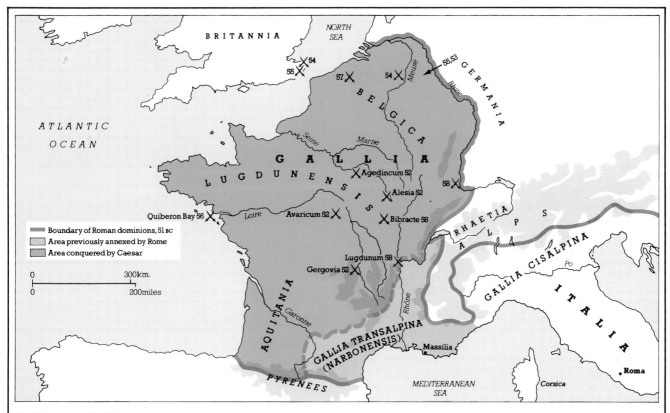

Gallic wars (58–51 BC)

Caesar was governor of Cisalpine Gaul when he embarked on the Gallic wars in 58 BC. His victory over the Helvetii at Bibracte was the first of a series of conquests in successive campaigning seasons. The Nervii, most powerful of the Belgic tribes, were defeated in 57, and a revolt by the Veneti was put down at Quiberon Bay in 56. These earlier victories were a tribute to superior organization and generalship, but Caesar faced his greatest test in 52, when the Gauls united under Vercingetorix. The rebellion was broken by the siege of Alesia, and Gaul was subsequently organized as a Roman province.

Ultramontanism, which gave the papacy complete control over the Church throughout the world. Gallicanism was founded firstly on the assertion of the French monarch's independence from the papacy, secondly on the collaboration of clergy and secular powers to limit papal intervention within France, and thirdly on the superiority of an ecumenical council over the papacy. It figured in disputes dating back to the 13th century, and was authoritatively set out in the *Gallican Articles* (1682), approved by the assembly of the French clergy.

Gallic wars, Julius *Caesar's campaigns (58–51 BC), which established Roman rule over central and northern Europe west of the River Rhine (*Gaul). Crossing into Transalpine Gaul, Caesar repelled German tribes in the south and east, Belgae in the north, and Veneti in the west. He even crossed the Rhine to demonstrate Roman control of that crucial natural frontier. With speed and ruthlessness and helped by inter-tribal disunity he subdued the northern and western coasts. He twice (55 and 54 BC) invaded Britain, which was regarded as a Belgic refuge and threat to Rome. In the winter of 53–52 BC, *Vercingetorix rallied the central Gallic tribes in unusual unity. In a long and bitter war, Caesar defeated him and his successors, and he was executed. Caesar's war dispatches, *De Bello Gallico*, supply most of the information about these events.

Gao   *Songhay.

Gardiner, Stephen (*c.*1490–1555), Bishop of Winchester (1531–51, 1553–5). A protégé of Thomas *Wolsey, he assisted in the negotiations to secure *Henry VIII's divorce from Catherine of Aragon. He was made Bishop of Winchester (1531), and defended the royal supremacy over the Church, most notably in *De Vera Obedientia* (1535), but he was opposed to Protestantism. Under *Edward VI he was imprisoned (1548–53) and deprived of his see. *Mary I restored him to Winchester, made him Lord Chancellor, and relied on him until his death.

Gascony, a former duchy in south-western France. It was named for the Vascones who took it in the 6th century and it enjoyed a great degree of independence until 819 when it submitted to King Louis the Pious of France. It came to Henry II of England on his marriage to *Eleanor of Aquitaine. Rebel nobles were subdued by Simon de *Montfort but it was lost to the French in the *Hundred Years War. *Edward III regained it and it was recognized as an English possession by the Treaty of *Brétigny in 1360. The English were finally driven out by the army of Charles VII of France. Royal authority was only fully realized, however, when the House of Armagnac was defeated and Gascony was formally joined to the crown of France by Henry IV.

Gates, Horatio (1728–1806), American general, born in England. He fought under *Braddock and *Amherst in the *French and Indian War but thereafter supported the American cause. His victory at *Saratoga (1777) led the Conway Cabal of New England officers to plot for him to replace *Washington. His rout at *Camden (1780) ended his military career, though he later served under Washington.

Gaul, the Roman name for the lands of the *Celts in western Europe. The Gauls invaded northern Italy in the 4th and 3rd centuries BC. In 222 BC the territory south of the Alps was declared the Roman province of Cisalpine Gaul. The River *Rubicon formed part of the frontier with Italy proper. The area north of the Po was known as Transpadane Gaul, that south of the Po as Cispadane Gaul. Beyond the Alps the southern coast of modern France and its hinterland was known as Transalpine Gaul, or often simply as 'Provincia' (hence modern *Provence) after its annexation in 121 BC. Its capital was at Narbo. Julius *Caesar was given command of both the Gallic provinces in 59 BC. He extended Transalpine Gaul to the Atlantic, the English Channel, and the Rhine in the course of his *Gallic wars. Despite the last stand by *Vercingetorix, there was generally little unity of resistance. Transpadane Gaul was given Roman citizenship by Caesar in 49 BC and the whole Cisalpine Gaul was incorporated into Italy by *Augustus, thus ceasing to be a province. (Cispadane Gaul had been given Roman citizenship in 90 BC.) Augustus divided Gaul north of the Alps into Narbonensian Gaul under the Senate and Lugdunensian (Lyons) Gaul, Aquitania, and Belgic Gaul, under his own legates. Lyons was the venue of a provincial assembly of the 'Tres Galliae'. In the later 3rd century usurper emperors created a semi-independent 'Gallic empire' which served as a temporary buffer against Germanic invasions. Increasingly vulnerable as the heartland of the Western empire it was devastated by *Goths, *Huns, and *Vandals until it was finally ceded to the *Franks in the 5th century.

Gaunt, John of (1340–99), Duke of Lancaster (1362–99). He was born at Ghent (whence 'Gaunt'), the third surviving son of *Edward III of England. He was the classic example of the over-mighty subject, who had sufficient power to threaten the king, although he proved loyal to his nephew *Richard II. His father provided him with the vast inheritance of Lancaster by marrying him to the heiress Blanche of Lancaster in 1359; his second marriage, to Constance, elder daughter of King Pedro the Cruel of Castile, gave him a claim to the throne of Castile and León, and his third wife, Katharine Swynford, brought to prominence her family, the *Beauforts. He was particularly powerful in 1376–7, as Edward III approached death, and in the 1390s; by outliving such rivals as Thomas of Woodstock, he came to seem such a threat that at his death Richard II forbade his son Henry of Bolingbroke (*Henry IV) from entering into his inheritance.

John of **Gaunt** being entertained by John I of Portugal on the occasion of the king's marriage to John of Gaunt's daughter, Philippa of Lancaster. Among the children of this marriage was Prince Henry the Navigator, one of Gaunt's many notable descendants. (British Library, London)

Gautama Siddhartha   *Buddha.

Gaveston, Piers (c.1284–1312), Earl of Cornwall. He was a *Gascon who was brought up in the English royal household as the foster-brother of the future *Edward II, and he exploited his infatuation for him. Edward gave him the earldom of Cornwall in 1307, and appointed him Regent of England (1307–8). The enraged English barons called for his banishment; Edward twice complied (1308, 1311), but Gaveston returned and in 1312 was killed by the Earl of Warwick.

Geneva, a Swiss city situated at the south-western corner of Lake Geneva. It is first mentioned as a settlement of the Celtic Allobroges in the 1st century BC; it later fell to the Romans and became the seat of a bishop. From the mid-12th century to 1401 the bishops effectively ruled the city and region, but from 1401 to 1536 the dukes of Savoy had effective control. In 1535 the citizens rejected the authority of both bishop and duke and accepted the Protestant religion. They turned Geneva into an independent self-governing city-state. In the 16th, 17th, and 18th centuries this 'Protestant Rome' became a haven for religious refugees (who often added considerably to its economic prosperity), and a training-ground for *Huguenot ministers.

Genghis Khan (c.1162–1227), founder of the *Mongol empire. At a time of rapidly shifting alliances among the people of the steppe, he gradually subjected other Mongol tribes. In 1206 a *kuriltai* (assembly) of all the tribes proclaimed him Genghis Khan, Universal Ruler. He proceeded to complete his conquest of the Xi Xia kingdom in Central Asia (1205–9), followed by the *Jin in northern China (1211–15), and the Kara Khitai (1211) and Khorezm in Turkistan (1219–21), thus extending his lands up to the Black Sea. Before he died he divided his empire into four *khanates.

His dynamic leadership and genius for organization largely account for the Mongols' swift rise to power. He was adaptable and had the ability to learn from others. He brought unity to the Mongol peoples and gave them a code of laws supreme over even the khan. An alphabetic script was devised for the Mongol language. He employed a Chinese *Liao prince to establish a system of administration and taxation, which later helped in the Mongol conquest of China. As a commander he created a disciplined military machine. His bowmen could kill at 180 m. (200 yards). Used to living on horseback and supplying themselves from plunder, they moved at incredible speed, once covering some 440 km. (270 miles) in three days. A master of strategy, he used espionage to outwit his opponents and terror to weaken their wills. Against disunited agrarian and commercial states he acted with overwhelming might and ruthlessness.

Genoa, an important port on the coast of north-west Italy. The port and fort established in the 5th century BC were captured by the Romans in the 3rd century BC. After the Roman empire collapsed, Genoa fell to the *Lombards in 634, and was repeatedly sacked by the *Saracens, but recovered its fortunes by the 10th century to become an independent republic and a seafaring and trading power. Influence and colonization spread to Sicily, Spain, North Africa, and the *Crimea. Civil disorder was reduced after a popular coup in 1257 and

Genghis Khan, founder of the Mongol empire, was noted for his brilliant generalship. This Persian manuscript illustration from an account by Rashad ad-Din shows dismounted Mongol archers firing their composite bows against cavalry in a battle against Jelal ad-Din. (Bibliothèque Nationale, Paris)

the office of *doge was created in 1339. Rivalry with neighbouring powers brought the defeat of Pisa in 1284, and persistent fighting with *Venice, which the Genoese nearly captured in 1380. They also contributed many mercenaries, particularly crossbowmen, to foreign armies. The advance of the *Ottomans weakened Genoese control of their colonial possessions, and its importance as a major European power diminished.

Genseric (or Gaiseric) (c. AD 390–477), King of the *Vandals from 428. An *Arian Christian, he was ousted by the *Goths from his lands in Spain and took his entire nation of 80,000 across to North Africa in 429. He besieged Hippo just after the death of *Augustine. Carthage fell ten years later and from there he declared an independent kingdom. Three times he defeated the Roman armies. With a fleet he took Sicily and Sardinia. Rome nicknamed the campaign the 'Fourth *Punic War'. In 455 he took Rome and indulged in a fortnight of looting. Fleets sent against the Vandals in 457 and 468 were defeated, and, at his death Genseric was in possession of all of his conquered territories.

Geoffrey of Monmouth (c.1100–55), Bishop of St Asaph in Wales and the author of a chronicle, the *Historia Regum Britanniae* ('The History of the Kings of Britain') (c.1136). In this work King *Arthur was projected as a national hero defending Britain from the Saxon raiders after the departure of the Roman armies. Much of the material was taken from a 9th century Welsh writer, *Nennius.

George I (1660–1727), Elector of Hanover (1692–1714) and King of Great Britain and Ireland (1714–27). His mother Sophia (1630–1714), a granddaughter of *James I, and her issue were recognized as heirs to the throne

of England by the Act of *Settlement (1701) which
excluded the Roman Catholic *Stuarts. He succeeded
peaceably to the throne on Queen *Anne's death in
1714, and the *Jacobite rebellion a year later helped to
unite the country behind him. He had little sympathy
for British constitutionalism, the need to accept the
limitations of Parliament and ministers, and he disliked
England, spending as much time as possible in Hanover.
But he developed a good command of English, despite
his German accent, and his unswerving support for
*Walpole from 1721 helped to consolidate the supremacy
of the Whigs. He ruled without a queen, having divorced
his wife, Sophia Dorothea, in 1694.

George II (1683–1760), King of Great Britain and
Ireland (1727–60), and Elector of Hanover. He resented
his father, *George I, because of his treatment of his
mother, Sophia Dorothea. George's own marriage, to
Princess *Caroline of Ansbach, was very successful, and
through her he learned to accept *Walpole as his Prime
Minister. He had a fiery temper and was intolerant in
his dealings with others: he insisted on the execution of
Admiral *Byng in 1757, and was always on bad terms
with his son, Prince *Frederick Louis. He was the last
English king to lead his troops in battle, at Dettingen in
1743 in support of the Empress *Maria Theresa during
the War of the *Austrian Succession. He disliked *Pitt
the Elder and endeavoured to keep him out of office; his
final acceptance of Pitt in 1757 paved the way for British
success in the *Seven Years War.

George III (1738–1820), King of Great Britain and
Ireland and of dependencies overseas, King of Hanover
(1760–1820). He was the first Hanoverian ruler to be born
in Britain. The son of *Frederick Louis, he succeeded to
the throne on the death of his grandfather *George II,
with strong convictions about a monarch's role acquired
from *Bolingbroke and *Bute. He was a devoted family
man and a keen patron of the arts, building up the royal
art collection with impeccable taste. He disliked the domination of the government by a few powerful Whig families
and preferred to remain above politics, which was a major
reason for the succession of weak ministries from 1760 to
1770. He was against making major concessions to the
demands of the American colonists, and he shared with
many Englishmen an abhorrence of the American aim of
independence. He suffered from porphyria, a metabolic
disease which causes mental disturbances, which manifested itself briefly in 1765 when plans were made for a
regency council, and for several months in 1788–9 his
illness was so severe as to raise again the prospect of a
regency. Although his political interventions were fewer
than have often been alleged, his interference did bring
down the *Fox–North coalition in December 1783. Increasing reliance on *Pitt the Younger reduced his political influence, although Pitt always had to take it into
account. When the King refused to consider *Catholic
emancipation in exchange for Pitt's Act of Union, Pitt
resigned (1801). In 1811 increasing senility and the onset
of deafness and blindness brought about the regency of
the profligate Prince of Wales, which lasted until he succeeded as George IV.

Georgia, Russia, a mountainous region of the western
Caucasus, a distinctive state since the 4th century BC. In
the 3rd century AD it became part of the *Sassanian

A portrait of **George III** by Johann Zoffany, 1771. It shows
the king wearing a general officer's coat, with the ribbon
and star of the Garter. (Royal Collection)

empire but the Persians were expelled c.400. The 12th
and 13th centuries saw territorial expansion and cultural
achievement cut short by Mongol destruction. Revival
was likewise curtailed by the ravages of *Tamerlane
and national decline was confirmed by the decision of
Alexander I to split the kingdom between his three sons.
After some two and a half centuries of partition, the
western half being under *Ottoman rule, the eastern
under Persian, the area was reunited by the Russian
conquest of 1821–9.

Georgia, USA, a colony and state on the southern
Atlantic coast, founded in 1732 with English parliamentary support both as a bulwark against the Spanish
in *Florida and as a new start for English debtors. The
trustees under *Oglethorpe established tightly controlled
settlements at Savannah (1733) and elsewhere, prohibiting slavery, rum, and land sales, and encouraging
silk and wine production. With the relaxation of this
regime in the 1750s and immigration from Europe and
other colonies, it slowly prospered. Georgia was a strong
centre of loyalism during the War of *Independence,
and did not join the Union until 1782.

Germantown, battle of (4 October 1777). When the
British occupied Philadelphia during the American War
of *Independence the greater part of their army camped
at Germantown, north of Philadelphia. After the defeat at
*Brandywine and the British occupation of Philadelphia,
*Washington attempted a surprise counter-attack on the
main British camp. Bad weather and bad co-ordination
resulted in American defeat, heavy losses, and withdrawal
to winter quarters at *Valley Forge.

Germany, a country in central Europe. It was originally
occupied by Teutonic tribes who were driven back across
the Rhine by Julius *Caesar in 58 BC. When the Roman

empire collapsed eight Germanic kingdoms were created, but in the 8th century *Charlemagne consolidated these kingdoms under the *Franks. The region became part of the *Holy Roman Empire in 962, and almost 200 years later was invaded by the Mongols. A period of unrest followed until 1438 when the long rule of the *Habsburgs began. The kingdom, now made up of hundreds of states, was torn apart during the *Thirty Years War and when it ended in 1648, the Elector of Brandenburg-Prussia emerged as a force ready to challenge Austrian supremacy. However, the multiplicity of small German states achieved real unity only with the rise of the German confederation in the 19th century.

Ghana, an ancient kingdom in what is now east Senegal, south-west Mali, and southern Mauritania. As early as c.800 al-Fazari called Ghana 'the land of gold'. Local tradition claims that there were twenty-two princes ruling there before *Muhammad, and twenty-two after, but little is known of this period. The capital was at Koumbi Saleh, with a population of Africans and *Berbers. In 990 the King of Ghana took the Berber kingdom of Audaghost, and so took control of the gold and salt caravan trade. The capital was in two parts, one of Muslim merchants and scholars, with the royal court 8 km. (6 miles) away. It fell to the *Almoravids in 1054, to Sosso in 1203, and finally to *Sundjata in 1240. The name was adopted by the former British colony of the Gold Coast when it became independent in 1957.

Ghaznavid, *Turkish dynasty whose founder, Sebuktegin, was appointed governor of Khurasan by the Persian Samanids and established his own power in Ghazna, Afghanistan, in 962. His successors extended their realm into Persia and the Punjab, but after the reign of *Mahmud (969-1030) it fragmented under *Seljuk pressure, although the dynasty clung on in Lahore until 1186, when it was extinguished by the Ghurids, founders of the *Delhi sultanate.

Ghent, Pacification of (1576), an alliance forged during the *Revolt of the Netherlands. It enshrined the agreement of the Catholic and *Calvinist Netherlands to oppose Spanish rule and to call for the removal of imperial troops. Owned by Austria from 1714, Ghent was siezed by the French in 1792, and was incorporated into the kingdom of Belgium in 1830.

Gibbon, Edward (1737-94), British historian. He was inspired to write his *The History of the Decline and Fall of the Roman Empire* (1776-88) by a visit to Rome in 1764. He travelled widely, and entered politics briefly, and his six-volume work took several years to complete. On publication it was attacked by contemporaries for suggesting that Christianity played a significant role in the decline of Rome. Although Gibbon's work has been modified by later research, it remains a masterpiece of historical writing.

Gibraltar, a town at the north-west end of the Rock of Gibraltar which dominates the passage from the Mediterranean to the Atlantic. A low sandy area attaches it to the Spanish mainland. It was captured by the Moors in 711 and their leader who fortified it gave it its name, Jabal al-Tariq (Mount of Tariq). It became Spanish in 1462, but in 1704 during the War of the *Spanish Succession the fortress surrendered to Sir George Rooke, in command of a large Anglo-Dutch fleet, and it was retained by Britain at the Treaty of *Utrecht. Spain made repeated attempts to recapture it, notably in 1779 when Gibraltar withstood a long siege by powerful Franco-Spanish forces.

Gilbert, Sir Humphrey (c.1539-83), English navigator. He had an active and distinguished career as a soldier in France, Ireland, and the Netherlands, but his real interest was in the discovery of a north-west passage to China, and in 1576 he published his *Discourse* on the subject. In 1583, with the help of his half-brother Sir Walter *Raleigh, he set off with five ships for *Newfoundland where he established the first English colony in North America. Fierce storms were met on the return voyage, and the smallest ship of the company, in which Gilbert was sailing, went down off the Azores with all hands.

Gildas (c.500-70), monk and British historian. He is best known as the author of a polemical work (c.550) *De excidio et conquestu Britanniae* ('The Ruin and Conquest of Britain') in which he attacked the British for their wickedness. In spite of its rhetorical tone and historical inaccuracies it is the only substantial written source for the condition of Britain during a crucial period. It recorded the British victory over the Saxons at *Mount Badon.

gild *guild.

Gilgamesh, a legendary king of the *Sumerian city-state of *Uruk in southern Mesopotamia some time in the first half of the 3rd millennium BC, and hero of the Gilgamesh epic, one of the best-known works of ancient literature. The epic, which recounts Gilgamesh's exploits in his ultimately unsuccessful quest for immortality, contains an account of a flood that has close parallels with the biblical story of Noah.

Girondin, a member of a French political party whose main exponents came from the Gironde region. The Girondins were closely associated with the *Jacobins in the early days of the *French Revolution. Their radical leaders, *Brissot, Marguerite Guadet, and Marie-Jeanne *Roland, quickly gained ascendancy in the Legislative Assembly and, after the king's powers had been abolished, dominated the republican government of the Convention set up in September 1792. They held power at a critical time and were responsible for provoking the wars with France's enemies. The eventual failure of these wars led not only to the king's execution but also to the downfall of the party and the introduction of the Reign of *Terror. The crisis was brought about after the desertion of Charles-François Dumouriez, the commander-in-chief of the army, to the Austrians in April 1793. The party tried to re-establish itself by appealing to the moderates and arresting *Hébert. When this failed an immense mob expelled them from the Convention and later more than twenty of their ministers were arrested and guillotined.

gladiator, literally 'swordsman', a slave or prisoner trained to fight other gladiators, wild beasts, or condemned criminals for the entertainment of the people in ancient Rome. Gladiatorial combat originated as a ritual

A detail from a mosaic pavement in a Roman villa at Torrenova, c.300 BC, of a group of **gladiators** variously armed and in combat. (Galleria Borghese, Rome)

at funerals of dead warriors in Etruria and was introduced to Rome in 264 BC. Gladiators belonged to four categories: the Mirmillo, with a fish on his helmet, and the Sammite, both heavily armed with oblong shield, visored helmet, and short sword; the Retiarus, lightly clad, fighting with net and trident; and the Thracian, with round shield and curved scimitar. Thumbs up or down from the crowd spelt life or death for the loser. Formal combat degenerated into butchery watched by huge crowds. Women and even the physically handicapped sometimes fought. Despite popular antagonism, the emperors *Constantine and *Theodosius outlawed the combats.

glebe, land belonging to a parish church used to support its priest. The size of glebes varied enormously from 1 ha. (2 acres) to a few hundred hectares; priests might afford to engage labourers or be obliged to work the land themselves, and they could sub-lease part or all of the land. Since the glebe was a freeholding, the lord of the manor could not demand labour duties of the priest, although this immunity was not always observed.

Glencoe, Massacre of (13 February 1692). A massacre for political reasons of members of the Macdonald *clan in Scotland. By failing to swear allegiance to *William III by 1 January 1692, the rebellious clan Macdonald was technically guilty of treason; their chief's delaying tactics had been compounded by bad weather, and his oath was six days late. The Campbells, hereditary enemies of the Macdonalds, undertook to destroy them, which they attempted after enjoying Macdonald hospitality for twelve days. The clan chief and more than thirty of his followers were murdered, but the rest (about 300) escaped. Although the king probably did not order the atrocity, he did little to punish the perpetrators.

Glendower, Owen (Owain Glyndŵr) (c.1354–c.1416), Welsh national leader. Of princely lineage, he fell out with the *Lancastrian English rulers of north Wales. In September 1400 he was proclaimed Prince of Wales, and he became leader of a revolt that spread throughout Wales. The castles built by *Edward I proved to be

crucial and it was *Henry IV's control of these, rather than the failure of Glendower's alliances, that ultimately led to the end of the rebellion in 1408. Glendower remains the national hero of Wales.

Glorious Revolution, the bloodless English revolution of 1688–9. The term covers the specific events of *James II's removal from the throne and his replacement by *William and Mary, and in a more general sense it marked the end of Stuart attempts at despotism, and the establishment of a constitutional form of government.

From his accession in 1685, James II's actions aroused both Whig and Tory concern. In defiance of the law he appointed Roman Catholics to important positions in the army, the church, the universities, and the government. He claimed the right to suspend or dispense with the laws as he pleased, and his two *Declarations of Indulgence suspended penal laws against Roman Catholics and dissenters. He arrested seven bishops and brought them to trial for objecting to the illegality of the second Declaration of Indulgence. The bishops were found not guilty of seditious libel, a verdict which discredited the king. The birth of a son to the king in 1688 appeared to ensure the Roman Catholic succession and provoked leading politicians of both the main parties to invite the king's son-in-law William of Orange to England. William landed with a Dutch army in Devonshire in November. James's army refused to obey its Catholic officers, his daughters deserted him, and he was allowed to escape abroad. Parliament asked William and Mary to take over the vacant throne. A revival of arbitrary government was made impossible by the *Bill of Rights of 1689, which offered the crown to the new monarchs. James II landed in Ireland with French troops (March 1690), besieged *Londonderry, and was defeated at the battle of the *Boyne (July 1690). He returned to exile in France. The Act of *Settlement of 1701 provided for the Protestant succession.

Gloucester, Humphrey, Duke of (1391–1447), a younger son of *Henry IV. From 1422 he was guardian of his infant nephew *Henry VI, but his claim to be Regent of England was rejected by the House of Lords, and especially Cardinal Henry Beaufort, who distrusted him. He was popular with the House of Commons and the citizens of London, and came to be seen as a threat by Henry VI. In 1441 his wife, Eleanor of Cobham, was imprisoned on a charge of treason and heresy, and his own death six years later was generally accounted murder; within a few years the *Yorkists began fostering the myth of him as the 'Good Duke'.

Goa, a district on the west coast of India, Portuguese colony from 1510 to 1961. Little is known of its early history, under Hindu rule, until its seizure by Muslims in 1327. The Portuguese governor, *Albuquerque, captured it in 1510, and expanded the settlement to form an entrepôt for Indian, and eventually south-east Asian trade. It became the administrative centre of Portugal's Asian empire. When Portuguese power declined elsewhere in Asia, Goa remained as an enclave of European influence.

Godfrey of Bouillon (Godefroi de Bouillon) (c.1060–1100), French Crusader and ruler of *Jerusalem. He was a prominent leader of the First *Crusade and financed a large part of the expeditionary force. Having distinguished himself at the siege of Jerusalem, he was elected Advocate of the Holy Sepulchre—in effect, king—a position which his brother, *Baldwin I, was to assume on his death. His victories against the Muslims confirmed the Crusaders' hold on Palestine and were celebrated in the medieval song cycle, the *Chansons de Geste*.

Godolphin, Sydney, 1st Earl of (1645–1712), English statesman who gave loyal service to *Charles II, *James II, and Queen *Anne. Although he maintained close links with the Jacobites, he served *William III until he quarrelled with his colleagues in 1696. He was Queen Anne's Lord Treasurer for most of her reign, and in spite of being himself a Tory played an important part in excluding Tories from her government and in securing the Act of *Union with Scotland in 1707. Godolphin's fortunes were linked with *Marlborough's: he financed the duke's campaigns in the War of the *Spanish Succession, but when Marlborough lost favour and the Tories regained influence in 1710 his career ended.

Godwin (or Godwine) (d. 1053), Earl of Wessex. He was the father of *Harold II of England. He arranged the accession of both *Harold I (1040) and *Edward the Confessor (1042). When his daughter Edith married Edward (1045) Godwin's dominance in English politics seemed assured. Edward, however, countered his influence by relying on Norman advisers and when Godwin rebelled (1051) he exiled him together with his son Harold. They invaded England in 1052 and Edward was forced to reinstate Godwin and his family.

Golconda, a ruined city west of Hyderabad city, on India's Deccan plateau. From 1518 to 1687 it was the capital of the Muslim Qutb Shahi sultanate, whose rulers came to power at the expense of the *Bahmani Deccan sultanate. Their control extended from the lower Godavari and Krishna valleys to the eastern coast, but was ended in 1687 by Mogul conquest. Many of the royal tombs and palaces remain in a good state of repair.

Golden Horde, the name given to the *Tartars of the Mongol *khanate of the Western Kipchaks (1242–1480). The word 'horde' derives from the Mongol '*ordo*', meaning a camp. 'Golden' recalls the magnificence of Batu Khan's headquarters camp. In 1238 Batu, a grandson of *Genghis Khan, invaded Russia with a Mongol-Kipchak force. He burned Moscow and in 1240 took Kiev. After a sweep through eastern Europe he established his camp at Sarai on the Lower Volga. Khan of a region extending from Central Asia to the River Dnieper, he claimed sovereignty over all Russia but, apart from demanding tribute in money and military contingents, interfered little with the Russian princes, who in general avoided trouble by co-operating. The destruction of Kiev led to the rise of a more northerly, forest-based Russian civilization, and it was from Moscow that resistance to the Horde started.

Defeat by *Tamerlane in 1391 seriously weakened the Horde. Independent khanates emerged in the Crimea and Kazan. In 1480 the power of the Tartars was broken by *Ivan the Great.

Gondomar, Diego Sarmiento de Acuña, Conde de (1567–1626), Spanish diplomat. He achieved notoriety as ambassador to England (1613–18 and 1620–2), when he made himself one of the most influential members of *James I's court, much to the chagrin of zealous English Protestants. He was largely responsible for the execution of Sir Walter *Raleigh, and he tried to interest the king in a royal Spanish wife for the Prince of Wales. When his unpopularity in England reached its peak in 1622, he was recalled to Spain.

Gordon riots, anti-Catholic riots in London in 1780. They were led by Lord George Gordon (1751–93), who strongly objected to parliamentary moves towards *Catholic emancipation. His followers terrorized London for a fortnight. Prisons were broken open, property damaged, and many people killed before order was restored. by the military. Gordon was acquitted of high treason, but was later convicted of libel and died in Newgate Prison.

Goths, Germanic tribes who overran the Western Roman empire. Originally from the Baltic area, by the 3rd century AD they had migrated to the northern Black Sea and the Lower Danube. The eastern group on the Black Sea were known as *Ostrogoths, the western settlers on the Danube in Dacia were known as *Visigoths. In the

A contemporary print of the storming of Newgate Prison during the **Gordon riots**. The mob set fire to the jail and set the prisoners free amid cries of 'No Popery'.

4th century the Visigoths settled within the Roman empire under treaty. They expanded their sway over the West, being finally responsible for the fall of the Western empire. Their most famous leader was *Alaric. The *Ostrogoths allied themselves to the *Huns and in the 5th century under *Theodoric established their kingdom in Italy. In Spain the Visigothic kingdom survived until it was overrun by the Muslims in 711. Ulfila (c.311–83) translated the Bible into Gothic and was responsible for the conversion of the Visigoths to *Arianism. The building style of the 13th century onwards with exaggerated pointed arches and complex embellishment was called 'Gothic' or 'barbarous' by Italian artists of the Renaissance, an unfavourable comparison with Roman and Romanesque simplicity. 'Gothic' script was long used for the printing of German.

Gotland, an island in the Baltic which has been inhabited since the Stone Age. During the *Viking period it had extensive commercial contacts and its capital, Visby, was a town of the *Hanseatic League from the 11th to the 14th century. It was of strategic importance in the Baltic; in 1570, by the terms of the Treaty of Stettin, Denmark obtained control of the island, but since the Treaty of Brömsbro (1645) it has been a Swedish possession.

Gracchus, Gaius Sempronius (c.158–121 BC), Roman tribune 123 and 122 BC, brother of Tiberius Sempronius *Gracchus. He introduced extensive reforms to alleviate poverty and curb senatorial corruption and abuse of power. Innovations included the introduction of a corn subsidy and the foundation of an overseas colony in the territory of Carthage. He lost popularity when it became clear that he wanted to extend the franchise to Latins and Italians. In 121 a violent incident occurred among his following and the Senate empowered the consul to take strong action. Gaius and his supporters were hounded and killed.

Gracchus, Tiberius Sempronius (168–133 BC), Roman tribune 133 BC. He introduced a bill to repossess illegal holdings of the 'public land' and redistribute this recovered land among the needy. He bypassed the *Senate and deposed a colleague who attempted to veto him. His disregard of the Senate's authority provoked a violent reaction and when he sought re-election and massed his supporters, the extremists in the Senate, claiming to uphold law and order, charged the crowd and killed many including Tiberius himself.

Grafton, Augustus Henry Fitzroy, 3rd Duke of (1735–1811), British statesman, Prime Minister (1768–70). He was first Lord of the Treasury until *Pitt the Elder's resignation from office. Although he favoured a conciliatory policy towards the Americans, he was overruled in cabinet. He handled *Wilkes's return from exile badly, and was subjected to ferocious personal attacks by *'Junius'. He held office again under *North in 1771 and under *Shelburne in 1782, but played little part in politics thereafter.

Granada, a province in south-eastern Spain, probably of Roman origin, with the city of Granada as its capital. Granada was conquered by Moorish invaders and survived as a Muslim state long after virtually all the rest of Spain had been seized by the Christians. The Alhambra, an old Moorish citadel and royal palace, gives an indication of the splendour of Moorish civilization in Granada. It finally submitted easily to *Ferdinand and Isabella in 1491. By 1609 the Jews and Moriscos (Muslim converts to Christianity) had been expelled and the *Inquisition was at work in the province. Granada never regained the influence it had enjoyed in Muslim hands.

Granby, John Manners, Marquis of (1721–70), British army officer. He became a hero during the *Seven Years War. He was made commander-in-chief of the British army in 1766, but was subjected to bitter attacks by *'Junius' for political reasons. Unnerved by such savage criticism, and in declining health, he resigned most of his public offices in 1770, and died in debt.

Grand Alliance, War of the * Nine Years War.

Grand Remonstrance (1641), a document drawn up by opposition members of the English *Long Parliament, indicting the rule of *Charles I since 1625 and containing drastic proposals for reform of church and state. Although it passed the House of *Commons by just eleven votes, and swords were first drawn in the Commons over the question of its printing, many saw it as a vote of no confidence in the king. It drove Charles into his disastrous attempt to arrest its prime movers, including John *Pym, an act of force which further alienated opposition Members of Parliament.

Grand Siècle, the age of *Louis XIV (1643-1715), the period of France's greatest magnificence, when it replaced Spain as the dominant power in Europe and established its cultural pre-eminence. This pre-eminence was not surprising: France was under strong political control, agriculturally fertile, and its population of some nineteen to twenty millions was far greater than that of any other European state. The genius of *Richelieu as chief minister (1624-42) had established the authority of the monarchy and achieved a far greater degree of internal unity for France than was possessed by its rivals. Europe was impressed by the splendours of the court of *Versailles. French military predominance was won by the brilliant victories of *Condé and *Turenne and the creation of the first modern standing army. The frontiers of France were strengthened by the acquisition of Artois, Alsace, and the Franche-Comté. French fashions in dress were copied everywhere, as were the elegant products of French craftsmanship. The works of the French classical writers Racine, Molière, and La Fontaine were accepted as models, and French became the polite language of Europe, the language spoken by German princes and Russian nobles, the language in which international treaties and learned books were written. The splendour of the *Grand Siècle*, based as it was on heavy taxation of the poorest classes, and a commitment to expensive military campaigns, gave way after the king's death to the more turbulent climate of the 18th century.

Grand Tour, a leisurely journey through Europe, often lasting several years, made by young Englishmen in the 18th and 19th centuries. The sons of the aristocracy, often accompanied by a tutor, completed their education by enriching their knowledge of classical art and of European society. The eventual destination was Italy, specifically Naples and Rome, where there were well-established

The **Grand Tour** offered young English gentlemen of the 18th and 19th centuries the opportunity to inspect the glories of classical Italy. The painting, dated 1750, of which this is a detail, shows a group of 'connoisseurs' in Rome, seemingly as interested in each other's modish dress and conversation as in the Roman ruins they have come to admire. (Yale Center for British Art, Paul Mellon Collection)

colonies of expatriate painters, architects, and connoisseurs. As well as purchasing antique sculpture these patrons bought contemporary Italian paintings, including portraits of themselves, with which to adorn their houses. The wealth of Greek and Roman statuary and Italian drawings and paintings in the *country houses and museums of Great Britain are the legacy of the Tour.

Granville, John Carteret, 1st Earl of (1690–1763), English statesman. He was at first Tory in sympathy, but he welcomed the Hanoverian succession and was employed as a diplomat by *Stanhope who recognized his talent. *Walpole saw him as a potential rival and sent him to Ireland as Lord Lieutenant (1724). As the opposition to Walpole grew, Carteret became a leading member of it, and on Walpole's fall in 1742 he dominated the new ministry. Yet he relied too exclusively on royal support, and in 1744 he was ousted by the *Pelhams who persuaded George II to dismiss him. He remained in politics, but without power.

Grattan, Henry (1746–1820), Irish statesman, a champion of Irish independence. He was born and educated in Dublin, where he trained as a barrister and entered the Irish Parliament in 1775. A brilliant orator, he led the movement to repeal *Poynings' Law, which made all Irish legislation subject to the approval of the British Parliament. After considerable agitation the British government yielded and repealed the Act (1782). He also strongly opppposed the Act of *Union (1801), which merged the British and Irish parliaments. In 1806 he

became member for Dublin in the British House of Commons and devoted the rest of his life to the cause of *Catholic emancipation.

Gravettian culture *Upper Palaeolithic.

Great Awakening, an American revivalist movement, which was a response to the growing formalism and *Arminianism of early 18th-century American Christianity. Though revivals began in New Jersey in 1719, Jonathan *Edwards's preaching, and the resultant conversions in the 1730s gave it widespread recognition and influenced the *Wesleys. George *Whitefield's mission (1739–41) won many converts from Pennsylvania to Maine, but his followers Gilbert Tennent and James Davenport precipitated schisms in both Congregational and Presbyterian churches, which also affected colonial politics. In Virginia, Samuel Davies led revivals (1748–53) among the 'New Side' Presbyterians. Baptists and Methodists also embraced the new movement. By questioning established authority, founding new colleges, and revivifying evangelical zeal, it helped to prepare the revolutionary generation in America.

Great Britain *Britain, Great.

Great Plague (1664–5), a disastrous epidemic, mainly confined to London and south-east England. Bubonic plague had recurred at intervals since the Middle Ages, but there had been no serious outbreak for thirty years and its violent reappearance was not expected. About a fifth of London's population of almost half a million died. Business in the city came to a standstill. The court and all those able to move into the countryside prudently did so, as the disease was less virulent there. It reached the village of Eyam in Derbyshire, however, in a box of infected clothes, and in order to prevent its spread the villagers agreed to isolate themselves: nearly 300 of the population of 350 died. The *Royal Society conducted

The **Great Plague** of 1664–5, as this contemporary woodcut shows, sent those Londoners who had the means to do so fleeing from the city. Thousands took refuge from the epidemic in rural areas. (Ashmolean Museum, Oxford)

post-mortems to try to establish the cause of the plague, but without success. At the height of the epidemic plague pits were dug to receive the dead, and hand-carts were taken from house to house, collecting the bodies. The *Fire of London in the following year destroyed many of the close-packed slums in which the plague flourished, and after 1665 the disease disappeared from London.

Great Schism, a term used to describe two breaches in the Christian Church. The Great or *East–West Schism (1054) marked the separation of the Eastern (Orthodox) and Western Christian churches. The Great Schism of 1378–1417 resulted from the removal of the papacy from Italy to France in 1309. Feuds among the Italian cardinals and their allies among the Italian nobility led to Pope Clement V (1305–14) moving the papal residence from Rome to Avignon in southern France. French interests came to dominate papal policy and the popes, notorious for their luxurious way of life, commanded scant respect. An attempt to return the papacy to Rome was followed by schism as two rival popes were elected by the cardinals, Urban VI by the Roman faction and Clement VI by the French faction. The period of popes and rival *antipopes lasted until the Council of *Constance (1417) elected Pope Martin V of the Roman party and deposed his French rival. The division of the papacy discredited the Church and was criticized by those demanding reform, notably *Wyclif.

Great Wall of China, a fortification built across northern China as a protection against the nomadic tribes of Mongolia and Manchuria. By the 3rd century BC some Chinese frontier states had built defensive walls. The first Qin emperor *Shi Huangdi ordered that these walls should be joined together and extended, using the forced labour of vast gangs of men and women, many of whom perished. The course of his wall was from the Gulf of Liaodong 2,250 km. (1,400 miles) across mountain, steppe, and desert to southern Mongolia. The wall was extended by the *Han to Yumen (Jade Gate) in Gansu province in order to facilitate their expansion into Central Asia. Thereafter much of the wall was rebuilt or reconstructed on different routes, notably by the *Northern Wei, the *Sui, and lastly by the *Ming. The

wall as seen today is largely of Ming construction. It is approximately 7.5 m. (25 ft.) high and 3.75 m. (12 ft.) wide at its top and is constructed of earth, with stone facings on the eastern sections. The wall acted as a demarcation line between the steppe and cultivated land. If adequately manned it could delay raiding parties, but frequently the steppe-dweller was able to ride through its undefended gates. The borders of *Qing China were far to the north of the wall, which ceased to have any military significance after 1644.

Greece, a mountainous country in south-east Europe. Its history begins c.2000–1700 BC with the arrival in the mainland of Greek-speaking peoples from the north. There followed the *Mycenaean civilization which flourished until overthrown by the *Dorians at the end of the 12th century BC. After an obscure period of history (the Greek 'Dark Ages') the city-state (*polis) emerged.

In the early 5th century the Greeks repulsed Persian attempts to annex their land. *Athens and *Sparta were

A Dutch engraving of the **Great Wall of China** from *China monumentis illustrata* (1667) by Athanasius Kircher. Horses demonstrate the width of its elevated roadway and an elephant conveys the impressive dimensions of a gateway. The wall incorporated watchtowers as signalling posts to warn of raiders from Central Asia, while the gateways allowed traffic to pass between civilized China and the uncultivated steppe beyond.

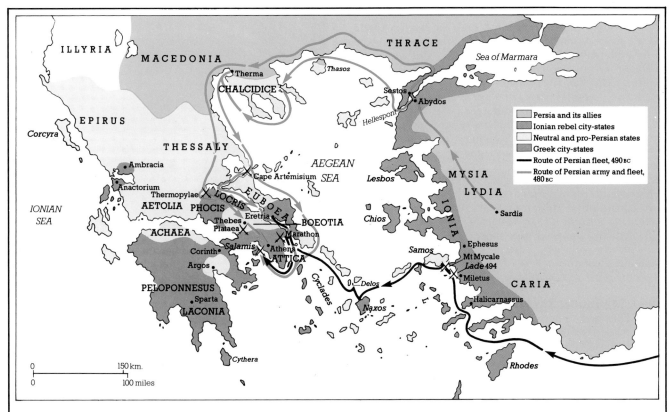

Greek-Persian wars (499–479 BC)

The revolt of the Greek cities in Ionia against the Persian empire was eventually put down, but it demonstrated to the Greeks that resistance to Persia was not hopeless. Their defeat of Xerxes' punitive expedition at Marathon further inspired resistance; it also ensured a renewal of the Persian attack (delayed until 480 BC by a revolt in Egypt). The Greek victory over the Persian fleet at Salamis broke the lines of supply to the Persian land forces, and prepared the way for the destruction of Mardonius' army in 479.

now the major sea and land powers respectively, and after a prolonged struggle it was Sparta who by 404 had crushed Athens and destroyed the Athenian empire in the *Peloponnesian War. In the 4th century *Thebes toppled Sparta, but Greece as a whole was soon forced to bow before an outside conqueror—*Philip II of Macedonia. After the death of his son, *Alexander the Great, the Greek world was dominated by the Hellenistic kingdoms with the cities of Greece playing comparatively minor parts in the power struggle. Then Rome intervened in the *Macedonian wars, until the year 146 BC saw the defeat of the *Achaean League, the sacking of Corinth, and the final incorporation of Greece into the Roman empire. Later it was part of the *Byzantine empire, but fell under the control of the Ottoman Turks in 1460.

Greek city-state   *polis.

Greek–Persian wars, conflicts which dominated the history of the eastern Mediterranean in the first half of the 5th century BC. In 499 BC the Greek cities of Ionia in Asia Minor revolted from the Persian empire. With some short-lived support from Athens and Eretria, they captured and burnt the important city of Sardis, but gradually the Persians regained control, the Greek fleet being finally crushed at Lade in 494. In 490 a Persian expeditionary force sailed across the Aegean. The capture of Eretria—the first goal—was achieved after a week-long siege and with help from Eretrian traitors. The Persians then landed in Attica but after a defeat at *Marathon they were forced to withdraw to Persia.

In 480 a much larger invasion force threatened Greece, advancing along the northern and western shores of the Aegean. A small Greek army and a large Greek fleet were positioned respectively at *Thermopylae and Artemisium, but despite vigorous fighting on land and sea the Greeks were forced to withdraw to the Isthmus of Corinth. With central Greece lost, the Athenians evacuated their city, while the Greek fleet, at *Themistocles' urging, lured the Persians into battle off *Salamis. In these narrow waters the Greek warships had the advantage and won a decisive victory which caused the Persian king *Xerxes to withdraw to Asia. Mardonius, his second-in-command, remained to continue the campaign with the army. In 479 Greeks and Persians met at *Plataea. The Greeks were eventually successful, the Spartans and their Tegean allies ensuring victory when they overcame the élite Immortals (the Persian royal bodyguards) and killed Mardonius. Meanwhile a Greek fleet was winning another great victory off Mycale in Asia Minor. Soon afterwards some of the Greeks formed the *Delian League to be the instrument by which they would continue the war against the Persians.

Greek religion, the religion of the ancient Greek world. It was polytheistic, involving the worship of several gods and goddesses. The most important deities were the sky-god Zeus (ruler of Olympus), his wife Hera (goddess of marriage), Poseidon (god of sea and earthquakes), the virgin goddess Athene (learning and the arts), Apollo and his sister Artemis (sun and moon, the one patron of music and poetry, the other of chastity and hunting), Hephaestus (fire and metalwork), Aphrodite (love and beauty), Ares (war), Demeter (crops), Hestia (hearth and home), and Hermes, the messenger of the gods. Although all were revered, different cities had different individual gods as their special patrons. Apollo's shrine at *Delphi was recognized throughout the Greek world. Despite the efforts of poets and philosophers the Greek gods never lost their essentially anthropomorphic character, and Greek religion largely lacked that insistence on high standards of personal morality which is associated with Christianity, Judaism, and Islam, and had no developed concept of an 'afterlife'. Its influence was weakened by some of the more mystical aspects of Near-Eastern religion, especially in the period after Alexander's conquest of Asia Minor and Egypt. It finally fell victim, first to Christianity, and then to Islam.

Green Mountain Boys, American patriots from Vermont. They were guerrilla fighters organized in 1771 by Ethan *Allen to defend their settlements against New York claims. Named after Vermont's mountain chain, they came to support the patriot cause, assisting in the capture of *Ticonderoga (1775), the battle of *Bennington, and Burgoyne's surrender at *Saratoga (1777).

Greene, Nathanael (1742–86), American general and Rhode Island Quaker. He led his colony's troops at the siege of Boston (1775–6) and fought at Long Island, Trenton, Brandywine, Germantown, and Monmouth. In command in the south (1780–1) he waged a brilliant hit-and-run campaign against *Cornwallis, eventually recapturing most of the region.

Greenland, a large island, lying mainly within the Arctic Circle. It was sighted in the early 10th century by the Icelander Gunnbjörn Ulf-Krakuson, but it was not until 985–6 that it was settled by *Erik the Red. Climatic shifts in the North Atlantic rendered the southwestern region habitable, and here Erik founded Brattahlid (modern Qagssiarssuk), known as the 'Eastern Settlement'; to the north Gardar, the 'Western Settlement', was established near Julianehaab. Farming was difficult but was supplemented by fishing and the export of furs and walrus ivory. Christianity came with Leif Ericsson c.1000. A chapel was eventually replaced by twelve parish churches, a monastery and nunnery, and a cathedral at Gardar by the 12th century. At its height the population reached c.3,000, but the *Black Death caused it to decrease in the 14th century. The settlements persisted into the 15th century, but only in a poverty-stricken state until they disappeared.

Gregory I, St (the Great) (540–604), Pope (590–604). When he became pope Italy was in a state of crisis, devastated by floods, famine, and Lombard invasions, and the position of the Church was threatened by the imperial power at *Constantinople; it was owing to Gregory that many of these problems were overcome. He made a separate peace with the Lombards in 592–3, and (acting independently of the imperial authorities) appointed governors to the Italian cities, thus establishing the temporal power of the papacy. One of his greatest

A fresco of **Gregory I** by Nicolò Pizzolo. He worked to restore and promote the prestige of the Western Christian Church. (Chiesa degli Eremitani, Padua)

achievements was the conversion of England to Christianity, by St *Augustine. Throughout his papacy he effectively opposed the double assault on the Church from paganism and the *Arian heresy. His interest in music led to developments in the plain chant which bears his name—the Gregorian Chant.

Gregory VII, St (Hildebrand) (c.1021–85), Pope (1073–85). He argued for the moral reform of the Church and that the Christian West should be united under the overall leadership of the papacy. The latter was opposed by many secular rulers and the prolonged struggles which followed have come to be known as the *Investiture contests. His most formidable opponent was the Holy Roman Emperor Henry IV. When in 1077 he submitted to the pope at *Canossa papal supremacy seemed nearer. However, Henry's submission was merely a tactical one and he later attacked Rome itself, forcing the pope to retreat to Salerno in southern Italy, where he died. He urged celibacy of the clergy and opposed *simony.

Gregory of Tours, St (538–94), a churchman and historian in the court of the Frankish king Childebert II. He was educated by his uncle, St Gall, and resisted attacks on the church during the reign of King Chilperic. Of his many works the best known is his vast *Historia Francorum* ('History of the Franks') a source of early *Merovingian history. He was canonized soon after his death.

Grenada, a West Indian island, discovered by Columbus in 1498. Colonized by the French governor of Martinique in 1650, it passed to the control of the French crown in

1674. The island was conquered by the British during the *Seven Years War and ceded to them by the Treaty of Paris (1763). An uprising in 1795 against British rule, supported by many slaves, was put down the following year.

Grenville, George (1712-70), British statesman. He became Prime Minister in 1763. He was largely responsible for the government's mishandling of the *Wilkes affair and also for the effort to raise revenue in America which resulted in the *Stamp Act of 1765. George III came to dislike him and some of his fellow politicians objected to his authoritanism, although his budgets were well-received by backbenchers in Parliament. His failures led to his dismissal from office in 1765.

Grenville, Sir Richard (1542-91), English naval commander. He became Member of Parliament for Cornwall (1571), led the unsuccessful expedition to colonize *Roanoake planned by his cousin Sir Walter *Raleigh, and supplied three ships to the force assembled against the Spanish Armada. He died after an epic battle off the Azores, during which his ship *Revenge* held out for fifteen hours against a powerful Spanish fleet.

Gresham, Sir Thomas (c.1519-79), English cloth merchant, financier, diplomat, and the founder of the Royal Exchange. His greatest skill lay in negotiating loans for the English government, especially in Antwerp after 1551. He worked with William *Cecil, providing him with valuable information from the Continent, and exerting considerable influence over Elizabethan economic policy. He founded Gresham College in 1579 as a venue for public lectures and the *Royal Society grew from these meetings at Gresham's house. ('Gresham's Law', which states that 'bad money drives out good', was not in fact formulated by him.)

Grey, Lady Jane (1537-54), the 'Nine Days' Queen' of England in July 1553. As a descendant of *Henry VII's younger daughter Mary, she had some claim to the English throne, and her father-in-law, John *Dudley, persuaded *Edward VI to name her as his successor. *Mary I ousted her easily, and she was beheaded after her father had incriminated her further by participating in *Wyatt's Rebellion.

Grotius, Hugo (1583-1645), Dutch jurist and scholar. He distinguished himself in both literary and diplomatic fields. In 1613 he supported his patron *Oldenbarneveldt in the dispute between the Arminians and the Counter-Remonstrants, for which he was arrested by Maurice of Nassau, tried, and sentenced to life imprisonment. His wife helped him to escape and he took refuge in Paris, entered Sweden's diplomatic service, and was Swedish ambassador to France from 1634 to 1645. His best-known work, *De Jure Belli ac Pacis* ('Concerning the Law of War and Peace', 1625), is generally considered to be the first definitive book on international law.

Guangzhou (Canton), the capital of Guangdong province, China. It lies on the Zhu Jiang (Pearl River). Under the *Tang Arab and Jewish traders had quarters there and enjoyed extraterritorial rights, not being subject to Chinese laws. From the 16th century it was a trading port for European vessels. After 1757 it was the only port in China open to Europeans, who were restricted to thirteen 'factories' (trading posts) outside its walls, where they brought tea and silk, and after c.1800 sold increasing quantities of opium.

Guatemala, a country in Central America. In prehistory it was culturally linked to the *Yucatán peninsula and witnessed the rise of pre-Maya and *Maya civilizations. In the northern and central lowlands arose the great, classic Maya cities such as Tikal, Uaxactún, Altar de Sacrificios, Piedras Negras, Yaxhá, and Seibal; in the southern highlands were the cities of Zacualpa, Kaminaljuyú, Cotzumalhuapa, and others. They had political and economic connections with each other, and with prehistoric cities in southern and central Mexico, such as *Teotihuacán and Monte Albán (in Oaxaca). Spanish *conquistadores arrived in 1523, seeking new American conquests, and the region soon became the Audiencia (a high court with a political role) of Guatemala, under the viceroyalty of *New Spain.

Guelph, a member of a faction originating in the German Welf family, who were dukes of Saxony and Bavaria. The Welfs were the traditional opponents of the *Hohenstaufens in Germany and Italy (where they were known as Guelphs and the latter were known as the Ghibellines). In the 12th century the Guelph leader was *Henry the Lion and he tended to support the papacy against the aspirations of the Holy Roman Emperors. Guelph support was in the major Italian towns and cities. Their rivals were the imperial party and their strength came mainly from the great aristocratic families. In local feuds, no matter what the cause, the antagonists came to associate themselves with one or other of the opposing families whose names continued to be used for many years after the original disputes were forgotten.

Guesclin, Bertrand du (c.1320-80), French army commander and Constable of France from 1370. He attracted attention at the siege of Rennes (1356-7), and was promoted by the regent, *Charles V, to the office of Constable of France. He fought in campaigns against the English, and, from 1366-9 against Spain where he was defeated and captured at the battle of Najera in 1367. It was his conduct of the war against the English which helped Charles recover his kingdom.

Guicciardini, Francesco (1483-1540), Italian statesman and historian. As Florentine ambassador to Aragon (1512-14), and then in the service of the papacy (1515-34), he showed outstanding administrative ability, and he also became a prolific political writer. In 1536, back in Florence, he began his monumental *Storia d'Italia* ('History of Italy'), which covered the years from 1494 to 1534. Although he died before completing the final revision, it stands as the most objective contemporary history of the country during the period of Italy's wars with France.

guild, an association of townspeople formed to provide mutual protection of trading practices. Guilds may have developed in Syria and Egypt or in Roman and Byzantine trade associations (*collegia*) but are more likely to have arisen in western Europe in the 7th century. In England they are found in the Anglo-Saxon family associations (frithgilds) which protected members' interests, including

Apprentices exhibit their woodworking skills before the master of their **guild**, hoping to become members themselves and free either to serve a master of their own choosing or to set up in business on their own. Their training would have lasted for several years. (British Library, London)

those of trade. These early benevolent associations of Europe were usually based in towns, were often religious in character—perhaps dedicated to a saint—and engaged in charity and local administration. Religious guilds, mainly devoted to devotional, charitable, and social activities, remained important in English towns and parishes throughout the Middle Ages. From the early 11th century merchants and traders combined to regulate trade. The merchant guilds they formed controlled markets, weights and measures, and tolls, and negotiated *charters granting their towns borough status. They maintained the charitable work of the earlier religious guilds. However, their monopolistic character forced the small crafts and trades to form their own associations, craft guilds, before the end of the 12th century. Each craft had its own guild which set quality standards and evolved a hierarchy consisting of master, journeymen, and apprentices (serving for up to twelve years). Guilds declined from the 16th century, being unable to adapt to the emergence of new markets.

guillotine, the instrument used to inflict capital punishment by decapitation during the *French Revolution. A similar device had been used in Europe since the Middle Ages and had fallen into disuse when Dr Guillotin (1738–1814) suggested its reintroduction. After satisfactory tests on dead bodies it was erected on the Place de Grève in 1792. 'La Guillotine' was used extensively during the *Terror, accounting for 1,376 victims between 10 June and 27 July 1794.

Guiscard, Robert (c.1015–85), *Norman warrior. He was the son of Tancred de Hauteville, and with his brother Roger established himself in southern Italy. In 1053 they defeated the forces of Pope Leo IX, securing Apulia and Calabria. Pope Nicholas II, enlisting Norman aid against the *Byzantines, gave him Sicily, though it was not finally conquered till 1090. Excommunicated by Pope Gregory VII for his attack on Benevento, he nevertheless fought for him against the invading *Henry IV of Germany. The brothers sacked Rome in 1084, driving Henry out.

Guise, a branch of the ducal house of Lorraine that rose to prominence in 16th-century France. **Claude de Lorraine** (1496–1550) was created duke in 1528; he had distinguished himself in a number of French military victories, including Marignano (1515). **Francis** (1519–63), his son and heir, became the most effective commander in the armies of Henry II. He was active throughout the 1550s, capturing Calais from the English (1558) and helping to bring about the Peace of Cateau-Cambrésis in 1559. His brother **Charles** (1524–74) became Cardinal of Lorraine in 1550, and his sister **Mary** (1515–60) married James V of Scotland and was the mother of *Mary, Queen of Scots.

In 1559, on the accession of Francis II, the Catholic Guise family was the most influential in France. Its dealings with the *Huguenots and *Bourbons (1559–62) led directly to the outbreak of the *French Wars of Religion. Francis was assassinated in 1563. His son **Henry** (1550–88), the third duke, fought in the third and fourth wars, and was one of the instigators of the *St Bartholomew's Day Massacre. In 1576 he took the lead in organizing the *Holy League, but Henry III had him assassinated in 1588, when he was being put forward as a possible heir to the throne. His brother, **Charles** (1554–1611), kept the Guise and extremist Catholic causes alive until 1595, when he submitted to *Henry IV. The Guise ducal line died out in 1688.

Gujarat, a region of India, consisting of the Kathiawar peninsula on the north-western coast plus a narrow hinterland. *Indus civilization sites indicate early urban settlement, followed by absorption into numerous Hindu and Buddhist empires from as early as the *Mauryas (3rd century BC). Its name originated from a Hun tribal dynasty, the Gurjaras (8th–9th century). After earlier contact with Arab traders, its era of Muslim rule began in 1298 with the invasion of the *Khalji Delhi sultan, Ala ud-Din. A strong independent Muslim sultanate, based on the new capital city of Ahmadabad, ruled from 1411 until 1573, when Mogul invasion again reduced Gujarat to provincial subordination. *Maratha expansion in the mid-18th century brought the area under the Marathas, before the region was absorbed into the *East India Company's Bombay presidency.

Gunpowder Plot, a Catholic scheme to murder *James I of England and his Parliament at the state opening on 5 November 1605, to be followed by a national Catholic uprising and seizure of power. The plotters, *recusants led by Robert Catesby, saw violent action as the only

The conspirators in the **Gunpowder Plot**, a detail from a contemporary German print. Their leader, Robert Catesby, is second from the right; next to him is Guy Fawkes. They were all all hung, drawn, and quartered in January 1606, and their severed heads were subsequently displayed on poles.

way to gain toleration for English Catholics. They were subsequently disowned by the majority of their fellow religionists, who had little sympathy for the conspiratorial tradition established by *Ridolfi, *Throckmorton, and *Babington. It has been suggested that Robert *Cecil manufactured the plot, in order to discredit the Catholic cause. Cecil learned of the plot through Lord Mounteagle, a Catholic peer. On the eve of the opening, Guy Fawkes (1570–1606) was discovered in the cellar under the House of Lords on guard over barrels of gunpowder. The other plotters were overcome in the Midlands after brief resistance. Fawkes and seven others, including Sir Everard *Digby were tried before *Coke and executed in January 1606. Immediately afterwards, the penal laws against Catholics were stiffened, and an Oath of Allegiance imposed, but to the chagrin of many Puritans and Anglicans, enforcement of the new legislation soon became sporadic. Bonfires, fireworks, and the burning of 'guys' still mark 5 November in Britain.

Gupta, a dynasty ruling from Pataliputra in north-east India from the mid-4th to the mid-6th century AD. The origins and rise of the family are uncertain, but the era is usually dated from the accession of Chandra Gupta I about AD 320. During the reigns of his son, Samudra Gupta (c.330–80), and grandson, Chandra Gupta II (c.380–415), military conquests achieved direct sway over most of the Indo-Gangetic plain, and suzerainty over much of central and eastern India. Failure to establish tight administrative control over this far-flung empire, and Hun attacks on the northern heartland which began during the next two reigns, caused the collapse of Gupta power by the mid-6th century.

At its height the Gupta empire experienced stable political and economic conditions in which a 'golden' or 'classical' age in religion, philosophy, literature, and architecture was born. Both Hinduism and Buddhism flourished during this period as attested by a Chinese

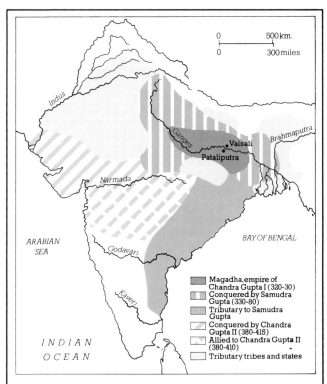

The Guptas: 320–415

The Guptas exercised their influence over the greater part of India during the 4th century AD. The dynasty was founded in the Ganges valley state around Pataliputra (Patna) by Chandra Gupta I and its influence was roughly doubled by the conquests of his successor Samudra Gupta. In the distant provinces Gupta authority was diffuse and short-lived, but their cultural influence endured.

Buddhist monk, Fa Xian, who travelled through northern India in the early 5th century.

Gustavus I Vasa (1496–1560), King of Sweden (1523–60) and founder of the Vasa dynasty. He fought against the Danes in 1517–18, but was successful only after 1520, thanks to financial and naval backing from the city of Lübeck. His election as king ended the 126-year-old Union of *Kalmar by which Sweden had been subordinated to Denmark. He created a national army of volunteers and built an efficient navy. He also modernized the economy, and in 1527 broke with the Roman Catholic Church for mainly political reasons. In 1544 the Swedish crown was made hereditary in the Vasa family.

Gustavus II Adolphus (1594–1632), King of Sweden (1611–32). He was the grandson of *Gustavus I and is generally recognized as Sweden's greatest ruler. His partnership with the Chancellor, *Oxenstierna, bore fruit in important reforms in the government, the armed forces, the economy, and education. His reign was notable for the absence of friction between crown and aristocracy.

Abroad he inherited three Baltic struggles: the Kalmar War with Denmark (1611–13); the Russian War (1611–17); and the intermittent conflict with Poland. The successes achieved by his mobile, highly motivated, and disciplined forces impressed Cardinal *Richelieu, who negotiated the Treaty of Altmark (1629) between Sweden and Poland, so that the Swedes could be released for action in the *Thirty Years War. Leaving the domestic government in the hands of Oxenstierna, Gustavus crossed to Germany in 1630 and proceeded to turn the tide of the war against the imperial forces. He was a devout Lutheran, and his war aims grew more ambitious as his invasion prospered. Originally intending to prevent the Catholic Habsburgs from dominating the Baltic, by 1632 he was pursuing grand imperial designs of his own. He was killed in action at *Lützen.

Gutenberg, Johann (c.1398–1468), German craftsman and printer, credited with the introduction of moveable type in *printing. In 1438 he contracted with three partners to develop printing techniques. In 1450 he raised money from a Mainz merchant, Johann Furst, and was able to produce his celebrated Bible. Furst later sued him for debt, and seized and used his printing facilities. His later career is obscure, although he seems to have continued printing.

Gwalior *Sindhia.

Gwyn, Nell (1650–87), English comic actress, and mistress of *Charles II of England. She sold oranges at Drury Lane Theatre before taking up acting. Charles made her his mistress in 1669, and she bore him two sons. She was always popular with the people, proudly proclaiming, 'I'm the Protestant whore', when a mob attacked her coach in mistake for that of his Roman Catholic rival. Charles II's deathbed instruction 'Let not poor Nelly starve' was faithfully carried out by his brother James II, who gave her a pension of £1,500 a year.

Haakon IV Haakonsson (the Old) (1204–63), King of Norway (c.1220–63). His reign was troubled by internal dissensions and he had Earl Skule executed in 1239. Iceland and Greenland were added to the Norwegian crown but control of the *Hebrides was lost. This followed his defeat by *Alexander III of Scotland in the decisive battle at Largs in 1263.

Habsburg (or Hapsburg; also called the House of Austria), the most prominent European royal dynasty from the 15th to the 20th century. Their name derives from Habichtsburg (Hawk's Castle) in Switzerland, built in 1020. The founder of the family power was Rudolf I, who was King of the Romans (1273–91) and conqueror of Austria and Styria, which lands he bestowed on his two sons in 1282, beginning the family's rule over Austria which lasted until 1918. Habsburg domination of Europe resulted from the shrewd marriage policy of Maximilian I (1459–1519), whose own marriage gained the Netherlands, and that of his son, Philip, which brought Castile, Aragon, and the Spanish New World possessions. His grandson, Emperor *Charles V (1500–58), ruled the whole Habsburg empire, almost achieving the family motto '*Austriae est imperare orbi universo*' ('Austria is destined to rule the world').

On Charles's death his brother Ferdinand (1503–64) ruled Germany and Austria, while his son, *Philip II (1527–98), ruled the Spanish inheritance. The Protestant *Reformation and the the Turkish invasion of eastern Europe forced the two branches to co-operate in preserving European Catholicism throughout the Counter-Reformation and the *Thirty Years War (1618–48). The long struggle weakened the family despite the addition of Portugal and Hungary to their territory. In 1700 the Spanish line became extinct and in the subsequent War of the *Spanish Succession (1703–13) the Spanish inheritance passed to the Bourbons. The Austrian Habsburgs (after 1740 the House of Habsburg-Lorraine) flourished again under *Maria Theresa (1717–80) and her son *Joseph II (1741–90). The Habsburgs ended the Napoleonic wars with the loss of the Austrian Netherlands and the title *Holy Roman Emperor.

Hadar, a site in the Afar region of Ethiopia where early human fossils dating back to 3.2–2.9 million years ago have been found. The finds were attributed to a new species of *australopithecine, *Australopithecus afarensis*, but this has been challenged by some scientists who say that they belong to another species, *Australopithecus africanus*, or even that two or more forms are represented at the site. The best-known of the fossils is 'Lucy', a skeleton nearly half complete. She was in her twenties when she died, with a height of only 1.1–1.3 m. (3.6–4.3 ft.) and a weight of about 27 kg. (60 lb.). *Oldowan stone tools from Hadar, dated about 2.5 million years, are the earliest known human artefacts.

Hadrian (Publius Aelius Hadrianus) (c. AD 76–138), Roman emperor (117–38). He was born in Spain, and

Habsburgs

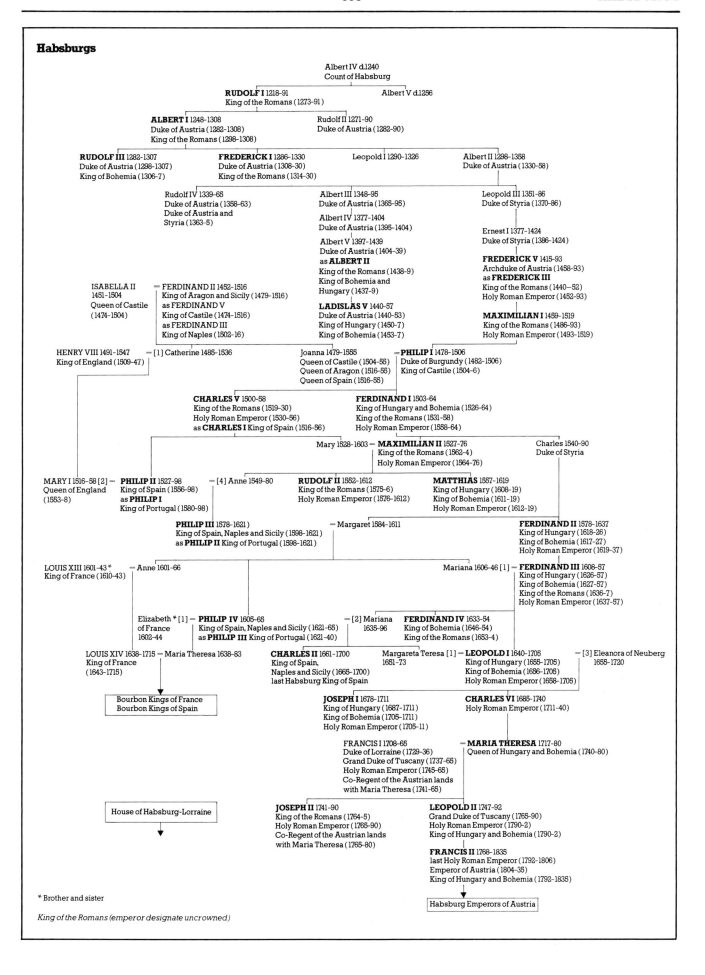

Albert IV d.1240
Count of Habsburg

RUDOLF I 1218-91   Albert V d.1256
King of the Romans (1273-91)

ALBERT I 1248-1308   Rudolf II 1271-90
Duke of Austria (1282-1308)   Duke of Austria (1282-90)
King of the Romans (1298-1308)

RUDOLF III 1282-1307   **FREDERICK I** 1286-1330   Leopold I 1290-1326   Albert II 1298-1358
Duke of Austria (1298-1307)   Duke of Austria (1308-30)   Duke of Austria (1330-58)
King of Bohemia (1306-7)   King of the Romans (1314-30)

                   Rudolf IV 1339-65   Albert III 1348-95   Leopold III 1351-86
                   Duke of Austria (1358-63)   Duke of Austria (1365-95)   Duke of Styria (1370-86)
                   Duke of Austria and
                   Styria (1363-5)   Albert IV 1377-1404
                                                Duke of Austria (1395-1404)   Ernest I 1377-1424
                                                                                Duke of Styria (1386-1424)
                                                Albert V 1397-1439
                                                Duke of Austria (1404-39)   **FREDERICK V** 1415-93
                                                as **ALBERT II**   Archduke of Austria (1458-93)
                                                King of the Romans (1438-9)   as **FREDERICK III**
ISABELLA II   = FERDINAND II 1452-1516   King of Bohemia and   King of the Romans (1440—52)
1451-1504   King of Aragon and Sicily (1479-1516)   Hungary (1437-9)   Holy Roman Emperor (1452-93)
Queen of Castile   as FERDINAND V
(1474-1504)   King of Castile (1474-1516) **LADISLAS V** 1440-57
                   as FERDINAND III   Duke of Austria (1440-53)   **MAXIMILIAN I** 1459-1519
                   King of Naples (1502-16)   King of Hungary (1450-7)   King of the Romans (1486-93)
                                                King of Bohemia (1453-7)   Holy Roman Emperor (1493-1519)

HENRY VIII 1491-1547   = [1] Catherine 1485-1536   Joanna 1479-1555   = **PHILIP I** 1478-1506
King of England (1509-47)   Queen of Castile (1504-55) Duke of Burgundy (1482-1506)
                                                         Queen of Aragon (1516-55)   King of Castile (1504-6)
                                                         Queen of Spain (1516-55)

                   CHARLES V 1500-58   **FERDINAND I** 1503-64
                   King of the Romans (1519-30)   King of Hungary and Bohemia (1526-64)
                   Holy Roman Emperor (1530-56)   King of the Romans (1531-58)
                   as **CHARLES I** King of Spain (1516-56) Holy Roman Emperor (1558-64)

                                        Mary 1528-1603 = **MAXIMILIAN II** 1527-76   Charles 1540-90
                                                         King of the Romans (1562-4)   Duke of Styria
                                                         Holy Roman Emperor (1564-76)

MARY I 1516-58 [2]   **PHILIP II** 1527-98   = [4] Anne 1549-80   **RUDOLF II** 1552-1612   **MATTHIAS** 1557-1619
Queen of England   King of Spain (1556-98)   King of the Romans (1575-6)   King of Hungary (1608-19)
(1553-8)   as **PHILIP I**   Holy Roman Emperor (1576-1612) King of Bohemia (1611-19)
                     King of Portugal (1580-98)   Holy Roman Emperor (1612-19)

                   PHILIP III 1578-1621)   = Margaret 1584-1611   **FERDINAND II** 1578-1637
                   King of Spain, Naples and Sicily (1598-1621)   King of Hungary (1618-26)
                   as **PHILIP II** King of Portugal (1598-1621)   King of Bohemia (1617-27)
                                                                                              Holy Roman Emperor (1619-37)

LOUIS XIII 1601-43 *   = Anne 1601-66   Mariana 1606-46 [1] = **FERDINAND III** 1608-57
King of France (1610-43)   King of Hungary (1626-57)
                                                                                    King of Bohemia (1627-57)
                                                                                    King of the Romans (1636-7)
                                                                                    Holy Roman Emperor (1637-57)

         Elizabeth * [1] = **PHILIP IV** 1605-65   = [2] Mariana   **FERDINAND IV** 1633-54
         of France   King of Spain, Naples and Sicily (1621-65) 1635-96   King of Bohemia (1646-54)
         1602-44   as **PHILIP III** King of Portugal (1621-40)   King of the Romans (1653-4)

LOUIS XIV 1638-1715 = Maria Theresa 1638-83   **CHARLES II** 1661-1700   Margareta Teresa [1] = **LEOPOLD I** 1640-1705   = [3] Eleanora of Neuberg
King of France   King of Spain,   1651-73   King of Hungary (1655-1705)   1655-1720
(1643-1715)   Naples and Sicily (1665-1700)   King of Bohemia (1656-1705)
                                              last Habsburg King of Spain   Holy Roman Emperor (1658-1705)

                                              JOSEPH I 1678-1711   **CHARLES VI** 1685-1740
┌─────────────────────────┐   King of Hungary (1687-1711)   Holy Roman Emperor (1711-40)
│ Bourbon Kings of France │   King of Bohemia (1705-1711)
│ Bourbon Kings of Spain │   Holy Roman Emperor (1705-11)
└─────────────────────────┘
                                              FRANCIS I 1708-65
                                              Duke of Lorraine (1729-36)
                                              Grand Duke of Tuscany (1737-65)
                                              Holy Roman Emperor (1745-65)
                                              Co-Regent of the Austrian lands
                                              with Maria Theresa (1741-65)

┌─────────────────────────┐   **JOSEPH II** 1741-90   = **MARIA THERESA** 1717-80
│ House of Habsburg-Lorraine │   King of the Romans (1764-5)   Queen of Hungary and Bohemia (1740-80)
└─────────────────────────┘   Holy Roman Emperor (1765-90)
                                              Co-Regent of the Austrian lands
                                              with Maria Theresa (1765-80)   **LEOPOLD II** 1747-92
                                                                                Grand Duke of Tuscany (1765-90)
                                                                                Holy Roman Emperor (1790-2)
                                                                                King of Hungary and Bohemia (1790-2)

                                                                                FRANCIS II 1768-1835
                                                                                last Holy Roman Emperor (1792-1806)
                                                                                Emperor of Austria (1804-35)
                                                                                King of Hungary and Bohemia (1792-1835)

                                                                    ┌──────────────────────────────┐
                                                                    │ Habsburg Emperors of Austria │
                                                                    └──────────────────────────────┘

* Brother and sister

King of the Romans (emperor designate uncrowned)

Housesteads in Northumberland is the best preserved of the fortress bases built along **Hadrian's Wall** to house infantry and cavalry troops. From the wall, built on high ground, the garrison manning such bases had a commanding vantage point over the approach of any would-be attacker.

became *Trajan's ward after his parents died. He pursued a successful military career, married Trajan's niece, and was adopted as his heir. Extensive travel characterized his reign after 117. The western empire was covered 121–6 and the eastern parts 128–34. *Hadrian's Wall was commenced during his visit to Britain 121–2. The *Bar Cochba revolt broke out after his visit to Palestine and his plan for rebuilding a Romanized *Jerusalem. He was a patron of the arts and the ruins of his villa at Tibur (modern Tivoli) remain.

Hadrian's Wall, a defensive fortification in northern Britain. It was built AD 122–6 after a visit to Britain by Emperor *Hadrian. It is 117 km. (73 miles) long, a stone barricade with a turf section in the west. Large fortress bases, mile-castles, and signal towers marked its length. A road ran along it to the south. Defensive ditches accompanied it on both sides. The wall was damaged several times by the Picts, and was finally abandoned in 383. Long stretches of the wall still stand.

Hainaut, a province in south-western Belgium. Originally inhabited by a tribe called the Nervii, it became part of the Roman empire in 57 BC. It changed hands several times and in the 11th century passed to Flanders and then in 1275 to Count John of Holland. In 1345 it came to the Wittelsbach family and in 1433 went to Duke Philip the Good of Burgundy. *Louis XIV divided it between France and Austria, but in 1794 the Austrian part was absorbed by France.

Haiti, a Caribbean country that occupies the western one-third of the island of *Hispaniola. Hispaniola was discovered by Columbus during his first voyage to the New World, and became a Spanish colony in the 16th century. French corsairs settled on the western part of the island in the 17th century and Spain recognized the

French claims to the area in 1697 in the Treaty of Ryswick. Known as Saint Domingue in the 18th century, it became a rich source of sugar and coffee for the European market. African slaves replaced a decimated Indian population and by the end of the 18th century the population of Haiti was predominantly black.

Halidon Hill, battle of (19 July 1333). A site near Berwick-on-Tweed, on the border between England and Scotland saw a major victory for Edward *Balliol over the nationalist Scots. Balliol, crowned King of Scotland in 1332, had been driven out of Scotland, but this victory, which was won with the help of English archers supplied to him by *Edward III, regained him his kingdom—at the price of doing homage for it to Edward III.

Hallstatt, the site of a famous prehistoric cemetery in Austria which has given its name to the culture of the early *Iron Age (c.750–450 BC). At first cremation was the rule, as were flat or low graves, though later the tumulus or raised *barrow became standard. Then as iron became common, interment was used as well as cremation, and the quality of the geometric-style pottery degenerated. It was superseded by the *La Tène culture.

Hamilcar Barca (d. c.229 BC), Carthaginian general and father of *Hannibal and *Hasdrubal. He commanded the Carthaginian forces in the later part of the first of the *Punic wars and negotiated the peace of 241 BC. When the mercenaries in Carthaginian service rebelled, Hamilcar, along with his rival Hanno, defeated them. In 237 he went to Spain, and brought the southern and eastern areas under Carthaginian control.

Hamilton, Alexander (1755–1804), American statesman and fiscal innovator. He organized an artillery company and saw action in campaigns round New York, before becoming Washington's private secretary and aide-de-camp (1777–81). He married into a prominent New York family. He served in the Continental Congress (1782–3) and was a delegate to the *Constitutional Convention (1787), where he advocated a strong central government. He wrote more than half of the *Federalist Papers*. As Secretary of the Treasury (1789–95) he devised the American fiscal programme (1790), recommending the treasury to accept old securities at face value in exchange for new bonds, the assumption by the federal government of Revolutionary War debts, the passage of a protective tariff, and the establishment of a Bank of the United States. This 'Hamiltonian System' was supported by the *Federalists and opposed by *Madison and *Jefferson, helping to precipitate the formation of the Democratic Republican Party. Hamilton resigned from the cabinet (1795) and returned to New York to practise law. In 1800 he thwarted Aaron Burr's ambitions to become President and in 1804 was challenged by Burr to a duel; he was killed by Burr at Weehawken, New Jersey.

Hamilton, James, 3rd Marquis and 1st Duke of (1606–49), Scottish nobleman, a supporter of the *Cavalier cause. *Charles I appointed him the king's commissioner in Scotland in 1638, but despite negotiating with the *Covenanters, he was unable to avert the *Bishops' wars. Charles kept faith with him on the outbreak of the *English Civil War, but his negotiations in Scotland came to nothing in 1643, when he was expelled for refusing to sign the *Solemn League and Covenant. His attempt to revive the Cavalier cause in 1648 ended in defeat at *Preston, and his execution.

Hammurabi (d. 1750 BC), Amorite king of *Babylon. He greatly extended the lands he had inherited until they stretched from the Persian Gulf to parts of Assyria. He was much more than a warrior, however, encouraging agriculture, literature, and intellectual pursuits, and drawing up a code of laws which was inscribed on a column found at Susa in 1901.

Hampden, John (1594–1643), English politician and Parliamentarian. He played a leading part in the opposition to *Charles I's arbitrary government. In 1627 he was imprisoned for refusing to pay the 'forced loan' imposed by Charles to finance his unpopular foreign campaigns. Ten years later he was prosecuted for refusing to pay *Ship Money. As a member of the *Long Parliament he was prominent in the impeachment of *Strafford, and a close ally of John *Pym. In 1642 he survived the king's attempt to arrest him (*five members), and was appointed to the Committee of Safety to organize the parliamentary *English Civil War effort. He died of wounds received in action.

Hampton Court Conference (1604), a meeting in which the new king of England, *James I, presided over an assembly of bishops and Puritans. The 'Millenary Petition' presented by the Puritans in 1602 had listed church practices offensive to them and had asked for reform in the *Anglican Church. Most of their demands were refused, although it was agreed to produce a new translation of the *Bible, the Authorized Version of 1611.

Han (Western Han, 202 BC–AD 8; Eastern Han, AD 25–220), Chinese dynasty established at the overthrow of the *Qin by the rebel peasant Liu Bang. Invoking the glories associated with this dynasty the Chinese still distinguish themselves from other *Mongoloid peoples by calling themselves the Han.
 When Wudi ('the Martial Emperor') (141–87 BC), halted *Xiongnu onslaughts, Chinese armies penetrated

Hammurabi standing before the Babylonian sun god Shamash to receive the symbols of justice. Carved on a 222-cm. (87 in.) basalt stele c.1700 BC, the relief surmounts an inscription of Hammurabi's laws which are the most comprehensive form of his code that is known. (Musée du Louvre, Paris)

A terracotta model of a house from a **Han** dynasty tomb near Kaifeng. The ornamental brick construction is in the form of a gateway guarded by soldiers and gives a good impression of the light-walled, timber-framed house architecture of the period. The roof is tiled by a method invented 400 years earlier. (British Museum, London)

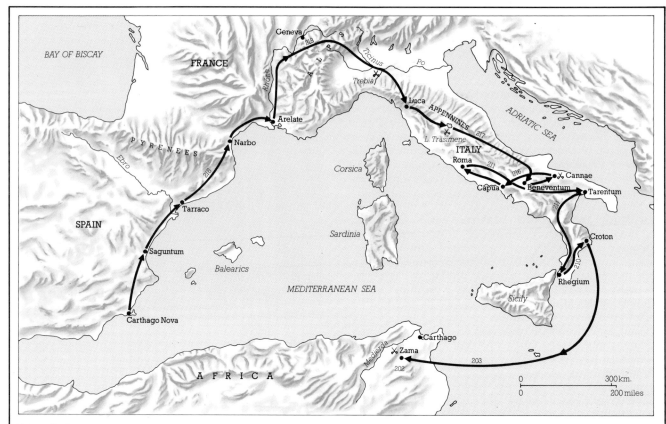

Hannibal

Hannibal marched north from Carthago Nova (Cartagena) into France in 218 BC and astounded the Romans by crossing the Alps, with an army of troops and elephants, and descending into Italy. The Alpine crossing, accomplished in a mere fifteen days, is regarded as one of the greatest feats of ancient warfare. Despite several devastating victories over the Romans,

Hannibal was not strong enough to attack Rome itself, and his march on the city in 211 was merely a feint to divert the Romans from their siege of Capua. In 203 he was recalled to Carthago (Carthage). The Second Punic War ended in 201, following Hannibal's defeat at Zama (202).

deep into Central Asia. Some marched over 3,200 km. (2,000 miles) west of the Western Han capital Chang'an (now Xi'an). Envoys seeking alliances against the Xiongnu also travelled far west, returning with information about the *Roman empire. Camel trains taking out silks, bringing back jade, and 'heavenly' horses larger than those then known in China, travelled the *Silk Route. Much of south China and *Annam was conquered, though the southern Chinese were not assimilated until later, and Han rule was established over part of *Korea. Gradually *Confucianism was accepted as the state philosophy.

In AD 8, the court was oppressed by economic problems and torn by strife. Wang Mang (33 BC–AD 23), chief minister of a boy-emperor he had placed on the throne, seized power and established the Xin (Hsin) dynasty (AD 8–23). He attempted to allay peasant discontent by redistributing landholdings. But the Taoist-inspired 'Red Eyebrows' rebellion in Shandong in AD 18 and renewed Xiongnu invasions brought his downfall.

In AD 25 a Han prince Liu Xiu set up a court in Luoyang. Soon Chinese armies again penetrated central Asia and temporarily subjugated the lands east of the Caspian Sea. The period was noted for its artistic achievement and for advances in technology. The first Buddhist missionaries arrived in China. By the end of the 2nd century the ambitions of empresses and *eunuchs

and the Taoist Yellow Turbans rebellion led to near collapse and the growth of regional armies. The last Han emperor abdicated, and China again became divided during the period of the *Three Kingdoms.

Hancock, John (1737–93), American Revolutionary leader. He was a radical Boston merchant who supplemented his inherited fortune by smuggling, and he came into open conflict with British customs officers when his sloop *Liberty* was seized (1768). He was a generous backer of the patriot cause, helped to organize the *Boston Tea Party, and as president of the Second *Continental Congress was the first to sign the *Declaration of Independence. He was first governor of independent Massachusetts.

Hannibal (247–183 or 182 BC), Carthaginian general, an outstanding military tactician and leader. He accompanied his father *Hamilcar to Spain in 237, and helped him establish a province there. He was himself granted supreme command in Spain in 221 and adopted an aggressive policy towards the Romans. His eight-month siege of Saguntum in 219 precipitated the second of the *Punic wars, and in 218 he marched over the Alps, into northern Italy, though many of his elephants and troops died during the arduous journey. He inflicted three crushing defeats on the Roman forces, at Trebia

(218), Lake Trasimene (217), and *Cannae (216). Despite winning over many of the southern Italian communities, central and northern Italy remained largely loyal to Rome. He was unable to break Rome's dogged resistance and gradually the tide of war turned against him, until in 203 he was recalled with his army to Africa. The following year he was defeated at Zama by *Scipio Africanus. His programme of political reforms in Carthage c.196 provoked his enemies to complain to Rome, and he fled abroad. He spent time at the courts of Antiochus the Great and Prusias of Bithynia, before committing suicide in 183-2.

Hannibal was one of the great generals of history. His military genius lay in his ability to use cavalry and infantry in combination, and to inspire deep loyalty in a mercenary army.

Hanover, House of, the family of sovereigns of Great Britain and Ireland from George I to Victoria (1714–1901). The dynasty was named after the city of Hanover, the capital of Lower Saxony in Germany. In 1658 Sophia, daughter of Elizabeth of Bohemia and granddaughter of James I of England married Ernest Augustus, Duke of Brunswick-Lüneburg, who subsequently became an Elector of Germany (1692), taking Hanover as his title and capital city. Their son became *George I, the first Hanoverian King of Great Britain in 1714. Hanover's territories included the important towns of Göttingen and Hildersheim and their defence was an important factor in British foreign policy in the 18th century.

Hanseatic League, an association of north German cities, formed in 1241 as a trading alliance. Cologne had enjoyed special trading privileges with England and was joined by other traders. An agreement between Hamburg and Lübeck (1241), and a Diet of 1260 marked the origin of the League. The towns of the League dealt mainly in wool, cloth, linen, and silver. In the later Middle Ages the League, with about 100 member towns, functioned as an independent political power with its own army and navy. It began to collapse in the early 17th century and only three major cities (Hamburg, Bremen, and Lübeck) remained in the League thereafter.

Hapsburg   *Habsburg.

Harappa, a village in south Punjab, Pakistan, an important *Indus civilization site. The term 'Harappan culture' is sometimes used to identify common features found at other sites along the Indus valley where urban civilization flourished c.2500 to 1500 BC. Harappa was the first such site to attract the attention of archaeologists. Excavations have revealed a city, built c.2500 BC, which was second in size and splendour only to *Mohenjo-daro. Important constructions were a granary and two cemeteries.

Hardy, Thomas (1752–1832), Scottish radical leader, a champion of parliamentary reform. He moved to London in 1774, where he became a shoemaker. In 1792 he founded the *London Corresponding Society, whose aim was to achieve universal manhood suffrage. The country was at war with Revolutionary France and the government became alarmed at the Society's growing influence. In 1794 Hardy was arrested on a charge of high treason but at the subsequent trial he was acquitted.

Harley, Robert, 1st Earl of Oxford and Mortimer (1661–1724), English statesman. He entered Parliament as a Whig, and during the early years of Queen *Anne's reign served variously as Speaker of the House of Commons and Secretary of State. He abandoned the Whigs and used the influence of his cousin Abigail Masham, a lady-in-waiting, to undermine *Marlborough's standing with the queen. In 1710 he headed the new Tory government whose greatest achievement was the Peace of *Utrecht (1713). In the subsequent power struggle among the Tories just before Anne's death in 1714 he lost to *Bolingbroke, and the Whig administration of *George I imprisoned him and began impeachment proceedings against him. He was released two years later, but took no further part in public affairs.

Harold I (c.850–933), first King of all Norway (872–933). A series of battles with minor kings culminated in his decisive victory at Hafrsfjord. He then succeeded in bringing the Orkney and Shetland islands, together with much of northern Scotland, into his kingdom and forced out many Vikings who went on to conquer Iceland and land in western Europe.

Harold II (c.1020–66), King of England (1066). He was the second son of Earl *Godwin of Wessex, whom he succeeded in 1053. The Godwin family had great political ambition. Harold's sister Edith was married to *Edward the Confessor, and his brother Tostig was Earl of Northumbria (1055–65). Exiled after an abortive attempt to intimidate the king, Harold and his father returned (1052) to dominate political affairs in England. Harold succeeded *Edward the Confessor in 1066, despite the Norman claim that Edward had designated Duke William as his heir and that Harold had recognized William's right. Tostig, who had been dispossessed of his earldom, raided the south-east coast before joining the invasion by Harald Hardrada of Norway in northern England. Harold defeated them at *Stamford Bridge and then marched 402 km. (250 miles) south to meet William's invasion at the battle of *Hastings, where he died.

Harrington, James (1611–77), English philosopher and political theorist. He sympathized with republicanism, as his work *The Commonwealth of Oceana* (1656) showed, but he was a friend of *Charles I and briefly shared his imprisonment with him. Even so *Charles II had him arrested in 1661 for alleged conspiracy. His ideas are said to have influenced the makers of the American Constitution, especially in his insistence on a written constitution and a two-chamber legislature.

Harsha (c.590–647), Buddhist ruler of a large empire in north India (c.606–47). He dominated the entire Gangetic plain, and also parts of the Punjab and Rajasthan, but was repulsed from the *Deccan. He allowed conquered rulers to keep their titles in return for tribute, and he is considered an enlightened and talented ruler. He organized Buddhist assemblies, established charitable institutions, and patronized learning, particularly poetry. His reign is well documented, notably by his celebrated court poet, Bana, and by a Chinese Buddhist pilgrim, Xuan Cang.

Harun al-Rashid (literally, 'Aaron the rightly guided') (c.763–809), the fifth *Abbasid caliph (786–809). Under

The battle of **Hastings**, illustrated in the Bayeux Tapestry, which depicts the main events surrounding Duke William's progress towards the throne of England. The English defended a hilltop position against the Norman attacks, but ill-disciplined pursuits of retreating enemy soldiers proved costly and Harold's death was the final blow. (Musée de Bayeux)

him Baghdad reached its greatest brilliance, partly thanks to the ability of the Persian *viziers of the house of Barmak (until their fall in 803). He was a competent commander and a patron of learning and the arts. His court and capital provided the setting for many of the stories in the *Thousand and One Nights*. His division of the empire among his heirs led to conflict after his death (809-18). According to French chronicles, there was an exchange of embassies between Harun and *Charlemagne.

Hasdrubal (d.207 BC), Carthaginian general and son of *Hamilcar Barca. He remained in Spain when his brother *Hannibal invaded Italy in 218. Despite mixed fortunes, by 211 he had established Carthaginian power as far north as the River Ebro. Defeated at Baecula by *Scipio Africanus in 208, he extricated most of his troops and marched to Italy in an attempt to join forces with Hannibal but was intercepted by two Roman armies and defeated and killed at Metaurus in 207.

Hastings, battle of (14 October 1066). A battle fought at Senlac, inland from Hastings (south-east England) between the English under *Harold and an invading army under Duke William of Normandy (*William I). Harold heard the news of the Norman invasion after his defeat of Harald Hardrada at *Stamford Bridge, near York, and immediately marched southwards with his troops. The English resisted the Norman attack throughout a long day's fighting but the Norman cavalry and crossbowmen were superior to the English soldiers, fighting on foot and armed with axes. Harold was killed, traditionally by an arrow piercing his eye, and William, the victor, marched towards London.

Hastings, Warren (1732-1818), first governor-general of *Bengal (1774-85). He consolidated *Clive's conquests in north-eastern India, but was afterwards impeached by his enemies. He worked his way up from a clerkship in the *East India Company's service, gaining experience in both commerce and administration, and a deep and sympathetic insight into Indian culture. Although well equipped for the governorship, he faced insurmountable difficulties when enemies on his council continually vetoed his policies. A quarrel with Philip Francis ended in a duel; both survived and Francis carried home tales of Hastings's alleged malpractices. Among the subsequent impeachment charges were the waging of unjustified wars, extortion from Indian rulers and their families, and the judicial murder of an Indian moneylender who had threatened him. After a seven-year trial he was finally

The trial of Warren **Hastings** caused considerable controversy in England. In this cartoon by James Gillray, Hastings, the 'Saviour of India', (on a camel) is assailed by (from left to right) Edmund Burke, Lord North, and Charles James Fox, who instituted the attack upon him in the House of Commons in 1788.

acquitted in 1795, and lived on to see his policies vindicated by the Company's success in India.

Scholarly opinion now holds that Hastings, although not entirely blameless, did not wittingly exceed the limits expected of a colonial administrator of his day. Indeed his constructive plans for improving the administration of Bengal, and his scholarly interest in India, make him in retrospect one of the most admirable of the first generation of British rulers of India.

Hatshepshut (c.1540–c.1481 BC), the daughter of Thutmose I of Egypt. After the death of her half-brother and husband, Thutmose II, the young *Thutmose III succeeded, but she soon replaced him as the effective ruler and reigned until her death twenty years later. As well as furthering her father's building program at *Karnak, she had a magnificent temple constructed at Deir al-Bahri.

Hausa, the people of northern Nigeria. The original Hausa states, which include *Kano and *Zaria, were for many years the vassals of *Bornu. Muslim missionaries seem to have come in the 14th century, but during the reign of Muhammad Rumfa of Kano (1463–99) the celebrated divine al-Maghili is said to have introduced the *shariah* (the Muslim code of law), *Sufism, and a body of constitutional theory. The Hausa states were conquered by the *Songhay in 1513 and by the *Fulani in the early 19th century.

Hawke, Edward, 1st Baron (1705–81), British admiral. He won fame by his great victory over the French at Finisterre in 1747. During the *Seven Years War he blockaded the French Atlantic fleet at Brest, and when it broke out in 1759 he destroyed it at the battle of *Quiberon Bay, thus effectively cutting France's communications with its Canadian colonies. Hawke then retired from active service.

Hawkins, Sir John (1532–95), English seaman. He made the first ever slave-trading voyage from Africa to the West Indies in 1562. His third such voyage (1567–9) ended in disaster, when Spaniards attacked his ships at San Juan de Ulua, Mexico, thus exacerbating Anglo-Spanish tension. He became a Member of Parliament (1572), and succeeded his father-in-law as Treasurer of the Navy (1577). He was largely responsible for creating the fleet which defeated the *Spanish Armada (1588), was third in command after Lord Howard of Effingham and his own kinsman, Sir Francis *Drake, and was knighted at sea during the battle. He died on an expedition to the Spanish West Indies.

Hawkins, Sir Richard (1560–1622), commander in the Elizabethan navy serving against the *Spanish Armada (1588). In 1593 he left England with the intention of surveying eastern Asia, where he hoped to establish an English trading empire. On the way, he plundered Valparaiso in Spanish America, and was held by the Spaniards until a ransom was paid in 1602.

Hébert, Jacques René (1757–94), French journalist and Revolutionary. He became a prominent member of the Cordeliers Club (an extreme Revolutionary club) in 1791. The following year he was a member of the Commune of Paris and was arrested in 1793 after his

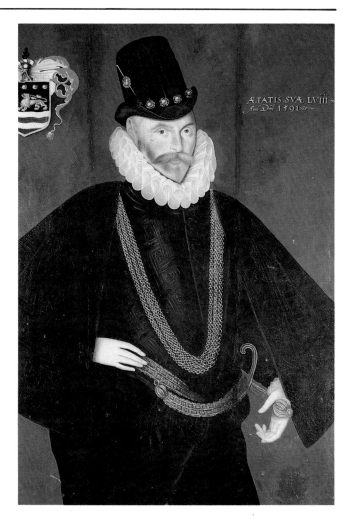

A portrait of Sir John **Hawkins**, by an unknown artist, 1591. Hawkins made a fortune from his slave voyages before assuming command of the English navy in the years leading up to the Spanish Armada. (City Museum and Art Gallery, Bristol)

violent attacks on the *Girondists. Although he was always popular with the mob, he and his followers, the Hébertistes, came into conflict with *Robespierre when he organized 'The Worship of Reason'. This substitute for the worship of God led to his arrest and execution in March 1794.

Hebrides, a group of about 500 islands off the west coast of Scotland. They comprise two archipelagos, lying parallel to the mainland—the Outer Hebrides (Lewis, Harris, the Uists, and Barra) and the Inner Hebrides (Skye, Mull, Islay, and Jura). The largest few dozen islands have been inhabited since c.3800 BC, settled by Picts and then, from the 3rd century, by Scots. Except in the *Viking era, their way of life changed little until the 19th century, when large-scale sheep farming and subsequent clearances of crofters caused depopulation and widespread deprivation.

hedonist, a believer in the doctrine that pleasure is the chief good or the most worthwhile aim. Hedonism (from the Greek *hēdonē*, 'pleasure') has been professed in various forms throughout the ages. The Cyrenaics, a minor school of philosophers who flourished in the 4th century BC, saw the satisfaction of immediate, sensual pleasures as the single dominating factor in life. A more subtle and complex theory of hedonism was professed by the *Epicureans, though their philosophy has often been misunderstood. It came under great pressure from the rise of Christianity, and doctrines of asceticism.

hegira (Arabic, *hijra*, 'exodus', 'migration', or 'breaking of ties'), *Muhammad's secret departure from *Mecca in 622, accompanied by *Abu Bakr, to live among the people of Yathrib, later *Medina, thus founding the first Muslim community. Under the second caliph, *Umar, this key event in the history of Islam was chosen as the starting-point for the Muslim calendar.

Helena, St (*c.*250–*c.*330), wife of Emperor Constantius Chlorus and mother of the Roman emperor *Constantine, the first Christian emperor. When he became emperor he made her empress dowager. In *c.*312, aged over 60, she was converted to the Christian faith and thereafter was noted for her piety and acts of charity. She died while on *pilgrimage to Palestine and was buried in Rome. Legend associates her with the finding of the wooden cross on which *Jesus Christ was crucified.

Heliopolis ('Sun City'), the Greek name for the Egyptian city of Iunu or Onu, important as the centre of the worship of Ra, the sun god, whose symbols were the pyramid and the obelisk. The temple there was the largest in Egypt after that of Amun at *Karnak. The temple has not survived, but many obelisks have, most notably two erected by Thutmose III, the so-called Cleopatra's Needles now in London and New York.

Hellenistic civilization, the result of the adoption of the Greek language and culture by non-Greeks. (Hellas, an area of southern Thessaly, was synonymous with Greece from the 7th century BC.) It has come to refer specifically to the civilization which arose in the wake of the conquests of *Alexander the Great. The many cities founded by him and his successors were the centres for a fusion of Greek and 'barbarian' ways of life, with *Alexandria in Egypt becoming the literary focus of the Mediterranean world. An important element in the diffusion of the new culture was the development of a common Greek dialect, *Koine*.

Hellespont, the strait that joins the Aegean Sea with the Sea of Marmara, and which separates Europe from Asia. It has long been a key strategic point. Ancient Troy stood at its western end. King Xerxes crossed it with his Persian army over a bridge of boats (*c.*481 BC); it was also crossed by Alexander the Great in 334 BC. Control of it in 405 BC enabled the Spartans to cut off corn supplies to Athens from the Black Sea area and thus bring the *Peloponnesian War to a conclusion. In course of time it became known as the Dardanelles.

Hellfire Club, a notorious English society which met in the ruins of Medmenham Abbey in Buckinghamshire. It was founded by Sir Francis Dashwood in 1745, and its members reputedly indulged in debauchery and in the mocking of organized religion by the performance of blasphemous 'black masses'. In the late 18th century its membership included many politicians, the most famous of whom were *Wilkes, *Bute, and Sandwich.

helot, an inhabitant of ancient Greece forced into serfdom by conquering invaders. Helots were used as agricultural labourers and in domestic service. The Messenians, subjected by *Sparta, greatly outnumbered the Spartan citizens, and fear of their rebellion caused the city to keep them under ruthlessly tight military control.

Helvetii, Celts who migrated from southern Germany to south and west of the Rhine in the 2nd century BC. In 102 BC they joined the Cimbri and Teutones in invading Italy and were defeated by Emperor *Marius. Under pressure of Germanic migrations they attempted a mass migration into Roman *Gaul in 58 BC. Julius *Caesar drove them back. *Augustus incorporated their territory into Belgic *Gaul. Overrun in the 5th century by a succession of Alemanni, *Franks, Swabians, and Burgundians their name is preserved in the formal name for *Switzerland—the Helvetic Confederacy.

Helvétius, Claude Adrien (1715–71), French philosopher. He was a wealthy French financier whose book, *De l'Esprit*, published in 1758 was condemned as immoral by both the church and *Parlement, and was publicly burned. Its theme was that self-interest is the mainspring of all human activity and should be utilized by government for the promotion of public interest. This doctrine gave the philosophical movement the foundation for a practical programme: the well-being of the human race could be secured by enlightened government without changing human nature. Helvétius was exiled to the country and finally went to England in 1763.

Hengist and Horsa, *Jutish brothers, leaders of the first Anglo-Saxon invasion of England. According to *Bede, they were invited by King *Vortigern to help reinforce British resistance to the raiding Picts and Scots (*c.*449). The *Anglo-Saxon Chronicle* claimed that Hengist and Horsa were joint kings of Kent and that when Horsa was killed (455) in battle, Aesc, the son of Hengist (d. *c.*488), succeeded him.

Henrietta Maria (1609–69), French princess, Queen consort of *Charles I of England from 1625. At first she was eclipsed in her husband's affection by *Buckingham, but after his death (1628) she became closely identified with the king's politics and tastes. Her Catholicism made her unpopular and heightened public anxieties about the court's religious sympathies. As the Puritan opposition strengthened she began to intervene in politics. The rumour that Parliament was to impeach the queen drove Charles to attempt to arrest five Members of Parliament (*five members), a contributory cause of the *English Civil War. From 1644 she lived mainly in France.

Henry I (the Fowler) (*c.*876–936), Duke of Saxony (912–36), and King of Germany (919–36). He was elected king by the Franks and Saxons in 919 and was the first king of the Ottonian dynasty. He developed a system of fortified defences against Hungarian invaders whom he defeated in 933 at the battle of the Riade. Despite challenges from his nobility, he laid a strong foundation for his son *Otto I.

Henry I (1068–1135), King of England (1100–35). He was the fourth and youngest son of *William I. When

Normandy led to the accession of Henry's nephew *Stephen and a period of anarchy and civil war between Stephen and Henry's daughter, *Matilda.

Henry II (1133–89), King of England (1154–89). He succeeded King *Stephen, who by the treaty of Winchester (1153) had recognized Henry as his heir. As the son of Geoffrey, Count of Anjou, Henry also inherited Normandy, Maine, Touraine, Brittany, and Anjou (this last title making him the first *Angevin king of England). His marriage to *Eleanor of Aquitaine (1152), the repudiated wife of Louis VII of France, brought Henry even greater estates in France so that his kingdom stretched from northern England down to the Pyrenees. These territorial gains were reinforced by the homage of Malcolm III of Scotland (1157) and by his recognition as overlord of Ireland (1171).

Henry's immediate task on becoming king was to end the anarchy of Stephen's reign. He dealt firmly with barons who had built castles without permission, and undertook the confirmation of *scutage (1157), an overhaul of military obligations in a review of feudal assessments (1166), and the introduction of a law that his subjects should equip themselves for military service (1181). He initiated a number of important legal reforms in the Assizes of *Clarendon (1166) and of Northampton (1176), and in his reign the land-law was developed to meet the needs of a more complex society. His reign, however, also saw rebellions led by his sons (1171–4) and the murder of the Archbishop of Canterbury, Thomas à *Becket (1170), a crime of which Henry was later absolved by Pope Alexander III (1172).

Henry III (1207–72), King of England (1216–72). He succeeded his father *John at the age of 9. During his minority (until 1227) England was managed by William Marshal, the first Earl of Pembroke, Peter des Roches, Bishop of Winchester, and Hubert de Burgh. Henry's personal rule soon proved his general incompetence as king, his preoccupation with aesthetic pursuits, including the rebuilding of Westminster Abbey, and his preference for foreign advisers and favourites. He received an early warning of baronial frustration when a rebellion broke out (1233–4) led by Richard Marshal, the third Earl of Pembroke. In 1258, one of the king's French favourites, Simon de *Montfort, led the English barons to draft a series of reforms (the Provisions of Oxford). While appearing to accept these, Henry sought to recover his independence. The ensuing civil war (1264–7) led to the temporary control of England by de Montfort following his victory at Lewes (1264). Although Henry recovered control after the battle of Evesham (1265), where de Montfort was killed, he became increasingly dependent upon his son, the future Edward I.

Henry IV (1366–1413), King of England (1399–1413). He was the only legitimate son of John of *Gaunt, and would have inherited vast estates on his father's death (in 1399) had *Richard II not banished him. He retaliated by invading England and forcing Richard to yield both the estates and the crown of England. Henry's position as king was not a strong one. He needed the support of the church (which caused him to be a persecutor of *Lollards), the nobility (who dominated his councils), and the House of Commons (which resented his frequent requests for money). Until 1408 he had to

A contemporary formal portrait of the young Queen **Henrietta Maria** by an unknown artist. The background is thought to be by Hendrik van Steenwyck the Younger, c.1635. (National Portrait Gallery, London)

his brother *William II died, Henry seized the treasury at Winchester and was crowned three days later in London, while his elder brother Robert was still on Crusade. Although Robert received the duchy of Normandy in compensation and exacted an annual pension of £2,000 (1101), Henry invaded Normandy in 1106. Robert was defeated at Tinchebrai and imprisoned at Cardiff Castle until his death in 1134. Louis VI of France exploited the situation in Normandy but in two campaigns (1111–13, 1116–20) he failed to take the duchy.

A determined ruler, Henry clashed with Archbishop *Anselm over his claim to appoint bishops (lay *investiture). He improved royal administration, particularly in the *Exchequer, and extended and clarified the judicial systems. His law code, the *Leges Henrici Primi*, embodied much that had survived from Anglo-Saxon law. Unfortunately the death by drowning (1120) of his only legitimate son William whilst journeying to England from

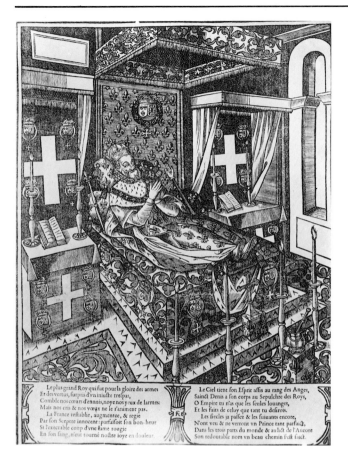

Le plus grand Roy qui fut pour la gloire des armes
Et des vertus, surpris d'vn iniuste trespas,
Comble nos cœurs d'ennuis, noye nos yeux de larmes:
Mais nos cris & nos vœux ne le s'animent pas.
La France restablie, augmentée, & regie
Par son Sceptre innocent: parfaisoit son bon-heur
St l'execrable coup d'vne ame rouge
En son fang, n'eut tourné nostre ioye en douleur.

Le Ciel tient son Esprit assis au rang des Anges,
Sainct Denis a son corps au Sepulchre des Roys,
O Empire tu n'as que les seules louanges,
Et les faits de celuy que tant tu desirois.
Les siecles ja passez & les suiuants encore,
N'ont veu & ne verront vn Prince tant parfaict,
Dans les trois parts du monde & au lict de l'Aurore
Son redoutable nom vn beau chemin fest fauct.

Henry IV of France on his death bed in 1610. He was stabbed to death by Francois Ravaillac, a Catholic fanatic.

deal with the rebellions of Owen *Glendower and the *Percys, and for the remaining five years of his short life he was in poor health.

Henry IV (1553–1610), King of Navarre (1572–89) and first *Bourbon King of France (1589–1610). He was educated as a Calvinist, mainly under the supervision of his mother, Queen Jeanne d'Albret of Navarre. As a leading Protestant nobleman, he was recognized by *Coligny as the nominal head of the *Huguenots. In 1572 he married Marguerite de Valois, daughter of Catherine de *Medici, and a few days later narrowly escaped with his life during the *St Bartholomew's Day Massacre by professing himself a Catholic. He was kept a virtual prisoner at court until 1576, when he escaped, resumed his Protestant beliefs, and played a major part in the subsequent *French Wars of Religion.

He fought the War of the *Three Henrys in a successful bid to secure his succession to the French throne, professed himself a Catholic again in 1593, and brought peace to France at last in 1598. By the Edict of *Nantes he established political rights and some religious freedom for the Huguenots. With the aid of his Huguenot chief minister, *Sully, he reconstructed his war-torn kingdom, while continuing to pursue an anti-Habsburg foreign policy. He is sometimes regarded as the founder of the absolute, centralized regime of later 17th-century France, but the breakdown of royal authority after his assassination by a Catholic fanatic showed that his power had depended on his own personal popularity.

Henry V (1387–1422), King of England (1413–22), eldest son of *Henry IV. He was a skilful military leader, with experience gained in campaigning against Owen *Glendower, and took advantage of French weakness by claiming the French crown (1413) and then invading France (1415), where he won the magnificent victory of *Agincourt. In 1417 he invaded France again, and again fortune favoured him. He was able to conclude the very favourable Treaty of Troyes (1420) by which he was to succeed to the French crown at Charles VI's death and meantime to marry Charles's daughter, Katherine of Valois. However, he died only fifteen months later. He was a popular hero, as celebrated in Shakespeare's play *Henry V*, and restored civil order in England in addition to his French campaigns.

Henry VI (1422–71), King of England (1422–61, 1470–1). He inherited the throne from his father *Henry V when only 9 months old. During his minority, the court was divided by a power struggle between *Humphrey, Duke of Gloucester, and the *Beauforts, which led to the Wars of the *Roses. A pious and withdrawn man, he suffered from insanity from 1453. His marriage to *Margaret of Anjou brought him a son, Edward, in 1453, who displaced *Richard, Duke of York as heir to the throne. Richard claimed it for himself in 1460 and

A portrait of **Henry V**, King of England, by an unknown artist. Henry renewed the Hundred Years War soon after coming to the throne and in 1415 won a resounding victory at Agincourt. He died prematurely, leaving an infant son to succeed him. (National Portrait Gallery, London)

when he was killed at *Wakefield, his son Edward seized it in March 1461 and was crowned Edward IV. Henry was later captured, and spent the rest of his life in the Tower of London apart from the brief period when *Warwick restored him to the throne (1470-1), but Edward then defeated Warwick and reimprisoned Henry. When Henry's son, Prince Edward was killed at *Tewkesbury in 1471, Edward IV won back the throne and had him put to death.

Henry VII (1457-1509), King of England (1485-1509), the son of Edmund *Tudor and Margaret *Beaufort. Through his mother he was an illegitimate descendant of John of *Gaunt and so had a tenuous claim to the throne, but rivals' deaths in the Wars of the *Roses strengthened his position as the Lancastrian claimant and enabled him to oust *Richard III at the battle of *Bosworth Field. He established the Tudor dynasty on firm foundations. The children of his marriage to Elizabeth of York were married to foreign royalty (his son Arthur to *Catherine of Aragon, and his daughter Margaret to James IV of Scotland), while he built up the crown's financial resources so that he was not dependent on Parliament and was able to leave a considerable fortune to his son, *Henry VIII.

Henry VIII (1491-1547), King of England (1509-47). He was the second son of Henry VII and Elizabeth of York. He succeeded to the throne aged 18 and began his reign by executing Dudley and Empson, two of Henry VII's financial officers. From 1513 to 1529 Thomas *Wolsey managed affairs of state and diplomacy while Henry played the part of the Renaissance prince, preferring hunting and dancing to government.

From 1525 he turned against his wife *Catherine of Aragon because of her failure to provide him with male heirs. The pope's refusal to annul his marriage led to England's break with the *Roman Catholic Church. With the assistance of Thomas *Cromwell and a compliant, anticlerical Parliament (1529-36), legislation was passed to sever the English Church from papal jurisdiction and Henry became Supreme Head of the English Church (1534). He exploited the Dissolution of the *Monasteries for his own profit and used the revenues from the dissolution to pay for his military campaigns of the 1540s. But he remained conservative in doctrine, believing in Catholicism without the pope and retaining the title 'Defender of the Faith' granted him by the pope in 1521 for his treatise against *Luther. Meanwhile he married Anne *Boleyn and subsequently Jane *Seymour, *Anne of Cleves, Catherine *Howard and Catherine *Parr, of whom only Jane Seymour bore him a son, the future *Edward VI.

Little was achieved by his expensive wars with France and Scotland, but a powerful English navy was created. His attempts to capitalize on the struggles of *Francis I of France and *Charles V of Spain severely undermined the English economy.

Henry the Lion (1129-95), Duke of Saxony and Bavaria. He was the son of Henry the Proud of the German Welf family. He obtained his duchies with the help of Emperor *Frederick Barbarossa, expanding his territories and influence as far as the Baltic, and encouraging trade. Ambitious and powerful, he lost the emperor's favour and was banished in 1180. He was

A formal portrait of **Henry VIII**, painted by an unknown artist about five years before his death. Slim and athletic in youth, Henry grew fat and heavy when ill health prevented him from taking strenuous physical exercise. (Montacute House)

welcomed in England, as his second wife was the daughter of *Henry II and he became a power in *Angevin political life. By 1190 he had regained much of his former territory. His son, Otto IV, continued the challenge to the *Hohenstaufen.

Henry the Navigator (1394-1460), Portuguese prince, the third son of John I of Portugal and grandson of John of *Gaunt. He did not himself undertake any voyages of *exploration, but was the patron of a succession of Portuguese seamen who made voyages of discovery among the Atlantic islands and down the west coast of Africa, which led, after his death, to the discovery of the Cape of Good Hope and the sea route to India. He established a school at which navigation, astronomy, and cartography were taught to his captains and pilots, and constructed the first observatory in Portugal.

Henry, Patrick (1736-99), American patriot, born of Scots-Irish parents in Virginia. He attracted notice as the anti-establishment lawyer in the Parsons' Cause, in which he argued that the use of a royal veto (obtained by clergy wishing to reverse a new law requiring them to be paid in money, at a discounted rate, rather than tobacco) was improper as it was against the public good. As the author of the Virginia Resolves (1765) against the *Stamp Act, he became leader of the frontier radicals, declaiming in 1775: 'Give me liberty or give me death'. He was the first governor of the state and opposed the *Constitution on states' rights grounds, but in the 1790s became a convert to Federalism.

Heraclius I (575–642), Byzantine emperor (610–41). He came to power having ousted the usurper Phocas, and set about the reorganization of the empire and its army, driving back the Persians in Asia Minor and securing a peace treaty. He then turned his attentions to the north, coming to terms with the invading *Avars, and strengthening the frontier. He then devoted his energies to the study of religion, neglecting the empire and losing Syria and Egypt to the Muslims.

heraldry, the study of coats of arms worn for individual identification, and of the accessories of crests, badges, mottoes, and flags which accompanied them. Its origins are military. Soldiers in armour and helmets could not easily be identified in battle and so the practice evolved of displaying a sign or device on the shield and on the linen surcoat worn over the armour (from which the terms 'coat-of-arms' and 'court armoury' derive). The first heraldic designs may have been worn by the Crusaders, but their use became widespread in Europe in the 12th century. A similar system also emerged in Japan during the 12th century.

By the 13th century heraldry had so developed that it had its own terminology, based on Old French. Its colours are called 'tinctures' of which there are two metals—gold (*or*) and silver (*argent*)—and five colours—blue (*azure*), black (*sable*), green (*vert*), purple (*purpure*), and red (*gules*). In England heralds were formed into the College of Arms (1484). Scotland has its Court of the Lord Lyon (1592).

Herculaneum, a small coastal town near Naples, in Italy. Lying near the volcano Vesuvius, it was the first to be buried in the eruption of AD 79 which also claimed *Pompeii. Over 20 cm. (8 in.) of ash fell on 24 August. This was followed by blast clouds and glowing avalanches of lava early the following day. Many of the population are now thought to have been caught in flight on the beach, buried, like the town, under 20 m. (over 60 ft.) of volcanic mud 0.5 km. (0.25 mile) from the present coastline. It was discovered accidentally by a well-digger in 1709. Systematic archaeological excavation of the site began only in the 19th century.

heresy, belief in a doctrine held to be false by the Christian Church. During the Middle Ages it was believed to be necessary to follow the one 'true' religion, which provided the only guarantee of salvation and afterlife. Consequently those who came to believe that orthodox teaching was inadequate or wrong risked being declared heretics. Since the Church sought to maintain the unique validity of its declared doctrine conflict was inevitable.

The early Church condemned Gnostics in the 2nd century and the *Arians and *Nestorians in the 4th century. The *Iconoclasts were condemned at the Council of *Nicaea in 787. The condemnation of the *Cathars in southern Europe led to the *Albigensian Crusade. Later dissatisfaction with orthodox teaching led ultimately to *Protestantism. The *Inquisition from its earliest days upheld the Church's doctrine and became responsible for the rooting out of unorthodoxy.

Hereward the Wake (11th century), a semi-legendary Anglo-Saxon outlaw best known for his resistance to the Norman conquerors of England and as the leader of a rebellion against them (1070–1). He supported a Danish raid on the monastery of Peterborough in retaliation for the recent appointment of a Norman abbot. Subsequently he set up camp on the Isle of Ely in eastern England where he was joined by Morcar, the English Earl of Northumbria. Although Morcar later surrendered, Hereward escaped capture; his exploits inspired many legends.

hermit (from the Greek word for 'desert'), someone who for religious reasons takes up a solitary life. The first Christian hermits, from the late 3rd century onwards, were most numerous in and around Egypt; some of them were highly influential and were visited by pilgrims. St Basil the Great (*c*.330–79) prepared a monastic rule that is still followed in the Eastern Church. The eremitic way of life influenced European monastic orders such as the *Carthusians and *Carmelites.

Herod, dynasty of Idumaean Jews who ruled in Palestine with the support of Rome. Its founder, Antipater, who was appointed governor of Judaea in 47 BC by Julius *Caesar, was murdered in 44 BC. Herod the Great was appointed 'King of the Jews' in 40 BC and he ruled Judaea (37–4 BC) with ruthlessness and skill, maintaining an unaccustomed peace. His encouragement of Greek culture was resented by Jewish nationalists and the end of his reign was marked by violent palace feuds. After his death, the kingdom was divided between three of his sons. Archelaus ruled Judaea, Idumaea, and Samaria (4 BC–AD 6); he was deposed and his territory put under a Roman governor. Herod Antipas ruled Galilee and Peraea (4 BC–AD 39); *John the Baptist criticized him for marrying his brother's wife *c*.AD 27. After religious riots, Caligula exiled him. Philip ruled the territory east of Galilee (4 BC–AD 34). By 41 Herod Agrippa I, a grandson of Herod the Great who had been educated at the imperial court in Rome, had, with the help of *Claudius, reunited these territories which he ruled as king (41–4). His son Herod Agrippa II, the last of the dynasty, only ruled in northern Palestine (50–*c*.93). He helped the Romans crush the *Jewish Revolt (66–70).

Herodotus (*c*.490–80 to *c*.430–20 BC), Greek historian, accorded the title the 'Father of History'. He wrote a comprehensive nine-book account of the *Greek–Persian wars. The scope of his work, which included an account of the earlier history of the Persian empire and a long digression on Egypt, reflects both his inquiring mind and his extensive travels; he visited places as far apart as Athens, Babylon, Egypt, and the Black Sea. Information was elicited from people encountered on his journeys and from his own observations, though he may often have accepted too readily what he heard, failing to subject these accounts to critical scrutiny. Nevertheless, his *History*, full of information about the contemporary Persian and Greek worlds, was an outstanding achievement.

Hexham, battle of (15 May 1464). A battle of the Wars of the *Roses in which a *Yorkist force led by John Neville, Earl of Montagu, captured Henry *Beaufort, 3rd Duke of Somerset, three miles from the town of Hexham, in Northumberland. Somerset was beheaded on the field of battle and many of his followers were executed soon afterwards, while Neville was rewarded with the estates of the *Percy family and the earldom of Northumberland.

Hezekiah (d. 687 BC), King of Judah (715-687 BC). When he came to power, Judah was a vassal state of the *Assyrian empire, and with the leaders of neighbouring states he was involved in a number of planned rebellions. The prophet *Isaiah spoke against these, but eventually Hezekiah did rebel and was heavily defeated in 701, when *Sennacherib invaded, the land was devastated, and only Jerusalem escaped destruction. The Bible describes his work of religious reform, destroying local shrines and various cult objects, and attempting to suppress the worship of local gods. The reform was short-lived, pagan practices being re-introduced after his death by his son and successor Manasseh.

Hideyoshi (1536-98), Japanese warrior. He continued *Oda Nobunaga's work of unifying the country that had been fragmented by the feuds between *daimyo. Between 1582 and 1591, by a mixture of military strategy and skilful diplomacy he broke their power. Mistrustful of the power of Buddhist monks, for a time he encouraged Catholic missionaries but later savagely persecuted Chris-

A portrait of the Japanese warrior, **Hideyoshi**. Despite his humble birth he achieved power as a result of his outstanding military ability.

tians in Nagasaki. He built castles, carried out land surveys, and disarmed peasants. His ambition was to conquer China, and when in 1592 *Korea, a vassal state of China, refused passage to his troops, his army numbering 200,000 captured Seoul and advanced north until *Ming armies forced him to retreat. The Koreans routed him at sea. A second campaign was abandoned when Hideyoshi died. He appointed *Tokugawa Ieyasu a guardian of his son, Hideyori.

hieroglyphs, the signs used for formal inscriptions in ancient *Egypt. They were devised c.3000 BC and were used, mainly for religious and monumental purposes, until the late 3rd century AD. They were representational in design, and were held to have near-magic powers. For practical purposes the structure of the script, as well as its form, was much too clumsy, and even cursive or 'long-hand' versions, the hieratic and demotic scripts, required long professional training to learn. Hieroglyphs have played an important part of archaeological research (*Rosetta Stone), extending written history by 2,000 years beyond classical times. The term has been applied more loosely to other complicated but ornamental scripts used by the *Minoan civilization, the *Hittites, and the *Maya.

highwayman, a robber who plagued Britain's main roads during the 17th and 18th centuries. There had been thieves and footpads from Anglo-Saxon times, but improved roads, more frequent travelling, and the growth of coaching inns made rich pickings for mounted thieves who sometimes worked in gangs in collusion with innkeepers. Their cry of 'Stand and deliver' was in reality much less romantic than is generally shown in fiction; the fear of being identified made them only too willing to murder their victims. Some, such as Swift Nick Levison, hanged at York in 1684, Dick Turpin (1705-39), and Jack Sheppard (1702-24), became folk heroes, but in fact highwaymen were always dangerous, and their appearance in central London in broad daylight in the mid-18th century provoked vigorous efforts to stamp them out. By the early 19th century the menace of highwaymen had been largely overcome.

Hinduism, the name given to the religious beliefs and social customs that originated in India. These are a complex mixture of elements from the Indo-European culture brought by the *Aryan invasions of 1500 BC onwards, and from the indigenous pre-Aryan and Dravidian societies. The religion of the early period is called Vedic, from the texts and rituals of the *Veda, which is called 'revelation' (sruti) and retains a very important place in Hinduism. However, many other texts and sources have become significant, for example the *Bhagavadgita, ('The Song of the Lord'), believed to have been revealed to man by the god Vishnu in his form as Krishna.

Hindus have very diverse religious beliefs, but most varieties believe that man is trapped in a cycle of death and rebirth, and seek to escape from this transmigration into union with God, or with Brahman, the eternal principle behind all existence. This release is sought by a variety of ascetic and devotional paths, as well as by actions deemed appropriate to an individual's caste (right dharma). Caste, a form of social organization unique to Hindus, is distinguishable from other social hierarchies by its distinctively religious character, based on ideas of purity and pollution, which are particularly associated with marriage and rules concerning the cooking and eating of food. Hindus have traditionally believed it to be part of the divine nature of things, but scholars disagree about its origins. There are four 'estates' (*varna*), Brahmin, Kshatriya, Vaisya, and Sudra, subdivided into numerous castes (*jati*) and sub-castes.

hippodrome, a course on which the ancient Greeks and Romans held *chariot and horse races. The courses were U-shaped with a barrier down the centre. The competitors would race down one side and then back up the other. Spectators watched from tiered stands. *Olympia had an early example, and hippodromes were a typical feature of major Greek cities of classical and Hellenistic times. The one at *Constantinople held about 100,000 spectators and was the scene of fierce rivalry among partisan supporters. The Circus Maximus at Rome was modelled on the Greek hippodrome.

Hispaniola, the second largest island of the West Indies. Hunter-gatherer Ciboney Indians from South America had settled in the island by 5000 BC. Agriculture reached the island with the arrival (*c*.1000 BC) of Arawak Indians from north-western South America. From *c*. AD 200, under the influence of Mexico, a Taino Arawak culture, which included the growing of maize, the construction of ceremonial centres, and the worship of human and animal representation of spirits, developed and spread to other islands. In 1492 *Columbus landed on the island and in 1496 the town of Santo Domingo was established, the first European settlement in the Americas. Spaniards developed plantations at the eastern end of the island but lost control of western Hispaniola to France, who established the colony of Santo Domingue (*Haiti).

Hittite archers hunting lions from a chariot, a 9th-century BC limestone relief from the 'Lion' gate of Milid, near present-day Malatya, Turkey. Hieroglyphs decorate the top. Bas-reliefs found at Milid are among the best remaining examples of Hittite art. (Turkish National Museum, Ankara)

Exploitation and European diseases made the Taino Arawak virtually extinct by the 18th century.

history writing, the recording and interpretation of past events. It began with the retelling of legends handed down through oral traditions: the epic tales of *Homer were the poetic expression of oral history, while in the classical age of ancient Greece *Herodotus and *Thucydides wrote narrative histories of their own times. In China *Sima Qian (*c*.145–*c*.85 BC) is known as the 'Father of Chinese History'. The Roman historians, who include *Tacitus, *Livy, and *Suetonius, wrote works which served as models for later medieval and Renaissance historians. Within the Arab world al-Tabaric (838–923) wrote the *Annals*, a history of the world from its creation to 915, and Ibn Khaldun (1332–1406) the *Kitab a'ibar* ('Book of Examples'), a major history of Islam. In medieval Europe history was written by the literate clergy (*Bede) and was mostly confined to chronicles (*Anglo-Saxon Chronicle*, *Froissart). In the 15th and 16th centuries the Italian historians *Machiavelli and *Guiccardini wrote political analyses of the state and its rulers. The 18th century *Enlightenment injected a considerable measure of rationalism and scepticism into historical writing, producing such masterpieces as *Gibbon's *The History of the Decline and Fall of the Roman Empire*.

Hittites, the ancient people of Asia Minor who flourished from 1700 to 1200 BC. They were Indo-Europeans, probably from north of the Black Sea, and entered Anatolia towards the end of the 3rd millennium BC. They established Hattush as their capital and gradually extended their power through much of Anatolia and into Syria. Mursilis I even penetrated as far as Babylon *c*.1595 BC, but his death was followed by considerable internal discord. The high-point of Hittite rule was achieved under Suppiluliumas (*c*.1375–1335). In *c*.1285 the Hittites met the Egyptians at Kadesh in a famous but indecisive battle. The two great powers then made peace, but *c*.1200 the Hittite empire collapsed before the invasions of the *Sea Peoples. Hattush (near modern Boğazköy) has yielded an invaluable collection of Hittite records, written in *cuneiform.

Hobbes, Thomas (1588-1679), English philosopher and political theorist. After receiving his degree from Oxford University in 1608, he became a tutor to the Cavendish family, and travelled extensively in Europe with his pupils, meeting such thinkers as Galileo and Descartes. Hobbes's philosophy arose from a project of investigating the nature of matter, of man, and of society. He planned to write a three-part work of philosophy but it was his ideas on the state, published as *Leviathan* in 1651, that were most controversial. In it he argued that man is motivated entirely by selfish concerns, notably fear of death; the necessity of avoiding anarchy leads him to surrender his individual rights and accept the rule of an absolute monarch. This sovereign Hobbes called the Leviathan, the monster to whom all must submit in the interests of security: the alternative, popular government, leads to civil war. The Royalists regarded Hobbes's views as an inducement to *Cromwell to set himself up as an absolute ruler and the book was condemned by Parliament. None the less Hobbes returned from exile in Paris at the end of the English Civil War and accepted the *Restoration settlement.

Hohenstaufen, a German royal house, members of which held the throne of the Holy Roman Empire (1138-1254), the rivals of the *Hohenzollerns. The emperor Henry IV gave them the duchy of Swabia and in 1138 Conrad himself became emperor as Conrad III. The family provided many emperors, including *Frederick Barbarossa who attempted to build up German power in Italy. The relationship between the papacy and the Hohenstaufen was frequently acrimonious, resulting from their respective claims to land and personal powers. The empire grew to include Germany, northern Italy, and Sicily but proved too large to be managed in the face of papal and Lombard opposition, and the dynasty's last ruling member, Manfred of Sicily, was killed in battle in 1266.

Hohenzollern, a powerful German princely family whose roots can be traced back to the 11th century. From 1415 they ruled the electorate of *Brandenburg and the following century saw great expansion, Margrave Albert becoming grand master of the *Teutonic Knights in 1511. In 1614 the duchy of Cleves was acquired and in 1701 the Elector Frederick III of Brandenburg took the title of Frederick I of *Prussia.

Hojo, a branch of a powerful Japanese family, the Taira. After *Minamoto Yoritomo's death they provided regents for puppet *shoguns, nominated by themselves. From 1219 the regency was hereditary, and the country prospered under them until c.1300. They refused tribute to *Kublai Khan and executed his envoys. His two invasions, though failures, weakened Hojo power. Vassals the Hojo were unable to reward for their victories turned against them. From 1331 there was war between the regent's forces and those attempting to restore imperial rule under Go-Daigo. Their power ended (1333) when *Ashikaga Takanji, a Hojo vassal, defected to the emperor and another vassal took *Kamakura. The last regent and his family committed *seppuku* (ritual suicide).

Holkar, Indian *Maratha family of peasant origin that became rulers of one of the most powerful of the *Maratha confederacy of states. The family's founder, Malhar Rao

Holkar, rose through military service to the Peshwas to establish by 1766 virtually independent control of the Malwa region. The family's control was consolidated in the late 18th century, but succession disputes and quarrels with other Maratha chieftains offset the gains. During the reign of Jaswant Rao Holkar (1797-1811) British expansion destroyed Holkar claims in north India. After defeat in 1804 the Holkars had to accept British protection, remaining a Princely State until 1947.

Holland *Netherlands.

Holy League, the name given to several European alliances formed during the 15th, 16th, and 17th centuries. The League of 1511-13 was directed against French ascendancy in Italy. The coalition was organized by Pope Pius II, and included England, Spain, Venice, the Holy Roman Empire, and Switzerland. It succeeded in its initial aims, but then there was squabbling over strategy and the hard-pressed French were able to conclude separate peace treaties with each member. The Holy League of 1526 was formed against emperor *Charles V by France, the papacy, England, Venice, and Milan. It achieved little in the subsequent war, and *Francis I of France made peace with Charles at Cambrai in 1529.

The French Holy League of 1576, also known as the Catholic League, was led by the *Guise faction during the *French Wars of Religion. Henry III ordered its dissolution in 1577, but it was revived in 1584 to play a major part in the War of the *Three Henrys (1585-9). Its power waned after *Henry IV accepted Catholicism in 1593.

The Holy (or Catholic) League of 1609 was a military alliance of the German Catholic princes, formed at the start of the War of the Jülich Succession (1606-14). During most of the *Thirty Years War its forces served the imperial cause, with *Tilly as its principal commander.

Holy Roman Empire (962-1806), the empire in the West, conceived as a replacement for the *Carolingian empire and in the Christian imperial tradition of *Constantine. The creation of the medieval popes, it has been called their greatest mistake; for whereas their intention was to appoint a powerful secular deputy to rule Christendom, in fact they generated a rival.

From *Otto I's coronation in 962, the empire was always associated with the German crown, even after it became a *Habsburg (Austrian) title in the 15th century. The empire consisted of duchies, counties, and bishoprics owing formal allegiance to the emperor, but there were often clashes between the emperor and the nobility (*elector), especially as the emperor's claims to sovereignty over Italy resulted in frequent absences from Germany. Emperor Henry IV (ruled 1084-1106) had a long struggle with the Church over the right to appoint bishops and senior clergy (*investiture). Emperor Frederick I (ruled 1155-90) made a sustained attempt to bring Italy and the papacy under military subjugation but was finally defeated at the battle of Legnano in 1176. Open warfare broke out between the *Guelphs (allies of the pope) and the Ghibellines (the imperial party). After the death of Emperor *Frederick II in 1250 German imperial power, both in Italy and within Germany, waned.

The empire became virtually hereditary, ruled by successive members of the same princely family. The first

Holy Roman Empire:
emperors including dates of reign

Saxon dynasty
Otto I, the Great, 936-73. Crowned Emperor of the Romans 962
Otto II, 973-83. Added 'Roman' to the title
Otto III, 983-1002. Crowned 996
Henry II, 1002-24. Crowned 1014

Salian or Franconian dynasty
Conrad II, 1024-39. Crowned 1027
Henry III, KoR[1] 1039-46, HRE[2] 1046-56. Crowned German King or KoR 1028 in his
    father's lifetime. Crowned HRE 1046
Henry IV, 1056-1106. Crowned 1084. Investiture Contests with Pope Gregory VII
    Rivals during this period: Rudolf of Swabia, 1077-80, Hermann of Luxemburg,
    1081-93, and Conrad of Franconia, 1093-1101
Henry V, 1105-25. Crowned 1111, Diet of Worms 1122
Lothair II, Duke of Saxony, 1125-37. Crowned 1133

Hohenstaufen dynasty
Conrad III, KoR 1138-52. Never crowned
Frederick I, (Frederick 'Barbarossa'), KoR 1152-5, HRE 1155-90. Crowned 1155.
    Lombard League formed 1167
Henry VI, KoR 1169-91, HRE 1191-7. Crowned 1191. Feud in Italy between papal and
    imperial parties
Philip of Swabia, KoR 1198-1208. Never crowned
Otto IV of Brunswick, 1198-1212. Crowned 1209
Frederick II, KoR 1212-20, HRE 1220-50. Crowned 1220
    Rivals: Henry Raspe, 1246-7, and William of Holland, 1247-56. Neither crowned.
Conrad IV, KoR 1237-54. Never crowned

The Great Interregnum 1254-73
Decline in power of HRE. Rivals: Richard of Cornwall, KoR 1257-73 and Alfonso X of
    Castile, 1257-72. Neither crowned

Habsburg, Luxemburg and other dynasties
Rudolf I (Habsburg), KoR 1273-91. Never crowned but recognized by the Pope 1274
Adolf of Nassau, 1292-8. Never crowned
Albert I (Habsburg), KoR 1298-1308. Never crowned
Henry VII (Luxemburg), 1308-13. Crowned 1312
Louis IV (Wittelsbach), 1314-46. Crowned 1328. Electors vote to choose HRE without
    papal intervention 1338
    Rival: Frederick of Habsburg, co-regent 1325-30
Charles IV (Luxemburg), 1346-78. Crowned 1355. Golden Bull which regulated
    election procedure 1356
    Rival: Günther of Schwarzburg, 1347-9
Wenceslas (Luxemburg), 1378-1400. Crowned 1376. Great Schism 1378-1417
Rupert (Wittelsbach), 1400-10. Never crowned
Sigismund (Luxemburg), 1410-37. Crowned 1433
    Rival: Jobst of Moravia, 1410-11

Habsburg dynasty
Albert II, KoR 1438-9. Never crowned
Frederick III, KoR 1440-52, 'HRE of the German people' 1452-93. Change of title.
    Last Emperor to be crowned in Rome
Maximilian I, KoR 1486-93, 1493-1519. Never crowned. Italian Wars 1494-1559.
    Reichsregiment established
Charles V, KoR 1519-30, HRE 1530-56. Last Emperor to be crowned by the Pope
    (at Bologna). Peace of Augsburg 1555
Ferdinand I, KoR 1531-58, HRE 1558-64
Maximilian II, KoR1562-4, HRE 1564-76
Rudolf II, KoR 1575-6, HRE 1576-1612
Matthias, 1612-19. Thirty Years War 1618-48
Ferdinand II, 1619-37
Ferdinand III, KoR 1636-7, HRE 1637-57. Peace of Westphalia 1648
Leopold I, 1658-1705. League of Augsburg 1686. War of the Grand Alliance 1688-97.
    Treaty of Ryswick 1697
Joseph I, 1705-11
Charles IV, 1711-40

Interregnum 1740-2
War of Austrian Succession 1740-8

Charles VII (Wittelsbach-Habsburg), 1742-5
Francis I (Lorraine), 1745-65. Treaty of Aix-la-Chapelle 1748.
    Seven Years War 1756-63. Treaty of Paris 1763

Habsburg-Lorraine dynasty
Joseph II, KoR 1764-5, HRE 1765-90
Leopold II, 1790-2
Francis II, 1792-1806. French Revolutionary Wars 1792-1802. Confederation of the
    Rhine and renunciation of the title of HRE 1806

1. KoR = King of the Romans (emperor designate uncrowned)
2. HRE = Holy Roman Emperor

Holy Roman Empire: its extent *c.*1100

By the end of the 11th century the Holy Roman Emperor was
the crowned ruler of Germany, Burgundy, and northern
Italy, in addition to his imperial role as temporal leader of
Christendom, and the empire was close to its height in terms
of cohesion and extent. The degree of power he exercised
varied, however, at different times. Within Germany there
was constant rivalry between the emperor and the rulers of
the margraviates, duchies, and kingdoms.

*Habsburg emperor was Rudolph I (from 1274) and the
empire remained in Habsburg hands until its dissolution.
King Charles I of Spain was crowned as the Holy Roman
Emperor *Charles V in 1530. His empire included
Spain, Germany, the Netherlands, Sardinia, and Sicily,
together with newly conquered territory in the Americas
(*New Spain). But the *Reformation and the *Thirty
Years War saw challenges to the power of the Catholic
Habsburgs and the emperors suffered a loss of prestige
and power during a series of wars against *Louis XIV.
In the 18th century *Prussia under *Frederick II emerged
as the leading German power, and, in 1806, the imperial
crown was surrendered to Napoleon. The empire was not
revived after his downfall.

Homer, traditionally the author of the *Iliad* and the
Odyssey, the greatest of the early Greek epic poems. The
date when they achieved their final form is disputed, but
may have been as late as the 7th century BC. They stand
as the final flowering of a long oral bardic tradition
which told the stories and legends of the heroic age of
Greece, roughly the end of the *Mycenaean period; but
they also contain elements which are later than that. The

Iliad concentrates on one series of connected episodes during the siege of *Troy. The *Odyssey*, which may be by a different author, tells of the wanderings of one of the Greek leaders at Troy, Odysseus, during his protracted and adventurous return to his kingdom of Ithaca. As the earliest surviving literary record, they are of great use in helping to reconstruct the early history of Greece, for which they provide much incidental evidence; but such evidence is often contentious and has to be treated with great circumspection. By about 500 BC, prose histories began to be written, superseding the older medium of verse.

Hominids, members of the family Hominidae, including our own species *Homo sapiens*, our presumed forebears *Homo erectus* and *Homo habilis*, and forms believed to be closely related called collectively the *australopithecines. Many scientists now also include the African great apes— the two chimpanzees and gorilla—in the human family too, rather than grouping them with the more distantly related Asian apes, the orang-utan, gibbon, and siamang. The traditional way of grouping the large apes (chimpanzees, gorilla, and orang-utan) is in their own family, Pongidae. Estimates of the date of divergence of the ape and human lineages vary. The Asian apes probably branched off 8–12 million years ago and the African apes 10–5 million years ago.

Homo erectus ('upright man'), the presumed predecessor of our own species, *Homo sapiens*, who lived in Africa and Asia and possibly in Europe. This hominid was larger than the *australopithecines and *Homo habilis* and its brain approached the size of a modern human's. However, the facial bones remained relatively massive and the skull was long and low. One of the hallmarks of this species was a teardrop-shaped stone tool flaked on both sides, the *Acheulian handaxe, which was more specialized than the *Oldowan tools of *Homo habilis*. *Homo erectus* was the first member of the human lineage to control and use fire and this, and perhaps clothing, may have contributed to its spreading so widely from its place of origin in tropical East Africa. It had evolved presumably from *Homo habilis*, by 1.6 million years ago. By around 1 million years ago or not long before these hominids are presumed to have begun their travels that took them as far as China (*Peking Man) and Indonesia (*Java Man). The last representatives disappeared 400,000–200,000 years ago.

Homo habilis ('handy man'), the name given to a group of human fossil remains found at *Olduvai Gorge in Tanzania in the early 1960s, and now known from other eastern African sites, especially *Koobi Fora in Kenya and perhaps also from southern Africa. They date from about 2.5 million to 1.6 million years ago. Although similar in size to the contemporary *australopithecines, their brains were larger, their faces more human-like, and they very likely evolved into *Homo erectus*. They were probably the first makers of stone tools—simple pebble and flake artefacts collectively called the *Oldowan industry.

Homo sapiens ('wise man'), our own species, that evolved from *Homo erectus* by 400,000–200,000 years ago. By this stage the brain had enlarged to the modern size, the skull bones had become less heavy, and the back of the head was rounded. The next development is obscure as the *Homo sapiens* lineage apparently split into two main lines, one leading to the *Neanderthals (*Homo sapiens neanderthalensis*), the other to fully modern people (*Homo sapiens sapiens*). The latter development took place gradually during the past 125,000 years. Anatomical and genetic evidence support the idea that this happened in Africa, but it is possible that there was at least one parallel development in the Far East. In the Middle East, anatomically modern humans had appeared by around 50,000 years ago; in Europe, they came slightly later, and more abruptly around 35,000 years ago. The earliest modern Europeans are often called *Cro-Magnons. It is not known what part, if any, the Neanderthals played in these Middle Eastern and European developments. Very likely they were not our direct ancestors but they may have interbred with modern people entering Europe from Africa via the Middle East.

With the evolution of modern people came marked advances in tool technology, rapid increase in population, the grouping of social activities in dwellings, and the first appearance of art; the cultural period called the *Upper Palaeolithic had begun. These Upper Palaeolithic people almost certainly had a spoken language. With population growth came the colonization of new territories, which seems to have begun soon after the origin of fully modern humans. People had reached New Guinea and Australia from Indonesia by at least 40,000 years ago and developed *Australoid characteristics in isolation there. The timing of the first settlement of the New World is more controversial. It was probably before 15,000 years ago but there is little firm archaeological evidence so far for an earlier colonization. Genetic, linguistic, and anatomical evidence of modern *Amerindians, however, is increasingly suggesting that the first entry into North America occurred between 40,000 and 30,000 years ago.

Honduras, a Central American country whose native inhabitants are mestizo Indians. One of the lieutenants of Hernan *Cortés, Francisco de las Casas, founded the first settlement, the port of Trujillo, in 1523. Honduras was attached administratively to the captaincy-general of *Guatemala throughout the colonial period.

Hood, Samuel, 1st Viscount (1724–1816), British admiral. He served in the Seven Years War, the American War of Independence, and the French Revolutionary wars. He entered the navy in 1741, was promoted post-captain in 1756, and in 1759 captured a French frigate after a fierce action. In 1780, with the rank of rear-admiral, he went to the West Indies as second-in-command to Lord *Rodney. There he displayed masterly tactical skills and took a prominent part in the defeat of the French fleet near Dominica two years later. During the French Revolutionary wars he commanded the British fleet in the Mediterranean.

Hooker, Richard (*c.*1554–1600), English theologian. He enjoyed a reputation as a controversial London preacher, but in 1591 he took a country parish and there wrote *The Laws of Ecclesiastical Polity* (1594–7). In it he attacked the Puritan idea of basing all human conduct upon the scriptures alone, and he gave the Church of England its first systematized Anglican theology and written justification. Hooker's pragmatic emphasis on reason influenced Tudor statesmen as well as churchmen.

Hopewell cultures, a group of related cultures of the eastern USA, dating c.500 BC–AD 500. They are known for their conical burial mounds, most highly developed in Ohio, and often associated with earthworks in various shapes, the most famous being the Great Serpent Mound of Adams County, Ohio, 213 m. (700 ft.) in length.

hoplite, a citizen-soldier of the cities of ancient Greece. Each man had to provide his own formidable armour (Greek, *hopla*)—2.7 m. (9 ft.) spear, short sword, large round shield, breastplate, and greaves (shin-pads). They fought in the close-packed *phalanx formation, and were extremely effective when operating in the plains of Greece. However, over rough terrain they were vulnerable to fast-moving light infantry. The professional hoplites of *Sparta were pre-eminent in classical times until their defeat by the Thebans in 371 BC.

Horn, Filips van Montmorency, Graaf van (or Hoorn) (c.1524–68), Flemish soldier and statesman. He had a long record of distinguished service to both Emperor *Charles V and *Philip II of Spain, but as a member of the regency council in the Netherlands (1561–5), he followed a similar opposition course to that of his colleague, Lamoral *Egmont. In 1566 he aligned himself with the Calvinists at Tournai, but then obeyed the regent's command to return to Brussels. Late in that year, he rejected *William the Silent's plan for armed resistance to the Spaniards, and withdrew to his home in Weert. In 1567 *Alba found him out, had him convicted of treason and heresy by the Council of Troubles, and he was beheaded.

Hospitaller   *Knight Hospitaller.

Hotspur   *Percy.

Howard, Catherine (c.1521–42), Queen consort of *Henry VIII of England from 1540. She was the king's fifth wife, her marriage in 1540 the result of the ambitions of her Catholic relatives, the dukes of Norfolk (the Howard faction). After the king's marriage to *Anne of Cleves failed, they engineered the downfall of Thomas *Cromwell and promoted the 19-year-old Catherine. Protestant enemies of the Howards subsequently accused her of infidelity and she was beheaded. After her death the Howards continued to promote Catholicism. Thomas Howard, 3rd Duke of *Norfolk (1473–1554) was imprisoned by *Edward VI but was favoured by *Mary I. Thomas Howard, 4th Duke of Norfolk (1536–72) was involved in the *Ridolfi Plot and was executed.

Howard, John (1726–90), British philanthropist and reformer. He first experienced prison conditions when captured by the French in 1756 while on his way to help the survivors of the Lisbon earthquake. Later, as high sheriff of Bedfordshire in 1773, his prison responsibilities aroused his interest in prison conditions. He published several works on the subject and secured some important prison reforms through Parliament. His name is preserved in the Howard League for Penal Reform.

Howe, Richard, 4th Viscount and Earl (1726–99), British naval officer. He served with distinction in the American War of *Independence and the French Revolutionary wars. He entered the navy in 1739, and gave early proof of his abilities during the *Seven Years War. He was promoted vice-admiral in 1775 and put in command of the North American station in the following year to enter into peace negotiations with the rebellious American colonists. When these proved fruitless his fleet helped the British army to capture New York and Philadelphia. In 1782 he raised the siege of Gibraltar after defeating a combined Franco-Spanish force. He brought his career to a triumphant conclusion with his victory over the French fleet off Ushant in 1794 in the Battle of the 'Glorious First of June'.

His brother, **William, 5th Viscount Howe** (1729–1814), joined the army in 1746 and, after winning rapid promotion, took part in the successful assault on *Quebec in 1759, leading the march to the *Plains of Abraham. Although, like his brother, he sympathized with the American colonists, he was made supreme commander of the British forces in North America in 1776 and inflicted a series of defeats on *Washington's army. In 1778 he resigned his position and returned home.

Hoysala, an Indian dynasty, first Jain and later Hindu, which ruled the south Deccan c.1006–1346, from a capital at Dwarasamudra (near Mysore). Beginning as marauding hill chieftains, they finally established sway over a region corresponding to the later *Mysore state. Visnuvardhana (c.1110–41) expelled the powerful *Cholas from the Deccan, and during the reign of his grandson, Ballala II (1173–1220), they became a force to be reckoned with in south India. However, their strength finally depended on the acquiescence of Hindu rivals whom they antagonized by rendering assistance to Muslim invaders. On their downfall in the mid-14th century, their territories fell to the *Vijayanagar empire. The Hoysalas were patrons of learning and the arts, notably architecture and sculpture.

Hsia   *Xia.

Hsiung-nu   *Xiongnu.

Hudson's Bay Company, a company chartered in 1670 to Prince *Rupert and seventeen others by *Charles II to govern and trade in the huge area of the Canadian north-west, called Rupert's Land, which drained into Hudson Bay. Although huge profits accrued from the *fur trade, the company was, until 1763, threatened by competition and military attack from the French. From 1787 there was occasionally murderous conflict with the North-West Company over control of the fur trade until the two companies amalgamated in 1821.

hue and cry, the term given to the practice in medieval England whereby a person could call out loudly for help in pursuing a suspected criminal. All who heard the call were obliged by law to join in the chase; failure to do so would incur a heavy fine and any misuse of the hue and cry was also punishable. The system was regularized by Edward I in the Statute of Winchester (1285), which rationalized the policing of communities. The obligation on the public to assist the police in the arrest of a suspect has survived in principle to the present day.

Huguenot, the name given in the 16th and 17th centuries to a French Protestant who followed the beliefs of *Calvin. By 1561 there were 2,000 Calvinist churches in France and the Huguenots had become a political faction that seemed to threaten the state. Persecution followed and during the *French Wars of Religion the Huguenots fought eight civil wars against the Catholic establishment and triumphed when, by the Edict of *Nantes in 1598, *Henry IV gave them liberty of worship and a 'state within a state'. Their numbers grew, especially among merchants and skilled artisans, until they were again persecuted. The centre of their resistance in 1627 was *La Rochelle, which the *Richelieu government had to besiege for over a year before capturing it. In 1685 the Edict was revoked; many thousands of Huguenots fled to England, the Netherlands, Switzerland, and Brandenburg, some settling as far away as North America and the Cape of Good Hope. All these places were to benefit from their skill in craftmanship and trade, particularly as silk-weavers and silversmiths.

humanism, a philosophical and cultural movement forming part of the 15th-century European *Renaissance. Humanists were originally Christian scholars who studied and taught the humanities (grammar, rhetoric, history, poetry, and moral philosophy) by rediscovering classical Latin texts, and later also Greek and Hebrew texts. They came to reject medieval *scholasticism, and made classical antiquity the basis of western Europe's educational system and cultural outlook. Among their ranks can be numbered Petrarch, *Guicciardini, and *Machiavelli. They had no coherent philosophy, but shared an enthusiasm for the dignity of human values in place of religious dogma or abstract reasoning.

The invention of *printing enabled the movement's ideas to spread from its birthplace in Italy to most of western Europe. Thomas *More, *Erasmus, and John Colet all contributed to the humanist tradition. Its spirit of sceptical enquiry prepared the way for both the *Reformation and some aspects of the *Counter-Reformation.

Humayun (1508-56), the second *Mogul Emperor of India (1530-40, 1554-5). His name means 'fortunate', yet after ten years of precarious rule he was driven into exile in Persia, recovering his empire only shortly before his death in an accident. His reign is significant mainly because of the introduction of Persian influences into India when he returned from exile accompanied by Persian scholars and artists. Persian became the court language, and pockets of *Shiite religious influence grew up in India.

Hume, David (1711-76), Scottish philosopher and historian. He published his *Treatise of Human Nature* in 1739. It provoked accusations of atheism, which lost him a hoped-for professorship at Edinburgh University. During the War of the *Austrian Succession he became secretary to General St Clair and was later sent on diplomatic missions. Between missions he wrote his famous *History of Great Britain*. Service at the British embassy in Paris brought him into contact with *Rousseau, whom he befriended, and after further government service he returned to Edinburgh in 1769. He enjoyed a great reputation as a historian in his own day but is now remembered as a philosopher. His philosophy was man-centred, an attempt to rationalize moral issues and to undermine by argument the idea of a religion based on the necessity for the universe to have a creator. His

other writings made an important contribution to the development of economic theory, and in their political discussion they influenced the makers of the American Constitution.

Humphrey of Gloucester *Gloucester, Humphrey, Duke of.

hundred, an administrative subdivision of an English *shire between the 10th century and the Local Government Act (1894) which established District Councils. Hundreds were probably based upon units of 100 hides. (A hide was a measure of land, calculated to be enough to support a family and its dependants, ranging from 25 to 50 ha. (60–120 acres) according to locality.) They did not exist in every shire. Their equivalents in the *Danelaw were wapentakes, in Kent lathes, in Yorkshire ridings, and in Sussex rapes. The hundred court of freeholders met once a month to deal with military defence, private pleas, tax levies, and to prepare indictments for the royal justices. The hundred bailiff served the sheriff's writs and the constable maintained law and order.

Hundred Years War, a war between France and England which stretched over more than a century between the 1340s and 1450s, not as one continuous conflict but rather a series of attempts by English kings to dominate France. The two key issues were the sovereignty of Gascony (the English king was Duke of Gascony and resented paying homage for it to the kings of France), and *Edward III's claim, through his mother, to the French throne, following the death of the last *Capetian king. Rivalry over the lucrative Flanders wool trade and provocative French support for the Scots against England also contributed.

In 1328 Philip of Valois was crowned King of France and his subsequent confiscation of *Aquitaine (1337) provoked Edward's invasion of France (1338). The English won a naval battle at Sluys (1340) and major military victories at *Crécy (1346), Calais (1347), and *Poitiers (1356), where *Edward the Black Prince captured and later ransomed Philip's successor John II. In 1360 the Treaty of Brétigny gave Edward considerable territories in France in return for abandoning his claims to the French throne. The French gradually improved their position and in the reign of Edward's successor, his grandson, *Richard II, hostilities ceased almost completely.

The English retention of Calais and Bordeaux, however, prevented permanent peace, and English claims to France were revived by *Henry V (invoking *Salic Law). He invaded Harfleur and won a crushing victory at *Agincourt (1415), followed by occupation of Normandy (1419) and much of northern France. The treaty of Troyes (1420) forced Charles VI of France to disinherit his son, the dauphin, in favour of the English kings. However, following Henry V's early death (1422) the regents of his ineffectual son *Henry VI gradually lost control of conquered territory to French forces under the leadership of *Joan of Arc. The English were defeated at Orleans (1429) and by 1450 France had conquered Normandy and much of Gascony; Bordeaux, the last English stronghold, was captured in 1453. This effectively ended the war and thereafter the English retained only Calais (until 1558). The English were forced to turn attention to internal affairs, notably the Wars of the

*Roses and gave up all claims to France. In France the virtual destruction of the nobility saw the *Valois monarchy emerge in a strong position.

Hungary, a country in central Europe. The Roman provinces of Pannonia and Dacia were overrun by Germanic tribes in the *Dark Ages and then conquered by *Charlemagne. By 896 elected *Magyar Arpad leaders ruled and Hungary emerged as the centre of a strong Magyar kingdom in the late Middle Ages. A Mongol invasion devastated the population in 1241 and the Arpad line ended in 1301. Thereafter, the crown was usually passed to a foreigner. The advance of the *Ottoman empire threatened, especially after the battle of Nicopolis in 1396, when Sigismund, King of Hungary, was defeated by the Turks. John Hunyadi (d.1456) and his son *Matthias Corvinus brought revival, but in 1490 the Jagiellons gained the throne, and, in 1515, a *Habsburg claim arose. The disastrous defeat of the Hungarian king, Louis II, at *Mohács (1526), led to the partition of Hungary between the Habsburgs and the Ottomans although *Transylvania retained its independence. By 1711 all of Hungary had come under Habsburg rule and remained part of the Habsburg empire until 1919.

Huns, pastoral nomads famed for their horsemanship, who entered history *c.* AD 370 when they invaded south-

A **Huron Indian**, his body painted in preparation for a hunting expedition or feast, a watercolour by John White, *c.*1585. (British Museum, London)

Hundred Years War

	Dates	The English	The French
Phase I	1337	Edward III (Duke of Aquitaine) has control of French coastal territories of Guienne and Gascony	
			Philip VI tries to extend his rule over the whole of France and declares Edward's duchy of Aquitaine forfeited
		Edward sends letter of defiance to Philip and claims French throne	
	1338	Edward invades France	
	1340	Edward assumes title 'King of France'	
		English win major naval battle at Sluys, gaining control of the Channel	
	1346	English win major victory at Crécy	
	1347		Calais surrenders to Edward III
	1347-56	Frequent truces. The Black Death subdues any significant military tactics by either side	
	1350		Philip VI dies; succession of John II
	1356	The Black Prince wins victory at Poitiers; John II captured, but Edward III fails to take advantage	
Phase II	1360		Brétigny treaty temporarily ends war: considerable territories transferred to Edward's rule on the agreement that he gives up his claim to the French crown
	1364		John II dies; succession of Charles V France becomes stronger
	1369	After intermittent fighting and violation of Brétigny treaty, Edward III reassumes title 'King of France'	
			Charles V reclaims Edward's French possessions; fighting resumes
	1377	Edward III dies; succession of Richard II. England now only holds Calais and part of Gascon coast	
	1380		Charles V dies; succession of Charles VI
	1396	Peace of Paris: Richard II marries Charles VI's daughter – 28-year truce	
	1399	Deposition of Richard II; succession of Henry IV	
	1410		Civil War (Burgundians v. Orleanists) weakens France
Phase III	1413	Henry IV dies; succession of Henry V	
	1415	Henry V renews claims to France; captures Harfleur; wins the great battle of Agincourt	
	1417	Henry begins systematic occupation of Normandy, helped by alliance with John the Fearless, duke of Burgundy	
	1420	Having gained considerable strength, Henry V marries daughter of Charles VI and by Treaty of Troyes becomes heir to French crown	
			Paris occupied by English
	1422	Henry V dies	Charles VI dies
		Henry VI (aged 10 months) rules France **N** of the Loire	
			Charles VII rules France **S** of the Loire
	1429		Joan of Arc leads a French revival and raises siege of Orléans
			Charles VII crowned King of France at Rheims
	1430		Joan of Arc captured by Burgundians
		Henry VI maintains control in Paris	
	1431	Joan of Arc burned	
	1435		Charles VII gains alliance with Burgundians by Treaty of Arras
	1436		Charles recaptures hold on Paris
	1450-3		Charles reconquers Normandy (1450), Guienne (1451) and Bordeaux (1453)
	1453	England only holds Calais. Weakened and distracted by Wars of the Roses, England ceases attempts to conquer France	

eastern Europe and conquered the *Ostrogoths. In 376 they drove the *Visigoths into Roman territory and early in the 4th century themselves advanced west, driving the *Alans, *Vandals, and others west into Gaul, Italy, and finally Spain. Under *Attila (434–53) they ravaged the Balkans and Greece, but a defeat was finally inflicted on them in 451 at the *Catalaunian Fields by the Romans and Visigoths under the command of Aetius. However that did not prevent them penetrating and plundering Italy the following year. Two years after the death of Attila they were decisively defeated near the unidentified River Nedao, and thereafter ceased to be of historical significance. The White Huns occupied Bactria and territory west towards the Caspian Sea. They vigorously attacked the power of the *Sassanian empire, defeating and killing Peroz in 484, but then moved south to establish an empire in northern India at the expense of the *Guptas.

Huron Indians, a confederacy of five Iroquoian-speaking tribes of North American Indians with a farming economy who lived in large villages in southern Ontario when first encountered by Europeans in the 17th century. The name Huron was given them by the French. In 1609 they were met by Samuel de *Champlain, who thereupon drew the French into a conflict between the Hurons and the *Iroquois to the south, particularly the *Seneca. By the 1620s the Hurons had become important suppliers of furs to the French. From 1635 Jesuits founded several missions among them, two of which were destroyed in Iroquois raids of 1648–9. Those Hurons who were not killed dispersed into other tribes to the west or settled near Quebec.

Huss, John (c.1372–1415), Bohemian religious reformer. A preacher in Prague and an enthusiastic supporter of *Wyclif's views, he aroused the hostility of the Church, was excommunicated (1411), tried (1414), and burnt at the stake. By his death he was acclaimed a martyr and his followers (*Hussites) took up arms against the *Holy Roman Empire and inflicted a series of dramatic defeats on the imperial army.

hussar, a soldier of a light cavalry regiment. Hussars were originally mounted troops raised in 1458 by *Matthias Corvinus, King of Hungary, to fight the Turks. As good light cavalry was scarce, other countries soon developed their own hussars: *Frederick the Great proved the superiority of Prussian hussars over those of Austria during the War of the *Austrian Succession. Britain hired hussars from several German states in the 18th century and sent them to America where they were much hated by the patriots.

Hussites, followers of John *Huss (c.1372–1415), the Bohemian religious reformer who was condemned for heresy and put to death by the Council of *Constance in 1415. Even before his death Huss had a large following in various parts of his native country *Bohemia and his execution at the stake sparked a nationwide protest in Bohemia, notably in the signing of a solemn protest by 452 nobles on 2 September 1415. Later the Hussite movement split into two main parties: the moderate Utraquists and the Taborites (named after Mount Tabor, their fortified stronghold), who held more extreme theological and social views. The Taborites were eventually

defeated by an alliance of Utraquists and Roman Catholic forces at the battle of Lipany in 1434. Most of the Utraquist demands were granted by the Church at the Compactata of Prague in 1436. Some groups of Hussites survived until today, under various names, but the movement was largely overtaken by the Reformation in the 16th century.

Hyde, Edward *Clarendon.

Hyder Ali (1722–82), sultan of *Mysore, south India (1761–82). Born in relative obscurity, he managed to supplant the Hindu ruler of Mysore. However, he soon had to face the expanding *East India Company armies, aided by Indian allies. In wars with the Company (1767–9 and 1780–4) he proved himself one of its most formidable obstacles, winning, with some French mercenary assistance, a series of remarkable victories. But heavy defeats in 1781 persuaded him that further struggle was pointless, and he urged his son, *Tipu Sultan, to seek terms with the Company.

Hyksos ('rulers of foreign lands'), invaders, probably from Palestine, who ruled Lower Egypt and part of Upper Egypt from c.1674 BC. Their power lasted until c.1550 BC, when they were overthrown by a rebellion started by the Egyptians of Thebes. The introduction of the horse and chariot was attributed to them, but for the most part they seem to have deferred to the native culture and gods. Egyptian remained the official language.

A **Hussite** army, from a manuscript illustration showing the priest Jan Siska at the head of a force made up of peasants armed mainly with flails, pitchforks, and clubs. The Hussites made strenuous efforts to counter the crusades sent against them at the instigation of the papacy. (Universitäts Bibliothek, Göttingen)

I

Ibarra, Francisco de (1539-75), Spanish explorer credited with the exploration and colonization of Nueva Vizcaya, the huge area of Mexico which today encompasses most of the states of Durango, Chihuahua, and part of Sonora. With funds provided by his uncle, who had made a fortune in the Zacatecas silver mines, Ibarra, as governor of Nueva Vizcaya, opened up much of the northern frontier of *New Spain in the 1560s before returning to Mexico City.

Iberians, early inhabitants of southern and eastern Spain. In the north of the peninsula they intermingled with the Celts to produce the Celtiberian tribes. From the 9th century BC they came into increasing contact with *Carthage, and Iberian and Celtiberian mercenaries were a vital element of the armies with which Carthage fought the Greek cities of Sicily and the Romans.

Iceland, a North Atlantic island, near the Arctic Circle. It was conquered by the *Vikings between 874 and 930. Its capital, Reykjavik, was founded, and the country was governed by some thirty-six chieftains, who met periodically in the Althing, an official assembly. A lawspeaker was appointed, and, in 1005, a Supreme Court. Authority, once derived from the pagan priests and temples, changed with conversion to Christianity in c.1000 to a partnership of church and Althing. In 1262 Iceland passed to Norway and, in 1380, with Norway to the Danish crown. The Lutheran religion was established in the 16th century.

Iceni, a pre-Roman tribe of eastern England. By the time of the Roman invasion of AD 43 they were part-Romanized and had come under the rule of the dynasty of *Cunobelinus. Their ruler Prasutagus was a treaty ally of Rome until his death in AD 60. The treaty was broken by Rome and his widow *Boudicca led the tribe in a revolt which was brutally suppressed.

Iconoclastic controversy (from the Greek, 'breaking of images'), a movement within the Eastern Christian Church in the 8th and 9th centuries which opposed the veneration of icons, both religious paintings and statues, which iconoclasts condemned as idolatry. Emperor Leo III banned such veneration in 726 and despite popular antagonism the decision was confirmed by Constantine V in 753. In the seventh ecumenical council of 787 at *Nicaea Empress Irene overturned the decrees but they were again enforced under the emperors Leo V, Michael II, and Theophilus. Veneration of icons was finally restored in 843 and the practice survives today in the Eastern Orthodox Church.

Ife, a holy city in Oyo Province, south-west Nigeria. For the Yoruba people it is the legendary birthplace of all mankind. It was the capital of a kingdom already established by the 11th century, and by the following century its craftsmen had produced the terracotta sculptures and bronze heads for which it is famous. Ife began

A bronze statue of an oni, or king, of **Ife**, cast during the 14th or 15th centuries. The ruler is shown in official regalia, bearing a ram's horn and sceptre and wearing beaded accoutrements and decorations.

to lose political influence among the Yoruba during the 16th century as the power of *Oyo increased.

Île de France, the name given to the centre of the Paris basin where the rivers Marne and Ouse join the River Seine. It is well-populated and fertile, supplying Paris with vegetable and dairy produce. As the original duchy of France it was the cradle of the French monarchy: in 987 Duke Hugh *Capet was chosen as king and his domains were the nucleus of the ever-growing crown lands. In the 15th century it became a province which was partitioned after the French Revolution.

Ilium *Troy.

Illyria, an area on the eastern coast of the Adriatic, north-west of Greece. In ancient times it was inhabited by a number of related Indo-European tribes from c.1000 BC. They caused considerable trouble to neighbouring Epirus and Macedonia, though *Pyrrhus annexed their southern territory in the 290s BC. Subsequently they reasserted their power southwards, the threat they posed to shipping eventually causing the Romans to declare war and suppress their ambitions (228-227 and 219). They supported the Romans against Macedonia at

Cynoscephalae (197), but after supporting King Perseus of Macedonia at Pydna (168) their territory was divided into three. The Roman province of Illyricum was disturbed by rebellion in AD 6, which *Tiberius had suppressed by AD 9. It was later divided into two provinces, known as Dalmatia and Pannonia.

Imhotep, architect, astrologer, and chief adviser to King Zoser (*fl.c.*2700 BC) of Egypt. To him are credited the first temple at Edfu and the first of the *pyramids, the so-called Step Pyramid at Saqqara. Within a century of his death he was viewed as semi-divine and believed to have healing powers, and *c.*525 BC he was elevated to the status of a full god. The Greeks identified him with their god of healing, Asclepius, and his temples attracted many suppliants.

impeachment, a prosecution brought by the English House of Commons, usually against a minister of the crown, and tried by the House of Lords. Dating from the 14th century, it was most frequently used in the 17th century against *Charles I's supporters and ministers including *Buckingham, *Strafford and *Laud. It was last used in England against Viscount Melville (1806).

Impressed ware cultures, the name linking groups of people who spread round virtually the whole coastline of the western Mediterranean, where they represent the first *Neolithic farmers, from 6000 BC. The characteristic pottery consists of round-bottomed bowls decorated with impressed designs, particularly using the crinkled edge of the cockle shell, the so-called Cardial ware. Material comes from both caves and open villages. By the 4th millennium the culture had split into many regional variants.

Inca, the pre-Columbian Indian people of western South America. They comprised Quechua-speaking tribes round *Cuzco (their capital), who formed a state contemporary to, and eventually superseding that of *Chimú. Sixteenth-century records indicate that the ruling dynasty was founded *c.* AD 1200 by Manco Capac, but real

---

The Incas

The Inca empire expanded rapidly in the 15th century from its capital, Cuzco. Expansion began under Pachacuti and his son, Topa Inca, who extended the empire to the north during his father's reign, and to the south after his own accession. The empire was consolidated under Huayna Capac, who founded a second capital at Quito, and divided the empire between his two sons, Atahualpa and Huáscar, at his death. Atahualpa invaded the southern half of the empire, which was led by Huáscar, and defeated him, before himself being overthrown by the Spanish conquistador, Pizarro.

The history of the Incas

*c.*1200	Ruling dynasty founded by Manco Capac, first Lord Inca
1250	Inca culture in and about Cuzco valley. A small group, the Incas vie for supremacy with their neighbours, the Chanca, Colla, Lupaca and others in the southern Andes
1300	Rise of the Chimu empire, rivals to the Incas
1350	Inca expansion begins
1390	Chimu empire extends over 960km. (600 miles)
1437	Viracocha, 8th Lord Inca, continues Inca expansion outside Cuzco area. Cuzco besieged by Chanca tribe
1438	Chancas defeated by Inca army led by Yupanqui, son of Viracocha. Proclaimed 9th Lord Inca, he takes the name of Pachacuti
1450	Pachacuti enlarges Inca empire, completing conquest of Titicaca basin
1463	Pachacuti wages war against Lupaca and Colla tribes. Quechua, the Inca language, is established. Pachacuti entrusts control of the army to his son Topa
1466	Chimu empire overrun by Inca troops commanded by Topa
1471	Topa Inca becomes 10th Lord Inca. Era of road-building
1485	Topa Inca conquers all of Chile to Maule River. South coast of Peru and north-west Argentina also conquered in his reign
1493	Huayna Capac becomes 11th Lord Inca. Quito founded as second capital, a decision which later leads to civil war between Huayna's sons Atahualpa and Huascar
1498	Huayna Capac extends Inca territory beyond Quito into Columbia. Andean highway completed
1513	Balboa discovers Pacific
1519	Atahualpa (last independent Inca) takes part in military campaigns
1525	Death of Huayna Capac. Civil war breaks out between Huáscar (crowned 12th Lord Inca) and his brother Atahualpa who dominates the north
1532	Huáscar defeated. Francisco Pizarro begins his conquest of the Inca empire for Spain and takes Atahualpa captive
1533	Atahualpa executed for crimes against the Spaniards
1535	Inca empire completely subjugated by the Spaniards

Above: Quito, the Inca capital, and its court, as it would have appeared in the 16th century, drawn *c.*1620 by the chronicler Guaman Poma de Ayala.
Right: Despite the size of the Inca empire, and the absence of wheeled vehicles, communications were swift and efficient. There was a good system of roads, along which messengers running in relays, armies, and llamas carrying goods were all able to travel freely.

------ Inca empire under Pachacuti and Topa Inca (1438-71)

—·— Inca empire under Topa Inca (1471-93)

— — Inca empire under Huayna Capac (1493-1525)

━━━ Greatest extent of Inca empire

expansion did not take place until 1438, forming an empire stretching from northern Ecuador, across Peru, to Bolivia and parts of northern Argentina and Chile by 1525 (some 3,500 km., 2,175 miles, north to south). Three important rulers carried out these conquests and the development of the imperial administration: Pachacuti (1438–71), *Topa Inca (1471–93), and Huayna Capac (1493–1525). After Huayna Capac civil wars broke up the empire of his son *Atahualpa just before Spanish troops led by Francisco *Pizarro landed on the coast in 1532. Atahualpa was captured in 1533 and killed shortly thereafter. In the same year Pizarro captured Cuzco, and by 1537, after the defeat of Manco Capac, most of the empire had been subdued by Spain.

The Sapa Inca, 'Son of the Sun', ruled by divine right and was worshipped as a god in his own lifetime. Under him was a vast administrative bureaucracy which regulated a complex system of regional capitals (for example, *Quito), agriculture, food collection and re-distribution, craft production, and roads and bridges. An efficient army, including a messenger system and strategically placed fortresses (for example, Machu Picchu), kept control; and rebellious populations were transferred wholesale to other parts of the empire. Although writing was unknown, records were kept on *quipus*, sets of cords of different colours and thicknessess tied with a system of coded knots.

Inca technology was of a high standard and included specialized factories and workshops producing ceramics, textiles, and metal artefacts, with fine decoration, in-corporating many regional styles. Architecture included accurately fitted stone masonry. Agriculture was based on systems of hillside terracing and included the potato, quinoa, and maize, and the guinea pig (for food), domestic dog, llama, and alpaca. Religion was centralized, local gods being respected but secondary to the Sun cult as the divine ancestor of the ruling dynasty and Viracocha, the creator god.

Independence, American War of

Independence, American War of, also known as the American Revolution (1776–83), the American revolution against British rule. It was triggered by colonial re-sentment at the commercial policies of Britain and by the lack of American participation in political decisions that affected their interests. Disturbances such as the *Boston Tea Party (1773) developed into armed resistance in 1775 (for example at *Lexington and Concord and *Bunker Hill), and full-scale war, with the *Declaration of Independence in 1776. Britain, fighting 3,000 miles from home, faced problems of supply, divided command, slow communications, a hostile population, and lack of experience in combating guerrilla tactics. America's disadvantages included few trained generals or troops, a weak central authority unable to provide finance, intercolonial rivalries, and lack of sea power. The French Alliance (1778) changed the nature of the war. Though France gave only modest aid to America, Britain was thereafter distracted by European and West and East Indian challenges.

The course of the war can be divided at 1778. The first, northern, phase saw the British capture of New York (1776), their campaign in the Hudson valley to isolate New England culminating in defeat at *Saratoga (1777), and the capture of Philadelphia (1777) after the victory of *Brandywine. The second phase switched British attentions to the south, where large numbers of

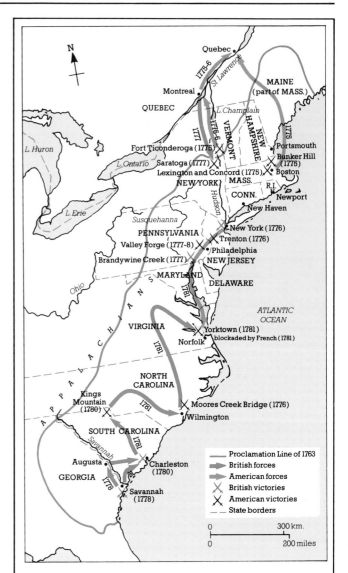

American War of Independence (1776–83)

The British strategy of attempting to break up the union of the colonies was initially successful, when Howe captured New York and forced Washington to retreat to Pennsylvania. However, Burgoyne's surrender at Saratoga raised American morale and persuaded the French to make an alliance with them. Having failed to cut off New England, the British began a southern campaign. Over 5,000 Americans surrendered at Charleston, but Cornwallis was trapped at Yorktown and, denied reinforcements by the French blockade, was forced to admit defeat.

Loyalists could be recruited. Philadelphia was re-linquished (1778) and *Washington camped at West Point to threaten the British headquarters at New York. After *Clinton's capture of Charleston (1780), *Cornwallis vainly chased the Southern Army under *Greene before his own exhausted army surrendered at *Yorktown, Virginia (October 1781), effectively ending hostilities. Peace was concluded at *Paris (1783).

Despite frequent victories, the British did not destroy Washington's or Greene's armies and could not break the American will. America's success has been depicted as influencing the French Revolution (1789) and subsequent revolutions in Europe and South America.

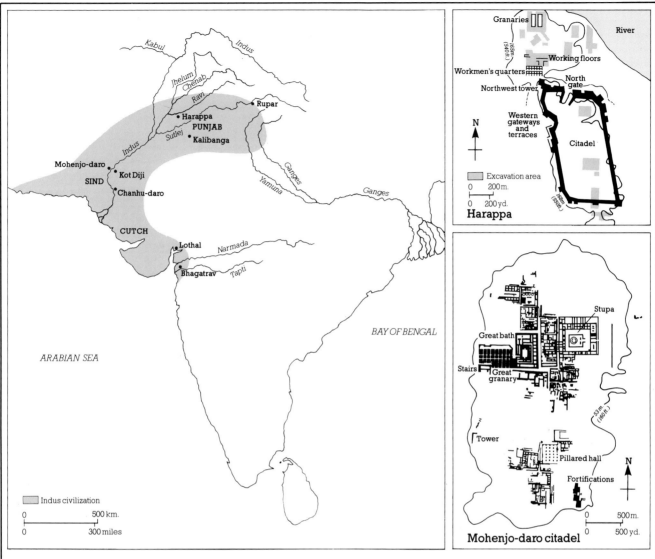

Indus civilization

Like other early civilizations in Egypt and Mesopotamia, the Indus civilization flourished in a region of low rainfall nourished by a great river. It covered a very wide area and the discovery of new sites is constantly extending the known range of its cultural influence. The citadels of Mohenjo-daro and Harappa were both built on artificial mounds on riverside sites. Major buildings were substantial and the system of sanitation and drainage with extensive brick culverts, one of the hallmarks of the Indus civilization, implies careful planning under close state control.

India, the greater part of the subcontinent of South Asia. Inhabited from an unknown date by Dravidian peoples, the *Indus civilization sites, dating from c.2500 BC, indicate one of the world's earliest urban cultures. It was destroyed c.1500 BC, possibly by the *Aryan invasions. The next 1,000 years saw the evolution of the religious and social systems which remain characteristic of the *Hindus. Regional kingdoms rose and fell under Hindu, and later Buddhist, dynasties, but mastery over the entire subcontinent was rarely achieved. The *Mauryan empire (c.325–185 BC), was the first all-India empire, only the southern tip remaining outside its influence. After its disintegration, internecine struggles between local powers remained the characteristic pattern.

Waves of invasion from from Central Asia from the 11th to the 16th century resulted in Muslim control over the north and the Deccan, and the evolution, through immigration and conversion, of India's largest minority.

Only in a few areas, notably the *Rajput states and *Vijayanagar, was Hindu political power maintained. Rule by the *Moguls (1526–1857), who claimed most of the subcontinent, marked the height of Indo-Muslim civilization. On their decline European trading powers were poised to take advantage of the power vacuum and the renewal of internecine struggle. Victorious over its French rival, the English *East India Company laid the basis in the 18th century for the subsequent hegemony of the British Raj, the third example of successful all-India imperialism.

Indo-Greek dynasties, rulers of parts of north-west India from the 3rd to the 1st century BC. In 326 BC *Alexander the Great's army had invaded India and explored the Indus valley. His direct impact was slight, but in the next century Greek commanders, who already held *Bactria in Central Asia again crossed the Indus, this time to establish power in the Punjab. Their successors were driven out of India after 200 years, but the 'Yavana' (Greek) rulers had an important impact on art and architecture, astrology and medicine. Evidence for Greek activity in India comes mainly from their inscribed coins.

Indonesia *Dutch East Indies.

indulgence, the cancelling by the Christian Church of the temporal punishment still owed for sins after they have been forgiven. The idea was found in various forms in the early Church, but as a widespread doctrine it dated from the 11th century, especially with the granting of indulgences to those who went on *Crusades. They could be obtained by saying certain prayers or by performing specified good works, such as helping the needy, taking part in a Crusade, or giving money to churches. It was through the connection with money payments that indulgences were most open to abuse, and that they became a focus for criticism by Martin *Luther and others at the Protestant *Reformation.

Indus civilization, a highly developed urban civilization in the lower valley of the River Indus, in South Asia, which flourished c.2500–1500 BC. Archaeological excavation, which began in the 1920s, is still being carried out and many important questions remain unanswered. City life seems to have ended abruptly c.1500 BC, possible causes being flooding, alteration of the course of the Indus, overpopulation, or an *Aryan invasion.

Among more than seventy sites now excavated, those at *Mohenjo-daro, *Harappa, Kalibangan, and Lothal seem particularly important for revealing a civilization based on the use of bronze, copper, and stone tools, and also city planning, including granaries, baths, drains, and straight streets. There is evidence of contact with Mesopotamia, but the script, known by the discovery of seals which were probably used in trade, has not yet been deciphered. The nature of religious worship is also uncertain, although stone and terracotta figures, as well as motifs on the seals, suggest links between the Indus cults and later post-Aryan Hindu religious concepts. The civilization provides evidence of an important stage in India's past, and of one of the world's first highly developed urban communities.

Industrial Revolution, a term to describe the change in the organization of manufacturing industry which transformed Britain from the 18th century, then other countries, from rural to urban economies. The process began in England as a result of a combination of economic, political, and social factors, including internal peace, the availability of coal and iron ore, and the availability of capital. *Steam power replaced the use of muscle, wind, and water power, and factories were built for the mass production of manufactured goods. A new organization of work known as the factory system increased the division and specialization of labour. The textile industry was the prime example of industrialization and created a demand for machines, and for tools for their manufacture, which stimulated further mechanization.

Coalbrookdale by Night, a painting by de Loutherbourg of an early centre of the **Industrial Revolution** in Shropshire. The blast furnaces established there by Abraham Darby produced iron more efficiently and cheaply by using coke instead of charcoal. (Science Museum, London)

Improved transport was needed, provided by canals, roads, railways, and the skills acquired were exported to other countries. It made Britain the richest and most powerful nation in the world by the middle of the 19th century.

Simultaneously it radically changed the face of British society, leading to the growth of large industrial cities, particularly in the Midlands, the North, Scotland, and South Wales, as the population shifted from the countryside, and causing a series of social and economic problems, the result of low wages, slum housing, and the use of child labour.

Innocent III (1160–1216), Pope (1198–1216). As pope he was an active and vigorous reformer of the Church. He reasserted control over the *Papal States and was acknowledged as the overlord of *Sicily. In Germany he asserted the pope's right to choose between two rival candidates for the imperial crown; he eventually supported *Frederick II's claims provided that he did homage for Sicily. He intervened in English affairs, excommunicating King *John for refusing to recognize Stephen *Langton as Archbishop of Canterbury and declaring *Magna Carta void; he also attempted to curb the independence of *Philip II of France.

Innocent's overriding concerns were crusades against heresy, and church reforms. He took steps to improve the quality of the clergy and ordered a crusade against the *Albigensians in southern France. He supported the disastrous Fourth *Crusade of 1204, but undeterred by its failure, remained zealous in support of the crusading effort. The Fourth *Lateran Council, which he summoned in 1215, resulted in seventy decrees for reform and prepared the ground for a new Crusade, though Innocent died of fever the following year. His reign marked the climax of the medieval papacy.

Inquisition, an ecclesiastical court established *c*.1232 for the detection and punishment of heretics, at a time when sectarian groups were threatening not only the orthodoxy of the Catholic religion but the stability of contemporary society. The Inquisition came into being when *Frederick II issued an edict entrusting the hunting-out of heretics to state inquisitors; Pope Gregory IX claimed it as a papal responsiblity and appointed inquisitors mostly drawn from the Franciscan and Dominican orders. He had previously ordered the Dominicans to crush the *Albigensians (1223). Those accused of heresy who refused to confess were tried before an inquisitor and jury and punishments were harsh, including confiscation of goods, torture, and death. The Index (a list of books condemned by the Church) was issued by the Congregation of the Inquisition in 1557. The *Spanish Inquisition was a separate organization established in 1479 by the Spanish monarchy with papal approval.

intendant, an agent of the French king under the *ancien régime*. The office was developed as an emergency measure to counter disobedience during the 1630s, building on an earlier practice of sending royal officials from the central councils on tours of inspection in the provinces. Under *Richelieu and *Louis XIV their authority was extended into every sphere of administration, and they became the principal link between the central government and the provinces. They supervised local courts, oversaw the tax system, and kept the crown informed about the political and economic situation in their *généralités* (administrative units). The office was abolished at the Revolution, but many of the same functions were later performed by the *préfets*.

Intolerable Acts   *Coercive Acts.

Investiture, the formal act whereby senior clergy receive their offices with their attendant properties and revenues. In the Middle Ages when kings 'invested' bishops with their bishopric, it looked as if bishops were receiving their spiritual powers as well as their property from the semi-sacred figure of the king. In the 11th century radical churchmen questioned this view of royal authority and lay investiture was denounced as a corrupt practice that shackled the church to the world. Under Pope *Gregory VII (1073–85) the increasingly bitter disputes over the claims of church and state erupted into the Investiture Conflict, itself part of a wider movement for church reform that swept Western Europe. Conflict was especially bitter in Germany where Emperor *Henry IV, as most exalted lay ruler, had most to lose.

Pope and emperor denounced each other in propaganda campaigns, manoeuvred for diplomatic advantage (*Canossa), and fought to depose each other. Civil war broke out in Germany until peace was established at the Concordat of Worms (1122) between Pope Calixtus II and Emperor Henry V whereby the king abandoned investiture of churchmen and thus claims to spiritual power, but managed to retain much practical control over the appointment of bishops. This was the first great conflict between pope and emperor in the Middle Ages and laid the foundations for papal claims to full authority over the Church.

Iran   *Persia.

Iraq   *Mesopotamia.

Ireland, an island of the British Isles to the west of Great Britain. It was inhabited by Goidelic *Celts as early as the 6th century BC, by Bretonnic Celts from the 3rd century, and by *Picts from the 1st century AD. It was never conquered by the Romans, who gave it the name Hibernia, and Irish raiders frequently attacked Britain and Gaul as the Romans withdrew. Its Celtic society, ruled by numerous petty kings and chiefs, lasted into the medieval period. The *Celtic Church which evolved from the 4th century, was based on this feudal model, with bishops subordinate to the heads of land-owning communities. Viking raids and domination by Viking settlers from the late 8th century ended at Clontarf (1014), when the High King *Brian Boru decisively defeated the Scandinavians. From 1169 Anglo-Norman barons invaded and seized land, until in 1172 *Henry II of England was formally acknowledged as King of Ireland.

For the next 400 years English monarchs made successive attempts to subdue Irish resistance to their rule. Bitterness increased with the imposition of Protestantism by *Henry VIII and the establishment of 'plantations' of English and Scottish settlers in Ulster under *Elizabeth I. In 1599 the queen sent *Essex to put down a revolt led by the Earl of *Tyrone and the Irish were defeated in 1601. Religious oppression and settlements continued and, with the extortionary policies of *Charles I, led to the Irish Rebellion, which was not finally put down until Oliver Cromwell led the Parliamentary army to Ireland (1649). He further extended Protestant settlement. After the *Glorious Revolution (1688) the 'Protestant Ascendancy' was further confirmed. In 1782 Henry *Grattan's party in the Irish Parliament secured the repeal of *Poynings' Law and limited independence, but the rising of the *United Irishmen (1798) helped bring about the Act of Union (1800), under which Ireland became part of Great Britain and forfeited its own parliament.

Irene (c.752–803), Byzantine empress (780–802). After her husband Leo IV died (780) she ruled jointly with her son, Constantine VI, until 790 when he banished her. She soon returned, intrigued against him, had him blinded and imprisoned and ruled as emperor (not empress) until again exiled (802) to Lesbos. She strongly opposed the iconoclasts (*Iconoclastic controversy). For her zeal in this cause she was canonized by the Greek Orthodox Church.

Ireton, Henry (1611–51), English *Roundhead commander and politician. After fighting at Edgehill (1642) and Marston Moor (1644), he was instrumental in pushing through the *Self-Denying Ordinance (1645). In 1646 he married *Cromwell's daughter Bridget, and in 1647 opposed the *Levellers' extreme demands at the Putney debates. He figured prominently in the moves which led to *Pride's Purge and the execution of *Charles I. After serving as Cromwell's second-in-command on his Irish campaign (1649–50), he was made Lord Deputy of Ireland. He died of plague.

Iron Age, the period of prehistory distinguished technologically by the use of iron. This was first mastered on a large scale by the *Hittites in Anatolia between 1500 and 1200 BC, and spread to the Aegean, and thence to south-east and central Europe and Italy. The spread was slow across Europe, as it only gradually replaced bronze. In Africa bronze had never been adopted, so the Iron Age there immediately followed the Stone Age. In America, iron was not discovered before being introduced from Europe.

Iron Mask, Man in the   *Mattioli.

Ironsides   *New Model Army.

Iroquois, generally used to refer to the North American Indian tribes of the precolonial period who inhabited present-day upper New York State; more specifically, to the confederacy of Five Nations—*Mohawk, Oneida, Onondaga, Cayuga, and *Seneca—to which the Tuscarora were added in 1720. Other Iroquoian tribes include the *Huron, Susquehanna, Neutral, and Erie.

In the *fur-trade rivalry that formed in the 17th century between the Dutch and English at Albany, New York, and the French at Montreal, the Iroquois were tied to Albany. They attempted to control the beaver lands in the north-west by raiding French settlements on the St Lawrence and (in Seneca raids in 1648–9) the Hurons, who lived near Georgian Bay in south-eastern Ontario, virtually destroying them as a nation. In 1784 the Iroquois, who had been allies of the British against the French and later of the rebelling colonies, were

An **Iron Age** mirror of engraved bronze from an ironstone quarry at Desborough, England. Dating from the La Tène period, about the beginning of the Christian era, the mirror is decorated in typical curvilinear style. (British Museum, London)

rewarded with a large land grant and moved to the Grand River (Ontario), led by Joseph Brant, and the Bay of Quinte (Desoronto, Ontario).

Isabella I (the Catholic) (1451–1504), Queen of Castile (1474–1504). She united her kingdom with that of Aragon by her marriage with its king *Ferdinand V in 1469, retaining sole authority in Castilian affairs. She was noted for her Catholic piety, encouragement of the *Spanish Inquisition, and her intolerance towards Jews and Mus-

The destruction of the **Iroquois** on the shores of Lake Champlain, which was named after the French explorer, an engraving from his *Voyages du Sieur de Champlain* (1613), Paris.

lims. She patronized Spanish and Flemish artists, and supported the exploration of the *Americas.

Isabella of France (1292–1358), Queen of England (1308–27). She was the daughter of *Philip IV of France, and was married in 1308 to *Edward II. She bore Edward four children, including the future *Edward III, between 1312 and 1321, but seems to have become alienated from him in 1322, resenting the influence of the *Despensers. In 1325 she became the mistress of Roger *Mortimer in Paris, and in September 1326 they invaded England, imprisoning and then ordering the murder of Edward II. They effectively ruled England until Edward III captured them both in October 1330; Isabella was pensioned off with £3,000 a year.

Isaiah (8th century BC), Hebrew prophet, who preached during the reigns of Jotham, Ahaz, and *Hezekiah. Isaiah's message was that the safety of Judah was in God's hands and the king should trust him and not rely on foreign allies. He advised Hezekiah to acknowledge Assyrian power and not ally with Egypt; when Judah was invaded by Israelites and Syrians in 735 BC, and by the Assyrians in 710 BC, and again in 703–701 BC, Isaiah promised that faith in God would guarantee the people's deliverance. The passages in the prophecies ascribed to him heralding the coming of a future king were taken by later Christian writers to refer to *Jesus Christ.

Islam, the religion preached by the Prophet *Muhammad. Islam asserts the oneness of God and the inevitability of judgement and lays upon believers five major obligations—to accept the basic creed ('I testify that there is no god but Allah and that Muhammad is the prophet of Allah'); to offer prayers five times a day; to pay *zakat*, a charitable levy; to fast during daylight in the month of Ramadan; and, once in a lifetime, to undertake the *hajj*, the pilgrimage to the Ka'aba at *Mecca, believed by the Muslims to have been built by *Abraham. These observances, as well as a code governing social behaviour, were given to Muhammad as a series of revelations and were written down in the *Koran.

Muhammad's preachings were subsequently elaborated to produce a corpus of religious law covering every contingency of life, from family affairs to crime, and from commercial contracts to good manners, which provided a stable social framework in harsh and troubled regions for a thousand years after his death. *Arab conquests spread the faith from Spain to India within a century of Muhammad's death, and subsequently, by conquest or the peaceful example of merchants and *Sufi mystics, it has spread throughout Africa and south and south-east Asia.

Ismaili, the name of a branch of *Shiite Islam which recognizes seven rather than twelve imams (spiritual leaders). Ismail, the eldest son of the sixth imam, Jafar al-Sadia (d.765) was disinherited and most Shiites recognized his brother Musa al-Kazim as imam. Ismailis regarded Ismail as the seventh and last imam and expected that he would soon return as the Mahdi ('expected one') to overthrow existing corrupt governments and establish justice on earth. The *Fatimid dynasty (909–1171) promoted their beliefs in Egypt and Syria but attempted no mass conversions among the *Sunnite majority. The Ismailis developed the idea that the religious precepts have a secret inner meaning, passed from *Muhammad to Ali and from him to later imams, who could thus instruct the ignorant. This encouraged major contributions to Islamic philosophy, though doctrinal differences led the movement to split into various sub-groups, such as the *assassins, Druzes, and Khojas.

Ismail I Safavi (d. 1524), first ruler of the Safavid dynasty in *Persia (1501–24). His ancestor Safi ud-Din (1252–1334) was a *Sufi holy man and founder of the Safaviyya, the mystic brotherhood after which the dynasty is named. Supported by Turcoman tribesmen, he established his rule over Persia, extending it into *Kurdistan and driving the *Uzbeks from Khurasan in the north-east. His expansion was checked in the west by the *Ottoman sultan Selim I at Chaldiran in 1514. His most enduring achievement was to convert his realm from *Sunnite to *Shiite Islam.

Israel, the name first appearing in the Old Testament of the *Bible as an additional name given by God to the ancestor of the Hebrews, Jacob. It probably means 'he that strives with God'. The Hebrews who occupied *Palestine in 12th and 11th centuries BC called themselves 'the children of Israel', and their tribal divisions were named after the traditional sons of Jacob. These names are nearly always twelve in number, and include Reuben, Simeon, Levi, Judah, Issachar, Zebulun, Dan, Naphtali, Gad, Asher, Joseph, and Benjamin. Joseph is sometimes subdivided into Ephraim and Manasseh.

Israel was also the name given to the kingdom over which *Saul, *David, and *Solomon ruled in Palestine, between c.1020 and c.922 BC. Thereafter it was the name of the northern Hebrew kingdom, which seceded from the Jerusalem-centred kingdom of Judah, and which covered the territory of all the tribes except Judah (and parts of Benjamin). This kingdom was overthrown by the Assyrians in 721 BC. After the return of the Jews from *Exile in Babylon the name Israel was used more generally of the Jewish nation.

Italy, a country in southern Europe. It had come under *Etruscan, Greek, and Celtic influence before it was united in c.262 BC under *Roman rule. In the 5th century it was overrun by the barbarian *Goth and *Lombard tribes. In 775 *Charlemagne conquered the north and it became part of the Carolingian empire, while the south was disputed between the Byzantine empire and the Arab conquerors of Sicily. By the 12th century city-states had emerged in northern and central Italy and the south united under first Norman and then, in 1176, Spanish control. The 14th century was a time of great commercial activity, followed by the *Renaissance period. The country, now divided between five major rival states, came under first Spanish (1559–1700) and then, after the Treaty of *Utrecht in 1713, Austrian domination. In 1796–7 Italy, having been used to maintain the balance of power in Europe, was invaded by *Napoleon and hopes of independence and unification re-emerged. The exclusion of Austria and partial unity were shortlived and after 1815 the 'old order' was restored. Modern Italy was created by the nationalist movement of the mid-19th century.

Ithaca, a small island off the north-west coast of Greece. Mycenaean remains testify to the importance of the island in early times, which was due in part to the fact that it was a staging-post for Corinthian trade with Sicily and southern Italy. By classical times it had declined, never again to regain its significance. At the beginning of the 16th century it was virtually uninhabited and Venetian traders had to encourage resettlement.

Ivan III (the Great) (1440–1505), Grand Prince of *Muscovy (1462). He was responsible for extending the territories of Muscovite Russia, becoming independent of the *Tartars and subjecting the principalities of Livonia and Lithuania. Introducing a legal code in 1497, he claimed the title of 'Ruler of all Russia', reorganized and reduced the independence of the nobility, and built up a class of new, loyal, dependent officials. Influenced by contemporary Italy and Byzantium, he claimed leadership of the *Orthodox Eastern Church. His authority subsequently declined as alcoholism, conspiracy, and succession problems diminished his effectiveness.

Ivan IV (the Terrible) (1530–84), Grand Prince of Muscovy (1533–84), the first ruler to assume the title of Tsar (Emperor) of Russia (1547). He had a violent and unpredictable nature, but his nickname is better translated as 'Awe-inspiring' rather than 'Terrible'. From 1547 to 1563 he pushed through a series of legal and administrative reforms. He also continued to expand

A 16th-century icon of **Ivan IV**. Russia was at war for most of Ivan's reign. He extended his rule to Kazan and Astrakhan, and conquered Siberia in 1582. He also established diplomatic and trade links with England. (National Museum, Copenhagen)

An engraving of **Ivan III**, who freed Russia from the Tartars by ending payment of tribute to Khan Ahmed of the Golden Horde in 1480. No longer a vassal of the khan, Ivan was able to enter the arena of European diplomacy as an independent sovereign.

Russian territory although his campaigns against the Mongols and in Siberia were more successful than those in the west. In 1564 he entered on a reign of terror, caused partly by his deteriorating mental condition, and partly by his determination to wrest power from the *boyars. He used a special body of civil servants, the *oprichniki*, to break the power of the nobility. Shortly before his death, he provoked further turmoil for Russia by killing his gifted son and heir, Ivan, and though another son, Fyodor, succeeded him, power soon fell into the hands of his favourite, *Boris Godunov.

J

Jacobin, the most famous of the political clubs of the *French Revolution. It had its origins in the Club Breton which was established after the opening of the *States-General in 1789, and acquired its new name from its headquarters in an old Jacobin (Dominican) monastery in Paris. Its membership grew steadily and its carefully prepared policies had great influence in the *National Assembly. By August 1791 it had numerous affiliated clubs and branches throughout the country. Its high subscription confined its membership to professional men who, at first, were not distinguished by extreme views. By 1792, however, *Robespierre had seized control and the moderates were expelled. The club became the focus of the *Terror the following year, and in June was instrumental in the overthrow of the *Girondins. Its success was based on sound organization and the support of the *sans-culottes. It was closed after the fall of Robespierre and several attempts to reopen it were finally suppressed in 1799.

Jacobite, a Scottish or English supporter of the exiled royal house of *Stuart. The Jacobites took their name from Jacobus, the Latin name for *James II, who had been deprived of his throne in 1688. Their strength lay among the Highland *clans of Scotland, whose loyalty was personal; the weakness of Jacobitism was that it failed to win over the Tories in England, who might have made it a more powerful and dangerous movement. The Jacobites were politically important between 1688 and 1745. The *Fifteen and the *Forty-Five were their major rebellions, but neither succeeded and after 1745, with the government's suppression of the clans, Jacobitism ceased to have a firm political base.

Jacquerie, a rebellion of French peasants in northern France (May–June 1358), named after 'Jacques Bonhomme', the aristocrats' nickname for a French peasant. A leader, Guillaume Karle (or Cale) emerged, and a bourgeois revolt in Paris helped the movement. The *Black Death, the French defeat at *Poitiers, the ravages of brigands, feudal burdens, and governmental demands for extra fortification work were all contributary causes. Castles were demolished and looted; but the rebellion was short-lived, collapsing after the execution of Karle and the massacre of a mob at Meaux.

Jagiellon, a Polish dynasty which reigned in Lithuania, Poland, Hungary, and Bohemia in the 14th, 15th, and 16th centuries. The family gained prominence under Jagiello, Grand Duke of Lithuania (c.1345-1434), who became King of Poland, as Ladislas I, in 1386. A descendant, Ladislas III, King of Poland (1434), governed Hungary as Lazlo I (1440) but was killed by the Turks at Varna in 1444. His brother, Casimir IV of Poland, fought the Teutonic Knights and gained the throne of Bohemia for his son, Ladislas II, and in 1490, as Lazlo II, of Hungary. His son was defeated and killed at the battle of *Mohács (1525). The line died out with Sigismund II Augustus, King of Poland (1548-72).

Jahangir (1569-1627), Emperor of India (1605-27), whose contribution to *Mogul greatness was artistic rather than military. His name means 'holder of the world', but control over the huge empire he inherited was threatened by court quarrels and by the dominance of his Persian wife, Nur Jahan. The fame of the Mogul empire had reached Europe during the reign of his father, *Akbar, when Jesuit missionaries reached Agra. In Jahangir's time there were visits by English ambassadors. He was criticized for addiction to alcohol and opium, but his own artistic interests encouraged a new naturalism in the work of his court painters.

Jainism, an Indian religion whose recorded history goes back to the 6th century BC, but which is certainly of greater antiquity. Vardhamana Mahavira (c.599-527 BC) is revered as the twenty-fourth *Tirthankara* or prophet, who like his near contemporary, the *Buddha, protested against the dominant *Hindu ritualistic cults.

Jainism teaches a strict moral code centred on *ahimsa* (avoidance of injury to any living creature) and a path of spiritual training through austerity, leading to *nirvana* (absorption into the supreme spirit). While monks and nuns adhere to this rigorously, lay people follow the 'three jewels' of right faith, knowledge, and conduct. The concepts of *karma* (destiny) and *nirvana* reflect the Hindu environment, but Jain thinkers have given them a distinctive interpretation. Belief in a creator god is rejected. The two main sects are named from the dress of their monks, Svetambara ('white-robed') and Digambara ('sky-clad' or naked). Jainism never spread beyond India.

Jaipur, former state in Rajasthan, north-west India. It was a leading *Rajput state from the 12th century. The city of Jaipur, built in 1727 by Raja Saiwai Jai Singh II

The Hawa Mahal (Palace of the Winds) in the city of **Jaipur** in Rajasthan, built in 1799 for the Maharaja Sawai Pratap Singh. Delicate tracery decorates the ornate façade behind which the ladies of the zenana (the apartments to which they were confined) watched public events in the street below.

(*c.*1699-1743) is renowned for its outstanding pink-coloured palace. Its rajas held off Mogul interference by advantageous bargaining, but in the 18th century fell victim to *Maratha, and then British, expansion, finally acknowledging British paramountcy in 1818.

Jamaica, the third largest island in the West Indies. Originally inhabited by Arawak Indians (*Hispaniola), it was discovered by Columbus in 1494 and settled by the Spanish in 1509. In 1655 it was captured by the British and prospered as a *buccaneer base. The importation of slaves to work on sugar-cane *plantations made Jamaica the leading sugar producer of the 18th century.

James I (1394-1437), King of Scotland (1406-37). He was the third son of *Robert III, and was shipwrecked and held captive in England from 1406 until his ransom was arranged in 1423. His rule was firm and effective, particularly once he had arrested most of his opponents who had governed Scotland during his absence, but his policy of reducing the powers of the nobility and making more use of the Scottish Parliament ultimately led to a reaction. In February 1437 he was killed by Sir Robert Graham, leader of a conspiracy against him.

James I (1566-1625), King James VI of Scotland from 1567, King of England (1603-25). He succeeded *Elizabeth I of England, since she had never married and the *Tudor dynasty was ended. He was the son of *Mary, Queen of Scots and Henry, Lord Darnley. As King of Scotland he survived several plots and assassination attempts, while he strengthened the power of the crown over Parliament, Kirk (Church of Scotland), and sectarian religious groups, and fostered good relations with England. As King of England, he lacked the shrewd judgement of his predecessor, his reign being marked by several errors of policy. He angered the Puritans by refusing to hear their demands at the *Hampton Court Conference and by his insistence on the maxim 'no bishop, no king' to counter their demands for the abolition of bishops, which put an end to their hopes of reform. His court was tainted by sexual and financial scandal and although *Cecil attempted reform, the king's promotion of *Buckingham from 1618 led him into costly and extravagant schemes which alienated Parliament. Although learned, he was tactless in his handling of Parliament, insisting repeatedly on his prerogatives as king. However the unsettled financial and religious settlement was his legacy from Elizabeth and it was an achievement that his reign was largely peaceful.

James II (1430-60), King of Scotland (1437-60), son and successor of *James I. During his minority successive earls of Douglas vied for power, and when he began to reign in his own right the Douglases remained a threat to his authority. He improved the courts of justice and regulated the coinage. He was killed by an accidental cannon blast while leading a siege against Roxburgh Castle, held by the English.

James II (1633-1701), King of England and Scotland (1685-8), the second son of *Charles I. As Duke of York he was Lord High Admiral in the second and third Anglo-Dutch wars, during which the Dutch settlement of New Amsterdam was captured and renamed New

A contemporary portrait of **James I**, King of England attributed to Juan de Critz, *c.*1605. James was given a sound education by his tutors who included George Buchanan, the foremost Scottish man of letters of his day, with a European reputation. James's writings included *Demonology*, (1597), a treatise which advocated severe measures against witchcraft, and *A Counterblast to Tobacco* (1604). (Loseley House, Guildford)

York in his honour (1664). He became a Roman Catholic and married Mary of Modena, also a Roman Catholic, in 1673, resigning as admiral in that year under the *Test Act; attempts were made to exclude him from the succession during the years 1679-81, but on the death of *Charles II he ascended the throne without opposition. *Monmouth's rebellion came early in his reign, and the *Bloody Assizes which punished it were resented. Within three years of his accession he had provoked the widespread opposition leading to the *Glorious Revolution which replaced him on the throne by William and Mary. He died in exile in France.

James III (1452-88), King of Scotland (1460-88), son and successor of *James II. He came to the throne aged

9 and after his mother died in 1463 his rule was challenged by members of his family and their supporters from the nobility. He did not take control of his kingdom until 1469. In 1479 he imprisoned his brother, Alexander Stuart, Duke of Albany, who later led a rebellion against him. He was defeated and killed in battle by his own nobility, supported by his son (who succeeded him as James IV), at Sauchieburn, close to Bannockburn.

James IV (1473–1513), King of Scotland (1488–1513). He was an able and popular king, and was successful in restoring order to Scotland, quelling an uprising of discontented nobles, and improving the prosperity of the kingdom. He made peace with England by the Treaty of Ayton (1497–8), and married *Henry VII's daughter Margaret Tudor, which brought the *Stuart line to the English throne in 1603. He led his country into disastrous defeat against an army of Henry VII and was himself killed at *Flodden Field.

James V (1512–42), King of Scotland (1513–42). He succeeded his father James IV at the age of only 2 months. His mother, her husband, Archibald Angus, Earl of Douglas, and John Stuart, Duke of Albany, struggled for control of the kingdom during his minority. When he came to power he began to ally himself with France against his uncle *Henry VIII of England, and made diplomatic marriages to Madeleine, daughter of *Francis I of France (1536), and on her death to Mary of *Guise. His only child was a daughter, who succeeded him as *Mary, Queen of Scots.

Jamestown, Virginia, the first permanent English settlement in North America, founded by an expedition despatched by the Virginia Company from London in 1607. The site 64 km. (40 miles) up the James River was chosen mainly for strategic reasons, but the surrounding malarial swamp and typhoid infected river water caused endemic illness. Only John *Smith's determination saved the new outpost. Jamestown was the seat of the colonial assembly from 1619; it was burned in *Bacon's Rebellion, after which Williamsburg became the effective capital.

Janissaries (Turkish, *yeni cheri*, 'new troops'), an élite corps of slave soldiers, bound to the service of the *Ottoman sultans. They were originally raised from prisoners of war, but from the time of *Bayezid I (1389–1403) they were largely recruited by means of the *devshirme* ('gathering'), a levy of the fittest youths among the sultan's non-Muslim subjects. Having been converted to Islam, most, after intensive training, served as foot soldiers, while the ablest passed into civil administration. Decimation in the great wars against Persia and Austria (1578–1606) lowered the traditionally high quality of the intake and opened the corps to Muslims. They exercised a powerful role in political life until their abolition in 1826.

Jansen, Cornelius Otto (1585–1638), Dutch bishop, the founder of the Catholic school of theology known as Jansenism. After studying at Louvain and Paris, he became the director of a newly founded college at Louvain (1617). In 1626–7 he publicly opposed the Jesuits at Madrid, and in 1636 he was made Bishop of Ypres. His ideal was the reform of Christian life through a return to St *Augustine, and in 1628 he began to write

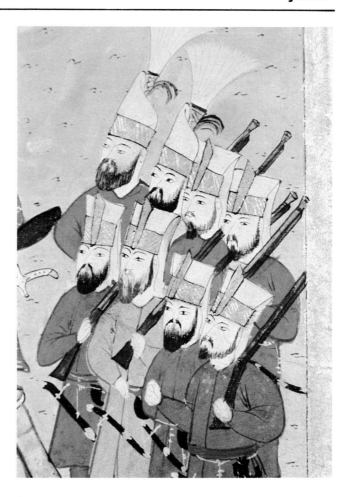

Detail from a manuscript portraying a group of **Janissaries** belonging to the Ottoman emperor Suleiman. These bearded Muslim warriors carried wide-muzzled muskets as well as the traditional swords.

his *Augustinus*, published posthumously in 1640. This work was condemned by the Sorbonne (1649) and then by Pope Innocent X (1653). Jansen himself insisted upon his Catholic orthodoxy, but a number of his tenets resembled those of *Calvin. Jansenists suffered persecution in France during most of the 18th century, but they were tolerated in the Netherlands, and in 1723 the Dutch Jansenists nominated a schismatic bishop of Utrecht. The best-known Jansenist was the French writer Blaise *Pascal.

Japan, a country chiefly comprising four large islands lying between the Sea of Japan and the Pacific Ocean. Originally inhabited by native Ainu, the Japanese themselves are thought to be descendants of people who migrated from various areas of mainland Asia. By the 5th century AD the *Yamato clan loosely controlled much of Japan and began to establish imperial rule. The developing state was much influenced by Chinese culture. *Buddhism was introduced in the 6th century and, after a brief conflict, coexisted with the Japanese religion, *Shinto. In the 7th century Prince *Shotoku was partially successful in establishing an administrative system based on that of *Sui China. However, by the 9th century the *Fujiwara family had gained control over the imperial court and its power was undermined.

The growing strength of feudal lords and of Buddhist

monasteries resulted in civil war for most of the 12th century, the ultimate victor being *Minamoto Yoritomo, who in 1192 became the first *shogun and established a military administration. From then effective power lay with the shogun rather than the emperor. Yoritomo's *Kamakura shogunate was replaced in 1333 by the *Ashikaga shogunate, but its rule was one of prolonged civil strife. In the late 16th century three warriors, *Oda Nobunaga, *Hideyoshi, and *Tokugawa Ieyasu broke the power of the feudal lords (*daimyo), and Ieyasu's *Tokugawa shogunate provided stable but repressive rule until the restoration of the emperor in 1868.

Europeans had begun to trade with Japan in 1542 and Catholic missionaries, including Matteo *Ricci, made numerous converts. The Tokugawa shogunate excluded all foreigners in 1639, except for a few Dutch and Chinese at Nagasaki, and proscribed Christianity. During the 18th and 19th centuries the wealth and power of merchants began to increase and Japan extended its influence over the northern island of Hokkaido.

Java, one of the Sunda islands, now the heartland of Indonesia. In the early centuries AD Hindu principalities emerged. About 700 Buddhism briefly displaced Hinduism in central Java; Borobudur, said to be the largest Buddhist temple in the world, a lasting monument to this period, was constructed by the Sailendra dynasty that ruled an area around Jogjakarta. Hinduism then re-established itself under a succession of kingdoms, the last of which was *Majapahit (1298–1520). During the 16th century Islam started to spread throughout Java. In 1619 the Dutch captured Jakarta on the north coast, renamed it Batavia after the Batavii, the Celtic tribe who lived in the Netherlands in Roman times, and made it the headquarters of the *Dutch East India Company. In 1628 and 1629 Sultan Agung (1613–46), the ruler of the kingdom of Mataram, besieged it unsuccessfully and the Dutch consolidated their position. By the mid-18th century the whole island was under direct or indirect Dutch rule.

Java man, name given to fossilized bones found at Trinil on the Solo River in Java. Originally classified as *Pithecanthropus erectus* ('erect ape-man'), these remains are now included within the species *Homo erectus. Their date is uncertain but is probably 750,000–500,000 years ago. Subsequent finds of hominids, probably also *Homo erectus*, in Java were made at Sangiran and Modjokerto. Some of these may be rather older (up to 1.3 million years ago) than the Trinil remains.

Jay, John (1745–1829), American statesman and jurist. He was a member of the first and second *Continental Congresses, became Chief Justice of New York, also a member of Congress, Minister to Spain (1780–2), and a member of the delegation that negotiated peace with Britain (1783). He was a conservative Federalist. He was Secretary of Foreign Affairs 1784–90, before becoming Chief Justice of the USA (1789–95). As special envoy to England he negotiated the Jay Treaty (1794) to settle outstanding differences resulting from the War of Independence. It enforced the terms of the Peace of *Paris (1783) and ordered the British to leave their trading post in the north-west of America. The British lost control of the lucrative *fur trade and ceded a share in the trade with the West Indies to the Americans.

Jefferson, Thomas (1743–1826), third President of the USA (1801–9). He was a delegate to the House of Burgesses (1769–75) and to the *Continental Congress (1775–6). He drafted the *Declaration of Independence, was active in Virginia during the War of *Independence, and was governor of the state (1779–81). A slave owner who favoured gradual emancipation, he never felt able to implement such a policy. After service as US minister to France (1785–9), he was *Washington's first Secretary of State. His opposition to *Hamilton's economic policies led to his resignation in 1794. He later became leader of the Democratic-Republican Party and was Vice-President under John *Adams before becoming President in 1801. His administration was marked by retrenchment and reduction in the scale of government itself, but also by the Tripolitan War, which ended tribute payment to Barbary pirates, the Louisiana Purchase, the Lewis and Clare expedition (the first overland expedition to the Pacific Coast of America), and the Embargo Act in defence of the US neutral rights. Jefferson, who believed in the virtues of an agrarian republic and a weak central government, has been a uniquely influential figure in the evolution of the American political tradition.

Jeffreys of Wem, George Jeffreys, 1st Baron (1645–89), English judge and Lord Chief Justice. He presided at the trials of the *Rye House conspirators, of Titus Oates after the *Popish Plot, and of Richard *Baxter, but he is chiefly associated with the *Bloody Assizes (1685). Contemporary reports of his brutality may have been prejudiced, but he certainly browbeat witnesses and his sentencing of the 80-year-old Alice Lisle to be burnt for treason caused widespread revulsion. At the accession of William III he was imprisoned and died before proceedings were taken against him.

Jehangir *Jahangir.

Jenkins's Ear, War of (1738–41), a war between Britain and Spain which broke out as a result of Britain's trade with South America. The Peace of *Utrecht, which gave the British South Sea Company a limited trade monopoly with the Spanish American colonies, caused general friction, but the main trouble-makers were illicit traders who defied both the Company and the Spanish government. In 1737 British merchants were protesting at a tightening-up of Spanish control. The Spanish government and *Walpole both wanted peace but Walpole's enemies made it an excuse to attack him: a merchant captain named Jenkins was produced to tell a story of torture and the loss of an ear. Popular clamour was such that Walpole consented reluctantly to declare war. Admiral *Vernon captured Porto Bello and France sent two squadrons to the West Indies. In 1740 the war merged into that of the *Austrian Succession.

Jeremiah (b. c.640 BC), Hebrew prophet. He was called to prophesy during the reign of King *Josiah. As Babylonian power increased, he maintained that resistance was useless and that the fall of Jerusalem was inevitable, views popular neither with the king nor with the people. After the fall of Jerusalem in 586 BC, he remained in the city until taken to Egypt by Jewish dissidents. His messages, always intensely personal, were preserved and edited by his scribe-secretary Baruch.

Jericho, an ancient city, now in Israel. A well-watered oasis near the Jordan river-crossing at the head of the Dead Sea, it was of strategic importance, located at the junction of the trade routes of antiquity. It was occupied from c.9000 BC and is one of the oldest continuously inhabited cities in the world. The principal mound, one of the best known of all Near Eastern *tells, accumulated over 15 m. (50 ft.) of deposit, even though the later occupation levels, from 2000 to 500 BC, have been swept off the summit by erosion. The most interesting layers are of the pre-pottery *Neolithic period c.7000 BC, when Jericho was already a walled settlement of some 4 ha. (10 acres). Little remains of the late Bronze Age period, the probable date of its destruction by Joshua recorded in the Old Testament of the Bible.

Jeroboam I (d. c.901 BC), first king of the northern kingdom of *Israel (c.922–c.901 BC). After *Solomon's death, his successor *Rehoboam failed to gain the support of the northern part of Israel, which seceded under the leadership of Jeroboam. He then extended and strengthened the kingdom taking advantage of Syrian and Assyrian weakness. He made the kingdom prosperous, but the prophet Amos spoke against his oppression and injustice. He was also criticized for encouraging the sanctuaries at Dan and Bethel to rival the Temple at Jerusalem.

Jerome, St (Eusebius Sophronius Hieronymus) (c.342–420), Christian writer and translator of the Bible into Latin. During a period spent as a hermit in the Syrian desert he learned Hebrew. From 382 he was Papal Secretary for two years during which he began his Bible translation. When the pope died he settled in Bethlehem where he completed his translation, known as 'The Vulgate' ('popular tongue'), and founded a religious house. He was a vigorous disputer, attacking among others Pelagius (*Pelagianism) and *Augustine. He advocated asceticism, notably celibacy.

Jerusalem, a holy city to *Jews, *Christians, and *Muslims. It was originally a Jebusite settlement, captured by *David c.1000 BC. Solomon's Temple, the central shrine of Judaism, destroyed by Nebuchadnezzar in 586 BC, was rebuilt in 516 BC and even more magnificently by *Herod the Great. It was razed by the Romans in AD 70 and in 135 they built the city of Aelia Capitolina on its site. St *Paul regarded it as the home of the original Christian congregation, headed until his death in 62 by James, the apostle of *Jesus Christ. Constantine marked its significance by building the Church of the Holy Sepulchre (c.335) over the supposed tomb of Christ. Muslim rule from 637 was symbolized by the Dome of the Rock, the city's holiest Muslim shrine, built in 691. The Christian knights of the *Crusades controlled the city from 1099 to 1187 as the chief city of the Latin Kingdom of Jerusalem. There was always a small Jewish community and Jewish pilgrimage to the Wailing Wall, a surviving part of the Temple, continued.

Jervis, John, 1st Earl of St Vincent (1735–1823), British admiral. He served with *Keppel and *Howe during the American War of *Independence. At the outbreak of the French Revolutionary War in 1793 he captured Martinique and Guadeloupe in the West Indies. His greatest victory was his destruction of the Spanish

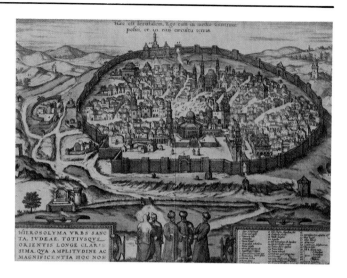

A coloured engraving of the walled city of **Jerusalem** in 1575 by F. Hogenberg and G. Braun, Cologne. It shows all the main sights of the city including Solomon's Temple (14), the Holy Sepulchre (15), and Calvary (16).

fleet at Cape St Vincent in 1797, where Nelson was one of his captains. He was First Lord of the Admiralty from 1801 to 1804.

Jesuit, a member of the Society of Jesus (SJ), a Roman Catholic religious order founded by St Ignatius *Loyola (1534), and approved by Pope Paul III (1540). Its original aim was to convert Muslims to Christianity but it acted to foster reform within the Roman Catholic Church, especially in the face of the challenge presented by the Protestant *Reformation.

By the early 17th century Jesuits had contributed incalculably to the success of the *Counter-Reformation, distinguished themselves as educationists, and established missions in India, China (where Matteo *Ricci gained the confidence of the Ming emperor, Wanli), Africa, and South America; but before the end of the century they had run into serious opposition within the Roman Catholic Church. The Dominicans, Jansenists, and others accused them of laxity in their ethical code and condemned their incorporation of local customs into their Christian teachings. Their opponents secured their expulsion from France (1764) and Spain (1767), and in 1773 they were suppressed by Pope Clement XIV. Jesuits continued to teach in Germany and Austria and also survived in England. The order was formally restored by Pope Pius VII in 1814.

Jesus Christ (c.6 BC–c. AD 30), the central figure of *Christianity, believed by his followers to be the Son of God, a person both human and divine. The Gospels of the *Bible are the main sources of information about Jesus. According to them, Jesus was born at Bethlehem to Mary, by tradition a virgin, in the reign of Augustus Caesar. He was brought up at Nazareth in Galilee and received a traditional Jewish education. He may have been a carpenter, the trade of Mary's husband, Joseph. About AD 27 he was baptized in the River Jordan by *John the Baptist and shortly thereafter started his public ministry of preaching and healing (with reported miracles). Through his popular style of preaching, with the use of parables and proverbs, he proclaimed the imminent approach of the Kingdom of God and the ethical and religious qualities demanded of those who were to enjoy it (summarized in the Sermon on the Mount). His interpretation of Jewish law did not reject ceremonial observances but placed them in subordination to the fundamental principles of charity, sincerity, and humility. From among his followers in Galilee he selected twelve *disciples to be his personal companions and to teach his message.

His preaching brought him into conflict with the Jewish authorities. In the knowledge of this he travelled to Jerusalem where he was betrayed by Judas Iscariot, one of his disciples, and condemned to death by the Sanhedrin, the highest Jewish court. He then appeared before the Roman governor, Pontius *Pilate, and he sentenced Jesus to death by crucifixion. His followers claimed that after three days the tomb in which his body had been placed was empty and that he had been seen alive in a glorified but recognizable form. Belief in his resurrection from the dead spread among his followers, who saw in this proof that he was the Messiah or Christ ('anointed one' in Hebrew and Greek respectively), the fulfilment of the hopes of Israel (as recorded in the Old Testament) and of all men. His followers started to form Christian communities around Jerusalem from which grew the *Christian Church.

Jewel, John (1522-71), Bishop of Salisbury, one of the key figures in the English Protestant church. He came from a Devonshire family, and came into contact with Protestantism at Oxford University. In the reign of *Mary I, he fled abroad in 1555, and then returned at the accession of *Elizabeth. Consecrated bishop in 1560, his best-known work was the *Apology for the Church of England* (1562) in which he justified the *Anglican Church as a true church, not a mere compromise between Catholicism and extreme Protestantism.

Jewish Revolt, a serious nationalist uprising led by the *Zealots against Roman rule (AD 66-70). *Vespasian invaded Palestine with 60,000 troops. By 68 the rebels were confined to the Jerusalem area. After Vespasian's succession in the civil wars of 68-9, his son Titus besieged the city, which fell district by district. Prisoners were taken to Rome and used as slave labour to build the *Colosseum. The last insurgents held out at *Masada until 73. In 132-5 *Bar Cochba led another revolt, which ended with the sack of Bethar and Cochba's death.

Jews, people descended from the inhabitants of Israel, whether residing in Palestine or, as part of the *diaspora, practising *Judaism elsewhere. During the *Exile (597-538 BC) following the Babylonian conquest, their religion developed from a sacrificial temple cult into an elaborate code for daily living which became the basis for communal identity. The revolt of the *Maccabees in 167 BC showed their determination to preserve that identity, which survived the Roman destruction of the Temple in AD 70 and of *Jerusalem in AD 135.

Dispersed throughout the Roman empire, they suffered much discrimination at the hands of its Christian successors and often welcomed *Arab conquests which brought greater toleration. In Spain and Cairo they prospered both materially and intellectually. In Christendom, they were free to engage in usury, a sin for Christians, and were herded into ghettos. Though they were tolerated for their usefulness, they suffered periodic persecution. Having been expelled from Spain in 1492, the Sephardim, speaking Ladino, a Spanish dialect, found refuge in north Africa and the Levant. The Ashkenazim, speaking Yiddish, a variant of German, established themselves in north-west and eastern Europe, greatly assisting the economic development of Germany and Romania. Secularization of political life in western Europe enabled them to gain civil liberties from 1789 onwards.

jihad, usually translated from Arabic as 'holy war', literally 'struggle'. One of the basic duties of a Muslim, prescribed as a religious duty by the *Koran and by tradition, is to struggle against external threats to the vigour of the Islamic community and also against personal resistance to the rules of divine law within oneself. Jihad in theory is controlled by the strict laws of war in Islam, which prescribe conditions under which war may be declared, usually against an enemy who inhibits the observance of the faith. In practice it has often been used by ambitious Muslim rulers to cloak political aims with religious respectability. Famous jihads include the early *Arab conquests, resistance to the Crusades, and the conquests of the Hausa reformer, Uthman dan Fodio in northern Nigeria in the early 19th century.

Jin (Chin) (1126-1234), a dynasty that governed Manchuria, part of Mongolia, and much of northern China. It was founded by the Juchen, nomad huntsmen, who came from around the Amur and Sungari rivers. They were ancestors of the *Manchus. When the Northern *Song set out to overthrow the *Liao, to whom they were tributary, they allied with the Juchen, hoping to play off one alien people against another. The latter, however, once having conquered the Liao, sacked the Song capital, Kaifeng, in 1126. The Song retreated south, establishing their new capital at Xingsai (Hangzhou).

The Juchen were in time tamed by their Chinese subjects, who far outnumbered them. Their frontier with Southern Song was stabilized. Jin emperors studied the Chinese classics and wrote poetry in Chinese. Their nomad vigour was sapped by a sedentary life. By 1214 much of their territory, including Beijing, their central capital, was in *Genghis Khan's hands. The dynasty survived, ruling from Kaifeng, until a final Mongol onslaught twenty years later.

Jin, Western   *Western Jin.

Joachim of Fiore (c.1132-1202), Italian mystic. He was the abbot of the Cistercian monastery of Corazzo in Italy, but left in 1189 to found a new reformed community at San Giovanni, in Fiore, which received papal sanction in 1196. From this developed the small but for a time very influential Florensian order. Joachim's work, *Concordia Novi*, claimed that there are three ages of creation: that of God the Father was past, that of the Son was drawing to a close, and that of the Holy Spirit was imminent.

Joanna the Mad (1479-1555), nominal Queen of Castile (from 1504) and of Aragon (from 1516). She was the daughter of *Ferdinand and *Isabella of Spain and the wife of *Philip IV of Burgundy. After Philip's death in 1506 she became insane. In 1509 she retired to Tordesillas, accompanied by Philip's embalmed corpse.

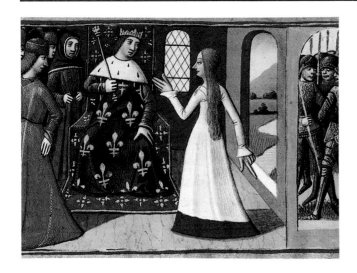

A manuscript illustration of **Joan of Arc** in the presence of Charles VII taken from the *Vigiles de Charles VII*. (Bibliothèque Nationale, Paris)

Her sons later became the emperors *Charles V and Ferdinand I.

Joan of Arc, St (*c*.1412-31), French heroine, sometimes called the 'Maid of Orleans'. The daughter of prosperous peasants, from the age of 13 she heard mysterious voices which she identified as the urgings of Saints Michael, Margaret, and Catherine. They insisted she help the French forces who, in 1429, were engaged in defending Orleans against English troops. A local commander was convinced of her claims as was the dauphin. Given command she carried out her mission of relieving Orleans and crowning the dauphin, Charles VII, at Reims Cathedral. This was a turning point in the *Hundred Years War. The jealousy of rival advisers, and doubts over her military skills, reduced her influence. Wounded and defeated at Paris, she was captured by the Burgundians in 1430, sold to the English, and tried as a witch. She was burnt at Rouen in May 1431, but an ecclesiastical court reversed the guilty verdict in 1455 and she was canonized in 1920.

Jodhpur, a district in north-west India, from 1459 to 1947 the capital of the largest *Rajput Princely State. Formerly known as Marwar, the state had existed since the early 13th century, but reached its peak after the construction in 1459 of a new capital, named after its founder, Raja Rao Jodha. Jodhpur occupies a magnificent site, its red sandstone fortress commanding the city from a rocky outcrop. Despite Mogul invasions in the 16th and 17th centuries, its rajas maintained their independence, but fell under *Maratha control in the 18th century.

John (1167-1216), King of England (1199-1216). He was the youngest son of Henry II, and succeeded his brother Richard I, supplanting Prince Arthur of Brittany. He was nicknamed John 'Lackland' because, unlike his elder brothers, he did not receive lands from his father. With Richard's accession (1189) John received the earldom of Gloucester, but during Richard's absences on Crusade intrigued against him with *Philip II of France.

John's reign was marked by the effects of spiralling inflation, which forced him to exploit all possible means of raising the royal revenues. The first crisis of the reign was the loss (1206-14) to Philip II of royal and baronial lands in Normandy, Anjou, Maine, and Brittany. The second crisis arose from his refusal to recognize *Langton as Archbishop of Canterbury, which led to John's excommunication and England being placed under papal interdict (1208-14). The final crisis reflected baronial discontent with John's financial policies and arbitrary style of government. John's opponents presented him with a formal statement of his obligations in the form of *Magna Carta (1215). His repudiation of it led to the first *Barons' War which only ended with his death.

John I (the Great) (1357-1433), King of *Portugal (1385-1433), the illegitimate son of Pedro I. He led nationalist sympathizers against the *Castilians, who supported Queen Eleanor, his rival. Defeating their invasion he went on to ally with England (through his marriage to Philippa, daughter of John of *Gaunt) and encourage African and Western exploration. His reign was the start of Portugal's period of colonial and maritime expansion; his sons, including *Henry the Navigator, consolidated his achievements.

John II (the Perfect) (1455-95), King of Portugal (1481-95). He destroyed his rival, the Duke of Braganza in 1483, and crushed a conspiracy led by his brother-in-law in 1484. He sponsored African and American exploration, and negotiated the Treaty of Tordesillas, which divided the lands of the New World between Spain and Portugal.

John III (the Pious) (1502-57), King of Portugal (1521-57), grandson of John II. He married *Charles V's sister and Charles married John's sister to create the *Habsburg claim to the Portuguese throne. He encouraged Portuguese settlement in Brazil, claimed the Moluccas, traded with Siam, and in 1535 conquered the Indian island of Diu. He was a fanatical Catholic noted for his sponsorship of the *Inquisition and the *Jesuits, who went as

King **John** of England hunting a stag in a royal forest, a manuscript illustration. Hunting was one of his favourite pastimes. (British Museum, London)

missionaries to Brazil. Agricultural decline and falling population overshadowed the end of his reign.

John IV (the Fortunate) (1605-1656), King of Portugal (1640-56). The founder of the Braganza dynasty, he expelled a Spanish usurper and proclaimed himself king in 1640. He defeated the Spanish at Montijo in 1644, drove the Dutch out of Brazil in 1654, and generally restored Portugal's international position.

'John Bull', a character invented by John Arbuthnot in *The History of John Bull* (1712), which was soon regarded as representative of the typical patriotic Englishman, and even of England itself. He was much used by caricaturists in the late 18th century in various (not always flattering) situations. He gained a new lease of life from the cartoonists of *Punch* magazine who made him so universally known that he was often used as a label for anything British.

John of Austria, Don (1545-78), Spanish general and admiral. He was an illegitimate son of Emperor *Charles V, educated in Spain, and recognized as his half-brother by *Philip II. He was appointed commander-in-chief of the Spanish navy in 1568, and organized the suppression of the Revolt of the Moriscos (1569-70). He commanded the *Holy League fleet which defeated the Turks at *Lepanto (1571), and went on to conquer Tunis (1573). His career began to founder after his posting to the Netherlands as governor-general (1576), at a critical point in the *Dutch Revolts. His impatience with negotiation led to the resumption of hostilities in 1577-8; and meanwhile he entertained grandiose schemes of marrying *Mary, Queen of Scots and replacing *Elizabeth I on the English throne. He died of typhus before he could effect these plans.

John of Salisbury (c.1115-80), English philosopher and theologian. He studied under Peter *Abelard and, after a short period at the papal court, was appointed secretary to Archbishop Theobald of Canterbury (c.1148).

A series of satirical sketches entitled *John Bull's Progress*, drawn by the cartoonist James Gillray and published in 1793. The drawings show the domestic consequences of **John Bull** (and therefore England) going to the wars. (British Museum, London)

He continued as secretary to Theobald's successor Thomas à *Becket, went with him into voluntary exile in France (1164-70) during Becket's dispute with *Henry II of England, and was with Becket on the day of his murder (1170). In 1176 he became Bishop of Chartres. He was the author of a number of works, the most important being the *Matalogicon* and the *Policraticus*, which outlines his ideal society.

John of the Cross, St (1542-91), Spanish Christian mystic and poet. He helped St *Teresa of Avila to found the Order of the Discalced *Carmelites. A Carmelite monk himself, he was ordained priest in 1567. His collaboration with St Teresa led to friction with the Calced Carmelites, and in 1577 he was imprisoned at Toledo. He escaped, and subsequently attained high office in the Discalced Carmelite order. His rigorously intellectual verse has been called the finest poetic expression of Spanish Christian mysticism. He was canonized in 1726.

John the Baptist, St (d. AD 28-30), Jewish preacher and prophet. In c.27 AD he began preaching against social injustices and religious hypocrisy, directed mainly to the *Pharisees and *Sadducees. As part of the Jewish renewal movement, he baptized his followers with water, and he also baptized *Jesus Christ, at the start of Jesus' own public ministry. John's denunciation of *Herod Antipas for marrying his brother's wife, Herodias, led to his imprisonment, and he was beheaded possibly at the request of Herod's stepdaughter Salome.

Johnson, Sir William (1715-74), *fur trader and superintendent of Indians in northern New York, who came from Ireland to the Mohawk valley as an estate manager in 1738. His honesty and geniality won the trust of the *Iroquois, who gave him huge tracts of land. His intimacy with the Indians was a vital element in the *French and Indian War, and after his victory at *Lake George (1755) he was knighted and appointed Superintendent of Indian Affairs (1756). He was influential at the *Albany Congress and in the promulgation of the *Proclamation Line.

Johore, a Malay sultanate founded by Sultan Mahmud, the last sultan of *Malacca before it was captured by the Portuguese in 1511. He tried unsuccessfully to retake Malacca a number of times but was finally defeated in 1526. Thereafter Johore was in frequent conflict with *Acheh, an emerging north Sumatran state. After Johore assisted the Dutch capture of Malacca in 1641, it established its rule over much of the southern Malay peninsular. However, an argument over a royal marriage led to its defeat by the Sumatran state of Jambi in 1673. Its capital moved to Rhio (on the island of Bintang) and a much weakened state only continued with support of *Bugis mercenaries.

joint-stock company   *chartered company.

Joinville, Jean de (1224-1317), French historian and courtier. He was a friend and confidant of *Louis IX, accompanying him on the Seventh *Crusade (1248-54). He advised against the Eighth Crusade and did not join it. His *Histoire de St Louis* was principally an account of the Seventh Crusade, written with humour and sincerity.

Jones, John Paul (1747–92), American admiral. He was born in Scotland and went to sea aged 12. He was employed in the slave trade and rose to the rank of mate, but his killing of a mutinous sailor led to his settling in America in 1773. He joined the Continental navy in 1775, and as a captain commanded a privateer in the West Indies (1776–7). He carried the news of *Saratoga (1777) to France, his base for a series of breath-taking exploits round the northern coasts of Britain. He spiked the guns of his old home port, Whitehaven, and captured HMS *Drake*. In a spectacular night battle off Flamborough Head (1779) his shattered *Bonhomme Richard* forced HMS *Serapis* to surrender. His own response to a demand to surrender, 'I have not yet begun to fight', made him a popular hero in America and France. After serving as admiral in the Russian navy against the Turks (1788–9), he died in Paris.

Joseph II (1741–90), Holy Roman Emperor (1765–90). He was co-regent of Austria with his mother *Maria Theresa from 1765 and sole ruler 1780–90. Intelligent, dedicated to the principles of the *Enlightenment, and hoping to improve the lives of his subjects, he was nonetheless autocratic and too hasty. Attempts at major reforms brought many parts of the empire close to revolt. His most lasting achievements were the edicts in 1781 granting toleration to Jews and Protestants. He abolished serfdom and curtailed the privileges of the nobles. In German affairs he was outmanoeuvred by *Frederick II of Prussia and his alliance with *Catherine of Russia led to a disastrous war against the Turks.

Josephus, Flavius (b. *c.* AD 37), Jewish historian. Born into a priestly family, a visit to Rome in AD 64 left in him a lasting impression of Rome's power and culture.

A portrait of **Joseph II**, by an unknown artist, painted five years after his death in 1795. The Austrian emperor is painted amid the imperial regalia representing the diversity of the lands he ruled. (Niederösterreichisches Landesmuseum, Vienna)

A portrait of the American admiral John Paul **Jones** by Charles Wilson Peale. (Independence National Historical Park Collection, Philadelphia)

Although hostile to the nationalist extremists, he was given command in Galilee in the Jewish revolt against *Nero, but after his capture in 67 went over to the Romans. Settled in Rome, he became a Roman citizen and was given a pension. His most important works were his histories of the revolt and its antecedents (*Bellum Judaicum*), and of the Jewish people from the Creation to the reign of Nero (*Antiquitates Judaicae*), both of them works of high quality and importance. The date of his death is unknown.

Josiah (649–609 BC), King of *Judah (640–609 BC) in succession to his father Amon. At the age of 17 he undertook a major reform of worship in and around Jerusalem, suppressing the worship of local gods, closing down outlying shrines, and making the Jerusalem Temple the sole centre of worship. Because of his concern for the Jewish faith, he is described in the Bible as a model king, the last good king of Judah. He died at the battle of Megiddo, defeated by the Egyptians.

journeyman, a qualified artisan working for someone else. Journeymen were workers (paid daily) who had served their apprenticeship and were not yet in a financial position to set up as masters. The late medieval craft *guilds, by restricting the number of masters without limiting the number of apprentices, caused increasing numbers of discontented and sometimes unemployed journeymen, and in the 16th century it was found necessary in England for Parliament to pass legislation to compel masters with apprentices to employ journeymen. This legislation was no longer enforced by the 18th century, if it ever had been, and the Industrial Revolution with its factory system and demand for unskilled labour spelled doom for the journeyman. Associations of journeymen were the earliest trade unions, as distinct from guilds, in both Britain and America, one of the longest-lived being the Federal Society of Journeymen Cordwainers in Philadelphia.

Juan-Juan   *Avars.

Judah (or Judaea), the southern part of ancient Palestine. After *Solomon's death the kingdom of Judah resisted repeated invasions before falling to *Nebuchadnezzar of Babylon, who destroyed its capital, *Jerusalem, in 586 BC. When the Jews returned from *Exile in Babylon, the land they occupied was named Judaea. *Seleucid desecration of the Temple in 167 BC prompted the revolt of the *Maccabees. Independence lasted until the Roman conquest of 63 BC. In AD 135 the area was absorbed into Roman Syria.

Judah Ha-Nasi (135–c.220), 'the Prince', patriarch of Judaea. He is best remembered for organizing the compilation of the *Mishnah*, the first comprehensive statement of Jewish religious law. He won the favour of Roman emperors, forged closer links with the Jews of the *diaspora, and worked to raise the status of the Hebrew language.

Judaism, the religion of the *Jews. According to Hebrew scriptures, the Jews owe their foundation to *Abraham, who, through a covenant (agreement) with their God, was promised the land of *Israel for himself and his descendants. The teachings of *Moses, recorded in the Torah (the first five books of the *Bible), provided instruction in religious belief and moral matters. Many of the characteristics of Judaism, including the idea of one all-powerful God emerged during the Babylonian *Exile when the importance of teaching and prayer was emphasized. Jewish communities became widely scattered (the *diaspora) after the destruction of the Temple at Jerusalem (the central sanctuary of Judaism) in AD 70 by the Romans. Leaders of these communities (rabbis) encouraged a greater loyalty to the Torah and its daily demands. During the following centuries various generally accepted interpretations of scriptures were collected and then written down in the *Talmud, and Judaism revolved around the study of the Torah and the Talmud. Jewish communities were centred on the synagogue and they lived by their own laws (as far as any state permitted) and kept to their own calendar.

Judas Maccabaeus   *Maccabees.

Jugurtha (c.156–104 BC), King of *Numidia. He had contested the throne with his two brothers following the death of his adoptive father Micipsa in 118. A Roman commission divided the kingdom between them, but Jugurtha attacked and killed one brother, Adherbal, in 112. Armed intervention by Rome followed disturbances, but Jugurtha resisted stubbornly until *Sulla, the legate of *Marius, induced Bocchus of Mauretania to hand the Numidian king over. After Marius had celebrated a triumph in Rome, Jugurtha was executed.

Julian (the Apostate) (Flavius Claudius Julianus) (c.331–63), Roman emperor 360–3. He was born at *Constantinople, a nephew of *Constantine. After receiving a classical Greek education he gave up Christian belief for *Neoplatonism, earning himself his posthumous insulting title. He blamed *Christianity for sapping Rome's traditional strengths. Reversing the Edict of *Milan, he reopened temples and restored paganism as the state cult in place of Christianity, but this move was reversed after his death on campaign against the Persians.

Julius II (Guiliano della Rovere, 1443–1513), Pope (1503–13). He strove to restore and extend the *Papal States and to establish a strong independent papacy. He crushed Cesare *Borgia and sponsored the League of *Cambrai and the *Holy League against France in 1510. Before the end of 1512 the French were forced to leave Italy and several new territories were added to the papacy's holdings. When Julius died he was mourned as the liberator of Italy from foreign domination.

Politics and war dominated his reign and he devoted little time to church reform. He did, however, send missionaries to India, Africa, and America, and was a noted artistic patron, commissioning work from Michelangelo, Raphael, and Bramante.

'Junius', the pen-name of an unknown British writer of seventy political letters, published in the London *Public Advertiser* between 1769 and 1772. The letters were informed, satirical, and vicious, and directed their attack mainly against the Duke of *Grafton, particularly on behalf of *Wilkes. The letter to *George III, a model of studied impertinence, made 'Junius' famous. To this day there is no certainty of the writer's identity—Sir Philip Francis seems the most likely, but many politicians and authors have been credited with the letters.

jury, trial by, the system in criminal and civil justice whereby twelve sworn (Latin, *jurati*) men and women determine the truth from disputed evidence. Its development was exclusive to England. Initially, jurors were regarded as sworn witnesses to the accuracy of a claim. In the Anglo-Saxon 'compurgation' an accused person could be cleared simply on the sworn word of twelve neighbours to his good character. The Assize of Clarendon (1166) instructed jurors to present their evidence and suspects before the king's justices. Trial by jury was adopted in civil cases when the Lateran Council (1215) forbade the trial by *ordeal, and it became compulsory for certain criminal cases under the Statute of Westminster (1275).

Justinian I (482–565), Byzantine emperor (527–65). Throughout much of his reign his troops were engaged in a defensive struggle against Persia in the east and a successful war against the barbarians in the west. Believing that they had lost their initial vigour, he hoped to revive the old Roman empire. His general, *Belisarius, crushed the Vandals in Africa (533) and the Ostrogoths in Italy (535–53), making Ravenna the centre of government. His greater claim to fame lay in his domestic policy in which he was strongly influenced by his powerful wife, *Theodora. He reformed provincial administration and in his *Corpus Juris Civilis* he codified 4,652 imperial ordinances (*Codex*), summarized the views of the best legal writers (*Digest*), and added a handbook for students

K

Trial by **jury**, as illustrated in a miniature from an English law treatise of the early 15th century. It shows the Court of the King's Bench. At the top are five judges; the jury is shown at the left, possibly being sworn in by an usher at the table. The prisoner stands in the dock, his feet shackled, while other prisoners await trial in the foreground.

(*Institutes*). A passionately orthodox Christian, he fought pagans and heretics. His lasting memorial is the Church of St Sophia in Constantinople.

Jutes, a Germanic tribe who invaded Britain in the 5th century AD under *Hengist and Horsa, according to tradition. Their name is preserved in Jutland and Juteborg. According to *Bede (and he is supported by archaeological evidence) they occupied the Isle of Wight, the Hampshire coast, and Kent, the former land of the Belgic Cantii with its capital at Canterbury. Here they permitted *Augustine's mission to re-Christianize Britain. In the 11th century the New Forest in Hampshire was still known as 'Ytene' ('of the Jutes').

Kabir (1440–1518), Indian mystic and poet, who preached the unity of all religions. There is uncertainty about his origins, but it is believed that he was adopted by a Muslim weaver after being abandoned by his Brahmin mother. His borrowings from both Hinduism and Islam resulted in the preaching of a new mystic path, the Kabirpanth, and the founding, by one of his disciples, of the *Sikh religion. His mystical poems, stressing the oneness of God, together with his rejection of caste, have made Kabir one of the most popular religious figures in his country's history.

Kalmar, Union of (1397). The joining together of the crowns of Denmark, Sweden, and Norway. Margaret I (1353–1412), daughter of the King of Denmark and wife of Haakon VI of Norway (d.1387) defeated (1389) the King of Sweden and persuaded the Diets of Denmark, Norway, and Sweden to accept Eric of Pomerania, her grandnephew, as king. He was crowned (1397), the beginning of the Union of Kalmar, though Margaret herself ruled the three kingdoms until her death. The union was dissolved by *Gustavus I of Sweden in 1523.

Kamakura, a city south-west of Tokyo, the headquarters of Japan's first *shogunate. There, near his estates and remote from court influence, *Minamoto Yoritomo set up a military administration, the *bakufu* (tent government). After his death the *Hojo family acted as regents of the shoguns. The Kamakura shogunate (1192–1333) saw the emergence of the *samurai and the organization of military power. The bronze Daibutsu (Great Buddha) (1252), 15.8 m. (52 ft.) high, showed the popularity of Buddhism at that time, *Zen Buddhism particularly appealing to the warrior class.

Kanem-Bornu, the names of two successive major African states in the Lake Chad region between the 11th and the 19th centuries. Ethnically and linguistically the peoples were mixed. They include Arab, Berber, as well as other African elements, and were mostly Muslims. An Islamic sultanate of Kanem, ruled by the Seyfawa family, existed by the 11th century, which, under Dunama (1221–59), came to extend from Fezzan and Wadai to the Niger, and included Bornu. Following civil wars this empire collapsed in 1398, but a member of it created a new state of Bornu with N'gazargamu as capital, and Kanem as a province. Idris Aloma (ruled 1571–1603) was the most powerful of the Bornu rulers; he introduced firearms into the army and Bornu reached the peak of its power under his rule. A long period of stability followed until 1808, when the *Fulani sacked N'gazargamu. Muhammad al-Kanemi, a leading chief, restored the titular kings, retaining effective power himself. The last Mai, or titular king, was executed in 1846.

Kangxi (K'ang-hsi), (1654–1722), second *Qing Emperor of China (1662–1722). He extended the Qing empire by a series of military campaigns, subduing opposition to Manchu rule in southern China (1673–81), incorporating *Taiwan into China for the first time in 1683, and personally leading a campaign into Outer *Mongolia (1693). He opened certain ports to overseas traders and the Treaty of *Nerchinsk (1689), established diplomatic contact with Russia. In 1692 he permitted Catholic missionaries to make converts and he employed *Jesuits to teach astronomy and mathematics. He was renowned for his tours of inspection of China and for his sponsorship of scholarship, including a history of the Ming dynasty and an encyclopedia of literature.

Kano, a city in northern Nigeria. According to tradition, it was founded in the 10th century, and probably became Muslim in the 14th century. It was one of the seven *Hausa city-states, and was an important trading and commercial centre. In the 16th century the Muslim teacher al-Maghili made it famous, teaching law and mysticism. In the 15th century it was probably subject to *Bornu, and in the 16th it came under *Songhay, but it was always ruled by an indigenous vassal.

Kant, Immanuel (1724-1804), German philosopher. He spent his whole life in East Prussia: from 1770 he was professor of logic at Königsberg University in spite of some difficulties with the authorities due to the divergence of his teaching from orthodox Lutheranism. His most important book is *The Critique of Pure Reason* (1781). He shared the 18th-century absorption with the study of the nature of human reason and the measure of its infallibility when it was considered as man's sole support in the pursuit of truth. Kant disagreed with *Hume and the sceptics and maintained that in spite of some errors man could achieve fundamental knowledge by the use of reason. His *Critique of Practical Reason* (1788) dealt with ethics. His writings had widespread influence.

Karnak, the religious centre of ancient *Thebes, situated on the east bank of the Nile, where the great temple of Amun was constructed. This complex of buildings, the work of some 2,000 years, includes the Hypostyle Hall with 134 columns each *c.*24 m. (79 ft.) high. It was begun

A view of the complex at **Karnak**, with its sacred lake in the foreground. The ruined columns of the Hypostyle Hall are in the centre, while to the right is one of the many obelisks. The first pylon (originally a tall, tapered gateway tower) is on the left.

by Ramesses I and completed by Seti I and *Ramesses II. A road lined with statues of sphinxes linked the site to nearby Luxor.

Kassites *Babylon.

Kaunitz, Wenzel Anton, Count von (1711-94), Austrian diplomat and statesman. As Chancellor (1753-92) he controlled foreign policy under Empress *Maria Theresa and Emperor Joseph II. Convinced that Prussia was Austria's most dangerous enemy his main diplomatic feat was to reverse (1756-7) long-standing European alliances although two constant factors remained: British opposition to France and Austria's alliance with Russia. But when, to safeguard Hanover, Britain allied with its former enemy, Prussia, Kaunitz negotiated an alliance with France, thus isolating Prussia on the Continent. Although *Frederick the Great's ambitions were not fully checked, Kaunitz was a leading negotiator of the Treaty of Paris (1763).

Kempis, Thomas à (1379-1471), *Augustinian canon and religious writer. He was the author of the *Imitatio Christi*, an influential work which emphasized the need for asceticism, and reacted against the worldliness of the 15th century Catholic establishment.

Kenilworth, siege of (June–December 1266). An episode during the *Barons' wars when *Henry III attacked Kenilworth Castle, refuge of the *Montforts and their supporters. The Dictum of Kenilworth (31 October 1266) asserted the king's powers over the barons and offered inducements to peace by allowing them to recover their confiscated lands: the beseiged earls finally surrendered in December.

Kenneth I MacAlpine (d. *c.*859), King of Scotland (*c.*843-858). He united the Scots and Picts to form the kingdom of Scotia (*c.*843), having succeeded in *c.*841 as King of Dalriada in the Highlands. In *c.*848 he moved the relics of St *Columba to Scone, where the kings of Scotland were crowned.

Kenneth II (d. 995), King of Scotland (971-95). In return for recognizing the lordship of King *Edgar of England, Kenneth received Lothian two years after his accession in 971. He was murdered by Constantine III

A French engraving of Thomas à **Kempis**, the Augustinian canon and author, who was named after his birthplace of Kempen, near Düsseldorf. His real name was Hammerken.

who, in turn, was killed by Kenneth III (d. 1005), whose brief reign of civil wars from 997 ended with the accession of *Malcolm II.

Keppel, Arnold Joost van, 1st Earl of Albemarle (1669–1718), Dutch soldier who entered the service of *William of Orange and accompanied him to England in 1688. He served William in various capacities in England, but returned to Holland after William's death and fought with distinction at the battles of *Ramillies (1706) and *Oudenarde (1708).

Kerala, a region in India which occupies the narrow Malabar coastal plain between the Western Ghats and the Arabian Sea. Its name evokes an ancient kingdom, Keralaputra, to which there are references in *Asoka's inscriptions (3rd century BC). From early times trade developed with distant parts of the world. The dominant rulers up to the 5th century AD were the Cheras, but the region subsequently fragmented into separate kingdoms and was never again united. However the evolution of the Malayalam language maintained some cultural unity.

The arrival of Vasco *da Gama at Calicut (1498) began an era of European dominance of the spice trade. Cochin, which developed as a Portuguese port, was conquered by the Dutch in 1663. By the late 18th century British influence predominated, signalled by the annexation of Malabar District (1792). In 1795 the

southern princely state of Travancore accepted British protection.

Kett's Rebellion (July–August 1549), an orderly English peasant protest against the profiteering and *enclosures of local Norfolk landlords. Led by Robert Kett, a well-to-do tradesman, 16,000 small farmers encamped outside Norwich, and eventually gained control of the city. By their disciplined self-government, the rebels aimed to impress the authorities and shame the local magnates. The rebellion was suppressed by forces under John Dudley (later Duke of *Northumberland) who routed the rebels at Dussindale on 27 August. Kett and his brother William were among those executed.

Khalji, a Muslim dynasty of Turkish origin which seized power in northern India in 1290. Its three kings successively ruled the *Delhi sultanate for the next thirty years. Ala ud-Din (1296–1316), the second sultan, was the most successful. His armies held off Mongol threats, subdued large parts of Rajasthan and Gujarat, then carried Islam to Madurai in the extreme south of the subcontinent. Their object was pillage rather than permanent empire, yet Khalji expansion began a new era of Muslim penetration of Hindu southern India. On Ala ud-Din's assassination the dynasty declined, to be replaced in 1320 by the *Tughluq dynasty.

khanate, region ruled by a khan (a Mongol-Turki word meaning a supreme tribal leader elevated by the support of his warriors). On *Genghis Khan's death in 1227 his empire was divided into four parts, each ruled by one of his descendants. By the mid-13th century the *Mongol empire consisted of four khanates; the khanate of the Western Kipchaks (the *Golden Horde); the khanate of Persia, whose ruler was called the Il-khan; the khanate of Turkistan (the White Horde of the Eastern Kipchaks), and the khanate of the Khakhan in East Asia. The three khans were subject to the Khakhan (the Great Khan), but were generally resentful in their relations with him. After the death of *Kublai Khan (1294) the Khakhan's authority was nominal. In 1368 the Mongols were driven out of China and by c.1500 all four khanates had

An engraving by P. Schent of 1702 of the busy port of Cochin in **Kerala**, a major trading station on the spice route. (National Maritime Museum, London)

disappeared. A number of lesser khanates emerged; the khanates of Kazan, Astrakhan, the Crimea, Khiva, Bukhara, Tashkent, Samarkand, and Kokand. These long presented a threat to the communities surrounding them. One by one all were absorbed by Russia. The last to fall was Kokand (1876).

Khandesh, a medieval Muslim kingdom occupying a narrow strip of territory between the Satpura and Ajanta hills in the Tapti valley of western India. From 1382 to 1599 it was an independent sultanate, but following conquest by *Akbar it became a province of the Mogul empire. Under subsequent Maratha and British rulers it was absorbed into larger units, and the region, a cotton-growing area, is now part of Maharashtra state.

Khmer, a subject of the Hinduized kingdom of Funan, based in the Mekong valley in the 1st century AD. In the 6th century the Khmers overthrew their rulers, whose title 'Kings of the Mountain' was taken over by the Khmer kings. The Khmers were Buddhists, but their kings, with two exceptions, remained Hindus. Early in the 9th century Jayavarman II expelled Javanese invaders, re-united his country, and instituted the cult of the god-king. He and his successors ruled from *Angkor or its vicinity. They built cities and temples with forced labour and constructed canals and reservoirs which supported a rice-growing economy. They engaged in savage wars against their neighbours. At the end of the 12th century under Jayavarman VII they were at the height of their power. He defeated *Champa, extended Khmer territory, and rebuilt Angkor Thom, providing rest-houses and hospitals.

During the 13th century the Khmers were converted from Mahayana Buddhism, which accepted the idea of god-kings, to Hinayana Buddhism, which rejected such concepts. Oppressed by forced labour and decimated by wars, they became unwilling to build temples for god-kings; the complex irrigation systems broke down; the Thais of *Siam, former tributaries, attacked Angkor. After repeated defeats the Khmers abandoned Angkor in 1431 and made the village of Phnom Penh, further from the Siamese frontier, the new royal capital of Cambodia. The kingdom lost further lands to its stronger neighbours of Siam and *Annam in the 18th century.

Khosrau I (d. AD 579), King of Persia (531–79). His reign, after a long period of turbulence, marked the highest point of *Sassanian power. He restored royal authority over the army, bureaucracy, and lower nobility, reformed taxation, and restored defences and public works. He invaded Byzantine Syria in 540 and took Antioch. In 565, in alliance with the western Turks, he destroyed the Hephthalite empire on his eastern frontier. He also annexed *Yemen and died during negotiations with Byzantium over his invasion of Mesopotamia.

Khosrau II (d. AD 628), King of Persia (590–628) He succeeded to the throne after the deposition of his father, Hormidz. After being unseated by a coup, he accepted Byzantine aid to regain his throne in return for most of Armenia. This he recovered, with Edessa and Caesarea, in 610. In 611 *Jerusalem was taken and several thousand Christians massacred. In 616 he simultaneously invaded Egypt, captured Ankara, and besieged Constantinople. A Byzantine counter-attack drove him back to Ctesiphon,

where he was assassinated. He overtaxed the resources of his empire, which fell to *Arab conquest within a decade of his death.

Kiakhta, Treaty of (1727), a treaty between Russia and China signed at Kiakhta, a town in Russia immediately north of the Mongolian frontier. A border was agreed between Siberia and Mongolia, China losing a large amount of peripheral land to Russia. Trade in Chinese silks, tea, and porcelain and Russian furs was permitted but limited to Kiakhta. The Russians were also allowed to send language students to Beijing and to build a church for them there.

Kidd, William (c.1645–1701), Scottish privateer. He was based as a shipowner and trader in New York from 1689 to 1699. After service in the Caribbean in *King William's War and anti-privateering ventures for New

The sack of Kiev, capital of **Kiev Rus**, by a Mongol army in 1240. The bloody massacre, which left much of the city destroyed and most of its people killed, is shown in this illustration from the city's medieval chronicle. A traveller to Kiev six years later reported that only 200 houses were still standing.

The remains of a mosque at **Kilwa**. Although now almost completely overgrown, such buildings once caused Kilwa to be described, by the much-travelled Arab writer Ibn Battuta, as 'the most nobly built city on earth'.

York, he was commissioned by the crown to put down piracy in the Indian Ocean and attack French shipping. Though he obeyed the latter order, he joined up with other pirates in Madagascar and ravaged the Malabar coast of India (1687-9). He was arrested on his return to America and hanged in London in 1701.

Kiev Rus, the historical nucleus of Russia. Kiev was probably founded in the 6th or 7th century, the centre of a feudal state ruled by the Rurik dynasty from the 9th to the 13th century. About 878 Igor advanced along the Dnieper River from Novgorod and made Kiev capital of the Varangarian-Russian principality. As the oldest established city it is known as 'the mother of Russian cities' and also 'the Jerusalem of Russia' as the first centre of the Greek Orthodox Church in Russia.

Killiecrankie, battle of (27 July 1689). A narrow densely wooded pass near Pitlochry in Scotland was the site where John Graham of Claverhouse, Viscount Dundee, led the first *Jacobite attempt to restore *James II to the English and Scottish thrones. He overwhelmed the inexperienced forces of General Mackay, who lost 2,000 dead and 500 taken prisoner, but Dundee was killed at the moment of victory. The Highlanders were subsequently unable to follow up their success.

Kilwa, an island off the Tanzanian coast. It was a sultanate founded by Arabs in c.957, and by 1200 it had a monopoly of the *Zimbabwe gold trade. Its agents reached *Malacca. After 1500 the Portuguese ruined its economy, and in 1587 the Zimba, marauders from the Zambezi valley further reduced the population. Kilwa recovered briefly; in the 17th century ivory was traded there; in the 1770s French slave-traders used it as a base.

King George's War (1744-8), American component of the War of the *Austrian Succession, which saw the capture of *Louisburg on Cape Breton Island by a combined British Navy–New England force under William Pepperell in 1745. A subsequent campaign against the St Lawrence valley was abandoned. The fortress was returned to France in exchange for Madras in 1748.

King Philip's War (1675-6), a North American Indian rising which resulted from encroachments on Indian lands in New England. It was led by Metacomet (or King Philip), chief of the Wampanoag, whose lands were in southern Massachusetts, but Mohawks of the *Iroquois Confederacy devastated frontier settlements in northern and western interior New England as well. Before Philip was betrayed and killed by *Church in 1676 near Kingston, Rhode Island, Indians had raided within 32 km. (20 miles) of Boston and one out of every ten adult males in Massachusetts had been killed.

King William's War (1689-97), a North American frontier war between the French and the English and their Indian allies, which was a colonial adjunct to the War of the League of *Augsburg in Europe. The two main theatres were the northern coast and the Upper Hudson–Upper St Lawrence valleys. In 1690 Sir William Phipps's New England expedition sacked Port Royal in Acadia, but an intercolonial campaign against Quebec and Montreal ended in disaster. *Frontenac organized Abuski raids on English outposts in Maine and successfully intimidated the *Iroquois. Both sides lacked resources for full-scale war, and assistance from Europe was thwarted. The war was ended by the Treaty of Ryswick (1697) and a truce in Maine (1699).

knighthood, the special honour bestowed upon a man by dubbing (when he is invested with the right to bear arms) or by admission to one of the orders of chivalry. In England the emergence of knighthood was slow (the Anglo-Saxon word *cniht* means 'servant'). In the late 11th and early 12th centuries, knights were the lowest tier of those who held land in return for military service. During the 12th century their economic and social status improved, as society became more complex, and the market in free land developed. They became involved in local administration, and the new orders of knights which emerged in Europe in the aftermath of the *Crusades helped to give them a distinct identity. First to appear were the military orders of the *Knights Hospitallers (1070), the Knights of the Sepulchre (1113), and the *Knights Templars (1119). Their potential for military colonization was best realized by the German Order of

The investiture of a **Knight Hospitaller**, a woodcut illustration from a history of the order published in Germany in 1496. Here the new knight kneels before the Grand Master. (Order of St John)

the *Teutonic Knights (1190) which pushed eastwards on the frontiers with Poland and acquired Prussia for itself. The Order of the Livonian Knights gained similar successes along the Baltic. The Order of the Garter (1348) was England's first and most important, followed by the Order of the Bath (1399). France created the Order of the Star (1352), and *Burgundy the Order of Golden Fleece (1429).

Knight Hospitaller, a member of a military religious order, formally the Knights Hospitallers of St John of Jerusalem, so called after the dedication of their headquarters in Jerusalem to St John the Baptist. From 1310 they were known as the Knights of Rhodes, from 1530 the Knights of Malta. They began in the 11th century with Muslim permission to run a hospital for sick pilgrims in Jerusalem, and were made a regular order when the city fell in 1099 to the First *Crusade. They adopted a black habit bearing a white eight-pointed (Maltese) cross. Under the first Master their function became primarily military and spread to western Europe. They followed the *Augustinian rule and were divided into three classes: knights, chaplains, and serving brothers. When they were driven out by *Saladin they went to Acre, only to be expelled a century later when Cyprus became their headquarters. In 1310 they captured the island of Rhodes and retained it till 1522. Given the island of Malta by Emperor *Charles V they held it, having fought off the assaults of the Turks, until it finally fell to *Napoleon I. By this time the order had lost its former influence. Some members moved to Russia where Paul I was made Grand Master. His death in 1801 led to a period of confusion. The English branch of the order was revived in the 1830s and today cares for the sick.

Knight Templar, a member of a military religious order properly called the Poor Knights of Christ and of the Temple of Solomon, founded in 1118 by Hugh de Payens, a knight of Champagne in France. He and eight companions vowed to protect pilgrims travelling on the public roads of the Holy Land (*Palestine). At the Council of Troyes (1128) approval was given to their version of the *Benedictine rule. They quickly became very influential, attracting many noble members and growing in wealth, acquiring property throughout Christendom. When Jerusalem fell in 1187 they moved to Acre together with the *Knights Hospitallers and great rivalry and hatred developed between the orders. In 1291 when Acre also fell, they retreated to Cyprus. In Cyprus their great wealth enabled them to act as bankers to the nobility of most of Europe and this affluence attracted much hostility, in particular that of *Philip IV of France. In 1307 they were charged with heresy and immorality. Though some of the charges may have been true, envy of their wealth seems to have been the reason for their persecution. They were condemned, their wealth confiscated, and the order suppressed. The Grand Master and many others were burned at the stake.

Knossos, the leading city of the *Minoan civilization, situated a few miles inland from the north coast of the island of Crete, excavated by the British archaeologist Sir Arthur Evans from 1900 onwards. The ancient city was dominated by a palace built originally (*c*.2000 BC) on the remains of a Neolithic settlement. The palace was destroyed *c*.1700, probably *c*.1550, and again shortly afterwards as a result of the massive explosion of the volcano at Thera (modern Santorini). The city seems to have been taken over by Mycenaean invaders *c*.1450, and the palace suffered another destruction *c*.1375. Inhabitation of Knossos continued, but its power was broken—none of the magnificent frescos which were such a feature of the palace date from after this destruction. Although it never regained its lost glory, it remained one of the leading cities of *Crete, an often bitter rival of Gortyn. It was captured and destroyed by the Romans in 68–7 BC.

Knox, Henry (1750–1806), American general, who commanded the Continental Artillery during the War of *Independence. In 1775 he hauled the guns from Ticonderoga 124 km. (200 miles) through the wilderness to Boston, forcing British evacuation. He was *Washington's right-hand man throughout the War, and after it organized the Veterans' Society of the Cincinnati (1783). A Federalist, he was Secretary of War to the Confederation and under President Washington (1785–94), but his scheme for a national militia was thwarted.

Knox, John (*c*.1513–72), Scottish Protestant reformer. He played the key role in the establishment of the Scottish Kirk or Church of Scotland. By 1546 he had fallen under the influence of the Protestant preacher George *Wishart. His association with the assassins of Cardinal *Beaton led to his capture by the French and nineteen months of forced labour. In 1551 he became chaplain to *Edward VI of England, but on the accession of *Mary I (1553) he fled to Frankfurt, then Geneva, where he ministered to Protestant refugees from Britain and met John *Calvin.

His *First Blast of the Trumpet against the Monstrous Regiment* (meaning 'rule') *of Women* (1558) alienated him from *Elizabeth I of England, but she approved of his return to Scotland (1559), to lead the Protestant anti-French faction. Appointed minister of St Giles, Edinburgh, he was closely involved in drawing up the *Scots Confession* (1560) and *First Book of Discipline* (1560). He was a fierce opponent of *Mary, Queen of Scots.

Kongo, a kingdom in Central Africa, established south of the River Congo by 1300 which became one of the most powerful kingdoms in the region. The Kongo people traded over long distances, exploiting iron and salt mines. On Loanda Island they had a monopoly of *nzimbu* shells, which provided a local currency. It was the first African kingdom after Ethiopia to be converted to Christianity, by Portuguese missionaries in the 16th century. The Portuguese also brought the slave trade, which encouraged civil wars and weakened the Kongo kingdom by the mid-17th century.

Koobi Fora, a large area in northern Kenya where early *hominid remains (from over 160 individuals), have been found, including the remarkably complete '1470 skull' found in 1972. The dating of the hominids and tools was once controversial but it now seems clear that most of the finds date to between 2 and 1.4 million years ago. The '1470 skull' is evidence of a large-brained hominid in East Africa at an early date in human evolution. Scientists still question its affinity but most now agree that it is an early representative of *Homo*, possibly *Homo habilis*, and was the maker of the *Oldowan tools of the same age (about 2 million years). By

The throne room of the Minoan palace at **Knossos**. Dated to the 15th century BC, the throne itself is made of alabaster. Decoration in the throne room is typical of the period. Minoan fresco artists revelled in depicting flora and fauna; here the creatures displayed are wingless griffins. The frescos are copies of the originals which were in position when the room was excavated. Craftsmen and artisans were accorded favourable status in Minoan society.

John **Knox**, an engraving from *Icones* (1580) by Theodore Beza. A scholar and teacher by inclination, Knox became the driving force behind the Scottish Reformation.

1.6 million years ago, *Homo erectus* with advanced Oldowan tools called the Karari Industry appeared; *Acheulian tools came a little later.

Köprülü, an influential family of able administrators who through the office of *vizier dominated *Ottoman affairs for half a century. Muhammad (1656-61) crushed internal discord and bolstered the war effort against Venice. His son Ahmed (1661-76) ended the conflict successfully, acquiring Crete, and also won Podolia from Poland. Mustafa (1689-91) died in a counter-offensive in the lengthy war (1683-99) against Austria. Hüseyn (1697-1702) ended that conflict by the Peace of Carlowitz (1699), ceding many territories. By their energy the Köprülü had arrested the decay of the empire, but their demise saw it enter its long decline from greatness.

Koran, the holy book of *Islam, composed of the revelations which came to the Prophet *Muhammad during his lifetime from c.610 to his death in 632. Written in Arabic, the revelations are grouped into 114 units which are known as suras. They touch upon all aspects of human existence, from the doctrinal messages of Muhammad's early career in *Mecca to those concerning social organization and legislation, which were communicated while the Muslim community was based in *Medina (622-30). Considered to be the direct word of God, the Koran is held by Muslims to be untranslatable, although versions or interpretations in many other languages are available. Memorization of the Koran forms the basis of a traditional Islamic elementary education.

Korea, a country in north-eastern Asia. It is deeply indebted to Chinese culture but its people are racially and linguistically distinct. The earliest Korean state was Choson (Morning Calm) (3rd century BC) in northern Korea. Chinese influence, already strong, increased following conquest and colonization by the Western *Han. Meanwhile the states in southern Korea acted as a cultural bridge between the mainland and Japan. In the 1st century AD the kingdom of Koguryo appeared in the north and by the 4th century Buddhism had reached Korea. Koguryo repulsed Chinese attacks but was overrun when the Silla kingdom, established in the south c.350, allied itself with the *Tang (668). The Silla became a tributary state of China and ruled a unified Korea for 200 years. A period of civil war ended with the supremacy of the *Koryo kingdom (918-1392). After Mongol incursions the Koryo allied with them, but following the fall of the *Yuan dynasty in 1392, Yi Song-gye, supported by the *Ming, seized power.

Under the Yi dynasty (1392-1910) Korea was greatly influenced by Ming China. A new capital was built at Seoul, and Confucianism largely displaced an increasingly degenerate Buddhism. From the late 15th century the administration was weakened by factional disputes. The Japanese invasion, led by *Hideyoshi in 1592 took six years to repel and further weakened the state. Shortly thereafter the *Manchus invaded and in due course Korea became a vassal state of China. The Yi dynasty continued only through the support of the *Qing. Unwilling to consider change they were more devoted to traditional Confucianism than the Chinese themselves.

Koryo, a Korean kingdom that gave its name to the whole country. From 986 its kings ruled a united

Kublai Khan, wearing a splendid black and white fur cape, leading his handsomely dressed band of Mongol huntsmen across the desert. A c.14th-century painting on silk attributed to Lui Guandao. (National Palace Museum, Taipei)

Korea from their Chinese-style capital, Kaesong. Chinese influence was strong in the administration of the kingdom and Buddhism flourished. A period of disorder in the 12th century was checked after 1196 by military families with powers similar to those of the Japanese *shogunate. Tributary to the *Song, Koryo also had to pay tribute to the *Liao and *Jin. After 1231 the Mongols repeatedly invaded and despoiled Koryo, which later depended entirely on *Yuan support. After the overthrow of the Yuan, a Koryo general, Yi Song-gye, seized Kaesong and in 1392 established the Yi dynasty.

Kosciuszko, Tadeusz Andrzej Bonawentura (1746-1817), Polish soldier and statesman. He volunteered for service during the American War of *Independence after serving for two years in the Polish army. He distinguished himself at *Yorktown and was rewarded with American citizenship and a pension. He returned to Poland and played a leading part in the war of 1792.

He then spent three years in Leipzig before returning to lead the Polish forces against the powers attempting further partition of Poland. He was defeated and wounded in October 1794. He moved to Paris four years later but refused Napoleon's offers of command. He spent the final years of his life in Switzerland.

Kossovo (or Kosovo Polje, the Field of Blackbirds), the site (in what is now Yugoslavia) of a decisive battle (28 June 1389) when the troops of the *Serbs were defeated by an *Ottoman force. The Serbs enjoyed early success and Sultan *Murad I was killed before battle commenced. His son *Bayezid took command, and won the battle. Victory opened the way for Turkish invasion of central Europe. A second battle, between Hungarians and Turks took place nearby in 1448 at which *Murad II defeated the Hungarians.

Kublai Khan (1214–94), first *Yuan Emperor of China (1279–94), a grandson of *Genghis Khan. Elected Khakhan (Great Khan) in 1260 on the death of his brother Mangu, he completed the conquest of China. In 1276 Hangzhou fell and in 1279 the Song fleet was defeated off southern China. Contrary to Mongol practice, he did not entirely lay waste the country. In 1267 he made Khanbaligh (Beijing) the Mongol capital. His rule, according to Chinese historians, was harsh, though *Marco Polo, who served him from 1275 to 1292, described it as mild. He followed Chinese precedents of government, though civil service examinations were temporarily abandoned. Many foreigners, particularly Muslims from Central Asia, were appointed to civil and military posts. A curfew and much spying enforced obedience. The Grand Canal, linking north with central China was reconstructed, post roads were built, and food stored for periods of shortage. Zhangzhou (Zayton), exporting silk and porcelain, was the busiest port in the world. His military adventures in Annam, Champa, Burma, and Java had limited success. Campaigns against Japan (1274 and 1281) failed dismally.

Kurdistan, homeland of the Kurds, a pastoral people. It is centred on the Zagros and Taurus mountains and is at present divided between Turkey, Iraq, Iran, Syria, and Soviet Armenia. The area was conquered by the Arabs in the 7th century and converted to Islam and became successively part of the *Seljuk, *Mongol, and *Ottoman empires. The Kurdish Shaddadids ruled Armenia from 951 to 1174 and the Ayyubids, a dynasty established by *Saladin, ruled Egypt and Syria from 1169 to 1250.

Kush (or Cush)   *Nubians.

Labourers, Statute of (1351), a statute passed after a large part of the English population had died of the *Black Death. It followed an ordinance of 1349 in attempting to prevent labour, now so much scarcer, from becoming expensive. Everyone under the age of 60, except traders, craftsmen, and those with private means, had to work for wages which were set at their various pre-plague levels. It was made an offence for landless men to seek new masters or to be offered higher wages. The statute was vigorously enforced for several years and caused a great deal of resentment; it was specifically referred to in the *Peasants' Revolt of 1381.

Laetoli, a site in northern Tanzania some 40 km. (25 miles) south-west of *Olduvai Gorge, where a human-like trail of footprints was found. With a date of 3.75–3.6 million years ago, they are the earliest known human footprints and show that upright stance and walking on two legs were developed at this early time in human evolution. A small collection of *australopithecine hominids, mostly jawbones and teeth, is also known from Laetoli. These hominids, along with those from *Hadar in Ethiopia, have been attributed to a new species, *Australopithecus afarensis*.

Lafayette, Marie Joseph, marquis de (1757–1834), French soldier and aristocrat. He was a youthful enthusiast for the American War of Independence and joined the struggle there in 1777 as a volunteer. He was

The French soldier and aristocrat, the marquis de **Lafayette** in military uniform, an engraving made at the time of the French Revolution which he actively supported in its early days. (Bibliothèque Nationale, Paris)

befriended by *Washington and fought at *Brandywine, endured *Valley Forge, and fought again at Monmouth (1778). In 1779 he returned briefly to France, where his wealth and court connections won the promise of naval reinforcements. In command in Virginia in 1781 he helped to besiege *Yorktown.

He was an early supporter of the French Revolution in the *States-General and the *National Assembly, but his liberal humanist republicanism was swept aside by radical *Jacobinism. He fled in 1792 and was imprisoned by the Prussians and Austrians. He retired to private life under Napoleon but was active as a liberal spokesman under the restored monarchy (1815–30). A lifelong advocate of liberty, endowed with great charm, his visit to America (1824–5) brought him fame and popularity.

laissez-faire, a phrase denoting government abstention from interference with individual action. '*Laissez faire et laissez passer*' was the maxim of the physiocrats (economists) in 18th century France who argued for non-interference of the state in economic matters, individualism as opposed to collectivism, and free trade. The policy of non-interference was advocated by Adam *Smith and was applied in 19th-century Britain.

lake dwelling, a house of the kind built since prehistoric times either beside or over shallow water in lakes, rivers, and swamps. Probably not all were raised on piles, as in Thailand, Borneo, and elsewhere today; some may have been on man-made mounds on or near the shore. Even so, the site gave access to water for drinking, washing, refuse disposal, fishing, fowling, transport, and defence, and the wet land was good for some crops. Such sites were used on all continents and round the lagoons of the Pacific islands. In the British Isles they are called crannogs; an example is a 1st-century village excavated near Glastonbury. Other examples were lake-side villages in *Neolithic Switzerland and *Tenochtitlán.

Lake George, battle of (8 September 1755). An engagement in the *French and Indian War fought 80 km. (50 miles) north of Albany, New York. New England and New York militia and Iroquois commanded by William *Johnson defeated a French force and halted their advance. This victory offset *Braddock's defeat of the summer.

Lally, Thomas Arthur, comte de, baron de Tollendal (1702–66). He was a French general, the son of an Irish Jacobite, Sir Gerard O'Lally. From 1756 he was an able commander of the French army in India during the *Seven Years War. He had to give up the siege of Madras through lack of supplies, and surrendered Pondicherry after being defeated by the British commander Eyre Coote in January 1760. This put an end to the French empire in India, and on his return to France he was tried for treason and executed after two years imprisonment. *Voltaire worked with his son to get him posthumously vindicated, and on his death-bed recorded his pleasure that their efforts had succeeded, for the condemnation was declared unjust in 1778.

Lamaism, a form of *Buddhism practised in *Tibet and *Mongolia. (Lama, 'Superior One', is the name given to its higher clergy.) It is a fusion of Bon, the native animist religion of Tibet, with Mahayana Buddhism introduced from north-west India in the 8th century by the scholar Padmasambhava. In times of disorder monks built fortress monasteries and by the 13th century spiritual and temporal power had fused. In the 14th century the Red Sect, whose lamas wore red robes, was discredited though not entirely displaced by the new Yellow Sect. Red Sect lamas were not celibate, sons succeeding fathers as abbots of monasteries. The Yellow Sect, by contrast, demanded of its lamas a life of celibacy and poverty. The head of the Yellow Sect, the Dalai ('All-Embracing') Lama, based in Lhasa, became Tibet's priest-ruler. Below him was the Panchen Lama, based in Shigatse. When either of these lamas died a child believed to be his reincarnation succeeded him.

Lambert, John (1619–83), English major-general. He rose to prominence as a *Roundhead officer during the *English Civil War. He accompanied *Cromwell as second-in-command on the invasion of Scotland (1650). He entertained high political ambitions and was chiefly responsible for drafting England's first written constitution, the *Instrument of Government* (1653). He supported Cromwell loyally (1653–7), but then resigned all his commissions when his own path to power seemed blocked. In 1662, after the *Restoration, he was tried for treason, and spent the rest of his life in captivity.

Lancastrian, a descendant or supporter of John of *Gaunt, Duke of Lancaster. The Lancastrians held the throne of England as *Henry IV, V, and VI (their badge was a red rose). In the Wars of the *Roses, a series of battles for the throne fought with the *Yorkists from 1455 onwards, the Lancastrians suffered a major reverse in the displacement of Henry by *Edward IV in 1461; they took refuge in France, and under the leadership of *Margaret of Anjou invaded England and succeeded in restoring Henry to his kingdom in October 1470. Henry's rule was ended after a few months by the Yorkist victories at *Barnet and *Tewkesbury; most of the remaining Lancastrian leaders died in the latter battle. Yet the Lancastrian party was ultimately successful, for it then supported Henry Tudor, who in 1485, by his victory at *Bosworth Field, became king as *Henry VII.

Lanfranc (c.1010–89), scholar, teacher, and Archbishop of Canterbury (1070–89). He was born in Italy, and set up a school at Avranches, Normandy (1039). He studied as a monk at the abbey of Bec, Normandy (1042), becoming its prior (1046) and making it into one of the finest schools in Europe, whose pupils included *Anselm and Theobald, both future archbishops of Canterbury. Lanfranc's association with *William I began with his negotiation of papal approval for William's marriage while he was Duke of Normandy (1053) and continued after the conquest of England. Lanfranc sought to reform the English church and to unite it under Canterbury, but he also recognized the king's right to intervene in church affairs. He supported *William II in the rebellion of 1088.

Langton, Stephen (d. 1228), Archbishop of Canterbury (1207–28). He was one of the main figures in the drafting of *Magna Carta (1215). Langton's appointment as archbishop was bitterly opposed by King *John, who agreed to it (1213) only after England had been placed under papal interdict. Langton sought to mediate between

John and his barons but was suspended by Pope Innocent III and summoned to Rome to explain his actions. He returned to Canterbury in 1218 and was responsible for a revision of the relations between church and crown.

Languedoc, a former province in south-east France. It was colonized by the Romans, and overrun by Visigoths in the 5th century, and later settled by the *Carolingians. It suffered during the *Albigensian Crusade and passed to the French crown in 1271. Traditional local independence, language (*langue d'oc*), and culture survived, and its *Parlement*, founded in 1443 was second in importance only to that of Paris. A 16th-century Protestant stronghold, its towns fostered the *Camisard rebellion.

Lansdowne, William Petty Fitzmaurice *Shelburne.

L'Anse aux Meadows, an important archaeological site in northern Newfoundland, the only confirmed *Viking settlement in the Americas. Excavations have uncovered foundations of turf-walled houses unlike any dwellings built by indigenous cultures, but identical to Viking structures in Iceland and Greenland. Norse construction is proved by associated finds, including a spindle-whorl, a stone lamp, a bronze ring-headed pin, iron nails, and a smith's hearth. Radio-carbon datings from the structures range from AD 700–1080, concentrating in the late 10th century, thus corresponding to the period of Norse *Vinland sagas. Identification with *Leif Ericsson, however, is purely speculative.

Laos, a country in south-east Asia. It has a variety of inhabitants, some akin to the *Khmers, others, the Lao, akin to the Thais. The Lao, originating in southern China, were driven south by *Mongol pressure. In 1354, following a period of Khmer rule, they set up the Buddhist kingdom of Lanxang ('Million Elephants') which for a time was very powerful. Harassed by neighbours and by its tribal minorities, Lanxang in 1701 split into Vientiane in the north and Luang Prabang in the south. By the early 19th century there were several states in what had been Lanxang.

Laozi (Lao-tzu, 'Master Lao'), a probably mythical Chinese philosopher, long honoured in China as the founder of *Taoism. He is said to have worsted *Confucius, reputedly his junior, in debate. The *Daodejing* (*Tao Te Ching*), which dates from about the 3rd century BC (about 300 years after Confucius), was attributed to him. Taoists later claimed he was an immortal who left China for India, where he converted the *Buddha to Taoism.

La Rochelle, a port in south-west France. As part of the county of Poitou it was controlled by the kings of England (1154–1224) until it was captured by the French. It was a *Huguenot stronghold in the *French Wars of Religion, and was besieged by Catholic forces in 1573. Attacked again by Cardinal *Richelieu (1627–8), it was the scene for a disastrous English intervention under the Duke of *Buckingham, (1629). Its inhabitants were finally besieged and starvation forced the Huguenot surrender. Decline followed the Revocation of the Edict of *Nantes and its transatlantic trade suffered from the French loss of Canada in 1763.

La Salle, Robert Cavalier, sieur de (1643–87), French explorer in North America. He went to Canada in 1666 and, after exploring the Great Lakes, in 1682 he descended the Mississippi River to its mouth, took possession of the whole valley, and named it Louisiana in honour of his king, *Louis XIV. On his return to Paris he was given authority to govern the whole region between Lake Michigan and the Gulf of Mexico. He set out with four ships for the Gulf but could not find the mouth of the Mississippi. The ships separated and he and his men landed on the Texas shore and attempted to find the way overland. Eventually the men became mutinous and murdered him, and it was not until twelve years later that the French established a settlement in Louisiana.

Las Casas, Bartolomé de (1474–1566), Spanish missionary priest, the 'Apostle of the Indies'. He was a Dominican friar who criticized the conquest and exploitation of the Indians of the Spanish colonies. He had himself participated as a settler in *Hispaniola in 1502 and as a member of the expedition to Cuba in 1511–12. His change of heart came in 1514, and for the next fifty years he travelled between the colonies and Spain advocating humanitarian reform, especially the abolition of Indian slavery, and writing books arguing the equality of Indians as subjects of the king. The most famous work was his *Brief History of the Indies* (1539). In 1542 his campaigning led to the New Laws to protect Indians in Spanish colonies. Their effectiveness, however, was limited by the opposition of the conquistadores.

La Tène, an archaeological site near Lake Neuchâtel, Switzerland, which has given its name to the Celtic culture of the late *Iron Age. It began c.450 BC,

A bronze openwork harness plaque of the **La Tène** culture, with typical circular compass decoration. Both abstract and realistic flower patterns can be seen in the solid and cut-away parts of the plaque, which was discovered in a French chariot grave dating from the 5th century BC. (British Museum, London)

superseding *Hallstatt, when the Celts came into contact with Greek and Etruscan civilization. It lasted, with various developments, until the 1st century BC, when most of the Celts came under the aegis of the Roman empire. The characteristic style was established early— s-shapes, spirals, and circular patterns—with clear debts to Greek and Etruscan motifs. Its finest work is a remarkable mixture of abstract and figurative, of animal and vegetable representations. The many finds bear witness to the warlike nature of the Celts—long iron swords, decorated scabbards, belts, shield bosses, and the like. Nevertheless, archaeology has also revealed more domestic artefacts—hammers, sickles, and plough shares.

Lateran Council, one of the synods or meetings of senior churchmen held at the Lateran Palace in Rome. The First in 1123 was to bring to an end the *investiture contests and to specify which aspects of life should be governed by the church and which by the secular authorities. The Second was called in 1139 to clarify doctrine and to heal the schism which had been caused by the activities of the antipope Anacletus II. In 1179 the Third Council condemned *simony, and regularized papal elections.

The Fourth Lateran Council (1215) is known as the 'Great Council' and was called by Pope *Innocent III. It condemned the *Albigensian heresy and clarified church doctrine on the Trinity, the Incarnation, and transubstantiation. The Fifth Council from 1512 to 1517 condemned heresy and additionally revoked the *Pragmatic Sanction of Bourges by which Charles VI of France had claimed authority over the church.

Latimer, Hugh (c.1485-1555), English Bishop of Worcester. He was appointed to his bishopric in 1535 after *Henry VIII had broken with the pope, but resigned over the passage of the Act of Six Articles (1539) which limited the introduction of *Reformation doctrine. Best known for his vigorous Protestant preaching, he also spoke out against social injustice during *Edward VI's reign. On *Mary I's accession (1553), he refused a chance to escape abroad before being arrested for his 'heretical' views. Along with *Cranmer and Nicholas Ridley (1500-55), Bishop of London, he was made to take part in a theological disputation at Oxford in 1554. Condemned for heresy, he, Cranmer, and Ridley were burned at the stake in Oxford (1555), providing the Anglican Church with three of its most celebrated martyrs.

Latvia, a region on the Baltic Sea. Originally inhabited by Lettish peoples, it was overrun by the Russians and Swedes during the 10th and 11th centuries, and settled by German merchants and Christian missionaries from 1158. In the 13th century the *Hanseatic League forged commercial links, while the *Teutonic Knights and German bishops imposed feudal overlordship. With Estonia it became part of Livonia in 1346, was partitioned under *Ivan IV of Russia, and came under Lutheran influence in the *Reformation. It then fell to the Poles, and in the 17th century to the Swedes. Their rule lasted till 1721, when parts again reverted to Russia, the remainder succumbing in the partitions of *Poland.

Laud, William (1573-1645), Archbishop of Canterbury. The son of a Reading clothier, he was ordained in 1601. He became president of St John's College, Oxford

A portrait of William **Laud** after Sir Anthony van Dyck (c.1636). (National Portrait Gallery, London)

in 1611, and attached himself to the party of *Buckingham at court. From early in his career he sought to return to some pre-Reformation practices in the Anglican Church rituals. Promotion to the bishopric of St David's (1621) and further preferment followed, despite James I's fears about his 'restless spirit' and the antagonism that his doctrinal attitudes would arouse. He did not adhere to the prevailing Calvinist belief in Predestination and was suspected of wanting to promote Catholicism. As Bishop of London (1628), he became one of Charles I's closest advisers, and as archbishop (1633), he supported Charles's absolute rule. A considerable opposition party developed as Laud and *Henrietta Maria were suspected of plotting together. Resistance came first from Scotland, where his attempt to 'Anglicanize' the Kirk (Church of Scotland) provoked the *Bishops' wars (1639-40). Arrested and impeached by the Long Parliament, he was found innocent of treason by the Lords, but then executed by Act of Attainder. His dogmatic policies were instrumental in precipitating the *English Civil War.

Law, John (1671-1729), Scottish financier based in France. He was an exiled Scotsman who believed that increased circulation of paper money and proper organization of credit would bring prosperity. The Regent of France, Philippe, duc d'*Orléans, facing financial crisis, allowed him to set up a state bank and a trading company, the 'Compagnie du Mississippi', to trade with *Louisiana, and later gave it a monopoly of overseas trade. In January 1720 Law became Controller-General of Finance, but in December 1720 a wave of speculation brought his system to an end. The episode, which brought fortune to a few and ruin to many, is similar to the *South Sea Bubble.

Lebanon, a south-west Asian country with a coast on the Mediterranean. Much of it formed part of *Phoenicia, including the important trading towns of Tyre, Sidon, Byblos, and Arvad, which retained their importance under Roman rule. Mount Lebanon was a refuge for persecuted minorities such as the Christian Maronites, who settled there from the 7th century AD, and the Muslim Druze, who occupied the southern part of the mountain from the 11th century. After the Arab conquest during the 7th century Arab tribesmen settled in Lebanon. Successive governments in the region usually left the people of the mountain to manage their own affairs and contented themselves with exercising authority on the coastal plain. During the course of the late 18th century the balance in Lebanon was upset by the expansion of the Maronite community, its educational development, its tighter ecclesiastical organization, and its political hegemony on the mountain.

Lechfeld, battle of (955). A site near Augsburg, in Germany, was the scene of a major battle when the forces of the Holy Roman Emperor *Otto I defeated the *Magyars, putting an end to their westward expansion and the long period of harassment which they had inflicted on the German empire.

Legions   *Roman legions.

Leicester, Robert Dudley, 1st Earl of (c.1532–88), courtier and favourite of Elizabeth I. The queen honoured him with offices and lands, and it was rumoured that he hoped to marry her. However, the convenient death of his wife, Amy Robsart, in mysterious circumstances, tarnished his attempted wooing. He was given command of a military force to the Netherlands in 1585, to aid the Dutch in their revolt from the rule of Spanish Catholic monarchs. He was not an able commander and was recalled to organize land troops to fight the Spanish Armada, but died the same year.

Leif Ericsson   *Eric the Red.

Leisler, Jacob (1640–91), American rebel, who migrated from Germany to *New Amsterdam in 1660. Leisler and other merchants resented English control of the colony from 1664. As a militia officer in the *Glorious Revolution in New York, he led the humbler residents against both *James II and the aristocratic patrons (lords of the manor) and assumed the governorship in 1689. When Governor Slaughter arrived from England in 1691 Leisler resisted and was captured; he was tried for treason and hanged. His execution precipitated factional conflict for a generation.

Lenclos, Anne (called Ninon de Lenclos) (1620–1705), famous French courtesan. She was clever and well-read and delighted in conducting philosophical discussions in her Paris salon which was frequented by men of letters including Racine and Molière. She was a disciple of Montaigne and a professed free-thinker in religious matters.

Leo I, St (the Great) (c,390–461), Pope (440–61). He established the authority of the papacy by defending orthodoxy in regions far distant from his authority in Italy, most notably in Spain, Gaul, Africa, and the East.

The *Pelagians and *Manicheans were particular threats. At *Chalcedon in 451 he obtained agreement between Eastern and Western churches on defining for Christian believers the relationship of God the Father to the Son. The following year he persuaded the barbarian *Attila to leave Italy. In 455 he saved Rome from *Vandal destruction.

Leo III (c.750–816), Pope (795–816), succeeding Hadrian I, who had tried to maintain papal independence by supporting the Byzantine emperor against the power of *Charlemagne. Leo reversed this policy, acknowledging the temporal suzerainty of Charlemagne. Accused of perjury and adultery by Hadrian's relatives, he was waylaid and imprisoned but escaped to Charlemagne at Paderborn. After sending him back safely, Charlemagne went to Rome (800), heard the accusations, and acquitted him. On Christmas Day in St Peter's, apparently unexpectedly, Leo crowned Charlemagne as Emperor of the West, establishing the precedent that only a pope could crown an emperor.

Leo IX, St (1002–54), Pope (1048–54). An able church reformer, he enlisted like-minded churchman to assist him including Hildebrand (later Pope *Gregory VII). His chief concerns were *simony and clerical celibacy; at the Easter synod of 1049 celibacy was enforced on all clergy. Attempting to establish papal control in southern Italy, Leo's forces were defeated by the Normans and he was made prisoner. His interference in south Italy in areas claimed by the *Byzantine empire led to the *East–West schism of 1054 when the Patriarch of Constantinople

A portrait of Anne **Lenclos**, known as much for her intelligence as her beauty, by an unknown 17th-century artist. (Musée de Versailles)

was excommunicated. He died soon after being released from prison and, for his work in restoring the prestige of the papacy, was declared a saint.

León, a province in northern Spain, once an independent kingdom. It was captured from the Iberians by the Carthaginian general *Hannibal in 217 BC and later became the Roman province of Lusitania. It was conquered by the *Visigoths in the 5th century and fell to the *Moors in the 8th century. It was freed from Muslim control by the Asturians and united with *Asturias and Galicia in the 10th century. *Ferdinand I of Castile seized it and it was finally united with Castile in 1230. The armies of the kingdom were important in the reconquest of Spain from the Moors.

Leonidas (d.480 BC), King of Sparta. He won immortal fame when he commanded a Greek force against the invading Persian army at the pass of *Thermopylae. He held the pass long enough to make possible the naval operation at Artemisium (*Greek–Persian wars). When counter-attacked he remained behind with 300 Spartans and 700 Thespians, and died fighting—an action which allowed his allies to escape.

Leopold I (1640-1705), Holy Roman Emperor (1658-1705). Originally intended to enter the Church, he was ill-equipped to rule, and relied heavily on his ministers and generals. His long reign nevertheless saw a major revival of Habsburg power, particularly after the *Ot-

The battle of **Lepanto** (1571) by an unknown artist. The Turks had over 270 galleys while the Holy League fleet under Don John comprised 200 oared galleys and six Venetian galleasses, which had both oars and sails. The galleasses with their heavy broadside guns and concentrated fire contributed significantly to the League's victory. (National Maritime Museum, London)

toman attack on Vienna in 1683 was repulsed by an army led by King John *Sobieski of Poland. The subsequent eastern campaigns brought the reconquest of Hungary, confirmed by the Peace of Carlowitz (1699). Increasing resentment at *Louis XIV's intervention in German affairs also allowed him to re-establish imperial leadership in Germany, an important contributory factor to the coalitions which inflicted heavy defeats on France between 1689 and 1713. After the removal of the Ottoman threat Vienna became a major European capital.

Lepanto, battle of (7 October 1571), a great sea-battle near the northern entrance to the Gulf of Corinth. On one side were the *Ottomans, who were seeking to drive the Venetians out of the eastern Mediterranean. On the other were the *Holy League forces of Venice, Spain, Genoa, and the papacy, under Don *John of Austria. Despite the Ottomans' superior number of *galleys, the League won the battle. Lepanto was the last naval action fought between galleys manned by oarsmen and the first major Turkish defeat by the Christian powers, but it was not followed up, and had little long-term effect on Ottoman power.

Levant, a general term for the countries bordering the Mediterranean from Egypt to Turkey, sometimes more particularly applied to Syria and Lebanon. It has been a trading coast since Phoenician days, and saw the main confrontations between Muslims and Christians at the time of the *Crusades. The Levant Company was an English *chartered company with a monopoly of trade with the Levant region. It was formed in 1592, on the amalgamation of the Turkey Company and the Venice Company. Cloth was the staple item of trade from England; currants, silk, oils, wines, and cottons came from the Levant. The *Anglo-Dutch wars (1652–74) disrupted the trade, as did later competition from the *East India Company. It ceased trading in 1821.

The Retreat

From Concord to Lexington of the Army of Wild Irish Asses Defeated by the Brave American Militia

A cartoon of the retreat from the battle of **Lexington and Concord**. The British are shown as the 'Wild Irish Asses' of the caption to the drawing, burning houses and barns and making off with plunder. Their mounted commander was Major John Pitcairn (second from right), later killed at Breed's Hill.

Levellers, English radicals of the mid-17th century. The Levellers were led by John *Lilburne, William Walwyn, and John Wildman, and their early strength lay with the London poor. By 1647 they had won considerable support among the lower ranks of the *New Model Army. In that year Leveller 'Agitators' were elected from each regiment to participate in the Putney debates, with *Cromwell and the Army Grandees, in an attempt to resolve disagreements. Their political programme, embodied in documents like the *Agreement of the People*, was less radical than that of the True Levellers or *Diggers. It demanded the abolition of the monarchy and House of Lords, sovereignty for the people, manhood suffrage, social reform, liberty of conscience, and equality before the law. Exasperated by the conservatism of the Grandees and Parliament, they mutinied in 1647 and 1649. By May 1649 both the civilian and military wings of the movement had been broken.

Lexington and Concord, battle of (19 April 1775), the first engagement of the American War of *Independence. When General *Gage learnt that patriots were collecting military stores at Concord, 32 km. (20 miles) north of Boston, he sent a force of about 800 men to confiscate them. Forewarned by Paul Revere and others, the troops were met by 70 militia (the *minutemen) at Lexington. It is not known who fired the first shot, but in the ensuing skirmish eight minutemen were killed. The British marched on to Concord and confiscated some weapons, but retreated to Boston the same day, harried by patriots, who inflicted over 250 casualties on them.

Liao (945–1125), a dynasty which ruled much of *Manchuria and a small part of north-east China. It was founded by the Qidan (Khitan) tribesmen of Tungus stock, whose homeland was around the Liao River in Manchuria. From their name came the Russian *Kitai* and Marco Polo's Cathay. Qidan strength lay in cavalry. In 1004 their frontier with the Northern *Song was stabilized. Gradually they adopted Chinese habits, accepting the teachings of *Confucius, holding civil service examinations, and performing the customary Chinese rites. Overthrown by the Juchen, nomad huntsmen who founded the *Jin dynasty, some of them migrated westward to become the Kara Khitai of Central Asia.

liberty, an area or individual enjoying a special privilege of freedom from royal jurisdictions. Liberties of all kinds abounded in the Middle Ages: a ruler could grant privileges to people or places, and these would then be enforceable for ever, cutting across other laws and customs. In England the word 'liberty' was usually taken to mean a territorial area held by some lay or ecclesiastical magnate; most of these liberties dated back to Anglo-Saxon times. The greatest liberties were the palatinate franchises of the bishops of Durham and the dukes of Lancaster.

Libya, a North African country consisting of the provinces of Cyrenaica, *Tripolitania, and Fezzan. During most of its history it has been inhabited by Arab and *Berber nomads, only the coastlands and oases being settled. Greek colonies existed in ancient times, and later under the Romans; under the Arabs the cultivated area lapsed into desert. The region came under Turkish rule in 1561, but their writ hardly ran beyond the coast, which served as a *corsair base.

Lilburne, John (*c.*1614–57), English political radical, the leader of the *Levellers. He was gaoled for smuggling Puritan pamphlets into England in 1638. Released by

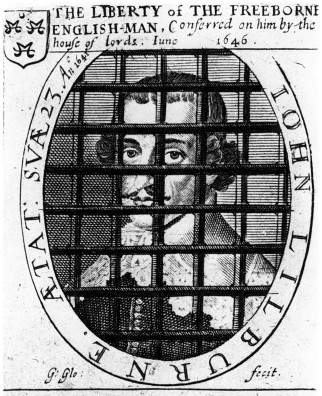

THE LIBERTY of THE FREEBORNE ENGLISH-MAN, *Conferred on him by the house of lords: Iune 1646*.

IOHN LILBURNE

ÆTAT. SVÆ 23. An 1641

G: Glo: fecit.

Gaze not vpon this shaddow that is vaine,
But rather raise thy thoughts a higher straine,
To GOD (I meane) who set this young-man free
And in like straits can eke deliuer thee.

A portrait of John **Lilburne**, leader of the Levellers, taken from a propaganda pamphlet of 1646 published by Richard Overton. Although dubbed 'Freeborn John', Lilburne spent much of the period 1645–57 in prison, as the satirical title and addition of bars to this engraving imply.

the Long Parliament, he fought as a *Roundhead in the English Civil War, rising to the position of lieutenant-colonel, but resigned in 1645 in opposition to the *Solemn League and Covenant. 'Freeborn John' always spoke and wrote about the rights of the people, rather than those of kings or parliaments. He attacked in turn all constituted authorities, and suffered frequent imprisonment for his opinions. He became a member of the *Quakers shortly before he died.

livery company, one of the London city companies which replaced the medieval craft *guilds, so called on account of the distinctive dress worn by their members. The liverymen constituted the freemen of the City of London, indirectly responsible for electing the mayor as well as the aldermen, while several of their companies, such as the Goldsmiths and Merchant Taylors, played an important role in the regulation of their trade and had monopoly powers within London. Since the 17th century there have been nearly 100 London livery companies but few have retained any importance other than as social and charitable institutions.

Livonia, a historic region of Russia, corresponding to nearly all of modern Latvia and Estonia. The Livs or

Livonians were a Finno-Ugric Baltic tribe, absorbed by the Letta and Estonians during the Middle Ages. The Knights of the Sword (the Livonian Order) conquered and Christianized the region during the 13th century. They subsequently emerged as the dominant estate within the Livonian confederation, which also consisted of ecclesiastical states and free towns. By 1558 rivalries within the order had combined with peasant discontent and religious disunity to render Livonia vulnerable to foreign intervention.

The Livonian War (1558–83) was a protracted struggle between Russia, Poland, Sweden, and Lithuania for control of the territory. (It overlapped with the *Seven Years War of the North.) By the peace treaties of 1582–3, *Ivan the Terrible of Russia renounced his claims, thus acquiescing in the seizures of Livonian land by Russia's three rivals; but after the Great *Northern War (1700–21), Russia gained Sweden's considerable holdings in the region, and then absorbed the Polish sections as a result of the 18th-century partitions of Poland.

Livy (Titus Livius) (BC 59–17 AD), Roman historian. He came from Padua to Rome where he was tutor to *Claudius. He wrote a history of Rome in 142 books; 35 survive in full, the rest in summary and fragments. He narrated dramatic history following the *annals, reconstructing speech and blending fact with tradition and legend. In part his work was a eulogy of the return to Roman ideals and values in the 'new age' of *Augustus.

Llywelyn (the Great) (d. 1240), Prince of Gwynedd (north Wales), the most powerful ruler in medieval Wales, his authority over other Welsh leaders being confirmed by the Treaty of Worcester (1218). Although married (1205) to Joan, the illegitimate daughter of King *John, Llywelyn took advantage of the political confusion in England to extend his influence over South Wales. He also had close ties with the *marcher lords.

Llywelyn ap Gruffydd (d. 1282), Prince of Wales. He encouraged the further development of feudalism in Wales and was recognized as Prince of Wales by *Henry III in the Treaty of Montgomery (1267). Forced into two disastrous wars (1277 and 1282) against *Edward I, Llywelyn was killed near Builth.

Locke, John (1632–1704), English philosopher, one of the most influential of English political thinkers. A friend of *Shaftesbury, he fell under suspicion of treason and fled to Holland in 1683, returning after the *Glorious Revolution (1688), of which he became the principal theoretical defender. He was employed from time to time by *William III, but spent his last years in writing and in defending his often controversial views.

As a philosopher he was an empiricist, insisting in his *Essay Concerning Human Understanding* (1690) that knowledge can be based only upon human experience, and that there is no innate or revealed knowledge. He argued vigorously for toleration of dissenters, although not of atheists or Roman Catholics. His political views, in his *Treatise on Civil Government* (1690), insist on the superiority of natural over man-made law, reject the Divine Right of Kings, and exalt the theory of private property. He developed arguments against taxation without representation, and believed in a mixed form of constitution with a balance of power: the makers of the

American *Constitution were much influenced by Locke's ideas. His treatise, *Some Thoughts Concerning Education* (1693), was also influential.

Lodi, a family of Afghan origin whose rule over northern India (1451–1526) marked the last phase of the *Delhi sultanate era. Their founder, Bahlul (1451–89), who already had a strong base in the Punjab, took advantage of *Sayyid weakness to seize power in Delhi. He and his two successors extended power eastwards through Jaunpur to the borders of Bengal and threatened Malwa to the south.

Sikander (1489–1517) consolidated his father's gains but was also renowned as a patron of poets, musicians, and other scholars. However, attempts by his successor, Ibrahim (1517–26), to secure greater centralization alienated many local governors. In retaliation, Daulat Khan, governor of the Punjab, invited the ruler of Kabul, the Mogul prince, *Babur, to invade India. His defeat of the Lodis in the ensuing battle at *Panipat (1526), resulted in the destruction of the dynasty and the establishment of the *Mogul empire.

Lollard, originally a follower of John *Wyclif, the name was later applied vaguely to anyone seriously critical of the Church. Lollards probably owed their name to the Dutch word *lollaerd*, meaning a mumbler (of prayers). Lollardy began in the 1370s as a set of beliefs held by Oxford-trained clerks who were keenly interested in Wyclif's teachings on papal and ecclesiastical authority; in an age unsettled by war and threatened by disease (especially the *Black Death), it also appealed to other educated sectors of society. They attacked clerical celibacy, *indulgences, and pilgrimages. *Richard II, who was himself an opponent of calls for ecclesiastical egalitarianism, none the less retained in his household some knights known to favour Lollardy. The nobility abandoned it only when *Henry IV came to the throne and backed Archbishop Arundel in a vigorous persecution of Lollards; further reaction against it, among the gentry, also resulted from the abortive Lollard uprising attempted by Sir John Oldcastle in January 1414. Thereafter, Lollardy's appeal seems to have been limited to craftsmen, artisans, and a few priests in the larger towns.

Lombardy, a region of central northern Italy, named after the Lombards, a tribe from the Danube region who arrived there in the 6th century having been driven south by the *Avar tribe. Though at first they were allies of the Byzantines, they later captured Ravenna from them. They were converted to Catholicism and formed a powerful kingdom until *Charlemagne conquered it in the late 8th century. By the mid-12th century the cities of the area (e.g. Milan) had become economically and politically powerful and formed the Lombard League to challenge *Frederick Barbarossa's ambitions in the area; he was defeated at the battle of Legnano (1176). Later the region was ruled by Spain, France, and Austria before becoming part of Italy in the 19th century.

London Corresponding Society, an organization founded in London in 1792 by Thomas *Hardy to agitate for universal manhood suffrage. It was the first real working-class political movement to appear in Britain—its members were mainly artisans of humble origin—and it rapidly established contact with similar societies in other

A portrait of the English philosopher John **Locke**, painted c. 1685 and attributed to Sylvester Brownover. (National Portrait Gallery, London)

towns, for example the Sheffield Society of Constitutional Information. The aim of the societies was to circulate letters and pamphlets and to initiate orderly debates on reform proposals. *Paine's writings were a stimulus to the popular societies. When there was talk of a national convention the government, then at war with Revolutionary France, became alarmed. The London Corresponding Society was believed to have been involved in the *Nore mutiny and in 1798 all the Society's committee members were imprisoned without trial and in 1799 the Society itself was suppressed.

Londonderry, a city on the River Foyle, in Northern Ireland. It grew up from the monastery founded by St Columba in 546 and was frequently attacked by the Norsemen. In 1613 it was granted to the city of London for colonization, taking the name of Londonderry and becoming staunchly Protestant. It suffered in the Irish rising of 1641, and in 1688–9 its Protestant defenders were besieged for 105 days by the deposed *James II of England; the city did not give in and its watchword 'No surrender!' dates from the siege.

Long Parliament (1640–60), the English Parliament called by *Charles I after the *Bishops' wars had bankrupted him. Led by John *Pym, by August 1641 it

The expulsion of the Rump from the English **Long Parliament** by Oliver Cromwell in 1653, from a contemporary Dutch print.

had made a series of enactments depriving him of the powers that had aroused so much opposition since his accession. These reforms were intended to rule out absolutism for the future, and were eventually incorporated in the Restoration settlement, and again during the *Glorious Revolution. The Parliament was also responsible for the execution of the king's advisers *Laud and *Strafford. Without its Cavalier members, the Long Parliament sat on throughout the *English Civil War, since it could be dissolved only with its own consent. Serious divisions emerged between the Presbyterian and Independent members, culminating in *Pride's Purge (1648). The remnant, the Rump Parliament, arranged the trial and execution of Charles I, and the establishment of the *Commonwealth (1649). *Cromwell ejected the Rump by force in 1653, but it was recalled after his son's failure as Lord Protector in 1659. In the next year General *Monck secured the reinstatement of those members 'secluded' by Pride. Arrangements for the Convention Parliament were made, and the Long Parliament dissolved itself in March 1660.

Longshan (Lung Shan), the Dragon Mountain of Shandong province in China, which gave its name to the later *Neolithic culture of the lower Huang He (Yellow River) valley. Its dates lay between 2500 and 1700 BC. It developed from the *Yangshao culture and, with bronze beginning to appear, was in turn ancestral to the *Shang civilization. Its economy was based primarily on millet, harvested with polished stone reaping knives, and on pigs, cows, and goats. Its distinctive pottery was the first in the Far East to be made on the fast wheel, and was kiln-fired to a uniform black colour.

longship, a term applied generally to all *Viking ships but properly reserved for warships. They were built usually of fir planks, and differed from the vessels of the Angles, Saxons, and Frisians in having a massive vertical keel of oak instead of a shallow horizontal one; this enabled them to carry a mast and sail. The clinker-built construction of overlapping planks secured by clench nails conferred great strength with flexibility, and the hulls were waterproofed with tar, seams between the planks being caulked with wool and hair. Later examples were over 46 m. (150 ft.) long and could carry hundreds of warriors who were also rowers. Longships were of extremely seaworthy design, and the addition of sails made very long voyages feasible, while the shallow draught meant that raiders could penetrate far inland by river. The violent expansion of the Norse peoples was dependent on the skilful use of such vessels.

Lord Appellant, one of five nobles who 'appealed' (accused) certain of the leading friends of *Richard II of England of treason in November 1387. Thomas of Woodstock, Duke of Gloucester, Richard Fitzalan, Earl of Arundel, Thomas de Beauchamp, Earl of Warwick, Henry *Bolingbroke, Earl of Derby, and Thomas Mowbray, 3rd Earl of Nottingham were opposed to the king's policy of peace with France. His position had been weakening for some months, and he was forced to summon a parliament to be held in February 1388; at this, the 'Merciless Parliament', the Archbishop of York and others of his friends and their associates were accused of treason or were impeached. Four men, including a former Chief Justice, were executed. Richard bided his time, gradually restoring his authority until in 1397 he was able to arrest the surviving Lords Appellant and have them accused of treason.

Lord Ordainer, a member of a committee chosen in March 1310 by the English lords who were opposed to *Edward II. They had forced Edward to agree to the appointment of this committee of twenty-one lords, with full power to reform both his household and realm; he had infuriated many peers by his infatuation with *Gaveston, and they wished to make the crown's officials answerable to Parliament. The resultant ordinances were drawn up by August 1311, and the Ordainers enforced them until Edward won the battle of Boroughbridge (1321) and at the Parliament of York in 1322 annulled all the ordinances and reasserted his authority.

Lords, House of, the upper chamber of the British *Parliament. It began as a form of the medieval kings' Great Council. In the 13th and 14th centuries, as the councils gave way to parliaments, the Lords evolved into a separate body which, together with the House of *Commons, presented bills to the crown for enactment as statutes. The immense individual importance of many peers did not prevent them gradually losing to the Commons the right to levy taxes on the king's behalf, and from the 16th century the Commons were the more powerful political force. The House of Lords was abolished in 1649 and revived in 1660. It was put on what is still its constitutional basis vis-à-vis the crown and the House of Commons by the *Glorious Revolution 1688–9. By the 18th century it had lost its power to amend money bills.

In the late 14th century the Commons developed the process of impeachment, but the Lords have otherwise retained their judicial function, exercised by the Lord Chancellor and Law Lords, of acting as the ultimate court of appeal.

Lord Lieutenant, an English magnate, originally commissioned to muster, administer, and command the militia of a specified district in times of emergency. *Henry VIII was the first to appoint them, and in 1551 during Edward VI's reign there were attempts to establish them on a permanent basis. From 1585 it became usual for every shire to have a lieutenant, and deputy lieutenants, and by the end of the 16th century they assumed additional roles, exercised on behalf of the sovereign, including the appointment of magistrates. They lost their military responsibilities in the army reforms of 1870-1, but still represent the crown in the counties. (The term was also used to denote the British viceroy of Ireland.)

Lorenzo the Magnificent   *Medici, Lorenzo de.

Lorraine, a former duchy in north-eastern France. After the death of *Charlemagne, under the terms of the Treaty of *Verdun in 843 the duchy passed to Emperor Lothair I. As Lotharingia it was divided between Charles the Bald and Louis the German in the Treaty of Mersen of 870. In 911 it passed to the Frankish kingdom, then back to the German in 923 as the duchy of Lorraine. In the 10th century it was divided into Upper and Lower parts of which only the former retained the name, continuing as a duchy until 1736. Stanislaus Leszczynski, formerly King of Poland, was given title to it in 1737, but it reverted to the French crown on his death in 1766.

Louis I (the Pious) (778-840), King of the Franks, German emperor (813-40), a son of *Charlemagne. He administered the empire well but failed to organize his succession, thus jeopardizing the unity of the Frankish kingdom. He had four sons, the youngest by his second wife, and tried originally to settle the empire and its title on the oldest. This plan faltered as father and sons fought for control, causing Louis to lose the throne briefly in 833.

Louis I (the Great) (1326-82), King of Hungary (1342-82), and of Poland (1370-82). He succeeded his father, Charles I in Hungary and his uncle, Casimir III, in Poland. He fought two successful wars against Venice (1357-8, 1378-81), and the rulers of Serbia, Walachia, Moldavia, and Bulgaria became his vassals. Under his rule Hungary became a powerful state, though Poland was troubled by revolts. His daughters Mary and Jadwiga succeeded him in Hungary and Poland respectively.

Louis IX, St (1214-70), King of France (1226-70). He succeeded his father, Louis VIII, and worked effectively to stabilize the country and to come to terms with the English who maintained territorial claims in France. Henry III of England was forced to acknowledge French suzerainty in the disputed region of Guienne. He had a profoundly religious nature and built the Sainte-Chapelle in Paris, to house holy relics brought from Constantinople. Prompted by his recovery from a severe illness he raised the Seventh Crusade, directed against Egypt, and sailed in 1248. After initial successes he was captured by Sultan Turanshah and only released upon payment of a ransom in 1250. His involvement in the Crusades was recounted by *Joinville. He later mounted another Crusade to Tunis where he died. He was canonized by Pope Boniface VIII in 1297, his sanctity conferring immense prestige on the Capetian dynasty.

Louis XI (1423-83), King of France (1461-83). He was exiled for plotting against his father, Charles VII, but succeeded to the throne in 1461. As king he imposed new taxes, dismissed his father's ministers, and attempted to curb the powers of the nobility. They retaliated by

forming a coalition against him which waged the 'War of Common Weal'. *Charles the Bold of Burgundy led a group including the Duke of Brittany, the Duke of Bourbon, and Louis XI's brother, Charles of France, supported by some lesser magnates, clergy, and a few towns. The battle of Montlhéry (July 1465) ended in stalemate, and Louis was able to gain the upper hand after Charles the Bold was defeated by the Swiss in 1477. He pursued a successful policy of territorial acquisition and centralization: by the time of his death only the duchy of Brittany remained largely independent.

Louis XI established firm government but nonetheless bequeathed a troubled legacy at his death. The minority of his son, Charles VIII, saw further outbreaks of noble discontent and attempts by the dukes of Brittany to undermine the monarchy.

Louis XIV (1638-1715) King of France (1643-1715). On his father's death in 1643, his mother *Anne of Austria became regent and *Mazarin chief minister. Louis survived the *Fronde, was proclaimed of age in 1651, and married the Infanta Maria Theresa of Spain in 1660. He took over the government on Mazarin's death in 1661 and embarked on a long period of personal rule.

Domestic policy was aimed at creating and maintaining a system of absolute rule: the king ruled unhampered by challenges from representative institutions but with the aid of ministers and councils subject to his will. The

Louis XIV, the 'Sun King', at the establishment of the Académie des Sciences and the foundation of the Observatory in 1667. Detail from a painting by Henri Testelin. (Musée de Versailles)

States-General was not summoned, the *Parlement* largely ignored, the great nobles were generally excluded from political office, and loyal bourgeois office-holders were promoted. Jean-Baptiste *Colbert expanded the merchant marine and the navy, and encouraged manufacturing industries and trade, though he largely failed in his attempts to improve the tax system. In the provinces the *intendants* established much firmer royal control. The French army became larger and more efficient; in his later years Louis was able to put between 300,000 and 400,000 men into the field. The greatest victories came in the earlier years, when the generals *Turenne and *Condé were available to take command. Victories were won in the War of *Devolution and the Dutch War, with the French frontiers strengthened by a series of strategic territorial gains, reinforced by the fortifications of *Vauban. The *Nine Years War and the *War of the Spanish Succession saw France hard-pressed as Europe united to curb Louis's aggressive policies; after 1700 France suffered a series of crushing defeats. The country was seriously impoverished by the burden of taxation.

Religious orthodoxy was strictly imposed, particularly after the Revocation of the Edict of *Nantes (1685) and the forced conversion of the *Huguenots, at least 200,000 of whom illegally fled the country. Within the Catholic church *Jansenists, Quietists, and other deviants were also persecuted. On the positive side, the achievements of the reign in literature and the arts based on the court at *Versailles have given it the name Le *Grand Siècle*. There was, however, a marked decline in these fields during the later part of the reign, and at his death Louis XIV left a series of political, economic, and religious problems of his great-grandson, *Louis XV.

Louis XV (1710–74), King of France (1715–74), a great grandson of *Louis XIV. During his minority Philippe, duc d'Orléans was regent, followed by Cardinal *Fleury. After Fleury's death in 1743 Louis decided to rule without a chief minister, but he proved to be a weak king who reduced the prestige of the French monarchy both at home and abroad.

At the age of 15 Louis married Marie Leszczynska, daughter of the King of *Poland, and France intervened in the War of Polish Succession, gaining the duchy of Lorraine in 1766. In foreign affairs France was involved in almost continuous warfare; in the *War of the Austrian Succession, in alliance with *Frederick II of Prussia until hostilities were concluded at *Aix-la-Chapelle. The *Seven Years War saw France and Austria fighting Prussia and Great Britain but with little success. The Treaty of *Paris (1763) marked the loss of most of France's overseas territories.

In domestic policy Louis XV was influenced by a succession of favourites and mistresses, including Madame de *Pompadour and Madame *du Barry, on whom he lavished enormous amounts of money. The extravagance of the court and the high cost of war absorbed all of France's resources and efforts to rationalize the tax system failed. The *Parlement* of Paris secured the suppression of the *Jesuits in 1764 but otherwise failed to achieve reforms. The members of the *Parlement* were banished and a compliant *Parlement* appointed in their place in 1771. The reign saw the aristocracy and the wealthy bourgeoisie prosper, though the country was close to bankruptcy. The king's failure to solve his financial affairs left an insolvent government for his successor, *Louis XVI.

Louis XVI (1754–93), the last King of France (1774–92) before the *French Revolution. Weak and vacillating, unwisely advised by his Austrian wife, *Marie Antoinette, he could neither avert the Revolution by supporting the economic and social reforms proposed by *Turgot and *Necker, nor, lacking all understanding of popular demands, become its popular leader. To meet the situation he summoned the largely aristocratic Assembly of Notables (1787) which achieved nothing and then (1789) the *States-General, which had not been called for 175 years. This marked the start of the Revolution. The royal family was forcibly brought back from *Versailles to Paris (October 1789) and their attempt to flee the country was stopped at *Varennes (1791). Thereafter they were virtually prisoners. The monarchy was abolished (September 1792), and Louis was guillotined (January 1793). His wife was executed six months later.

Louisburg, a French fortress on the southern coast of Cape Breton Island, Canada, built in 1720 after the ceding of Acadia (*Nova Scotia). As a threat to New England fishermen and shipping and to Nova Scotia, it was captured by Pepperell's combined operation (1745) during *King George's War. To the disgust of the colonists it was exchanged for Madras in 1748, but it was again taken by *Amherst and *Boscawen in 1758 prior to *Wolfe's attack on *Quebec.

A portrait of Ignatius **Loyola**, Spanish ecclesiastical reformer and founder of the Jesuits. Attributed to Juan de Roelas, it was painted c.1622, the year of Loyola's canonization. (Museo Provincial, Seville)

Louisiana, a French American colony round the delta and up the valley of the Mississippi, discovered by *La Salle's expedition from Canada in 1682 and named after Louis XIV. Its capital, New Orleans, was founded in 1718 as part of a chain of French forts and trading posts ringing British settlements in North America. The name Louisiana was extended to cover French claims on the Gulf Coast and in the west. In the *French and Indian Wars it was a serious threat to southern British colonies and West Indian possessions, especially after some 2,000 expelled Acadians from *Nova Scotia swelled its population in 1755. It was ceded to Spain in 1762, returned to France in 1800 and sold to the United States in 1803.

Low Countries   *Belgium, *Netherlands.

Lower Canada   *Quebec.

Loyola, Ignatius, St (1491–1556), Spanish ecclesiastical reformer, founder and first general of the Society of Jesus (the *Jesuits). Born into a noble family, he attached himself to the court of Ferdinand II of Aragon. His military career was ended by a leg-wound received while fighting for Navarre against France (1521). During his convalescence he underwent a spiritual transformation. He spent almost a year in prayer and penance (1522–3), and wrote the first draft of his *Spiritual Exercises*, an ordered scheme of meditations on the life of Jesus Christ and the truths of the Christian faith. After a pilgrimage to Jerusalem in 1523 he attended the University of Paris (1528–35). There he collected a band of like-minded followers, who worked through the *Exercises*. In 1534 he and six others took vows of poverty, chastity, and obedience to the pope, and Pope Paul III recognized their 'Society of Jesus' as an order of the Church in 1540. By the time of his death there were over 1,000 Jesuits in nine European provinces as well as those working in foreign missions.

Luba, a Bantu kingdom founded *c*.1500, lying north of Lake Kisale in modern Zaïre. The founders, the Balopwe clan, came from further north, and imposed their sovereignty over existing chiefdoms. Some of the Luba moved eastwards *c*.1600, and founded a kingdom among the Lunda; from there a large number of small chiefdoms proliferated stretching from eastern Angola to north-eastern Zambia, and making previously existing chiefdoms their vassals. The largest was the kingdom of Mwata Yamvo; others were the Bemba in north-eastern Zambia, Kazembe in the Luapula valley, and Kasanje in central Angola. They all paid tribute to the central kingdoms, but the organization was decentralized, and the kings served as settlers of disputes between communities. These were occupied not only by agriculture, but also in mining and trade in copper and salt against European goods obtained from the Portuguese. Kazembe was the richest and most important.

Luther, Martin (*c*.1483–1546), German theologian, the principal initiator of the Protestant *Reformation in Europe. He attended Erfurt University (1501–5) before joining the Augustinian Friars. Ordained priest in 1507, he took the chair of biblical theology at Wittenberg University in 1512. In evolving his doctrine of justification by faith alone, he challenged the hierarchy of the Catholic Church over such issues as the roles of the papacy and

A portrait of Martin **Luther** by Lucas Cranach. Cranach the Elder lived in Wittenberg and was a friend of Luther. This is one of several studies he made of the Protestant reformer. (Leipzig Museum and Art Gallery)

the priesthood, and the necessity of certain sacraments and observances. At the Diet of *Worms in 1521 he defended his doctrines before *Charles V, but he was excommunicated by the pope and made an outlaw of the Holy Roman Empire. Yet his challenge to the Catholic Church had been well publicized; his many pamphlets, including *To the Christian Nobility of the German Nation* (1520), had not fallen on deaf ears. By mid-century, a host of German and Scandinavian rulers had severed their links with Rome and set up new 'Lutheran' churches in their territories.

Luther's original intention had not been to split Christendom in this way, but he reluctantly accepted the inevitable when the Catholic establishment proved resistant to his proposals for reform. His fear of anarchy and his conservative political attitudes led him into acrimonious dispute with more radical Protestants such as the *Anabaptists. He spent the last twenty-five years of his life under the protection of Elector Frederick the Wise of Saxony. He married Katherine von Bora (1525) and had six children. His translation of the Bible into German provided the foundation of the literary language of northern Europe.

Lützen, battle of (16 November 1632), a battle during the *Thirty Years War between the Protestant forces under *Gustavus Adolphus and Bernard, Duke of Saxe-Weimar, and imperial Catholic troops commanded by *Wallenstein. The Protestant attack was delayed by foggy conditions and the numerically superior imperial forces came close to victory. Gustavus was killed but Bernard eventually secured a victory for the Protestants.

Luxemburg, a country in Europe. It was occupied by the Romans, then by the *Franks in the 5th century, passing to the counts of Luxemburg in the 11th century. The duchy of Luxemburg was created in 1354. Seized by *Burgundy in 1443, it passed to the *Habsburgs in 1477 and to *Spain in 1555. The French occupied it in 1684–97, and again during the Napoleonic wars.

Lydia, a territory in western Asia Minor. It derived much prosperity from trade as it lay athwart the two main roads which linked the coast to the interior. The first minting of coins is ascribed to it in the 8th century BC. It reached its apogee of independent power under *Croesus, thereafter entering the Persian empire. It later came under the control of the Seleucids and Pergamum, before being incorporated into the Roman province of Asia in 133 BC. Under Diocletian it became a separate province. Then, as previously, Sardis was its principal city.

Lysander (d. 395 BC), *Spartan admiral and statesman. He did much to bring about the defeat of Athens in the *Peloponnesian War. He revived Peloponnesian fortunes largely as a result of the friendship which he struck up with *Cyrus, son of the king of Persia. With his financial support he increased and improved the Peloponnesian fleet, and eventually sealed the Athenian fate with victory at *Aegospotami (405) and the subsequent blockading of the *Hellespont. Nevertheless, his high-handedness and lack of restraint made him many enemies at Sparta and elsewhere. He was killed on campaign in Boeotia.

Macao (Aomen), a Portuguese territory on the Zhu Jiang (Pearl River) estuary, near Guangzhou (Canton), China. Portuguese traders and missionaries established themselves there in 1557 with permission of the local authorities and called their settlement City of the Name of God in China. From its foundation until the late 17th century it had a flourishing trade with Japan. After the *Shimabara rebellion in 1637 Japanese Christians sheltered there. It became the staging-post and refuge for Europeans trading with or trying to enter China.

Macassar (or Makasar; now Ujung Pandang), a city in southern Sulawesi. After *Malacca fell to the Portuguese in 1511 many Malays transferred their business there. By 1600 Chinese, Indians, Arabs, and Javanese crowded its markets. The conversion of its people to Islam c.1600 added religious zeal to their commercial activities. After 1600 Dutch policies in the *Moluccas made Macassar increasingly attractive to smugglers and monopoly breakers. By 1630 its empire extended over much of Sulawesi, eastern Borneo, and many neighbouring islands. In 1667 the Dutch besieged Macassar and achieved its complete submission. Dutch overlordship and monopoly of trade were established.

Macbeth (c.1005–57), King of Scotland (1040–57). He was the grandson of Malcolm II and cousin to Duncan I, whom he challenged for the Scottish throne, defeating and killing him in battle near Elgin (1040). Previously he had been Mormaer (Earl) of Moray and had married Gruoch, a granddaughter of Kenneth III. Both Macbeth and his wife generously supported the Church and in 1050 he went on pilgrimage to Rome. He was killed in the battle of Lumphanan, Aberdeenshire, by Duncan's son *Malcolm III.

Maccabees, a Jewish dynasty founded by Judas Maccabeus (the Hammerer). In 167 BC the Syrian king Antiochus IV plundered the Temple in Jerusalem, set up an altar to the Greek god Zeus, and proscribed Jewish religious practices. A Jewish revolt began, led by Mattathias, an elderly priest, and guerrilla tactics were used against the Syrians. When Mattathias died in 166, his second son, Judas, assumed leadership. After a series of successful encounters with Syrian forces he retook the Temple area in 164 and cleansed the Temple in a ceremony that has from that time been commemorated annually as the feast of Hanukkah. Judas died in 160 and his brothers continued the struggle until independence from the Syrians was achieved, the third brother, Simon, becoming high priest, governor, and commander. The conquests and forced conversions of later rulers caused much discontent, and the dynasty ended with the arrival of the Romans in 63 BC.

Macdonald, Flora (1722–90), Scottish Jacobite. She gave invaluable assistance to Prince Charles Edward (the Young *Pretender) after the collapse of the *Forty-Five Rebellion. He was disguised as her maid-servant, and she travelled with him, so that he was able to escape from Scotland to France (June 1746). She was arrested, but released in 1747.

Macedonia, a region in northern Greece, linking the Balkans and the Greek peninsula. It was settled by migrating tribes during the Neolithic period. Perdiccas I established a kingdom in the southern plains c.640 BC, but nearly three centuries passed before *Philip II added mountainous Upper Macedonia to it. Under him the kingdom became the leading power in Greece and under *Alexander the Great the leading power in the known world. Macedonia had geographical advantages which helped Philip to dominate Greece. The fertile plains were good for horse breeding and cavalry training.

It was prominent in the wars which followed Alexander's death. Upper Macedonia came briefly under *Pyrrhus' control, but Antigonus II (ruled 276-239) wrested it back and restored the kingdom's strength. However, it was unable to match the expanding power of Rome, and the *Macedonian wars resulted in its being first partitioned into four republics and then (146 BC) turned into a Roman province. Large numbers of Slavs settled there in the 6th and 7th centuries AD. It then entered the Bulgarian empire until its collapse in 1014, and came under Turkish rule in 1389.

Macedonian wars, conflicts fought between Rome and Macedonia in the 3rd and 2nd centuries BC. In the first war (211-205) Philip V was opposed by an alliance of Rome, Aetolia, and Pergamum, but with Rome also deeply involved in the second of the *Punic wars he was able to force Aetolia to accept terms, and then to agree favourable ones with Rome itself. But war broke out again (200) and this time Philip was defeated decisively at Cynoscephalae (197). Philip's son Perseus came to the throne in 179, and set about winning influence and friends in Greece. This caused Roman suspicion, the outbreak of a third war, and another Roman victory, this time at Pydna in 168. Macedonia was divided into four republics. In 149-148 Andriscus, claiming to be a son of Perseus, attempted to set himself up as king but was defeated and Macedonia became a Roman province.

Machiavelli, Niccolò di Bernardo dei (1469-1527), Italian statesman and political theorist, one of the outstanding figures of the *Renaissance. By 1498 he had become Secretary and second Chancellor to the republic of Florence, and on his diplomatic missions (1499-1508) he encountered some of the most powerful political figures of the age. After the *Medici restoration in Florence (1512), he was excluded from public life, and he retired in some frustration to produce works on the art of war and on political philosophy. In *Il Principe* ('The Prince') (1513) he addressed a monarchical ruler, offering advice which was intended to keep that ruler in power. 'Machiavellian' has since been applied to one who uses deception and opportunism to manipulate others, but this does a considerable injustice to Machiavelli's far subtler view of the relationship between ethics and politics. In his less widely read *Discourses on the First Ten Books of Titus Livius* (1513-21) he advocated a republic with a mixed constitution modelled on that of ancient Rome, and stressed the importance of an incorruptable political culture and a vigorous political morality.

In 1520 he was appointed chief historiographer of

A portrait of the Italian statesman **Machiavelli** by Santi de Tito. After holding high office he was dismissed by the Medicis on suspicion of conspiracy, but was subsequently restored to some degree of favour. (Palazzo Vecchio, Cancelleria)

Florence. After the Medici fell again in 1527, his hopes of securing a position in the new republic were dashed, and he died soon afterwards.

Madagascar, an island off south-east Africa. The people are of Indo-Melanesian and Malay descent, mixed with some Bantu, Arabs, Indians, and Chinese. The time of arrival of different groups is controversial. Arab traders were probably visiting by the 10th century if not earlier. In 1500 a Portuguese sea captain, Diego Dias, stumbled on the island, calling it Sao Lourenço. Marco Polo had already named it Madagascar from hearsay knowledge, and this name endured. In the following centuries Dutch, English, and Portuguese vessels made frequent visits, and the French set up trading centres. Many of these were used as pirate bases. By the beginning of the 17th century a number of small Malagasy kingdoms emerged, and later the Sakalawa, from the west of the island, conquered northern and western Madagascar, but their kingdom disintegrated in the 18th century. The Merina people of the interior were later united under King Andrianampoinimerina (ruled 1787-1810).

Madeira, the name of the main island and of a group of three others in the Atlantic Ocean, belonging to Portugal. It was discovered in the 14th century by Portuguese colonists. They chiefly cultivated sugar and vines. Later the Portuguese were joined by Dutch, Spanish, and Italian immigrants, by Jews and Moors expelled from Spain, and yet later by slaves from Africa.

Madison, James (1751-1836), fourth President of the USA (1809-17). He helped draft the Virginia State

Constitution, served in the *Continental Congress (1780–3), and was the author of *Memorial and Remonstrances* (1784) in which he opposed taxation to support religious teachers and secured *Jefferson's bill for religious freedom. He played an influential role in the Constitutional Convention (1787) and was author of twenty-nine of the *Federalist Papers*; he also proposed the *Bill of Rights (1791). He became leader of the Democratic-Republican Party. Serving as Secretary of State under Jefferson (1801–9), he was involved in disputes with France and England over America's right to neutrality. After his election as president he lost popularity through his bad leadership in the War of 1812, which New England called 'Mr Madison's War'. Despite his dislike of a powerful central government and fiscal system, he signed the bill incorporating the Second Bank of the USA and introduced the first protective US tariff.

Magadha, a centre of ancient regional kingdoms, and intermittently of extensive empires, corresponding to the Patna and Gaya districts of modern Bihar in north-eastern India. Its position on the middle Ganges explains the economic and strategic advantages which drew successive rulers there from the 6th century BC to the 7th century AD. Among them were the *Mauryas (c.325–185 BC) and *Guptas (mid-4th–6th century AD). After Muslim invasions in the 12th century Magadha ceased to exist as a separate region.

Magdalenian *Upper Palaeolithic.

Magellan, Ferdinand (c.1480–1521), Portuguese navigator. On Portuguese service in the East Indies in 1509–12 he explored the Spice Islands (*Moluccas), but in 1517 offered his services to Spain to undertake a voyage to the same islands by a westward route. He left Spain with five ships in 1519 and sailed down the coast of South America and through the long and tortuous strait that now bears his name. When they emerged into a peaceful new ocean to the west, he named it the Pacific. It took more than three months of hardship before they reached the Philippines. Magellan was killed in a skirmish between the islanders, but one of his ships, the *Vittoria*, reached Spain by the Cape of Good Hope, thus becoming the first vessel to circumnavigate the world.

Magna Carta, the document which the English barons, aided by Stephen *Langton, forced King *John to seal at Runnymede on 15 June 1215. It was a charter of 61 or 63 clauses (the final clause is sometimes subdivided into three) covering a wide range of issues. There were clauses dealing with weights and measures, fish weirs, and foreign merchants. The powers of sheriffs were restricted, the liberties and privileges of boroughs preserved. The crown agreed not to interfere with the rights of the church, nor to levy *scutage or any special tax without the consent of the king's council, nor to demand from the barons more than was due according to feudal practice and convention. Clause 39 guaranteed the right of trial to all freemen, and clause 40 justice to everyone. The charter was safeguarded by clause 61 which set up a group of 25 barons empowered to take up arms against the king if he failed to observe its conditions. John sought and obtained papal condemnation of the charter on 18 June 1215, which led to the first *Barons' War four months later. Although the charter was often violated by

medieval kings, it came to be seen as an important document defining the English Constitution.

Magyar, a member of the race speaking a Finno-Ugric language, whose ancestors came from an area round the River Volga in Russia. Under Prince Arpád, they entered what became *Hungary in the 9th century. They harassed the German kingdom but were finally defeated and repulsed by Otto I at the battle of *Lechfeld. Pope Sylvester crowned Stephen as the first king of their country in 1000 and he established unity and introduced Christianity. He was canonized after his death.

Mahican (or Mohican), a tribe of North American Indians who inhabited the Hudson River valley. They shared many traits with the *Iroquois to the west. European contact began when Henry Hudson sailed up the Hudson River in 1609. Beaver and other pelts were exchanged for beads and knives, and regular trade became centred round the Dutch Fort Orange (later Albany). As middlemen for other tribes they jealously kept the *Mohawk Iroquois at bay until their control was finally broken in wars of 1662–9. Thereafter epidemics and dispersal by Dutch, then English and American, colonial pressures relegated them to an increasingly lesser role through the 18th century, and, like the *Delaware, they were eventually officially relocated to midwestern reservations in the 19th century.

Mahmud of Ghazna (969–1030), Muslim ruler of the *Ghaznavid dynasty of Afghanistan and Khurasan (999–1030). He led seventeen raids into northern India in the name of Islam. Sapping Hindu power in the process, he paved the way for Muslim conquest of the subcontinent. He also extended his power into Transoxania, Persia, and Mesopotamia.

Maimonides, Moses (Mosheh ben Maymun) (1135–1204), Jewish physician and philosopher. He was born

A scene depicting the arrival of the **Magyars** in Hungary from a manuscript chronicle, the *Képes Krónika*. Despite the army's presence in this heavily fortified region, the inclusion of peaceful groups of women and cattle in a flower-strewn landscape seems to belie the reputation of the Magyars as the scourge of Europe in the 9th and 10th centuries.

in Córdoba in Spain, but was driven out by the persecution of the conquering *Almohad dynasty. He settled in Egypt in 1165, where he wrote two major scholarly works, the *Mishnah Torah*, a massive codification of rabbinical law and ritual, and a *Guide for the Perplexed* which, attempting to reconcile faith and reason, caused much controversy among orthodox Jews.

Maine, the north-easternmost state of the USA, named after Queen *Henrietta Maria's French property. Discovered by Sebastian *Cabot (1496) and settled by Sir Ferdinando Gorges and his heirs in the early 17th century, it was infiltrated by settlers from Massachusetts who contended for control. It was purchased by Massachusetts in 1677, and its outposts were devastated by Abnaki Indians in alliance with the French (1687–99). Maine prospered after the American War of *Independence and became an independent state in 1820.

Majapahit, a Hindu empire based in the fertile valley of the Brantas River in eastern *Java, which flourished between 1293 and the latter part of the 15th century. It experienced its 'golden age' under its last great ruler Hayam Wuruk (1350–89) whose reign is extolled in an epic poem, *Nagarakertagama* (1365). This poem claimed an empire for Majapahit covering much of peninsular Malaya, Sumatra, Borneo, Sulawesi (Celebes), Bali, and other islands, though its control in the more far-flung areas must have been weak. Its chief minister, Gajah Mada (d. 1364), codified its laws and is said to have bequeathed a more centralized administration. After it was partitioned between Wuruk's sons there was decline. By c.1527 the last remnants of Majapahit's authority had been extinguished and many of its royal family had fled to Bali.

Malacca (or Melaka), a Malayan port commanding the strait connecting the Indian and Pacific Oceans. It was founded by a refugee Hindu prince from *Srivijaya in 1402; he was later converted to *Islam and took the title Iskandar Shah. Following visits by the Chinese admiral *Zheng He it became a Chinese vassal, thus securing itself against Thai and Javanese attacks. By 1500 it was an entrepôt where merchants traded in spices from the *Moluccas, Chinese silk and porcelain, camphor from Borneo, teak from Burma, Indian cloths, and woollens from Europe. During the 16th century the merchants of Malacca greatly assisted the spread of Islam in south-east Asia. It was captured by *Albuquerque in 1511 and the Portuguese repulsed attacks from *Acheh, *Johore, and Java. In 1641 it fell to the Dutch, aided by Johore, but with their headquarters in Java, the Dutch regarded it as of little importance.

Malaya (now West Malaysia), a region in south-east Asia. Originally inhabited by Negritos, it was later settled by people from *Yunnan. By c. AD 400 Indian influence was apparent in Kedah and Perak in north-western Malaya. Later *Srivijaya, *Majapahit, and *Siam exerted some control. It remained relatively backward until, following *Malacca's foundation in 1402, Islam was introduced. After Malacca fell to the Portuguese in 1511, sultanates were established in *Johore, Pahang, and Perak. Manangkabau Malays from Sumatra settled in Negeri Sembilan and *Bugis in Selangor. The Dutch, influential after they took Malacca in 1641, came under pressure following the British occupation of Penang in 1786. Following the British acquisition of Singapore in 1819, the Dutch withdrew from Malaya in 1824.

Malcolm III (Canmore) (c.1031–93), King of Scotland (1058–93). Malcolm was brought up at the English court of *Edward the Confessor after *Macbeth murdered his father, Duncan I. With English assistance he defeated Macbeth in 1054, and killed him in 1057. Malcolm married (1068) Margaret, the granddaughter of *Edmund Ironside. His support of Saxon exiles (including *Edgar the Atheling) fleeing from the Normans, led to an invasion (1072) by *William I and Malcolm's homage to him. Tension continued, and Malcolm was killed at Alnwick while on his fifth invasion of England.

Maldon, battle of (August 991). A major battle between the English under the leadership of Brhtnoth and Danish raiders led by Anlaf. The battle is the subject of a short but moving poem written by an eyewitness, in which the heroism of the defeated English is celebrated.

Mali, an empire in the upper Niger region of West Africa, established in the 13th century. The founder, *Sundjata, conquered the remains of the *Ghana empire c.1235–40 with his army of Malinke soldiers. Mali soon controlled the rich trade across the Sahara and became a major supplier of gold. The empire reached its peak in the early 14th century under *Mansa Musa, who established an efficient administration. The Muslim traveller Battuta visited Mali in 1351–2 and gave a detailed account of the court and trade. However, by then the empire was beginning to decline. In 1335 *Songhay became independent of Mali and by the 15th century had conquered the rest of the empire.

Malplaquet, battle of (11 September 1709). The site in north-east France close to the Belgian frontier of *Marlborough's last victory over the French in the War of the *Spanish Succession. Marlborough's invasion of France in 1709 was an attempt to make *Louis XIV agree to the allies' harsh peace terms. Though Malplaquet was a victory, the losses of the combined forces of England and the Holy Roman Empire exceeded those of France and the invasion attempt was abandoned.

Malta, an island in the Mediterranean roughly midway between Tunisia and Sicily. It was settled, possibly as long as six thousand years ago, during the *Neolithic era. In historic times it was a Carthaginian centre, falling to Rome in 218 BC who named it Melita. The Byzantine empire controlled it until 870 when it was conquered by Muslim Arabs. The *Normans captured it and included it in their kingdom of *Sicily, but having been recovered by Muslim forces it finally fell to the Spanish kingdoms of Aragon and Castile and thence to Spain itself. Under Emperor Charles V it was given to the *Knights Hospitallers (1530), and they defended it against Turkish attacks and fortified and enriched it. They were eventually expelled by Napoleon I of France (1798) and the island was taken by the British in 1800.

Mameluke (from the Arabic *mamluk*, 'possessed', a synonym for 'slave'), the name for two Egyptian dynasties. The Bahri Mamelukes (1250–1390) were Turks and Mongols recruited as slave bodyguards by the Ayyubid

The variety of **Mameluke** dress and racial types shown in this German manuscript reflects the Mameluke division into Circassian and Arab-Turkish dynasties. (Forschungsbibliothek Gotha)

sultan al-Salih (1240-9), himself a Turk, and were stationed in barracks beside the Nile. The Bahri recruited the Burji (1390-1517), likewise as bodyguards, stationing them in the citadel. They were chiefly Circassians. Captured in childhood as slaves, the Mamelukes were trained in every branch of warfare and had an exacting academic education. Though theoretically they were elected as sultans, in fact they challenged their rulers and in some cases usurped their offices. Succession was hereditary with sons succeeding fathers. Their rule extended over Egypt and Syria (including the present Israel, Jordan, Lebanon, and western Arabia). There was an elaborate court, and a highly organized civil service and judiciary. Active encouragement of trade and commerce brought great prosperity throughout their dominions, as is witnessed by the splendid monuments which they built in Cairo and elsewhere. Their external trade reached across Africa as far as Mali and Guinea, and throughout the Indian Ocean as far as Java. In 1517 the *Ottoman Turks captured Cairo and overthrew the Mamelukes. As Turkish power waned they re-established themselves as rulers. *Napoleon defeated them in 1798 and they were brought down by Muhammad Ali in 1811.

Mameluke sultanate (Slave sultanate), a series of Muslim kings based in Delhi who ruled a north Indian empire from 1206 to 1290. The founder, Qutb ud-Din Aibak (ruled 1206-11) had risen in the service of Muhammad of Ghor, the previous ruler of Delhi. One of Delhi's most famous monuments, the Qutb Minar, a carved red sandstone minaret, was built during his reign. His son-in-law and former slave Iltutmish (1211-36) consolidated his hold on the Punjab, Bengal, and Rajputana, conquered Sind, and expanded south as far as the River Nerbudda. Having survived Mongol threats in the mid-13th century, the sultanate reached its peak under Ghiyas ud-Din Balban (1266-87). Soon afterwards Delhi fell to the *Khaljis (1290).

Manchuria, now called Dongbei (East-North), a region lying north-east of the *Great Wall of China. Between 200 BC and AD 900 China exercised little influence over its various nomadic peoples except in the extreme south. A succession of dynasties established by different tribes then ruled it—for example the *Liao and the *Jin, both of whom extended their empires into China. By the early

17th century the *Manchus, from whom its name derives, were in control. From 1644, with China under Manchu rule, it was part of the *Qing empire.

Manchus, descendants of the Juchen, founders of the *Jin dynasty. Vassals of the *Ming, their base was north of the Liaodong Peninsula in Manchuria. After 1582 their chief, Nurhachi (1559-1626), made alliances with neighbouring tribes, built a strong castle, and imported Chinese technicians and advisers. Everyone—tribesman, captive, serf, or slave—was registered under a distinctive banner, making possible an efficient system of taxation and military control. In 1616 Nurhachi took the title of emperor and in 1625 made Shenyang, renamed Mukden, his capital. When he died he had built his bannermen into a nation. His son, Abahai, campaigned extensively in Korea, Mongolia, and northern China. Twice he attacked Beijing. He ordered his people to call themselves Manchus, a name of obscure origin, and in 1636 proclaimed the Da Qing (Great Pure) dynasty. Eight years later Nurhachi's grandson became the first *Qing emperor of China. While the Manchus adopted many aspects of Chinese life, they remained segregated from them, intermarriage with Chinese was forbidden, and they had separate quarters in all Chinese cities. During the 19th century segregation began to break down and in the 20th century they have merged into the mass of the Chinese people.

mandarin (Portuguese *mandarim*, from the Sanskrit *mantrin*, 'counsellor'), a name given by Europeans to a senior official in imperial China. From the Song dynasty (AD 960), officials were recruited predominantly by examination in the Confucian classics. (Since the *Han dynasty (206 BC) examinations had been used within the civil service.) Study for the examinations lasted many years and success brought status not only to the candidate but also to his family. There were nine grades of mandarin.

Manfred (1231-66), King of Sicily (1258-66). The illegitimate son of the Holy Roman Emperor Frederick II, he ruled in Italy on behalf of Conradin, his half-brother, and with support from the Saracens took the Kingdom of the Two *Sicilies (1257). He was excommunicated by Pope Alexander VI but invaded papal territories in Tuscany. He was again excommunicated by Pope Urban IV who gave his crown to Charles I of Anjou, and he was finally defeated and killed at the battle of Benevento.

Manichaeism, the teaching of Manes (c.216-76), a Persian influenced by *Mithraism, *Christianity, and Gnosticism. He taught a dual principle of Good and Evil in conflict, symbolized as Light against Darkness, God against Satan. He counselled asceticism for an elect group following the teaching of the Jewish prophets, *Jesus Christ, *Buddha, and himself. *Zoroastrians drove him into exile in India, flayed him alive, and crucified him. His followers were condemned by *Diocletian, though this did not prevent their influence spreading to Rome and Africa by the 4th century. The sect survived in Chinese Turkistan until the 10th century and influenced various heresies in medieval Christianity.

manor house, the home of the lord of an estate in medieval times. As well as housing the lord—or his

Manor house

first floor plan

a) Boothby Pagnell, Lincolnshire
*c.*1180

ground plan

b) Little Wenham Hall, Suffolk
*c.*1270

c) Penshurst Place, Kent
*c.*1340

ground plan

d) Great Chalfield, Wiltshire
*c.*1480

ground plan

e) Compton Wynyates, Warwickshire
*c.*1500

Early manor houses were fortified buildings, centred on a great hall, with their living quarters on the first floor. The stone manor at Boothby Pagnell (a) is one of the earliest surviving examples, while Little Wenham Hall (b) is the earliest extant brick manor. Penshurst Place (c) is similar in design, with an additional wing for service rooms. Great Chalfield (d) retained the great hall as an important feature but the house has a more domestic appearance. By early Tudor times the need for defence had disappeared and is no longer apparent in the design of Compton Wynyates (e).

resident *bailiff—the house was the administrative hub of the feudal estate. Throughout Europe, manor houses varied considerably in size and design, depending upon what materials were locally available and how much fortification seemed necessary. In France, and elsewhere in battle-scarred Europe, defensive considerations predominated until the 17th century. Rectangular, fortified tower-houses within walled and moated enclosures were familiar sights on the landscape. In England, the move toward more luxurious accommodation began earlier. Already by the 14th century, interiors were being divided up into private living apartments and service rooms, rather than being dominated by the traditional great halls. By the 17th century the manor house had evolved into the *country house.

manorial system, the social, economic, and administrative system (also called seigneurialism) which emerged in Europe in the 5th century from the chaos and instability which followed the collapse of the Roman empire. Farmers sought the protection of powerful lords and in return surrendered certain rights and control over their lands. Gradually a system of obligations and service emerged, especially relating to manorial agrarian management, and set down in records called custumals.

The manor consisted of demesne land (private land of the lord) and tenants' holdings. Tenants were free or unfree (villeins), rank being determined by personal status or the status of their land. Not all manors had this balance of demesne, free land, and unfree land. In addition, meadow land for grazing livestock was available to all, and thus known as common land. Access to woodland for timber and grazing of pigs might be a further facility. The lord presided over the manorial court and received money or labour services from his tenants regularly (week work) or seasonally (boon work). A tendency in the 12th century for labour services to be commuted to cash rents was reversed after *c*.1200, when inflation encouraged landlords again to exact services in kind. Labour shortages following the *Black Death (1348) when Europe's population fell from 80 million to 55 million, enclosures, tenant unrest, and rebellions such as the *Peasants' Revolt (1381) effectively ended the manorial system in England by *c*.1500.

Mansa Musa (ruled 1307–37), the most celebrated of the rulers (kankans) of *Mali, chiefly because of his spectacular pilgrimage to *Mecca in 1324. He caused a sensation in Cairo, with his 500 slaves and 80–100 camels carrying gold. In his absence one of his generals acquired Gao, the capital of the neighbouring *Songhay state for him. He returned from Mecca with the Andalusian poet-architect Es-Saheli, who built the palace and Great Mosque of *Timbuktu. He greatly expanded the commerce and prosperity of Mali, and gave encouragement to Islamic learning and culture.

A Catalan map of 1375 showing **Mansa Musa**, emperor of Mali (bottom right), that traces the trade routes of northern Africa. Until the mid-13th century, two-thirds of the world's gold supply came from Africa, much of it from Mansa Musa's dominions. (Bibliothèque Nationale, Paris)

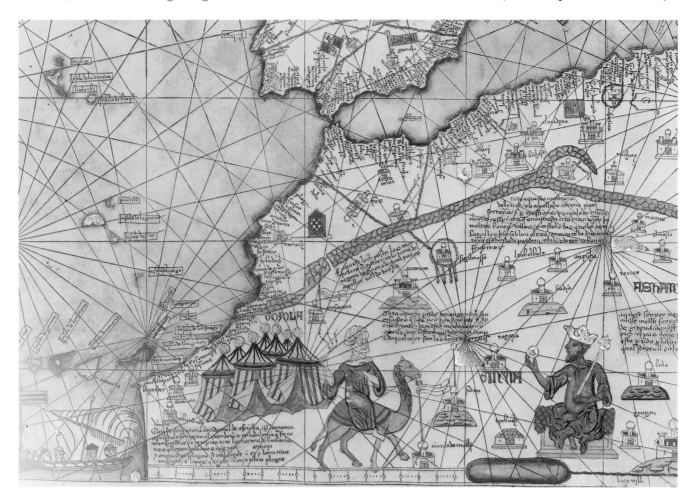

Manzikert, battle of (1071). At a site near Lake Van in Turkey *Seljuk Turks, recent converts to *Islam, routed a Byzantine force, capturing its leader, Emperor Romanos Diogenes IV. Alp Arslan, the Seljuk leader released him, but the defeat left Anatolia open to Turkish invaders and the weakening of Byzantine control which was indirectly to provoke the First *Crusade.

Marat, Jean Paul (1743-93), French Revolutionary leader. He practised as a doctor in London, where in 1773 he published a *Philosophical Essay on Man*. This criticized *Helvétius's view that science was unnecessary for a philosopher. He returned to Paris to edit the radical journal *L'ami du peuple*. He was a leader of the extremist Cordelier Club and became a key figure in the *French Revolution. His attacks on those in power led to his brief exile in 1790 but in 1792 he was elected to the Assembly where he opposed the *Girondins. Their defeat in May 1793, a triumph for Marat, led to his murder by Charlotte *Corday.

Maratha, a Hindu warrior people of western India who in the 17th and 18th centuries led a military revival against Muslim expansion. They rose to prominence under the inspired leadership of *Sivaji, who, after victories against the *Moguls, established a Maratha kingdom in 1674. Their great age was the early 18th century when, after a temporary collapse, they benefited from Mogul decline to sweep over the north and central Deccan. They seemed poised for all-India mastery, but failure in 1761 of their bid to take Delhi (in the battle of *Panipat) was followed by increasing internal disunity. Authority had passed from Sivaji's line to a Brahmin family based at Pune, who as hereditary *peshwas* (ministers) struggled to hold the dissident chiefs together. Rivalry among these 'confederates', notably the *Sindhia, *Holkar, Bhonsla, and Gaekwar families, prevented a united stand against expanding British power. By 1818, besides considerable loss of territory, the surviving chiefs had been forced to accept British protection over their Princely States.

Marathon, a plain in north-east *Attica, scene in 490 BC of a battle in which the Athenians, with their *Plataean allies, defeated the invading Persians. The 11,000 Greeks used tactics which prevented encirclement of their flanks and inflicted some 6,400 casualties. They then marched swiftly back to Athens, thereby deterring a sea-borne Persian attack on the city itself. The marathon race is named after Pheidippides, who ran to summon Spartan help after the news of the Persian landing, and is celebrated for covering the distance of 240 km. (150 miles) in two days. The modern marathon race is 42.195 km. (26 miles, 385 yards).

marcher lord, a holder of land (a lordship) which lay on the border (or march) between countries. The term applied to Italy (marche), Germany (mark), and in England along its borders with Scotland and Wales. The marcher lords on the Welsh border were particularly powerful. Between 1067 and 1070 three large marcher lordships were created, based on Chester, Shrewsbury, and Hereford. They extended their power into Wales despite Welsh resistance in the 12th century led by Owain Gwynedd and Rhys ap Gruffydd. About 140 lordships and sub-lordships were created, the most powerful being

The waterfront at Venice, from a manuscript dated 1338. **Marco Polo**, with his father and uncle, set sail from Venice to Acre on the first stage of their journey east. St Mark's Church with its four bronze horses, brought to Venice in 1204, is clearly visible on the left. (Bodleian Library, Oxford)

Glamorgan, Pembroke, and Wigmore. The crown sought to deal with them by first setting up the Council of the March in Wales (1472) and then abolishing them in the second Act of Union (1543). Their lands were attached to the six existing and the seven new Welsh shires and to the English border counties.

Marco Polo (*c.*1254-1324), Venetian traveller. Between 1271 and 1275 he accompanied his father and uncle on a journey east from Acre into central Asia along the *Silk Route, eventually reaching China and the court of *Kublai Khan. He entered the service of the khan and travelled widely in the *Mongol empire during the next seventeen years. In 1292-5 the three Venetians returned home by sea, calling at Sumatra and southern India before reaching Persia and making the last part of their journey to Venice overland. Three years later Marco Polo was captured by the Genoese in a sea battle, and during his imprisonment he dictated an account of his travels. His book was widely read and stimulated European interest in the East and its riches.

Marcus Aurelius, (AD 121-80), Roman emperor (161-80) and *Stoic philosopher. He was born of Spanish parents, and adopted as *Hadrian's grandson. A student of literature, philosophy, and law, his succession to the imperial title in 161 plunged him into military activity on frontiers in the Balkans, Dacia, Pannonia, and Syria. His 'Meditations', largely written on campaign, survive with some of his letters. While an admirer of *Epictetus, he persecuted the Christians.

Margaret, Maid of Norway (*c.*1283-90), Queen of Scotland (1286-90). She was the daughter of Erik II, King of Norway, and the granddaughter of *Alexander III of Scotland. She became Queen of Scotland at the age of 3, although six guardians were appointed to govern the kingdom during her minority. Edward I of England proposed a marriage between Margaret and his son Edward, the first English Prince of Wales. Her death by drowning when crossing the North Sea from Norway to Scotland brought to an end the dynastic House of Canmore (since 1057), and led to a dispute (1291-2) over the succession involving thirteen claimants. Edward I of England judged in favour of John *Balliol.

Margaret, St (*c*.1046-93), Queen consort of *Malcolm III of Scotland from 1069. She was descended from the House of Wessex, being the daughter of Edward the Atheling and the granddaughter of Edmund Ironside. A pious woman, Margaret encouraged the Scottish Church to adopt the Roman form of Christianity. She refounded Iona and brought the Benedictine monastic order to Scotland. She was canonized in 1250.

Margaret of Anjou (1430-82), Queen of England. Her marriage to *Henry VI in 1445 ensured a truce in the war between England and France. Henry's weakness caused the *Lancastrian party to centre on his indomitable wife; in February 1461 she won the second battle of *St Albans but by her hesitation lost the chance to keep Henry on the throne, and she had to flee to Scotland and thence to France. Except for the few months that Henry regained the throne in 1470-1, she spent most of the rest of her life in her native Anjou.

margrave (German, *Markgraf*, 'count of the mark'), a governor appointed to protect vulnerable areas known as marks in the Holy Roman Empire. They were the equivalent of English marches (*marcher lord) and were usually frontier territories. Charlemagne introduced this office, which in the 12th century became hereditary, and the title came to rank equally with a prince of the empire.

Maria Theresa (1717-80), ruler of the Habsburg dominions (1740-80). She was the daughter of Charles VI, and succeeded to the Habsburg lands in Austria, Bohemia, Hungary, the southern Netherlands, and north Italy in 1740 according to the *Pragmatic Sanction. Though only 23, she was a woman of great courage and determination, and rallied her peoples to defend her territories in the War of the *Austrian Succession and the *Seven Years War. She lost Silesia but gained Galicia in the First Partition of Poland. Her husband Francis of Lorraine became Holy Roman Emperor in 1745, and on his death in 1765 her son *Joseph II succeeded and became co-regent with his mother until her death.

She was a benevolent and practical woman and chose able ministers, notably *Kaunitz, Haugwitz, and Chotek. Hungary remained strongly independent but Austria and Bohemia were brought under more centralized control, local government and taxation were reorganized, and important steps were taken in army and educational reform. Her improvements were gradual but proved more enduring than the more radical changes imposed by her son.

Marie Antoinette (1755-93), Queen of France. She was the daughter of *Maria Theresa and *Francis I of Austria, and was 14 when her arranged marriage to the Dauphin of France took place. Her brief popularity as *Louis XVI's queen from 1774 was sacrificed by her frivolous and extravagant behaviour, associated with the *Petit Trianon*, a small country house near Versailles, given

A portrait of Empress **Maria Theresa** by an unknown artist. A central figure in European politics of the 18th century, she involved Austria in two major wars, the War of the Austrian Succession and the Seven Years War. (Kunsthistorisches Museum, Vienna)

by Louis XVI to the queen and used for parties. Throughout the difficult days of the *French Revolution she displayed remarkable courage and dignity. Several attempts were made to free her, but her close relationship with Austria and support for an allied invasion led to her trial and execution.

Marius, Gaius (157-86 BC), Roman general and seven times consul. He was a commander in the Numidian campaign and concluded the war by capturing *Jugurtha in 105. From 104 to 101 he fought the Cimbri and Teutones who threatened Italy and defeated them in two major battles. For the African war he had broken with precedent in recruiting men into the army of no property, and created a professional army. In Gaul he introduced rigorous training and established the cohort and the century (*centurion) as the essential units of the legion. His soldiers were nicknamed 'Marius' mules' because they carried everything on their backs. In 88 he attempted to deprive *Sulla of his command against Mithridates. When Sulla marched on Rome to maintain his position, Marius fled. After Sulla's departure for the East, he captured the city with his partisans and massacred his enemies. He died just after entering on his seventh consulship in 86.

Mark Antony (Marcus Antonius) (83-30 BC), Roman general. He had served with *Caesar at the end of the *Gallic wars. As tribune in 49 he defended Caesar's interest in the Senate as civil war loomed. He was present at *Pharsalus, and represented Caesar in Italy. His offer of a crown to Caesar was refused. After Caesar's murder he took the political initiative against the assassins, and delivered the funeral speech. *Octavian, however, was Caesar's designated heir and hostility arose between the two. During Antony's struggle for ascendancy over the Senate led by *Cicero, he was denounced in the 'Phillippic' orations and defeated at Mutina by the forces of the consuls and Octavian. He was then reconciled with Octavian, and together with Lepidus they formed the Second Triumvirate, disposed of enemies including Cicero and defeated the 'Liberators', Brutus and Cassius, at *Philippi in 42.

Antony received the government of the eastern Mediterranean and began (42) his liaison with Cleopatra. Although a powerful ally she cost him much support at Rome. Their marriage, Antony's fifth, was illegal in Roman law. In 34 he declared Caesarion (Cleopatra's son allegedly by Caesar) as Caesar's heir in Octavian's place and divided the east among his family. War followed. After *Actium he committed suicide in Egypt.

Marlborough, John Churchill, 1st Duke of (1650–1722), English general, one of the most outstanding strategists in English military history. He showed military ability in his early campaigns and political sagacity in his marriage to Sarah Jennings, the favourite of Princess *Anne. *James II appointed him second-in-command against *Monmouth's Rebellion in 1685, but he failed to support James's Roman Catholic aims and deserted him in 1688 to welcome *William of Orange. When William's wife Mary quarrelled with her sister Anne, Churchill fell from royal favour, but began to recover his influence after Mary's death in 1694.

War seemed inevitable over the *Spanish Succession by 1701 and the king appointed Churchill commander-in-chief of the English forces. Anne became queen in 1702 and he received a dukedom and became commander-in-chief of the allied armies. His brilliant campaigns in Europe led to a series of spectacular victories: *Blenheim (1704), *Ramillies (1706), *Oudenarde (1708), and *Malplaquet (1709), but the increasingly heavy casualties brought criticism at home. The Tory revival in 1710 led to his dismissal in the following year. He resumed his old offices on the accession of George I in 1714, but two strokes in 1716 incapacitated him and ended his career.

His wife, Sarah (1660–1744), was as loyal and devoted to her husband as she was quarrelsome and vindictive to everyone else, including her own children. Her influence over Queen Anne was undermined by her cousin Abigail Masham, working for *Harley, and Sarah was dismissed by the queen when the Whigs fell in 1710. After the duke's death Sarah supervised the completion of Blenheim Palace, the country house given by the nation to her husband as a reward for his victories. In 1742 she published an account of her long and stormy public life.

Marprelate tracts, satirical English pamphlets signed by the pseudonymous 'Martin Marprelate', which appeared in 1588–9. They featured scurrilous attacks on Anglican bishops, and were the work of Presbyterians who wished to discredit the episcopacy. *Elizabeth I, angered by them, prompted a search for the secret presses on which they were printed. Star Chamber prosecutions of leading ministers followed. Having appeared at a time when Presbyterian fortunes were already at a low ebb, the tracts probably served to discredit the movement still further with the public.

Marston Moor, battle of (2 July 1644), a decisive victory for the *Roundheads and Scots during the *English Civil War. The *Cavalier general Prince Rupert had pursued them to Marston Moor, Yorkshire, after his relief of York. They attacked him unexpectedly in the evening, and Cromwell's disciplined cavalry routed the Royalist troops. The Cavaliers lost perhaps 3,000 men through casualties, and 4,500 prisoners. After the encounter few northern fortresses held out for the king.

martyr, originally the term for a legal 'witness', came to denote anyone who died for his or her Christian beliefs after Stephen, the first Christian martyr, was stoned to death in Jerusalem (c.35). Martyrs' graves became shrines, from the Roman *catacombs to that of St Alban outside Verulamium (modern St Albans). After the last persecution in the early 4th century their remains were either transferred to or marked by new churches. Relics were prized, and services were celebrated over a martyr's remains. Later centuries provided countless martyrs, including the Protestant and Catholic martyrs of Tudor England and missionaries sent abroad from the expanding of European empires.

Mary, Queen of Scots (1542–87), Queen of Scotland (1542–67), the daughter of *James V of Scotland and Mary of *Guise. She was betrothed to the future *Edward

Mary, Queen of Scots, with her son, the future James I of England (born June 1566). Known as the Duff Ogilvie portrait, the painting is by an unknown artist. Mary's height and red-gold hair made her a striking figure. (Metropolitan Museum of Art, New York)

VI of England in 1543. Cardinal *Beaton's veto led to war with the English, and the Scottish defeat at *Pinkie (1547). Mary was then sent to the French court, where she received a Catholic upbringing under the supervision of her Guise uncles. She married the dauphin Francis (1558), who succeeded to the French throne in 1559 and died in 1560. By 1561 she had returned to Scotland, and had also proclaimed herself the rightful queen of England, as granddaughter of *Henry VIII's sister, Margaret Tudor.

She had to adapt to the anti-monarchical, anti-Catholic, anti-French atmosphere of Reformation Scotland. Her ill-considered marriage to *Darnley (1565), although it produced a son, the future *James VI and I, also led to the murders of *Rizzio and of Darnley himself, and the scandalous marriage to *Bothwell. The subsequent rising of the Scottish lords resulted in her military defeat and flight to England. There she threw herself on the mercy of *Elizabeth I, who kept her confined in various strongholds until her death. Wittingly or not, she was involved in a number of Catholic conspiracies against Elizabeth, figuring in the scheming behind the *Northern Rising as well as the *Ridolfi and *Throckmorton Plots. Her implication in the *Babington Plot (1586) provided enough damaging evidence for a commission to find her guilty of treason. For years Elizabeth had turned a deaf ear to Protestant pleas to execute this fellow monarch. Even now she delayed signing the death warrant, and then disclaimed responsibility for the execution of Mary at Fotheringhay.

Mary I (1516–58), Queen of England and Ireland (1553–8), the only surviving child of *Henry VIII and Catherine of Aragon. During her parents' divorce proceedings, she was separated from her mother (1531), never to be reunited. She was banished from court, declared illegitimate, and barred from the throne before being restored to the succession in 1544.

During the reign of her half-brother *Edward VI she clung tenaciously to her Catholic faith. Then she outmanoeuvred Lady Jane *Grey to win the throne, and appeared to enjoy considerable public support, despite being the first ruling queen since Matilda. Many people had remained loyal to the old Catholic religious forms, and there was little opposition to her reversal of Edward VI's Protestant legislation, but her projected marriage to the future Philip II of Spain (1554) provoked *Wyatt's Rebellion. She proceeded with the marriage, which turned out to be unhappy and childless.

After 1554 she relied increasingly on Reginald *Pole for guidance in the reversal of Henry VIII's Reformation, except for the Dissolution of the *Monasteries, and the revival of severe heresy laws. Between 1555 and 1558 nearly 300 Protestants were executed including *Cranmer, Ridley, and *Latimer, earning her the name 'Bloody Mary'. She also lost popularity through her foreign policy. In 1557 Philip dragged England into the final phase of the European Habsburg–Valois struggle and England lost Calais, its last outpost on the Continent.

Mary II (1662–94), Queen of England, Scotland, and Ireland (1689–94). She was the daughter of *James II by his first wife Anne Hyde. She married *William of Orange in 1677, and in 1688–9 supported her husband against her father during the *Glorious Revolution. She always deferred to her husband and refused to become queen in 1689 unless he was made king. Her popularity both in the *United Provinces and in England enabled William III to trust her with the administration of England during his frequent absences abroad. Her lack of children and her quarrel with her sister *Anne, the successor to the throne, saddened her last years.

Maryland, USA, a state on the central east coast round the northern end of Chesapeake Bay. It was founded as a proprietary colony by the Roman Catholic Lord Baltimore in 1632 and named after Queen *Henrietta Maria. It became a major tobacco producer. Despite its Toleration Act (1649) it suffered religious uprisings from Puritan settlers in the 1650s and a revolution in 1689, after which it became a royal colony, though in 1715 it reverted to being a proprietary colony. In the American War of *Independence, Maryland was responsible for forcing other states to cede their western lands to the national government.

Masada, a mountain fortress in the *Judaean desert, 395 m. (1,300 ft.) above the western shores of the Dead Sea. Between 37 and 31 BC *Herod the Great strengthened fortifications possibly dating from the 2nd century BC and added two palaces, a bath-house, and aqueducts. In AD 66 Masada was seized from its Roman garrison by the Zealots, an extremist Jewish sect, who held it until AD 73 when, after a two-year siege, it fell to the 15,000 men of the Tenth Legion. The 1,000 defenders, with the exception of two women and five children, committed suicide rather than surrender.

Mason–Dixon line, dividing line surveyed (1765–8) to determine the disputed southern border of Pennsylvania with Maryland by Charles Mason and Jeremiah Dixon. As the dividing line between a slave and a free colony, its name was later used to designate the westward boundary between slave and free states, 36° 30′ N, laid down in the Missouri Compromise (1820).

An aerial view of **Masada** from the north-east, showing clearly the defensive qualities of the mountain-top fortress which enabled it to withstand a lengthy siege. The ruins of Herod the Great's summer palace can be seen in the foreground.

Massachusetts, USA, a New England colony and state founded in 1630 by the Puritan Massachussetts Bay Company. Persecution and economic depression in England drove some 20,000 people to emigrate in the 1630s, including leading Puritan ministers and gentry. They were granted a charter authorizing trade and colonization. The first governor, John *Winthrop, chose Boston as capital and seat of the General Court, the legislature, and established a strict congregational regime. Massachusetts fought a delaying battle against royal interference, but its charter was revoked in 1684 and direct government substituted. After 1689 it became a royal colony in which Pilgrim *Plymouth was incorporated. It played a leading role in the American War of *Independence, and after *Shays's Rebellion (1786), in which state troops had to defend a federal arsenal, its élite clamoured for a new federal constitution to strengthen central government. The Massachusetts constitution became a pattern for later American states.

Mather, Cotton (1663–1728), American Puritan clergyman. The son of Increase Mather, he became his father's clerical colleague in 1685. Immensely learned, often pedantic, he published nearly 500 works on theology, history, political questions, science, medicine, social policy, and education. Although he did not support the trials of witches at *Salem, his *Memorable Providences relating to Witchcraft and Possessions* (1689) helped to stir emotions. After being passed over as president of Harvard, he helped found Yale University (1701), advocated Puritan involvement in social welfare, and championed smallpox inoculation. He was the first American-born member of the *Royal Society.

Mather, Increase (1639–1723), the son of Richard Mather (1596–1669), who had helped define Congregational orthodoxy in 1648. He became a Boston minister in 1664 and married the daughter of John Cotton. He was a conservative president of Harvard College (1685–1701) but as colonial agent in London (1688–92) he negotiated a liberal royal charter for the state of Massachusetts. On his return he helped end the *Salem witch trials. He was a forceful preacher against 'declension' (spiritual decline) as well as a prolific author, and was the foremost minister of his generation.

Matilda (or Maud) (c.1102–67), the only daughter of *Henry I of England and, after the death (1120) of his heir, William the Atheling, his only legitimate child. Married (1114) to Henry V, Emperor of Germany, Matilda returned to England after his death (1125) to be recognized by the English barons (1127) as Henry's successor. However, after her unpopular marriage to Geoffrey, Count of Anjou (1128), the barons accepted Henry's nephew *Stephen as king (1135). Matilda's invasion of England (1139) and the defeat and capture of Stephen at Lincoln (1141) proved but temporary successes. She returned (1148) to Normandy but lived to see her son succeed Stephen as *Henry II.

Matthias I (Corvinus) (1443–90), King of Hungary (1458–90). He was the son of the Hungarian leader and hero John Hunyadi. He was 18 on his father's death and for the first few months of his reign he was under the control of a regent, his uncle. He had to repulse a military threat from Emperor Frederick III, and fight the Turks,

An engraving of Cotton **Mather**, the most celebrated New England Puritan minister.

before he was officially crowned in 1464. His reign saw almost continuous warfare; his military successes were based on army and fiscal reforms. In 1468 he accepted an overture from the papacy to lead a crusade to challenge the *Hussites in Bohemia; meanwhile he continued to wage war against the Turks who remained a constant threat. After the death of King George of Bohemia (1471) Matthias was successful over Bohemia, and the Peace of Olomuc (1478) granted him extensive territories and the (shared) title of King of Bohemia. In 1477 his armies moved into Austria and in 1485 he beseiged and captured Vienna. As well as administrative reforms, he also codified the law, founded the University of Buda, and encouraged the arts and learning. At the time of his death he was lord of an empire that dominated south-central Europe but his successes were short-lived, as the Jagiellon dynasty came to power.

Mattioli, Ercolo Antonio (the Man in the Iron Mask), one of history's mystery figures, the hero of a legend largely created by *Voltaire. The unknown prisoner incarcerated by *Louis XIV in Pignerol and later in the Bastille is thought by some to have been Count Mattioli, an agent of the Duke of Mantua, who had deceived the king over a secret treaty to purchase the strategic fortress of Casale. Others suggest that he was a brother or son of Louis XIV or even *Fouquet. The prisoner who died in the Bastille in 1703 wore a velvet mask.

Maud *Matilda.

Mauryan empire (c.325–185 BC), the first empire in India to extend over most of the subcontinent. The dynasty was founded by Chandragupta Maurya (c.325–

297 BC), who overthrew the *Magadha kingdom in north-eastern India. He established his capital at Pataliputra, then expanded westwards across the River Indus, annexing some trans-Indus provinces deep into Afganistan from *Alexander's Greek successors. His son, Bindusara (c.297–272 BC), moved south, annexing the Deccan as far as Mysore. Although the third emperor, *Asoka (c.265–238) soon renounced militarism, his reign marked the high peak of Mauryan power, for his humane rule permitted the consolidation of his father's huge empire. On his death decline quickly set in, and the dynasty finally ended with the assassination of Birhadratha (185 BC) by the founder of the subsequent Sunga dynasty.

At its height the empire was a centralized bureaucracy organized round the king. It was divided into four provinces, each headed by a prince, and revenue was drawn from the land and from trade. Royal patronage of Buddhism appears to have ceased on Asoka's death. Sources for the extent and character of the empire include the account of Megasthenes, a Greek envoy to Chandragupta's court, and the rock and pillar inscriptions of Asoka's reign. Examples of sculpture that survive indicate that the fine arts flourished during this period.

Maximilian I (1459–1519), King of the Romans (1486–93), *Holy Roman Emperor (1493–1519). By marrying Mary, daughter and heiress of *Charles the Bold (1477) he added the duchy of Burgundy (which included the Netherlands) to the *Habsburg lands, thus earning the enmity of France. He defeated the French at the battle of Guinegate (1479) but the Habsburg–Valois rivalry continued in the Netherlands and Italy.

In 1490 he drove out the Hungarians, who, under *Matthias Corvinus, had seized much Austrian territory, and by the Treaty of Pressburg (1491) he was recognized as the future king of Bohemia and Hungary. After repulsing the Turks in 1493, he turned to Italy where war was waged between French and Habsburg troops until 1516. He was at a military disadvantage since the German princes refused to finance his campaigns and, despite allying with England against France, he was forced to cede Milan to France and Verona to the Venetians, and to sign the Treaty of Brussels with Francis I in 1516. He was also forced to grant the Swiss independence from the Holy Roman Empire in 1499.

Dynastically he had great success; his son Philip's marriage to the Infanta Joanna (daughter of *Ferdinand and Isabella) united the Habsburgs and Spain, and his grandson's marriage to the daughter of the King of Bohemia and Hungary secured his inheritance to those lands. In Germany Maximilian's attempt to impose centralized rule on the princes and cities was resisted, since they were determined to remain self-governing. His achievements were in increasing the Habsburg territories far beyond Germany, notably by linking it to Spain, and thus to Spain's empire in the Americas.

Maya, Central American Indians, the dominant cultural and linguistic group in southern Mexico, Guatemala, and the Yucatán peninsula until the 15th century. Their area of influence is divided by archaeologists into a southern region—the Guatemalan Highlands; a central region—the Petén (central Guatemala), and regions to east and

The Maya

Although best known for their architecture, the Maya were also expert craftsmen in stone and pottery (despite the absence of the potter's wheel). These pieces all come from the late classic period of Maya civilization, from the 7th to the 10th century. (a) Terracotta figure of a nobleman, about 17.5 cm. (7 in.) high, from the island of Jaina. (b) Jade mosaic mask, with eyes of mother of pearl and obsidian, from Chiapas, c.692. Containing over 200 pieces this was a death mask placed over the face of an important person at burial. (c) Ceremonial plate decorated with a fish, emblem of the fish-god Xoc, from Tikal. (d) Polychrome earthenware jar painted with a scene from mythology, 16.5 cm. (6½ in.) high, from Guatemala.

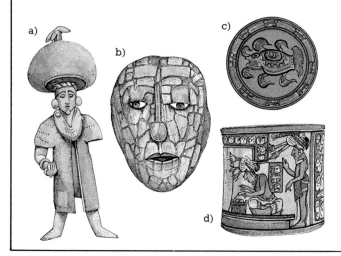

Mayan civilization extended over a very large area, including southern Mexico, Guatemala, Belize and western parts of El Salvador, and Honduras. It was not a unified empire but consisted of hundreds of cities, each with its own distinctive character and artistic style and its own government. It was well advanced in art and science.

west called the Southern Lowlands; and a northern region—the Yucatán peninsula—called the Northern Lowlands. They built cities throughout these regions: in the period c. AD 300-800 the Southern Lowlands were dominant with sites like Tikal, Uaxactún, and Palenque; the emphasis later shifted to the Northern Lowlands, at sites like Chichén Itzá, Uxmal, and Dzibilchaltún. In the later period (c. AD 1200-1450) *Mayapán became the most powerful city until it was overthrown in AD 1441 and several smaller Maya states were established. In the highlands Kaminaljuyú and San Antonio Frutal were the earliest important cities, while Zacuelu, Utatlán, and Iximché (forming the Quiché Maya state) were later more important. All the regions had political and trade connections with the cities of southern and central Mexico.

Significant features of Maya culture include their use of hieroglyphic writing, still only partly deciphered; codices; the Calendar Round, a 52-year ritual cycle combining a solar and a sacred calendar, and a 'Long Count' calendar in which absolute dates were calculated from a zero date (corresponding to our 3113 BC); ceremonial plazas with steeply stepped pyramids, temples, and vast multi-roomed palaces decorated with several different artistic styles through the ages; the ritual ball game; use of the corbel or false arch; and their polychrome painted pottery.

Mayapán, the most powerful *Maya city of Yucatán in the late prehistoric period. In c. AD 1200 its ruler, Hunac Ceel, conquered its principal rival, Chichén Itzá, and dominated politics and trade until a coalition of other cities, led by the city of Uxmal, defeated and sacked it in AD 1441. Its circuit wall, over 9 km. (5.6 miles), enclosed an area, of 4.2 square km. (1.6 square miles) with nearly 4,000 structures inside.

Mayflower *Pilgrim Fathers.

Mazarin, Jules (1602-61), French statesman and cardinal. An Italian by birth, he acted as an envoy for Pope Urban VIII. From 1631 to 1639 he acted as France's unofficial representative at Rome, and then joined *Richelieu's French service permanently. After the deaths of Richelieu (1642) and Louis XIII (1643), he became France's chief minister during the minority of *Louis XIV, deriving his authority mainly from his close relationship with the queen mother, Anne of Austria. He secured good terms for France at the Treaty of *Westphalia (1648), but his decision to continue the war against Spain led to the upheavals of the *Fronde (1648-53). He survived that crisis, then brought the Spanish War to a successful conclusion at the Treaty of the Pyrenees (1659). His continuation of Richelieu's policies enabled him to bequeath to Louis XIV the most powerful kingdom in Europe.

Mecca, capital of the Hejaz in Saudi Arabia, the birthplace of the Prophet *Muhammad, and the holiest city of Islam. Lying in a narrow valley in an arid region, it nevertheless prospered from trade and from the cult associated with its central shrine, the Ka'aba. Muhammad's teachings were strongly critical of his native city, and his life was crowned by the incorporation of pilgrimage to the Ka'aba into Islam. Although it has retained its sacred character ever since, the city soon lost its commercial significance, its prosperity resting henceforth on the *pilgrimage. It was sacked in 930 by the Qarmatians, a radical *Ismaili sect, and fell under *Ottoman suzerainty in 1517.

Medes, an Indo-European people who occupied Media, an area south-west of the Caspian Sea. It seems to have been Phraortus (c.675-653 BC) who established the Median empire. His son Cyaxares (625-585) extended it and, in alliance with Babylon, conquered the Assyrians, capturing *Nineveh in 612. He was succeeded by Astyages, who added various Babylonian territories to his domain before being overthrown by a vassal, *Cyrus the Great, who established the Persian empire of the *Achaemenids.

Medici, Catherine de (1519-89), Queen consort of *Henry II of France. She was the daughter of Lorenzo de Medici and was married to the duc d'Orléans (later Henry II) in 1533. On the accession of her second son, Charles IX, to the throne of France in 1560, she acted as his regent and then as his adviser until his death in 1574. The outbreak of the *French Wars of Religion saw her abandon an initial policy of conciliation in favour of an alliance with the Catholic *Guise faction against the *Huguenots, which led to the *St Bartholomew's Day Massacre (1572), in which hundreds of Protestants were killed. After the accession of her third son, Henry III, she still ruled the court and tried once more to reconcile Catholics and Protestants, but was trusted by neither faction.

Medici, Cosimo de (1389-1464), Florentine banker, the first member of the Medici family to rule Florence.

A detail from Benozzo Gozzoli's fresco, *Journey of the Magi* (1459), depicting Lorenzo de **Medici** as one of the Magi. (Palazzo Medici-Riccardi, Florence)

In Florence the struggle for power between rival patrician families was intense and Cosimo was expelled from the city in 1433 before triumphing over his rivals in 1434. The basis of his wealth was the Medici bank and he managed it prudently and expanded the family's financial dealings into other areas of commerce. In c.1455 he owned a company for the manufacture of silk, two companies for the manufacture of wool, and a bank (all in Florence), and branch banks in Geneva, Bruges, London, Avignon, Rome, Milan, Pisa, and Venice. He was a keen patron of the arts.

Medici, Lorenzo de (Lorenzo the Magnificent) (1449-92). Aged 20 he became joint ruler of Florence with his brother Giuliano. In 1478 the brothers were the targets of a plot organized by the rival *Pazzi family and the pope: Giuliano was killed but Lorenzo survived. His main concern was the promotion of his family, and he was rewarded by seeing his second son become Pope Leo X. He was a collector of antiquities and was Michelangelo's first patron.

Medina, western Arabia, the second holiest city of Islam. Originally an oasis settlement called Yathrib, it welcomed *Muhammad after his *hegira from Mecca in 622 and became the first Muslim community. It grew rapidly until 661, and became known as Madinat al-Nabi—the city of the Prophet. When the *Umayyads shifted the capital to Damascus, Medina declined and came under *Ottoman control from 1517. The tombs of Muhammad and the caliph *Umar are both in Medina.

Medina Sidonia, Alonso Pérez de Guzmán, Duke of (1550-1619), Spanish nobleman. He succeeded to his title in 1555. In 1588 *Philip II charged him with responsibility for commanding the *Spanish Armada against England. He begged to be relieved of the commission, on grounds of ill health and inexperience, but without success. He organized his fleet with great efficiency, and was by no means exclusively to blame for the Armada's eventual failure. He subsequently served both Philip II and Philip III.

megalith, literally a large stone. The practice of building with these (presumably primarily for purposes of ostentation) occurred in such diverse places as Inca Peru, ancient Egypt, and Easter Island. The term is generally reserved for large blocks built into tombs and other monuments in western Europe in the *Neolithic to *Bronze Age, c.4000-1500 BC. They were once thought to have been derived from a single source, but further study and close dating suggest that that is too simple a view, and that many areas were involved.

While many monuments consist of separate stones raised on end as *menhirs, stone circles, and avenues, the same technique was often used in walling chambers. Roofs could be of horizontal capstones to make the so-called dolmens, or of oversailing courses of slabs, which are known as corbelled vaults. The largest block recorded is the capstone of the tomb at Browneshill, County Carlow, Ireland, estimated to weigh 100 tonnes.

Mehmed II (Muhammad II, the Conqueror) (1430-81), *Ottoman sultan (1451-81). He was frustrated while ruling briefly (1444-6) during the retirement of his father, *Murad II, but backed expansionist factions on coming

Among the **megalith** structures of the Neolithic period was the dolmen. Browneshill dolmen, in Country Carlow, Ireland, is one of the largest in Europe. Two of its smaller supports have collapsed, although the enormous capstone still remains. The true dolmen type was confined to northern Europe and was originally designed for a single burial.

to power, and by 1453 had achieved the long-standing strategic objective of the Ottomans of taking *Constantinople and thus uniting the European and Asian parts of their empire. Ceaseless campaigns brought further gains in the Balkans, consolidated control of Asia Minor, and took Otranto in Apulia, but failed to wrest Rhodes from the *Knights of St John. He modernized his forces by equipping them with firearms and artillery, and created the institutional framework of the developed Ottoman state.

Melanchthon, Philip (1497-1560), German religious reformer. He became professor of Greek at Wittenberg in 1518, and came under the influence of *Luther, whose teachings he helped to systematize. After 1521 he assumed an even more prominent role in the Reform movement: his *Loci Communes* (first published in 1521) was the first ordered presentation of Reformation doctrine; he took part in the Diet of Speyer (1529) and the Colloquy of Marburg (1529) and was largely responsible for drawing up the *Augsburg Confession (1530). A conciliatory man influenced by Christian *humanism, he was particularly active in reforming Germany's educational system. In 1537 he signed the Schmalkaldic Articles (a statement on doctrine drawn up by Luther) with the reservation that he would accept the papacy in a modified form.

Melanesians, peoples of New Guinea, New Caledonia, and neighbouring island groups. With their dark wavy (to kinky) hair, they were once assumed to be closely related to the *Negroids of Africa. Recent work on blood groups now links them more closely to the neighbouring *Australoids and *Mongoloids of East Asia. There are no fossil remains to document their early story. They had apparently developed subsistence farming by the early date of 7000 BC. The pig was the principal domesticated animal and fishing was an important part of their economy. In New Guinea, elaborate rituals developed, until recently including head-hunting and cannibalism.

Memphis, the capital city of Egypt throughout the Old Kingdom (c.2700-2200 BC). Tradition attributes its foundation to King Menes who supposedly united Upper and Lower Egypt, on whose boundary near the head of the Nile delta the city stands. It was the centre of worship of the god Ptah. The most impressive remains from the site are two colossal statues of *Ramesses II and an alabaster sphinx.

Mencius (Mengzi, Meng-tzu) (*c*.372–289 BC), Chinese philosopher. His impact on China was second only to that of *Confucius and he consistently championed Confucius's school. He argued that an individual's innate goodness would be revealed in the right environment. By pithy examples he demonstrated that a ruler who saw to the welfare of the state would attract all people to his sway; 'all under Heaven' would obey him. He argued that rebellion against oppressive rule is justified. Like Confucius he went from court to court but failed to win high office. His teachings later became part of the Chinese classics.

Mendoza, Antonio de (1490–1552), member of an aristocratic Spanish family, appointed first viceroy of Mexico by *Charles V. He served in that important position from 1535 to 1550 and provided the new colony with resolute leadership. During his administration he improved relations between Spaniards and Indians, fostered economic development, especially in mining, and lent his support to important educational initiatives for both the Spanish and Indian populations. In 1551 *Charles V named him viceroy of Peru but he served in Lima for only a year before his death.

menhir, a standing stone. The practice of raising *megaliths on end spread widely in western Europe in the 3rd to 2nd millennia BC, and occasionally elsewhere. They could stand isolated, or in avenues as at *Carnac, or in circles in Britain. In some areas round the central Mediterranean they were given human features by the addition of carved breasts, shoulders, eyes, or weapons. Whether they were to commemorate the dead, to serve as places of worship or congregation, as route-markers, or as astronomical observation points remain matters for debate.

Mennonite, a member of a Protestant sect that evolved in Friesland in the 16th century. Mennonites take their name from Menno Simons (1496–1561), a Dutch ex-priest who became an *Anabaptist in 1536. He shepherded the dispersed Anabaptists of northern Europe into congregations, which were soon named after him. Persecutions drove many Mennonites from Holland and Switzerland where they were initially strongest. The first emigration to the USA took place in 1663.

mercantilism, a term used in the 18th century to describe the previous century's practices and beliefs in economic matters. It aimed at exploiting natural resources fully to promote exports and limit imports. Mercantilists believed that the possession of gold or 'bullion' was all-important and countries without a source of precious metal must obtain it by commerce. Trading was controlled by government-backed companies, tariffs were imposed, and trade wars such as the *Anglo-Dutch wars were fought. Mercantilism is particularly associated with *Colbert, who hoped to strengthen France by improving its public finance, though in reality prolonged warfare meant higher taxes which weakened industrial prosperity. Later supporters of free trade (*laissez-faire) opposed the mercantilist theory that the volume of trade is fixed and that to increase one's share one must lessen that of others.

Merchant Adventurer, originally any English merchant who engaged in export trade. A trading company of Merchant Adventurers was incorporated in 1407 and flourished in the 15th and 16th centuries. It derived from loosely organized groups of merchants in the major English ports who sold cloth to continental Europe, especially the Netherlands. They acquired royal *charters in cities such as Bristol (1467) and London (1505) and in their European settlements. They became dominant in England's foreign trade, ousting their rivals, the German merchants of the *Hanseatic League. Until 1564 their principal continental market was in Antwerp, the commercial capital of north-western Europe; from 1611 they made Hamburg their foreign centre, but their main base had long been London. They were the forerunners of the great *chartered companies, and declined in importance in the 18th century.

Merchant Stapler (or member of the Company of the Merchants of the Staple), any English merchant who traded in wool which passed through the *Wool Staple (at Calais in France from 1363), the fixed place for its marketing in continental Europe from the 13th to 16th century. The setting up of the Staple (which had between twenty–six and thirty-eight members) had the effect of keeping down the price paid for wool in England, and thus of encouraging the rise of the English cloth industry and its merchants, such as the Clothworkers and *Merchant Adventurers, but until well into the 16th century the Merchant Staplers were pre-eminent in English overseas trade as a result of their monopoly in the trading of wool.

Mercia, the kingdom of central England formed in the 7th century largely from the settlements made by the Angles in the 5th century. The original communities located in the valley of the Trent had been settled by tribes from the Fenland in the east and from the Humber estuary to the north. This loose tribal confederation was first welded into a kingdom in the 7th century through the military and political skills of its rulers *Penda and his son Wulfhere. The borders of Mercia were extended to meet Wales in the west, Wessex in the south (including London), the eastern coastline, and the northernmost regions. In the 8th century, during the reigns of Ethelbald and *Offa, Mercian supremacy was recognized throughout England south of the Humber and by *Charlemagne and the caliph *Harun al-Rashid. In the 9th century Mercia collapsed dramatically and was partitioned between the *Danelaw and Wessex (877).

Meroë, an ancient city on the Nile *c*.200 km. (125 miles) north of Khartoum. It was the capital of the Kush kingdom from the early 6th century BC until *c*.AD 320–5 when it seems to have fallen to conquerors from *Axum. Excavation has revealed extensive remains of the ancient city which include palaces, temples, houses, streets, baths, and, not far distant, pyramids. It was an important centre of iron-working as early as *c*.500 BC.

Merovingians, a dynasty of kings of the *Franks, named after Mérovée (d.458), the grandfather of *Clovis.

The Merovingians were warriors rather than administrators, few of them showing any interest in government. After the death of Dagobert I in 638, power passed from the kings to the 'mayors of the palace'. The mayor of the palace was originally the head of the royal household and came to represent royal authority in the country, administer the royal domains, and command the army in the king's absence. The most notable mayors of the palace were Pepin (ruled 687–714), *Charles Martel (ruled 714–41), famous for his victories over *Saracen invaders from Spain, and *Pepin, the father of *Charlemagne, who in 751 deposed the last of the Merovingian kings and usurped the throne.

Mesolithic ('Middle Stone Age'), the transitional period between the *Palaeolithic and *Neolithic ages. Its people were the hunting and gathering groups that existed about 10,000 years ago as the climate became warmer at the end of the last Ice Age. The term is most applicable to western Europe, where Mesolithic hunting societies continued to exist contemporaneously with Neolithic farming groups further east.

Mesopotamia (from the Greek, 'between rivers'), the land which lies between the rivers Euphrates and Tigris in the Middle East. By the end of the 4th millennium BC a number of city-states had been established by the *Sumerians in the south where agriculture (helped by irrigation), trade, and industry developed. The first Mesopotamian empire was forged by Sargon of *Akkad c.2350 BC and lasted approximately 150 years. The second was centred on the Sumerian city of *Ur c.2150–2000. Thereafter the area came under the influence of Amorites from Canaan, a new empire being established by Sumuabum in 1894 BC with its centre in *Babylon. Northern Mesopotamia had by this time been occupied by the *Assyrians. Both powers subsided with the appearance of more invaders from the north in the second half of the 2nd millennium BC, but the Assyrians re-emerged to create a great empire (744–609) which extended as far as Egypt. This was shattered by the Chaldean dynasty of Babylon, which also established a regime which extended to the Mediterranean until conquered in 539 by *Cyrus the Great. From then on Mesopotamia came under the aegis of several major empires—the *Achaemenids, the *Seleucids, the *Parthians, and the *Sassanians. Arab conquest (635–7), and then a century of intra-Muslim rivalry with Syria, culminated in the *Abbasid dynasty's building of a new capital, Baghdad. The area was devastated by Mongols in 1258 and 1401, and eventually became a loosely ruled part of the *Ottoman empire. Mesopotamia is of cardinal importance as one of the first great *civilizations of antiquity. Here urban life and a written language first developed, and excavation has revealed the area's artistic legacy.

Methodist, a follower of the evangelical Protestant religious movement founded by John *Wesley in the 18th century. He organized his followers into 'societies' with class-meetings, regular attendance at which was a prerequisite of membership. His doctrine of universal salvation brought him into dispute with the *Calvinists, and the intense emotionalism of his meetings, with the difficulty of subjecting his societies to ecclesiastical control, caused conflict with the *Anglican Church. His consecration of Dr Thomas Coke (1784) to superintend the

A statue from the city of Mari in **Mesopotamia** which dates from c.2800–2400 BC. An inscription on the shoulder of this government official reads 'Statue of Ebih-il . . . to Ishtar, he has consecrated it'. The stone figure was discovered in the temple at Mari in 1934. It is dressed in the typical 'kaunakes' skirt of rough, hairy material. (Musée du Louvre, Paris)

societies in America and to ordain ministers led to a definite breach which widened with Wesley's death in 1791, when the Methodists formally separated from the Anglican Church. The movement spread rapidly both in Britain and in America.

Mexico, the Central American country immediately south of the USA. In prehistory it formed the greater part of ancient Mesoamerica, within which arose a succession of related civilizations which shared many cultural traits: socio-political organization based on cities; ceremonial plazas of pyramids, platforms, and temples; similar deities; calendrical systems; long-distance trading; and the ritual ball game. Some of these were the *Olmec (Gulf Coast), *Maya (Yucatán), *Teotihuacán (Central Valley), Zapotec and *Mixtec (Oaxaca), *Toltec (North Central), western cultures, and *Aztecs.

The conquest of the Aztec empire by *Cortés was complete by 1521, and *New Spain became the first Spanish American viceroyalty, eventually including all of ancient Mesoamerica, northern Mexico, the Caribbean, and most of the south-western USA. A rigid colonial administration, including repression and exploitation of the native population, lasted for the next 300 years. As many as 90 per cent of the Indians had died of European-introduced diseases by the early 17th century, and thereafter their numbers only slowly increased.

Michael VIII Palaeologus (c.1225–82), Byzantine emperor (1259–82). He usurped the throne of the young emperor of *Nicaea, John IV, for whom he had been regent. He then recovered *Constantinople and returned the Byzantine capital there, and was crowned Byzantine emperor. He attempted to reconcile the two branches of the church at Constantinople and Rome, though a voluntary union between them agreed at the Council of Lyons (1274) was for immediate political purposes and only lasted until 1289. His great rival was Charles I of *Anjou and Naples; after the *Sicilian Vespers (1282) Angevin power was reduced and Michael's successor, Andronicus, withdrew from the union with Rome, no longer needing the pope as an ally.

Micronesians, inhabitants of the Mariana, Caroline, Kiribati, and Marshall island groups and neighbouring islands in the Pacific Ocean, nearly 3,000 islands in total. Archaeological and linguistic evidence suggests that prehistoric people had moved into this area from islands in south-east Asia and from Melanesia by at least 4000 years ago.

Middle Ages, the period in Europe from c.700 to c.1500 (though this is not a period for which precise dates can be given). The decline of the Roman empire in the West and the period of barbarian invasions in the 5th and 6th centuries (*Dark Ages) was followed by the emergence of separate kingdoms and the development of forms of government. The coronation of *Charlemagne in AD 800 is held to mark the end of anarchy and the revival of civilization and learning. England, under *Alfred, similarly saw the encouragement of learning and the establishment of monastic houses. Territorial expansion by *Vikings and *Normans throughout Europe in the 9th and 10th centuries, initially violent and disruptive, led to their assimilation into local populations. Trade and urban life revived.

The High Middle Ages (12th and 13th centuries) saw a growth in the power of the papacy which led to clashes between the pope and secular rulers over their respective spheres of jurisdiction. The creation of new monastic orders encouraged scholarship and architecture. The obsession with *pilgrimage to holy shrines was the impetus behind the *Crusades, in which thousands of Christian knights went to Palestine to fight the Muslims and convert them to Christianity. Society was organized on a military basis, the *feudal system in which land was held in return for military service. But although war dominated this period, it also saw the growth of trade (notably the English wool trade), the foundation of *universities, and the flowering of scholarship, notably in philosophy and theology (*scholasticism). Gothic art and architecture had its finest expression in the cathedrals built from the 12th century.

During the 13th and 14th centuries various factors combined to cause social and economic unrest. The *Black Death, and the *Hundred Years War between France and England, resulted in a falling population and the beginnings of *anticlericalism. In the 15th and 16th centuries the *Renaissance in Italy marked a new spirit of sceptical enquiry and the end of the medieval period.

Milan, a province and city of northern Italy. Originally an Etruscan settlement, in 222 BC it became the Roman town of Mediolanum and developed into the prosperous second city of the early Roman empire. It gave its name to the Edict of Milan issued by *Constantine in 313 which recognized Christianity and gave universal religious toleration. Towards the end of the 4th century its governor *Ambrose became its most famous bishop. After *Charlemagne defeated the Lombards the city which had briefly functioned as capital of the old Western empire was ruled by a patriarch or prince-bishop. In the 16th century it passed to Spain.

Miletus, one of the leading Ionian cities of Asia Minor. It established a number of colonies on the Hellespont and Black Sea coasts in the 7th and 6th centuries BC, and traded widely. It continued to thrive even after coming within *Croesus' sphere of influence, and in 499 led the revolt of the Ionians against Persia. After its final defeat in 494 it was razed, and never thereafter recovered its former power. It revolted from the Athenian empire in 412, but then came under Persian control. The city's economic decline was hastened by the silting up of the harbour while it was part of the Roman empire.

militia, a military force composed of citizens, enlisted or conscripted in times of emergency, usually for local defence. In England, it developed during the Middle Ages from the Anglo-Saxon *fyrd. Its forces were usually raised by impressment (forcible recruiting), and until the 16th century it was supervised by the local sheriff. Then the Lord Lieutenants were given the responsibility until the later 19th century. Control of the militia was disputed in 1642 between Charles I and Parliament. Parliament prevailed, but at the Restoration the militia was again placed under royal command. It declined in importance as a standing army emerged in the late 17th century, and the same applied in Europe. In colonial America, however, they were the only form of defence. As the *minutemen, they played a significant part in the Revolution. The Militia Act (1792) required every free, able-bodied, white, male citizen between the ages of 18 and 45 to be enrolled in the militia by local and state authorities. The law went unenforced, and during the next century voluntary militias developed into the National Guard, local organizations supervised, armed, and paid by the federal government.

Minamoto Yoritomo (1147–99), Japanese general, founder of the *Kamakura shogunate. His family, like their rivals, the Taira, had risen to prominence when imperial factions called for military support. In 1160 the Taira, having slain his father, placed him under the surveillance of their *Hojo kinsmen. Taira power reached its zenith in 1180 when Taira Kiyomori made his infant grandson, Antoku, emperor. However, by 1185 Yoritomo was master of Japan. The victories that swept him to power were won largely by his younger brother, Yoshitsune, later enshrined in history and legend as a tragic hero. After destroying the Taira's Inland Sea bases he annihilated their fleet on 25 April 1185 at Dan-no-ura at the southern tip of Honshu. Antoku was drowned.

In 1192 another child-emperor appointed Yoritomo as the first *shogun, whereupon he set up his military administration in Kamakura, which effectively became the central government of Japan. Yoshitsune and other

Minamoto Yoritomo, a 14th-century copy of a portrait in colour on silk attributed to the 12th-century artist Fujiwara no Takanobu. The art of the realistic depiction of individuals in portraits developed during the turbulent Kamakura shogunate founded by Yoritomo. (British Museum, London)

Minamoto rivals had already been killed on Yoritomo's orders, but his supporters were given estates and were to become the basis of the *daimyo. On Yoritomo's death Hojo Tokimasa, whose daughter had married Yoritomo, made himself regent. By 1219 Yoritomo's own line was extinct.

Minden, battle of (1 August 1759), a battle in the *Seven Years War. A French army seized the town of Minden, a German city guarding access to Hanover, but was surrounded by a large force of British, Hanoverian, and Hessian troops under Prince Ferdinand of Brunswick. On 1 August the French were severely defeated and Hanover was saved. This was the only pitched battle in Europe in which British troops were involved during the *Seven Years War.

Ming (1368-1644), the last dynasty of native-born Chinese rulers. It was founded by Zhu Yuanzhang, who in 1368 drove the Mongol *Yuan dynasty from Beijing. Orphaned during a famine, he learned to read as a Buddhist novice and for a time begged for a living. As Emperor Hong Wu (Extremely Martial) he invaded

Mongolia, brought *Yunnan for the time being under effective rule, and obliged *Korea to pay tribute. After *Yongle's reign (1403-24) the Ming abandoned expansionist policies. A hundred and fifty years of peace were marred only by the depredations of Chinese and Japanese pirates and brief Mongol incursions. Meanwhile the administrative system established by the Yuan was improved, for example by the appointment of provincial governors and governors-general. Great public works were undertaken. From 1517 Portuguese and other Europeans, traders and missionaries, appeared on the coast. The Portuguese were permitted to settle in *Macao whilst Matteo *Ricci and other Jesuits were allowed into Beijing. During this period southern Chinese began settling in south-east Asia, where their presence has remained significant.

The invasion of Korea by the Japanese military ruler *Hideyoshi in 1592 threatened China. Although the Ming successfully resisted any incursion into China, it destabilized the country. The *Manchus began to make attacks on Beijing. Banditry became rife in the provinces, and the bureaucracy fell into disorder. Pressure of population on fertile land brought famine and discontent. A rebellion by a bandit, Li Zicheng, which started in Shaanxi Province, cost a million lives. In 1644 Li occupied Beijing and the last Ming emperor hanged himself. The *Qing dynasty followed.

Minghuang (or Xuanzong) (684-762), *Tang Emperor of China (712-56). He came to the throne when the Tang were at their zenith. An army he dispatched to prevent the Tibetans allying with the Arabs crossed the Pamirs and reached the Hindu Kush. In 745 the emperor took as his consort Yang Guifei, formerly his son's concubine, and neglected his imperial duties. In 751 the Arabs defeated the Chinese at the Talas River, resulting in the loss of earlier Tang gains in Central Asia. In 755 An Lushan, a general Yang Guifei favoured, rebelled and took Luoyang. The court fled from Chang'an, Yang was executed, and the emperor abdicated. The revolt, which dragged on until 763, was suppressed only by calling in Uighur troops from Central Asia.

Minoan civilization, the first great European civilization, so-called after the legendary King Minos. It developed in Crete from c.3000 BC, its two leading cities being *Knossos and Phaestus. The evidence of early Minoan pottery suggests a possible influx of settlers from the east and strong trade contacts with Egypt during the early Minoan period (c.3000-2000), though it is not clear whether the Minoan civilization originated with immigrants or grew from the previous Neolithic culture. Minoan civilization enjoyed its greatest prosperity c.2200-1450, due largely to the Minoans' control of the sea, which for the most part rendered major land defenceworks unnecessary. The large Minoan palaces were of complex design, each centred on a large courtyard, with many staircases, smaller courtyards and rooms for cult worship. Magnificent frescos adorned the walls. Pottery, metalworking, gem-engraving, and jewellery-making reached high artistic standards. Bull-leaping seems to have been part of religious or magical rites. The palaces at Knossos and Phaestus suffered destruction c.1700—either through war or as a result of an earthquake—but were rebuilt. A further destruction occurred c.1500, and soon after that the massive eruption of the volcano on Thera

Minoan civilization: detail from painted scenes of a religious procession on a carved limestone sarcophagus, from Aghia Triada, near Phaistos, Crete. One of the long sides shows men bringing offerings to the tomb. On the left (shown here) women carry libations to pour into a vase. Behind them a man plays the lyre. (Heraklion Museum, Crete)

(Santorini) which overwhelmed much of Crete.

*Mycenaean invaders may have taken over Knossos *c.*1450. Within a century it had suffered its last major destruction. That period was a prosperous one for the city, to judge from the grave finds and the continued occupation of the palace. Subsequently, however, the indications are of a civilization under pressure from outside, with communities forsaking the coast for safer sites in the mountains, though Knossos continued to be occupied.

The early Minoan period used a form of pictorial writing which *c.*1800 was superseded by the still undeciphered Linear A. This in turn was superseded by Linear B, which represented an early form of the Greek language. It was a sophisticated society with the beginning of a palace bureaucracy that used written tablets.

minuteman, an American Revolutionary militiaman ready at a minute's notice to take up arms in defence of his property or country. Minutemen distinguished themselves in local, short-term skirmishes and guerrilla actions like *Lexington and Concord, but proved so unreliable in long campaigns and pitched battles that George *Washington turned to long-term recruits. The Second Amendment to the US *Constitution, guaranteeing the right to bear arms, is said to owe its enactment to the 'minuteman philosophy'.

Mirabeau, Honoré Riqueti, vicomte de (1749–91), French orator and statesman. Until 1789 he led a life of violent excesses, and was often in gaol and also exiled.

He was the author of numerous political pamphlets, and when the *States-General was summoned in 1789 he was elected as a delegate from Aix-en-Provence, for the Third Estate, not as a noble. At the royal session of 23 June *Louis XVI disregarded the *Tennis Court Oath and ordered the delegates to deliberate separately from the nobles and clergy. When they were ordered to leave the hall Mirabeau declared: 'We are here by the will of the people and will not leave our seats unless forced by bayonets'. He hoped for the establishment of a constitutional monarchy, allowing some power to the king, in which he hoped to play a major part as chief minister. He was the dominating figure in the events of 1789–91; in 1790 he established secret communications with the court and tried to advise the king but was opposed by the queen. He died in April 1791.

Miranda, Francisco de (1750–1816), South American revolutionary. He was a Creole born in Caracas, who received a commission in the Spanish army and fought against the English in Florida and the West Indies during the American Revolution. When that conflict ended with the independence of the thirteen colonies he dedicated himself to the advancement of the same cause on the South American continent. From 1783 until his death he was the most active promoter of the idea of Spanish American independence. More successful as a political propagandist than as a military leader, he was ultimately captured by Spanish forces and died in a Cádiz prison.

Mississippi cultures, a North American tradition of interrelated cultures in the central and lower Mississippi

A 13th-century stone effigy tobacco pipe of the **Mississippi cultures**. Such elaborate artefacts were used in ceremonials which demanded the practice of ritual smoking. The figure shown here probably represents some form of ritual sacrifice. (Museum of the American Indian, New York)

valley from *c.* AD 700 to 1700. Three principal new features distinguish it from the preceding *Hopewell cultures. Most famous are its huge politico-religious centres of pyramidal, flat topped, earthen temple mounds, which were part of special religious practices, known as the *Southern cult. They were surrounded by the wattle-and-daub houses of farmers. Inspiration for these ceremonial centres was ultimately derived from the cultures of Mexico (Mesoamerica), but exactly how is unclear. Famous sites include Cahokia (Illinois), Aztalan (Wisconsin), and Macon (Georgia), with vast mound complexes, the last two fortified. A second feature was the much decreased importance of burial mounds, and a third was the appearance of completely new pottery styles, also showing indirect Mesoamerican influence.

Mithras, the central figure of a cult popular among Roman soldiers of the later empire. Scholars are divided as to whether there is real continuity between this cult and the reverence for 'Mithra', a creator sun-god, shown in much earlier scriptures of *Hinduism and *Zoroastrianism. The Roman cult focused on secret rituals in cave sanctuaries devoted to sculptures of Mithras killing a cosmic bull. Initiates underwent severe tests which demonstrated the cult's concern with the soldierly virtues of courage and fortitude. Women were excluded. Mithraism flourished along the empire's frontiers—the rivers Danube, and Rhine, and Britain—but finally succumbed to the challenge of Christianity.

Mithridates VI (Eupator) (120–63 BC), King of *Pontus in Asia Minor, which he led to its period of greatest power. He brought under his control the northern coast of the Black Sea and expanded his domains within Asia Minor. In 88 BC he swept through the Roman province of Asia, and then advanced into Greece, but his armies were crushed at *Chaeronea and he made peace with *Sulla in 85. The second and third Mithridatic wars followed before *Pompey inflicted a final defeat on him in 66. He failed to re-establish himself in the Crimea and ordered one of his guards to kill him.

Mixtec, a people of the mountainous regions of Oaxaca, Mexico. Several of their historical codices survive, from which their dynastic history can be traced back to AD 692, including their famous king Eight-Dear-Tiger-Claw.

By *c.* AD 1000 they had formed a loose confederation of city-states and in the 14th century began to infiltrate the valley of Oaxaca where they sometimes fought, sometimes mixed with, the Zapotec culture there.

Mochica (Moche), a culture developed *c.* AD 200–700 in the Moche and Chicam valleys of Peru. It was the earliest major civilization on the northern coast, which expanded into adjacent valleys, but was eventually eclipsed by the Huari culture to the south. Its rulers built huge pyramids of adobe bricks, known as Huaca del Sol (Temple of the Sun) and Huaca de la Luna (Temple of the Moon), south of Trujillo, which were painted with polychrome murals. They also constructed extensive irrigation works and fortifications round their ceremonial centres. Its craftsmen mass-produced pottery bottles and bowls, including fine-quality water jars with 'stirrup' spouts, painted with a variety of religious, military, and everyday scenes and its metal-smiths produced cast, alloyed, and gilded artefacts.

Model Parliament, the English Parliament summoned by *Edward I (November 1295) and subsequently idealized as the model for all parliaments since it was supposed to be truly representative of the people. In addition to earls (7 attended), barons (41), archbishops, bishops, abbots (70), heads of religious houses, two knights from each shire, two representatives from every city or borough, Edward called representatives of the lower clergy (one from each cathedral chapter, two from each diocese). The 'model' was hardly effective. Knights and burgesses did not attend regularly until the mid-14th century. Representatives of religious houses disappeared at the Reformation. The lower clergy preferred their own parliament, Convocation.

Mogadishu, a city on the East African coast, now the capital of the Somali Republic. It is first mentioned in history in the 10th century as having the monopoly of the gold trade of *Zimbabwe, which it lost to *Kilwa in the 12th century, though it remained an important

Detail from a page of the *Codex Zouche-Nuttal*, an elaborate painted manuscript of the Mixtec, who employed a form of picture writing to record details of their history and religion. (British Library, London)

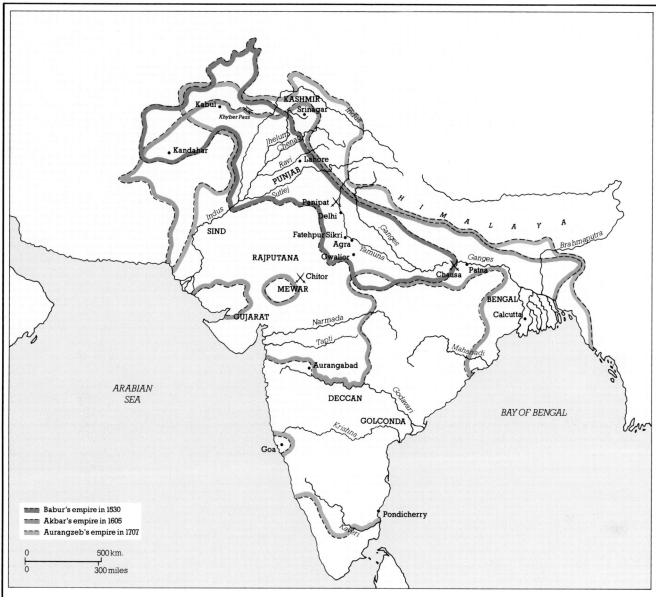

The Moguls: 1526–1707

Babur's victories at Panipat (1526) and elsewhere established the basis of the Mogul empire, but his conquest was essentially military and it was left to his grandson, Akbar, to extend and consolidate Mogul rule in the latter part of the 16th century.

Although the empire reached its greatest extent during Aurangzeb's reign (1659–1707), his authority was seriously threatened by the time of his death and the Moguls' power declined steadily in the 18th century.

commercial port. It was visited by the Chinese admiral *Zheng He *c.*1432. In the Middle Ages a federation of leading families governed it under a sheikh, with a council of viziers and emirs. This Fakhr al-Din dynasty was succeeded in the 16th century by the Muzaffarids. Its prosperity depended on its relations with the nomads of the hinterland, with whom it was frequently at war.

Mogul (or Mughal), a Muslim dynasty of mixed Mongol and Turkish descent which invaded India in 1526, expanded over most of the subcontinent except the extreme south, and ruled in strength until the early 18th century. The first emperor was *Babur (1483–1530). He was succeeded by a line of remarkable emperors: *Humayun, *Akbar, *Jahangir, *Shah Jahan, and *Aurangzeb. They created a strong administration for the

rapidly growing empire, and the official attitude of conciliation towards the majority Hindu population encouraged religious harmony. Culturally, the introduction of the Persian language and Persian artistic styles led to a distinctive Indo-Muslim style in miniature painting and architecture, the legacy remaining today in the tombs and palaces of Delhi, and Agra, and several other cities of India and Pakistan.

Internal and external pressure accelerated the weakening of central power during the 18th century. Rival court factions undermined the position of less capable rulers, allowing provincial governors to seize local power. The abandonment of conciliatory religious policies encouraged a resurgence of Hindu power, notably among the *Marathas. Hostile invasions from central Asia revealed the hollowness of the dynasty's claim to all-India hegemony, so that by 1803, when Delhi fell to the *East India Company, all real power had already been lost. For another half-century they enjoyed a 'twilight era' as nominal 'kings of Delhi', dependent on British goodwill, but in 1857 the last Mogul king was exiled and the title abolished.

Mohács, a town and port on the River Danube, in Hungary. It was the site of two important battles: in 1526 the Hungarian Louis II encountered the invading forces of the Ottoman sultan *Suleiman II there, and lost decisively. Louis's death led to civil war and the domination of most of Hungary by the Ottomans. In 1687 a second battle was fought there and an army under Charles of Lorraine routed the Ottomans.

Mohawk, a North American Indian tribe, the easternmost of the Five *Iroquois nations, living in eastern New York, west of the *Mahicans. They were the first Iroquois to experience the impact of European trade goods, and as trade rivalry developed between tribes were the prime movers in establishing the Iroquois League. The effectiveness of Iroquois control over trade routes increased as warfare changed with the use of guns, obtained from the English and Dutch from c.1640. After the destruction of Huronia (1647-9) peace treaties were signed with the French in 1653, 1667, and 1701 at Quebec. Jesuit missionaries converted many in the late 17th and 18th centuries and English missionaries began work among them in 1704. Many moved to new villages near Montreal in the 1670s and other villages on the St Lawrence in the mid-18th century. In the American War of *Independence most sided with the British and as a result any who had remained in New York were driven out in 1777 to join their relatives in Canada, where most remain today.

Mohenjo-daro, the most important site so far excavated for the understanding of the *Indus civilization, which flourished c.2500-1500 BC. It is situated near the right bank of the River Indus, in *Sind province, Pakistan. It is about a mile square, comprising a citadel mound on the west, and a larger lower city, laid out on a grid pattern, to the east. In the citadel is a brick bath, about 2.5 m. (8 ft.) deep, which was possibly used for ritual bathing. Other civic buildings included a large granary and an oblong hall which may have been a temple.

Mohican *Mahican.

Moluccas (or Maluku), a group of islands in eastern Indonesia. They include the Bandas, Ternate, Tidore, and *Amboina—the fabled Spice Islands, the source of cloves, nutmeg, and mace. In the 16th century they fell under Portuguese control. Not long converted to Islam, the islanders found in religion a focus of opposition to the Portuguese and later to the Dutch. Francis *Drake visited Ternate in 1579. In 1599 the Dutch made their appearance and by 1666 they had subjugated all the islands. They exercised a tight monopoly, systematically destroying spice trees when over-production threatened. During the French Revolutionary and Napoleonic wars the British twice occupied the Moluccas.

Mombasa, East African seaport. It was settled by Arabs in the 8th century and became an autonomous city-state and a centre for trade in ivory and slaves. It is mentioned by al-Idrisi, the Moroccan geographer, in the 12th century as a settlement of iron-workers and hunters. Vasco *da Gama visited it in 1498, and the Portuguese admiral, Francisco d'Almeida, sacked it in 1505. The Turks tried to take it in 1589, and the Portuguese therefore built Fort Jesus, transferring their trading centre there. They lost it, and all the coast, to the sultanate of Oman in 1698. Omani administration of Mombasa was in the hands of the Mazrui family until 1837, when control passed to *Zanzibar.

Monasteries, Dissolution of the (1536-40), the systematic abolition of English monasticism and transfer of monastic property to the Tudor monarchy, part of the English *Reformation. Thomas *Cromwell, *Henry VIII's vicar-general, pointed the way ahead by commissioning the *Valor Ecclesiasticus* (1535), a great survey of church wealth, and by sending agents to investigate standards within the religious houses. An Act of Parliament (1536) dissolved monasteries with annual revenues of under £200. This provoked an uprising, the *Pilgrimage of Grace. In its aftermath, Cromwell forced certain abbots to surrender larger houses to the king. Another Act (1539) confirmed all surrenders that had been, and still were to be, made and monastic lands passed to the Court of Augmentations of the King's Revenue, a state department. Resistance was minimal. By 1540 all 800 or more English houses were closed. Eleven thousand monks, nuns, and their dependants were ejected from their communities, most with little or no compensation.

The Dissolution had a number of consequences apart from the immediate wholesale destruction of monastic buildings and the despoliation of their libraries and treasures. The nobility and gentry benefited financially from the distribution of former monastic lands, which were used to form the basis of new private estates, and the laity gained a monopoly of ecclesiastical patronage which survived for the next three centuries. The termination of monastic charity and the closure of monastery schools stimulated the introduction of the *Poor Law system and the foundation of grammar schools.

monastery, a community of monks living by prayer and labour in secluded, often remote, locations. Monasticism is common to most religions. *Buddha (c.563-c.483 BC) founded a monastic order (*sangha*) and a code of discipline which is still used and was spread by missionaries throughout Asia. The Essenes, a Jewish messianic sect (2nd century BC) founded a remote community by the Dead Sea. The 6th century revival in China of *Taoism, with its emphasis on individual salvation, prompted the founding of many monasteries on the Buddhist model; however, from the 8th century elements of worship were taken somewhat indiscriminately from both religions and from *Confucianism. Numerous *Hindu orders had also emerged by that period, but Islam did not fully develop a monastic organization until the *Sufis formed the Rifaite and Mawlawite brotherhoods in the 12th century.

Christian monasticism originated in Egypt and Syria in the remote hermit communities of the Desert Fathers (particularly St Antony). After St Pachomius (c.292-c.346) had founded the first organized community at Tabennisi in Egypt, Christian monasticism spread rapidly

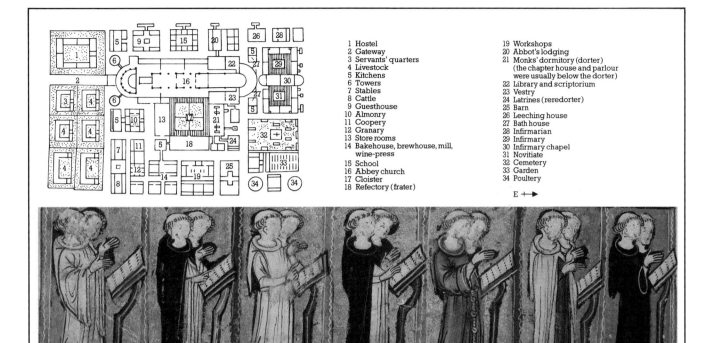

Monastery

Top: Plan of a Benedictine monastery based on the great 9th-century abbey of St Gall, Switzerland. Centred on the cloisters and the chapter house (on the south side of the church), western Christian monasteries were planned as self-sufficient communities, able to cater for the material and

spiritual needs of both monks and travellers. Monasteries usually grew their own food on land within their walls.
Bottom: Religious orders from a 14th-century psalter (left to right): Cistercians, Dominicans, Premonstratensians, Austin friars, Franciscans, Carmelites, and Benedictines.

Key to plan:

1 Hostel	19 Workshops
2 Gateway	20 Abbot's lodging
3 Servants' quarters	21 Monks' dormitory (dorter)
4 Livestock	(the chapter house and parlour
5 Kitchens	were usually below the dorter)
6 Towers	22 Library and scriptorium
7 Stables	23 Vestry
8 Cattle	24 Latrines (reredorter)
9 Guesthouse	25 Barn
10 Almonry	26 Leeching house
11 Coopery	27 Bath house
12 Granary	28 Infirmarian
13 Store rooms	29 Infirmary
14 Bakehouse, brewhouse, mill, wine-press	30 Infirmary chapel
15 School	31 Novitiate
16 Abbey church	32 Cemetery
17 Cloister	33 Garden
18 Refectory (frater)	34 Poultery

E ⟶

in Eastern Europe through the Rule of St Basil (*c*.330–79), with its emphasis on a repetitive liturgy, and in Western Europe with St *Benedict's Rule of carefully regulated worship and labour. In the 10th century the great orders of monasteries evolved. *Cluny, in France, (founded 909) built a series of 'daughter houses' which extended throughout Europe, all under the direct control of the powerful abbot of Cluny. The *Cistercians (founded 1098) also built monasteries in Europe and England, though these foundations enjoyed a semi-autonomous position and were only subject to the direct influence of the abbot of Citeaux at an annual council. Other orders were the Carthusians (1098), the Premonstratensians (1120), and the Gilbertines (1131).

Monasteries were an integral part of medieval life, as centres of pilgrimage, hostelry, medical care, and learning. Many of the poets and chroniclers of the Middle Ages were monks; monastery libraries housed the classical texts and biblical manuscripts which were the basis of contemporary scholarship. Monastic architecture combined the skills of mason and sculptor: the form of monastic building became more or less standardized in the 11th century, following the Benedictine model.

Monck, George, 1st Duke of Albemarle (1608–70),

English general, admiral, and statesman. He began his career as a Royalist and was taken prisoner during the *English Civil War; but he was given a command by Parliament and later completed the suppression of the

Royalists in Scotland. In the first *Anglo-Dutch War he fought three naval battles before returning to Scotland in 1654. He was trusted by *Cromwell, but after Cromwell's death he acted to secure the restoration of *Charles II in 1660, and he received many honours. Monck was placed in charge of London during the *Great Plague (1665) and *Fire of London (1666).

money, metal coins or paper notes given and accepted when buying and selling goods. In primitive societies goods were exchanged directly in a barter system, a method that depended upon having suitable goods to exchange and hindered the development of specialization in trade. As trade expanded goods began to be exchanged for portable objects that represented an item of fixed value, for example an ox. Initially these objects included cowrie shells, stones, and precious metals. Coins of fixed values issued by a government first appeared in the 8th century BC both in Lydia, in Asia Minor, and in China. Early Lydian coins were made of electrum, a mixture of gold and silver, but under *Croesus they were made of gold and of guaranteed value. Britain's first coins, of silver and copper, appeared in the 1st century AD. Thereafter until recently most coins have been made of gold and silver, with copper and bronze used for coins of low value. The style of a head on the obverse and a symbol on the reverse was very widely used from the 4th century BC onwards. Both clipping (cutting small pieces off the coin) and counterfeiting were common practices

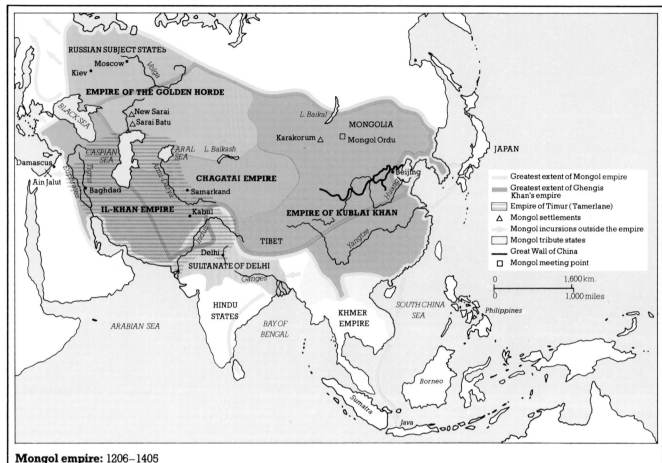

Mongol empire: 1206–1405

The Mongol empire expanded dramatically from 1206 under Genghis Khan, and after his death, under his four sons between whom the empire was divided. Total domination of Europe and Muslim Asia was probably prevented only by the defeat at Ain Jalut (1260), which halted the advance in the west, and by the

succession disputes after the Great Khan's death in 1259. Tamerlane, the last great Mongol conqueror, founded a new empire in the 14th century, at the expense of the Chagatai and Golden Horde khanates, the last remnants of the old empire, but his empire barely survived his death in 1405.

until the introduction of milled edges in 1724 made this more difficult. Banknotes were first issued by banks who undertook to pay the sum of money that appeared on the note from their deposits of gold. The *Bank of England issued notes from its foundation in 1694 and banknotes are now the principal form of money in circulation.

Mongol empire, an empire founded by *Genghis Khan early in the 13th century. Loosely related nomadic tribesmen who lived in felt huts (yurts) and subsisted on meat and milk—and fermented mares' milk (koumiss)—were united for the first time under his leadership. From Mongolia they swept out to Asia and eastern Europe. Splendid horsemen and archers, their onslaught was difficult to resist. Khakhans (Great Khans) elected from among Genghis's descendants continued his conquests. Central Russia, Poland, Hungary, Bulgaria, and Romania were overrun, but following the death of the Khakhan Ogodei in 1241 the Mongols withdrew to attend an election in their capital, Karakorum, in Mongolia. However, the *Golden Horde remained in control of Russia. In 1245 an advance towards *Mesopotamia began. In 1258 Hulagu, Genghis's grandson, sacked Baghdad, but was defeated by the *Mamelukes at Ain Jalut (1260). The conquest of China, begun under

Genghis, was completed sixty-five years later under *Kublai Khan.

Campaigns were notable for the death and destruction they left behind. But once the empire was established a great peace, the *Pax Mongolica*, descended. Travellers, of whom *Marco Polo was only one, passed to and fro between the *caravanserais of the empire. There was a steady flow of trade—and ideas and technology—between men of many lands and faiths. For a time *Nestorians flourished in the steppe areas and parts of northern China. After Kublai shifted the capital to Khanbaligh (now Beijing), it became increasingly difficult to maintain the Khakhan's authority over remote parts of the empire. Quarrels over succession, corrupt and incompetent administration, and revolts accelerated disintegration. After 1300 the *khanates were fully independent. By 1368 the Mongols were driven out of China and in 1372 a Chinese army burned Karakorum. *Tamerlane and *Babur both claimed descent from Genghis Khan.

Mongolia, an immense thinly populated steppe, divided by the Gobi Desert into Inner and Outer Mongolia. Although named after the *Mongols, up to the 12th century they only controlled a small area near the sources of the Orkhon River, and other nomad tribes, such as the Merkit and Naiman, held greater power in the

Eastern steppes. In the 13th century, however, the Mongols swept out to create the *Mongol empire. In the 16th century they were converted to *Lamaism. During the 17th century the *Manchus won control of Inner and then of Outer Mongolia.

Mongoloids, one of the major human racial groups, distributed widely through Asia from the Caspian Sea eastwards. They are also found on many islands off the Asian mainland and, as Inuits and Aleuts, in northern Canada and in Greenland. The *Amerindians, the other indigenous peoples of America, are also sometimes grouped with the Mongoloids. The prehistory of this racial group is unclear. The distinctive physical characteristics of Mongoloids could have developed in isolated populations in the Far East over a long period of time, perhaps many tens of thousands of years, independently of developments in Africa and Asia. In this scheme, which is not widely accepted today, there was a steady evolution from *Peking Man to Mongoloid *Homo sapiens* in China. Alternatively, Mongoloids developed from a modern human stock that migrated to the Far East from Africa via western Asia and intermixed with indigenous populations. In this latter scheme, Mongoloid characteristics developed relatively recently in human evolution. In historic times, Mongoloids became more widely distributed with the rise of the *Mongol empire spreading to northern India, Persia, and Central Asia.

monk *monastery.

Monmouth's Rebellion (1685), an insurrection in south-west England against *James II, led by the Duke of Monmouth (1649–85), illegitimate son of *Charles II. Monmouth had failed to gain any political benefit from the crises of his father's reign other than popularity, which was due more to his Protestantism and lenient treatment of rebels than to his rather meagre talents. The Duke of Argyll led a revolt in Scotland against James and persuaded Monmouth to launch a rebellion in the south-west. He landed at Lyme Regis in Dorset, and was proclaimed king at Taunton, but could muster only limited support. He failed to take Bristol and, with forces inferior in training, experience, and equipment to the king's army, was routed at *Sedgemoor. Monmouth was captured a few days later and executed; his followers were harshly punished by the *Bloody Assizes.

Monomatapa *Mwene Mutapa.

Mons, inhabitants of *Burma, akin to the *Khmers. They early established themselves in the Irrawaddy delta, adopted an Indian script that spread throughout Burma, and were converted to Theravada *Buddhism brought by missionaries from South India and Ceylon. By 825 they had established themselves in Lower Burma and founded the city of Pegu. In the 11th century they were conquered by *Pagan, but after its collapse retained a shadowy rule in the south until overthrown by the *Toungoo in 1539. They in turn overthrew the Toungoo in 1752 but in 1757 lost their independence to Alaungpaya, the founder of the Konbaung dynasty. Many of the Mons were killed; others fled to Siam.

Montcalm-Gozon, Louis-Joseph, marquis de (1712–59), French general. He fought in the Rhineland,

Bohemia, and Italy before his appointment as commander in Canada in the *French and Indian War (1756). Though initially victorious at Fort Oswego (1756), Fort William Henry (1757), and Ticonderoga (1758), he was surprised by *Wolfe's daring attack on Quebec in 1759 and defeated on the *Plains of Abraham (1759).

Montespan, Francoise Athénais, marquise de (1641–1707), mistress of *Louis XIV (in succession to Mademoiselle de la Vallière) from c.1667. She had several children by the king and employed as their governess the future Madame de Maintenon. The king was tiring of her by the late 1670s, and her influence was finally destroyed by her implication in the *affaire des poisons*, when she was accused of involvement in black magic.

Montesquieu, Charles de Secondat, baron de la Brède et de (1689–1755), a distinguished French lawyer and writer. His *Lettres Persanes* (1721) contained a searching criticism of political and religious institutions as told by two imaginary travellers from Persia. In *De l'Esprit des Lois* (1748) he analysed different forms of government, believing that they should relate to the nature of each society but that all should establish the rule of law. He admired the British Constitution without fully understanding it, and carefully described the separation of powers into executive, judicial, and legislative which influenced the makers of the *Constitution of the United States. He contributed to the *Encyclopédie* and was one of the great figures of the *Enlightenment.

Montezuma (more correctly Moctezuma II Xocoyotzin) (c.1480–1520), the ninth *Aztec ruler (1503–21). His priests foretold the arrival of *Cortés in 1519, as the return of the legendary god-king, Quetzalcóatl, from the east. This caused a certain hesitation and indecision on his part, which Cortés exploited by taking him hostage and forcing him to parley with his people. Because he offered to pay tribute to Spain, they deposed him in 1521 and attacked the Spaniards. In the melée which followed he was wounded and died three days later.

Montfort, Simon de, Earl of Leicester (c.1208–65), the leader of the English barons against *Henry III.

Montezuma addressing his Aztec subjects from the palace rooftop in Tenochtitlán, a 17th-century painting by an unknown artist. (Private Collection on loan to British Embassy, Mexico City)

Born in Normandy, he came to England (1230) and successfully claimed the earldom of Leicester (1239). His marriage to Eleanor (1240), Henry's sister, and appointment as lieutenant of Gascony (1248-52) appeared to secure his position when Henry subjected him to a five-week trial on uncorroborated charges against his Gascon rule. He was acquitted, but left Gascony to return to England. His support of reforms incorporated in the Provisions of Oxford (1258) led Simon into the *Barons' War against the king. Between his victory at Lewes (1264) and death at Evesham he was effectively ruler of England. In that period borough representatives attended Parliament for the first time.

Montrose, James Graham, 5th Earl and 1st Marquis of (1612-50), head of an ancient Scottish family. He fought for the *Covenanters, but fearing their control of both church and state later joined *Charles I in England. In 1644 he was appointed lieutenant-general of Scotland and commissioned to reconquer the country. His force was small, but in a brilliant campaign he won six battles, entered Glasgow, and summoned a Parliament. Support for the Royalists then faded, however; Montrose was utterly defeated in a surprise attack and escaped abroad. Returning in 1650 to raise an army for *Charles II, he was betrayed, captured, and hanged by the Presbyterians in Edinburgh.

Moor, a term commonly used by European authors from the Middle Ages on as a synonym for *Saracen. The Spanish and Portuguese used it of the Muslim inhabitants of North Africa and of Spain; and then, by extension, applied it to Muslims as distinct from pagans in Africa, and from Hindus in India. It is used especially in Spain for its Arab conquerors, who landed first in 710. In 711, under *Tariq ibn Zaid, a *Berber army swept the country, and the *Visigoth kingdom collapsed. The Moors reached the Pyrenees, and crossed into France. Their defeat near Poitiers in 732 marked the limit of their expansion. The new state of al-Andalus (Andalusia) was consolidated by the *Umayyad amir Abd al-Rahman ibn Mauwiya in 756, succeeded by a caliphate in 912 which lasted until 1031. Seville was the first capital, but was superseded by Córdoba in 717. Following the collapse of the caliphate the state dissolved among the Reyes de Taifas, or 'Party Kings' (1031-85), while the Christian kingdoms further north expanded southwards. But in 1090 new North African invaders, the *Almoravids, who were primarily fighting men, made themselves supreme. They gave way in 1145 to another Berber group, the *Almohads, whose collapse early in the 13th century gave the Christians their opportunity, and by 1235 only the kingdom of *Granada remained, covering most of the present Andalusia. Torn by factions, it could not withstand the united monarchy of *Ferdinand and Isabella who brought an end to Moorish rule in Spain in 1491.

The Moorish period in Spain was the zenith of Islamic culture in learning, law, poetry, art, and architecture. The Great Mosque of Córdoba and the Alhambra of Granada were among their supreme creations.

Moravia, an area of central Czechoslavakia. Occupied during the 4th century by Celtic and Germanic tribes, it was conquered by the Romans, and as the Roman empire declined, came under *Avar, and subsequently

Moors at the defence of the island of Majorca (1229), when it was captured by the Aragonese. This fresco detail was painted some fifty years later. 'White' as well as 'black' Moors are among the horsemen. Europeans considered as Moors both those of mixed Spanish, Arab, and Berber origins, and their slaves from south of the Sahara. (Palacio Berenguer de Aguilar, Barcelona)

*Magyar domination. It passed to *Bohemia in 1029. Under *Hussite influence in the 15th century, a moderate form of the religion survived despite persecution. Gaining from the influence of 16th-century *Anabaptists, it evolved as the Moravian Brethren, whose beliefs survived, often in exile, notably in America, after the persecutions following the Bohemian rebellion in the *Thirty Years War. The *Habsburgs ruled Moravia from 1526.

More, Sir Thomas (1478-1535), Lord Chancellor of England (1529-32), English statesman and scholar. He trained as a lawyer and entered Parliament in 1504, becoming Speaker in 1523. By then he had made a European reputation for himself as a leading *humanist. His *Utopia* (1516) bitterly attacked contemporary political and social evils. He was appointed Lord Chancellor by *Henry VIII in 1529 and remained in this office until 1532, a stern campaigner against Protestant heresy. His resignation was probably hastened by reservations over the king's divorce from Catherine of Aragon and the government's anti-papal stance. In 1534 he refused to acknowledge Henry's supremacy over the English Church. After a trial at which he defended himself, he was executed as a traitor. He was canonized in 1935.

Sir Thomas **More** as Lord Chancellor, a sketch by Hans Holbein the Younger. Holbein was given an introduction to More by Erasmus on his first visit to England in 1526. (Royal Collection)

Morgan, Sir Henry (c.1635-88), Welsh *buccaneer, the scourge of Spanish settlements and shipping in the Caribbean between the 1660s and 1680s. Although he had semi-official employment as a privateer, he was little more than a pirate. Among his exploits were the capture and ransom of Porto Bello (1668), the sacking of Maracaibo (1669), and the taking of Chagres and Panama (1670-1). Although knighted and appointed lieutenant-governor of Jamaica in 1674, he continued to encourage piracy and lawlessness. He was disgraced in 1683 but restored just before his death.

Morocco, a North African country. By the 5th century BC *Phoenicians had stations on the coast, when the Carthaginian admiral, Hanno, passed the Straits of Gibraltar, and perhaps reached the Gulf of Guinea. A kingdom of Mauritania was formed in northern Morocco in the 4th century BC; the Romans made it the province of Mauritania Tingitana, based on Tangier. Vandals from Spain occupied it from 428, but the Berbers controlled the interior even when the Byzantines recovered the coast in 533. It did not come under Arab control until Musa ibn Nusayr's conquest c.705. Under Byzantium the puritanism of the Berber character had been manifested in the *Donatist heresy; under Islam a similarly austere movement, Kharijism, arose. True Arab domination was brief, and Berber dynasties emerged, Idrisids (788-974), *Fatimids (909-73), *Almoravids (1056-1147), *Almohads (1145-1257), Merinids (1248-1548), and finally the *Sharifian dynasties from 1524 until the present. Having defeated the Portuguese at *Al-cazarquivir (1578), Morocco itself attempted colonial expansion, defeating the *Songhay empire with the help of firearms in 1591, but ruling it inefficiently.

Mortimer, Roger, 8th Baron of Wigmore and 1st Earl of March (1287-1330), a member of a family of *marcher lords of medieval England. While living in exile in France he won Edward II's wife *Isabella of France as his mistress. In September 1326 Roger and Isabella invaded England and forced Edward to abdicate in favour of her son, the young *Edward III, leaving them as *de facto* rulers of England. Roger's pro-France foreign policy and his financial greed made him unpopular, and in October 1330 Edward III and Henry, Earl of Lancaster, captured him and had him executed.

The family was later restored to favour and **Edmund** (1352-81), 3rd Earl of March, married Philippa, only child of Lionel, Duke of *Clarence; in the early 1370s he and *Edward the Black Prince led the so-called constitutional party against the court party of John of *Gaunt. His son **Roger** (1374-98), the 4th earl, was heir-presumptive of Richard II, but was killed while fighting for Richard in Ireland.

Mortimer's Cross, battle of (2 February 1461). An important battle in the Wars of the *Roses, which took place at a site near Wigmore in Herefordshire. The *Yorkist forces led by Edward, Earl of March, won a major victory over the Earl of Wiltshire's Lancastrian army. It was not negated by the victory of Queen *Margaret of Anjou's army at St Albans less than a fortnight later, for Margaret could not win over the Londoners; on 4 March they acclaimed Edward as King *Edward IV.

Morton, John (c.1420-1500), English ecclesiastic. He was trained as an academic lawyer; the patronage of Archbishop Bourchier led to his becoming a master in *Chancery and a member of the king's council. A *Lancastrian, he lost all in 1461 and fled to France; after the battle of *Tewkesbury (1471) he joined the *Yorkists and was favoured by *Edward IV, becoming Bishop of Ely (1479). He then supported Henry Tudor and soon after Henry's accession (as *Henry VII) was made Archbishop of Canterbury (1486) and Chancellor of England (1487), as well as a cardinal (1493). His assistance to Henry in tax gathering has made him known traditionally as the author of 'Morton's Fork'; his two-pronged argument that the rich could afford to pay and the frugal must have savings.

Moses (13th century BC), one of the patriarchs of *Israel. According to the Old Testament of the *Bible, he was born in Egypt and, having escaped a massacre of Hebrew male children, was brought up by one of the pharaoh's daughters. After much opposition from the pharaoh, he led the Jews away from bondage in Egypt (the *Exodus). During the journey he was inspired to write down the Law and Ten Commandments that give religious and moral principles based on the Hebrew conception of God. He died in Moab, within sight of the 'promised land' of *Canaan.

Mossi, West African peoples of the plains and valleys between the upper Black and White Volta rivers. They reached there apparently in the 10th century. In the

Moses receiving the Law on Mount Sinai, a manuscript illustration from the Grandval Bible. The lower portion of the picture shows the prophet presenting the Law to the Israelites on his return. (British Library, London)

11th century they formed three kingdoms, Yatenga, Wagadugu, and Gurma. In the 12th and following centuries, by skilful organization and the use of cavalry, they ruled a wider area, administered by 393 *nabus* under a *morho-naba*, or emperor, at Wagadugu. In the 16th century they fell to *Mali and *Songhay, but Yatenga and Gurma later managed to recover independence, and continued into the 19th century as sovereign states.

mound-builders, American Indian tribes in Ohio and Illinois who, from c.1000 BC, erected richly furnished circular burial mounds, resembling the European *barrows in shape and function. In the *Mississippi basin some centuries later (c. AD 700), larger and more complex mounds, presumably for ceremonial gatherings, were raised to support temples. Though most were rectangular, and frequently built in large groups as at Cahokia in Illinois, there are more bizarre ones like the 400 m. (1,300 ft.) long snake mound in Adams County, Ohio. These may have been influenced from Mexico.

Mount Badon (according to *Gildas), the site of a major battle or siege between the Britons and the invading Saxons c.500. *Nennius, 300 years later, claimed that King *Arthur fought there. The British victory gained for them a brief period of peace, perhaps for half a century. Although the site of the battle is not known, a possible location is the Iron Age hill fort of Badbury Rings near Wimborne, Dorset. Similarly, the date can only be ascribed tentatively to around the year 500.

Mousterian, a prehistoric culture typically associated with *Neanderthals and the Middle *Palaeolithic in Europe. Tools of this type are named after the Neanderthal cave site of Le Moustier in the Dordogne, France but are now known from throughout Europe from about 130,000 years ago to about 30,000 years ago and disappear with the emergence of anatomically modern humans and the start of the *Upper Palaeolithic. They are also found in North Africa and the Near East. Characteristically, Mousterian tools were made from flint. Compared with the earlier *Acheulian tools, there was a much greater emphasis on flakes, which allowed a larger range of tools to be produced. Small, neat handaxes and stone and bone projectile points were among other typical Mousterian tools.

Mozambique, a country in south-eastern Africa, known to medieval Arab geographers as Musambih. According to the Arab historian al-Masudi it was already exporting gold from mines in the interior of what is now *Zimbabwe in the 10th century. Merchants from *Mogadishu had a monopoly for a time, though it was taken over by *Kilwa in the 12th century. The Portuguese sacked the port of Sofala in 1505, and built a new town as the seat of a captaincy to control the gold and other trade. The present city of Moçambique was begun with a fort in 1508 as a refreshment station on the way to *Goa. The first inland settlements at Sena and Tete were Arab trading towns, from which they made contact with the *Mwene Mutapa and other hinterland rulers until the 19th century.

Mughal *Mogul.

Muhammad (c.570-632), the Prophet of *Islam. He was born in *Mecca of the Hashemite clan of the tribe of Quraish and orphaned in infancy. He won the name 'al-Amin', 'the trustworthy', and married his employer, a wealthy widow, Khadija. Religious contemplation led him to a vision in 610 and to the revelations which were subsequently compiled as the *Koran and established the framework of Islam. From 613 he preached openly against idolatry and the social evils of his day, proclaiming the oneness of Allah, 'the God', and the inevitability of judgement. The death of Khadija and of his protective uncle Abu Talib in 619 exposed the Prophet and his followers to the hostility of the Meccans and led ultimately to his flight from Mecca (the *hegira) of 622 and the establishment of the first community of Muslims at *Medina. A lengthy period of sometimes violent struggle followed, ending in the capitulation of Mecca, the purgation of the Ka'aba of idols, and the submission to Islam of most of the tribes of the Arabian peninsula. The Prophet's sudden death in 632 led to the establishment of the *caliphate and indirectly inaugurated the tide of *Arab conquests. Muhammad never claimed to be more than God's messenger, but generations of *Sunnite Muslims have taken his example, in word and deed, as recorded in the *Hadith*, as an unchallengeable model for godly living.

mummy, the body of a person or animal embalmed or otherwise treated to ensure its preservation, a practice associated especially with ancient Egypt. It seems to be connected with a belief in life after death, the body being preserved so that the soul could return to it. Artificial methods of preservation were introduced during the Old Kingdom, but the most complex method of embalming was perfected in Dynasty 21 (*c*. 1100–950 BC). Fine examples of mummification are the bodies of the pharaohs Seti I and Ramesses II. It continued into the Roman period, until Christianity gradually made it obsolete. Local conditions also encouraged mummification among the *Incas.

Murad I (c.1326–89), *Ottoman sultan (c.1362–89). He consolidated his empire's hold on Asia Minor by marriage alliances and outright purchase and rapidly extended its Balkan territories, taking Adrianople in 1362, Macedonia after the battle of Cirnomen (1371), and Sofia and Nish in the 1380s. A Serbian counter-offensive was defeated at the first battle of *Kossovo, in which he was killed.

Murad II (*c*.1403–51), Ottoman sultan (1421–51). He overcame early opposition to his claim to the throne and, after significant reverses, routed a Hungarian-led 'crusade' at the battle of Varna and the second battle of Kossovo (1448). He also made the *janissaries a basic pillar of the Ottoman state.

Muscovy, a state, centred on Moscow which first became a distinct territory in the 13th century. Cut off from the West by the 13th century *Tartar conquest, and by its subscription to *Orthodox faith, it evolved into a princedom under Ivan I (1328–41). Major reconquest of

The second **mummy** of the pharaoh Tutankhamun, from the 18th Dynasty of the New Kingdom (*c*.1550–1300 BC), one of four miniature coffins, all of the same form. Each bore an internal organ of the king—liver, stomach or spleen, lungs, and intestines—removed from the body during embalming.

former Russian territory ensued under *Ivan III, and under *Ivan IV (the Terrible). The state surmounted setbacks in the so-called 'Time of Troubles' (1605–13) and, after 1613, under a new dynasty, the *Romanovs, the state of Muscovy was the nucleus of a unified country.

Muscovy Company, an English *chartered company, incorporated in 1555, with a monopoly of Anglo-Russian trade. It was formed by London merchants, after Richard Chancellor had made contacts with Muscovy in searching for a north-east passage to Asia (1553–4). The trade was largely in cloth and firearms from England, and in naval stores from Muscovy. After the *Restoration it was reorganized as a regulated company, and its monopoly was ended in 1698.

musketeer, a soldier who aimed from the shoulder a large-calibre, smooth-bore firearm which was called a musket. Inefficient hand cannons had been used in Europe during the 14th century, and matchlock 'arquebuses' were subsequently used, rather haphazardly, in battle. In the mid-16th century Spanish troops pioneered the use of the more powerful, more accurate *mosquete* (musket). They also evolved complementary battle tactics. Effective as these weapons were, infantrymen still needed forked stands as props for aiming and firing; and since they were slow to load, pikemen had to be included in battalions to protect musketeers from enemy cavalry charges. The 17th-century development of the bayonet eventually removed the need for pikemen. Wheel-lock and flintlock muskets also became practical for military use at this time.

Muslim *Islam.

Mutiny Act (1689), English legislation concerning the enforcement of military discipline primarily over mutineers and deserters. The *Declaration of Rights (1689) had declared illegal a standing army without parliamentary consent. To strengthen its control over the

Mycenean civilization: the so-called Mask of Agamemnon, recovered from shaft grave V at Mycenae by Heinrich Schliemann in 1876. Convinced that he had discovered the graves of the Greek heroes of Homer's *Iliad*, Schliemann had actually unearthed burials from an earlier Bronze Age Mycenean civilization, spanning the 16th century BC. (Athens Museum)

army this Act was enforced for one year only, theoretically giving Parliament the right of an annual review. In fact there were years (1689–1701) when it was not in force and both army and navy long retained their close connection with the sovereign. Only when the crown ceased to pay for the upkeep of the army did Parliament's annual review become effective.

Mwene Mutapa, the title taken by Mutota, founder of the *Rozvi empire in East Africa. It has been translated as 'master pillager' and as 'lord of the conquered mines'. The Portuguese corrupted it to Monomotapa, and used it as a name for the empire. The founder died *c*.1470, and was succeeded until 1480 by his son Matope, a great conqueror who took possession of the greater part of eastern Central Africa. On his death his son Changa seized most of the empire from the designated heir, proclaiming himself *Changamire. The dynasty lasted until the end of the 18th century.

Mycenaean civilization, Greek culture which dominated mainland Greece from *c*.1580 BC to *c*.1120 BC, when the invading *Dorians destroyed the citadels of Mycenae and Tiryns. Another important Peloponnesian centre was Pylos, and Mycenaean influence spread as far north as southern Thessaly. In *c*.1450 Mycenaeans seem to have conquered *Knossos in Crete, and traders travelled widely to Asia Minor, Cyprus, and Syria. It seems that they also sacked Troy *c*.1200, though the duration and scale of the expedition was doubtless exaggerated by the poet *Homer in his epic, the *Iliad*. Finds from the early period bear witness to considerable

wealth and a high artistic skill. At Mycenae, Tiryns, and Pylos have been excavated palaces, in each of which the central room is the *megaron*, rectangular in shape with one side open and columns supporting a roof; here stood the royal throne and the hearth.

The creations of Mycenaean artists indicate a sophisticated audience: frescos; inlaid bronze weaponry; gold death masks, utensils, and ornaments. Ivory imported from Egypt was a favourite medium, and pottery, of stylized design, reached a high standard of quality. The scenes, depicted in art, of war and hunting indicate the preoccupations of Mycenaean kings.

Mysore, a former Princely State in south-west India. In early times this region had formed part of successive Hindu kingdoms, including that of the Hoysalas (11th–13th centuries). In the 14th century Muslim expansion from the north led to the emergence of the great Hindu empire of *Vijayanagar whose power centred in this region. When Vijayanagar declined the Wadiya family was well placed to set up an independent kingdom. However, strong neighbours prevented consolidation, and in 1761 the Wadiya raja had to surrender power to an enterprising Muslim soldier, *Hyder Ali. He and his son, *Tipu Sultan, fought bravely, and with initial success, against *East India Company forces in four 'Mysore wars' (1767–9; 1780–4; 1797 and 1799), but their capital, Seringapatam, finally fell in 1799. The British annexed almost half of Mysore, but restored the core of the kingdom to the Wadiya rajas.

N

nabob, a term of derision applied in the 18th century to those *East India Company servants who had amassed fortunes in India, sometimes unscrupulously, which they then used for bettering their economic and social positions in England. The term is corrupted from the Persian title nawab, which originally designated governors administering Indian provinces for the *Mogul emperors. Rulers of some Muslim Princely States continued to use the title during the British Raj period.

Nadir Shah (1688–1747), ruler of Persia (1736–47) and scourge of central Asia and India. Of Turkish origin and until 1726 a bandit chieftain, he rose to prominence under the *Safavid shahs of Persia, acting as king-maker during an era of disputed succession. In 1736 when the infant shah died he seized the throne and immediately embarked on expeditions against neighbouring states. In 1739 he attacked Delhi, capital of *Mogul India, but retreated after slaughtering the citizens. Campaigns against Russia and Turkey followed, but military adventures were by then at the price of Persia's economic stability. His own subjects suffered as much as his enemies from his ruthless methods, and he was assassinated by his own troops.

Nanak (1469–*c*.1539), founder of the *Sikh religion. He was born in a village in the Punjab, north-western India, in the mercantile Khatri caste, and abandoned family and occupation to travel in search of religious inspiration. On his return to the Punjab he preached a new path to the orthodox Hindu goal of release from the cycle of rebirth and attainment of union with God. He practised a form of inward and disciplined meditation on the name of God, the hallmark of the new Sikh faith which soon spread in the Punjab. Many details of his life and travels remain uncertain, although hagiographical collections of

A contemporary Dutch engraving of Henry IV, King of France, signing the **Edict of Nantes** in 1598. (Bibliothèque Nationale, Paris)

anecdotes, called *Janam-sakhis*, are in circulation among the *Sikhs ('disciples'), who revere him as the first of their ten gurus (religious teachers).

Nantes, Edict of (1598), a decree promulgated by *Henry IV which terminated the *French Wars of Religion. It was signed at Nantes, a port on the Loire estuary in western France. The Edict defined the religious and civic rights of the *Huguenots, giving them freedom of worship and a state subsidy to support their troops and pastors. It virtually created a state within a state, and was incompatible with the policies of *Richelieu and *Mazarin and of *Louis XIV. The fall of the Huguenot stronghold of La Rochelle to Richelieu's army in 1628 marked the end of these political privileges. After 1665 Louis XIV embarked on a policy of persecuting Protestants, and in 1685 he revoked the Edict.

Naples, city on the Bay of Naples in south-east Italy. It was settled by Greeks from Chalcis and Athens, who submitted to Roman conquest in 328 BC. When Roman rule weakened it was invaded by the Goths, but revived under *Byzantine influence in the 6th century and survived as an independent duchy until 1139 when it was conquered by the *Normans. It became part of the Kingdom of the Two Sicilies and passed successively to the Angevins, the Aragonese, and from 1504 to Spain, becoming a key base for Spanish and *Habsburg power in their Italian disputes with the *Valois kings of France. It passed to the Austrians in the War of the *Spanish Succession, but was conquered for the *Bourbons in 1734. Napoleon gained it in 1799.

Napoleon I (Napoléon Bonaparte) (1769–1821), Emperor of the French (1804–14). Born in Ajaccio, he was a Corsican of Italian descent. He was educated in military schools in France and served in the French Revolutionary army. By the age of 26 he was a general, and placed in supreme command of the forthcoming campaign against Sardinia and Austria in Italy (1796–7). This provided him with some of the most spectacular victories of his military career and resulted in the creation of the French-controlled Cisalpine republic in northern Italy. In 1798 he led an army to Egypt, intending to create a French empire overseas and to threaten the British overland route to India. *Nelson, by destroying the French fleet at the battle of the *Nile (1798), prevented this plan. Bonaparte returned to France (1799) and joined a conspiracy by *Sieyès which overthrew the Directory and dissolved the First Republic. Elected First Consul for ten years, he was now the supreme ruler of France. During the next four years he began his long-lasting reorganization of the French legal system (Code Napoléon), administration, the Church, and education.

With the Treaties of Lunéville (1801) with Austria, and Amiens (1802) with Britain, France became paramount in Europe. In 1803 Britain again declared war on France, and Napoleon prepared to invade it. The ruthless execution (1804) of the duc d'*Enghien on suspicion of conspiracy provoked criticism throughout Europe. In the same year Napoleon crowned himself Emperor of the French. He created an imperial court and nobility around himself, while at the same time restricting the liberal provisions of the earlier revolutionary constitution. In 1804–5, a European coalition against Napoleon was formed, and his armies attacked it. He defeated the

Austrians at Ulm, occupied Vienna, and won his most brilliant victory over the combined Austrian and Russian forces at Austerlitz (1805). The naval victory by Nelson at the battle of Trafalgar (1805) led Napoleon to seek Britain's defeat by the introduction of the Continental System, which aimed to stop all trade between Britain and France and its allies on the continent of Europe.

In 1806 the *Holy Roman Empire was dissolved and Napoleon consolidated his domination of the continent. The difficulty of enforcing the Continental System, the ill-fated invasion of Russia (1812), and the set-backs of the Peninsular War (1807–14), all contributed to Napoleon's decline, and following his defeat in the battle of Leipzig (18 October 1813), his abdication in 1814. After a brief exile on Elba he returned, only to end his 'Hundred Days' rule by defeat at the battle of Waterloo (1815). He spent the rest of his life in exile on St Helena.

In 1796 he married Josephine de Beauharnais, whose failure to give him a son led to their divorce. In 1810 Napoleon married the Austrian princess, Marie-Louise; their only child, Joseph-François-Charles, crowned as the Roi de Rome, died aged 21.

Narva, battle of (30 November 1700). In 1700 at the beginning of the *Northern War the port of Narva, in Estonia, was the scene of the crushing defeat of the *Peter the Great of Russia by *Charles XII of Sweden. In 1704 Peter recovered the town.

Narváez, Pánfilo de (c.1478–1528), Spanish conquistador. He gained a reputation as a ruthless soldier in the conquest of Cuba in 1511; later, as governor, he watched his men slaughter some 2,500 unarmed natives. When he was sent to fetch the rebellious *Cortés in Mexico in 1520, most of his men deserted, and he lost an eye in the ensuing battle. Leading an expedition to Florida in 1528, he ignored advice from his lieutenant, *Cabeza de Vaca, divided his forces, and was lost at sea trying to return to Cuba in makeshift boats.

Naseby, battle of (14 June 1645), a decisive victory for the Parliamentary forces during the *English Civil War. The battle took place near Naseby in Northamptonshire, after *Charles I's storming and sacking of Leicester. Led by Fairfax and Cromwell, the *New Model Army outnumbered the *Cavalier force by about one-third. Prince Rupert's cavalry squandered an early advantage, as at *Edgehill, and after a bitter struggle the Roundhead forces proved superior, and the Cavaliers suffered extremely heavy losses.

Nash, Richard (1674–1762), known in later life as 'Beau' Nash, English dandy. In 1705 he moved to Bath when it was a declining health resort, attracted by its gambling. He soon became its most influential resident, organizing such improvements as street lighting and the suppression of duelling, and he was especially famous for defining and insisting upon correct dress. His influence turned Bath into a fashionable *spa town. New anti-gambling laws in the 1740s caused financial difficulties for Nash, but in 1758 the city corporation, in gratitude, granted him a pension of £126 a year.

Natchez, a North American Indian tribe of the middle Mississippi River region, especially important because their tribal organization survived long enough for it to be

A portrait of 'Beau' **Nash**, after William Hoare (c.1761), portrayed in more sober dress than might be expected of such a famous dandy. (National Portrait Gallery, London)

documented, giving a unique insight into the prehistoric *Mississippi cultures. Their chiefdom—two or more villages linked under the same authority—was organized into a complex social hierachy of nobility (comprising 'suns', 'nobles', and 'honoureds') and commoners (called 'stinkards'). Nobility were required to marry commoners, and descent was through the female line. Thus children of stinkard fathers took their mother's rank, while those of commoner mothers were one rank below their fathers.

An attack on Fort Rosalie in 1729 led to a strong French response, and many Natchez villages were destroyed. In 1731 450 Natchez captives were sold as slaves and most of the remainder joined the *Chickasaw.

National Assembly, the name taken by the Third Estate on 17 June 1789 when it failed to gain the support of the whole of the French *States-General. Three days later its members signed the *Tennis Court Oath. It was accepted by *Louis XVI the following month, having added Constituent to its title. In August it agreed upon the influential declaration of the *Rights of Man and the Citizen and two years later its constitution was accepted by the king. Its reorganization of local government into departments, although long lasting, had less immediate success. Renamed the Legislative Assembly (1791) and the National Convention (1792), it was dominated by the *Girondins and the *Jacobins before being replaced by the *Directory in 1795.

Natufian culture, a group of plant-food gatherers and gazelle-hunters in the *Levant in prehistory. At the end of the last glacial period (between 14,000 and 10,000 years ago), as the northern ice-sheets began to retreat, a period of increased rainfall in the region allowed wild cereals to grow. The people who took advantage of these opportunities are called Natufians, after a site discovered in the Wadi el-Natuf. Their remains have been found both in caves and on campsites underlying the mud brick villages of succeeding farming groups who had taken the step from collecting to deliberately sowing and cultivating cereals. *Jericho is one of the most important places where this transition can be seen.

Navaho, a North American Indian tribe who, like the related *Apaches, originated in western Canada. They moved southwards to the 'four corners' area of Utah, Colorado, Arizona, and New Mexico between the 11th and 15th centuries, displacing much of the native *Anasazi culture to the south by the end of the 13th century, and raiding the neighbouring Hohokam to the south. Spanish contact began in 1540-2, and from 1609 Spanish missions were working among them. The Navaho adopted horse-breeding and pastoralism from the Spanish, and learned Anasazi weaving skills and Spanish silver-working.

Navigation Acts, legislation enacted by the English Parliament to prevent foreign merchant vessels competing on equal terms with English ships. The earliest goes back to the reign of Richard II, but the most important was that of 1651, requiring that goods entering England must be carried in English ships or ships of the country where the goods originated. Its aim was to destroy the Dutch carrying trade, and it provoked the *Anglo-Dutch wars. The Acts applied to the colonies, and despite the impetus they gave to New England shipping, they were widely resented in America, where their increasingly strict measures against smugglers, and against the colonial manufacture of certain goods which competed with English products, made them a major grievance. They were modified in 1823, and withdrawn in 1849.

navy, a fleet of ships and its crew, organized for war at sea. In the 5th and 4th centuries Athens and Corinth relied on *triremes (galleys with three banks of oars) and high-speed manoeuvrable quinqueremes (five-banked galleys) were developed by the Macedonians. At the battle of *Salamis an Athenian fleet won a decisive victory over the Persians, established Greek control over the eastern Mediterranean, and the fleet remained the crucial basis of Athenian supremacy. The Roman empire, though essentially a land-based power, fought Carthage at sea in the First *Punic War, and gained control of the Mediterranean.

Navies were needed to protect trading vessels against pirates: the *Byzantine empire maintained a defensive fleet to retain control over its vital trade arteries. In England, King *Alfred created a fleet in the 9th century in defence against Scandinavian invasions. The *cinque ports supplied the English navy from the 11th to the 16th centuries and it was organized and enlarged under successive Tudor monarchs. The Italian city-states kept squadrons of galleys and adapted carracks (merchant ships) to defend their ports against the *Ottoman Turks and the battle of *Lepanto saw a Christian fleet decisively

beat the Ottomans. The 17th century saw naval re-organization in England under *Pepys, and the Dutch and French also expanded their fleets as trade and colonial expansion accelerated in the 18th century.

From the early Middle Ages the warship altered from being a converted merchant ship, modified by the addition of 'castles', fortified with land artillery, and manned by knights, into a specially armed vessel. By the 14th century ships were being fitted with guns and by the 16th century special warships were being built with heavy armaments. Success or failure in battle, however, was determined by tactical skill as all sailing ships were at the mercy of the wind, and it was the introduction of the steamship in the 19th century that revolutionized naval warfare.

Nazca (or Nasca), a culture developed on the southern Peruvian coast c.200 BC–AD 600, eventually eclipsed by the expansion of the Huari culture of central Peru in the 7th century. In the extremely dry environment its settlements and population remained modest, but its craftsmen produced a long sequence of pottery styles, with animal and human figures. It also produced large drawings of animals, abstract designs, and straight lines on the coastal plain, by clearing and aligning the surface stones to expose the underlying sand; their purpose is uncertain, but may have been religious.

Neanderthals, people who lived in Europe, the Near East, and Central Asia from about 130,000 to 30,000 years ago. No people with characteristic ('classic') Neanderthal features are known from Africa or the Far East. They are named after the Neander valley near Düsseldorf in Germany where part of a skeleton was discovered in 1856. Neanderthals flourished particularly during the last Ice Age and were adapted for living in cold environments. Called Homo sapiens neanderthalensis to distinguish them from fully modern people *Homo sapiens sapiens, their features included heavy bones, strong musculature, large brow ridges across a sloping forehead, and larger brains than those of fully modern people. Neanderthals were probably the first people to have burial rites. Flint tools of *Mousterian type are usually found with their remains and characterize the Middle *Palaeolithic period.

The part played by Neanderthals in later human evolution is controversial. One widely held view is that none were our direct ancestors. An alternative opinion is that at least some Neanderthal groups evolved into fully modern people. If they were an evolutionary dead end, their extinction 50,000–30,000 years ago remains a mystery (the date varies according to geographical locality). It is possible that they became too specialized and were supplanted by modern people migrating into Europe from elsewhere, probably Africa via the Middle East. Some interbreeding between Neanderthals and modern humans probably also took place.

Nebuchadnezzar II (d. 562 BC), a Chaldean who came to the throne of *Babylon in 605 BC, shortly after leading his father's army to victory over the Egyptians at Carchemish. He campaigned vigorously in the west, capturing Jerusalem for a second time in 587 and 'exiling' many of the Judaeans to Babylon. He instituted a major building programme in his capital, including massive city walls and the Hanging Gardens, one of the *Seven Wonders of the World.

Navy

The oar-holes were in balconies outside the hull, leaving a long handle inboard.

a) **Roman trireme**
early 3rd century BC

b) **Viking longship**
early 9th century AD

Viking ships were clinker-built, using individually shaped, overlapping planks.

c) **floating 'castle'**
early 14th century

d) **Mediterranean carrack**
late 14th century

e) *Revenge*
late 16th century

f) *Sovereign of the Seas*
mid 17th century

The Roman trireme (a) probably differed little from the earlier Greek trireme but little is known about the appearance of these ships. Well-preserved examples of Viking longships (b) have been found. They were exceptionally well designed and built, ideally suited for long sea journeys. The clumsy floating 'castles' (c) were intended as platforms for conventional troops rather than as naval fighting units. Carracks (d) were primarily merchantmen, adaptable for warfare. *Revenge* (e) and *Sovereign of the Seas* (f) were both prototype warship designs. *Revenge* was a lighter design of galleon; *Sovereign* was Britain's first 100-gun battleship.

Necker, Jacques (1732-1804), French statesman and financier. He was a Protestant banker from Geneva who was Director-General of Finances in France (1777-81). He analysed the weaknesses of the French taxation system and opposed some of the ideas of the physiocrats, such as free trade in grain. In 1788 he became chief minister. France was bankrupt and a *States-General was summoned and reforms introduced. News of Necker's dismissal in July 1789 angered the people and, with other rumours, led to the attack on the *Bastille. He was reinstated as Controller-General of Finances but finally resigned in September 1790.

Nefertiti (14th century BC), wife and queen of *Akhenaten, Pharaoh of Egypt. She was a devoted worshipper of the sun god Aten, whose cult was the only one permitted by her husband. She fell from favour, and was supplanted by one of her six daughters. She is known to posterity through inscriptions, reliefs, and above all a fine limestone bust which was found at ancient Akhetaton (modern Tell el-Amarna).

Negroids, the indigenous peoples of Africa south of the Sahara and their descendants in other parts of the world. Bantu-speaking Negroid pastoralists and crop-growers are traditionally believed to have spread from western to eastern and southern Africa during the past few thousand years but recent evidence suggests that Negroids speaking other languages were in other parts of sub-Saharan Africa much earlier. They may indeed have originated in southern, not western Africa, but this is controversial. Unquestionably, though, they gained knowledge of agricultural techniques and domesticated animals from northern parts of the continent. Negroids are extremely variable in appearance but they can be seen in their most typical form in West Africa; in the east, there has been much intermixing with Hamitic-speaking *Caucasoids (for example, Ethiopians and Egyptians); and in the south with the related hunting and gathering San (Bushmen) and the cattle-raising Khoikhoi (Hottentots). The pygmies of Central Africa are of Negroid stock but not the *Melanesians nor the Negritos of southern Asia.

Nehemiah (5th century BC), Jewish leader, the cupbearer to the Persian king *Artaxerxes I. He obtained leave to visit Jerusalem in 444 BC where, despite opposition from local officials and from Samaria, he supervised the speedy rebuilding of the city walls. In 432 he made a second visit to Jerusalem and introduced important moral and religious reforms. His firm action at a time of crisis probably saved the new state of Judaea from collapse and enabled *Ezra to undertake his important reforms of Judaism.

Nelson, Horatio, Viscount, (1758-1805), British admiral, whose brilliant seamanship twice broke the naval power of France. The son of a Norfolk rector, he entered the British navy at the age of 12 and became a captain at the age of 20. On the outbreak of war with France in 1793 he was given command of the battleship *Agamemnon* and served under Admiral *Hood in the Mediterranean. He lost the sight of his right eye during a successful attack on Corsica in the following year. In 1797 he played a notable part in the defeat of the French and Spanish fleets at the battle of *Cape St Vincent and was subsequently promoted rear-admiral. Later the same year

A portrait bust of the Egyptian queen **Nefertiti**. The bust, painted on limestone, c. 1355 BC, was found at Tell el-Amarna, Egypt, the site of the royal capital. Nefertiti was a loyal wife to Akhenaton, though she apparently lost favour or retired from public life some time before her death. (Staatliche Museen Preussischer Kulturbesitz, Berlin)

he lost his right arm while unsuccessfully attempting to capture Santa Cruz de Tenerife in the Canary Islands. In 1798, after pursuing the French fleet in the eastern Mediterranean, he achieved a resounding victory at the battle of the *Nile. While stationed at Naples he began his life-long love affair with Lady Emma Hamilton, the wife of the British ambassador. In 1801 Nelson was promoted vice-admiral and, ignoring a signal from his commander, Sir Hyde Parker, defeated the Danish fleet at the battle of Copenhagen. Following this engagement he was created a viscount. In 1803, after the renewal of war with France, Nelson was given command of the Mediterranean and for two years blockaded the French fleet at Toulon. When it escaped he gave chase across the Atlantic and back, finally bringing the united French and Spanish fleets to battle at Trafalgar in 1805. This decisive victory, in which Nelson was mortally wounded, saved Britain from the threat of invasion by *Napoleon.

Nennius, a Welsh monk and chronicler of the late 8th or early 9th century. He is best known for his *Historia Britonum* ('History of the Britons') based on earlier chronicles and compilations. It was a source for stories about King *Arthur and also of *Hengist and Horsa, and was used by *Geoffrey of Monmouth in the 12th century.

Neolithic ('New Stone Age'), a term used to describe the later part of the *Stone Age, characterized by polished stone axes and simple pottery. The Neolithic discovery of farming brought an end to the slow development of the hunting societies of the *Palaeolithic and *Mesolithic periods and initiated a time of rapid change that soon produced metalworking, cities, states, and empires. The term is thus best applied to the stone-using, farming populations of Asia and Europe, who used polished axes to clear the forests and cooked their grain in pottery vessels. The very first farmers, at sites like *Jericho, had not discovered pottery, and are called pre-pottery Neolithic.

Neoplatonists, followers of *Plotinus (c.205–70) and other thinkers of the school of *Plato in the 3rd century AD. They searched for an intellectually 'respectable' basis for the act of reasonable religious belief. God, described as the 'One' or 'Absolute', was the unifying factor making sense of Plato's two worlds—thought and reality, the mental or 'ideal' and the physical. Mystical experience brought man closer to the 'One'. The experience was known as 'ecstasy' or 'standing outside self'. *Julian turned to this doctrine from Christianity, *Augustine of Hippo from *Manichaeism.

Nepal, a mountainous country in southern Asia. Its first era of centralized control was under the Licchavi dynasty (c.4th–10th century). Buddhist influences were then dominant, but under the Malla dynasty (10th–18th century) Hinduism became the dominant religion. In 1769 a Gurkha invasion brought to power the present ruling dynasty. From their capital at Kathmandu they wielded absolute power over the indigenous Nepalese tribes. A hard-fought war with the British conceded some territory (Treaty of Kathmandu, 1816), and effective rule passed to a family of hereditary prime ministers, the Ranas.

Nerchinsk, Treaty of (1689), a treaty between Russia and China signed at Nerchinsk, a town in Russia. It was the first treaty China signed with a western power. Drawn up in Latin by Jesuits from the Chinese emperor *Kangxi's court, the treaty fixed the Sino-Russian frontier well to the north of the Amur River. Albazin, a fortress town the Russians had built on the Amur, was dismantled and rebuilt in the Western Hills near Beijing.

Neri, Philip, St (1515–95), Italian mystic, known as the 'Apostle of Rome'. He went to Rome in 1533, where he studied, tutored, and undertook charitable works. He established the Confraternity of the Most Holy Trinity (1548) for the care of pilgrims and convalescents. Ordained priest in 1551, he joined the ecclesiastical community at San Girolamo. His popular religious conferences there took place in a large room called the Oratory, built over the church nave. Those who met there, and participated in the devotional, recreational, and charitable activities, were thus called the Congregation of the Oratory or Oratorians (approved in 1575 by Pope Gregory XIII). Philip helped to secure papal absolution for *Henry IV of France from excommunication for his temporary Huguenotism (1595). He was canonized in 1622.

Nero, Claudius Caesar Drusus Germanicus (AD c.37–68), Roman emperor (54–68). He was adopted by *Claudius, who had married his own niece, Agrippina, Nero's mother. On Claudius' suspicious death in 54 he succeeded to the throne and poisoned Britannicus, Claudius' own son by Messalina. He proceeded to have his mother murdered, to compel his boyhood tutor and state counsellor *Seneca to commit suicide, and had his own wife Octavia executed. Another wife Poppaea died as a result of Nero's violence towards her. He was the first emperor to persecute Christians, many of whom were put to death. He saw himself as artist, singer, athlete, actor, and charioteer. Reputedly he set Rome alight in AD 64, hoping to rebuild it in splendour. The surviving remains of his 'Golden Palace', a massive edifice dominated by a statue of himself as the sun-god, and magnificent coinage show high aesthetic standards. Revolt

A bust of the Roman emperor **Nero**. The historians Suetonius and Tacitus recorded the details of his flamboyant career.

broke out in Palestine in 66 followed by an army rebellion in Gaul and he committed suicide.

Nestorian, a member of a sect of the Eastern Orthodox Church. Nestorians followed the teaching of the controversial Syrian Nestorius (d. *c.*451), who was appointed Bishop of Constantinople in 428 and exiled to Egypt in 431. He taught that *Jesus Christ was a conjunction of two distinct persons, one divine and the other human, in whom the human and the divine were indivisible. The implication of this doctrine was that Mary was not the mother of God but simply of Jesus the man. This attack on the popular cult of the Virgin Mary led Nestorius' followers to establish a breakaway church in Edessa. They were expelled in 489 and settled in Persia until they were wiped out by the 14th-century Mongol invasions. Missionaries had established other groups as far away as Sri Lanka and China.

Netherlands, a European country, situated on the North Sea, with Belgium to the south and Germany to the east. It is also called Holland, after one of the historic regions which make up the modern nation. The area was conquered as far north as the River Rhine by the Romans; the Franks and Saxons moved in during the early 5th century. After the collapse of the Frankish empire in the mid-9th century, there was considerable political fragmentation. Consolidation began under the 14th- and 15th-century dukes of *Burgundy, and in 1477 the whole of the Low Countries passed to the House of *Habsburg. In 1568 the *Dutch Revolts against Spanish Habsburg rule began. The independence of the *United Provinces of the Netherlands was finally acknowledged at the Peace of *Westphalia (1648). During the 17th century it was a formidable commercial power, and it acquired a sizeable *Dutch empire. It began to decline after the *Anglo-Dutch wars and the protracted wars against *Louis XIV's France.

Neville's Cross, battle of (17 October 1346). A major victory of the English over the Scots at a site close to the city of Durham, in northern England. *Edward III's northern barons, the Nevilles and the Percys, and the Archbishop of York, inflicted a severe defeat on the invading army of *David II and captured him. It did not gain Edward the kingdom of Scotland, but it cost David's people a very expensive ransom for the release of their monarch.

New Amsterdam   *New York.

New Brunswick, a Canadian province on the east coast, south of the St Lawrence estuary. The Bay of Fundy separates its southern coastline from *Nova Scotia and its western frontier adjoins *Maine. First settled in 1604 by the French, it formed part of Acadia until 1713 when it was ceded to Britain. Its forests, dissected by numerous rivers, were the hunting ground for fur traders and trappers. Small settlements of Scottish farmers in the 1760s were expanded by the arrival of large numbers of loyalists during the American War of *Independence.

Newcastle, Thomas Pelham-Holles, 1st Duke of (1693–1768), English statesman. Born Thomas Pelham, he inherited substantial estates from his uncle John Holles, whose name he added to his own (1711). He became Secretary of State in 1724, and devoted his political career to the management of the House of Commons by using the vast crown patronage at his disposal. He supported in turn *Walpole and his own brother Henry *Pelham. His own ministry of 1754–6 was a disaster, but his coalition with the Elder *Pitt, 1757–61, was a triumph and brought success in the *Seven Years War. His resignation in 1762 was accepted by George III, who disliked the domination of government by a few powerful Whig families, and he never enjoyed high office again. Thereafter his followers increasingly rallied to the Marquis of *Rockingham.

New England, a region of the north-eastern USA, comprising the six states of *Connecticut, *Massachusetts, *Rhode Island, *New Hampshire, *Vermont, and *Maine. First named by Captain John *Smith in 1614, it was granted to the Council for New England in 1620, from whom the *Pilgrims and the Puritan New England Company (later the Massachusetts Bay Company) obtained permission to settle. Connecticut, Rhode Island, New Hampshire, and Maine were Massachusetts offshoots. In 1643 the New England Confederation was formed to co-ordinate defence and Indian policy. James II centralized the government of the region into the Dominion of New England (1686–9). Though the region was predominantly agrarian, American merchants and fishermen quickly entered coastal, West Indian, and transatlantic trades. It was a leading patriot centre in the American War of *Independence, and was *Federalist in the early national period.

Newfoundland, an Atlantic offshore island of Canada in the Gulf of the St Lawrence. Claimed for England by John *Cabot in 1497, its rich cod banks and deeply indented coastline quickly attracted European fishermen. Humphrey Gilbert (1583) and Lord Baltimore (1621) attempted colonization, but by 1650 the settlers numbered only 2,000. France contested Britain's claims, and Newfoundland changed hands a number of times. France accepted British sovereignty there in the Peace of *Utrecht (1713) and in 1728 it became a chartered colony. All claims on Newfoundland by France ended with the Treaty of *Paris (1763), although France retained some fishing rights.

New France, the name given to French possessions in North America discovered, explored, and settled from the 16th to the 18th centuries. Its centres were Quebec (founded in 1608) and Montreal (founded in 1642) on the St Lawrence River. In 1712 New France stretched from the Gulf of St Lawrence to beyond Lake Superior and included *Newfoundland, Acadia (*Nova Scotia), and the Mississippi valley as far south as the Gulf of Mexico. It began to disintegrate when the Peace of *Utrecht was signed in 1713, when France lost Acadia, Newfoundland, and Hudson Bay, and ceased to exist as a political entity in 1763 under the terms of the Treaty of *Paris. *Louisiana, the last French colony on mainland North America, was sold to the United States in 1803.

As a major competitor for the fur trade and western lands, and as a threat to the British colonies to the south, New France became embroiled in the *French and Indian wars. Wolfe's victory on the *Plains of Abraham (1759) led to the end of French rule in 1763. Under British rule the territory east of the Ottawa River and

A map of **New France**, from Samuel de Champlain's *Les Voyages de la Nouvelle France Occidentale* (1632). The map shows the Gulf of St Lawrence, with Hudson Bay, or '*la mer du nort glacialle*', to the north-west. Champlain's detailed cartography shows the extent of French exploration, northwards into what is now Quebec and west around the Great Lakes. (National Map Collection, Public Archives, Canada)

north of the St Lawrence River became Lower Canada (1791), Canada East (1841), and the Province of Quebec (1867) in the Dominion of Canada.

New Granada, a Spanish colony which embraced much of northern South America. The *conquistadores had occupied the Caribbean coast of present-day Colombia by 1510, but it was not until 1536 that one of them, Gonzalo Jiménez de Quesada, led a two-year campaign into the Andes. By 1538 he had conquered the Chibcha Indians, looted their gold, and founded the city later known as Bogotá. Gold, silver, and precious stones from the mountainous regions of Antioquia and Cauca were exported through Cartagena, which was held to ransom by Sir Francis *Drake in 1585. In 1564 New Granada became a presidency subject to the viceroyalty of Peru. The distance to Lima created difficulties, however, and in 1717 it became a viceroyalty comprising Colombia, Ecuador, Panama, and Venezuela.

New Hampshire, a northern central *New England colony and state of the USA, named by John Mason who received a grant there about 1629 from the Council of New England. New Hampshire was mainly settled from *Massachusetts and *Connecticut and became a royal colony in 1679 after long controversies with Mason's

family. It was prized by the crown for its ships' timbers and was also attractive to land speculators. Its claims west of the Connecticut River, relinquished in 1782, became *Vermont in 1791.

New Jersey, a colony and state of the USA situated on the east coast between the Hudson and Delaware estuaries. Originally claimed by the Dutch, it became two English proprietary colonies in 1664, and a united crown colony in 1702. Its western part was closely linked with Quaker Pennsylvania and tended to loyalism, and union while the eastern part was settled by New Englanders and Scottish Presbyterians, supporters of independence. The new state was a battleground in the early stages of the American War of *Independence, with actions at Trenton (1776), Princeton (1777), and Monmouth (1778). Its delegates led the smaller states in the *Constitutional Convention (1787).

New Learning, the name given to the approach to education pioneered by 15th and early 16th century European *humanists. Originally its method was to apply new critical techniques to a close scrutiny of the *Bible and ancient Christian texts in order to reach a deeper understanding of Catholic Christianity. In England humanist thinking also influenced those advocating *Protestantism. This was partly due to the work of great exponents of the New Learning like Sir John Cheke (1514-57), the first professor of Greek at Cambridge University and tutor to *Edward VI and Nicholas Udall, headmaster of Eton and Westminster schools. Such men helped to establish a new educational ideal: the formal training of gentlemen to become valuable servants of the state. The foundation of their training was rigorous study both of the Bible and of classical literature.

New Model Army, the English *Roundhead force established by Parliamentary ordinance on 15 February 1645. A single army of 22,000 men, it was formed largely from the uncoordinated Roundhead forces of the first phase of the *English Civil War. Its first commander-in-chief was *Fairfax, with Philip Skippon commanding the infantry and, after the *Self-Denying Ordinance, Oliver *Cromwell in charge of the cavalry. Derided at first by the Cavaliers as the 'New Noddle Army', its men, regularly paid, well disciplined, and properly trained, became known as the Ironsides. Promotion was by merit. Resounding victories like *Naseby and *Preston won the war for the Roundheads. The army was inextricably involved in national developments until the Restoration. Religious and political radicalism quickly permeated its ranks, with *Leveller influence particularly strong between 1647 and 1649. The army was responsible for *Pride's Purge (1648), and formed the basis of government in the following years.

New Plymouth *Plymouth Colony.

New Spain, the name for Spain's colonial empire in north and Central America (*Spanish empire). The formation of the viceroyalty of New Spain began in 1518 with *Cortés's attack on the *Aztec empire in central Mexico. Following his destruction of Aztec power, Cortés erected a new capital at Mexico City and was named governor and captain-general of New Spain (1522). He and his lieutenants extended Spanish authority south into Salvador, Guatemala, and Honduras, and north into the Mexican hinterland. New Spain grew to encompass California, the American south-west, and Florida, although Spanish settlement in many areas was very limited. In the 18th century Spain's involvement in European wars had affected its colonial possessions. In 1763 it ceded Florida to Britain and received Louisiana from France, regaining the former in 1783, but being forced to return Louisiana to France in 1800.

New York, a state bordering Canada in the north-east of the USA. Its history dates from 1609, when Frenchmen first sailed down Lake Champlain and Englishmen in Dutch service sailed up the Hudson River. The Dutch established Albany as a fur-trading post and in 1625 founded their capital at the mouth of the Hudson on Manhattan Island, which they bought from local Indians for trinkets worth $24. They called it New Amsterdam; but in 1664 it was captured by the British and renamed New York for the Duke of York (later *James II). British colonists joined the Dutch settlers; the *Iroquois Confederacy prevented French interference; and landowners, merchants, and pirates alike prospered for a hundred years. When the issues leading to the War of *Independence arose, New Yorkers were divided. Many were loyalists, and the British, after capturing New York City in 1776, held it throughout the war. Even so, *Washington's triumphal entry into the city (1783) was welcomed tumultuously. His presidential inauguration took place there in 1789 and New York City was briefly (1789–90) the new national capital.

Nicaea, Councils of. The city of Nicaea (now Iznik, Turkey) was the scene of two councils of the Christian Church. The first Council (325) was summoned by the Roman emperor *Constantine and issued a statement of

New Model Army: these contemporary portraits of the chief commanders of the Parliamentary forces include Oliver Cromwell as Lieutenant-General. He is placed in the front rank of officers, behind the Earl of Essex. (Ashmolean Museum, Oxford)

orthodoxy against *Arianism, later known as the 'Nicene Creed'. The Second Council (787) was called by the Byzantine empress *Irene to end the *Iconoclastic controversy.

Nicaean empire (1204–61), the name given to the territories centred on Nicaea (now Iznik, Turkey). The empire was established by Theodore I (1175–1222) following the sack of Constantinople during the Fourth *Crusade and adopted the institutions of the Byzantine empire, including emperors and patriarchs. Resisting the pressures of *Seljuks and Crusaders, he successfully annexed lands from the Komnenoi of Trebizond. His son-in-law John III sustained this miniature empire in exile but the early death of his son, Theodore II (1254–8), enabled *Michael VIII to dispossess and subsequently blind and imprison the infant heir, John IV Lascaris, after retaking Constantinople in 1261.

Nicaragua, the largest Central American country. The first Spanish colonization was undertaken by Francisco Hernándes de Córdoba, who founded the towns of Granada on Lake Nicaragua and León on Lake Managua in 1524. Administratively part of the viceroyalty of *New Spain and the captaincy-general of Guatemala, the area grew slowly. It depended upon agriculture, which developed substantially in the 18th century.

Nicopolis, battle of (23 September 1396). The town of Nicopolis, on the River Danube was besieged by a force of Crusaders under *Sigismund of Hungary and John, son of Philip the Bold of Burgundy. A combined force of Hungarians and European knights had answered an appeal to relieve *Constantinople from Turkish attack. After initial success they were confronted by the *Turks under Sultan Bajazet I (ruled 1389–1402). The Crusaders were defeated and many of the knights were executed after the battle, although Sigismund escaped.

Nijmegen, Treaty of (1678). The Franco-Dutch War (1672-8) was terminated by the treaty signed at Nijmegen, in the Netherlands between France, the United Provinces, Spain, and the Holy Roman Empire; France gained substantially from the terms, which rationalized its frontiers.

Nile, battle of the (1 August 1798). A naval battle fought at Aboukir Bay on the Mediterranean coast of Egypt, in which a British fleet defeated a French fleet. The French admiral had anchored his fleet of thirteen vessels in the bay. He believed his ships to be safe from attack, but *Nelson, the British commander, was able partially to encircle the French fleet. Nine French ships were destroyed, including the flagship *L'Orient*. This conclusive victory established Nelson's prestige, destroyed *Napoleon's plans for leaving Egypt, and encouraged the signing of the second coalition against France.

Nine Years War (1688-97), also known as the War of the Grand Alliance, a conflict which resulted from French aggression in the Rhineland, and which subsequently became a power struggle between *Louis XIV of France and *William III of Britain. In 1688 when French armies invaded Cologne and the Palatinate, the members of the League of *Augsburg took up arms. Meanwhile William had driven *James II from the throne of England and in 1689 a Grand Alliance of England, the United Provinces, Austria, Spain, and Savoy was formed against France. The French withdrew from the Palatinate. James II, supported by French troops, was defeated in Ireland at the battle of the *Boyne. In 1690 the French navy won a victory off Beachy Head, but in 1692 was defeated at La Hogue, though their privateers continued to damage allied commerce. The French campaigns in north Italy and Catalonia were successful, but the war in the Spanish Netherlands became a stalemate as one lengthy siege succeeded another. William's one success was the retaking of Namur. The war was a severe defeat for France, despite a good military performance, because its financial resources were not equal to those of Britain and the United Provinces. Peace was finally concluded by the Treaty of *Ryswick.

Nineveh, an ancient city on the River Tigris. It was selected by *Sennacherib of Assyria (ruled 704-681 BC) as his capital and he instituted a major building programme there. It was sacked by the *Medes in 612 BC, and never thereafter regained its former prestige. Excavations have revealed much of the ancient city, including five of the fifteen gateways in the 12 km. (7-mile) wall, the royal palaces with magnificent sculptured reliefs, and a remarkable store of cuneiform tablets which are a valuable source of historical information.

Nizam Shahi, Muslim dynasty of Ahmadnagar in the north-western Indian Deccan from 1490 to 1637. Its founder was Malik Ahmad who revolted against the *Bahmanis and set up an independent kingdom centred on a new capital, Ahmadnagar, which took its founder's name. His main achievement was the conquest of Daulatabad (1499). He and his successors then engaged in almost constant warfare. After abandoning an alliance with the Hindu kingdom of *Vijayanagar they participated in the final destruction of that city in 1565. The dynasty presented spirited resistance to subsequent Mogul

A terracotta head of the Nigerian **Nok culture**, sculpted between 400 BC and AD 200 and standing 35 cm. (14 in.) high. Typically of well-fired clay, heads of Nok figures are usually tubular, although sometimes conical or spherical. They display features which seem to derive from woodcarving. (Lagos Museum, Nigeria)

encroachment into its territories, but had lost all independent existence by 1637.

Nok culture, a term for the people occupying northern Nigeria from *c*.400 BC to AD 200, important for two reasons. As the earliest known centre of iron-working south of the Sahara, it played a major part in passing on metallurgy; and its distinctive terracotta figurines are considered ancestral to the later statuary of *Ife.

Nonconformist (or Dissenter), a Protestant who did not conform to the disciplines or rites of the *Anglican Church. The term covers a number of groups including Catholic *recusants. The *Puritans wished to purify the Church from within while the *Presbyterians were specific in their demands for the replacement of organization by bishops for a system of elected elders. The separatists under Robert *Browne left the Anglican Church entirely. All Nonconformists were subject to penalties; the *Pilgrim Fathers emigrated to escape persecution. During the Civil War Nonconformists (especially *Congregationalists and *Baptists) fought on the Parliamentary side and the Restoration Settlement (1660) enacted harsh measures against all Nonconformist groups. The 1662 Act of Uniformity deprived them of freedom of worship and subsequent persecution led to a further exodus to North America. In 1681 Pennsylvania was founded as a refuge for *Quakers. The Toleration Act (1689) brought some improvements in England, but until the 19th century Nonconformists were debarred from holding political office.

Nootka Sound, a harbour on the west coast of Vancouver Island, Canada, lying between it and Nootka Island. The mouth of the sound was visited by Juan Perez in 1774 and by Captain Cook in 1778. In that year a fort was built by John Meares. The seizure of the fort by the Spanish in 1789 led to a controversy with Britain. Agreement was reached in the Nootka Convention of 1790, which opened the West Pacific coast to British settlement.

Nore mutiny (May 1797). A mutiny by sailors of the British navy stationed at the Nore anchorage in the Thames estuary. Encouraged by the earlier naval mutiny at *Spithead they demanded improvements in their conditions, the removal of unpopular officers, a greater share in prize money and, under the influence of their ringleader, Richard Parker, certain radical political changes. This time the Admiralty would make no concessions and eventually the mutineers surrendered. About nineteen men, including Parker, were hanged. Alarm at the mutiny probably contributed to the decisive defeat of Grey's parliamentary reform motion of May 1797.

Norman, an inhabitant or native of *Normandy, France, a descendant of a mixed Scandinavian ('Northmen') and Frankish people established there in early medieval times.

The area, secured by *Rollo in 912 from Charles III of France, was inadequate for settlement since inheritance laws left younger sons without territory; land hunger provided the impetus towards conquest and colonization. Under Duke William the Normans conquered England (1066), and later Wales, Ireland, and parts of Scotland as well as large areas of the Mediterranean. Their expansion southwards, led by a spirited adventurer Robert *Guiscard, was initially as mercenaries fighting the Muslims but they soon controlled much of Europe. In 1154, the year of *Roger II of Sicily's death and *Henry II's accession to the English throne, Norman power was at its height, witnessed in the highly efficient governments of Sicily and England which were renowned in Europe for their sophisticated legal and administrative systems.

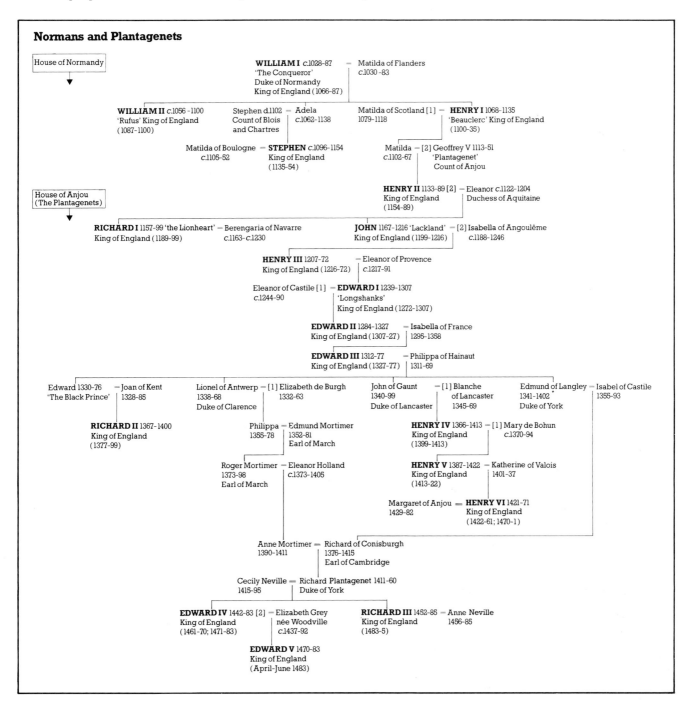

Normans and Plantagenets

Norman Conquest, the period which opened in 1066 with Duke William of Normandy's victory over the English at the battle of *Hastings. As *William I (1066–87) he established a military superiority over the English, rebellions were crushed (1067–71), and about 5,000 castles were constructed by the time of his death. England's frontiers were protected, first by *marcher lords and then through conquest. Ruthless attention to detail characterized the Norman approach to government. English institutions were either retained and developed (such as the treasury, the king's council, the king's peace, sheriffs, and the shire system) or replaced with Norman versions. Not all changes were popular with the English, who had already lost heavily in terms of status, land holdings, and public office. Taxation was heavier, forest laws were harsh and outside the common law. Norman efficiency produced the unique survey recorded in *Domesday Book (1086), though it owed a great deal to existing English records. The language of government and of the court was Norman French. England prospered commercially: its towns grew, as did its population. Many of these developments would probably have arisen without Norman rule—as is also true of the reorganization of the English Church under Archbishop *Lanfranc—but the rapid nature of these changes owed most to the Norman Conquest. In architecture the Norman style, characterized by rounded arches and heavy pillars was introduced after the Conquest.

Normandy, a former province in north-western France, originally the home of Celtic tribes, and part of the kingdom of *Clovis. It was in Neustria under *Merovingian rule and suffered from Viking invasions in the 9th century. The Viking *Rollo accepted it in 912 as a *fief from the French king, who was powerless to prevent its falling to the Vikings, and the present name derives from those invaders. They accepted Christianity and adopted the French language but Norman expansion meant that their power rivalled that of the French kings.

In 1066 Duke William of Normandy conquered England, becoming *William I. The duchy was recovered for France by Philip Augustus in 1204, but fell once more to England in the Hundred Years War. After the battle of Formigny in 1450 it was permanently reunited with France.

Norseman *Viking.

North American Indians, the original *Amerindian inhabitants of North America, who migrated from Asia from about 30,000 years ago. By the time of European colonization, the Indian population was probably under 900,000, mostly living along the coasts rather than in the barren interior. They lived in small villages which, except in the south-west, were organized round hunting, with agriculture a secondary activity. The overall social organization was that of the tribe, and warfare between tribes was endemic. Conflict with British and French settlers in the north-east forced the Indians inland, as did clashes with the Spanish in the south-west. The acquisition of horses from Europeans increased the number of nomadic Indians on the Great Plains. The major tribes are usually divided geographically, North-eastern Woodland (for example, *Algonquin, *Delaware, *Iroquois), South-east (e.g. *Cherokee, *Choctaw, *Creek),

Great Plains (e.g. *Blackfoot, *Comanche, *Dakota), Desert-west (e.g. *Apache, *Pueblo, *Navaho), Far west (e.g. Paiute),Pacific North-west (e.g. Chinook), and Mountain or Plateau (e.g. Nez Percé).

North, Frederick, Lord (1732–92), British statesman, Prime Minister (1770–82). A statesman of an impeccable *Whig background, he was first given a ministerial post by *Newcastle in 1759. He was an able Chancellor of the Exchequer under *Grafton and he became Prime Minister in 1770, restoring stability after a decade of frequent changes. He kept taxes low and avoided expensive foreign ventures, and he tried to cool American passions by withdrawing all but one of the *Townshend Acts, which imposed import duties on the colonists. He refused to be provoked by incidents such as the Boston Massacre, but his divided Cabinet would not allow the *Boston Tea Party to go unpunished, and North was led into a situation which made the American War of *Independence inevitable. Defeat in the war brought his resignation in 1782. He had been popular in the House of Commons as a peacetime Prime Minister and in the early years of the American War, but increasing frustration with failures in America and opposition propaganda (especially from the *Rockingham Whigs) led to suggestions that his ministry was being sustained in power by excessive crown patronage and influence. This led to growing demands for financial reform which was implemented after his resignation by the Rockingham Whigs. He held office again briefly in the *Fox–North coalition of 1783, but failing eyesight from 1786 caused him to withdraw from active politics.

North Carolina, a colony and state of the USA on the east coast between Virginia and South Carolina. After Raleigh's ill-fated *Roanoke venture, official colonization did not begin until 1663 when *Charles II granted a charter to colonize the Carolinas to a syndicate of eight proprietors. The northern and southern parts of Carolina developed separately and the colony was divided into North and South Carolina in 1713. It endured recurrent unrest against proprietorial authority, including *Culpeper's Rebellion, until it became a royal colony in 1729. Scots-Irish and German pioneers settled inland, and many supported the *Regulator movement (1768–71) against the coastal planter oligarchy. After *Cornwallis's campaign through the back-country (1780–1), the state settled to a planter-dominated regime.

Northern Rising (November–December 1569), an English rebellion, led by the earls of Westmorland and Northumberland. Protesting loyalty to *Elizabeth I, the rebels aimed to remove 'new men' like William *Cecil from power, preserve the Catholic religion, and have *Mary, Queen of Scots declared heir to the English throne. They ventured south and took Durham before government troops intervened and ended the rising. At the queen's insistence, around 400 rank-and-file rebels were executed. Meanwhile a number of the leaders fled abroad or were allowed to buy themselves pardons.

Northern War (1700–21), a conflict between Russia, Denmark, and Poland on one side, and Sweden opposing them, during which, in spite of the victories of *Charles XII, Sweden lost its empire and Russia under *Peter the Great became a major Baltic power. In 1697, on the accession of the 15-year-old Charles, the Swedish empire included Finland, Estonia, Livonia, and territory in northern Germany. In 1700 Sweden was attacked by a coalition of *Augustus the Strong of Saxony and Poland, Denmark, and Russia, organized by the Livonian exile, Patkul. Charles forced Denmark to make peace, defeated Peter at *Narva, invaded Poland, and placed Stanislaus Leszczynski on the throne; he pursued Augustus into Saxony and in 1706 forced him to recognize Stanislaus as King of Poland. After a diplomatic mission from *Marlborough to persuade him not to invervene in the War of the *Spanish Succession, Charles turned east and by the summer of 1708 was deep into Russia and won

his last victory at Holowczyn. He then turned south but his ally Mazeppa, the hetman (leader) of the Dnieper Cossacks was defeated by Peter's general Menshikov. The winter was severe and he was defeated at *Poltava. He took refuge in Turkish territory for five years, trying to persuade the sultan to help him while his enemies attacked and his subjects fought desperate rearguard actions. Finally expelled by the Turks, Charles was killed in 1718 and his sister, Ulrica Leonora, began the peace negotiations which Charles had refused to consider. By the treaties of Stockholm (1719, 1720) Bremen and Verden were ceded to Hanover and most of Pomerania to Prussia. Finally, after further Russian naval successes,

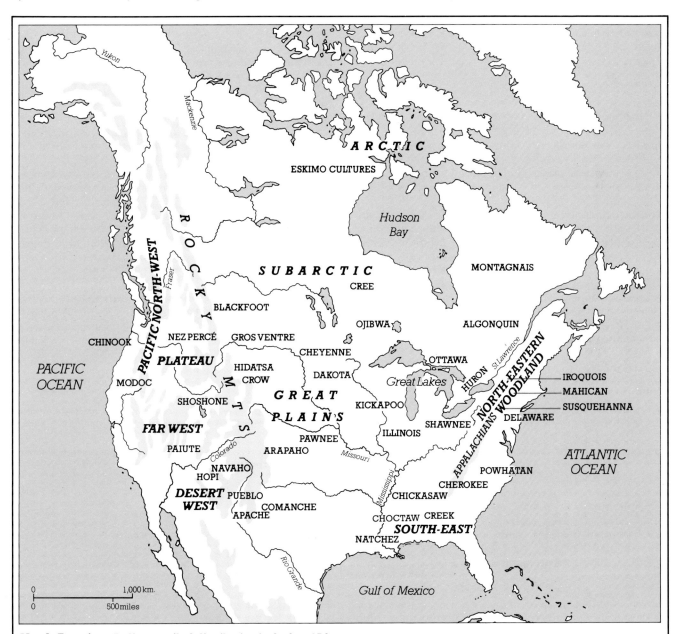

North American Indians: tribal distribution in the late 15th century

When the first European settlers arrived in North America, the Indian population was scattered widely throughout the whole subcontinent. There were numerous different tribes and an extraordinary diversity of languages and cultures. Some cultural similarities can, however, be found between tribes living in areas of similar geography and climate. For example, in the wooded north-west coastal area, with its hospitable climate and plentiful food supplies, tribes lived a settled life with a well-developed culture; on the central plains the majority of tribes were nomadic, living by hunting the buffalo; while in the fertile south-east an agricultural and trading economy was firmly established by the 14th century.

the Treaty of *Nystadt which ended the war gave Russia Sweden's Baltic provinces.

Northern Wei (Toba Wei) (386–535), one of the dynasties that ruled north China after the empire disintegrated following the collapse of the *Western Jin in AD 311. It was founded by a Mongol group, the Xianbi. Its rulers availed themselves of Chinese administrators and encouraged intermarriage with Chinese. Luoyang, once the Eastern Han capital, became their capital. Land reforms aimed at building up a stable peasantry, and an enduring system of collective responsibility, were instituted. Its rule ended when tribal frontier garrisons opposed to its policies revolted.

Northumberland, John Dudley, Earl of Warwick, Duke of (c.1502–33), ruler of England on behalf of Edward VI (1551–53). He began his political career under *Henry VIII and was a member of Edward VI's Privy Council. As Earl of Warwick he sought to undermine *Somerset's power and was created Duke of Northumberland in 1551 shortly before ordering Somerset's imprisonment and execution. The new government was committed to radical Protestantism and produced a new Prayer Book (1552), supported by a new Act of Uniformity prescribing penalties for not attending Church services. In 1552 *Cranmer produced the Forty-Two Articles, a comprehensive statement of Protestant doctrine which formed the basis of the *Thirty-Nine Articles (1571). Though posthumous tradition condemned Northumberland as an evil schemer, his regime promoted stability. He terminated the unsuccessful wars against France and Scotland initiated by Somerset and introduced several important financial reforms. Northumberland brought about his own downfall by trying to ensure the succession of Lady Jane *Grey. He was executed and *Mary, Henry VIII's daughter, succeeded to the throne.

Northumbria, territory in northern England, the kingdom of the Angles formed by the merger of the kingdoms of Bernicia and Deira (604). It extended from Yorkshire to the Firth of Forth. The greatest period of its history was the 7th century when, under a succession of powerful kings (including *Oswald and *Oswy) it appeared that Northumbria might create a united England out of the many Anglo-Saxon kingdoms. However, the death of Egfrith at the hands of the Picts in the battle of Nechtansmere (685) revealed the precarious nature of Northumbrian control, which collapsed before the threat of the kings of *Mercia. In the 7th and 8th centuries Northumbrian learning and monasticism were pre-eminent, developed by such figures as St Wilfrid, St Cuthbert, and *Bede. It attracted the attention of Danish raiders: the monastery of Lindisfarne was plundered by them in 793 and 875. Northumbria became part of the Danish kingdom of York (876) before being conquered by Wessex in the 10th century and reduced to an earldom in the following century.

Northwest Territory, the area in North America between the Great Lakes and the Ohio and Mississippi rivers, which formed the first national territory of the Indiana, Michigan, Wisconsin, and eastern Minnesota. French explorers and trappers arrived there in the 17th century. It was acquired by Britain in 1763 and transferred to the USA in 1783. The charter grants of individual states had already been transferred to Congress.

The Northwest Ordinance of 1787 was a major achievement of the Confederation. It laid down procedures for government of the territories, survey, division and sale of lands, and the democratically willed acquisition of statehood, on levels of total equality with the original states. Slavery was forbidden in the area, the major difference with the Southwest Ordinance of 1790. The Ohio Company with large areas of cheap congressional land was a major colonizer, but Indians still sympathetic to the British endangered frontier settlement.

Norway, a country in northern Europe. It was inhabited in prehistoric times by primitive hunting communities. Rivalry between chiefs, and the desire for land provoked excursions by the Norwegian *Vikings as far as England, Greenland, and Iceland. Political organization strengthened under Harald Fairhair (c.900) and under *Olaf I, who brought Christianity. *Olaf II furthered the work of Christian conversion, but was killed in a battle with the Danes. Danish rule (1028–35) followed, and thereafter civil war, and, in 1066, an unsuccessful expedition to assert Harald Hardrada's claim to the English throne. The reign of *Haakon IV brought order and, from 1254, Norway traded with the *Hanseatic League. In 1397 the Union of *Kalmar brought Norway, Sweden, and Denmark together under a single monarch. Danish rule resulted in conversion to *Lutheranism. The Union was dissolved in 1523, though Norway was ruled by Danish governors until 1814. A growing fishery and an important timber export trade to Great Britain increased prosperity in the 18th century.

Nostradamus (Michel de Notredame) (1503–66), French astrologer and physician. He first won fame for his innovative use of medicines and pioneer treatments during plague outbreaks at Aix and Lyons (1546–7). His *Centuries* of rhyming prophecies (1555), decidedly

An engraving of Catherine de Medici consulting the magic mirror of the French astrologer and physician, **Nostradamus** in which, it is claimed, she saw the future of France up to the French Revolution.

A battle between the province of **Novgorod** and the north-east Russian principality of Suzdal, from a late 15th-century painting. Prayers offered by the priests appear to be answered as the attackers are driven back from the citadel, under the protection of an icon of the infant Jesus Christ and his mother Mary. (Museum of Russian Art, Moscow)

apocalyptic in tone, caught the contemporary imagination. An enlarged second edition was dedicated to Henry II of France in 1558. He was appointed physician-in-ordinary to Charles IX on his accession in 1560, and his prophecies continued to excite speculation and controversy. In 1781 they were condemned by the Roman Catholic Congregation of the Index (*Inquisition).

Nova Scotia, one of the Atlantic provinces of Canada, a peninsula off the south-east coast adjacent to *New Brunswick. It was claimed by the French in 1603, and their settlement at Port Royal became the centre of Acadia. Britain contested their claim, naming the peninsula Nova Scotia in 1621, and launched several invasions (1613, 1654, 1690) before capture in 1710 and cession in 1713. The French retained Cape Breton Island, immediately adjacent, and built *Louisburg, which was captured in 1745 but returned in 1748. In the *French and Indian War, some 6,000 Acadians were forcibly deported to British colonies to the south, some ending up in Louisiana, others eventually returning. The second capture of Louisburg (1757) and the Treaty of Paris (1763) secured Nova Scotia in British hands. Settlement was vigorously encouraged and during the War of Independence large numbers of loyalists emigrated there.

Novgorod, a province and capital in north-west European Russia. It was allegedly the site of the founding of a Russian state under Prince Rurik in 826. It was forcibly Christianized in 989 and became self-governing in 1019. It survived the invasion of the *Golden Horde but submitted to their suzerainty in 1238. The *Swedes were beaten off in 1238 and the *Teutonic Knights in 1242, but a struggle with *Muscovy ensued in which *Lithuanian aid was sought, and which ended with submission to *Ivan III in 1478. *Ivan IV inflicted a massacre there in 1570, and it was occupied by the Swedes from 1611 to 1619. It retained political and commercial importance until the building of St Petersburg under *Peter the Great.

Nubians, a people who live chiefly in Egypt and the Sudan, between the First and Fourth Cataracts of the River Nile. Their recorded history begins with raids by Egyptians c.2613 BC, when their country was called Kush. Then a Nubian dynasty ruling at Napata from c.920 BC conquered all Egypt. The Nubian Shabaka ruled as King of Kush and Egypt with Thebes as his capital, but Assyrians forced Taharka, his successor, to withdraw (680–669). After several further struggles, the Nubians drew back to Napata, and c.530 BC their capital moved to *Meroë. The dynasty continued to AD 350, when Aezanas of *Axum destroyed it; its three hundred pyramids remain. Nubia was converted to Christianity in c.540. Three Christian kingdoms emerged, but in 652 an Egyptian army conquered that at Dongola, granting

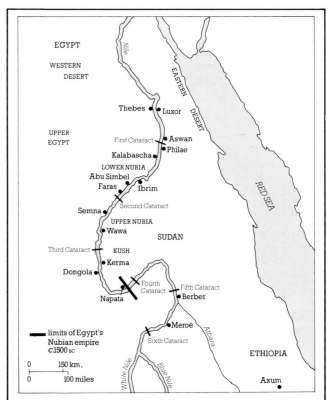

Nubia

The Romans gave the name Nubia to the region of the Nile valley extending roughly from the First Cataract in Egypt to the confluence of the Nile. In the 10th century BC its rulers became pharaohs of Egypt, and Napata, its capital, was briefly the centre of the ancient world.

peace for an annual tribute of slaves, and at the end of the 13th century *Mamelukes took the north. The southern kingdom remained until the 16th century, when the *Funj kingdom of Sennar absorbed it.

Numidia, the area to the south and west of Carthage in North Africa which was originally inhabited by the Numidae or Nomads. Its inhabitants were renowned in antiquity as horsemen, and the cavalry of its king Masinissa played a crucial part in *Scipio Africanus' defeat of *Hannibal at Zama in 202 BC. Later *Jugurtha led the Numidians against Rome, a struggle that ended with his capture in 106 BC. Numidia was ruled as a client kingdom of Rome before being incorporated into the province of Africa. It was the centre of the *Donatist movement in the 4th century AD.

nunnery, the building which houses a community of religious women. The word 'nun' is reserved for those living under strict vows of poverty, obedience, and chastity. Religious orders for women date from the 4th century and by the 11th century all communities lived lives devoted to prayer, reading, and work (spinning and weaving). From the 16th century sisterhoods were founded for active work, such as teaching and nursing. All communities were abolished in England at the Reformation but some were refounded in the 19th century.

Nupe, a kingdom in West Africa, now part of Nigeria. According to tradition, it was founded in the 16th century by Tsoede, who died in 1591 aged 120; he built Gbara and conquered much territory. Raba, the capital, became highly prosperous. It was largely dependent on wars to collect prisoners to sell as slaves. In the early 19th century the *Fulani asserted overlordship, leading to bitter struggles for independence.

Nystad, Treaty of (1721), the final treaty of the *Northern War. It was signed at Nystad in south-west Finland. Under this treaty Sweden recognized *Peter the Great's title to Estonia, Livonia, Ingria, Kexholm, and part of Finland and so lost its Baltic empire.

Oates, Titus   *Popish Plot.

Octavian   *Augustus.

Oda Nobunaga (1534–82), Japanese warrior. He overthrew many powerful *daimyo in an attempt to end the disorder of the *Ashikaga period. He was assisted by *Hideyoshi and *Tokugawa Ieyasu—and by the fact that many of his troops were armed with muskets, introduced by the Portuguese after 1542. In 1568 he entered Kyoto and in 1573 drove out the *shogun. He began 'sword hunts' to disarm much of the population and organized land surveys. When he was assassinated by one of his followers, firm rule had replaced political chaos in most of central Japan.

Odoacer (433–93), Gothic chieftain who became the first Germanic ruler of Italy. He and his troops were part of the Roman army and when Romulus Augustus became emperor (476) he led an uprising and deposed him, making Ravenna his capital. This event marked the end of the *Roman empire. His troops later overran Dalmatia, threatening the power of Zeno, the Eastern emperor, who encouraged *Theodoric, King of the Ostrogoths, to besiege Ravenna. Odoacer surrendered (493) on promise of retaining half of Italy but at a banquet was murdered by Theodoric, who became sole ruler.

Offa (d. 796), King of *Mercia (757–96). His kingdom was the midland region of Saxon England whose western boundary with Wales is marked by the great dyke he constructed (784–6). His overall supremacy extended to the kingdoms of Kent, Sussex, and Wessex. Offa, anxious to extend his European reputation, re-established direct contact between England and the papacy, and regarded himself as an equal of *Charlemagne. He codified the laws of Mercia, introduced a uniform silver penny which was to remain the basis of English currency until the 13th century, and stimulated its economy, particularly the cloth trade.

Oglethorpe, James Edward (1696–1785), English general, philanthropist, and founder of *Georgia. He entered Parliament (1722) after serving under Prince

A silver penny dating from the reign of **Offa**, king of Mercia. The coinage introduced by Offa bore the king's name and title and formed the basis of English currency for centuries. (British Museum, London)

*Eugène. Espousing the cause of imprisoned debtors, he led a group of philanthropists who became trustees of the new North American colony of Georgia in 1732. He founded Savannah (1733) and governed the colony on paternalistic lines, encouraging immigration of persecuted Protestants from Europe and of former soldiers. His military preparedness saved the colony from Spanish attack in 1742, after which he returned to England, where he fought against the *Forty-Five rebellion.

Olaf I (Tryggvessön) (969–1000), King of Norway (995–1000). He was exiled when his father was killed and he was brought up in *Novgorod in Russia. Viking expeditions took him to Iceland, England, and Ireland where he was converted to Christianity. In 995 he was accepted as King of Norway, and established his new religion. He was drowned at sea after a battle with the combined Danish and Swedish fleets.

Olaf II, St (Haraldsson) (c.995–1028), King of Norway (1015–28). He was converted to Christianity and continued the work of conversion begun by *Olaf I but his attempts at reform provoked rebellion and he was killed in a battle with rebel and Danish forces. Canonized as a saint, he is honoured as Norway's national hero.

Oldenbarneveldt, Johan van (1547–1619), Dutch statesman and lawyer who played a key role in the *Dutch Revolts. He was a Calvinist and keen supporter of *William the Silent and helped to negotiate the Union of Utrecht (1579). After 1586, as leader of the Estates of Holland, he managed to impose unity on the diverse political, economic, and religious interests of the *United Provinces. He negotiated a twelve-year truce with Spain in 1609, but after this his political differences with Maurice of Nassau became part of a bitter internecine quarrel between two rival schools of Calvinism, and eventually Maurice had Oldenbarneveldt tried and executed on a charge of treason.

Oldowan, the oldest tradition of human toolmaking, named after the simple stone tools found at *Olduvai Gorge but now known from many other early human occupation sites in Africa. The oldest certain stone tools, from the *Hadar and Omo regions in Ethiopia and from eastern Zaïre, were made between 2.5 and 2 million years ago, probably by *Homo habilis. Usually, the toolmaker started with a large cobble, probably picked out of a stream bed, and flaked it with a hammerstone into the required shape. The detached flakes were also trimmed and put to use. Several distinct types of Oldowan tools were made and were probably used for different tasks. A more advanced tradition, Developed Oldowan, occurs also at Olduvai Gorge around 1.5 million years ago, the maker probably being *Homo erectus.

Olduvai Gorge, a forked cleft 50 km. (31 miles) long and up to 100 m. (330 ft.) deep that cuts down through ancient lake and river beds into the Serengeti Plain in northern Tanzania, that contains an unsurpassed record of human evolution spanning the past 2 million years. Research carried out since 1935 has produced large quantities of important early human fossils, representing at least three hominid species, together with animal bones and stone tools.

The fossil-rich deposits are numbered I–IV, followed by three later sequences. From Bed I (2–1.7 million years ago) have come *australopithecines, including the famed *'Zinjanthropus' (Australopithecus boisei), and also *Homo habilis, probably the maker of the *Oldowan tools found in these deposits. Australopithecus boisei has also been found throughout Bed II (1.7–1.15 million years ago); in the lower part, the species is associated with Homo habilis, in the upper part with *Homo erectus, the maker of *Acheulian handaxes. Acheulian tools and Homo erectus continue into Bed III (1.15–0.8 million years ago), and Bed IV (800,000–600,000 years). Later evidence of human activity continues up to 15,000 years ago.

Olmec, the ancient Indian people of the southern gulf coast of Mexico who were the earliest culture in Mesoamerica (Mexico and northern Central America) to build large ceremonial centres, with huge carved stone heads of imported basalt and serpentine. Three principal sites successively dominated their culture, but overlapped in dates of occupation, between c.1500 and 400 BC: San Lorenzo, La Venta, and Tres Zapotes. Their art style stressed jaguar figures and a 'baby-face' image, and was spread widely, including Central Mexico and as far south as El Salvador and Costa Rica.

Olympia, a site in western Greece. It was the location of the most important shrine to the god Zeus (*Greek religion). An oracle of Zeus was there and every four years a festival of competitive games was held there in his honour. The large complex of religious and secular buildings which grew up survived until the Roman

An **Olmec** figure, carved in jade, of a jaguar spirit. Olmec art made wide use of the jaguar symbol, which was associated with Tlaloc, the god of rain and fertility. (Dallas Museum of Fine Art)

emperor *Theodosius decreed the destruction of all pagan sanctuaries. Archaeologists have recovered much, however, most notably many of the sculptures of the temple of Zeus (5th century BC), and also the workshop where the gold and ivory statue of Zeus was created by the Athenian Phidias in the 430s BC.

The first Olympiad is dated 776 BC, though the games were said to have begun before then. The date 776 has become a conventional one marking the 'beginning' of archaic Greece. Originally local, before long they began to attract Greeks from much further afield. They were greatly expanded from a one-day festival of athletics and wrestling to, in 472 BC, five days with many events, including horse and chariot racing, wrestling, and a race in full armour. Nevertheless, the most prestigious event continued to be the short foot-race over one length (c.200 m.) of the stadium. Winners received olive wreathes as their prizes, and they brought much glory to their cities, who might reward them with privileges. They might also commission choral poetry in celebration, such as the victory odes of Pindar. The games began and ended with sacrifices and feasting. Theodosius stopped the games in AD 393.

Oman, on the Arabian peninsula, a trading outpost of Mesopotamia, settled by Arabs in the 1st century AD. It was conquered for Islam in the 7th century. Expelling Portuguese incursions by 1650, the Omanis created a maritime empire with possessions as distant as *Mombasa and *Zanzibar and trade contacts with south-east Asia. By 1754 Ahmad ibn Said had expelled Turkish invaders and founded the sultanate that still rules Oman.

Omar *Umar.

Omri (9th century BC), King of Israel (c.876–c.869 BC). After the death of *Jeroboam I (c.901 BC) five kings had ruled ineffectively over the northern kingdom. Omri, an army general, founded his own dynasty in northern Israel. His main achievement was to establish his capital, *Samaria, where he built a lavish palace and temple as a symbol of political and religious unity for the northern kingdom. Omri expanded the power of northern Israel, aided by Egypt's temporary weakness and Syria's distraction by the Assyrian revival. Alliances with the Phoenicians were reinforced by his son *Ahab's marriage to Jezebel. The Omri dynasty ruled until c.842 BC.

oracle, place at which people consulted their deities for advice or prophecy. There were many of these in the ancient Greek world, most notably at *Delphi, Didyma on the coast of Asia Minor, Dodona in Epirus, and *Olympia. The most famous non-Greek oracle was that of the Egyptian Ammon at Siwah oasis in the Sahara, identified by the Greeks with Zeus and consulted by *Alexander the Great in 331 BC. Apollo was the god most favoured as a giver of oracles though many other deities presided over oracular shrines. At the most primitive shrines the god's reply was elicited through the casting of lots and the interpretation of signs. At healing oracles—for instance that of Asclepius at *Epidaurus—the god's reply came in the form of a dream. At Delphi an entranced priestess conveyed the divine message. Consultations usually concerned religious matters or where support was sought for political or military actions.

Oracle bones are animal bones which were used in divination by the *Shang kings of ancient China (c.1600 BC onwards). In many cases the question and/or answer has been written on the bone in pictographic figures.

Orange, the name of the ruling house (in full Orange-Châlons) of the principality centred on the small city of Orange, southern France. The city grew up around its Roman monuments, which include a semicircular theatre and a triumphal arch. In the 11th century it became an independent countship, and from the 12th century its rulers were vassals of the Holy Roman Emperor and came to style themselves 'princes'.

After 1530 the related house of Nassau-Châlons succeeded to the title, and in 1544 of William of Nassau-Dillenburg became Prince of Orange and subsequently, as *William the Silent, *statholder in the Netherlands. His younger son, Maurice of Nassau (1567–1625), assumed the military leadership of the *Dutch Revolts in 1584. Until the late 18th century the Orange dynasty continued to play a major part in the politics of the *United Provinces. The principality itself was conquered by Louis XIV (1672) and incorporated into France by the Treaty of Utrecht (1713), but the title of Prince of Orange was retained by *William III (1650–1702), who became King of England in 1689.

ordeal, a painful technique used in the early Middle Ages to determine the guilt or innocence of suspects by divine intervention, conducted and supervised by the Church. Ordeal by fire required suspects (usually freemen) to carry hot irons, or to walk blindfold and barefoot through red-hot ploughshares or over heated coals. If

A contemporary portrait of the Anglo-Irish army commander, James Butler **Ormonde**, after Lely, c.1665. (National Portrait Gallery, London)

they emerged unhurt or their wounds healed within three days they were innocent. Immersion of a hand or arm in boiling water was another method. Suspects (often witches) thrown into cold water were deemed guilty if they floated. Trial by blessed bread was a test for priests, for it was assumed guilty clergy would choke on hallowed food. Ordeals were repudiated by the Church in 1215. They were not particular to Europe; tribes in east Africa and Madagascar practised similar tests.

Orissa, an area in eastern India once known as Kalinga. In 261 BC it was conquered by the Buddhist emperor *Asoka, but local dynasties subsequently ruled until its absorption by the *Chola empire (1023). Its golden age is identified with the Ganga dynasty, founded in 1076 by Anantavarma Codagangadeva. Despite Muslim incursions, Hindu rule was upheld until the 16th century, when an Afghan invasion was followed by absorption into Akbar's *Mogul empire (1568). *Maratha expansion brought a short return to Hindu rule in the 18th century.

Orkneys, an archipelago of over 70 islands off north-east Scotland. Colonized first by the Picts (c.200 BC), they were already Christian when the first Norse colonists settled in the late 8th century. The Norse earls of Orkney extended their authority over the Shetlands, Caithness, and Sutherland, and their exploits are recorded in the *Orkneyinga Saga* (c.900-1200). The islands passed to the Scottish crown in 1472 as part of the marriage contract between *James III and Margaret of Norway.

Orléans, duc de, a title borne by younger princes of the French royal family from the 14th century. Charles VI of France bestowed the duchy of Orléans on his brother Louis (1392). Louis's grandson became *Louis XII of France. His great-grandson became *Francis I and the Valois-Orléans ended with the death of *Henry III in 1589. Philippe (1674-1723) of the second Bourbon-Orléans branch of the family became Regent of France in 1715 during the minority of *Louis XV. His great-grandson Louis Philippe Joseph ('Philippe Égalité') succeeded to the title in 1785. He was a supporter of the French Revolution from its beginnings, and in June 1789 he organized the forty-seven nobles who joined the Third Estate. In 1792 he voted for the death of the king but his eldest son, afterwards King Louis-Philippe, had him arrested, with all the remaining Bourbons, accused of conspiracy, and guillotined (1793).

Ormonde, James Butler, 12th Earl and 1st Duke of (1610-88), Anglo-Irish nobleman and army commander. He fought for *Charles I against the Irish rebels in 1641-3 and became Lord Lieutenant of Ireland. On Charles's death he proclaimed *Charles II as King of Ireland, but was defeated by *Cromwell and forced into exile. He served again as Lord Lieutenant throughout most of Charles II's reign. His loyalty to the crown was severely tested by *James II's arbitrary actions, some of which he bitterly opposed.

Ormonde, James Butler, 2nd Duke of (1665-1745), Anglo-Irish army commander. A firm Tory, he opposed the Hanoverian succession, and when George I dismissed him from the post of captain-general in 1714 he became a *Jacobite, was threatened with impeachment, and fled to France. Storms prevented him reaching England for

the *Fifteen Rebellion, and after he had settled in Spain he was given command of an abortive Jacobite expedition in 1719.

Orphic mysteries, a religious cult of ancient Greece, prominent in the 6th century BC. It was believed to have been established by the mythological hero Orpheus, who was able to charm all nature with his music. It was based on the belief that there was a mixture of good (divine) and evil in human nature and that the evil should be destroyed by initiation into the Orphic mysteries and by moral purification. Initiates purified themselves and adopted ascetic practices, such as abstaining from meat. It sank to the level of a superstition in the 5th century, though the profound thoughts which underlay it were perceived by *Plato.

Orthodox Church, the 'Eastern' or 'Greek' Christian Church which separated from the Western Church in the 9th century. The Orthodox Church developed from the church of the Byzantine empire. Originally the churches of Rome and Constantinople were in agreement (seven ecumenical councils were convened between 325 and 787 to establish accord on doctrinal matters) and Constantinople acknowledged the seniority of Rome. From the 9th century areas of conflict multiplied, cul-

The kneeling figure of **Osman I** in a painted Turkish miniature from a 16th-century manuscript illustrating the dynasty of the Ottoman sultans. (Topkapi Museum, Istanbul)

minating in the excommunication of the Patriarch of Constantinople by the pope in 1054, the date which marks the final rift (*East-West Schism). Following the separation the Orthodox Church expanded, most significantly into Russia.

The Orthodox rite differs in several respects from the Roman Catholic rite, principally in the matter of church sacraments. Parish priests are generally married (Catholic priests are not allowed to marry) and veneration of icons (holy pictures) plays an important part in worship.

Osman *Uthman.

Osman I (c.1258-1326), founder (c.1300) of the *Ottoman empire. He was the leader of the famed 'four hundred tents', who turned his band of Turkish nomads from raiding to permanent conquest. In 1290 he declared his independence from the *Seljuk Turks and expanded his territory by capturing minor Christian kingdoms in north-western Turkey.

ostracism, a method of banishment in ancient *Athens. At a stated meeting each year, the Athenian assembly voted whether it wanted an ostracism that year. If the vote was affirmative, an ostracism was held two months later. Every citizen who so wished then wrote a name on a sherd of pottery ('*ostrakon*'), and provided that at least 6,000 valid 'ostraka' were counted, the man with the most against him had to leave Attica for ten years, though he was allowed to enjoy any income from his property in Attica while absent. Ostracism often functioned as effectively a sort of 'general election', constituting a 'vote of confidence' for the policies of the most powerful rival of the man ostracized. Such trials of political strength were most notable in the ostracisms of Themistocles (c.471), Cimon (c.462), and Pericles' rival Thucydides (443). Ostracism was not resorted to after 417 or 416.

Ostrogoths, the eastern *Goths on the northern shores of the Black Sea. They became vassals of the *Huns whose westward migration displaced them and under *Attila they were defeated by Roman and barbarian allied armies on the *Catalaunian Fields, AD 451. Forty years later *Theodoric established a kingdom in Italy. After the murder in 533 of Theodoric's daughter, who was the regent of Italy, *Justinian's general *Belisarius twice invaded and defeated them, and the Ostrogothic kingdom was crushed by Narses in 552.

Oswald, St (c.605-42), King of *Northumbria (633-42). He gained the Northumbrian kingdom after his victory near Hexham over the Welsh king *Cadwalader (633) who had in the previous year invaded Northumbria and killed Oswald's uncle· *Edwin. Following that invasion Oswald was in exile on Iona where he became a Christian. He arranged for missionaries from Iona led by St Aidan (635) to convert his people. Oswald was killed in battle by the pagan king *Penda of Mercia.

Oswy (or Oswiu) (d. 670), King of *Northumbria (651-70). He succeeded his brother Oswald as ruler of the Northumbrian kingdom of Bernicia in 642 and incorporated the other kingdom, Deira, after arranging the assassination of his cousin Oswin (651). Although initially Northumbria was still in the control of *Penda, king of the neighbouring Mercia, Oswy was able to defeat

him (655) and establish himself as overlord of England. Oswy's support of St Wilfrid at the Synod of *Whitby (664), called to consider the rival claims of the Roman and Celtic forms of the Christian Church, was critical in gaining the decision in favour of Rome.

Otis, James (1725-83), American lawyer and patriot from Massachusetts. He first achieved fame by his attack on Writs of Assistance (search warrants) in 1761, basing his case on natural rights philosophy. Motivated partly by jealousy against the conservative Thomas Hutchinson, lieutenant governor of Massachusetts (1758-71), he rapidly became the leading orator, pamphleteer, and political spokesman of the patriot cause. After receiving head injuries in a fight with a British customs officer (1769) he became insane, but recovered sufficiently to fight at *Bunker Hill.

Ottawa Indians, an Algonquian-speaking group of North American indians living round northern Lake Huron, encountered by Samuel de *Champlain in 1615. Some were dispersed eastwards after 1649 when the *Iroquois defeated the *Huron, whose cultures theirs resembled. Thereafter they relocated their villages round the Great Lakes until the 17th, 18th, and early 19th centuries, when they finally settled on reservations round Lakes Michigan, Huron, and Erie, and in Oklahoma.

The head of the German king and Holy Roman Emperor **Otto I** from a statue in Magdeburg Cathedral.

Otterburn, battle of (5 August 1388). Otterburn, a village in northern England near the border with Scotland, was the scene of a victory for Scottish forces over the English. The Earl of Douglas captured Henry *Percy the Younger (Hotspur) and jeopardized English control of the north for years to come, in an exploit that was immortalized in the *Ballad of Chevy Chase*.

Otto I (the Great) (912–73), Holy Roman Emperor (936–73). He succeeded his father Henry I, who had established the Saxon (or Ottonian) dynasty on the German throne and so extended royal influence as to make his son's election a virtual formality. Otto I consolidated his position by defeating the independent dukes of south Germany in a three-year war and thereafter keeping most of the great duchies under direct royal control. He went on to restore the empire to its full power, destroying the *Magyar threat to the eastern frontier at the battle of *Lechfeld in 955, conquering Bohemia, Austria, and northern Italy, and bringing the German church into alliance with the throne despite the opposition of the German nobility. In a series of expeditions into Italy, Otto also established imperial supremacy over the papacy, a process highlighted by his coronation by the pope in 962 and his deposition of papal incumbents who failed to follow his wishes. His reign was also noted for the flowering of culture and learning in the so-called 'Ottonian Renaissance', fostered by the imperial court.

Otto III (980–1002), King of Germany and Holy Roman Emperor. He had ambitions of recreating the glory and power of the old Roman empire with himself as leader

Otto III, grandson of Otto I and a successor as Holy Roman Emperor, portrayed in state, surrounded by secular and religious dignitaries. This portrait miniature from the *Reichenau Gospels*, dated about AD1000, exemplifies the cultural flowering of the 'Ottonian Renaissance' (Bayerische Staatsbibliothek, Munich)

of world Christianity aided by a subservient pope. His short life was torn between periods of intense religious devotion, living for a year in a monastery, and dreams of secular power wielded from Rome where he lived in oriental seclusion in the palace which he had had built. To control the papacy he installed in turn his cousin, Gregory VI (996–99) and his tutor Sylvester II (999–1003) as popes but this did not prevent rebellions in Italy.

Ottoman empire, the Islamic empire, established in northern Anatolia by *Osman I at the end of the 13th century. It expanded north, west, and south, inspired by the spirit of *jihad and chiefly at the expense of the *Byzantine empire, the capital of which, *Constantinople, was taken by *Mehmed II in 1453. The army was the basis of the Ottoman expansion, its senior ranks being recruited from the subject Christian families of the Balkans in a five-year levy, the devshirme. Boys were forcibly recruited, converted to the Muslim faith, educated in the palace schools, and entered into the Sipahis (cavalry) and *Janissaries (infantry) of the army; they also staffed the civil service. At its height, under *Suleiman I, the empire exercised suzerainty over territories stretching from the Austrian border to Yemen and from Morocco to Persia. The Turkish fleet under the notorious pirate *Barbarossa, ravaged the coasts of Spain, Italy, and Greece, twice defeating the Italian admiral *Doria, before being beaten at *Lepanto (1571). Internal decline set in as civil service and army appointments became hereditary and the rule of each new sultan was marked by rivalry and bloodshed as there was no legally accepted line of succession. It was reinvigorated by the able services of the *Köprülü family in the late 17th century, but its conservatism condemned it to military backwardness in the face of Europe's industrializing challenge. (See map on page 266.)

Oudenarde, battle of (1708). At the Flemish city of Oudenarde *Marlborough and his Dutch and Austrian allies defeated the French army in the War of the *Spanish Succession. It was Marlborough's third great victory and led to the capture of Lille. In 1745 during the War of the *Austrian Succession the French took the town and dismantled the fortifications.

Outremer ('Beyond the Sea'), the Frankish name given to the Latin Kingdom of *Jerusalem established in Palestine after the First *Crusade (1096–9). On Christmas Day 1100 *Baldwin I was crowned as its first king. He ruled over a narrow strip of coastal territory with no clear eastern boundaries. The kingdom depended for its survival on sufficient settlers coming regularly from Europe. However, Muslims united against the invaders and Jerusalem fell to them in 1187.

Overbury, Sir Thomas (1581–1613), English courtier, essayist, and poet. He was a close associate of Robert Carr, a favourite of *James I of England. He opposed the rising influence of the *Howard faction, but Carr, now Earl of Somerset, fell in love with Lady Frances Howard. Despite Overbury's protestations, Carr persuaded James to arrange an annulment of Frances's marriage to the 3rd Earl of Essex. By the end of 1613 Carr and Frances were married, and Overbury was imprisoned in the Tower of London for displeasing the king. He died there. In 1615 it was discovered that he

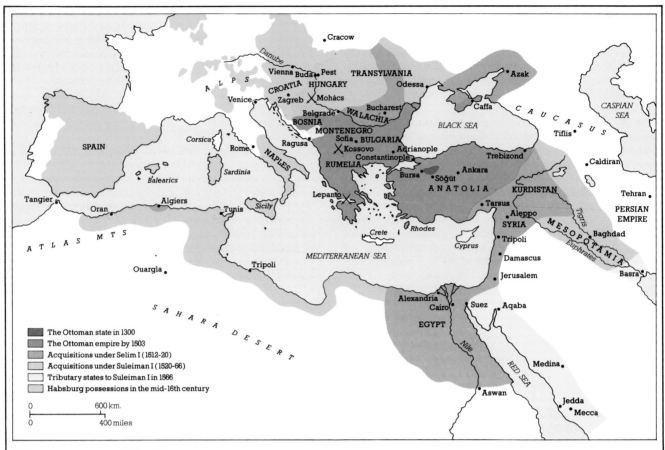

Ottoman empire: *c.*1300–1571

The Ottoman state in 1300 was one of many small Turkish states in Anatolia, but it expanded rapidly during the 14th century, absorbing Bursa (1326), Adrianople (Edirne) (1361), and, following the battle of Kossovo (1389), much of the Balkan peninsula. The expansion was briefly checked by Tamerlane at Ankara (1402), but began again with the accession of Mehmed (1413). It reached its peak under Selim I, who defeated the Mamelukes of Egypt and Syria, and Suleiman I, whose victory at Mohács brought in much of Hungary. The defeat at Lepanto marked the beginning of the empire's slow decline.

had been poisoned. Amid hysterical accusations of adultery and witchcraft, the Carrs were publicly tried and sentenced to death for the crime, but James pardoned them, and kept them in the Tower until 1622. The episode badly damaged the reputation of both king and court, especially in the eyes of *Puritans.

Oxenstierna, Axel Gustaffson, Count (1583–1654), Swedish statesman. He entered the service of the Swedish state in 1602 and joined the Council of State in 1609. *Gustavus Adolphus appointed him Chancellor in 1612, and for the next twenty-two years the two men worked together. An administrative reformer of note, Oxenstierna also made possible the reconciliation of the Swedish aristocracy to the monarchy. He negotiated the Truce of Altmark with Poland (1629), joined Gustavus in Germany in 1631, and after the king's death in 1632, was responsible for directing Swedish policy throughout the *Thirty Years War. He was also the effective ruler of Sweden from 1636 to 1644, when Queen *Christina reached her majority. His subsequent relations with her were not always harmonious. On his death, his son Erik succeeded him as Chancellor.

Oyo, the capital of Oyo Province, Nigeria, and seat of the Alafin of Oyo, titular head of all the Yoruba people.

The original Yoruba capital, Old Oyo, was settled probably in the 10th century, and rose to prominence because of its rich agricultural resources and advantageous trading position. In the 16th century Alafin Oronpoto organized a cavalry force, giving it a military dominance over other Yoruba states. In two wars, 1724–30 and 1738–48, Oyo took the kingdom of *Dahomey and came into contact with European merchants on the coast, with whom it traded in slaves. By 1817–18 the state had declined as a result of internal dissensions and civil wars, and Old Oyo was destroyed by the Fulani.

P

Pagan, a city on the Irrawaddy River and former state in central *Burma. It was originally a centre to which the Burmans, a people originating in Yunnan province, south-west China, migrated in the 9th century AD. In the 11th century they brought much of Burma under their rule. With their conquest of the *Mon people, Theravada *Buddhism spread to Pagan and in time throughout Burma. When *Kublai Khan, having conquered Yunnan, demanded tribute from Pagan, his envoys were executed. Five Mongol campaigns culminated in the occupation of Pagan (1287–1301), which was already weakened by Mon and *Shan revolts.

Paine, Thomas (1737–1809), British writer and political theorist. He lost his post as an excise officer in 1774 after agitating for higher wages in the service. He went to America with a testimonial from Benjamin *Franklin, and soon became involved in the political controversies which led to American independence. His pamphlet *Common Sense* (1776) influenced the American *Declaration of Independence and provided the arguments to justify it. He returned to England in 1787, and in 1791 published the first part of *The Rights of Man*, a reply to *Burke's *Reflections on the Revolution in France*. The radical views expressed in the second part of *The Rights of Man* (1792) stimulated the growth of the *London Corresponding Society but alarmed the British government. He was threatened with arrest and fled to France, where he was immediately elected to the Convention. He supported republicanism, but showed courage in opposing *Louis XVI's execution; on the orders of *Robespierre he was imprisoned for a year, during which he completed *The Age of Reason* (1795), a provocative attack on Christianity. Only his American connection saved him from the guillotine, and in 1802 he returned permanently to America.

Pala, a dynasty of Bihar and Bengal in north-eastern India that ruled from the 8th to the 12th century. Gopala, the founder, was chosen king by the great men of the region. The reign of his successor, Dharmapala (c.770–810) marked the dynasty's apogee, after which power rose and fell intermittently until its final eclipse in the 12th century. The surviving Pala artefacts in both stone and metal are of a particularly fine decorative quality. An important development was the spread of *Buddhism to Tibet by missionaries from the University of Nalanda in Bihar, which received Pala patronage.

Palaeolithic ('Old Stone Age'), a term to describe the earlier part of the *Stone Age, characterized by flaked stone tools. Today, the term is usually broadened to include all human existence as a scavenger, hunter and gatherer, and maker of tools before and during the last Ice Age. (The term *Mesolithic was invented later to describe the hunting groups in Europe in the postglacial period.) Among the first tools would have been wooden digging sticks but these have not survived in the archaeological record. However, simple stone tools of *Oldowan type produced by *Homo habilis in eastern Africa up to 2.5 million years ago can be recognized. More advanced *Acheulian handaxes were made by *Homo erectus for more than a million years from 1.5 million years ago. Together, these two traditions in Europe make up the Lower Palaeolithic. The more varied stone tools of the Middle Palaeolithic, which began 150,000–125,000 years ago, were produced by *Neanderthal people and their contemporaries. More sophisticated stoneworking techniques that produced long, narrow blades struck from a stone core are associated with the appearance of fully modern populations (*Homo sapiens sapiens) about 50,000 years ago, and define the *Upper Palaeolithic in Europe.

Palatinates, two regions in Germany: the Rhenish or Lower Palatinate, on the Rhine, and the Upper Palatinate, since 1628 part of Bavaria. *Frederick I bestowed the title of Count Palatine on his half-brother Conrad, who held lands east and west of the River Rhine (the Lower Palatinate). From 1214 these lands were ruled by the Bavarian Wittlesbach dynasty whose own lands near Bohemia formed the Upper Palatinate. In 1356 the Counts Palatine were made Electors of the Holy Roman Empire.
    The Rhenish (Lower) Palatinate became a centre of the Protestant *Reformation in the 16th century and the choice of Elector *Frederick V as King of Bohemia led to clashes with Catholic Habsburg authority and the outbreak of the *Thirty Years War. After the battle of the White Mountain (1621) the Palatinates were partitioned, with Bavaria absorbing the Upper Palatinate and the Lower Palatinate passing to Frederick's heirs under the terms of the Treaty of *Westphalia. It was invaded and brutally devastated by *Louis XIV in 1688–9. In 1777 the two Palatinates were reunited.

pale, a 14th-century term for a distinct area of jurisdiction, often originally enclosed by a palisade or ditch. Pales existed in medieval times on the edges of English territory—around *Calais (until its loss in 1558), in Scotland (in Tudor times), and, most importantly, as a large part of eastern Ireland (from *Henry II's time until the full conquest of Ireland under *Elizabeth); the actual extent of the Irish pale depended on the strength of the English government in Dublin. *Catherine the Great in 1792 made a Jewish pale in the lands she had annexed from Poland: Jews had to remain within this area, which ultimately included all of Russian Poland, Lithuania, Belorussia, and much of the Ukraine.

Palestine, an area on the east coast of the Mediterranean, also known as the Holy Land, where the kingdoms of *Judah and *Israel were in biblical times. The word 'Palestine' derives from Palaistina or Philistia, the land of the *Philistines.
    By 2000 BC the *Canaanites had settled in the area, though they were subsequently confined to the coastal strip later known as Phoenicia by the Hebrews and Philistines. In time the Hebrews predominated, but after the death of *Solomon they split into two—the kingdoms of Israel and Judah. Israel was overthrown by the *Assyrians in 722 BC, and Judah suffered a similar fate at the hands of *Nebuchadnezzar II, who destroyed Jerusalem and transferred many inhabitants to Babylon in 586 BC. When the Chaldean dynasty of Babylon was overthown by *Cyrus the Great in 536, the Jews in

Babylon were allowed to return to Palestine where, under the leadership of *Nehemiah and *Ezra, Jerusalem was rebuilt and the nation restored. Independence was finally regained when the *Maccabees rebelled against the *Seleucids in 168 BC, but in 63 BC Pompey annexed the area for Rome. The *Herod dynasty ruled parts of the area with Roman approval until unrest (the *Jewish Revolt) caused the Romans to destroy Jerusalem in AD 70. Further trouble flared in 132 under *Bar Cochba.

Palestine was Christianized under Rome and Byzantium, but it fell to the Arabs in 641, and later came under the control of the *Crusaders, the *Mamelukes, and the *Ottoman Turks.

Pallava, a south Indian dynasty which maintained a regional kingdom along the Carnatic coast between the 4th and 9th centuries. There is uncertainty about their origins and early history before their emergence to power from a previously subordinate role in the *Deccan. They established their capital at Kanchi, and traded in Sri Lanka and parts of south-east Asia. In the late 9th century they lost their territories to their own feudal vassals, the *Cholas. Their artistic legacy is important. Apart from patronage of music, painting, and literature, some of the greatest south Indian temples were built during their rule, including the 'Shore Temple', carved from solid rock on the coast at Mahabalipuram.

palmer, a professional pilgrim of medieval England who had made the long and dangerous journey to the Holy Land (Palestine). As evidence of this pilgrimage, palmers displayed a palm leaf. The name was later applied more

The 'Shore Temple' at Mahabalipuram in southern India is one of the finest examples of **Pallava** architecture. Dedicated to the Hindu god Shiva and dating from about AD 700, the temple is built in typical South Indian style with a pyramid-shaped spire consisting of stepped storeys. At the apex of the pyramid is a solid dome topped by a pot and finial.

generally to include those who devoted all their time to pilgrimages and were dependent upon charity for their livelihood.

Palmyra, an ancient Syrian city which rose to prosperity in the 1st century BC by organizing and protecting caravans crossing the desert between Babylonia and Syria. It was probably incorporated within the Roman empire in AD 17, but rose to its greatest power in the 3rd century AD under King Odaenathus (d. 267) and his second wife, Zenobia. The latter at one point ruled Syria, Egypt, and almost all of Asia Minor, but by 273 *Aurelian had captured her and destroyed Palmyra. Its extensive remains testify to the mixture of Hellenistic and Parthian elements in Palmyran culture.

Panama, the southernmost country of Central America, visited in 1501 by the Spaniard Rodrigo de Bastidas. It was explored more thoroughly in 1513 by Vasco Núñez de *Balboa, the first Spaniard to see the Pacific Ocean. Portobello on the Caribbean coast served as the principal port for the trade of the viceroyalty of Peru. In the 18th century, Panama became part of the viceroyalty of *New Granada.

Pandya, a Tamil dynasty which ruled in the extreme south of India from the 3rd century BC to the 16th century AD. Little is known about their early history during an era of frequent warfare among competing south Indian dynasties. Between the 7th and 14th centuries they expanded into Sri Lanka and Kerala, and northwards into *Chola and *Hoysala territories. Their peak was reached during the reign of Jatavarman Sundara (1251–68). They were already weakened by family quarrels when their capital at Madurai was invaded in 1311 by the sultan of Delhi, Ala ud-Din *Khalji. Although they retained local power until the 16th century, they never again aspired to empire. *Marco Polo recorded the prosperity of their realm during its peak years in the late 13th century.

Panipat, a town in Haryana state, north India, 70 km. (50 miles) north-west of Delhi. It is historically important as the site of three battles which proved decisive for India's future. On each occasion armies moving out to defend the capital of Delhi clashed here with invaders approaching from the north-west. In the first battle of Panipat (21 April 1526) the *Mogul invader, *Babur, defeated the Afghan sultan of Delhi. The second battle (5 November 1556) marked *Akbar's victory over the Sur Afghans who then held Delhi, and initiated the spread of strong Mogul power in India. In the third battle (14 June 1761) an invading Afghan army ended *Maratha ambitions to fill the power vacuum at Delhi caused by the decline of the Moguls.

papacy, the office of the pope (Bishop of Rome), derives its name from the Greek *papas* and Latin *papa*, which are familiar forms of 'father'. In early times many bishops and even priests were called popes, but in the Western Church the word gradually became a title and restricted to the Bishop of Rome; Pope Gregory VII in 1073 forbade its use for anyone except the Bishop of Rome. The traditional enumeration lists 265 holders of the office, excluding *antipopes, beginning with St *Peter and reaching to the present holder John Paul II. The basis

Papacy: Pope John XXII, enthroned and crowned in the presence of his cardinals. This painting from an early 15th-century version of *The Travels of Sir John Mandeville* shows the pope receiving an official communication from a messenger of the Greek church. (British Library, London)

of papal authority derives from St *Peter's position of leadership among the twelve Apostles, given him by Jesus Christ, the early tradition that he came to Rome and was martyred there, and the belief that Christ wanted there to be successors to St Peter in the Church. In the early centuries of Christianity the Bishop of Rome was a fairly low-key figure, exercising authority over the wider Church only in emergencies. But gradually the papacy extended its claims to jurisdiction over the whole Church. These extended papal claims were a major cause of various churches breaking with Rome, notably the *Orthodox Church definitively in 1054, and the Protestant churches at the time of the *Reformation in the 16th century.

Papal States, lands in central Italy held by the Roman Catholic Church. A law of 321 allowed the Church to own land, but it was the 'Donation of *Pepin' (754), which promised Lombard lands conquered by the Franks to the pope and guaranteed their protection, which increased the papacy's temporal power. In the early Middle Ages papal control weakened and land was alienated as the papacy 'bought' allies in return for territory. In the 13th century the situation was reversed: the Church's holdings were greatly extended by Pope *Innocent III and in 1213 Emperor Frederick II confirmed and increased papal possessions. *Julius II restored and enlarged the temporal power of the papacy: at their greatest extent the Papal States included Romagna, Ferrara, Ravenna, and much of Tuscany. They were taken over by the kingdom of Italy in the 19th century.

papyrus, a document written on the paper made from Egyptian papyrus, a plant which thrived in the Nile valley. The majority of surviving papyri have been discovered in Egypt where the dry conditions have favoured their preservation. These varied documents—funerary, legal, administrative, and literary—have yielded much detailed information on life in ancient Egypt.

Paraguay, one of two land-locked nations of South America. It was part of Spain's Rio de la Plata territory from the founding of the capital Asunción in 1537. Paraguay was only sparsely settled by Spaniards and was dominated by Jesuit mission villages among the Guaraní Indians until their suppression in 1767.

pardoner, an agent of the Christian Church licensed to sell *indulgences. The rapid proliferation of pardoners in the 14th century meant that often they were not licensed. They might carry 'holy' relics to assist sales and they exploited the gullibility of people for their own profit. Pope Boniface IX ordered an enquiry into these abuses (1390), and the abuse of indulgences was one of Martin *Luther's main quarrels with the Roman Catholic Church. The sale of indulgences was forbidden by the Council of Trent (1563).

Paris, Peace of (1783), the treaty which concluded the American War of *Independence. It was mainly engineered by John *Jay and the Earl of *Shelburne, and was damaging to France and Spain, which regained only Florida. The peace recognized American independence, gave it north-eastern fishing rights, and attempted (unsuccessfully) to safeguard creditors, protect loyalists, and settle the frontier between Canada and the USA. These failures led to 30 years' friction, especially in the *Northwest territory, where Britain retained forts in retaliation, and led to the War of 1812.

Paris, Treaty of (1763), the treaty signed by Britain, France, and Spain which brought the *Seven Years War to an end. Britain did not fully exploit the world-wide successes it had enjoyed, as *Pitt had resigned and the Earl of Bute was anxious for peace. Under the terms of the treaty Britain gained French Canada and all the

The **pardoner**, a woodcut illustration printed in the 1484 and 1498 editions of Chaucer's *The Canterbury Tales* by William Caxton and Wynkyn de Worde respectively.

territory France had claimed to the east of the Mississippi. France ceded some West Indian islands, including St Vincent and Tobago, but retained the islands of Guadeloupe and Martinique. In India, France retained its trading-stations but not its forts. Britain gained Senegal in West Africa and Florida from Spain; it also recovered Minorca in exchange for Belle Isle. Spain recovered Havana and Manila, and France's claims in Louisiana west of the Mississippi were ceded to Spain in compensation for Florida, which became British until 1783. Britain was supreme at sea and, for the time being, dominated the east coast of North America.

parish, the smallest unit of ecclesiastical and administrative organization in England. In the 7th and 8th centuries regional churches ('minsters') were founded, staffed by teams of priests who served large 'parochiae' covering the area of perhaps five to fifteen later parishes. These were broken up during the 10th to 12th centuries as landowners founded local churches for themselves and their tenants, though it was only in the 12th century that the territories which these served crystallized into a formal parochial system. The *Taxatio Nicholai* (1291), an assessment of clerical incomes, probably understated the number of parishes at 8,085: a more likely figure would be 9,500. The majority of these were rural communities averaging about 300 persons. Probably no more than a few hundred parishes were located in towns: London had about a hundred, Norwich fifty, Lincoln forty, and Cambridge twenty. Since the 16th century parishes have performed an administrative role for the state although their boundaries and those of ecclesiastical parishes do not always coincide. In 1555 they were responsible for highways and in 1601 for organizing poor relief.

Parlement, the sovereign judicial authority in France, the chief being in the capital, *Paris. First established in the 12th century, it functioned as a court of appeal, and as a source of final legal rulings. There were also provincial *Parlements*, those of Paris, Toulouse, Bordeaux, Rouen, Aix, Grenoble, Dijon, and Rennes. Political importance derived from their power to register royal edicts, and to remonstrate against them. This power could be overridden by the king, either by order or by *lit de Justice* (a personal intervention). The introduction of hereditary office-holding (the *paulette*), in 1604, considerably enhanced the privileges and power of the *parlementaires*, who challenged the monarchy during the *Frondes. Revolt led to their suppression by *Louis XIV, but they were restored in 1718 by the regent, Philippe, duc d'*Orléans. They then led opposition to the monarchy, posing as defenders of liberty, but chiefly defending aristocratic privilege, and halting royal reforms. Their resistance culminated in the calling of a *States-General in 1789. The resulting *French Revolution revealed their selfish motives and they were suppressed in 1792.

Parliament, the supreme legislature in Britain, comprising the sovereign, as head of state, and the two chambers, the House of *Commons and the House of *Lords. Beginning in the 13th century as simply a formal meeting of the king and certain of his officials and principal lords, it became partly representative, as in Simon de *Montfort's Parliament (1265), which contained commoners (knights of the shire and burgesses of the boroughs) who were elected in their locality, and in Edward I's *Model Parliament (1295).

Until the 16th century, both chambers grew in importance *vis-à-vis* the crown, as it came to be accepted that their approval was needed for grants of taxation; *Henry VIII effected the English *Reformation through the long-lived Reformation Parliament (1529-36). Kings such as *Charles I tried to manage without summoning a parliament (1629-40), but by the 17th century the Commons had made themselves indispensable. Charles I had to call Parliament in 1640 in order to raise money, and Parliament, led by John *Pym, led the opposition to him. The Parliamentary side won the *English Civil War, and at the end of the *Commonwealth period it was the members of the House of Commons who negotiated the *Restoration of *Charles II (1660) and the accession of *William III and Mary (1688). The legislation enacted in the *Glorious Revolution of 1688-9 and the Act of *Settlement (1701) settled the relationship of crown, Lords, and Commons definitively and made clear the ultimate supremacy of the Commons.

Parr, Catherine (1512-48), sixth and surviving wife of *Henry VIII. She was the daughter of Sir Thomas Parr, controller of the king's household. Herself twice widowed, she became the king's wife in 1543. She skilfully managed him in his decline, helped his daughters *Mary and *Elizabeth to regain their places in the succession, and concerned herself with the education of Elizabeth and the future *Edward VI. Soon after Henry's death in

Parliament: a view of the House of Commons meeting in St Stephen's Chapel, Westminster, c.1715, a painting by Peter Tillemans. The central seated figure is the Speaker; members of the House sit on benches on either side of the Speaker's chair. At the bar in the foreground stand officers, one carrying the mace.

1547 she married Thomas, Baron Seymour of Sudeley, but died after bearing his daughter.

Parsi *Zoroastrian.

parson, originally a clergyman who held a church living, and was supported by its revenues. The term comes from the Latin word *persona* (person), which seems to have been used in this way since the 11th century. According to *Coke, it derived from the view of a parson as the legal 'person' who holds the property of God in the *parish. Since the 17th century, however, the term has been used to describe all clergymen, especially those in the Anglican Church.

Parthia, an ancient country of Asia, originally a *satrapy of the *Achaemenid empire, which later passed under the control of the *Seleucids. The Parthian empire traditionally began in 247 BC, when Arsaces I led his people in a rebellion against the Seleucids. By the end of the 2nd century BC they ruled from the River Euphrates in the west to Afghanistan in the east and north as far as the River Oxus. In 53 BC they defeated the Romans, led by *Crassus, at Carrhae. The Parthians never developed beyond being a military aristocracy, though their geographical position allowed them to capitalize on the trade between East and West. Their rule came to an end in AD 224 when Ardashir of Persia defeated Artabanus V in battle, thereby founding the *Sassanian empire.

Pascal, Blaise (1623-62), French mathematician, physicist, religious philosopher, and man of letters. As a precociously clever child of 12 he worked out Euclid's *Elements* for himself. He subsequently designed a form of mechanical calculator, and his study of hydrostatics led him to invent the syringe and the hydraulic press. From 1651 to 1654 he suffered from chronically bad health. Part of his work during these years was concerned with laying the foundations for the calculus of probabilities. Then he underwent a religious experience that caused him to frequent the *Jansenist monastery at Port-Royal. He helped the Jansenists in their struggles with the Jesuits by anonymously writing the *Lettres provinciales* (1656-7), which helped significantly to undermine *Jesuit prestige and authority. His *Pensées*, published posthumously in 1670, established his influential principle of intuitionism, which taught that God could be experienced through the heart and not through reason.

pasha, the highest official title of honour in the *Ottoman empire. Used after the proper name, it was personal rather than hereditary, and not given to men of religion and rarely to women. It is derived from the Persian *padshah* and was first used by the *Seljuks. Ottoman usage assigned it to *viziers and provincial governors.

patrician, a privileged landed aristocrat of early republican Rome. The patricians (or 'fathers') gathered after the expulsion of *Tarquin to guide the state. Supported by revered tradition they were hereditary members of the *Senate. They monopolized all magistracies and priesthoods but during the 'struggle of the Orders' with the plebeians they were forced to share power with them—in 367 BC the consulship was open to plebeians. Thereafter a 'plebeian nobility' arose which together with the patricians formed the ruling class. Their

A portrait by a contemporary artist of Blaise **Pascal**, the 17th-century French scientist and religious philosopher. (Musée de Versailles)

ranks were thinned and their influence waned in the late republic but the ancient names still carried prestige.

Patrick, St (c.390-c.460), patron saint of Ireland. He was the son of a Romano-British Christian family but was captured by pirates as a boy, and spent six years as a slave and herdsman in Ireland. He escaped back to his home in Britain and went to Gaul; he received a training for the priesthood. He returned to Ireland as a missionary bishop (c.435) and played a leading role in the conversion of that country to Christianity. He challenged the influence of the *Druids, converted the royal family, and founded the cathedral church at Armagh (444). His *Confession* (450) is the main source for information about his life. The cult of St Patrick gradually spread to Irish monasteries in Europe.

Paul I (1754-1801), Emperor of Russia (1796-1801). He was the disturbed son of the future empress *Catherine II and Peter III, and was greatly affected by the murder of his father during Catherine's *coup d'état* in 1762. After Catherine's death in 1796 he began a reign of frenzied despotism. He was obsessed by a fear of revolution and joined the coalitions against France. His attempts to violate the order of succession led to a conspiracy, which included his son, Alexander I, and in March 1801 resulted in his murder.

Paul, St (1st century AD), early Christian missionary, originally named Saul. He was born into a Jewish family resident at Tarsus and given a thorough Jewish education. He belonged to the religious party of the *Pharisees and was at first strongly opposed to *Christianity. He was dramatically converted to it (c. AD 33) after seeing a vision while on a journey by road to Damascus. About ten years after his conversion Paul began to travel as a missionary in Asia Minor and the Aegean. This work expressed his conviction that he was an apostle to the Gentiles (non-Jews). His preaching stressed that Gentiles who became Christians needed only faith in Christ in order to be 'justified' before God, and need not accept Jewish customs. Paul's radical understanding of the Christian message provoked hostility, and a riot against him on a visit to Jerusalem led to his arrest by the Roman authorities. He was eventually taken to Rome, where he is thought to have died a martyr's death (perhaps after a further period of freedom) c.64 AD. Several of Paul's letters to early Christian groups have been preserved in the New Testament of the Bible. Through them his influence on Christian life and thought has been greater than that of any other of the first Christians.

Pavia, battle of (24 February 1525). An engagement in the Habsburg–Valois wars, which involved the papacy and England supporting *Charles V against *Francis I. In October 1524 the French invaded Italy and took

A contemporary engraving of the battle of **Pavia**. The French army laying siege to the city was itself engaged by the 23,000-strong advancing Habsburg imperial force. Hastily attempting to encircle the town, the French cavalry came under fire from 1,500 Spanish arquebusiers, represented here by two soldiers in the space right of centre, who effectively won the battle. (Bibliothèque Nationale, Paris)

Milan, and the pope changed sides to join them. Then the town of Pavia, in Lombardy, saw a battle between the French and imperial armies. The imperial forces (23,000) defeated the French army (28,000), and Francis was taken prisoner. He was released (1526) and the wars in Italy continued.

Pawnee, a North American Indian tribe who were descendants of the *Mississippi cultures west of the Missouri River in Nebraska. By c. AD 1000 they were farmers living in permanent villages, practising seasonal hunting. Some adopted foot nomadism on the Great Plains in the 15th and 16th centuries, and in the late 17th and 18th centuries buffalo hunting became increasingly important as horses and, later, firearms were acquired from northern Mexico.

Paxton Boys, American rebels in Pennsylvania, who were Scots-Irish frontiersmen from settlements round Paxton. In 1763, threatened by *Pontiac's Rebellion and agitated by lack of colonial defence and political representation, they first massacred some Christian Indians and then marched on Philadelphia, where *Franklin managed to pacify them. They were symptomatic of a long-term antagonism between frontier and coastal settlers, having many similarities with *Bacon's Rebellion (1676) and the Carolina *Regulators (1768–71).

Pazzi Conspiracy (1478), an unsuccessful plot to overthrow the *Medici rulers of Florence. Their rivals, the Pazzi family, backed by Pope Sixtus IV, conspired

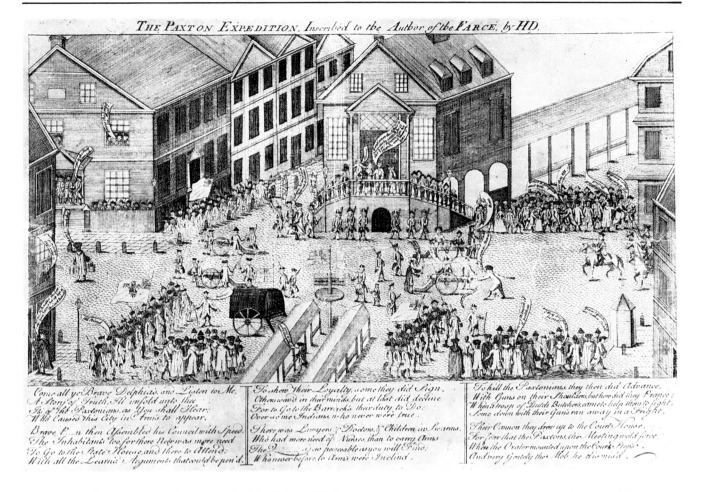

THE PAXTON EXPEDITION. Inscribed to the Author of the FARCE, by HD.

Come all ye Brave Delphia's, and Listen to Me,
A Story of Truth, I'll unfold unto thee,
So of the Paxtonians as You shall Hear,
Who Caused this City in Arms to appear.

Brave P...n then Assembled his Council with Speed,
The Inhabitants too for there Neer was more need,
So Go to the State House, and there to Attend:
With all the Learn'd Arguments that could be pen'd.

To show their Loyalty, some they did Sign,
Others wav'd in their minds, but at last did decline
For to Go to the Barracks their duty To Do:
Over some Indians, who never were true.

There was Lawyers & Doctors, & Children in Swarms,
Who had more need of Nurses, than to carry Arms
The Q.........y so peaceable as you will Find,
Who never before, to Arms were Inclind.

To kill the Paxtonians, they then did Advance,
With Guns on their Shoulders, but then did they Prance,
When a troop of Dutch Butchers came to help them to fight,
Some down with their Guns ran away in a Freight.

Their Cannon they drew up to the Court House,
For fear that the Paxtons, the Meeting would force,
When the Orator mounted upon the Court steps,
And very Gentely the Mob he dismist.

to murder Giuliano and Lorenzo de Medici at High Mass in Florence Cathedral and to sieze power. Although Guiliano was murdered as planned, Lorenzo escaped. The mob rallied to the Medici, and siezed and murdered the main conspirators.

Peasants' Revolt (1381), an English rebellion that had dramatic consequences as a result of the government's weak and delayed response to it. The upheaval after the *Black Death (1349) caused many *villeins, for whom economic prospects were improving, to resent their servile status the more keenly. The *Poll Tax of 1380 was the spark which caused peasants and others to rise in revolt in many parts of England in June 1381; under the leadership of Wat *Tyler and John *Ball, Kentishmen marched on London and joined men from Essex in destroying the property of lawyers, tax collectors, and other targets of their hatred. The complete overthrow of authority was only averted by the courage of the young *Richard II in riding out to meet the rebels on 15 June and satisfying them with the promise of pardons and other concessions. Once the danger was past, the government reacted sharply against the rebels and the teachings of the *Lollards with which they were associated.

Peasants' War (1524-6), a mass revolt of the German lower classes during the *Reformation. It began in south-west Germany and spread down the River Rhine and into Austria. Frustrated by economic hardships, the rebels were encouraged by radical *Protestant preachers to expect a second coming of Jesus Christ and the establishment of social equality and justice. They raided

An 18th-century print showing the settlement of the **Paxton Boys** march on Philadelphia; the six verses beneath describe the story of the Paxton expedition and the orator's 'gentle dismissal' of the mob. (New York Public Library)

The priest John Ball, a leader of the **Peasants' Revolt**, shown riding a farm horse at the head of a well-ordered and disciplined army. This miniature from a c.1460 manuscript of Froissart's Chronicles portrays both peasant rebels and their adversaries bearing the standards of England and St George. (British Library, London)

and pillaged in uncoordinated bands, driving *Luther to condemn them in his fierce broadsheet *Against the thieving and murdering hordes of the peasants* (1525). Luther also supported the army of the Swabian League under *Philip of Hesse, which helped to crush the main body of insurgents at Frankenhausen. Over 100,000 rebels were eventually slaughtered.

Peking Man, a popular name for the numerous remains of *Homo erectus pekinensis* (originally *Sinanthropus pekinensis*) found at *Zhoukoudian (Choukoutien) in China since 1927 and often used for all *Homo erectus* fossils from China. Besides those from Zhoukoudian, important remains are known from Hexian in Anhui province (about 250,000 years old) and Lantian in Shaanxi province (800,000–600,000 years old). The latter could be contemporary with *Java Man.

Pelagianism, the doctrine of the heretical British monk Pelagius (c.350–420). He rejected the pessimistic Christian view which saw mankind as totally depraved and without freedom of will. Instead he insisted that God could be reached, by human effort, through asceticism. His beliefs stemmed partly from a reaction to *Manichaeism, partly from the necessity for self-survival in the wake of the

collapse of the old imperial order. Condemned in 416, he was excommunicated in 418 and disappeared from record. Some of his works survive in fragmentary form. His influence survived particularly in Britain and Gaul which had grown increasingly independent of Rome. The issue later assumed a central importance in the dispute between Protestant and Catholic theologies at the *Reformation.

Pelham, Henry (1696–1754), English politician and statesman. He was the brother of the Duke of *Newcastle with whom he formed a powerful political combination from 1744, with Pelham providing the leadership and Newcastle the patronage and political stability. Pelham reduced the National Debt and in 1748 introduced a period of peace and prosperity by bringing the War of the *Austrian Succession to an end. His ministry also introduced legislation adopting the Gregorian *calendar. His death in office is often regarded as a major factor in producing the political instability of the 1760s.

Peloponnesian War, the war waged between Athens and Sparta and their respective allies between 431 and 404 BC. Its history was recorded in detail by *Thucydides and *Xenophon. According to Thucydides, the under-

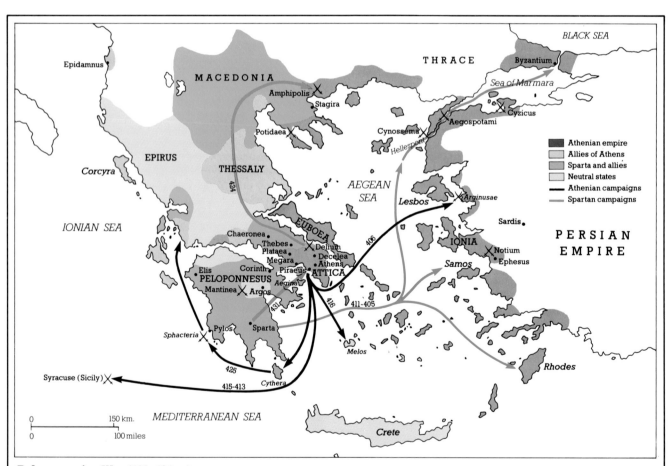

Peloponnesian War (431–404 BC)

Of the two chief contestants in the Peloponnesian War, one, Sparta, was essentially a land power, while Athens was a naval power. The Athenians were forced to abandon Attica in 431 BC, and it was regularly overrun by the Spartan army thereafter. Sparta was also more successful in gaining Persian aid, which was perhaps the decisive factor ultimately. However, Athens could not be defeated while it retained its maritime supremacy. Despite their disastrous losses at Syracuse, the Athenians inflicted several naval defeats on the Spartans between 411 and 406. Athens' surrender was forced only by the capture of its entire fleet at Aegospotami.

lying reason for conflict was the growth of Athenian power and the fear this aroused among the Spartans. Corinth was especially vigorous in pressing Sparta to declare war on Athens. Sparta invaded Attica with its allies in 431, but *Pericles had persuaded the Athenians to withdraw behind the 'long walls', which linked Athens and its port of Piraeus, and avoid a land-battle with Sparta's superior army. Athens relied on its fleet of *triremes to raid the Peloponnese and guard its empire and trade-routes. It was struck a serious blow by an outbreak of plague in 430 which killed about a third of the population, including Pericles. Nevertheless the fleet performed well and a year's truce was made in 423 BC.

The Peace of Nicias was concluded in 421 BC, but Alcibiades orchestrated opposition to Sparta in the Peloponnese, though his hopes were dashed when Sparta won a victory at Mantinea in 418. He was also the main advocate of an expedition to Sicily (415–413), aimed at defeating Syracuse, that ended in complete disaster for Athens. War was formally resumed in 413 BC; the fortification of Decelea in Attica by the Spartans and widespread revolts among her allies put pressure on Athens, who had lost much of its fleet in Sicily, was desperately short of money, and troubled by political upheaval. Nevertheless, and thanks in large part to Alcibiades, Athenian fortunes revived, with naval victories at Cynossema (411), Cyzicus (410), and the recapture of Byzantium (408). There was a further victory at Arginusae in 406. From then on, Persian financial support for Sparta and the strategic and tactical skills of the Spartan *Lysander tilted the balance. Sparta's victory at *Aegospotami and its control of the Hellespont starved Athens into surrender in April 404. An oligarchic coup followed immediately, supported by Sparta, and the reign of terror of the 'Thirty Tyrants', but democracy was restored in 403.

Penda (d. 655), King of *Mercia in the first half of the 7th century. He sought to establish the supremacy of Mercia over the other English kingdoms. He drove the King of Wessex into exile, killed the ruler of East Anglia in battle, and, in alliance with *Cadwalader, invaded *Northumbria. Edwin, its ruler, was defeated and killed in 633 and his successor *Oswald in 642. Although never a Christian, Penda allowed missionaries to preach in his kingdom. He was killed in battle by *Oswy, King of Northumbria.

Penn, William (1644–1718), Quaker founder of *Pennsylvania. He was granted the proprietary colony to satisfy debts owed by Charles II in 1681. Persecution and warfare in Britain and Europe ensured a high rate of emigration to his 'Holy Experiment'. Although a tolerationist and friend to Indians, Penn's original 'Frame of Government' was paternalistic. Settler discontent forced him to concede representative government by 1701. He was a tireless missionary and propagandist for Quakerism, but spent his later life in England, harassed by financial problems and suffering from apoplexy.

Pennsylvania, a colony and state of the USA regarded as a middle Atlantic state, though its only coastline is at its north-west corner on Lake Erie. Founded in 1681 by William *Penn, it grew rapidly under Quaker control, with efficient German farmers, still known as Pennsylvania Dutch (Deutsch), settling the rich farmlands of

A charcoal portrait sketch of William **Penn** by Francis Place, the only known authentic likeness. (Historical Society of Pennsylvania)

the coastal plain and Scots-Irish immigrants taming the frontier. By the 1770s Philadelphia was the largest city in the colonies, exporting grain in homebuilt shipping to the Caribbean and Europe. It was the national capital during the Revolutionary period, acting as host to the *Continental Congress (1774–87). In 1777 it was captured by the British and held for a year. It then hosted the Constitutional Convention (1787) and the new federal government (1790–1800).

Pepin, the name of three Frankish 'mayors of the palace' under *Merovingian rule who gave rise to the Carolingian dynasty. Pepin I of Landen was mayor of Austrasia, and his son Pepin II of both Austrasia and Neustria, the two most important parts of the Merovingian kingdom. Pepin III, the Short, was the grandson of the latter and son of *Charles Martel. He ousted the last Merovingian, Childeric III, in 751 and was crowned King of the Franks. A close ally of the papacy, he defended it from Lombard attacks and made the Donation of Pepin which was the basis for the *Papal States. He added Aquitaine and Septimania to his kingdom, which passed, on his death in 768, to *Charlemagne and Carloman.

Pepys, Samuel (1633–1703), English naval administrator and diarist. His connection with the navy started in 1660, the year he began his diary. In 1673 he became Secretary to the Admiralty Commission and a Member of Parliament. He worked hard for naval reform, but his closeness to the Duke of York led to his arrest for alleged complicity in the *Popish Plot. He was re-appointed in 1684 (he became president of the Royal Society in the same year), and during James II's reign he did his best to rebuild the navy which had again fallen into neglect. Pepys is remembered for his diary, an important document on contemporary life and manners, (including descriptions of the *Fire of London and the *Great Plague), written with engaging sincerity and humanity. It was written in a code of Pepys's own invention, which was not deciphered until 1825.

Pequot War (1637), a North American Indian war in New England, which arose from the murder of traders and settlers by the warlike tribesmen, hostile to white incursions. After several skirmishes, the Pequot camp at Mystic on the Connecticut coast near Rhode Island was surprised and fired, killing some 500 of the tribe. Poor colonist co-operation in this war led to the founding of the *New England Confederation (1643).

Percy, a family of *marcher lords of medieval England with lands in Northumberland. Henry de Percy (1341-1408), 1st Earl of Northumberland, was the first of the family to be of major importance in the defence of England's northern frontier. The earl's son, Sir Henry Percy ('Hotspur') (1364-1403), was a hero of the battle of *Otterburn. When Henry of Bolingbroke landed in the north of England in 1399, the earl and Hotspur helped assure him of the crown; they were well rewarded, but within four years their greed for more offices or money led them into open rebellion. Hotspur and his uncle Thomas, Earl of Worcester, were killed at *Shrewsbury in 1403. Five years later Earl Henry invaded England from Scotland, but he too was killed and his estates were forfeited.

Subsequently restored to their estates, later generations of the family resumed their role as guardians of the northern frontier and rivals of the Nevilles. The male line ended in 1670, but the earldom passed in the female line to Sir Hugh Smithson (1715-86) who took the name of Percy and in 1766 was created Duke of Northumberland.

Pergamum, an ancient city in what is now Turkey, some 20 km. (12 miles) inland from the west coast of Asia Minor. It developed into a major power during the 3rd and 2nd centuries BC under the Attalid dynasty. In particular, Attalus I (ruled 241-197), inflicted a severe defeat on the *Galatians and for a time wrested most of Asia Minor from the *Seleucids. He allied himself to Rome in the first two *Macedonian wars, and his pro-Roman policy was followed by his successors. Thus Eumenes II (d. c.160) helped to defeat Antiochus at Magnesia, and in accordance with Attalus III's will the kingdom was bequeathed to Rome in 133 BC. It became a province of Asia, and was soon eclipsed by Ephesus as the chief city of the region.

Attalid Pergamum was a brilliant centre of *Hellenistic civilization: its chief glories were its sculpture and its library, where parchment was developed in the 2nd century BC as more durable than papyrus for books.

Pericles (c.495-429 BC), Athenian statesman and general. Noted for his oratory, political acumen, and integrity, he dominated Athens throughout the third quarter of the 5th century BC. He supported Ephialtes in his attack on the *Areopagus, and after his death became champion of the *Athenian democracy, proposing pay for jurors, and other innovations. He was instrumental in strengthening and extending the *Athenian empire, sending out colonies, and personally leading the attack against rebellions at Samos and an expedition into the Black Sea, important as a source of corn. He originated a major building programme, of which the jewel was the Parthenon, the temple which dominated the *acropolis. When the spectre of war with the Peloponnesians threatened in the 430s, Pericles determined to resist their demands. After the *Peloponnesian War broke out, he persuaded the Athenians to abandon the countryside when the Spartans invaded and to rely on their fleet. He was briefly deposed from the generalship when plague shattered Athenian confidence, but was re-elected the following year (429). He died of plague soon afterwards.

Persepolis, the ceremonial capital of the *Achaemenid empire. A festival of tribute was held there each year, it was the burial place of the kings, and its treasury was a repository of enormous wealth. The city was captured, looted, and burnt in 331 BC by *Alexander the Great's troops. Excavation of the palaces—built by *Darius I and *Xerxes—and other buildings, while confirming the destruction which took place, has also revealed some magnificent examples of Achaemenid art and architecture, particularly the bas-reliefs.

A bas-relief on the ruined main staircase of the Apadana, or Audience Hall, at **Persepolis**, dating from the early 5th century BC and begun by Darius I. The men are guards, a Persian in the fluted hat and Medes in the round head-dresses.

Persia, a country in south-west Asia between the Caspian Sea and the Persian Gulf (now Iran). Early Persian dynasties included the *Achaemenids, whose rule ended with *Alexander the Great's defeat of Darius III, and the *Sassanians who were overthrown by the Arabs. Since the fall of the Sassanian empire in 642, it has been under the rule of Islam. Persians were prominent in the empires of their Arab, Seljuk, and Mongol overlords for nine centuries, until *Ismail I established a strong Persian state and converted the population to Shiite Islam. After *Abbas I Safavid power declined until the Qajars, ruling from Tehran, took power in 1796.

Persian wars   *Greek–Persian wars.

Peru, a country in the west of South America. It has had a succession of complex cultures and states from c.1000 BC: *Chavín in the central Highlands, from c.1000 BC, *Mochica on the northern coast, *Nazca on the southern coast, and *Tiahuanaco round Lake Titicaca in the Andes. Between c. AD 600 and 1000 Huari in the central Andes conquered a small 'empire', and the *Chimú state rose on the northern coast c. AD 1000. The *Incas were another such group, based round *Cuzco, who began their regional expansion c. AD 1200 and eventually conquered a vast empire stretching from Chile to Ecuador during the 15th century. *Pizarro's defeat of *Atahualpa in 1532 was followed by rivalry for control and led eventually to direct rule by the Spanish crown. Inca revolts continued for nearly fifty years. The vice-royalty, with its capital at Lima, attempted to placate the various factions but was not in reasonable control until the mid-16th century. Further Inca insurrections occurred in 1780, led by *Tupac Amarú, and in 1814.

Peter I (the Great) (1672–1725), Emperor of Russia (1682–1725). He became joint emperor with his imbecile half-brother Ivan at the age of 10: his half-sister Sophia was regent until Peter seized power in 1689. Russia (or Muscovy) was then shut off from Europe by Poland and the Swedish and Turkish empires, and Archangel was its only port. From boyhood Peter was fascinated by the sea and ships and with what he learned about Western skills from foreigners visiting Moscow. In 1697–8 he travelled to Germany, Holland, England, and Vienna; his chief preoccupations were learning the techniques of ship-building and recruiting foreign technicians and craftsmen. His attempts to Westernize his people were often crude and tactless, but he centralized the government under the senate and brought the church under state control, replacing the patriarch by a Holy Synod. A table of ranks was established, ennobling those who served the state in the higher grades, and requiring service from all. He improved the army and navy and founded St Petersburg as a capital modelled on European examples. He seized but later lost Azov on the Black Sea: in the *Northern War he defeated *Charles XII at *Poltava and at the treaty of *Nystadt gained the Baltic coastal provinces of Estonia, Livonia, and Ingria. His achievements for his country were marred by his barbaric cruelties, which included even the torturing to death of his own son in 1718.

Peter, St (d. c. AD 64), originally named Simon, the leader of the Apostles who followed *Jesus Christ. Jesus named him Cephas (Aramaic, 'rock'; Greek *petra*, 'rock')

to signify his key role in establishing the early Christian Church. After the death of Jesus, Peter was the undisputed leader of the Church, preaching, defending the new religious movement, and visiting newly established Christian communities. He was the first to accept Gentiles (non-Jews) into the Church but later disagreed with St *Paul over the admission of Gentiles. It seems certain that Peter spent the last years of his life in Rome and was probably crucified during *Nero's persecution of 64. The *papacy traces its origins back to Peter and the *Roman Catholic Church identifies him as the founder and first bishop of the church of Rome, although this cannot be proved.

Peter's pence, the name given to the tax paid annually to the papacy. First stated as compulsory in 787, it was levied in England from the 10th century at the rate of one penny per householder. It was revived by *William I as a single lump sum of £200 for the whole of England. It was abolished in England in 1534 during the Reformation.

Peter the Hermit (1050–1115), French monk, noted for his preaching in support of the First *Crusade. A charismatic figure, he raised up to 20,000 followers in France and Germany in 1096 but few were trained soldiers. They marched to Constantinople and crossed into northern Turkey where most were massacred or

An early 18th-century caricature of the attempts by **Peter I** to westernize the Russian people. After the revolt of the *streltsy* (imperial musketeers) in 1698, many of whom were Old Believers, Peter executed a large number of the *streltsy*, introduced compulsory shaving, and western uniforms. The barber in the illustration is trying to shave the Old Believer's beard.

enslaved: this disaster is known as the People's Crusade. Peter lived to participate in the seige of Antioch in 1097, and later became prior of an Augustinian monastery.

Petition of Right, 1628, a document drawn up by opposition members of the English Parliament, led by *Coke. It came at the time of *Charles I's wars against France and Spain, and the lengthy quarrel over *tunnage and poundage. It stated parliamentary grievances and forbade illegal unparliamentary taxation, the forced billeting of troops, the imposition of martial law, and arbitrary imprisonment. Charles did assent to the Petition but it was a limited parliamentary victory and did nothing to curb Charles's unconstitutional rule during the eleven years of government without Parliament.

Petra, an ancient city in modern Jordan. It was the capital of the Nabataeans (an Arab tribe) from the 4th century BC to the 2nd century AD. Its prosperity was derived from the caravan trade from southern Arabia, but declined after its annexation by the Romans in AD 106. The remains of the city, accessible only by a single narrow entrance cut through steep rocks, are extensive and spectacular, particularly the tombs carved in the pink rock of the hills.

petty sessions, *hundred courts held by a sheriff in 16th-century England. They were mainly concerned with maintaining order, but were also charged with enforcing terms of service on labourers, prosecuting grain profiteers, and other local economic regulations. They were distinct from the general or quarter sessions, to which they referred more serious criminal cases, and by the 19th century their jurisdiction had been enlarged by innumerable, usually minor, statutory offences.

phalanx, a military term denoting a formation of Greek infantry, especially of *hoplites. It is now popularly applied to the tightly packed formation of pike-armed infantrymen who were such an important part of the army of *Philip II and *Alexander the Great. Under their successors it became a less skilled, more cumbersome formation, the pike being extended from 4 to 6.5 m. (13 to 21 ft.) by the 2nd century BC. It failed to match the more flexible legions of Rome at Cynoscephalae (196 BC), Magnesia (189), and Pydna (168).

pharaoh, the term commonly used to describe the kings of ancient *Egypt. The Egyptians themselves only used it in this way from 950 BC onwards. It comes from the word for 'palace'. The pharaoh was thought of as a god, the son of Osiris ruling on earth, and acted as an intermediary between gods and men. He wielded immense power as the religious, civil, and military leader of the country. The pharaohs greatly increased Egypt's power and territory. *Thutmose I conquered much of *Nubia and campaigned as far east as the Euphrates. *Thutmose III defeated the powerful Mitanni, and strengthened Egyptian rule in Africa. *Ramesses II made peace with the *Hittites and his rule saw great prosperity. The most spectacular legacy of the pharoahs is their magnificent buildings as at *Karnak, *Thebes, and *Heliopolis.

Pharisee, a member of a religious party in ancient Israel which set great store by observance of every detail of the Jewish law. They tended to be rather isolated from

A gold coin of **Philip II** of Macedonia, with a laureate head of Apollo. Philip needed large amounts of coin to pay his troops and his basic unit of currency was the stater, often named a philippus, struck in 23-carat gold. (British Museum, London)

other Jews and came into conflict with *Jesus Christ whose compassion was often at odds with their dry legalism, as is clearly documented in the Gospels of the *Bible. They continued to have enormous influence on Jews—being more popular than the conservative, aristocratic *Sadducees—and on the development of Judaism after the destruction of Jerusalem in AD 70.

Pharsalus, battle of (48 BC), the scene of *Pompey's defeat by Julius *Caesar. After Caesar crossed the *Rubicon, Pompey retired to Greece to rally his forces. Caesar put down Pompey's support in Spain and then pursued him to northern Greece. Pompey's forces were routed in pitched battle, although he himself escaped.

Philip I (the Handsome) (1478-1506), King of Castile (1504-6). The son of *Maximilian I, he inherited *Burgundy in 1482, and took over from his father's regency in 1494. His marriage to *Joanna (the Mad) of Spain in 1496 brought the *Habsburgs a dynastic link with Spain. When Queen *Isabella died in 1504, the pair inherited Castile. His wife's increasing madness enabled him to assume considerable power as king consort. He died of fever soon after his arrival at Castile, his claims passing to his son *Charles V.

Philip II (382-336 BC), King of Macedonia (359-336 BC). He transformed an ineffectual and divided *Macedonia into a power which dominated the Greek world. His success was based upon exploitation of Macedonia's natural advantages and the highly professional army which he created around a core of pike-armed infantrymen, the *phalanx, and his excellent cavalry. He gradually extended his empire until, after the battle of *Chaeronea, it stretched from the Black Sea to the southern Peloponnese. When an assassin struck him down, he was preparing for the invasion of Persia, a project which his son *Alexander III inherited.

Philip II (Augustus) (1165-1223), King of France (1179-1223). One of the *Capetians, he succeeded his father Louis VII in 1180 and set about the restoration and expansion of his kingdom. Defeating the Count of

Flanders, and the Duke of Burgundy, he seized Artois and part of the valley of the Somme. He was obliged initially to accept the homage of King *John of England for Normandy, Aquitaine, and Anjou but later recovered Normandy, Anjou, Poitou, and the Auvergne for the French crown. He defeated John and the Holy Roman Emperor jointly at the battle of Bouvines (August 1214), leaving his country stronger and more united. After these military successes he devoted his energy to reforming the law and building and fortifying Paris.

Philip II (1527–98), King of Spain, Naples, and Sicily (1556–98) and, as Philip I, of Portugal (1580–98). He was the only legitimate son of Emperor *Charles V. He was married four times: to Mary of Portugal (1543), to *Mary I of England (1554), to Isabella of France (1559), and to Anne of Austria (1570), whose son by him was his successor as Philip III of Spain. After 1559 he never left the Iberian peninsula, and he ruled Spain and the *Spanish empire industriously during its 'golden age'. His strongly Catholic religious policies helped to provoke the *Dutch Revolts (1568–1648) and the Revolt of the Moriscos within Spain (1568–70). He generally subordinated his Catholic crusading zeal to more worldly considerations. His intermittent wars against the *Ottomans, his war with England (1585–1604), and his involvement in the *French Wars of Religion were all largely motivated by interests of state. His personal brand of absolute monarchy made things difficult for his successor, to whom he left a country economically crippled as a result of his military expenses.

Philip II, King of Spain, Naples, and Sicily, c.1556, one of many royal portraits executed by A. Sanchez Coello while in the service of Philip in Castile. (Museo del Prado, Madrid)

Philip IV (the Fair) (1268–1314), King of France (1285–1314). He inherited the throne from his father and strengthened royal control over the nobility as well as improving the law. Pope Boniface VIII resisted his claim to the right to tax the clergy but was imprisoned by a royal agent and died soon afterwards. The next pope, Clement V, was under the king's control and acquiesced in the removal of the papacy from Rome to Avignon in France, the beginning of seventy years 'captivity'. Coveting the wealth of the *Knights Templars, he seized much of their property after Pope Clement suppressed their order by royal command in 1313. He also persecuted the Jews and had their property confiscated.

Philip V (1683–1746), King of Spain (1700–46). He was Philip of Anjou, the younger grandson of *Louis XIV, and succeeded under the terms of the will of the last Habsburg king, Charles II. It was Louis XIV's acceptance of this will that plunged Europe into the War of the *Spanish Succession. Philip's claim was confirmed at the Peace of *Utrecht. Catalonia was deprived of its liberties when Barcelona surrendered to him in 1714. For the rest of his reign he was dominated by his second wife, Elizabeth *Farnese, whose ambitions for her sons influenced Spanish foreign policy.

Philip VI (1293–1350), King of France (1328–50), the first of the *Valois kings. His right to the throne was challenged by *Edward III of England and the *Hundred Years War began in 1337. His ill-fated reign also witnessed the *Black Death and war with *Flanders. Despite the ruinous expense of war and some disastrous military defeats, the government was strengthened during this period as an organized system of taxation evolved.

Philip of Hesse (1504–67), German prince. He played a leading role in establishing the *Protestant religion in Germany, and in asserting German princely independence against Emperor *Charles V. After becoming converted to *Luther's doctrines (1524), Philip turned Hesse into a Protestant state. He was active at the Diet of Speyer (1529) and a subscriber to the *Augsburg Confession (1530), and became a founder and leader of the Schmalkaldic League, an alliance of Protestant princes and cities (1531). He forfeited the loyalty of some of his followers by marrying bigamously (1540), and after his defeat in the *Schmalkaldic War he was imprisoned by Charles V. He lived to see Lutherans achieve equality with Catholics under the Peace of *Augsburg (1555).

Philippi, battle of (42 BC). At Philippi, a city of Macedonia in Greece, *Caesar's assassins, under *Brutus were defeated by the armies of *Mark Antony and *Octavian. Both Cassius and Brutus committed suicide after the defeat.

Philippines, an archipelago of about 7,000 islands in south-east Asia. The original Negrito inhabitants were largely displaced by waves of Malay peoples migrating from Yunnan province in south-west China after c.2000 BC. By AD 1000 the islands were within the south-east Asian trade network. By the 16th century Islam was advancing from Mindanao and Sulu into the central islands and Luzon. After Spaniards under *Magellan visited the islands (1521), Spanish seamen discovered how to return eastbound across the Pacific to

Mexico. In 1543 they named the islands after Prince Philip (later *Philip II of Spain). In 1564 Miguel de Legazpi, with 380 men, set out from Mexico to establish a settlement, Christianize the Filipinos, open up commerce with East Asia, and secure a share of trade in the *Moluccas. A settlement was made in 1565 at Cebu in the western Visayas, but the Spaniards moved their headquarters to Manila in 1571. Manila became the centre for a trade in Chinese silks with Mexico, in return for Mexican silver dollars. From there Spanish influence and control spread out through the Philippine island chain, particularly assisted by missionary activity. Christian outposts founded by Dominicans, Franciscans, and Augustinians grew into towns. Revolts against the harsh treatment of Filipinos by the Spanish were frequent, particularly in the 17th century. During the *Seven Years War the British occupied Manila for two years.

Philip the Bold (1342–1404), Duke of Burgundy (1363–1404). He was the fourth son of John the Good, King of France, and was created Duke of Burgundy in 1363. In 1369 he married Margaret, heiress of the Count of Flanders, and in 1380 he helped to quell a revolt by the Flemish burghers against the count, which finally ended with the massacre of 26,000 Flemings in 1382. On the death of the count in 1384 he inherited Flanders and proceeded to encourage commerce and the arts. During the minority of Charles VI (1380–88) Philip was virtual ruler of France and, when the king became insane (1392), Philip fought for power with Louis d'Orléans, the king's brother, a quarrel carried on after Philip's death by his son, John the Fearless.

Philip the Good (1396–1467), Duke of Burgundy (1419–67). His first act as Duke of Burgundy was to forge an alliance with *Henry V of England and to recognize him as heir to the French throne. He was a powerful ally: by the early decades of the 15th century his territories included Namur (acquired 1421), Holland and Zeeland (1428), Brabant (1430), Luxemburg (1435); the bishoprics of Liege, Cambrai, and Utrecht were under Burgundian control. Some were his by inheritance, others

An Assyrian alabaster carving of **Phoenicians**, traders *par excellence*, in their merchantmen. As well as an intriguing assortment of sea creatures, the carving also shows the typical carved horses' heads on the bows of the boats. (Musée du Louvre, Paris)

had come through marriage, purchase, or conquest and they combined to constitute a formidable 'state'. The Treaty of Arras (1435) released Philip from the duty of doing homage to the French king and rendered him virtually independent of royal control. However, the king of France succeeded in breaking the alliance between France and England and from 1435 France and Burgundy joined forces to wage war on England. The imposition of taxes on the Burgundians provoked a rebellion, led by Ghent, but the rebels (of whom 20,000 were killed) were defeated (1454). Under Philip the Burgundian court was the most prosperous and civilized in Europe. He founded an order of *chivalry, the Order of the Golden Fleece, and patronized Flemish painters.

Philistines, a non-Semitic people, originally a group of the *Sea Peoples, who settled in southern *Palestine in the 12th century BC. They established five cities—Ashdod, Askelon, Ekron, Gath, and Gaza. They gained control of land and sea routes and proved a formidable enemy to the Israelites, inflicting defeats on Samson and *Saul. King *David, however, gained decisive revenge, and from then on Philistine power declined until they were assimilated with the *Canaanites.

Phoenicians, Semitic-speaking people descended from the *Canaanites of the 2nd millennium BC. They inhabited a number of cities on the Levant coast most notably Byblos, Sidon, and Tyre. By 1000 BC they had invented the alphabet. As they had only a narrow, though fertile, hinterland, they turned to the sea and became the greatest traders of those times, penetrating as far afield as southern Spain and possibly even circumnavigating Africa c.600 BC. By 900 BC they were well known in the West, where they established many small trading posts; of these *Carthage was the exception, developing into a major city. Conflict with Assyria weakened the power of the cities of Phoenicia, and they subsequently came under the influence of other great powers—the *Babylonians, the *Achaemenids, *Alexander the Great, the *Seleucids, and *Rome. Their trade and industries included linen, metal, glass, wood, ivory, and precious stones.

Photius (c.810–c.893), Patriarch of Constantinople and defender of the independence of the Byzantine Church against the papacy. He was elected in 858 to replace the deposed Patriarch Ignatius, but the papacy refused to recognize him and he excommunicated the pope. Ignatius was later reinstated and Photius was restored in 886 by Emperor Leo VI and this time Pope John VIII, who urgently needed Byzantine naval assistance against the Moors, recognized him. The dispute, misleadingly known as the 'Photian Schism', was partly concerned with the presence of Latin missionaries in Bulgaria but principally concerned doctrinal matters.

Phrygia, the territory in north-west Asia Minor occupied by the Phrygians from Europe c.1200 BC. They clashed with the Assyrians c.1100 and again in the late 8th century BC. They were conquered by the invading Cimmerians from southern Russia in the early 7th century, and soon after fell under the sway of Lydia, and subsequently Persia, the Seleucids, and Pergamum. Most of Phrygia was incorporated into the Roman province of Asia in 116 BC, while the eastern sector became part of Roman *Galatia in 25 BC.

Pilgrimage to the Ka'aba at Mecca, from a Persian miniature. (Bibliothèque Nationale, Paris)

Picts (or 'Pictae'), 'painted people', those who lived beyond *Hadrian's Wall and threatened Britain in the 4th and 5th centuries. A serious invasion took place in AD 367. Traditionally there were northern and southern Picts. The former were invaded by *Scots from Ireland, the latter by the *Angles. As a result distinct Highland and Lowland cultures developed in Scotland. Kenneth MacAlpine united them into Scotland.

Pilate, Pontius (*fl.* 1st century AD), the Roman governor of Judaea (AD 26–36) who presided at the trial of *Jesus Christ. He sentenced Jesus to death by *crucifixion, the standard punishment for non-Romans found guilty of sedition, and his part in the trial is mentioned by *Tacitus and in the New Testament of the *Bible. Little is known about Pilate's life although Jewish writers recorded that he was corrupt and insensitive to Jewish customs.

pilgrimage, a journey, usually lengthy, to visit a religious shrine or site and undertaken as an act of religious devotion or penance.

Christian pilgrimage was initially made to sites connected with the life of *Jesus Christ in Palestine, and Muslim obstruction of this custom helped to provoke the *Crusades. In the Middle Ages, Christians had a deep reverence for sites connected with particular saints such as Rome (St *Peter and St *Paul), Compostella in northern Spain (St James), and Canterbury (associated

with Thomas à *Becket after his murder). Most pilgrims travelled in groups, like Chaucer's pilgrims in *The Canterbury Tales*, staying in hospices provided along the main pilgrimage routes. Protestants condemned pilgrimage both theologically and for its commercial debasement.

Pilgrimage to the Ka'aba at *Mecca, well established in pagan Arabia, was incorporated into *Islam, the detailed rites being based on *Muhammad's own practice. The *hajj* (Arabic, 'pilgrimage'), undertaken only in the twelfth month of the Muslim calendar, became highly organized, with special caravans and guides.

In India, pilgrimage was often linked to *Hindu festivals and associated with rivers, such as the Ganges, or sacred cities, like *Varanasi. In China and Japan mountains, such as Mount Tai and Mount Fujiyama were early favoured as sites for *Buddhist pilgrimage. Mendicants and scholars often adopted pilgrimage as a permanent way of life.

Pilgrimage of Grace (1536–7), the name given to a series of rebellions in the northern English counties, the most significant of which was led by Robert Aske, a lawyer. He managed, briefly, to weld together the

The departure of Puritans from Delft harbour in Holland to join the **Pilgrim Fathers** on the *Mayflower*, a contemporary painting by Adam van Breen, 1620.

disparate grievances of his socially diverse followers. The main causes of concern were the religious policies of Thomas *Cromwell, notably the Dissolution of the *Monasteries, although the rebels stressed their loyalty to *Henry VIII, and accepted his promises of pardon before dispersing. Severe retribution followed, as Henry authorized the execution of about 200 of those involved, including Aske.

Pilgrim Fathers, a 19th-century name for the 102 founders of the Plymouth plantation in America who had travelled from Plymouth in England in the *Mayflower* in 1620. Some of them had been persecuted as separatists from the Church of England, and the nucleus of them had fled from Scrooby in Nottinghamshire in 1608 to Holland. In 1618 they obtained backing from a syndicate of London merchants and permission to settle in Virginia. On the voyage, having no charter, they subscribed to a covenant for self-government, the Mayflower Compact. They made landfall at Cape Cod, Massachusetts in December and decided to settle there. Only half the party survived the first winter. William *Bradford left a vivid account of their experience. *Fur trading with the Indians provided a lucrative export, but the plantation grew slowly and was overshadowed by the huge Puritan influx into Massachusetts in the 1630s. In 1643 it joined the *New England Confederation, and in 1691 it was incorporated into Massachusetts. The Pilgrim Fathers played a relatively insignificant role in New England history, but as the region's first permanent settlers they are of symbolic importance in the European colonization of America.

pillory and stocks, wooden frames with holes for the head and hands of an offender in which he was exposed to public ridicule. They were a means by which local communities in the Middle Ages could punish minor offenders convicted in the manor or town court. Even in the time of the *Star Chamber (abolished 1641), a court could earn itself a bad name by sentencing men to stand in the pillory for what might be thought a political offence. The pillory was abolished in 1837.

Piltdown Man, supposed fossil remains 'discovered' by Charles Dawson in 1908–13 in gravels in Sussex, England. The association of a modern-looking skull, ape-like jaw, and extinct animal bones provided just the missing link in human evolution being keenly sought at the time. It was not until 1953 that '*Eoanthropus dawsoni*' was shown by scientific tests to be a forgery. It has not been clearly established who perpetrated the hoax.

Pinkie, battle of (10 September 1547), sometimes known as Pinkie Cleuch, fought to the east of Edinburgh, Scotland. A large but poorly equipped Scottish army was defeated by an English expeditionary force under the Duke of *Somerset, but the expedition failed in its purpose of engineering a marriage between *Mary, Queen of Scots and *Edward VI of England and only drove the Scots closer towards alliance with France.

pipe-roll, the financial account presented to the Exchequer by the sheriffs of England. The earliest surviving roll dates from 1130 and there is an almost unbroken series from 1156 to 1832. The rolls were compiled by the clerks of the treasury and included details of rents, leases, and other royal revenues. Their name originated with the practice of enrolling the records on a rod, or 'pipe'.

Piraeus, the chief port of Greece, located south-west of Athens. It superseded Phaleron as the harbour of Athens when *Themistocles had it fortified probably in 483–481 BC to protect the large fleet of *triremes which he had just had built. In 461–456 it was joined to Athens by the Long Walls, making a total circuit of 25.7 km. (16 miles) and thereafter became the power base and trading centre of the *Athenian empire. It was destroyed in 86 BC when *Sulla's troops laid siege to those of *Mithridates there.

pirate, a sea robber, operating wherever there has been unprotected merchant shipping to attack. Pirates are often described in classical literature, and during the Dark Ages the *Vikings committed piracy throughout European waters. The creation of the Spanish and Portuguese seaborne empires in the 15th and 16th centuries provided irresistible temptations. To Spain heroes like *Drake or *Gilbert were simply pirates, forerunners of the *buccaneers in the 17th-century Caribbean or Indian Ocean like Blackbeard Teach, *Morgan, and *Kidd, with their bases in Jamaica, the Bahamas, and Madagascar. As England's international trade grew, its merchantmen too became prey to French pirates from Dunkirk, Barbary *corsairs in the eastern Mediterranean or African Atlantic Coast, marauding from Algiers, Oran, Tunis, or Salli, and Far-Eastern sea-raiders in the China seas. Piracy generally declined by the middle of the 18th century, though incidents of piracy continued into the 19th and 20th centuries.

Pisistratus (*c*.600–527 BC), ancient Greek ruler. He became tyrant (unconstitutional ruler) of *Athens in 561 BC. Twice he was forced out of the city, but in 546, with the help of mercenaries, he re-established his tyranny, which he maintained until his death in 527. His rule was generally benevolent. He was conciliatory towards the nobles, yet improved the lot of the poor by encouraging agriculture (thus providing jobs for labourers) and instituting travelling judges. Under him and his sons, Athens thrived in both economic and cultural terms.

The English Prime Minister William **Pitt** the Elder, a painting from the studio of W. Hoare, c.1754. Hawk-nosed and with a tall, lean figure, Pitt's commanding presence dominated the House of Commons. (National Portrait Gallery, London)

Pitt, William, 1st Earl of Chatham (Pitt the Elder) (1708-78), British Prime Minister. He was a brilliant orator and parliamentarian. As a Whig opponent of *Walpole he earned the mistrust of *George II, but the *Pelhams forced the king to give Pitt a ministerial post in 1746. On Pelham's death he quarrelled with Newcastle, but the national emergency of the *Seven Years War drew them together in a coalition in 1757. For four years Pitt conducted the war, demonstrating mastery in world-wide strategy and the effective use of sea-power. While public opinion hailed him as a hero, his domineering and aloof nature exasperated his colleagues. By the end of the war Britain was everywhere victorious, but Pitt had already resigned following disagreements with his ministers over war with Spain. In 1764 and 1765 he was twice approached by *George III to form a ministry but declined. His ministry of 1766-8 was severely hampered by his increasing withdrawal from public affairs because of illness. This ill health dogged him in his later years. As an elder statesman he sympathized with the grievances of the Americans, but could not reconcile himself to independence for the colonies, as he showed in his last speech.

Pitt, William (Pitt the Younger) (1759-1806), British Prime Minister, second son of *Pitt the Elder. He became Chancellor of the Exchequer at the age of 23. His refusal to betray Lord *Shelburne began his long rivalry with *Fox, and on the overthrow of the Fox–North coalition in 1783, Pitt became Prime Minister, a post he was to hold until 1801 and again from 1804 to 1806. During the years of peace until 1793 Pitt was nurturing Britain's economic recovery from the American War of *Independence, and he showed his brilliance in his handling of finance. From 1793 he symbolized Britain's resistance to the French Revolution and Napoleon, but although he raised three European coalitions and gave subsidies to Britain's allies, victory eluded him. When George III refused to consider *Catholic emancipation in exchange for his Act of Union, he resigned in 1801. Pitt's sympathy with reform evaporated with the war; he suppressed British radicalism as readily as he suppressed Irish revolution, and although he called himself an independent Whig, his followers, after his early death, were to form the nucleus of the emerging Tory party.

Pius II (Aeneas Silvius Piccolomini, 1405-64), Pope (1458-64). Born of an impoverished noble family, he led a dissolute life as a poet but reformed, took holy orders in 1446, and became an outstanding *humanist. He became secretary to Felix V, the *antipope, from 1439, and an ecclesiastical diplomat, to be employed by Emperor *Frederick III as secretary and poet laureate. As pope he proclaimed a *Crusade against the *Ottoman Turks (October 1458) but a congress of Christian rulers summoned to Mantua in 1459 was a failure and despite repeated attempts to launch a Crusade the enterprise came to nothing. He had to face anti-papal movements in France and Germany and frequent quarrels with local rulers prevented him from carrying out his programme of reform. He died, bequeathing the problem of the *Hussite heresy and Turkish war to his successor.

Pius V, St (Antonio Ghislieri, 1504-72), Pope (1566-72). He was an austere Dominican, Bishop of Nepi and Sutri (1556), cardinal (1557), and Grand Inquisitor from 1558. As pope, he laboured to restore discipline and morality with a sometimes intransigent *Counter-Reformation zeal. His ill-timed Bull calling for the deposition of *Elizabeth I of England (1570) proved ineffectual, but he had more success in helping to organize the *Holy League, whose forces defeated the Turks at *Lepanto (1571). He was canonized in 1712.

Pizarro, Francisco (c.1475-1541), Spanish explorer and conquistador. He began his career as an explorer when he joined Vasco Núñez de *Balboa's expedition across Panama, which discovered the Pacific Ocean in 1513. He led his own expeditions in 1524-5 and 1526-7 to explore the Colombian, Ecuadorian, and Peruvian coasts. In 1532 he marched against the *Inca empire. At Cajamarca he met the emperor, *Atahualpa, imprisoned him, demanded a vast ransom, and then had him murdered upon learning of the Inca army marching to his rescue. In the march on *Cuzco, the Inca capital, battles were fought at Jauja, Vilcashuamáno, Vilcaconga, and Cuzco itself, which fell in 1533. Another contingent, under his half-brother Hernando Pizarro (c.1501-78), marched on Pachácamac. Thereafter, he consolidated his position by founding new settlements including Lima, and dividing the spoils of conquest among his supporters. A partner on his earlier expeditions, Diego de Almagro (c.1475-1538) felt cheated of his share, and in 1537 he seized Cuzco and imprisoned Hernando Pizarro, its governor. In 1538 Hernando escaped, defeating Almagro and executing him on the orders of Francisco. Almagro's

The Spanish explorer Francisco **Pizarro** supervising the collection of the Inca emperor Atahualpa's treasure amid apparently quarrelsome soldiers. The engraving also shows a priest in the background baptizing Indians, as more armed guards stand by.

son took revenge three years later by assassinating Francisco in Lima.

Pizarro, Gonzalo (c.1506–48), Spanish conquistador and brother of Francisco *Pizarro. He assisted his brother in the conquest of the *Incas and defended *Cuzco against the attacks of Manco Capac (1536–7). In 1538 he was appointed governor of Quito. The first viceroy of Peru, Blasco Núñez Vela, proclaimed the New Laws in 1542, drafted by *Las Casas, to protect Indian rights. Gonzalo became leader of conquistadores opposed to the New Laws and in the ensuing civil war defeated and executed the viceroy in 1546. He was offered a royal pardon by the new viceroy, but rejected it, and was beheaded in 1548 after his army had deserted him.

plague, a term used for a number of epidemic diseases, particularly linked with the infectious bubonic plague caused by fleas carried by rodents. The plague which struck Athens in 430 BC, causing the death of the great statesman Pericles and about a third of the population, cannot now be precisely identified. *Thucydides, who was one of those who caught the plague but recovered from it, described in vivid detail the despair, anguish, and lawlessness it caused.

It was from Central Asia that plague devastated the Roman world in the 3rd century, causing hundreds of deaths daily, and again in the time of Justinian (mid-6th century). The form known as the *Black Death swept through Europe during 1346–50: England may have lost as much as half its population, with severe long-term consequences. In 1665 London was afflicted by the *Great Plague, which carried off at least 70,000 victims. The following year, the *Fire of London destroyed in large part the insanitary slums which had made London such a fertile breeding-ground for the disease.

Plains of Abraham, battle of (13 September 1759). A plateau above the city of *Quebec, Canada, was the scene of a famous battle during the *French and Indian War. *Wolfe ferried British troops up the St Lawrence past the French fort. Having tricked the sentries into believing them to be French, the British climbed the precipitous Heights of Abraham. In the ensuing battle, the French garrison under *Montcalm was routed and Quebec surrendered. Both commanders died but the British victory ended French rule in Canada.

Plantagenet, English dynasty descended from the counts of Anjou in France and rulers of England from 1154 to 1485, when the *Tudor line began. The unusual name possibly arose from the sprig of broom plant, *Genista*, which Geoffrey (1113–51), Count of Anjou, wore on the side of his cap. It was Geoffrey's son Henry who became *Henry II (ruled 1154–89) of England and established the Plantagenet dynasty, although it is customary to refer to the first three monarchs Henry II, *Richard I (ruled 1189–99), and *John (ruled 1199–1216), as Angevins (descendants of the House of Anjou). The line was unbroken until 1399 when *Richard II was deposed and died without an heir. The throne was claimed by Henry Bolingbroke, Earl of Lancaster, Richard's cousin and the son of John of *Gaunt (Edward III's third son). In becoming king as *Henry IV he established the *Lancastrian branch of the dynasty which was continued by *Henry V (ruled 1413–22) and *Henry VI (ruled 1422–61, 1470–1).

The second branch of the family, the House of *York, claimed the throne through Anne Mortimer, the great-granddaughter of Lionel (Edward III's second son), who had married the father of Richard, Duke of York. The Yorkist claim succeeded when Edward, Richard's son, became *Edward IV (1461–70, 1471–83) and was in turn followed by *Edward V (ruled 1483) and Richard III (ruled 1483–5). In contending for the crown the Houses of Lancaster and York and their supporters resorted to civil war, known romantically as the *Wars of the Roses (1455–83), and to murder (of Henry VI, the Duke of Clarence, and Edward V). The Plantagenet line was ousted by *Henry VII.

plantation, a farming unit for the production of one of the staple crops in the southern colonies of North America and the British West Indies. Successful tobacco production after 1614 and the Virginia Company's emigration inducements of 20 ha. (50 acre) 'headright' land grants to each passage-paying settler and private estates ('particular plantations') for investors and company officers helped establish the landholding pattern. Between 1640 and 1660 Royalists emigrated to the Chesapeake and the Caribbean to grow tobacco and sugar. These two groups formed the dominant colonial plantation aristocracy. The pattern was copied in South Carolina where rice and indigo were raised. Small profit margins after 1600, credit-worthiness in Britain, access to navigable rivers, and influence with colonial land officers ensured that a tightly knit group of families monopolized staple production. With the increased availability of slaves, after 1650 in the Caribbean and 1700 on the mainland, larger areas could be cultivated and because tobacco was soil-exhaustive, planters acquired huge areas for future use. Robert 'King' Carter of Nomini Hall, Virginia, owned over 12,000 ha. (30,000 acres) in 1732. Successful West Indian sugar planters often returned to England, leaving their estates in charge of overseers. The larger plantations resembled small towns, with warehouses, smithies, boatyards, coopers' shops, wharves, schools, burial grounds, and slave quarters near the big house.

Plassey, battle of (23 June 1757). The village of Plassey in West Bengal, India, was the site of Robert *Clive's victory over Nawab Siraj ud-Daula, which opened the way for the *East India Company's acquisition of *Bengal. Clive had prepared Siraj's elimination by buying support from his enemies, notably by promising the nawabship to Mir Jafar, a rival. In the ensuing encounter, in which he was heavily outnumbered, Clive was aided both by good luck and by Mir Jafar's defection from Siraj. After the victory Mir Jafar was made nawab, with Clive as governor.

Plataea, a city in southern *Boeotia. It was helped by Athens in 519 BC against Thebes. Hence it alone of the Greek states helped the Athenians against the Persians at the battle of *Marathon (490). The Spartans arrived only after the battle. In 479 Plataea was the site of the crucial land-battle of the *Greek–Persian wars, when Mardonius was defeated. Its citizens were given refuge by Athens in 431 following a Theban attack, and a prolonged siege (429–427) led to the capture and the execution of the garrison who failed to escape.

Plato (c.429–347 BC), Greek philosopher, much influenced by *Socrates. After the latter's death in 399, he travelled widely before returning to *Athens where he founded the *Academy. Here he spent the last forty years of his life, devoting himself to teaching and writing. His *Dialogues* are the first unfragmentary body of Greek philosophical writing to have survived, and they had a powerful influence on later philosophers, notably Plotinus. His *Apology* provided a re-creation of the unsuccessful defence which Socrates offered at his trial in 399 BC. The early *Dialogues* reflected a serious questioning of current

A mosaic depicting the Academy at Athens where **Plato** taught mathematics and philosophy until his death in 347 BC. (Museo Nazionale, Naples)

philosophical thought. In them was displayed Socrates' renowned method of argument, whereby a protagonist's definition was demonstrated to be untrue by a logical series of questions and answers. In the later *Dialogues*, however, the figure of Socrates served more as a mouthpiece for Plato's developing ideas. Thus in the *Republic* he tackled the question of the perfect constitution, arguing that perfect unity would come not through democracy, but through communism. In the *Laws* he rejected the excesses of *Athenian democracy and argued that extremes of government were to be avoided.

He also developed a theory of knowledge based on comparing abstract ideas and physical realities which led him to conclude that the soul is immortal, argued in the *Phaedo*. His philosophy influenced the Romans, notably *Cicero and shaped the development of Christian theology and Western philosophy. His works were translated into Latin in the late 15th century and became central to *Renaissance scholarship and ideas.

plebs, the common people of ancient Rome, originally possibly the poor and landless. In the early republic they were excluded from office and from intermarriage with *patricians. The political history of the early republic reflects largely their increasingly organized claim for greater political participation, which was rewarded by the concession of eligibility for the consulship in 367 BC. During the 'Struggle of the Orders' in the 5th and 4th centuries BC the office of *tribune became a watchdog over the activities of the traditional *Senate. Their vote in *comitia was a 'plebiscite'.

Pliny (Caius Plinius Secundus) (23–79) (Pliny the Elder), Roman lawyer, historian, and naturalist. Of many works only the 37 volumes of his *Natural History* survive, a valuable quarry of ancient scientific knowledge blended with folklore and anecdote. He died while leading a rescue (and research) party on the stricken coastline near *Pompeii, during the eruption of Vesuvius.

Pliny (Gaius Plinius Caecilius Secundus) (61–112) (Pliny the Younger), a senator and consul, nephew of the Elder Pliny. A close friend of *Trajan and *Tacitus, he governed Bithynia. His 'Letters', which are really essays, give a detailed picture of the life-style adopted by wealthy Romans of his class. Other letters, written after 100, provide the only detailed accounts of the eruption of Vesuvius in 79 and the devastation of Campania of which he was a youthful eyewitness.

Plotinus (205–c.269), one of the first *Neoplatonic philosophers. He was apparently born in Egypt, though his name was Roman and his mother tongue was Greek. From 232 to 242 he studied in Alexandria, then attached himself to an expedition against Persia in the hope of learning more about Eastern thought. In 245 he went to Rome, where he established himself as a teacher, attracting a wide circle of adherents. At the age of 50 he began to write the *Enneads*. Although these were not published until well after his death (c.300–305), they established him as the greatest philosopher since *Aristotle and earnt him the title of founder of Neoplatonism.

Plutarch (AD c.46–126), Greek writer and philosopher of wide-ranging interests. His extant works include rhetorical pieces, philosophical treatises, and, most memorably, his *Parallel Lives*, paired biographies of famous Greeks and Romans. He was concerned to highlight the personal virtues (and sometimes vices) of his subjects, and the result has been an inspiration to later writers, most notably Shakespeare in his Roman plays.

Plymouth Colony, the first permanent *New England settlement, on the south-eastern Massachusetts coast in Cape Cod Bay. It was settled by the *Pilgrim Fathers in 1620 and grew slowly under the leadership of William *Bradford, profiting from the *fur trade with the Indians. Overshadowed after 1630 by *Massachusetts, it joined the New England Confederation in 1643. *King Philip's War (1675-6) began on its frontier. In 1691 it was incorporated into Massachusetts.

poaching, the unauthorized taking of game or fish from private property or from a place where hunting, shooting, or fishing are restricted. Poaching was pursued by the poor in search of free food, and by the 18th century poaching in the English royal forests was a well-organized and profitable activity. Poachers who were caught were harshly treated, particularly under the terms of the Black Act of 1723. The criminal conviction of poachers was bitterly resented since it was popularly held that the *Enclosure Acts had denied access to common land.

Pocahontas (*c*.1595-1617), North American Indian 'princess', the daughter of *Powhatan, Chief of the Indians in the *Jamestown region of Virginia. John *Smith claimed that she saved him from torture and death in 1607. To cement Anglo-Indian relations, she

The Virginian Indian 'princess' **Pocahontas** wearing European court dress, a portrait painted after 1616 from an engraving by Simon van de Passe. (National Portrait Gallery, Smithsonian Institution, Washington DC)

Ætatis suæ 21. Aº.1616.

Matoaks als Rebecka daughter to the mighty Prince Powhatan Emperour of Attanoughkomouck als Virginia converted and baptized in the Christian faith, and Wife to the worth Mr Tho: Rolff.

was married to John Rolfe in 1614 with the Christian name of Rebecca. He took her to England where she was presented at court, but she died of fever.

Poitiers, battle of (19 September 1356), a battle between the English and French during the *Hundred Years War. An English and Gascon force under *Edward the Black Prince was trapped by a superior French army while raiding. The English archers defeated the French and their king, John II, was captured. Revolt (the *Jacquerie) followed soon after.

Poland, a country in eastern Europe, on the Baltic Sea. It became an independent kingdom in the 9th century and was Christianized under Miezko I (962-92). Unity was imposed under Ladislas I (1305-33) and Casimir the Great, who improved the administration, and the country's defences, and encouraged trade and industry. *Jagiellon rule (1386-1572) culminated in the brief success of *Protestantism, and achievement in the arts and sciences. The 16th century saw Poland, after Lithuania was joined to Poland (1447, 1569), stretching from the Baltic to the Black Sea. However, the weakness of a hereditary monarchy took effect and despite the victories of John Casimir (1648-68) and John *Sobieski (1674-96), internal decline and foreign attack undermined Polish independence, and much territory was ceded to Sweden and Russia. Ravaged by the *Great Northern War and the *War of the Polish Succession, it lost its independence in the 18th century. From 1697 the Electors of Saxony took the title of king and partition between Russia, Austria, and Prussia followed in 1772. Brief resistance under *Kosciuszko resulted in two further partitions in 1793 and 1795, mainly for the benefit of *Catherine the Great's Russia, and Poland became effectively a protectorate of Russia.

Pole, Reginald (1500-58), English cardinal and Archbishop of Canterbury. He held a *Yorkist claim to the throne of England through his mother, the Countess of Salisbury. This high birth, combined with his devotion to Roman Catholicism, made him very important in the eyes of foreign rulers during the English Protestant Reformation. After 1532 he lived abroad, disenchanted with *Henry VIII's marital and religious policies. He was made a cardinal (1536), and urged France and Spain to invade England in the name of Catholicism. Henry revenged himself on Pole's relatives, executing his brother and his aged mother. In 1554 he returned to England. His task was to assist the new queen, *Mary I, in her *Counter-Reformation programme. As Archbishop of Canterbury he began to lay the foundations of a revived Catholicism, although he seems to have disapproved of Mary's persecution of Protestants, and his work did not survive after his death.

polis, the Greek word for city-state. The *polis* may have first emerged in the 8th century BC as a reaction to the rule of the early 'kings'. There were several hundred *poleis* in ancient Greece, many very small. Each consisted of a single walled town surrounded by the countryside; the territory might include villages. At its centre was the *citadel and the *agora. In the *Athenian democracy, which exemplified the *polis* in its highest form, power lay only in the hands of the citizen body, from which, for instance, females, resident foreigners, and slaves were excluded. Freedom, self-reliance, and autonomy were the ideals of the *polis*, but these aspirations were responsible for the innumerable wars between the Greek *poleis*. Even temporary unity in the face of a foreign invader, whether Persian or Macedonian, was very hard to achieve. The rise of the Hellenistic kingdoms at the end of the 4th century BC meant that the power of the *polis* was thereafter limited.

Polish Succession, War of the (1733–5/8), a contest between Russia and Poland on one side and France opposing them. It began after the death of *Augustus the Strong: Austria, Russia, and Prussia supported the candidature of his son, while the French supported Stanislaus Leszczynski, the father-in-law of *Louis XV. Stanislaus was elected but was driven out by Russian troops and Augustus III became king (1733–63). There was fighting in Italy between Austria and Spain, supported by France, and Austria was driven from south Italy. Negotiations began in 1735, though the final treaty was not signed until 1738. Naples and Sicily went to the Spanish Bourbon, Don Carlos; Austria retained Milan and Mantua and acquired Parma; Francis, Duke of Lorraine became Duke of Tuscany, and Lorraine went to Stanislaus (it was to come to France on his death); France accepted the *Pragmatic Sanction. The war, which began in Poland, chiefly affected Italy and France.

poll tax, a tax levied on every poll (or head) of the population. Poll taxes were granted by the English House of Commons in 1377, 1379, and 1380, but the third of these, for one shilling from every man and woman, was acknowledged as a cause of the *Peasants' Revolt, and no poll taxes have since been levied in England.

Poltava, battle of (1709). *Peter the Great's decisive victory over *Charles XII of Sweden in the *Northern War took place near the city of Poltava, near Kiev. Charles, with depleted forces, was besieging the town, but his army was destroyed and he and his ally, the Cossack leader Mazeppa, fled to Turkish territory.

Polybius (*c*.204–122 BC), Greek historian of the rise of Rome from the Second *Punic War. A man of noble birth and some political importance, he was taken to Rome with other Greek hostages after the Roman intervention in Greece by Aemilius Paullus, and joined a circle of intellectuals around his captor. He accompanied his other patron, *Scipio Aemilianus, at the sack of Carthage. His *Universal History* covered the years 220–145. In forty books he attributed Roman success to the *legions, fair administrators, and a balance of regal, aristocratic, and popular elements in the republican constitution. Only five books survive but his account of the war with Hannibal augments that of *Livy.

Polynesians, inhabitants of the islands of the central Pacific, from Hawaii to Easter Island and New Zealand. They show close affinity in physical features, language, and culture, and probably spread from a focal centre in the area of Samoa and Tonga within the past 3000–2000 years. The principal immigration of Maoris from the Marquesas into New Zealand, is dated to about AD 1350, though it was not the first. The origins of Polynesians are controversial. Some authorities, on the basis of their fair skin coloration, wavy hair, and stocky build relate them

The marquise de **Pompadour** by François Boucher, 1757, the most fashionable painter of the day. A favourite of the marquise, he painted several portraits of her. Here she is dressed in blue silk damask trimmed with lace. (National Gallery of Scotland, Edinburgh)

to the *Caucasoids; they have also been regarded as close to the *Melanesians. Probably the widest held view is that they are a distant offshoot of a *Mongoloid population in south-east Asia.

Pombal, Sebastião José de Carvalho e Mello, Marquis of (1699–1782), Portuguese statesman. He was made Minister of Foreign Affairs and War in Portugal on the accession of José I in 1750, and the king's indolence gave him control of the country. He regarded the dominance of the church as the chief reason for the retardation of Portugal. He suppressed the Jesuit missions established in South America, as part of the *Portuguese empire, and in September 1759 expelled the Jesuits from Portugal. The *Inquisition was brought under the control of the state. In 1755 when an earthquake devastated Lisbon, Pombal organized relief work and the rebuilding of the city. Though his reduction of ecclesiastical influence was held to be part of the *Enlightenment, Pombal had little interest in reform.

Pomerania, a territory around the River Oder with the Baltic to the north. Its name derives from a *Slav tribe which settled there in the 5th century. From 1062 to 1637 it enjoyed much independence, ruled by its dukes, but after the Peace of Westphalia in 1648 it was divided between Sweden and Brandenburg. In 1770 Prussia acquired most of Swedish Pomerania.

Pompadour, Jeanne Antoinette Poisson, marquise de (1721–1764), mistress of *Louis XV of France from 1745. She came from the world of wealthy officials and bankers, and was a lively witty woman, on friendly terms with the *philosophes* of the *Enlightenment. The people blamed her for the extravagance of the court and the disasters of the *Seven Years War, but her political influence has probably been exaggerated.

Pompeii, the largest settlement on the Bay of Naples to be destroyed by the eruption of the volcano Vesuvius

An illustration of the excavations at **Pompeii** published in 1777 in a work by the archaeologist and collector Sir William Hamilton. The enthusiasm of early excavators to hunt for antiquities caused much damage to otherwise well-preserved sites. (British Library, London)

in AD 79. An ancient town of mixed Etruscan, Greek, and Samnite background, it became fashionable and prosperous when it became a colony for veteran Roman soldiers. Dedicated to Venus, its population of 15–20,000 included Greeks, Jews, and Christians. An earthquake damaged many buildings in 62 AD. On 24 August 79 it was engulfed, shortly after Herculaneum, in three days of glowing avalanches of molten lava, and showers consisting of ash, gas, rock, and pumice. A blast cloud from Vesuvius was described by *Pliny the Younger, an onlooker, as 'like an immense pine tree'. *Pliny the Elder died leading a naval party into the burning ash and fumes. Local Jews and early Christians saw the eruption as judgement on immoral behaviour and a sign of the end of the world. The ruins of Pompeii were discovered in the 16th century and excavation work begun in 1748 still continues today.

Pompey (Gnaeus Pompeius Magnus) (the Great) (106–48 BC), Roman general and politician. His early military career was meteoric and brilliant. He welcomed *Sulla back to Italy with a private army and was dispatched abroad to fight *Marius. Later the Senate gave him special powers to combat Lepidus and Sertorius. On his return from Spain he and *Crassus, backed by their armies, obtained the consulship for 70, although Pompey was too young and had held none of the statutory offices. Tribunician bills gave him unprecedented powers against the pirates, whom he swept off the sea in only three months, and against *Mithridates. By his settlement of Asia and the annexation of Syria, he doubled the revenue of the treasury and vastly increased his personal fortune.

When a hostile Senate refused to ratify his acts and provide land for his army veterans, he was driven into the pact called the First Triumvirate with Crassus and Caesar, and married Caesar's daughter Julia. His immediate ambitions were satisfied by Caesar as consul in 59, but his relationship with Crassus was strained and he became jealous of Caesar's success in Gaul. Nevertheless he renewed the pact in 56 and obtained a second consulship with Crassus and the governorship of Spain with seven legions to be administered from Rome. In 52 he was made sole consul to deal with gang warfare and anarchy in Rome. Confident of his military superiority, he precipitated the civil war crisis in 49. He

finally engaged battle with Caesar at *Pharsalus and was defeated. He escaped to Egypt but was murdered by *Ptolemy's ministers who hoped to gain Caesar's approval.

Ponce de León, Juan (1460–1521), Spanish explorer. He sailed with *Columbus on his second voyage in 1493, and as deputy-governor of Hispaniola he founded the first settlement on Puerto Rico in 1508. From there in 1513 he sailed in search of the 'Fountain of Youth', an Indian legend, discovering Florida, which he though to be an island, and probably sighting the coast of Yucatán on his return voyage. He did not realize that he had landed on the mainland of North America, and left little description. In 1521 he returned to colonize Florida but was mortally wounded in a battle with Indians.

Pondicherry, a French colony (1674–1954), in southeast India, originally established by the *French East India Company. It provoked Dutch and British rivalry, and was captured several times in the 18th-century trade struggles. It survived the collapse of the French East India Company to remain the administrative centre for French interests in India throughout the British Raj era.

Pontiac (c.1720–69), leader of a North American Indian tribal confederacy, and chief of the Ottawa Indians, for many years allies of the French. After the French defeat in 1759 and British occupation of their forts, he managed to confederate many *Algonquin tribes, fearful of British expansion and intransigence. Spurred by religious enthusiasm, Ottawa, Ojibwa, Potawatomi, Wyandot, Shawnee, and Delaware tribesmen rose in a concerted frontier attack from the Great Lakes to Virginia in May 1763. Only Detroit and Fort Pitt held out and 200 settlers were killed, many in western Pennsylvania. British punitive expeditions weakened the confederacy, and in 1766 Pontiac made peace. He was murdered in 1769 near St Louis by hired Indian assassins.

Pontus, a kingdom in northern Asia Minor whose expansion came to pose a severe threat to Roman power. Pharnaces I captured the Black Sea port of Sinope in 183 BC, but was compelled to renounce many other gains. Mithridates IV and V adopted pro-Roman policies, the latter adding *Galatia and *Phrygia to his territories. *Mithridates VI greatly increased the power of the kingdom, but three wars with Rome led to the collapse of his regime, though it was not until AD 64 that eastern Pontus was incorporated into the Roman empire.

Poor Laws, legislation which provided the basis for organized relief and welfare payments, originating in England in the 16th century. They gradually reduced the charitable obligations placed upon ecclesiastical institutions, guilds, and other private benefactors in the Middle Ages. With the Dissolution of the *Monasteries an important source of charity had ended. Originally only those physically incapable were reckoned to deserve charity and severe measures were taken against ablebodied beggars. However, following local schemes initiated by the town corporations of London, Ipswich, York, Norwich, and Bristol, systematic poor relief evolved. A statute passed in 1576 recognized that men fit and willing to work might be genuinely unable to find employment and were in need of support. Three categories

Poor Laws: distribution of relief to the poor at Durham, c.1778. Elderly women are shown receiving bundles of food and also clay pipes. This drawing was made by S. H. Grimm during a tour of northern England with his patron, Dr Richard Kaye, a canon of Southwell and Durham cathedrals. Low wages and high unemployment made home relief customary at this time. (British Library, London)

of poor were subsequently recognized: sturdy beggars or vagabonds, regarded as potential trouble-makers, the infirm, and the deserving unemployed. Justices of the Peace listed the poor in each parish and appointed overseers to raise funds from parishioners sufficient to meet the cost of maintaining the poor and work was organized for able-bodied poor. Vagrants and vagabonds were sent to houses of correction which became mandatory in all shires from 1601. The government's measures helped to alleviate starvation during years of harvest failure, though there was famine in the northern counties during the 1590s. The Poor Laws demonstrate the recognition by 16th-century governments that a systematic scheme for the succour of the infirm and the unemployed should replace the policy of non-interference.

Popish Plot (1678), an alleged conspiracy by Roman Catholics to kill *Charles II of England and replace him as king by his Roman Catholic brother, James, Duke of York. The plot was invented by Titus Oates, an Anglican priest, who asserted that a massacre of Protestants and the burning of London were imminent. The plot achieved credibility because of *Shaftesbury's willingness to use Oates as a means to secure James's exclusion from the throne. A nationwide panic ensued during which more than eighty innocent people were condemned before Oates was discredited. He was punished for perjury, but survived to receive a pension from *William III.

Porteous Riots, disturbances in Edinburgh, Scotland, in 1736. Crowds had rioted during the hanging of a smuggler, and Captain John Porteous of the Edinburgh city guard tried to restore order by opening fire, killing several people. As a result he was condemned to death, but reprieved. An orderly crowd then attacked his prison, dragged him out, and lynched him on the day originally appointed for his execution; for this outrage Edinburgh was fined £2,000 and the Lord Provost was removed from office.

Portugal, a small European country on the western coast of the Iberian peninsula. It was settled by Celtic tribes after 1000 BC, and during Roman domination was known as 'Lusitania'. Periods of Gothic and Moorish control followed the collapse of the Western Roman empire, and Portugal struggled to develop a distinct identity until the papacy recognized the kingship of Alfonso I in 1179. In 1249 the Portuguese completed the reconquest of their country from the *Moors. Then, after a series of unsuccessful wars against Castile, peace was at last concluded in 1411, and under the ruling house of Avis (1385-1580) the vast overseas *Portuguese empire took shape. On the expiry of the Avis dynasty, *Philip II of Spain became king by force. The Spanish union lasted until 1640, when the native House of *Braganza was swept to power by a nationalist revolt. During the relatively peaceful and prosperous 18th century, close links were established with England. In the wake of the disastrous Lisbon earthquake (1755) the dynamic minister *Pombal exercised the powers of an enlightened despot.

Portuguese empire, the overseas territories accruing to Portugal as a result of the country's leadership of the first phase of European overseas expansion, beginning in the 15th century. Portuguese imperialism was stimulated by a scientific interest in maritime exploration, a desire to profit from the spice trade of the Orient, and a genuine determination to spread the Christian religion in non-Christian lands. The Portuguese empire came (by c.1530) to include the islands of Cape Verde, Madeira, and the Azores; a large part of *Brazil; fortress settlements in East and West Africa; continuous stretches of the coastlines of Angola and *Mozambique; Indian Ocean bases like Ormuz, *Goa, Calicut, and Colombo; and scattered Far Eastern posts including those in the *Moluccas, *Macao, the Celebes, *Java, and *Malacca.

The empire's wealth derived mainly from coastal entrepôts, and its representatives often had to face highly developed Muslim civilizations. Thus, except in Brazil, there was little conquest or colonization along the lines of the contemporary *Spanish empire. The Portuguese crown was slower than the Spanish to establish a bureaucratic system of administration, but from 1643 the Overseas Council performed a similar role to that of the Spanish Council of the Indies. At the colonial level, however, the various viceroys, governors, and captains-general retained considerable freedom of action. The empire enriched Lisbon, the court, and an increasingly foreign merchant community, but little of the new wealth was reinvested in the mother country. During the 17th century the *Dutch empire in Asia was assembled largely at the expense of the Portuguese.

Potemkin, Gregory (1739-91), Russian soldier and favourite of *Catherine II. He was a man of great energy and an able administrator who extended Russian rule in the south, carried out a series of army reforms, annexed the Crimea in 1783, and built a Black Sea fleet and a naval base at Sevastopol. In the war with the Turks he was made army commander and died in a year of Russian military victory.

Powhatan, chief of the North American Algonquin tribes of central coastal Virginia in 1607, when the English settlers arrived to found *Jamestown. Earlier contacts had been made in 1570-1 and 1588 by the

Spanish, and in 1584–6 by the English. During the early 17th century his position was strengthened by wars of expansion. His daughter *Pocahontas intervened to save the captured John *Smith and was herself held hostage by the English to force their policy on her father.

Poynings' Law (1494), an act of the Irish Parliament, properly called the Statutes of Drogheda, named after Sir Edward Poynings, Lord Deputy of Ireland (1494–5). By its terms, the Parliament was to meet only with the English king's consent, and its legislative programme had to be approved in advance by the English king. It was intended to bolster English sovereignty and destroy Yorkist influence. It soon became a major grievance to Irish parliamentarians, but it was not until 1782 that Henry *Grattan managed to have it repealed.

Praemunire, Statute of (1353), English anti-papal legislation. Like the Statute of Provisors (1351), it resulted from a nationalism and anti-papalism that was widespread in later 14th-century England and was designed to protect rights claimed by the English crown against encroachment by the papacy. It was a powerful weapon for the English king; it was used, for instance, to prevent Bishop Henry *Beaufort from becoming papal legate in England, and *Henry VIII several times resorted to it.

praetorian, originally a bodyguard for a Roman general or 'praetor'. In 27 BC *Augustus established nine cohorts of such troops in and near Rome, later putting two prefects in command. Expanded to twelve by *Caligula and sixteen by Vitellius the number of cohorts was fixed at ten by the end of the 1st century. They were an élite, better paid than legionaries, serving shorter engagements and with many privileges. They also became 'king makers' since their support was essential for gaining high political office. At least four prefects became emperor before *Constantine abolished them early in the 4th century.

Pragmatic Sanction, an imperial or royal ordinance issued as a fundamental law. The term was employed to denote an arrangement defining the limits of the sovereign power of a prince, especially in matters of the royal succession. The Pragmatic Sanction of Bourges, issued by the French clergy in July 1438, upheld the rights of the French church to administer its temporal property independently of the papacy and disallowed papal nominations to vacant benefices and church livings. In April 1713 the Habsburg emperor *Charles VI promulgated a Pragmatic Sanction in an attempt to ensure that all his territories should pass undivided to his children. By 1720 it was clear that his daughter *Maria Theresa would be the heiress, and Charles spent his last years in obtaining guarantees of support from his own territories and the major powers of Europe. On his death in 1740 the failure of most of these powers to keep their promises led to the War of the *Austrian Succession.

Prague, Defenestration of (1618). An act of rebellion by Bohemian Protestant nobles against Catholic *Habsburg rule. The 'Defenestration of Prague' was the ejection of two imperial representatives and a Secretary from a window of the Hradčany Castle in Prague. It precipitated the beginning of the *Thirty Years War and, following the Habsburg victory at the Battle of White Mountain (1620), near Prague, the city underwent enforced Catholicization and Germanization. In 1635, during the *Thirty Years War, the **Peace of Prague** reconciled the German princes to the emperor. In 1757 the city withstood a Prussian siege during the *Seven Years War.

The Defenestration of **Prague**, an engraving by Matthaeus Merian the Elder. The two imperial governors, Martinitz and Slavata, and their Secretary suffered the indignity of ejection, but no serious injury.

The Liberty of the Subject by the English caricaturist James Gillray, 1779, showing the terrorizing tactics employed by the **pressgang** and caricaturing the impressment of 'able-bodied' men.

prehistory, the period before written records, when the only source of evidence about early societies is archaeology. It thus covers an immense period of time, linking evolutionary biology, archaeology, and history. The subject begins with the study of early humans, and can be said to have lasted down to recent times in remote parts of the world. It is divided into the *Stone Age (*Palaeolithic, *Mesolithic, and *Neolithic), the *Bronze Age, and the *Iron Age. History, based on written records began *c.* 3000 BC in Egypt and Mesopotamia.

Presbyterian, a Protestant Christian who subscribes to the anti-episcopal theories of church government, and usually to the doctrines of John *Calvin. At the *Reformation proponents of the Calvinist system claimed that opposition to church government by bishops was not heretical, but rather a restoration of the original organization of the early Church as described in the Bible. In subsequent centuries Reformed and Presbyterian churches all over the world made various adaptations to the original Calvinist pattern, but they remained essentially similar. Government of Presbyterian churches is by elected representative bodies of ministers and elders. The first Presbyterian Church to be organized on a national basis was in 16th-century France; its members became known as *Huguenots, and they played a large part in provoking the French Wars of Religion. Reformed congregations contributed to the *Dutch Revolt, and once the Netherlands secured independence from Catholic Spain the Reformed Church became established there. Elsewhere in Europe, many congregations managed to survive the *Counter-Reformation. In 1628 a Dutch Reformed Church was organized on Manhattan Island. The first American Presbyterian Church was founded in Philadelphia in 1706.

pressgang, a detachment of sailors empowered to seize men for service in the British navy. The use of the pressgang had been sanctioned by law since medieval times but the practice was at its height in the 18th century. All able-bodied men were liable for impressment, although in fact the pressgangs confined their attention to the seaport towns where they could find recruits with suitable experience. The navy continued to rely on the pressgangs until the 1830s, when improvements in pay

and conditions provided sufficient volunteers. The system was also used to a lesser extent by the army but discontinued after 1815.

Prester John (Greek *presbyter*, 'priest'), a legendary Christian ruler to whom successive generations of Crusaders looked for help against the power of Islam. First mentioned in a 12th-century German chronicle, he was variously identified as a Chinese prince, the ruler of *Ethiopia, and, in a papal appeal of 1177, as 'illustrious and magnificent King of the Indies'.

Preston, battle of (17–19 August 1648), an encounter in Lancashire which effectively ended the second phase of the *English Civil War. On one side were the invading Scottish Engagers under *Hamilton. On the other was Cromwell's *New Model Army. The raw Scottish recruits, although greatly superior in numbers, were no match for the English veterans. Cromwell caught up with them at Preston and dispersed them in a series of running battles.

Prestonpans, battle of (21 September 1745). This battle on the Scottish coast 14.5 km. (9 miles) east of Edinburgh resulted in a famous Jacobite victory during the *Forty-Five Rebellion, when Bonnie Prince Charlie's untrained Scots met Sir John Cope's professional royalist forces and surprisingly routed them in little more than five minutes. The victory attracted many recruits to the Young *Pretender's standard and paved the way for his invasion of England.

pretender, one who puts forward claims to the thrones of others. Sometimes the claims are false, as were those of Lambert *Simnel and Perkin *Warbeck to the crown

Prince Charles Edward Stuart, the Young **Pretender**, in a portrait by Antonio David, displaying the youthful charm which gained his popular titles of 'Bonnie Prince Charlie' and 'The Young Chevalier'. (National Gallery of Scotland, Edinburgh)

of *Henry VII of England, and of *Pugachev during the reign of *Catherine the Great of Russia.

In England the Stuart Pretenders were excluded from the throne because of their religion. The Old Pretender, James Edward Stuart (1688-1766), was the son of the exiled *James II, and in the eyes of loyal *Jacobites became King of England on his father's death in 1701. He was a devout but unimaginative man, who failed to win the affection even of his followers. Two Jacobite rebellions, the *Fifteen and the *Forty-Five, were organized by his supporters to accomplish his restoration, but he arrived in Scotland when the Fifteen was virtually over, and he entrusted the leadership in the Forty-Five Rebellion to his son Charles Edward, the Young Pretender (1720-88). Charles Edward (Bonnie Prince Charlie) had youth and charm, and aroused loyal devotion to his cause, but the Forty-Five was a failure, and Charles's career after his miraculous escape from Scotland was an anti-climax, marred by moral and physical decline.

Pride's Purge (6 December 1648), an English army coup in the aftermath of the *English Civil War, in which Members of Parliament (the exact number is uncertain but it was more than 100) who wished to reach an agreement with *Charles I were forcibly excluded from the House of Commons by Colonel Thomas Pride, a Puritan army officer. The remaining members continued to sit in the Commons, forming the *Rump Parliament.

primogeniture, or 'the condition of being first born', the system whereby the father's estate descends to the oldest son to the exclusion of all other children. It was developed in western Europe and introduced in England in the late 11th century by Norman lawyers as a means of preserving intact the landed wealth of the *barons as the basis of their military service to the crown. As part of the *feudal system, primogeniture maintained the political and social status of the aristocracy. Although it was subsequently more generally extended, it never applied to personal or movable property; and where the practice had been for lands to descend to females, then such lands continued to be divided equally amongst the children. Other exceptions to the practice of primogeniture included burghs and the county of Kent, where an alternative system of inheritance (gavelkind) existed. Despite the Statute of Wills (1540), which permitted the disinheriting of an oldest son, primogeniture survived in England until 1926 and in Scotland until 1964, and it continues still to apply specifically to inheritances of the crown and of most *peerages.

printing, the mass production of text and illustrations by applying inked blocks or type to paper. Woodblock printing, where the image of each page is carved in wood, originated in China in the 8th century AD and the earliest known complete book is a Chinese Buddhist prayerbook of 868. Printing from carved woodblocks appeared in Europe in the 14th century. Movable type made of wood was developed in China, and later in Korea of metal,

Elizabeth I of England, aged about 67, on a stately royal **progress**, being carried to a wedding, surrounded by courtiers. The painting is attributed to Robert Peake, c.1600. (Sherborne Castle)

but this was unknown in Europe in the 1450s when *Gutenberg invented movable metal type. Printing spread quickly and by the 1480s nearly all European countries had printing presses. The development of printing meant that many copies of individual titles could be produced more quickly and cheaply than by scribes copying manuscripts. It also enabled ideas to be disseminated much more widely, frequently in vernacular languages rather than the Latin of scholars. *Erasmus and *Luther exploited printing's potential (in 1525 Luther published 183 pamphlets) and universities, lawyers, and doctors benefited from a much wider range of texts.

News and entertainment were spread by newspapers from the 18th century, and in the 19th century literature was increasingly made available through periodicals and journals as well as books. Encyclopedias and circulating libraries accelerated the diffusion of scientific and technical knowledge.

Prithviraj III (d. 1193), Hindu (Chauhan Rajput) King of Delhi, who died in a brave attempt to resist the establishment of Muslim power in northern India. Until the Muslim invasions he was preoccupied with defending his territories in Ajmer and Delhi against rival Hindu kings. Although he was victorious in 1192 in his first encounter with the Turkish invader, Muhammad Ghuri, he was defeated and killed by him in 1193, thus opening the way for the founding of the *Delhi sultanate. He has been immortalized in ballads and folk literature as a figure of romance and heroism.

privateer, a licensed sea-raider in time of war, who had government-issued letters of marque or reprisal to attack enemy shipping. Privateers were often employed in European wars in the 16th and 17th centuries by the English, French, and Dutch, and they later became common in the West Indies, North America, and the Indian Ocean during imperial conflicts. The Americans resorted to widespread privateering during the Wars of Independence and 1812, and the South followed this example during the Civil War. They were internationally abolished by the Declaration of Paris (1856).

Proclamation Line (1763), a declaration which prohibited American settlement west of the Allegheny Mountains in an attempt by the British government to regularize relations between frontiersmen and Indians after the *French and Indian War. The royal proclamation also established governments for captured territory and imposed royal control on Indian traders. It antagonized land speculators, pioneers, and war veterans who had been promised western land. It was partially amended by subsequent treaties with Indian tribes.

progress, royal, a journey around the kingdom, regularly taken by monarchs and their courts in the days of personal rule. When communications were poor, and regional control limited, progresses served to assert sovereignty and win loyalty. They also offered opportunities to hunt, to avoid the plagues that thrived in built-up cities, and to share the economic burden of maintaining the court among richer subjects. *Elizabeth I compelled her rich courtiers to entertain her and her retinue at their *country houses. Emperor *Charles V, with his widely scattered dominions, was by necessity a

'peripatetic monarch' *par excellence*. Monarchs who refused to make progresses ran the risk of forfeiting their subjects' obedience, as was partly the case with *Philip II and the *Dutch Revolt.

Protectorate, English (16 December 1653 to 25 May 1659), the rule of England established by Oliver *Cromwell. Unable to work with the *Barebones Parliament, Cromwell entrusted a council of army officers with the task of drawing up a new constitution. The resulting Instrument of Government made Cromwell Lord Protector, monarch in all but name, who would share power with a single House of Parliament elected by Puritans. Politically it was a failure. Cromwell could not work with his first Protectorate Parliament, so he divided England into eleven military districts ruled by army officers known as major-generals. This was so unpopular that he reverted to parliamentary rule through the second Parliament of the Protectorate in 1656. Although the Protectorate was successful in foreign policy and notable for religious toleration of all faiths other than Roman Catholicism, its stability depended on Cromwell's personal qualities. After his death in 1658 it did not take long for the army to remove Richard *Cromwell, his successor, bringing the Protectorate to an end in 1659 in preparation for the *Restoration of Charles II.

Protestant, a member or adherent of any of the Christian bodies that separated from the *Roman Catholic Church at the *Reformation. The term was coined after the imperial Diet summoned at Speyer in 1529, and derives from the 'Protestatio' of the reforming members against the decisions of the Catholic majority. These adherents of the Reformation were not merely registering objections: they were professing their commitment to the simple faith of the early Church, which they believed had been obscured by the unnecessary innovations of medieval Roman Catholicism. Since then, the term has been used to identify those who accept the principles of the Reformation, as opposed to Catholic or *Orthodox Christians. *Luther, *Zwingli and *Calvin founded the largest of the original Protestant branches, and there were other more radical groups, such as the *Anabaptists.

All the early Protestants shared a conviction that the *Bible was the only source of revealed truth and it was made available to all in vernacular translations. They also believed in the doctrine of justification by faith alone, and to the idea of the universal priesthood of all believers. They minimized the ceremonial aspects of Christianity, and placed preaching and hearing the word of God before sacramental faith and practice. Many Protestant sects and churches were formed, largely because the principle of 'private judgement' in the interpretation of the scriptures led to many shades of doctrine and practice. In England, members of both the established *Anglican Church and the various *Nonconformist churches were usually regarded as Protestants.

Provence, a region of south-east France with a Mediterranean coastline. It was named Provincia Gallica Transalpina by the Romans, who annexed it in the 2nd century BC, when it was occupied by Celtic tribes. It also contained Greek and Phoenician settlements and developed a distinct culture which it managed to retain, even through the chaos of Germanic invasions. As part of the kingdom of Arles it passed to the Holy Roman

Empire in the reign of Conrad II but continued to enjoy a large measure of independence, its language, Provençal, remaining substantially different from Middle French. It made a great contribution to the success of the First *Crusade, and its involvement continued throughout the crusading period. The *troubadours led a flourishing revival of secular literature and music. Its independence encouraged heresy and *Catharism took root, resulting in the suppression of the *Albigenses by the Church. In the 14th and 15th centuries it resisted *Angevin claims to sovereignty, but in 1480 René, the last king of Aix, died and it passed, via his nephew, to France.

Prussia, a former kingdom in north Germany. It was originally inhabited by pagan tribes who were conquered by the *Teutonic Knights in the mid-13th century. By the 15th century this Christianized region on the south-eastern coast of the Baltic Sea had come under Polish control. The lands east of the River Vistula (Royal Prussia) were ceded to the Polish crown, while the Teutonic Knights had to recognize Polish suzerainty over the rest (1466). The *Hohenzollern Albrecht of Brandenburg-Ansbach was chosen as High Master in 1511. After the rapid spread of Lutheranism in his territory, he secularized the Teutonic Order and in 1525 became hereditary ruler of ducal Prussia. In 1618 the duchy passed under the control of the Electors of *Brandenburg. By 1701 Brandenburg-Prussia had evolved into an ascendant Protestant power in Europe, and in that year its ruler became known as 'King in Prussia'.

The kingdom's most notable 18th-century ruler was *Frederick the Great. He and his successor Frederick William II added considerably to Prussia's holdings, but many of these were lost during the French Revolutionary and Napoleonic wars.

Prynne, William (1600–69), English Puritan pamphleteer, a fearless campaigner on religious, moral, and political issues. His most famous pamphlet, *Histriomastix* (1632), was an attack on stage-plays; he was tried before the Star Chamber for its implied criticism of Queen *Henrietta Maria, who was a devotee of plays and masques. He was sentenced to life imprisonment and cropping of the ears (1634). He continued to write anti-episcopal pamphlets and in 1637 the remaining parts of his ears were removed. The *Long Parliament freed him in 1640. Elected to Parliament himself in 1648, he was expelled at *Pride's Purge, and eventually supported the *Restoration.

Ptolemy, the Macedonian dynasty which ruled Egypt from 323 to 30 BC. Ptolemy I was an officer of *Alexander the Great who, after the king's death, was appointed *satrap of Egypt. He proclaimed himself king in 304, and by the time of his death in 283–2 he had established control over Cyprus, Palestine, and many cities in the Aegean and Asia Minor. The reigns of the Ptolemies who succeeded him were characterized externally by struggles with the *Seleucids for control of Syria, Asia Minor, and the Aegean; and internally by dissatisfaction and rebellion among the native Egyptians. Contact with the rising power of Rome came to a head during the reign of *Cleopatra VII whose liaison with *Mark Antony led ultimately to defeat at Actium, suicide, and the annexation of Egypt by *Octavian.

Ptolemaic Egypt was remarkable for its administrative, financial, and commercial infrastructure, for the most part developed by the first two Ptolemies. The discovery of many *papyri has enabled scholars to reconstruct a detailed picture of life in these times. All land except that owned by temples belonged to the state, which rented much of it out to the people. The state provided seed corn, but required that it be repaid in kind at harvest time. It also held monopolies on such goods as oil-producing crops, mining, and salt. Local and central registers of houses, people, and animals facilitated taxation and the close supervision of life that was a pervasive part of the Ptolemaic regime. It included the most advanced system of banking in antiquity, geared to obtaining the maximum income for the state, with little regard for individual Egyptians. Thus attempts to cheat the system and mass unrest were common.

Pueblo *Anasazi.

Puerto Rico, an island in the Caribbean, originally known as Boriquén, discovered by Columbus in 1493. Encouraged by tales of gold from the indigenous Arawak Indians, his companion, Juan *Ponce de León, was granted permission by the Spanish crown to colonize the island. In 1508 he founded the settlement of Caparra and in 1509 he was made governor. Caparra was abandoned and the settlement moved to nearby San Juan in 1521, to take better advantage of the bay for trading. By the end of the 16th century the Arawak were virtually extinct from European-introduced diseases and exploitation. In the 17th and 18th centuries the island remained important for its sugar and tobacco plantations, worked by imported black slaves, and as a key to Spain's defence of its trading interests in the Caribbean and Atlantic against France, Britain, and Holland.

Pugachev, Emelian Ivanovich (1726–75), *Cossack and leader of a massive popular uprising in 1773–4 against the rule of *Catherine II of Russia. Deserting the army, he won the support of discontented serfs, cossacks, miners, and recently conquered peoples like the Bashkirs and Tartars. He captured Kazan and established a court, claiming to be the assassinated Emperor Peter III. He promised the abolition of landlords, bureaucrats, serfdom, taxation, and military service and the restoration of traditional religion. His betrayal and execution were followed by the ruthless suppression of his followers.

Punic wars, the three wars fought in the 3rd and 2nd century BC between Rome and Carthage, so named from 'Poenicus' ('Dark skin' or 'Phoenician'), the term used to describe founders of the North African city. The contest was for control of the Mediterranean Sea. Rome emerged as victor from each war.

The First (264–241 BC) was fought largely at sea. Rome expanded its navy and took control of Sicily. Corsica and Sardinia were seized a few years later. *Hamilcar, father of *Hannibal led the defeated side. The Second (218–201) arose from Hannibal's invasion of Italy from Carthaginian bases in Spain via the Alps. He led a huge force including elephant squadrons. Rome suffered disastrous defeats, most notably in the mists by Lake Trasimene and at *Cannae. Italy was overrun by Hannibal but the Italian tribes did not rise against Rome. The strategy of the dictator *Fabius prevented further losses. In a long-drawn out series of campaigns Hannibal's extended lines of supply were threatened by defeats in Sicily and Spain and the brilliant generalship of *Scipio. *Hasdrubal, Hannibal's brother, was defeated on the Italian mainland in 207. By 203 Hannibal, who had no effective siege engines, was summoned to withdraw to Africa to defend Carthage itself, now threatened by Scipio. Pursued by Scipio he was defeated at Zama in 202 and the Carthaginians were forced to accept humiliating terms the following year. Spain was acquired as a provincial territory by Rome.

In 149 BC at a peak of its territorial expansion and at the insistence of *Cato, Rome intervened in an African dispute to side with Numidia against Carthage. In the Third War (149-146) the Younger *Scipio besieged and destroyed Carthage utterly, sowed the site with symbolic salt and declared Africa a Roman province.

Puritan, a member of the more extreme English Protestants who were dissatisfied with the *Anglican Church settlement and sought a further purification of the English Church from *Roman Catholic elements. Their theology was basically that of John *Calvin. At first they limited themselves to attacking 'popish' (Roman Catholic) practices—church ornaments, vestments, and organ music—but from 1570 extremists attacked the authority (episcopacy) of bishops and government notably in the Martin

A **Puritan** congregation listening to the preaching of the Lord Protector Oliver Cromwell. This Dutch print incorporates various scenes from the Civil War at the church windows as well as the spirit of Charles I, seen here holding a sword and book, on the pulpit. (British Museum, London)

*Marprelate tracts. However, James I resisted their attempts to change Anglican dogma, ritual, and organization, voiced at the *Hampton Court Conference. In the 1620s some emigrated to North America, but it was the policies of *Laud and Charles I in the 1630s that resurrected the Puritan opposition of the 1580s. The doctrine of Predestination (that God ordains in advance those who shall receive salvation) became a major source of contention between the Puritans, for whom it was a fundamental article of faith, and the Arminians who rejected it. Religion was a key factor leading to the outbreak of civil war in 1642. Puritanism was strong among the troops of the *New Model Army and in the 1640s and 1650s, with the encouragement of Cromwell, Puritan objectives were realized. After the Restoration they were mostly absorbed into the Anglican Church or into larger *Nonconformists groups and lost their distinctive identity. The term 'Puritan' covers many groups and attitudes and remains an area of historical debate.

Putnam, Israel (1718-90), American general, who served in the Connecticut militia throughout the *French and Indian War and against *Pontiac's Rebellion. He was a fervent patriot who joined the continental army at the start of the War of *Independence, and fought at *Bunker Hill. He served with Washington's army until 1779 when he became paralysed. Though a mediocre tactician, his personal bravery, patriotism, and geniality made him a national hero.

Pym, John (1584-1643), English politician. He entered Parliament in 1614, and by the 1620s was making his mark, especially as a manager of the impeachment of

*Buckingham (1626), and as a supporter of the *Petition of Right (1628). In the *Long Parliament his debating and tactical skills brought him great influence, and earned him the nickname 'King Pym'. He was the main architect of the reforming legislation of 1641, including the Acts of *Attainder against *Strafford and *Laud, and was responsible for having the *Grand Remonstrance printed and published. Pym was one of the *five Members of Parliament whom Charles I ill-advisedly tried to arrest in 1642. Once the *English Civil War began, he played a vigorous role on the Committee of Safety (1642) and in the year of his death engineered the *Solemn League and Covenant with the Scots.

pyramid, a large burial monument, especially characteristic of ancient Egypt. At first the pharaohs were buried in underground chambers over which were built rectangular *mastabas*; these were stone structures housing the food and accoutrements the pharaoh would need in the afterlife. The first pyramid was that constructed for King Zoser at Saqqara by *Imhotep c.2700 BC, the so-called Step Pyramid which has six enormous steps and is over 60 m. (197 ft.) high. A pyramid built at Meydum c.2600 was originally of similar design, but the steps were later filled in with limestone to produce the classical pyramid shape. Most of the best known pyramids date from the Old Kingdom (c.2700–2200 BC), though some were built during the eleventh and twelfth dynasties (c.2050–1750 BC). The pyramids of Khufu, Khafre, and Menkaure at Giza are a spectacular illustration of the skill of Egyptian architects—and of the state's ability to organize a large work-force. The Great Pyramid of Giza is estimated to have required a labour force equivalent to about 84,000 people employed for 80 days a year for 20 years.

In pre-Columbian Central America the pyramid was a stepped base for a temple. These were erected by the *Mayas, *Aztecs, and *Toltecs, for the most part between AD 250 and 1520. The Temple of the Sun in *Teotihuacán in Mexico is perhaps the most impressive.

Pyramids, battle of (21 July 1798). The decisive battle fought near the pyramids of Giza that gave *Napoleon control of Egypt. He took Alexandria by storm on 2 July, and then, with 40,000 men defeated a *Mameluke army of 60,000 led by Murad Bey. The victory enabled Napoleon to take Cairo and allowed France to control Egypt until its withdrawal in 1801.

Pyrrhus (319–272 BC), King of *Epirus in Greece (307–303, 297–272), first as a minor (307–303) and then after his return from exile in 297. After various campaigns against his neighbours, especially *Macedonia, he aided Tarentum against the growing power of Rome. At Heraclea (280) and Asculum (279) he was victorious, and he almost expelled the Carthaginians from Sicily, but because of his limited numbers of troops he could ill afford the losses which he suffered (hence the expression 'Pyrrhic victory'). After his campaign in Sicily and a third battle with the Romans he returned to Epirus. He was a fine tactician, especially in his use of elephants, and greatly admired by *Hannibal.

Qianlong (Ch'ien-lung) (1710–99), *Qing Emperor of China (1735–95). During his rule China reached its greatest territorial extent with campaigns undertaken in Turkistan (*Xinjiang), Annam, Burma, and Nepal. In 1757 he restricted all foreign traders to *Guangzhou (Canton), where they could trade only from November to March. He rejected in 1793 the requests of a British delegation led by Lord Macartney for an expansion of trade and the establishment of diplomatic relations. Towards the end of his reign his administration was weakened by corruption, financial problems, and provincial uprisings, notably of the *White Lotus Society in 1796. He was lauded by *Voltaire as a philosopher-king, patronizing the arts, writing poetry, and overseeing the compilation of literary collections (in which anything thought critical of the Qing was expunged). Four years before his death he abdicated in favour of his son.

Qin (Ch'in) dynasty (221–206 BC), China's first imperial dynasty. It was founded by Prince Zheng, ruler of the *Zhou vassal state of Qin. Unlike rival Chinese states, Qin used cavalry not chariots in battle and early adopted iron weaponry. It ensured a regular food supply by developing a system for land irrigation. Based in Shaanxi, it began to expand its territories c.350 BC. Under Zheng, it overthrew the Eastern Zhou and conquered (256–221 BC) all Zhou's former vassal states. Zheng then took the title Huangdi and is best known as *Shi Huangdi (First Emperor). He died in 210 BC and his dynasty was overthrown four years later. From that time, though China was sometimes fragmented, the concept of a united empire prevailed. From Qin is derived the name China.

Qing (Ch'ing) dynasty (1644–1912), the last dynasty to rule China. Its emperors were *Manchus. In 1644 a *Ming general, Wu Sangui, invited Manchu Bannermen massed at Shanhaiguan, the undefended eastern end of the Great Wall of China, to expel the bandit chieftain Li Zicheng from Beijing. The Bannermen occupied the city and proclaimed their child-emperor 'Son of Heaven'. Resistance continued for up to thirty years in south China. Chinese men were forced to braid their long hair into a queue or 'pigtail'. But Qing rule differed little from that of Chinese dynasties. It emphasized study of the Confucian classics and the Confucian basis of society. The civil service, half Manchu, half Chinese, continued to be recruited through an examination system based on knowledge of the classics. The empire of China reached its widest extent, covering Taiwan, Manchuria, Mongolia, Tibet, and Turkistan. The Qing regarded all other peoples as barbarians and their rulers as subject to the 'Son of Heaven', and were blind to the growing pressure of the West. Under *Kangxi and *Qianlong China was powerful enough to treat the outside world with condescension. Thereafter the authority of the dynasty was reduced and the last emperor abdicated following the Chinese Revolution in 1912.

Quaker, originally a term of contempt for members of

The Chinese emperor **Qianlong** carried by retainers to his open-air audience with the British delegation of Lord Macartney (shown right) in a watercolour by William Alexander, *c*.1792. Chinese hospitality awaits in the form of rest, music, and refreshment in decorated tents. (British Museum, London)

the *Nonconformist Society of Friends founded by the Englishman George *Fox. They rejected all manifestations of an organized church, believing that consecrated buildings and ordained ministers were irrelevant to the individual seeking God. They became convinced that their 'experimental' discovery of God—sometimes featuring trembling or quaking experiences during meetings—would lead to the purification of all Christendom.

By 1660 there were more than 20,000 converts, and missionaries were at work in Ireland, Scotland, Wales, and the American colonies. They continued to grow in number, despite severe penalization from 1662 to 1689 for refusing to take oaths, attend Anglican services, or pay tithes. After considerable debate, they evolved a form of organization, with regular monthly, quarterly, and annual meetings. This system essentially stands today, and any Quaker can attend any meeting.

In 1681 William *Penn founded the American Quaker colony of *Pennsylvania, and Quaker influence in the colony's politics remained paramount until the American War of Independence.

Quebec, a Canadian city and province. The city was founded in 1608 by Samuel de *Champlain on a near-impregnable plateau and served as the capital of *New France until it was captured in 1759. When New France ceased to exist in 1763 the Province of Quebec was created. The **Quebec Act** (1774) extended its boundaries at American expense and guaranteed religious tolerance and French Civil Law. This ensured loyalty during the War of Independence, and the American campaign to capture the city was repulsed (1775–6). In 1791 the province was divided into two, the predominantly French Lower Canada (now Quebec) and Upper Canada (now Ontario).

Queen Anne's War (1702–13), that part of the War of the *Spanish Succession fought in America. It saw a renewal of frontier warfare in New England with savage French and Indian attacks on outlying settlements. In 1710 the French lost Port Royal in Acadia and Acadia (known to the British as *Nova Scotia) came under British control. An attempt to capture Quebec the next year was prevented by storms. In the south, a South Carolinian expedition destroyed St Augustine, Florida, in 1702 and a retaliatory attack on Charleston (1706) was repulsed. In the Caribbean, St Christopher was captured from the French in 1702, but Guadeloupe held out in the following year. Thereafter only *privateers and *buccaneers remained active. The main British colonial gains at the Peace of *Utrecht were Nova Scotia, western Newfoundland, and St Christopher.

Queen's shilling (or King's shilling), a coin, the acceptance of which from a recruiting officer obliged the recipient to serve in the British army. Recruiting sergeants in the 18th century would ply likely young men with drink, and if they could persuade them in their alcoholic stupor to take the shilling, they were in fact accepting army pay. This ranked as a binding agreement from which escape was very difficult.

Quesnay, François (1694–1774), French physician and economist. He trained as a doctor and became physician to Louis XV; his writings on economics were not published until he was 60. He was the leader of the physiocrats, a group of economists who opposed *mercantilism and proclaimed *laissez-faire. They believed that land was the only source of wealth and that all taxation should be based on land; tax exemptions and all artifical restrictions on the circulation of wealth should cease, and in particular there must be a free market in grain; these changes could best be secured through a strong monarchy. A protégé of Madame de *Pompadour, his followers met at Versailles.

Quetzalcóatl, literally 'quetzal-bird snake'—hence 'feathered serpent', one of the chief gods of ancient Mesoamerica (Mexico and northern Central America). Quetzalcóatl was also a historical priest-king of the *Toltecs, and the official title of the *Aztec high priest. As a god he was known throughout Mesoamerica and

A stone image of the Aztec god **Quetzalcóatl**, in the form of a porter carrying maize. (Detroit Museum)

was called Kukulkán by the *Maya. Images and temples to him appear at early sites, such as as *Teotihuacán, but he was especially revered *c.*700–1520.

Quiberon Bay, battle of (1759). A site on the west coast of Brittany in northern France was the scene of a major British naval victory in the *Seven Years War. The French were planning an English invasion. In 1759 Admiral *Hawke was blockading Brest, but in November the French admiral Conflans broke out during a storm. Hawke pursued the French fleet and when most of the ships took shelter in Quiberon Bay he came through the dangerous shoals in a gale. The French lost eleven ships and 25,000 men, and the British only two ships. Any threat of a French invasion of Britain was over.

Quito, the northern provincial capital of the *Inca empire, linked by a long road to *Cuzco, the imperial capital. In 1530 it was governed by *Atahualpa, who used it as a base to seize the rest of the empire, defeating his brother and co-ruler, Huáscar. Under Spanish colonial rule it became a provincial capital, responsible to the viceroy; and in 1535 the church and monastery of San Francisco, built on the city's main plaza, became the first Christian foundation in South America.

R

Raikes, Robert (1735–1811), British printer, newspaper proprietor, and philanthropist. He took up the cause of prison reform early in his career, but his greatest achievement was the Sunday school he established at Gloucester in 1780 to teach reading and the church catechism to children who worked in factories during the rest of the week. This led to a national movement for the establishment of Sunday schools all over Britain.

Rajput (Sanskrit, 'son of a king'), the name of a predominantly landowning class, also called Thakurs, living mainly in central and northern India, who claim descent from the Hindu Ksatriya (warrior) caste. Many leading clans are of royal lineage, but others include cultivators of Sudra (menial) caste. Their clans are divided into four lines: Solar, Lunar, Fire, and Snake.

They are almost certainly descended from 6th- and 7th-century migrants from Central Asia. They became politically important from the 9th century when their chieftains gained dominance over the desert and hill area now called Rajasthan, between the north Indian plains and the *Deccan. Although they held off Muslim invasions between the 12th and 16th centuries, the *Moguls then penetrated their desert fortresses. However, Emperor *Akbar achieved Rajput acquiesence by allowing them to rule their conquered territories, and by drawing them into imperial service. This balance was destroyed, first by less tolerant Mogul rulers, and then by the *Maratha upsurge in the 18th century, which finally ended Rajput independence. When the Marathas fell victim to British arms, their Rajput clients also had to come to terms. However, the leading chiefs, including *Jodhpur, *Jaipur, and Udaipur, preserved considerable autonomy as Princely States under British rule.

Raleigh, Sir Walter (*c.*1552–1618), English explorer and courtier. He took part in two privateering expeditions to the West Indies and was then sent to Ireland to

The arrival of the English explorer Sir Walter **Raleigh** in Trinidad. Sick with fever, Raleigh remained there while five of his ships were sent on an ill-fated expedition up the Orinoco. (National Maritime Museum, London)

suppress a rebellion. On his return to England in 1581 he became a great favourite of *Elizabeth I, who showered honours and rewards on him, including land in Ireland where he introduced the potato as a crop. In 1585-91 he organized several colonizing expeditions to *Virginia, all ending in failure. His position as royal favourite was usurped by the Earl of *Essex in 1592, and he was imprisoned in the Tower of London and later exiled from the court. He set out in 1595 for Guiana in South America where the gold mines of *El Dorado were thought to lie. His account of the voyage, *The Discoverie of Guiana*, is one of the finest narratives of Elizabethan adventure. The death of Elizabeth I brought ruin to Raleigh. He was falsely accused of conspiracy against *James I, and in 1603 was again imprisoned in the Tower, where he remained for thirteen years. In return for his freedom he promised to discover a gold mine for the king on Guiana, but on condition that there should be no clash with the Spaniards, with whom England was now at peace. When his fleet reached the mouth of the Orinoco River, Raleigh remained at Trinidad, sending five small vessels up the river. They came unexpectedly on a Spanish settlement, fighting broke out, and Raleigh's son was killed. The expedition returned sadly home, and Raleigh was arrested and executed at the request of *Gondomar, the Spanish Ambassador.

Ramanuja (d. 1137), Brahmin (member of the *Hindu priestly caste) from southern India, whose teachings inspired the *bhakti* devotional school. He identified Brahman (the supreme soul) with the god Vishnu, whose worship he then encouraged on pilgrimages throughout India. His preaching that the visible world is real and not illusory, and that God should be worshipped devotedly, built a bridge between philosophy and popular *bhakti* religion.

Rambouillet, Catherine de Vivonne, marquise de (1588-1665), a French aristocrat who presided over the first of the *salons* which dominated the intellectual life of 17th-century Paris. The Hôtel de Rambouillet was at the height of its influence between 1620 and 1645 and was frequented by Corneille, Madame de Sévigné, and Bossuet. Great emphasis was laid on refinement and good taste expressed by the term *précieuse*, later mocked in Molière's comedy *Les Précieuses Ridicules*.

Ramesses II (the Great), Pharaoh of Egypt (ruled *c.* 1304-1237 BC). In the fifth year of his reign he fought the *Hittites at Kadesh where he managed to extricate himself from a perilous situation. In the twenty-first year of his reign the two powers concluded a peace treaty, and Ramesses married a Hittite princess. He also undertook campaigns against the Libyans. His reign, which marked a high point in ancient Egyptian history, was one of considerable prosperity, and included a substantial building programme, for example two temples cut out of the cliffs at Abu Simbel and the completion of Seti I's hypostyle hall at *Karnak and temple at Abydos.

Ramesses III, Pharaoh of Egypt (ruled *c.*1188-1156 BC). He successfully repelled three major invasions, two by the Libyans and one by the *Sea Peoples. Peace and prosperity followed, but the last years of his reign were marked by social unrest and an assassination attempt. He was the last Egyptian ruler to hold land in Palestine.

The seated figure of the pharaoh **Ramesses II**, carved in black granite. The pharaoh carries a shepherd's crook, the insignia of the god Osiris. (Torino Museum)

Ramillies, battle of (23 May 1706). A site in eastern Belgium between Namur and Louvain was the scene of *Marlborough's second great victory during the War of the *Spanish Succession over the French army under Villeroi. Marlborough duped his opponents into thinking his main attack was coming from the right and smashed through the French line from the left. The French losses were five times greater than Marlborough's. He went on to overrun much of Flanders and Brabant.

ransom, a sum of money paid for the release of a prisoner or for the restitution of property. The practice formed an accepted part of medieval warfare, and of medieval diplomacy. Knights who were *vassals of a lord were obliged to pay for the release of their lord, if he was captured in war, although in the late Middle Ages family and friends paid as well as a lord's estate. A suitable ransom would be negotiated and raised to secure eventual release. Needless massacre of prisoners, as after *Agincourt, aroused resentment among would-be captors. Notable ransom victims include John II of France and *Richard I of England.

Reason, Age of, a term generally applied to the 18th century, when it was optimistically assumed that all knowledge, encompassing the whole universe, its laws and functioning, was capable of rational (reasoned) discovery, explanation, and understanding. The medieval idea that knowledge was divinely revealed had been undermined in the 16th and 17th centuries by the empirical methods which led to a scientific revolution in astronomy, physics, medicine, and mathematics. As faith in divine revelation crumbled before 18th-century rationalism and logic, so too unquestioning religious conviction gave way to the search for specific proofs.

Rationalists often found religion unnecessary, though this attitude incurred clerical and state wrath.

Belief in the unlimited power of the human intellect gave rise to a self-confidence which became more concerned with the pursuit of happiness in this world rather than preparation for heaven in the next. These ideas found their expression in the work of the *philosophes*, most of whom were French, and among whom the *Encyclopédists were the most influential. Their political impact is to be found in the *Enlightenment. The so-called 'enlightened despots' included *Frederick the Great of Prussia, *Catherine I of Russia, and *Joseph II, Holy Roman Emperor, who were aware of the Enlightenment, often had personal links with individual philosophers such as Voltaire and Diderot, and attempted to put some enlightened ideas into practice within their states often by authoritarian means.

recusant, a term for those (usually Roman Catholics) who refused to attend *Anglican Church services, from the 16th century onward. Fines were imposed on them by Acts of Uniformity (1552 and 1559). Although *Nonconformists could be penalized for recusancy, the term was often used as an abbreviation of 'Catholic Recusants', distinguishing them from 'Church Papists', who were Catholics who attended Anglican services rather than pay the fines. The penal laws against Catholics were extended between 1571 and 1610, but were rarely enforced. They were systematically repealed in a series of Toleration Acts (*Catholic emancipation).

reeve, a local official in Anglo-Saxon and post-Conquest England. The most important, *shire reeves (sheriffs), administered royal justice and collected royal revenues within their shire. Manorial reeves organized the peasant labour force on estates and their duties were considerable. They received a money wage, grants of grazing land, and remission of rent and feudal dues. Although often of *villein status those reeves who contrived through annual re-election to make their office hereditary had considerably improved their economic condition by the 14th century, when Chaucer wrote of them in *The Canterbury Tales*.

Reformation, the 16th-century movement for reform of the doctrines and practices of the *Roman Catholic Church, ending in the establishment of Reformed or *Protestant churches. The origins of the movement can be traced back to the 14th century and attacks by *Lollard and *Hussite believers on the hierarchical and legalistic structure of the Church. The *Great Schism had also weakened papal authority, and there was widespread dissatisfaction with the papacy's worldliness, its financial exactions, and its secular involvement as an Italian territorial power.

The starting point of the Reformation is often given as 1517, when the German theologian Martin *Luther launched his protest against the corruption of the papacy and the Roman Catholic Church, but he was breaking no new controversial ground. In fact, most of the Reformation movements laid stress, not on innovation, but on return to a primitive simplicity. Luther's theological reading led him to attack the central Catholic doctrines of transubstantiation, clerical celibacy, and papal supremacy. He also called for radical reform of the religious orders. By 1530 the rulers of Saxony, Hesse,

Brandenburg, and Brunswick, as well as the kings of Sweden and Denmark had been won over to the reformed beliefs. They proceeded to break with the Roman church, and set about regulating the churches in their territories according to Protestant principles.

In Switzerland, the Reformation was led first by *Zwingli, who carried through antipapal, antihierarchic, and antimonastic reforms in Zürich. After his death the leadership passed to *Calvin, in whose hands reforming opinion assumed a more explicitly doctrinal and revolutionary tone. Calvinism became the driving force of the movement in western Germany, France, the Netherlands, and Scotland, where in each case it was linked with a political struggle. Calvinism was also the main doctrinal influence within the *Anglican Church. In Europe the reforming movement was increasingly checked and balanced by the *Counter-Reformation. The era of religious wars came to an end with the conclusion of the *Thirty Years War (1618–48).

Regulators (1764–71), American rebels from inland North Carolina who felt aggrieved at the political control of the aristocrats of the coastal region. They turned to violence when legal action failed to increase their representation and reduce their taxation. 'The Regulation', centred in Orange County, attacked magistrates and lawyers until overwhelmed by the militia under Governor Tryon at the battle of the Alamance. Such antagonisms between coastal and frontier settlers were not uncommon, as *Bacon's, *Culpeper's, and *Shays's rebellions demonstrated.

Regulus, Marcus Atilius (d. *c*.251 BC), Roman consul in 256 BC. He defeated the Carthaginian fleet and invaded Africa during the First *Punic War. He took Tunis but was defeated and captured. He was allowed to return to Rome on parole in order to negotiate peace terms involving the exchange of prisoners. He advised the Senate to refuse them, before returning to captivity. A posthumous tradition grew up that he returned to torture and execution.

Rehoboam (10th century BC), son of King *Solomon. He succeeded his father as King of *Israel (922–915 BC), but the people of the north, restive from the constraints of Solomon's rule, broke away and set up a new kingdom under *Jeroboam, after which Rehoboam continued as the first ruler of the southern kingdom of *Judah. His military efforts to regain control of the north were unsuccessful, partly because of an invasion of Judah by the Egyptian king, Shishak.

Reich *Holy Roman Empire.

relic, part of the mortal remains of, or an object closely associated with, a Christian saint or *martyr. In the Middle Ages the possession of relics greatly enhanced the prestige of a church or monastery and caused the shrine to become a place of pilgrimage. *Louis IX of France built the Sainte-Chapelle in Paris to house relics from *Constantinople. Abuses led the Fourth *Lateran Council to forbid their sale and decree that only those authenticated by the papacy might be venerated. Their use was attacked by the later Protestants, which led in 1563 to the Council of Trent confirming the doctrine of the veneration of holy relics simply as aids to devotion.

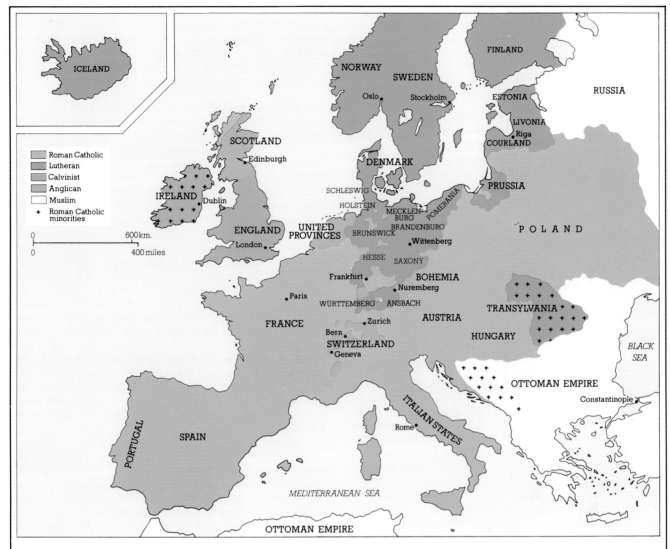

Reformation: Europe *c.* 1570

The adoption of the 'reformed' theology in Europe, promoted by Luther and Zwingli, spread rapidly from Zurich (1520) to the neighbouring towns and states of southern Germany, as well as to Scandinavia, and England. The movement was given further impetus in the 1540s under the leadership of Calvin. At the height of the Reformation Protestant churches had made substantial inroads into France and Poland and the theology

had been adopted by the majority of the German states. Roman Catholicism remained strong, however, and, of the leading powers, only England had broken with Rome. The Counter-Reformation was effective in checking the reforming movement, but Protestantism was nevertheless firmly entrenched, and the patterns of religious loyalty established in Europe in the 16th century would endure largely unchanged.

Religion, Wars of *French Wars of Religion.

Renaissance, the period in European civilization characterized by an interest in the classical past and a flowering in literature, art, and architecture. It began in Italy as a literary movement led by *humanist scholars, notably Petrarch (1304–74) and Dante (1265–1314). Poetry and prose was written in the vernacular instead of in Latin and the invention of printing contributed to the spread of ideas, particularly the new *humanist approach of *Erasmus to the Christian scriptures. In the 15th and 16th centuries the main centres of the movement were the cities of Florence, Venice, and Rome and the ducal courts of Mantua, Urbino, and Ferrara. In architecture Bramante, Palladio, and Alberti looked to classical examples for inspiration, while the artists Michelangelo, Raphael, Titian and others broke new

ground by introducing the human figure, naturalistically depicted, into their paintings. The invention of perspective also aided painters in their realistic depiction of nature. Important support for the arts came from the patronage of the *Medici in Florence and from a succession of popes including Julius II and Leo X.

The new spirit of enquiry extended to geography and cartography. The impulse to explore the world led to *exploration and the voyages of discovery of *Diaz, *da Gama, and *Magellan. The astronomers Copernicus, Kepler, and Galileo proposed new theories about the movement of the planets, and advances were made in biology, chemistry, physics, and medicine. The Flemish anatomist Vesalius wrote *De humani corporis fabrica* (1543), an influential anatomical treatise.

The Renaissance had far-reaching consequences including the colonization of new lands, and conceptual and technical advances in scholarship and the arts. The new spirit of enquiry also affected man's perception of the church and paved the way for calls for reform.

Restoration (1660), the re-establishment in England and Scotland of the Stuart monarchy by placing *Charles II, the exiled son of *Charles I, on the throne. Oliver *Cromwell had never succeeded in reconciling the royalists to the republican regime, and his rule had been based upon the strength of the army rather than popular consent. His death in 1658 undermined the *Protectorate, and his successor Richard *Cromwell was brushed aside by the army. The careful and deliberate actions of General *Monck in Scotland and England, and *Clarendon, in exile with Charles II, brought about the peaceful restoration in May 1660. Before he left Holland Charles's Declaration of *Breda promised forgiveness, reconciliation, and, in effect, a determination to work with Parliament. The Restoration was accompanied by the revival of the Church of England, and the growth of Cavalier fortunes, although those who had sold their estates to pay fines could not get them back, and a flourishing cultural and social life. The Restoration did not in fact restore the absolute authority of the Stuart monarchy, as Charles II was soon to discover.

Retz, Paul de Gondi (1613–79), French statesman and cardinal. The designated successor to his uncle, the Archbishop of Paris, he was extremely ambitious and wished to replace *Mazarin as chief minister of France. He was active in the first years of the *Fronde and was afterwards imprisoned. He escaped and fled to Rome to appeal to the pope, but without success. In exile he was a serious nuisance to *Louis XIV, and it was only in 1662 that he resigned his archbishopric as the price of being allowed to return to France. His memoirs give a spirited, though unreliable, account of the personalities and intrigues of mid-17th-century France.

Revere, Paul (1735–1818), American patriot, a silversmith and engraver of *Huguenot extraction living in

The years following the **Restoration** of 1660 brought widespread interest in cultural pursuits and scientific knowledge. This painting by Thomas Danckerts shows the royal gardener presenting Charles II with the first pineapple successfully cultivated in England. (Houghton Hall, Norfolk)

Boston. After service in the *French and Indian War he joined the *Sons of Liberty and published anti-British cartoons. Immortalized in Longfellow's account of his midnight ride before *Lexington and Concord (1775), he served in New England in the War of Independence, and died a wealthy merchant and manufacturer.

Rhode Island, USA, the smallest *New England colony and state, on the southern coast of the region around Narragansett Bay. Originally settled in 1636 by such dissidents from Massachusetts as Roger *Williams and Anne Hutchinson, it was chartered as a colony by Parliament in 1644 and became a Quaker and Baptist refuge. Its internal disunity was pacified under Samuel Cranston's governorship (1698–1727) and its eastern Narragansett country became a prosperous agricultural and horse-breeding area, while Newport and Providence thrived on trade, including slave-trading and rum distilling. It was occupied by British troops and ships during the War of *Independence (1776–9), but they withdrew before a French fleet arrived. After the American Revolution, it remained independent in its views. It did not send representatives to the *Continental Congress and had to be greatly pressured to ratify the *Constitution.

Rhodes, an island off southern Asia Minor. It was first settled by the *Dorians, who established three city-states—Ialysus, Lindus, and Camirus. These sent out colonies, endured tyrannies, and became subject to the Persians, before in the 5th century BC joining the *Athenian empire. In 412–411 they seceded and the subsequent war with Athens brought about a fusion of the three into the federal state of Rhodes. The island was well placed to benefit from trade, and the opening up of the Persian empire by *Alexander the Great meant that it became the most prosperous city-state in Greece. It survived a prolonged siege by Demetrius Poliorcetes in 305–304 and managed to sustain remarkably independent policies in the 3rd century. However ambivalent loyalties in the third *Macedonian War led Rome to declare Delos a free port, which drastically injured Rhodes' trade. Nevertheless it thrived, in a more modest way, under the Romans. Rhodes was part of the Byzantine empire until the capture of *Constantinople (1204). It was then held by *Genoa, and annexed by the emperor of Nicaea (*Nicaean empire). In AD 1309 the island was occupied by the *Knights Hospitallers of St John of Jerusalem, who were ejected in 1522 by *Suleiman I, and it became part of the *Ottoman empire.

Ricci, Matteo (1552–1610), Italian *Jesuit missionary. He was received at the court of the *Ming emperor Wanli in 1601, having arrived in southern China in 1583. He had made himself proficient in Chinese and always dressed in a Chinese scholar's robes. He interested the emperor in clocks brought from Europe, translated numerous books, among them the geometry of Euclid, into Chinese, and made a world map with China, the Middle Kingdom, at its centre. He established beyond doubt that China was Cathay, the land *Marco Polo had described. His tolerance and scholarship impressed influential Chinese, some of whom were converted. The emperor gave land for his tomb in Beijing.

Richard I (the Lionheart) (1157–99), King of England (1189–99). He was the third son of *Henry II and

A woodcut of 1496 showing the Knights Hospitallers fighting a naval action off **Rhodes**. The cross of St John flies on their flags and sails, while the defenders hurl rocks and shoot firearms at the attackers as they scale the fortress.

*Eleanor of Aquitaine. He was made Duke of Aquitaine at the age of 12, and in 1173 joined his brothers in their rebellion against Henry. Richard spent only six months of his life in England. Soon after his coronation he left with the Third *Crusade, and in *Palestine in 1191 he captured Acre and defeated *Saladin at Arsuf. The following year, after concluding a three-year truce with Saladin, he set out overland for England. In Vienna he was captured by Duke Leopold of Austria and imprisoned by Emperor Henry VI until in 1194 England paid a ransom of £100,000 (1194). During his absence his brother *John had allied himself with *Philip II of France against him. Within a few weeks of his return he began the military campaigns for the defence of Normandy against Philip which led eventually to his death whilst attacking the castle at Châlus. Richard's military exploits earned him the nickname *Coeur de Lion* (French, 'Lionheart'). However, his absence abroad led to a growth in the power of the barons, a problem inherited by *John.

Richard II (1367–1400), King of England (1377–99). He was the only son of *Edward the Black Prince, and the grandson of Edward III, whom he succeeded at the age of 10. In 1381 his courage helped prevent disaster in the *Peasants' Revolt, but in the next few years he had to face a more direct threat to his power from a group of magnates (*Lords Appellant) led by his uncle *Thomas of Woodstock, Duke of Gloucester. During the session of the Merciless Parliament (1388) they had Richard's chief supporters executed or imprisoned, and it was only in 1397 that he was able to strike back at them by punishing the Lords Appellant. His attempt to impose his personal rule upon England alienated support and enabled Henry of *Bolingbroke to seize the throne with comparative ease in 1399. Richard abdicated and died a few months later in prison. He was a sensitive man but temperamentally unbalanced and incapable of firm rule.

Richard III (1452–85), Duke of Gloucester (1461–85), King of England (1483–5). He was a younger brother of *Edward IV and the eleventh child of Richard Plantagenet, Duke of *York. Tudor propaganda, notably the biography by Thomas *More, portrayed him as a monster from birth, always a traitor to his own family, but as Duke of Gloucester he served Edward faithfully and was an able soldier and a capable administrator in northern England. Upon the accession of his young nephew *Edward V, he became Protector of England: the council over which he presided included his enemies, the Woodvilles, and he gained in popularity from striking at their power. His usurpation of the throne in June 1483 caused no outright hostility, but in all save the *Yorkist north of England there was revulsion when it came to be believed that he had had Edward V and his brother killed in the Tower of London. (When and how they died remains a mystery.) He had long expected a further invasion of England by the *Lancastrian Henry Tudor, but when a battle was fought at *Bosworth Field in August 1485 he was defeated and killed because he had lost the support of his army.

In Shakespeare's play *Richard III* he is portrayed as a villain and a hunchback, but there is no evidence to support the tradition that he was physically deformed.

A 15th-century illuminated manuscript showing **Richard II** imprisoned at Pontefract by Henry of Bolingbroke who succeeded him as Henry IV. Richard died in prison, probably from starvation, in 1400. (British Library, London)

Richard of York *York, Richard Plantagenet, 3rd Duke of.

Richelieu, Armand Jean du Plessis, duc de (1585-1642), cardinal, the chief minister of Louis XIII of France from 1628 to 1642, and architect of royal absolutism in France and of French leadership in Europe. He was consecrated Bishop of Luçon in 1607, and entered the council of the regent, Marie de Medici c.1616. He was temporarily ousted after Louis XIII's assumption of power, but had gained the king's confidence by 1624, and for the rest of their lives the two men worked together. On the domestic front he destroyed the political power and military capacity of the *Huguenots, and continued *Henry IV's policies of centralized absolutism. He alienated the Catholic 'Dévot' party, the high nobility, and the judicial hierarchy, but managed to survive a series of plots and conspiracies. His excessive taxation of the lower orders created an endemic state of revolt in many provinces. The money was needed to finance France's active anti-Habsburg foreign policy, and during the *Thirty Years War he subsidized the Protestant Dutch, Danes, and Swedes to fight against the Habsburgs before France declared war on Spain in 1635; he also supported anti-Spanish revolts in Catalonia and Portugal (1640). His successor was *Mazarin, whom he had trained to continue his policies.

Ridolfi Plot (1571), an abortive international Catholic conspiracy, intended to put *Mary, Queen of Scots on the English throne in place of *Elizabeth I. It took its name from Roberto di Ridolfi (1531-1612), a well-connected Florentine banker who had settled in England. The

Detail of a portrait by the Flemish painter Philippe de Champaigne of his patron, Cardinal de **Richelieu**. (Louvre, Paris)

English Catholics were to rise under Thomas Howard, 4th Duke of Norfolk. Then, with papal finance and Spanish military aid, Elizabeth was to be deposed in favour of Mary, who would marry Howard. Elizabeth's intelligence service uncovered the plot and its leading figures were arrested.

Riebeeck, Jan van (1610-77), governor of Cape Town (1652-62) and the founder of European settlement in South Africa. He was sent by the Chamber of Amsterdam to found a refreshment station at Cape Town for Netherlands vessels sailing to the East Indies, and arrived there on 6 April 1652 with 125 men. As commander he organized a simple government, and ruled for ten years. He followed *Dutch East India Company policy in regarding the station as an opportunity for profit rather than for colonization. Growing corn and vines and rearing cattle were permitted, but he discouraged trade with Africans. In 1662 he was transferred to *Malacca.

Rienzo, Cola di (1313-54), Italian popular leader. As spokesman for the Roman populace he attempted to lead a revolution in 1347. Taking the title of tribune, he sought fiscal, political, and judicial reform. Rejected by the papacy, and the powerful Orsini and Colonna lords, he was deposed in November 1347 and excommunicated. In 1350 he gained the support of the new pope, Innocent VI, who encouraged him to restore papal authority in Rome. He returned there in triumph in 1354 only to be killed by the mob. He was a symbol of Italian unity in the 19th century.

Rights of Man and the Citizen, a declaration of the guiding principles of the *French Revolution which was approved by the Constituent National Assembly in August 1789. *Jefferson, the American minister in Paris, was consulted and the Declaration was influenced by the American example as well as by the *Enlightenment. It set forth in clear language the principles of equality and individual liberty: 'Men are born and remain free and equal in rights'; 'No body of men, no individual, can exercise authority which does not issue expressly from the will of the nation'. Civil and fiscal equality, freedom from arbitrary arrest, freedom of speech and the press, and the rights of private property were affirmed. Later the Revolution denied many of these rights, but the declaration ensured an initial welcome for the French Revolutionary armies in many European countries and was the charter of European liberals for the next half-century.

Riot Act, an Act to prevent civil disorder passed by the British Parliament in 1715. *Jacobite disturbances following George I's accession alarmed the new Whig government, and gave the excuse for new legislation, which applied to any group over a dozen. The Act made it a serious crime for anyone to refuse to obey the command of lawful authority to disperse; thus the Act imposed upon the civil magistrates the dangerous duty of attending a riot, or a large meeting which might become riotous, and reading the Riot Act. Frequent use was made of the Act in the 18th century. Its use declined in the 19th century and it was repealed in 1911.

Rizzio, David (1533-66), Italian-born secretary and adviser to *Mary, Queen of Scots. He entered service at court in 1561 and by 1564 he had become her Secretary: he possibly arranged her marriage to his friend *Darnley. By March 1566 'Seigneur' David's arrogant monopoly of power, combined with fears of his being a papal agent, led to his assassination. Darnley, who suspected him of adultery with Mary, was involved in the plot.

Roanoke Island, the first English colony in North America, in Albemarle Sound off the north coast of North Carolina. A group of settlers financed by *Raleigh landed there in 1585, but were evacuated, just before relief arrived from England, by *Drake returning from the Caribbean. A second colonizing group was sent in 1587 under John White, who then had to go back to England to obtain further supplies. His return from England was interrupted by the Spanish Armada, and in 1590 he found the settlement empty. The 'lost colony' has been immortalized by White's paintings of local Indians and by Thomas Harriott's *Report* (1588).

Rob Roy (Robert MacGregor) (1671–1734), Scottish brigand. He was given a commission by *James II in 1688, which he used as an excuse to plunder his neighbours, both by cattle-stealing and blackmail. Ruined by debt, he waged virtual war on the Duke of Montrose whom he blamed for his misfortunes. He was trusted by neither side during the *Fifteen Rebellion, and despite a temporary reconciliation with Montrose in 1722 he was arrested in 1727 and escaped transportation only by the timely arrival of a pardon.

Robert I (the Bruce) (1274–1329), King of Scotland (1306–29). He had a successful reign, inheriting a contested throne in a country partly occupied by the English, and leaving a securely governed kingdom to his son, *David II. He was fortunate in that he was matched by an ineffectual English king, *Edward II, over whom he won an important victory at *Bannockburn in 1314. In 1322 Edward attempted a fresh invasion of Scotland, but Bruce outmanoeuvred him and then invaded England as far south as Yorkshire, nearly capturing Edward himself. In the Treaty of Edinburgh (1328), *Edward III recognized Bruce's title and Scotland's independence from England, though this was only a temporary lull in Anglo-Scottish hostilities.

The Seal of **Robert the Bruce**, inscribed '[RO]BERTVS DEO RECTORE REX SCOTTORVM' (Robert Under God's Governance King of Scots).

A contemporary portrait of **Robespierre** at his desk by Louis Léopold Boilly c.1784 painted some years before he was elected to the States-General. (Musée des Beaux Arts, Lille)

Robert II (1316–90), King of Scotland (1371–90). He inherited the throne from his uncle, *David II, fairly late in life; as Robert the Steward, he spent most of his active years in virtual opposition to David, leading the Scottish nationalists against the invading armies of *Edward III while David was in exile in France. Shortly after his accession Robert successfully concluded a treaty with *Charles V of France which reaffirmed the Franco-Scottish alliance.

Robert III (c.1337–1406), King of Scotland (1390–1406). He was the eldest son of *Robert II, and was christened John but assumed the name Robert on his succession. He had been severely injured by a horse-kick, so power was exercised by a regent—first his brother Robert, Earl of Fife, and then his son David, Duke of Rothesay.

Robespierre, Maximilien François Marie Isidore de (1758–94), French Revolutionary and 'incorruptible' leader of the *Jacobin Club. He trained as an advocate but resigned his post as a judge in 1782 and for a time lived the life of a dandy in Arras. He was a member of the *States-General called in 1789, and became an influential voice in the newly formed *National Assembly. Two years later he joined the Jacobin Club and began a long and bitter struggle with the powerful *Girondins. He was popular with the Paris Commune, though he opposed its violent methods. He argued for the king's execution and the following year he was instrumental in the overthrow of the Girondins and was elected on to the *Committee of Public Safety. He did not instigate the *Terror or its machinery, but supported moves to rid the Revolution of its enemies. By the end of March 1794 his power seemed secure, *Hébert and *Danton had been executed, and the Terror intensified. On 28 October, however, he was shouted down in the Convention, arrested, and executed the following day.

Robin Hood, a legendary figure, traditionally represented as the unjustly outlawed Earl of Huntingdon, fighting from Sherwood Forest in central England, the corrupt administration of King *John and his local officer the sheriff of Nottingham. The earliest literary evidence of Robin appears in William Langland's poem *Piers Plowman* (written *c*.1367-86). The fullest account of his exploits, given in the late medieval *Lytell Geste of Robyn Hode* (printed *c*.1495), locates him in Barnsdale, Yorkshire. This region coincides with that recorded in a Yorkshire pipe-roll of 1230 which mentioned an outlaw by that name. Robin's sympathies for the plight of the gentry, not just the peasantry, are best understood in the context of 14th-century disenchantment with royal justice: it has been suggested that the ballads actually describe events of *Edward III's reign.

Rockingham, Charles Watson-Wentworth, 2nd Marquis of (1730-82), British statesman and leader of the political faction known as the Rockingham *Whigs. Most of his supporters were originally followers of the Duke of *Newcastle, but from the mid-1760s they transferred their allegiance to Rockingham. He formed a ministry in 1765-6, which repealed the American *Stamp Act and the controversial cider excise. He and his supporters strenuously opposed Lord *North and the American War of *Independence, and argued for financial reforms, which the second Rockingham administration (1782) undertook. On his death in office the Rockingham Whigs split into further factions, of which the most important formed the basis of the new Whig party which was evolving at the end of the 18th century.

Rodney, George Brydges Rodney, 1st Baron (1719-92), British admiral. He gained his early naval expertise with *Hawke at Finisterre in 1747, and at Le Havre in 1759, where he destroyed the French flotilla poised to invade England in the *Seven Years War. His greatest victory was at the battle of *Les Saintes, 1782, in the West Indies, where he restored British supremacy at sea in the closing stages of the American War of *Independence.

Roger II (Guiscard) (*c*.1095-1154), King of *Sicily (1130-54). The *Norman expansion into southern Italy and Sicily was begun by the brothers Robert and Roger *Guiscard, initially in defiance of the pope, but subsequently with his grudging cooperation. The Treaty of Melfi (1059) empowered them to take south Italy from the Greeks and Sicily from the Muslims and by the time of Roger Guiscard's death, Sicily was in Norman hands. His son Roger II effectively ruled Sicily from 1113 but had to assert control over the anarchic Norman barons who threatened his rule, especially when backed by the pope in 1129. From 1130 Roger supported the *antipope's cause; despite excommunication by Pope Innocent II and internal revolt, he consolidated his power by 1140. He took Malta, Corfu, and many cities on the Greek mainland as well as controlling most of the land in North Africa between Tripoli and Tunis. In 1140 he issued a revised code of laws and in his later years he ruled one of the most sophisticated governments in Western Europe. His court at Palermo enjoyed a high artistic and scholarly reputation.

Rogers, Robert (1731-95), American frontier soldier.

He formed the Rogers Rangers, a force of 600 New England frontiersmen who fought with great bravery in the *French and Indian War at Lake George (1758) and the capture of Quebec (1759). He served in the relief of Detroit in *Pontiac's Rebellion. His loyalty to the Revolution was questioned and Washington had him arrested as a spy in 1776. Such treatment converted him into a loyalist, and in 1780 he went to live in London.

Roland de la Platière, Marie-Jeanne Philipon (1754-93), French Revolutionary, whose *salon* became the centre of the *Girondin party. Through her political influence her husband, Jean-Marie Roland (1734-93), was appointed Minister of the Interior in 1792. She helped him write a letter, read out in the council, which criticized the actions of *Louis XVI. He was dismissed in July 1792 but returned to power in August following the overthrow of the monarchy. However, the following year he was forced to resign and, as the power of the Girondins weakened, Madame Roland was arrested. She was executed on 8 November 1793 and two days later her husband, hearing of her death, committed suicide.

Rollo (*c*.860-931), leader of a band of *Vikings which invaded north-western France. In 912 as Duke Robert he accepted Normandy as a duchy from the French king Charles III, and was baptized, but remained quite independent of French authority. He married a French princess, gave parcels of land in Normandy to his followers, and began the long chapter of *Norman influence on Europe.

Roman Britain (AD 43-410), the period when most of Britain was part of the *Roman empire. Britain was first visited by the Romans under Julius *Caesar during the *Gallic wars. It was then the home of Gallic tribes and later a refuge for defeated allies of *Vercingetorix. *Claudius invaded Britain in AD 43, attracted by the island's minerals and grain. At first the Belgic tribes were subdued up to the *Fosse Way. The frontier was then extended into native Celtic territories and established by the building of *Hadrian's Wall. Native culture absorbed Roman ways: enlarged former tribal capitals adopted a Roman lifestyle. Army veterans settled there after discharge, as did traders, scholars, craftsmen, and soldiers from all parts of the Roman empire. Universally acknowledged Christian bishoprics were established. *Roman villas, *Roman roads, and titles abounded, but little Latin was spoken and the people remained essentially Celtic. In 406 and 409 the Britons rebelled against Roman rule. The Romans withdrew from Britain in 410. The period of Roman decline and the early history of the Saxon kingdoms remains obscure.

Roman Catholic Church, the part of the Christian Church (*Christianity) ruled by the Bishop of Rome (*papacy). In the early Church the Bishop of Rome, as the successor of St *Peter, exercised a limited measure of authority over all Christians. With the *East-West Schism however, his authority was restricted to the Western Church. At the *Reformation this authority was further restricted by the secession of the *Protestant churches. It was at this time that the term Roman Catholic Church came to be used commonly, initially by Protestants somewhat as a prejudiced term, to imply that the Roman Catholic Church was at most only one branch

Roman empire: its expansion to AD 117

Rome expanded from a small settlement in the 6th century BC to rule most of the known European world by the early 2nd century AD. Having established control over Italy by c.260 BC, Rome expanded to the east and west. By 133 BC all the territory formerly held by Carthage in the western Mediterranean and north Africa had been absorbed, together with Macedonia and Asia Minor in the east. Expansion continued for a further 150 years, the empire reaching its greatest extent in AD 117, under Trajan. His successor, Hadrian, attempted to consolidate the empire behind fixed frontiers, drawing back to the Euphrates in the east and building the famous wall in Britain. However, less than fifty years later the decline of the empire had begun.

of the Christian Church. But the term was also acceptable to many (Roman) Catholics, inasmuch as it asserted the primacy of the pope over Christians. Despite the reductions in its allegiance, the Roman Catholic Church remained much the largest of the Christian churches.

Roman civil wars, conflicts which afflicted the last century of the republic (88 BC–c.28 BC) and led to the inevitable institution of the unchallenged authority of one man, the Principate. Political life in Rome was unsettled from the period of *Sulla's dictatorship, and the *Catiline conspiracy (64–63 BC). Rivalry between the republican military leader *Julius Caesar and *Pompey began after the collapse of their alliance. Caesar defeated the Pompeian army in Spain at Ilerda (49) and Pompey himself at Pharsalus (48); he won further victories in Asia and Africa. *Cato's suicide (46) signified the collapse of the republican cause. On his return to Rome Caesar was made dictator and proceeded to virtual sole rule at Rome. His plans for safeguarding the empire by military expeditions against Dacia and Parthia were cut short by outraged republican traditionalists who murdered him in 44. Further civil wars followed. Initially *Octavian supported by the republican party struggled against *Mark Antony. In 43 Antony, Octavian, and Lepidus formed a coalition whose forces defeated the republicans led by Brutus and Cassius at *Philippi. Antony meanwhile joined forces with Cleopatra and was defeated by Oc-

tavian at *Actium. The Roman world was united under the sole leadership of Octavian, who annexed Egypt.

In AD 68 civil war broke out in the empire in the struggle for succession after *Nero's death. Galba was proclaimed emperor from Spain; he entered Rome in September but was murdered and succeeded by Otho; meanwhile Vitellius was proclaimed emperor in Germany and Otho committed suicide. *Vespasian then invaded Italy and took the throne, making 68–9 'the year of the four emperors'. This crisis period was followed by the settled rule of Vespasian.

Roman empire, strictly speaking the period when the Roman state and its overseas provinces were under the rule of an emperor, from the time of *Augustus (27 BC) until AD 476. It was divided in AD 375 by Emperor Theodosius into the Western and Eastern empires. The term is more commonly used of territories under Roman rule during both the republic and the empire.

The city of Rome gradually gained power from the time of the Tarquins (6th century BC), subduing the Etruscans, Sabines, Samnites, and Greek settlers, and by the mid-2nd century BC, controlled Italy. It came into conflict with *Carthage in the western Mediterranean and with the Hellenistic world in the east. Success in the *Punic wars gave Rome its first overseas possessions in Sicily (241), Spain (201), and north Africa (146), and the Macedonian Wars eventually left Rome dominant in

Macedonia, Greece, and parts of Asia Minor. Syria and Gaul from the Rhine to the Atlantic were added by the campaigns of *Pompey and Julius *Caesar and Egypt was annexed in 31 after the battle of *Actium. *Augustus planned to consolidate the empire within natural boundaries, but in AD 43 *Claudius invaded Britain. *Trajan, in 106, made Dacia a province in response to raids across the Danube, though it was abandoned in 270. His annexation of Mesopotamia was very brief (114–117).

This vast empire was held together by secure communication, and internal peace maintained by the army (*Roman legion). Fleets kept the sea safe for shipping and a network of *Roman roads, built to move troops quickly, facilitated trade, personal travel, and an imperial postal system. The development of a single legal system and the use of a common language (Latin in the west, Greek in the east) helped maintain unity. Roman cities flourished throughout the empire, assisted by efficient water and drainage systems. Roman influence and trade spread even further reaching India, Russia, south-east Asia, and through the *Silk Route, China.

The success of the empire also led to its downfall, its sheer size contributing to its collapse, exacerbated by power struggles and the invasion of land-hungry migrating tribes. Rome was sacked by the Visigoths in 410, Carthage was conquered by the Vandals in 455, and in 476 Romulus Augustulus, the last emperor of the Western empire was deposed. The Eastern empire (*Byzantine empire) lasted until 1453. *Gibbon's *The History of the Decline and Fall of the Roman Empire* (1776–81) is the classic account of the imperial disintegration.

Roman legion, a division of the army in ancient Rome. Legions evolved from the citizen militia which equipped itself in times of crisis for defence of the state. During the Second *Punic War *Scipio reorganized the battle array and improved the army's tactics. Under *Marius, with the recruitment into the army of men of no property, a professional army appeared and new training methods could be introduced. Ten cohorts with standards formed a named and numbered legion with an eagle standard. The cohorts, divided into six centuries commanded by a centurion, became the main tactical unit of the army. Cavalry and auxiliaries supported each regiment.

*Augustus established a standing army to man the frontiers of the empire. There appears to have been twenty-eight permanent legions, each having a number and an honorific title. *Severus added three legions; *Constantine increased the number but limited them to a thousand men to allow flexibility and to avoid mutiny. He also placed them under equestrian prefects instead of the traditional senatorial legates and placed a Christian symbol on their standards. On retirement a veteran in the early days earned a land grant in a 'colony' where he continued to act as a Romanizing and pacifying influence throughout the empire, but from the time of Augustus it was more useful for him to receive a payment. Many nevertheless settled in the area where they had served, thus effectively 'colonizing' it.

Roman religion, the religion of the Roman republic and empire. In its developed form it came to have much in common with *Greek religion, although it contained elements of Etruscan and other native Italian regional beliefs and practices. The Roman gods mirrored those of Greece: Jupiter = Zeus, Juno = Hera, Neptune = Posei-don, Minerva = Athene, Diana = Artemis, Mars = Ares, Mercury = Hermes, and so on; but it too had its array of minor deities. Romans also possessed domestic shrines of the spirits of the household (Lares and Penates). Religious observances and rites were very closely tied in with politics; from Julius *Caesar onwards, the practice of according divine honours and worship, even temples, to deceased rulers became common (though not all were deified after their deaths), and cults of 'Rome and the Emperor' spread through Italy and the provinces, largely to provide a public and visible focus of loyalty to the regime. Although *Christianity became the official religion of the empire from the early 4th century AD, pagan beliefs and practices proved tenacious in many areas, especially away from the cities, and in many places the Christian Church had often to take over Roman festivals and hallowed shrines or sites under a new guise.

Roman republic (Latin *respublica*, 'common wealth'), the political form of the Roman state for 400 years after the expulsion of *Tarquin. The rule of a sole monarch yielded to the power of a landed aristocracy, the *patricians, who ruled through two chief magistrates or consuls and an advisory body, the *Senate.

The city of Rome could operate as a 'public concern' as long as the small landed aristocracy managed the state. The simplicity of *Cincinnatus or *Cato was an ideal of statesmanship, leaders as ready as their citizen militia to return to their farms after holding power in a crisis. But, with overseas expansion, generals had to be given power to deal with problems abroad, most notably the *Scipios, *Marius, *Sulla, *Pompey, *Julius Caesar, *Mark Antony, and Octavian (*Augustus). They depended on the personal loyalty of their troops. This substantial independence threatened republican tradition with its corporate government and brief periods of high office for individuals in rotation. Eventually the generals simply ignored the law which required generals to lay down their commands on returning to Italian soil. The last of these commanders-in-chief, Octavian, achieved a settlement which appeared to combine republican institutions with personal military power. The Roman empire succeeded the republic.

Roman roads, a systematic communications network originating in the Italian peninsula joining Rome to its expanding empire. The *Appian Way was the first major stretch, leading into *Samnite territory. The Via Flaminia, constructed in 220 was the great northern highway to Rimini. For travellers landing from Brindisi the Egnatian Way continued overland through Greece and on to Byzantium. By the 1st century AD three roads crossed the Alps and the Domitian Way went from the Rhône valley to Spain. Every province had such roads which served military and commercial purposes. In Britain major highways fanned out from Londinium (London), some now known by their Anglo-Saxon names: *Watling Street and *Ermine Street. Designed with several thick layers they were drained by side ditches, and maintained by engineers.

Roman Senate, the assembly of the landed aristocracy and *patricians which originated in the royal council of the kings of Rome. Entry widened to include those of *plebeian origin by the late 4th century BC. A membership of six hundred established by *Sulla was standard

Roman villa

A Roman villa such as Chedworth in Gloucestershire would have been a large country mansion, surrounded by outbuildings and barns, and set in an estate. Chedworth was originally built in the early 2nd century AD, but it was rebuilt and extended until the end of the 4th century. It had an elaborate system of plumbing and underfloor heating and there were two bath suites, but relatively few living and sleeping rooms. A number of the rooms had mosaic floors.

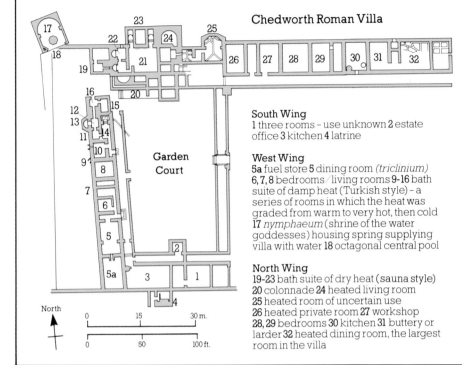

Chedworth Roman Villa

South Wing
1 three rooms – use unknown 2 estate office 3 kitchen 4 latrine

West Wing
5a fuel store 5 dining room (triclinium) 6, 7, 8 bedrooms / living rooms 9-16 bath suite of damp heat (Turkish style) – a series of rooms in which the heat was graded from warm to very hot, then cold 17 nymphaeum (shrine of the water goddesses) housing spring supplying villa with water 18 octagonal central pool

North Wing
19-23 bath suite of dry heat (sauna style) 20 colonnade 24 heated living room 25 heated room of uncertain use 26 heated private room 27 workshop 28, 29 bedrooms 30 kitchen 31 buttery or larder 32 heated dining room, the largest room in the villa

This detail is from a large Roman floor mosaic dating from c. 2nd–4th century AD, which was found at a villa at La Chabba in Tunisia. The mosaic is entitled 'The Triumph of Neptune and the Four Seasons'. The circular centre shows Neptune in a chariot and each corner depicts one of the seasons. Here the autumn fruits are being gathered. (Bardo Museum, Tunis)

although it rose to nine hundred in *Caesar's time. This advisory body consisted of hereditary (patrician) and life (conscript) members, the latter being ex-magistrates. It was summoned by the consuls as chief magistrates and passed decrees which were ratified by the people in assembly. It was expected that all magistrates would submit proposals to the Senate before putting them to the people. This procedure began to be flouted from the time of the *Gracchi onwards. Its power was real but informal, based on prestige and wealth. Even the emperors made at least the token gesture of consulting the 'Fathers'. Until the 3rd century AD all bronze coinage carried the mark 'By Consultative Decree of the Senate'.

Roman villa, originally a rural dwelling with its associated farm-buildings, forming the centre of an agricultural holding. Such holdings could vary from small farms chiefly dependent on family labour to large estates worked by slaves or tied-labourers who were effectively *serfs. From the 2nd century BC, villas of increasingly sophisticated design and elegance, usually constructed round a courtyard and often sited to enhance a landscape or command a vista, began to be built as suburban or country or seaside houses for the rich, some of whom owned several in different districts, the attendant estate-land being cultivated by tenants or under the supervision of an estate-manager (vilicus). The villa-type of dwelling and estate spread to the provinces, and some sixty villas are known in Britain. In post-Roman times the *manor-house and manor (and their equivalents elsewhere in western Europe) replaced the villa and its dependent estate.

Romanov, the ruling house of Russia from 1613 until the Revolution of 1917. After the Time of Troubles (1604–13), a period of civil war and anarchy, Michael Romanov was elected emperor and ruled until 1645, to be followed by Alexis (1645–76) and Fyodor (1676–82). Under these emperors Russia emerged as the major Slavic power. The next emperors established it as a great power in Europe: *Peter the Great (1689–1725) and *Catherine the Great (1762–96) were the most successful of these rulers.

Rome, a city on the River Tiber, the capital of Italy and bishopric of the pope. It was at the heart of the *Roman empire. Several traditions surround its foundation on seven low hills. One legend said that the twins, Romulus (after whom the city was supposedly named) and Remus, suckled by a wolf, began the first settlement on the Palatine Hill. The date 753 BC became accepted and is well supported by archaeological excavation. *Etruscan remains, dating from the *Tarquins (c.650–500 BC) have been discovered. The expansion of provinces under the republic and early empire brought wealth to Rome. *Augustus was said to have turned a city of brick into one of marble, and successive emperors added palaces, arches, columns, and temples. *Nero burnt much of it, hoping, it was said, to rebuild it and rename it after himself. As the empire declined, Rome was attacked by *Goths and *Vandals. By then it was politically overshadowed by *Constantinople, the capital of the Eastern Roman empire. During the Middle Ages Rome emerged as the seat of the papacy and the capital of Western Christianity. It became a centre of the *Renaissance and was largely rebuilt in the Baroque style

Wars of the Roses

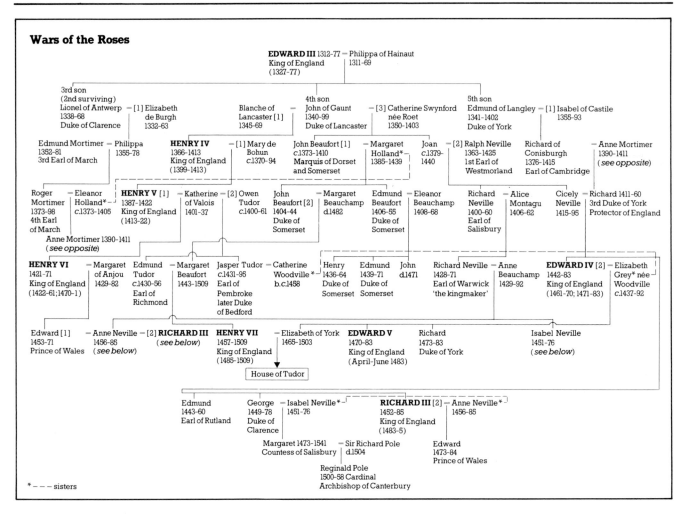

* – – – sisters

in the 17th century. The city was sacked in 1527 and again in 1798.

Root-and-Branch Petition (1640), a document drawn up by a London alderman, Isaac Pennington, calling for the abolition of episcopal government (government of the church by bishops), 'with all its dependencies, roots and branches'. The 15,000 signatures testified to the Puritan radicalism of some London citizens. A Root-and-Branch Bill was introduced into the *Long Parliament in May 1641, and was supported by *Pym. Eventually it was dropped in favour of more urgent measures such as the *Grand Remonstrance.

Roses, Wars of the (1455–85), a protracted struggle for the throne of England, lasting for thirty years of sporadic fighting. These civil wars grew out of the bitter rivalry between two aspirants to the throne—Edmund *Beaufort (1406–55), Duke of Somerset, of the House of Lancaster (whose badge was a red rose), and Richard, 3rd Duke of *York (whose badge was a white rose); the former was a close supporter of *Henry VI and *Margaret of Anjou, while Richard of York became their opponent. In 1455 Richard gained power by winning the first battle of *St Albans; a whole series of private enmities and disputes was now absorbed into a bitter and openly fought civil war. Richard of York was killed at the battle of *Wakefield (1460), and Henry VI's supporters, the *Lancastrians, won a further victory at the second battle of St Albans (February 1461), yet their hesitations

allowed Richard's son Edward to gain the throne a month later as *Edward IV, the first *Yorkist king of England. In September 1470 a Lancastrian invasion restored Henry VI to the throne (although power was effectively exercised by 'the kingmaker', Richard Neville, Earl of *Warwick), but in April 1471 Edward regained it by the victory of *Barnet. Most of the remaining Lancastrian leaders were killed at *Tewkesbury in May 1471, but the struggle ended only in 1485 when Henry Tudor defeated *Richard III at *Bosworth Field. *Henry VII married Edward IV's eldest daughter, Elizabeth of York, in order to unite the two factions. The wars weakened the power of the nobility and after a bid for the throne from Lambert *Simnel in 1487, there were no serious challenges to the *Tudor dynasty.

Rosetta stone, a piece of black basalt bearing inscriptions which provided the key to the deciphering of Egyptian *hieroglyphics. It was found in 1799 by a French soldier during Napoleon's occupation of Egypt, and contained three inscriptions, in Greek, in Egyptian demotic, and in Egyptian hieroglyphics. Comparative study of the three texts, which date from 196 BC, was undertaken by Thomas Young and Jean-François Champollion, the latter finally unlocking the secrets of hieroglyphics in 1821–2. The stone is housed in the British Museum in London.

Rosicrucian, a member of certain secret societies who venerated the emblems of the Rose and the Cross as

The **Rosetta stone**, found near Rosetta (Rashid) in Egypt, bears a decree drawn up in 196 BC by the priests of Memphis in honour of Ptolemy V (Epiphanes). On it can be seen the three inscriptions. The top 14 lines are in Egyptian hieroglyphics; the middle 32-line section is in Egyptian demotic; and the bottom portion of 54 lines is in Greek. (British Museum, London)

symbols of Jesus Christ's resurrection and redemption. Rosicrucians claimed to possess secret wisdom passed down from the ancients, but their origin cannot be dated earlier than the 17th century. The anonymous *Account of the Brotherhood* published in Germany in 1614 may well have launched the movement. It narrated the tale of a mythical German knight of the 15th century, Christian Rosenkreutz, who travelled extensively to learn the wisdom of the East, and then founded the secret order. Robert *Fludd subsequently helped to spread Rosicrucian ideas. In later centuries many new societies were founded under this name.

Roskilde, Treaty of (1658), named after a port in eastern Denmark. After the defeat of Frederick II of Denmark by Charles X of Sweden, this treaty expelled the Danes once and for all from the Swedish mainland: they surrendered Halland, Scania, Blekinge (provinces), and the island of Bornholm, and the Norwegian territories of Trondheim and Bohuslän.

Roundheads, the term applied to Puritans and Parliamentarians during the *English Civil War. It originated as a term of abuse, referring to the Puritans' disapproval of long hair and their own close-cropped heads. Roundhead strength during the Civil War lay mainly in southern and eastern England.

Rousseau, Jean-Jacques (1712–78), philosopher and writer, born in Geneva. He led an unsettled life, sometimes aided by benefactors (including *Hume), sometimes occupying humble situations, as footman or music-master. He came to the public notice by the works he wrote in which he expounded his revolt against the existing social order. Believing in the original goodness of human nature, he considered that the rise of property and human pride had corrupted the 'noble savage'. His most important work, the *Social Contract* (1762) expounded the theory that society is founded on a contract: the people sacrifice their natural rights to the general will in return for protection. These views influenced the American Declaration of Independence and were acclaimed during the French Revolution.

Royal African Company, a trading company to which *Charles II granted a monopoly charter in 1672. It built forts on the west African coast (the 'Gold Coast'), and trade was principally in slaves and gold. It lost its monopoly of trade with England by an Act of 1698; in 1750 it was succeeded by the African Company of Merchants. Its castles or forts were trading posts, fortified as much because of quarrels between Europeans as because of attacks from the local Fante chiefdoms, and also against pirates.

Royal Society, the oldest scientific society in Great Britain, founded in 1660. *Charles II granted it a royal charter in 1662. Its *Philosophical Transactions* (1665–) was the first permanent scientific journal. Among its earliest members were Boyle, Hooke, Pepys, Wren, and Newton,

A portrait by a contemporary artist of the French philosopher and writer Jean-Jacques **Rousseau** at Ermenonville, near Paris, where he moved shortly before his death in 1778. (Musée Carnavalet, Paris)

whose *Principia* was published in 1687 with the active encouragement of the Society. Among more literary members were Dryden, Evelyn, and Aubrey. Political and religious topics were excluded from its discussions and debates, and in 1848 the Society became wholly scientific: only those who had made a distinguished contribution to the sciences were eligible for election as Fellows of the Royal Society (FRS).

Rozvi empire, an empire in East Africa named from a Karanga clan of the Shona which established the authority of the *Mwene Mutapa in the 15th century. Probably originally spiritual leaders, and then military rulers, by 1480 the Rozvi occupied all of present-day Zimbabwe and Mozambique. After *c*.1500 the central and southern provinces broke away under *Changamire, while the ports were subject to *Kilwa. The Rozvi controlled gold-mining in the interior, and for a time successfully warded off Portuguese attempts to conquer them, but in 1629 the Mwene Mutapa acknowledged Portuguese suzerainty. The empire was finally broken up by the Ndebele in the 1830s.

Rubicon (literally 'reddish coloured'), in Roman times a stream in north-east Italy which flowed into the Adriatic and formed part of the border between the province of Cisalpine Gaul and Italy. Julius *Caesar in 49 BC crossed it, refusing to lay down his command of troops, and thereby committing an irrevocable act of treason, since this was taken to indicate a desire for personal power.

Rump Parliament, the remnant of the English *Long Parliament, which continued to sit after *Pride's Purge (1648). In 1649 it ordered *Charles I's execution, abolished both monarchy and House of Lords, and established the *Commonwealth. Its members were mostly gentlemen, motivated by self-interest, and its policies were generally unpopular. Oliver *Cromwell expelled the Rump in April 1653. Six year later it was recalled to mark the end of the *Protectorate; in 1660 the members excluded by Pride were readmitted, and the Long Parliament dissolved itself in preparation for the *Restoration of the monarchy.

Rupert, Prince (1619–82), son of Frederick V, the Elector Palatine, and Elizabeth, Queen of Bohemia, and the nephew of *Charles I of England. He grew up in the Low Countries, and saw military action during the Thirty Years War. In 1642 he joined the *Cavaliers in England, and rose to be commander of the royal army, taking part in all the critical engagements of the first phase of the *English Civil War. A daring, skilful cavalryman, he was none the less outwitted at Marston Moor and Naseby. After he surrendered Bristol (1645), Charles I dismissed him. Later he commanded privateers against Commonwealth shipping (1649–52), with diminishing success. Under *Charles II he was given naval commands in the Second and Third *Anglo-Dutch wars.

Russia, a country in eastern Europe and northern Asia. In the 9th century the house of Rurik began to dominate the eastern Slavs, establishing the first all-Russian state with its capital at *Kiev. This powerful state accepted Christianity *c*.985, but decline had set in long before the Mongols established their control over most of European Russia from the 13th to the late 14th century.

The principality of *Muscovy developed in the shadow of Mongol overlordship, and in the 15th century it emerged as the pre-eminent state. Gradually it absorbed formerly independent principalities such as *Novgorod (1478), forming in the process an autocratic, centralized Russian state. *Ivan the Terrible was the first Muscovite ruler to assume the title of Tsar (Emperor) of all Russia (1547). During his reign the state continued its expansion to the south and into Siberia. After his death a period of confusion followed as *boyar families challenged the power of Theodore I (ruled 1584–98) and Boris *Godunov. During the upheavals of the Time of Troubles (1604–13), there were several rival candidates to the throne which ended with the restoration of firm rule by Michael Romanov. The *Romanov dynasty resumed the process of territorial expansion, and in 1649 established peasant *serfdom. *Peter the Great transformed the old Muscovite state into a partially Westernized empire, stretching from the Baltic to the Pacific, which subsequently played a major role in European affairs in the 18th century. Under the empresses *Elizabeth I and *Catherine II, Russia came to dominate *Poland, and won a series of victories against the *Ottoman Turks. In 1798–99 the Russians were allied with Great Britain, Austria, Naples, Portugal, and the Ottoman empire to fight against *Napoleon.

Ruyter, Michiel Adrianszoon de (1607–76), Dutch naval commander. He served under Maarten *Tromp in the First *Anglo-Dutch War and was the rival of Cornelius Tromp in the Second and Third wars. His most daring coup was in 1667 when, knowing that the English fleet was laid up for lack of money, he sailed up the rivers Thames and Medway, remained in the Chatham dockyard for two days destroying many ships, and made off with the *Royal Charles*, the English fleet's newest ship.

Rye House Plot (1683), a conspiracy of Whig extremists who planned to murder *Charles II of England and his brother James, Duke of York, after the failure of attempts to exclude James, a Roman Catholic, from the succession. The conspiracy takes its name from the house in Hertfordshire where the assassination was to have taken place. Of those accused of conspiracy, the Earl of Essex committed suicide and Lord Russell and Algernon Sidney were condemned to death on the flimsiest of evidence.

Ryswick, Treaty of (1697), the treaty which ended the *Nine Years War. *Louis XIV agreed to recognize *William III as King of England, give up his attempts to control Cologne and the Palatinate, end French occupation of Lorraine, and restore Luxemburg, Mons, Courtrai, and Barcelona to Spain. The Dutch were allowed to garrison a series of fortresses in the Spanish Netherlands as a barrier against France. Strasburg and some towns of Lower Alsace were the only acquisitions made since the Treaty of *Nijmegen which France retained.

S

Sabines, a tribe native to the foothills of the Apennine hills north-east of Rome. According to tradition they joined the earliest settlement of Rome in the 8th century BC. Legend relates how the Romans abducted the Sabine women during a festival; an army was raised to take revenge but the women appeared on the battlefield with new-born babies and the two sides were reconciled. They became Roman citizens only after conquest.

Sacheverell, Henry (1674-1724), English divine and political preacher. In 1709 he preached two sermons which attacked the Whig government's policy of religious toleration, one of the principles of the *Glorious Revolution. The House of Commons condemned the sermons as

Some 3,000 Huguenots perished at the hands of Parisian Catholic mobs in the **St Bartholomew's Day Massacre**. The massacre spread from Paris into other areas of France and led to the resumption of the French Wars of Religion. (Bibliothèque Nationale, Paris)

seditious and Sacheverell was impeached. He attracted a popular following, the crowds shouting 'High Church and Sacheverell' in his support, and his sentence was a nominal one, a temporary suspension from preaching. The Sacheverell episode was important within a political context; the Tories relied on the message of 'The Church in danger' to attract support from the conservative Anglican squirearchy against the Whig opposition.

Sadducee, a member of a sect which was a rival to the *Pharisees in ancient Israel. The Sadducees were drawn mostly from the rich landowners, and were naturally conservative, but by the time of *Jesus Christ they had declined considerably from their previously dominant position. Nevertheless they still held a number of priesthoods and were a powerful voice in the *Sanhedrin.

Safavid *Ismail I

St Albans, in Hertfordshire, the site of two major battles in the Wars of the *Roses which were fought near the town. The first marked the opening of the wars, with Richard, 3rd Duke of *York, attacking the *Lancastrian forces. He won the battle, and with it control of England, on 22 May 1455. The second battle of St Albans, on 16 February 1461, was a Lancastrian victory, in which *Margaret of Anjou's forces outmanoeuvred Richard Neville, Earl of *Warwick, and released *Henry VI from captivity; but her indecision, enabled Edward to slip through to London and gain the throne as *Edward IV.

St Bartholomew's Day Massacre (23-4 August 1572), an event which marked a turning-point in the *French Wars of Religion. The Catholic *Guise faction prevailed upon Catherine de Medici to authorize an assassination of about 200 of the principal *Huguenot leaders. Parisian Catholic mobs used these killings as a pretext for large-scale butchery, until some 3,000 Huguenots lay dead, and thousands more perished in the twelve provincial disturbances that followed. *Coligny was killed in Paris, while *Henry IV of Navarre saved himself by avowing the Catholic faith. Catherine's reputation as a mediator was damaged by the massacre.

Saintes, Les, battle of (1782), a Caribbean naval battle fought off the coast of Dominica, between a British fleet under *Rodney and a French fleet under de Grasse. Five French ships were captured and one sunk. The engagement was notable because Rodney, taking advantage of the wind, broke the French line in two places, a manoeuvre later perfected by *Nelson. This victory somewhat offset the British defeat in the American War of Independence and re-established Britain's maritime supremacy.

St John, Order of *Knight Hospitaller.

Saint-Just, Louis Antoine Léon Florelle de (1767-94), French Revolutionary leader. He was elected an officer in the National Guard when the *French Revolution began. His great loyalty to *Robespierre led to his appointment to the *National Convention in 1792. In his first speech he condemned Louis XVI and was instrumental in the overthrow of the *Girondins *Hébert and *Danton. He was the youngest member of the *Committee of Public Safety, organized the *Terror, and

A Persian portrait (Fatimid School) reputed to be of
Saladin, Muslim warrior and Ayyubid sultan of Egypt.

carried out many missions to enforce discipline in the
revolutionary armies. He was executed with Robespierre.

Saladin (Salah al-Din, literally 'the Welfare of the
Faith') (1138–93), the Kurdish founder of the Ayyubid
dynasty (1169–93). He reunited Egypt, Syria, and Mesopotamia under one rule. Periodic hostilities with the
Crusader kingdom of *Jerusalem led in 1187 to his rout
of Christian forces under Guy of Lusignan and the
capture of Jerusalem and Acre, thus provoking the Third
*Crusade. The Crusaders eventually recovered Acre in
1191, when *Richard I also defeated Saladin at Arsuf
but, unable to recover Jerusalem, made a truce. Saladin's
dynasty fell to the Mongols in 1250, but his fame as a
shrewd and chivalrous commander was incorporated into
European as well as Muslim legend.

Salamis, battle of (480 BC). A naval battle fought
in the Aegean Sea during the Greek–Persian wars.
*Themistocles, the Greek commander, lured the Persian
fleet of *Xerxes, the Persian king, into the narrow waters
between the island of Salamis and the mainland. The
outnumbered but nimbler and expertly handled Greek
triremes took full advantage of the confusion engendered
by the confined space to win a victory which offset the
earlier reverses at *Thermopylae and Artemisium.

Salem witch trials, the trial and execution of nineteen
witches in 1692 at Salem, Massachusetts. The hunt for
witches began when girls in the minister's household
believed themselves bewitched. Panic about devil worship
and resultant accusations spread through surrounding
towns until fifty people were afflicted and 200 had been
accused. Increase *Mather and Governor Sir William
Phips managed to halt the witch craze in October 1692.

Salic law, the legal code of the Salian *Franks, which
originated in 5th century Gaul. It was issued by *Clovis
(465–511) and reissued under the *Carolingians. It
contained both criminal and civil clauses and provided
for penal fines for offenders. It also laid down that
daughters could not inherit land and was later used in
France and in some German principalities to prevent
daughters succeeding to the throne. The *Valois kings of
France used it against *Edward III of England, who
claimed the French throne through his mother.

Sallust (Caius Sallustius Crispus) (86–34 BC), Roman
historian. He was a supporter of Julius Caesar and later
governor of Numidia in north Africa. He grew extremely
wealthy, reputedly from extortion. In retirement he wrote
a history of the Numidian War fought by *Marius against
*Jugurtha and an account of the Catiline conspiracy.
Both survive but only fragments exist of his five-book
history of Rome from 78–67.

Samaria, the ancient city built by King *Omri which
was the capital of the northern kingdom of Israel after
it divided from the southern kingdom of Judah (centred
on Jerusalem) after the death of King *Solomon. In 723
BC it was captured by the Assyrians who deported most
of the inhabitants. The descendants of those who remained
and those who were imported were disliked by the Jews,
more for historical and racial than for religious reasons
since they worshipped the same God.

Samnite wars, a succession of wars fought between the
southern neighbours of Rome, the Sammites and the
Latins. The first war (343–341 BC) was brief, but the
second (326–304) was more protracted. Roman troops
experienced the humiliation of having to walk like slaves
under yoke of spears after their defeat at the Caudine
Forks. The *Appian Way was built in 312 to assist
communications between Rome and the war area. Gauls
joined against Rome in the third of the wars (298–290),
but were defeated. The Samnites were consistently hostile
to Rome. They helped *Hannibal in the Second *Punic
War and revolted for the last time in the Social War of
90 after which they became allies of *Marius. *Sulla
crushed them and devastated their homelands.

Samuel (11th century BC), Israelite leader and prophet.
He was the last leader (judge) of the tribes of *Israel
before the establishment of hereditary kingship. Ruling
during a period of *Philistine domination of Israel,
he rallied his people in opposition to them. He was
instrumental in creating the monarchy by anointing
*Saul as the first King of Israel. Samuel provided Saul
with prophetic advice until they had a disagreement over
his priestly duties, and he then anointed *David as the
next king.

samurai (from the Japanese, 'those who serve'), warrior
retainers of Japan's *daimyo (feudal lords). Prominent
from the 12th century, they were not a separate class
until *Hideyoshi limited the right to bear arms to them,
after which they became a hereditary caste. Their two
swords were their badge. Their conduct was regulated
by *Bushido* (Warrior's Way), a strict code which emphasized the qualities of loyalty, bravery, and endurance.
Their training from childhood was spartan. Their ultimate
duty when defeated or dishonoured was *seppuku*, ritual
self-disembowelment.

　Samurai were active during the *Kamakura shogunate and with their masters were responsible for the
strife of the *Ashikaga period. They served during
Hideyoshi's Korean campaigns and fought for and against
the *Tokugawa at Sekigahara. During the long Tokugawa

Above: armoured **samurai** warriors at the burning of the Sanjo Palace in Kyoto during the Heiji Insurrection of 1159. This detail from a 13th-century picture scroll shows scenes from the battle and belongs to the *senki monogatari* ('stories of military exploits') type of scroll. (Boston Museum of Fine Arts)

Below: the east gate of the 'Great Stupa' at **Sanchi**, *c.* 250 BC. The superstructure is supported by massive blocks, sculptured with figures of elephants. Between the cross-bars are symbolic scenes of the life of the Buddha. The voluptuous woman supported by a mango bough represents a tree spirit.

peace, with little to do except quell peasant revolts, many applied themselves to the pursuit of learning and the development of military skills, and many able samurai became administrators. When their regular incomes were ended by the Meiji government after 1868 some went into business.

Sanchi, one of the most important Buddhist sites in India, situated in Madhya Pradesh state, central India. During the 19th century a group of well-preserved stupas (shrines) and monasteries dating from the 3rd century BC were rediscovered at this site. The most renowned is the oldest, the 'Great Stupa', attributed to the Mauryan Buddhist emperor, *Asoka. It is surrounded by a stone railing and four gateways on which are carved episodes in the life of the *Buddha.

sanctuary, a sacred place recognized as a refuge for criminals. Such places existed in both Greek and Roman society and since 399 in the Christian world. In England a fugitive could claim refuge from immediate prosecution in a church or churchyard provided he agreed with the coroner to leave the realm by a specified port within forty days. Failure to do so would result in prosecution.

This right did not apply in cases of treason (1486). Towns having sanctuary privileges were restricted by Henry VIII to Derby, Launceston, Manchester, Northampton, Wells, Westminster, and York. In England criminals lost the right to sanctuary in 1623.

Sanhedrin (Greek, *synedrion*, 'council'), the term used by *Jews for their supreme court, headed by the high priest, before the fall of *Jerusalem in AD 70. It was probably founded around the 2nd century BC. Under the Romans its jurisdiction covered Palestinian Jews in civil and religious matters, though capital sentences required Roman confirmation.

Sankaracharya (*c.*700–50), Indian religious thinker who expounded and taught the Vedanta system of *Hindu religious philosophy, the most influential of the six recognized systems of Hindu thought. He was born in a Brahmin family in Kerala, south India, but renounced the world to travel all over India as a sannyasi (ascetic), having discussions with philosophers of various schools. His foundation of monasteries throughout India helped to spread his ideas, and he has remained one of the most influential teachers in the Hindu world. Within the Vedanta philosophy his teaching that Brahman (the supreme soul) and Atman (the human soul) are one, and that the visible world is *maya* (illusory), are the chief distinquishing marks of his school.

sans-culotte, a member of the ill-clad and ill-equipped republican 'army' of the early *French Revolution. Their name, meaning 'without knee-breeches', was chosen by the Parisian lower classes to distinguish themselves from the culotte-wearing upper classes. This mob played a key role during the *Jacobin dictatorship but their influence ceased with the fall of their leader *Robespierre in 1794, when the name itself was proscribed.

Saracen, a nomad belonging to tribes of the Syrian or Arabian deserts but at the time of the *Crusades the name used by Christians for all Muslims. In a surge of Islamic conquest they swept into the Holy Land (western Palestine), north into the Byzantine territory of Asia Minor, and westward through North Africa. Spain was conquered, together with most of the islands in the Mediterranean; they held Sicily from the 9th to the 11th century. Their expansion was halted by the Carolingians in France only with great difficulty. The Crusades against them, though initially effective, did not prove decisive in the long term, and they were not finally expelled from Spain until the 15th century.

Within their conquered territories they had a profound effect on cultural life, particularly in architecture, philosophy, mathematics, and religion. In religion they were often tolerant of local beliefs and customs. The lurid accounts of Saracen bloodshed must be offset by the financial advantages of their presence: Saracen gold, used to pay for European goods, invigorated the Frankish economy.

Saragossa (in Spanish, Zaragoza), a province and its capital city in north-eastern Spain. The city was built on the site of the Roman colony of Caesaraugusta. It was one of the first towns in Spain to receive Christianity, but fell to the Visigoths in the 5th century, and then to the Moors *c.*714. In 1118 it was seized by the Aragonese,

and subsequently prospered as the capital of Aragon. It declined in importance after the unification of Spain, but its citizens' heroic resistance to a long French siege (1808–9) caught the public imagination.

Saratoga campaign (1777), an operation devised by *Burgoyne to isolate New England from the other American colonies. He advanced south from Montreal expecting to meet Howe from New York and St Leger from Oswego at Albany and thus secure the line of the Hudson valley. Thanks to bad co-ordination, however, Howe chose this time to embark on his Philadelphia campaign. Burgoyne captured Ticonderoga but his advance was slowed by unsuitable equipment, lack of supplies, and guerrilla attacks. Defeated at *Bennington and Freeman's Farm, with no help forthcoming from Howe, he was halted, and forced to surrender to *Gates. This British defeat encouraged France to enter the war in 1778 and was a vital tonic to the American cause.

Sargon I   *Akkad.

Sassanian empire, an empire in the Middle East, founded *c.*224. Ardashir (ruled *c.*224–41) then overthrew Artabanus V, the last *Parthian king, in the name of vengeance for the last *Achaemenid king. The dynasty takes its name from his grandfather Sasan. Territorially the empire stretched from the Syrian desert, where Roman pressure was checked, to north-west India where the Kushan and Hephthalite empires, having restricted valuable trade routes, were eventually destroyed. Politically the empire fluctuated between centralization under strong monarchs like *Khosrau I (d. 579), who were served by the army and bureaucracy, and local control by great nobles. The religious life of the empire was dominated by *Zoroastrianism, established as the state cult in the 3rd century. Christians in Armenia and Transcaucasia survived persecution and, by breaking with the Byzantine Church in 424, threw off the suspicion of alien loyalties. The court at Ctesiphon provided a focus for a brilliant culture, enriched by Graeco-Roman and eastern influences, which enjoyed such pastimes as chess and polo. The closing years of the dynasty were overshadowed for the masses, however, by lengthy wars, which may explain the empire's rapid disintegration before the *Arab conquest of 636–51.

Satavahana, a dynasty which ruled the north-west *Deccan of India, probably from the late 1st century BC until the 4th century AD. Its greatest king, Gautamiputra Satakarni (AD 106–130), consolidated his hold over the north-west Deccan, and extended his sway from coast to coast. Under his successors, Satavahana power gradually declined and had been entirely lost by the early 4th century. The little that is known about the kingdom and its administration depends on inscriptions in the cave temples which are a significant feature of this era, and on numismatic evidence.

satrap, a provincial governor of the *Achaemenids, first established by *Darius I who divided his empire up into twenty satrapies. Although the satraps nominally owed allegiance to the king, the considerable power and autonomy vested in them fostered disloyalty and there were frequent uprisings, the most notable being that of 366–358 BC against *Artaxerxes II. *Alexander the

The martyrdom of **Savonarola** in Florence on 23 May 1498. This 15th-century painting by an unknown artist shows Savonarola and his two companions, Domenico and Silvestro, hanging over the fire in the Piazza della Signoria. (San Marco, Florence)

Great retained the system after his conquest, as did the *Parthians, but under the *Sassanian empire the term 'satrap' designated a less important figure.

Saul (11th century BC), the first King of the Israelites (c.1020–c.1000 BC). Following the *Philistines' capture of the Ark of the Covenant (the most sacred object of the Israelites) and the destruction of Shiloh and its ruling priesthood, Saul united the tribes of Israel in order to defeat the Ammonites at Jabesh-Gilead, after which he was crowned king. Initially supported by the prophet *Samuel, he invoked Samuel's anger by usurping some of his priestly duties. Saul's later years were dominated by *David's rise to power. David's military successes and friendship with his son Jonathan provoked Saul's jealousy, and he banished David. This further increased support for David, whom Samuel had secretly anointed king. When Saul and Jonathan were killed fighting the Phil-

istines at Mount Gilboa, David assumed leadership of the tribe of Judah.

Savonarola, Girolamo (1452–98), Italian Dominican friar noted for his vehement denunciations of religious and political corruption. As prior of the convent at San Marco in Florence, he attracted large audiences for his attacks on corruption in church and state in his Lent sermons of 1485-6. His prophecies of doom seemed vindicated when Charles VIII of France invaded in 1494. The feeble policy of the *Medici led to their expulsion, and Savonarola briefly led a democratic republic of Florence. The Duke of Milan, and the notoriously corrupt Pope Alexander VI opposed him, and, when the French withdrew in 1497, the excommunicated leader was isolated. Refusal to undergo ordeal by fire was followed by his trial, crucifixion, and burning. The Medici were restored to power in 1512.

Savoy, a region and former kingdom in north-west Italy. It was ruled by the Savoy dynasty for nine centuries, from 1003. It gained in importance through its strategic control of the Alpine passes, and from the family's ability to hold the political balance between the *Holy Roman

Empire and the *papacy, Spain, and France, and subsequently France and Austria. The power of Savoy in Piedmont increased from the 11th century. A strong military tradition was built up as the dynasty acquired the ducal title of Savoy (1416), and the royal title of King of Sicily (1713). Dominion over Sicily was exchanged in 1720 for control of Sardinia.

Saxe, Maurice, comte de (Count of Saxony) (1696–1750), Marshal of France and one of the best known military theorists of his age. He was an illegitimate son of *Augustus the Strong, and was half-German and half-Swedish. In the War of the *Austrian Succession he won a series of victories, including *Fontenoy, and gained control of most of the Austrian Netherlands, which strengthened France's position at the Treaty of *Aix-la-Chapelle (1748).

Saxons, Germanic tribes, possibly named from their single-edged *seax* ('sword'). Under pressure from the migrating *Franks they spread from their homelands on the Danish peninsula into Italy and the Frisian lands and engaged in piracy on the North Sea and English Channel between the 3rd and 5th centuries. They appear to have entered Britain, together with *Angles and *Jutes as mercenaries in the late period of the Roman occupation. By the 5th century their settlements had marked the beginning of *Anglo-Saxon England. Their name survives in Wessex ('West Saxons'), Essex ('East Saxons'), and Sussex ('South Saxons') in England, as well as in Saxony in Germany.

Saxony, a former duchy in north Germany. The *Saxons expanded into Europe during the *Dark Ages and conquered many lesser tribes, before they were defeated in battle by *Charlemagne and converted to Christianity. On the collapse of the Carolingian empire the Saxon duchy survived and expanded, reaching its greatest extent under *Henry the Lion, but the state was dismembered in 1180, only the part taken by Bernhard of Anhalt retaining the name. In 1423 it came to Landgrave Frederick the Warrior and, as an 'electorate' of the *Holy Roman Empire, remained with his descendants.

Sayyid dynasty, Muslim rulers of the Delhi sultanate in northern India (1414–51). They seized power from the *Tughluqs, but never equalled their predecessors' imperial pretensions. Rival neighbours soon threatened their claims even in the north, and in 1448 their last sultan abandoned Delhi, to be replaced three years later by the Afghan *Lodis. The name 'Sayyid' reflected the family's claim to be direct descendants of the Prophet Muhammad.

Scandinavia, the northernmost part of Europe, traditionally Denmark, Norway, and Sweden; a broader definition of the area also includes Finland, Iceland, and the North Atlantic islands which have come within Scandinavia's influence during the past thousand years.
In the *Viking age (c.800–1050), Scandinavia was an important centre of civilization, which sent colonists to Iceland, the North Atlantic islands, and Greenland, while Scandinavian rulers dominated much of England, Ireland, Normandy, Finland, and western Russia. This primarily sea-borne civilization also made contact with the shores of North America. By the early 11th century,

the national kingdoms of *Norway, *Denmark, and *Sweden were well-established. A period of further overseas conquest was followed by Scandinavian unification when in 1397 the rulers of Denmark set up the Union of *Kalmar. The secession of Sweden under *Gustavus Vasa in 1523 heralded a long period of disunion, despite common acceptance of reformed religion. Sweden's swift rise to European prominence in the 17th century owed much to the inspirational leadership of *Gustavus Adolphus and his Chancellor *Oxenstierna. It eclipsed Denmark–Norway as a Baltic power, and until the *Great Northern War (1700–21) held the balance of power in Europe as a whole. Thereafter it gradually declined in importance.

Scanian War (1676–8), a struggle between Denmark and Sweden for the latter's southernmost province of Scania. For centuries it was controlled by Denmark but was gained by Charles X of Sweden at the Treaty of *Roskilde in 1658. Christian V of Denmark invaded Scania and was welcomed by the population. Charles XI of Sweden fought back in a harsh campaign and finally won a pitched battle at Lund (3 December 1676); the victory was acknowledged in the Treaty of Lund (1679), and Scania was ceded to Sweden.

Schmalkaldic War (1546–7), a brief and indecisive phase in the struggle between the Roman Catholic emperor *Charles V and the *Protestant party within the *Holy Roman Empire. The defensive League of Schmalkalden was formed in the town of that name by Protestant states in 1531. It was led by *Philip of Hesse and John Frederick I of Saxony. The emperor was heavily committed elsewhere and did not come face to face with the League until 1546. Then he crushed the League with the help of Duke Maurice of Saxony, winning a notable victory at the battle of Mühlberg (24 April 1547).

scholasticism, the educational tradition of the medieval 'schools' (universities) which flourished in the 12th and 13th centuries. It was a method of philosophical and theological enquiry which aimed at a better understanding of Christian doctrine by a process of definition and systematic argument. The writings of Aristotle (translated from Greek into Latin by *Boethius) and of St *Augustine played a crucial part in the development of scholastic thought. Scholastics did not always agree on points of theology; *Aquinas and *Duns Scotus argued from different standpoints. Scholasticism declined in the later Middle Ages; in the 14th century the writings of *William of Occam challenged the scholastic position by stressing the opposition between faith and reason.

Scientific Revolution, the increase in knowledge and understanding of the universe which began in the 16th and flourished in the 17th century. At first it was a product of the *Renaissance spirit of enquiry which led astronomers like Copernicus (1473–1543), Kepler (1571–1630), and Galileo (1564–1642) to overthrow the medieval notion of the Earth as the centre of the universe, but not until the early 17th century did their theories gain wide acceptance.
The 17th century was an age of intellectual activity marked by great progress in science. Unbiased enquiry, shrinking from no conclusion merely because it was unorthodox and testing all conclusions by experiment

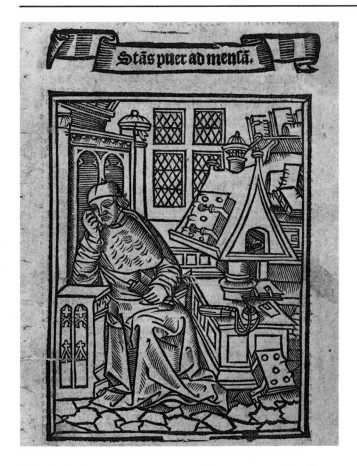

Scholasticism: a scholar deep in thought, a woodcut from
a book published in 1518. The study has a reading desk,
seat, and bookcase, and a pen case is ready for use.

and observation, characterized the work of leading
scientists. An early figure was Francis *Bacon (1561-
1626), not himself a scientist but a philosopher, who saw
theology as the study of God's mind and science as the
study of God's works by means of experiment. With the
emerging division of science into its various branches
progress was made in chemistry by Robert Boyle (1627-
91), in medicine by William Harvey (1578-1657), with
his discovery of the circulation of the blood, above all in
mathematics and physics by Isaac Newton (1642-1727),
whose brilliant mind, with its assurance that everything
in heaven or on earth was comprehensible in terms of
reason, affected all branches of western thought and
research. Scientific discovery forged ahead in all fields
including electricity, first described (1600) by William
Gilbert. Meetings of scientists to exchange information
had begun during the English Civil War (1642-49) but
their gatherings were given royal patronage by *Charles
II (1662) when the *Royal Society was founded. By the
18th century the application of science to industry,
with the invention of the steam engine and advances
in textile manufacture, marked the beginning of the
*Industrial Revolution.

Scipio, Publius Cornelius (236-184 BC), later entitled
'Africanus Major' for his part in defeating Carthage on
the African mainland in the Second *Punic War. After
the Roman defeats at Ticinus and Cannae he rallied the
aristocracy and the army. His tactics of aggression led to
campaigns in Spain and his landing in Africa in 204. He

later retired from public life after being accused by *Cato
of corruption in 184.

Scipio, Aemilianus Africanus Numantinus (185-
129 BC), grandson by adoption of Publius Cornelius
Scipio. He earned the titles 'Africanus' for his destruction
of Carthage in the Third *Punic War and 'Numantinus'
for his later capture of Numantia in Spain. Rapid
promotion gave him command of the war in Africa in
147. He blockaded Carthage and levelled the site in 146.
He died suddenly and suspiciously in 129. He was at the
centre of the intellectual group called the 'Scipionic
Circle', which included the historian *Polybius.

Scotland, a country in the north of *Great Britain, a
part of the United Kingdom since 1707. It was peopled
in the early Middle Ages by five different races: *Picts,
who by *c.*500 had retreated to the north-east highlands;
Britons, in the south-west; *Scots, from Ireland, in Argyll
and Galloway; *Angles, in the Lothians; and from *c.*800,
*Norsemen in the far north and west. These peoples spoke
different languages, and pursued an entirely agricultural
economy when not at war with each other. Scotland was
converted to Christianity in the 6th and 7th centuries by
such missionaries as St *Columba, but the whole country
was cut off from the rest of Europe by Norse encirclement.
   Malcolm *Canmore, its first king (1058-93), and his
queen *Margaret, established a dynasty that ruled the
country for over two centuries. They and their successors,
especially *David I, did much to civilize the land,
introducing burghs and initiating an urban economy,
*feudalism (in place of tribal law), and church reforms
(such as diocesan and parish organization).
   Entanglement with England resulted from *Edward
I's attempt to make Scotland subject to him, through
the *Balliol family. Intermittent warfare ensued for the
next 250 years; at times the Scots gained the upper hand,
as at *Bannockburn under the leadership of *Robert the
Bruce, but they were seriously defeated at *Halidon Hill
(1333), *Flodden (1513), and *Pinkie (1547).
   In 1560 the Scottish Parliament accepted John *Knox's
Confession of Faith. The Reformation caused Scotland to
turn away from France, its main ally since the early 14th
century, and to draw closer to England. James VI of
Scotland became *James I of England in 1603, and in
1707 an Act of *Union joined the Parliaments of England
and Scotland. Union brought great advantage to both
countries during Britain's imperial and economic ascendancy, and Scotland experienced a remarkable *Enlightenment in the 18th century.

Scots (or Dalriads), Celtic Irish settlers, in what is today
Scotland. In the early 6th century they settled Argyll
(Ar Gael) after two centuries of raiding the coasts of
Britain and Gaul. They overcame the northern *Picts in
the Highlands and introduced the Celtic Gaelic language.
The name 'Scotia' (Scotland), formerly a name for
Hibernia (Ireland) passed to the land of *Caledonia and
the territories of all the Picts.

Scottish Martyrs, a group of political reformers who
were persecuted for their beliefs during the period of
unrest in Scotland in the 1790s. In 1792 a Society for
the Friends of the People was formed to promote
parliamentary reform. The government reacted strongly
and Thomas Muir, a member of the society, was convicted

of treason. In October 1793 a meeting of the Society was broken up by force, three of its delegates being subsequently sentenced to long terms of transportation. A group of radical reformers calling themselves the United Scotsmen continued to meet in secret, but after further trials the movement broke up.

scutage, or escutage (from Latin *scutum*, 'shield'), the payment (usually 20 shillings) made by a knight to the English king in lieu of military service. *Henry II raised seven scutages between 1157 and 1187. *Richard I was tempted (1198) to turn scutage into an annual tax not necessarily connected with military needs. The barons' opposition to John's annual scutages (1201–6) was a factor in their revolt (1214) and was reflected in clause 12 of *Magna Carta (1215), which stated that the king was not to levy scutage without consent, except in recognized and reasonable cases.

Scythians, a group of Indo-European tribes which briefly occupied part of Asia Minor in the 7th century BC before being driven out by the Medes. They subsequently established a kingdom in southern Russia and traded with the Greek cities of the Black Sea, but in about the 2nd century BC were compelled to move into the Crimea by the related Sarmatian tribe. They were a nomadic people, famed for their horsemanship and their skill as archers. When *Darius the Great attempted to subdue them in *c*.512 BC, they successfully adopted a scorched-earth policy and *c*.325 BC they crushed a large detachment of Macedonian troops before making peace with *Alexander the Great. The graves of Scythian kings and nobles have revealed many objects of gold and bronze, which bear witness to outstanding technical and artistic skill.

Sea Peoples, a term used to describe the various groups of sea-borne invaders who attacked the countries of the eastern Mediterranean in the later 13th and early 12th centuries BC. They probably included Greeks, Sardinians, and Tyrrhenians, though their activities and identities are indistinctly known. They were instrumental in causing the collapse of the *Hittite empire *c*.1200 BC, and Egypt was attacked from the sea on several occasions. One group to settle successfully was the *Philistines.

Sedgemoor, battle of (6 July 1685), the decisive battle of *Monmouth's Rebellion. Monmouth was blocked in his retreat from Bristol by the army of *James II, commanded by Lord Feversham and John Churchill

Part of a political broadside depicting the defeat of Monmouth's rebellion by the forces of James II at the battle of **Sedgemoor** on 6 July 1685. Monmouth attempted to surprise the king's forces who were encamped on Sedgemoor but his men could not cross a broad ditch and were mowed down by the royal artillery. Monmouth fled but was taken two days later and beheaded on Tower Hill.

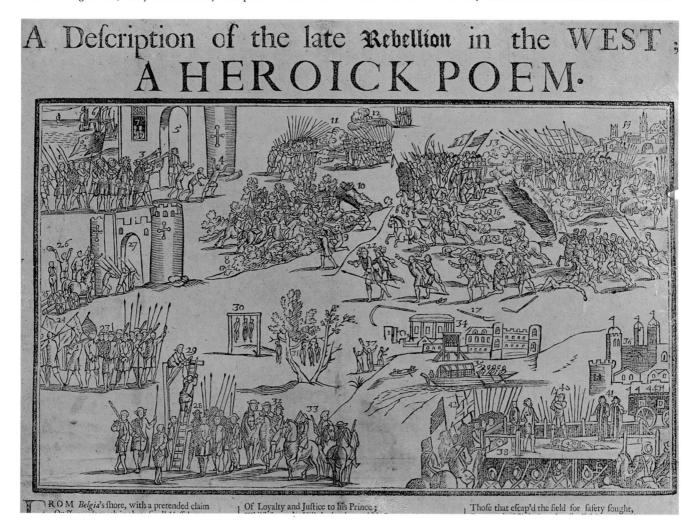

(later Duke of *Marlborough). Monmouth attempted a night attack to give his raw recruits some advantage over the professional royalist army, but his plans miscarried and Monmouth was defeated.

Selden, John (1584-1654), English lawyer, historian, and antiquary. Although not a Puritan, he used his knowledge of the law on Parliament's behalf in its conflicts with Charles I, and was repeatedly imprisoned. A member of most of the Parliaments after 1621, he was active in the impeachment of *Buckingham (1626), and helped to draw up the *Petition of Right (1628).

Seleucid, the Macedonian dynasty which ruled an Asian empire from 312 BC when Seleucus I, who had served *Alexander the Great, gained Media and Susiana to add to Babylon. His subsequent expansion was westwards: he occupied Syria, where he founded Antioch in 300 BC, and by defeating Lysimachus in 281, secured control of Asia Minor. Under his successors Syria and Asia Minor were lost and regained more than once, while in the east Bactria asserted its independence and the Parthian kingdom was established. Antiochus III (ruled 223-187 BC) forced both these kingdoms to acknowledge his lordship. He also recovered Syria and Palestine, but when he conquered Thrace and then invaded Greece, he came into conflict with Rome. He was defeated at *Thermopylae and Magnesia, and made peace in 188 on terms which excluded him from Asia Minor. Thereafter Seleucid power declined—Tyre regained its independence in 126 BC, as did other cities and chiefdoms—until in 64 BC Syria became a Roman province.

Self-Denying Ordinance (3 April 1645), an English parliamentary regulation which made all Members of Parliament resign their military commands. Oliver *Cromwell was determined to create an efficient national army controlled and paid from Westminster, rather than by the counties. The House of Lords amended the Ordinance, so that it was possible for certain Members of Parliament to be reappointed to the *New Model Army, enabling Cromwell to continue his military career as lieutenant-general under commander-in-chief Sir Thomas Fairfax.

Selim I (c.1470-1520), Ottoman sultan (1512-20), known in English as 'the Grim', though 'Relentless' better conveys the Turkish sense. Recalled from Crimean exile after an aborted attempt to ensure his own succession, he defeated and killed his rival brother Ahmed in 1513. In 1514, responding to Safavid-inspired subversion in Asia Minor, he crushed a Persian army at Chaldiran. Turning against the *Mamelukes, he next conquered Syria and Egypt and took the titles of *caliph and protector of the holy cities of Mecca and Medina.

Seljuk, a Turkish dynasty. Its early members rose to prominence as mercenaries, raising Turkish nomad troops to serve and ultimately to challenge the *Ghaznavids. By 1055, under the leadership of Tughrul Beg, they had entered Baghdad and subjected the eastern lands of the Muslim empire to their control, while maintaining the fiction of an *Abbasid caliphate and using the existing administrative system and such talented officials as the vizier Nizam al-Mulk. Reaching their apogee under Alp Arslan (1063-72) and Malik Shah (1072-92), they

disturbed the existing regional balance of power, crushing the Byzantines at *Manzikert, and, by interrupting the *pilgrimage to Jerusalem, indirectly provoking the First *Crusade. They were ultimately undermined by the turbulence of their own nomad troops among a settled population and by the rivalries of ambitious subordinates. Their decline in the early 12th century was, however, soon followed by the emergence in Asia Minor of the Seljuk sultanate of Rum, centred on Konya, under Kilij Arslan II (1155-92). Under Kaykobad I (1220-37) this regime achieved great splendour, but it suffered a crushing defeat at *Mongol hands at Kösedagh in 1243 and became a dependency of the Mongol Il-Khans of Persia until its extinction in 1308.

Semite, a speaker of a Semitic language. The Semites were farmers in Arabia who spread north and west to create some of the major empires of antiquity. The inhabitants of *Akkad who overthrew the Sumerians were Semites, as were the Assyrians, the Egyptians, the Aramaeans, the Canaanites, the Phoenicians, and the Hebrews.

Senate   *Roman Senate.

Seneca, a North American Indian tribe, the westernmost of the Five *Iroquois Nations in western New York. They became involved in the *fur trade and its wars, and French Jesuit missionaries began work among them in 1668. In the American Revolution they joined the British and in 1778 an American expedition destroyed their villages. New Christian missions began at the end of the 18th century, and simultaneously 'The Code of Handsome Lake', based on Iroquois traditions, was preached by Chief Handsome Lake, both beliefs gaining followers.

Seneca, Lucius Annaeus (4 BC-AD 65), a *Stoic philosopher and *Nero's tutor. Banished from Rome in 42 on suspicion of adultery with *Claudius' sister he was recalled in 49 to act as tutor to the 12-year-old Nero. On the young emperor's accession the post evolved into that of political counsellor. At first he and the *Praetorian Prefect were able to influence Nero. In 59 he found himself a reluctant accessory to the murder of Nero's mother Agrippina and composed Nero's explanation for the Senate. He chose retirement in 62. Three years later Nero accused him of treason and he was compelled to commit suicide. His surviving works include letters, dialogues, ethical treatises, and plays.

Sennacherib, King of the *Assyrian empire (704-681 BC). He was preoccupied with fighting rebels for much of his reign. In the west a campaign in 701 was largely successful in crushing discontent; Tyre and Jerusalem both defied capture, though the latter was compelled to pay a large indemnity. Babylonia was a source of more persistent discontent, and Babylon itself was finally destroyed and looted in 689 after a nine-month siege. Sennacherib carried out a major building scheme at *Nineveh, which included a palace and two city walls.

Serbia, a republic of Yugoslavia, formed from the former kingdom of Serbia. Slavic tribes from the Danube region won the area from Greeks and Romanized peoples in the 7th century. The first Slavic state of Rascia, under threat from Bulgaria in the 10th century, was obliged to accept the protection of the Byzantine empire, and the rival state of Zeta was supported by Rome. Bogomilism, a dualist heresy, later complicated the rivalry between the two major Christian churches which was finally resolved by acceptance of the *Orthodox faith. The 13th century saw exploitation of mineral resources and much trade, particularly with Venice. Under the rule of *Stephan Dushan (1331–55), the law was codified and the status of the serfs was regularized. Macedonia, Albania, and parts of Greece were annexed, but this expansion came to an abrupt end in 1371 and 1389 with disastrous defeats by the *Ottomans.

serf, an unfree peasant under the control of the lord whose lands he worked. As villeins or servants of a medieval lord they represented the bottom tier of society. They were attached to the land and denied freedom of movement, freedom to marry without permission of their lord, and were obliged to work on their lord's fields, to contribute a proportion of their own produce, to surrender part of their land at death, and to submit to the justice and penalties administered by their lord in the manorial court in the case of wrongdoing. The lord had obligations to his serfs (unlike slaves), most notably to provide military protection and justice.

Serfdom originated in the 8th and 9th centuries in western Europe and subsequently became hereditary. In much of Western Europe the system was undermined in the 14th century by the *Black Death and starvation

A group of **serfs** reaping corn on the lord's demesne under the direction of the reeve or bailiff, a decoration from the early 14th-century *Queen Mary's Psalter*. (British Library, London)

resulting from war which led to acute labour shortages. Commutation of their labour for cash meant that the lord became a rentier and the serf a tenant; in the *Peasants' Revolt in England (1381) the main demand was for the abolition of serfdom and the substitution of rent at 4 pence an acre for services. However, in the eastern regions of Germany and *Muscovy, the increased power of the nobility, and the development of absolutism led to consolidation of serfdom. It was formally abolished in France in 1789, but lingered in Austria and Hungary till 1848, and was abolished in Russia only in 1861.

Settlement, Act of, the name of several English Acts, that of 1701 being the most politically significant. It provided for the succession to the throne after the death of Queen *Anne's last surviving child, and was intended to prevent the Roman Catholic Stuarts from regaining the throne. It stipulated that the crown should go to James I's granddaughter, the Electress Sophia of Hanover, or her surviving Protestant heirs. The Act placed further limitations on royal power, and made the judiciary independent of crown and Parliament. On Anne's death in 1714, Sophia's son became Britain's first Hanoverian monarch as *George I.

Seven Wonders of the World, the most remarkable man-made sights of the ancient world, most commonly listed as the *pyramids of Egypt; the Hanging Gardens of *Babylon—a series of terraced roof-gardens attributed to *Nebuchadnezzar II; the statue of Zeus at *Olympia— a large statue overlaid with gold and ivory by the Athenian sculptor Phidias, c.430 BC; the temple of Artemis at Ephesus—built in the mid-6th century BC with assistance from *Croesus, burnt in 356 BC, rebuilt, and

finally destroyed by the Goths in AD 263; the mausoleum of Halicarnassus; the Colossus of *Rhodes—an enormous (over 30.5 m., 100 ft., high) bronze statue of the sun god erected c.292-280 BC by the Rhodians in their harbour to commemorate their repulse of Demetrius Poliorcetes in 305; the Pharos of Egypt—a huge lighthouse built by Ptolemy II c.280 BC on the island of Pharos outside the harbour of *Alexandria.

Seven Years War (1756-63), a wide-ranging conflict involving Prussia, Britain, and Hanover fighting against Austria, France, Russia, Sweden, and Spain. It continued the disputes which had been left undecided after the treaty of *Aix-la-Chapelle, and was concerned partly with colonial rivalry between Britain and France and partly with the struggle for supremacy in Germany between Austria and Prussia. Fighting had continued in North America with the *Braddock expedition. Each side was dissatisfied with its former allies and in 1756 *Frederick II of Prussia concluded the Treaty of Westminster with Britain. This made it possible for *Maria Theresa of Austria and her minister *Kaunitz to obtain an alliance with France (known as the 'diplomatic revolution') by the two treaties of Versailles in 1756 and

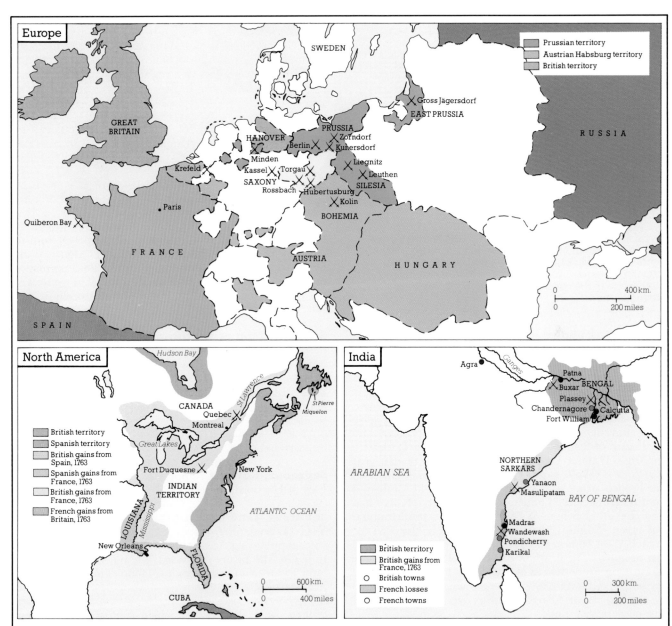

Seven Years War (1756–63)

The Seven Years War was both a battle for maritime and colonial supremacy between Britain and France and a wider European power struggle centred on competition between Austria and Prussia for Silesia. In Europe, Prussia retained Silesia and became a power of equal weight to Habsburg Austria, while the British victory at Quiberon Bay destroyed French naval ambitions. Overseas, the ascendancy of Britain as a commercial and colonial power and the relative decline of France were confirmed. In India, the British victory at Wandewash (1760) laid the foundations for its future domination of the subcontinent; in North America, British naval superiority paved the way for military victory against the over-extended French. Britain gained Canada and the French possessions east of the Mississippi, together with Florida, formerly Spanish.

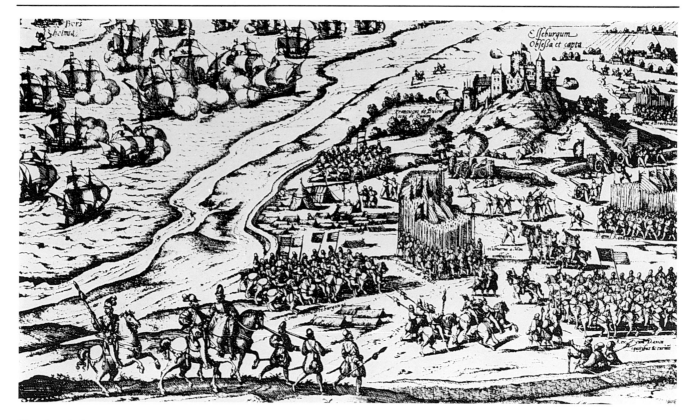

Naval and siege engagements during the **Seven Years War of the North**, portrayed here in a contemporary print. As Denmark and Sweden struggled for control of the vital sea lanes, their land forces fought exhausting battles but, in the event, these were to prove largely inconclusive.

1757; she was also allied with *Elizabeth of Russia. At first the advantage was with the French and Austrians, but in July 1757 *Pitt the Elder came to power in England and conducted the war with skill and vigour. In November Frederick II won his great victory of Rossbach over the French, and in December he defeated the Austrians at Leuthen. Frederick was hard pressed in 1758, but he defeated the Russians at Zorndorf and Ferdinand of Brunswick protected his western flank with an Anglo-Hanoverian army. 1759 was the British year of victories: *Wolfe captured Quebec, Ferdinand defeated the French army at *Minden, and *Hawke destroyed the French fleet at *Quiberon Bay. In India *Clive had won control of Bengal at Plassey, and in 1760 Montreal was taken. Admiral *Boscawen successfully attacked the French West Indies. In 1761 Spain entered the war and Pitt resigned. The death of Elizabeth of Russia eased the pressure on Frederick, as her successor Peter III reversed her policy. All were ready for peace which was concluded by the Treaty of *Paris in 1763: overall England and Russia were victorious.

Seven Years War of the North (1563-70), a bitter conflict resulting from a collision of the expansionist aims of Denmark and Sweden in the Baltic, and merging with the contemporary war in *Livonia. Frederick II of Denmark's troops achieved greater tactical successes on land, but the Swedes compensated with a series of victories at sea. In 1568 the insane Eric XIV of Sweden was deposed by an alliance of the nobility and replaced by John III. The new king swiftly sought peace from the equally exhausted Danes. By the Treaty of Stettin (1570) Denmark's grip on the entrance to the Baltic remained unbroken, but Sweden retained Estonia. The contest for the dominion of the whole Baltic was not resolved until the *Northern War.

Severus, Lucius Septimius (145-211), Roman emperor (193-211). A professional soldier, he became governor of Pannonia and was proclaimed emperor by his troops in 193. Rivals were eliminated and the *praetorians were disbanded and replaced with his own men. He adopted himself into the *Antonine dynasty. He persecuted the Christians in 203. During a long campaign on the *Caledonian border in Britain he died at the military headquarters in Eboracum (York).

Seymour, Edward   *Somerset, Edward Seymour, 1st Earl of Hertford, and Duke of.

Seymour, Jane (c.1509-37), Queen consort of *Henry VIII of England from 1536. She was the king's third wife, supplanting Anne *Boleyn. In 1537 Jane gave birth to a son, the future *Edward VI, but died twelve days later. Henry had been genuinely fond of her and mourned her death.

Sforza, an Italian family which rose to prominence in the 15th and 16th centuries. Muzio Attendolo (1369-1424) was one of the most powerful *condottieri of the period (his assumed name Sforza means 'force'). His illegitimate son Francesco (1401-66) was also a successful condottiere, whose armies were involved in a three-way war with the Milanese republic and Venice, after which he entered Milan in triumph as duke (1450), and thereafter governed ably.

Ludovico (1451-1508), known as 'Il Moro' (the Moor),

usurped the Milanese government in 1480. He helped Charles VIII of France to invade Naples (1494), but he was subsequently driven out of his duchy by Louis XII (1499). In 1512 his son Massimiliano (1493–1530) was restored to Milan with Swiss aid; *Francis I of France defeated him at Marignano (1515), and forced him to cede his dominions, granting him a pension of 30,000 ducats. Massimiliano's brother Francesco II (1495–1535) was restored by Emperor *Charles V in 1522, but his death marked the end of the male ducal line.

Shaftesbury, Anthony Ashley Cooper, 1st Earl of (1621–83), English statesman. He entered Parliament in 1640 as a royalist supporter, but changed sides in 1643, eventually becoming a member of *Cromwell's council of state. In 1660 he was one of the Commissioners of the Convention Parliament who invited *Charles II to return, and Charles rewarded him with the Chancellorship of the Exchequer. After *Clarendon's fall he became one of the *cabal, but was dismissed in 1673 because of his support for the *Test Act and his unwavering opposition to Roman Catholicism. He became leader of the opposition, and used the *Popish Plot to try to exclude the Roman Catholic James, Duke of York, from the succession (*Exclusion Crisis), but his political failure led him to flee into exile in 1682.

Shah Jahan (1592–1666), *Mogul Emperor of India (1628–58) whose outwardly splendid reign ended in imprisonment by his son and successor. He extended Mogul power, notably in the Deccan, and rebuilt the capital at Delhi. His buildings there and in Agra, notably the Taj Mahal, mark the high peak of Indo-Muslim architecture. His severe illness in 1657 caused a succession war between his four sons in which *Aurangzeb, the third son, killed his rivals, imprisoned his father in the Agra palace, and seized the throne. On his death Shah Jahan was buried with his favourite wife in the Taj Mahal.

Shaker, a member of a religious sect, so-called because of their uncontrolled jerkings in moments of religious ecstasy. They were a revivalist group and held many *Quaker views although they left the Quaker movement in 1747. Ann Lee, who joined in 1758, declared herself the female Christ and, inspired by visions, established a community near Albany in New York colony in 1774. The community prospered and Shakers gained a reputation in New England as skilled craftsmen. The sect reached a peak of about 6,000 members in the 1820s, but a decline set in after 1860.

Shang (c. 16th century to c. 11th century BC), China's first verified dynasty. It was authenticated in the 1920s after the discovery of oracle bones ('dragon bones') near *Anyang. Devoted to hunting and war, the Shang kings regulated their activities by consulting diviners. Scratches on the oracle bones, the earliest form of Chinese characters, proved to be their oracular writings. From these records have emerged the traditional names and sequence of the kings of this dynasty, who gradually extended their rule over the Huang He (Yellow River) plain. The Shang developed a complex agricultural society and saw the emergence of skilled artisans, most notably bronze casters.

Shankaracharya *Sankaracharya.

An early 17th-century portrait of the Mogul emperor **Shah Jahan**. The painting is inscribed in the margin in Shah Jahan's hand: 'a good portrait of me in my twenty-fifth year'. (Victoria and Albert Museum, London)

Shans, a people of *Burma. They are akin to people in *Laos and *Thailand, and originated in Yunnan province in south-west China, entering Burma about the 13th century AD. Based in the hills east of the Irrawaddy River, they lived under chieftains thought to have divine powers. There was deep enmity between Burmans and Shans, though the latter, except in very remote areas, adopted Burman culture and Theravada *Buddhism. A kingdom founded by a Burmanized Shan prince at Ava had considerable power from c.1360 until the unification of Upper Burma under the first *Toungoo dynasty in the late 15th century.

Sharifian, the name given to two *Moroccan dynasties, called *sharifs* (nobles) by reason of their descent from al-Hasan, son of *Muhammad's daughter Fatima. The *sharifs* of *Mecca, and others, claim similar descent.

The Sadian dynasty of *sharifs* originated in Sus in 1509, and speedily conquered all Morocco. Sharif Muhammad traded with Spain and England, a policy followed by his successors. In 1664, after a period of great confusion, a cadet branch of *sharifs* known as Filali supplanted them, and remain the ruling dynasty of Morocco.

Shays's Rebellion, an armed American uprising from August 1786 to February 1787 led by Captain Daniel Shays (c.1747–1825), a Massachusetts war veteran. Shays led a group of destitute farmers from western Massachusetts against the creditor merchants and lawyers of the seaboard towns. The Rebellion was caused by the Massachusetts legislature adjourning without hearing the petitions of debt-ridden farmers for financial help. The rebellion prevented the sitting of the courts; the state militia routed Shays's force, but he escaped and was later pardoned. The uprising won concessions, but it also boosted the campaign for an effective constitution and central government for the USA.

Shelburne, William Petty Fitzmaurice Lansdowne, 1st Marquis and 2nd Earl of (1737-1805), British statesman. He joined Chatham's ministry in 1766, but failed to build up a political following and was regarded with distrust by many of his colleagues. His ideas were often regarded as impracticable, especially his conciliatory scheme of 1767 for settling the American question. He opposed the American policies of Lord *North and succeeded Lord *Rockingham in 1782 as Prime Minister. He was responsible for settling the main outlines of the peace treaty between Britain and America, but before these could be concluded he was brought down by a combination of the supporters of *Fox and North in 1783.

sheriff (shire-reeve), the chief representative of the crown in the shires (counties) of England from the early 11th century, taking over many of the duties previously performed by ealdormen. Sheriffs assumed responsibility for the *fyrd, royal taxes, royal estates, shire courts, and presided over their own court, the Tourn. They abused these powers, as an inquest of 1170 showed, when many were dismissed. However, by c.1550 the office had become purely civil, as a result of the proliferation of specialist royal officials (Coroners, 1170, Justices of the Peace, 1361, *Lords Lieutenant, 1547).

Sheriffmuir, battle of (13 November 1715), the only major Scottish battle of the *Fifteen Rebellion. The *Jacobite army of 10,000 men, commanded by the Earl of Mar, met the much smaller loyalist force of the Duke of Argyll. Although the fighting was inconclusive, Mar was forced on to the defensive, and the chance of Jacobite success disappeared.

Sher Shah Suri (c.1486-1545), Emperor of northern India (1540-5). His short-lived seizure of power from the second Mogul emperor, *Humayun, made an important impact on Indian administration. An Afghan of humble origins, he had risen through military service to be well placed to take advantage of temporary Mogul weakness. After defeating Humayun twice (1539 and 1540) he made himself Emperor of Delhi, extending his control to Gwalior and Malwa. Before his death in battle he carried out innovations in the land revenue system and the army which were subsequently built on by the great Mogul administrator, *Akbar.

Shetlands, an archipelago of about 100 islands 160 km. (100 miles) off the north coast of Scotland. They were settled by the Norse people from the end of the 8th century. Controlled by the powerful earls of Orkney, the Shetlands eventually became part of the kingdom of Scotland in the marriage contract of *James III and Margaret of Norway (1472).

Shia *Shiite.

Shi Huangdi (Shih Huang-ti) (259-210 BC), first *Qin Emperor of China (221-210 BC). He became ruler of the state of Qin in 246 BC and declared himself emperor in 221 BC after overthrowing the *Zhou and their vassal states. He ordered that the frontier walls in northern China should be joined together and extended to make the *Great Wall and enlarged his empire into southern China. He could not accept the Confucian belief that an emperor should follow traditional rites, and so ordered the burning of all Confucian books, the banning of Confucian teaching, and the killing of scholars. In death as in life he was heavily guarded: close to his burial mound outside Xi'an stood an army of life-size pottery warriors and horses. The dynasty outlasted him by only three years, but he had imposed lasting unity on China by standardizing scripts, weights, and measures.

Shiite, the name of a *Muslim minority, who take their name from the Shiat Ali, the 'party of Ali', the cousin and son-in-law of *Muhammad. Ali and his descendants are regarded by Shiites as the only true heirs to Muhammad's position as leader of the faithful. Important differences of doctrine, ritual, and spiritual organization have developed between Shiism and Sunnism, notably with reference to the status of imams (religious leaders). Shiite tradition, emphasizing sacrifice, martyrdom, and belief in an inner hidden meaning of the Koran, has generated many different subgroups, such as the *Ismailis. Most, however, revere the sites associated with the deaths of Ali and his sons, Hasan and Husain, and the tenth day of the month of Muharram, the first in the Islamic calendar, which marks their martyrdom.

Shimabara, a peninsula near *Nagasaki, Japan. In the 17th century it was a Catholic stronghold. Its converts, regarded by the *Tokugawa as subversive, were persecuted, and in 1637 some 40,000 Christians rebelled against oppression and poverty. Having taken refuge in a castle, they held out for some months against the shogun's army, numbering 100,000, and a Dutch warship sent to his aid. They were virtually annihilated. Japan's few surviving Christians, risking torture and death, continued to practise their rites in secret.

Shinto, a Japanese religion that reveres ancestors and nature-spirits. The name means 'Spirits' Way'. Things that inspire awe—twisted trees, contorted rocks, dead warriors—are believed to enshrine *kami* ('spirits'). In early times each clan had its *kami*. With the supremacy of the *Yamato, its sun-goddess, enshrined at the temple at Ise, became paramount. The name Shinto was adopted in the 6th century AD to distinguish it from Buddhist and *Confucian cults, though most Japanese practised both Shinto and *Buddhism. Shinto offers no code of conduct, no philosophy. It stresses ritual purity—which may explain why the Japanese, to the amazement of early Western visitors, bathed frequently. At simple shrines worshippers rinse hands and mouth, bow, and offer food and drink.

In the 18th century scholars began stressing the emperor's descent from the sun-goddess. After the Meiji restoration in 1868, the government encouraged State Shinto, classed not as a religion but as a code requiring loyalty and obedience to a divine emperor. This was clearly distinguished from Sect Shinto with its simple rites and local festivals.

ship money, originally an occasional sum of money paid by English seaports to the crown to meet the cost of supplying a ship to the Royal Navy. *Charles I revived the tax in 1634, while he was ruling without Parliament. From 1635 he extended it to the inland towns, and raised up to £200,000 a year as a result. In 1637 John *Hampden was taken to court for refusing to pay and claimed that Charles needed Parliament's approval to levy such a regular tax. The judges decided by 7 to 5 in Charles's favour, but the narrowness of the victory encouraged widespread refusal to pay tax afterwards. The *Long Parliament made ship money illegal in 1641.

shire, the main unit of local administration in England. Shires evolved as territorial units in Wessex in the 9th century, replacing the Roman system of provinces. They were extended over a wider area of England by *Alfred the Great and his heirs as administrative and political units. The English shire system reveals many different evolutionary processes. Some were based on former kingdoms (Kent, Sussex, Essex); others on tribal subdivisions within a kingdom (Norfolk and Suffolk); others were created during the 10th-century reconquest of the *Danelaw or as territories centred on towns (Oxfordshire, Warwickshire, Buckinghamshire, and so on). England north of the River Tees was not absorbed into the shire system until the Norman Conquest when shires were re-styled 'counties'.

Shivaji *Sivaji.

shogunate, a Japanese institution under which government was in the hands of a *Sei-i dai-shogun* ('barbarian-conquering great general'). The shoguns exercised civil and military power in the name of emperors, who became figure-heads. In the 8th century generals with this title had been appointed during the wars against the native Ainu of northern Japan, but their commands were limited in time and purpose. The shogunate as a form of government originated with *Minamoto Yoritomo's appointment without any limit to his authority (1192). After he died the *Hojo regents took control of affairs, but in theory they remained subject both to the emperor and the shogun. During the *Ashikaga period the shoguns were independent of any other authority though their rule was ineffective. Under the *Tokugawa the power of the shogunate was decisive in national politics. This dual system of government ended when the imperial power was restored under the Meiji emperor in 1868.

Shotoku Taishi (574–622), Japanese prince. As regent for Empress Suiko, he set out at the age of 20 to convert a clan society into a centralized administration like that of China. He sent embassies to the *Sui, brought in Chinese artists and craftsmen, adopted the Chinese calendar, created a constitution, and instituted a bureaucracy based on merit. He promoted both *Buddhism and *Confucianism. After Shotoku's death, during the Taika ('Great Change') period (645–710), an imperial prince and a *Fujiwara initiated further reforms. Gradually, Chinese practices were adapted to Japanese conditions and a more centralized administration emerged.

Siam (now Thailand), a country in south-east Asia. The Thais, akin to the Shans and Lao, originated in the Yunnan province of south-west China. Their name means 'free'. *Mongol pressure accelerated their southward movement from Yunnan. They set up kingdoms in Sukhotai and Chiengmai, formerly under *Khmer rule, became Theravada *Buddhists, and adopted an Indian script. About 1350 Ayuthia became the capital of a new Thai kingdom which, after prolonged fighting, captured *Angkor in 1431. Ayuthia ruled much of Cambodia and at times Tenasserim and northern Malaya. Wars with *Burma, whose kings coveted Ayuthia's sacred white elephants, brought no lasting loss of Thai territory. Among Europeans who became active in Ayuthia the French were dominant. In 1684 Thai envoys presented *Louis XIV with elephants, rhinoceroses, and a letter engraved on gold. The Burmese finally destroyed Ayuthia in 1767. Under the leadership of General Taksin, the Burmese were expelled from Siam by about 1777. His successor, General Chakri (later Rama I) founded the Chakri dynasty and established Bangkok as his capital. The Chakri dominated much of *Laos and northern Malaya.

Sicilian Vespers, the name given to an uprising and massacre in Sicily which began at the time of vespers (the evening church service) on Easter Tuesday in 1282. It marked the end of the rule of the *Angevins in the island and of their dynastic ambitions in Italy. Charles I of Anjou had received the Kingdom of the Two Sicilies from Pope Urban IV in 1266 and to claim it had defeated the Hohenstaufen *Manfred, son of the Holy Roman Emperor Frederick II. His rule was extremely harsh, enforcing heavy taxation, and the French occupation was generally hated. Within a month all the French had been killed or forced to flee and the crown was later given to Pedro III of Aragon, who thwarted Angevin attempts at reoccupation, and who passed the crown to his son Frederick III of Sicily.

Sicily, an island in the Mediterranean Sea, south-west of Italy. In the 8th century BC Phoenicians founded trading posts in the west of the island, while Greeks colonized the eastern and southern coasts. From the 5th to the 3rd centuries BC conflict between *Carthage and the Greeks (led by Syracuse) was a feature of the island's history. Then in 264 BC the first of the *Punic wars between Carthage and Rome began, and by 210 BC the island was totally under Roman control, and remained so until it was occupied by the Vandals and then the Ostrogoths. *Belisarius took possession of the island for the Byzantine empire in AD 535, a rule which lasted until the 9th century when the Arabs won control after prolonged and ferocious fighting. Under their benevolent regime Palermo in particular flourished. Two centuries of Arab rule were ended by *Norman colonization. The Kingdom of the Two Sicilies was formed by *Roger II in 1139, when Sicily was linked to the kingdom of *Naples. Norman rule was followed by Surabian and *Angevin rule, the latter ended by the infamous *Sicilian Vespers in 1282. There then followed more than four centuries of rule by Aragonese princes and Spanish kings. Ferdinand II, King of the Two Sicilies, was overthrown by a French Revolutionary army in 1799 and seven years later Joseph Bonaparte was crowned King of Naples.

Sierra Leone, a country in West Africa. The Portuguese navigator Pedro de Cintra reached it in 1462, at about the time the Temne, its chief inhabitants, were reaching the coast. During the 16th and 17th centuries the *slave trade and piracy attracted many Europeans, including English, so that the coast has a very mixed population. In 1787 the Anti-Slavery Society bought the coastal territory from the local ruler as a haven for slaves found destitute in Britain. This failed, and in 1791 Alexander Falconbridge formed the Sierra Leone Company for the same purpose, landing the first colonists at Freetown in 1792. It became a crown colony in 1808.

Sieyès, Emmanuel Joseph (1748-1836), French statesman and abbot, one of the chief theorists of the Revolutionary era. He was vicar-general of the diocese of Chartres when he wrote his famous pamphlet, *What is the Third Estate?* (1788). The following year he became a member of the *National Assembly and played a major part in writing the constitution. He voted for the king's execution but refused to serve under the constitution of 1795. Four years later he was appointed to the *Directory but conspired with *Napoleon to overthrow it and set about producing the 'perfect' constitution. When this was modified by Napoleon he retired from public life and, after the Restoration, left France.

Sigismund (1368-1437), Holy Roman Emperor (1411-37), King of Hungary (1387-1437), of Germany (1411-37), Bohemia (1419-37), and Lombardy (1431-37), the last emperor of the House of Luxemburg. In 1396 he was defeated by the Turks at *Nicopolis but went on to acquire and secure a large number of territories and titles in a long and violent reign which featured warfare with the *Hussites, Venetians, and rivals for the thrones of Hungary and Germany. An orthodox Catholic, he acted severely against the Hussites, and put pressure on the pope to call a council at *Constance to end the Hussite Schism; Sigismund promised Huss safe-conduct to attend the council, which subsequently ordered his death.

Sikh ('disciple'), follower of the teachings of the Indian religious leader Guru *Nanak (1469-c.1539) and his nine successor gurus. Sikhism is rooted in *Hinduism, but stresses meditation on the name of the one God. Guidance is found in the *Adi Granth*, compiled in 1604 by Guru Arjun, which contains hymns compiled by the first five gurus. The orthodox Sikh wears 'the five ks': the *kesh* (unshorn hair and beard), the *kungha* (a hair comb), the *kuchcha* (shorts), the *kara* (an iron or steel bangle), and the *kirpan* (a sword or dagger). Clashes with the Moguls in the 17th century resulted in the execution of the fifth and ninth gurus. The tenth and last guru, Gobind Singh, in 1699 established the *khalsa* (military brotherhood) whose members are called *akalis*. A century later Ranjit Singh (1780-1839) set up a powerful Sikh kingdom in the Punjab, but its shortlived dominance was ended by British annexation in 1849.

Silk Route, a term loosely applied to the caravan routes linking China, Central Asia, and the Levant via the Pamir mountains, the Ferghana valley, Samarkand, and Merv. From the 1st century BC Chinese silks were exchanged for western bullion, much of the trade being through middlemen. Few merchants and only a small proportion of the goods ever travelled the whole distance. *Buddhism and *Nestorian Christianity came to China by this route and, from the 7th century, so did Levantine traders. The Mongol conquests of the 13th century gave the route new life, and *Marco Polo gave it added significance, until in the 15th century it was superseded by east-west trade by sea.

Sima Qian (Ssu-ma Ch'ien) (c.145-85 BC), Chinese official and historian. He was an official at the court of the *Han emperor Wudi. His *Historical Records* is a history of China from earliest times to the days of Wudi, and is a model for later dynastic histories. As well as ancient myths, he provided much source material, quoting inscriptions from old bronzes and imperial decrees from the archives. One section of the work, *Assorted Traditions*, has lively biographies of generals, poets, scoundrels, and court ladies. Angered by his defence of a general forced to surrender to the *Xiongnu, Wudi had him castrated. Although such punishment often led officials to commit suicide he decided to live on to complete his history.

Simnel, Lambert (c.1487-c.1525), English royal *pretender. Although he was actually a joiner's son certain *Yorkists pretended or believed that he was Edward of Warwick, son of the murdered George, Duke of *Clarence, and on 24 May 1487 he was crowned in Dublin as Edward VI. Next month he was brought to England, but the Yorkists were defeated and *Henry VII, showing mercy, gave him employment in the royal kitchens.

simony, a term derived from Simon Magus who, according to an account in the New Testament of the Bible, tried to buy spiritual powers from St Peter. It came to mean the purchase of any office or authority within the *Roman Catholic Church. The church's policy that its benefices should not be sold for money was often jeopardized because many secular lords claimed that they were theirs to dispose of as they wished. Wealthy families bought offices for their members and used them as a form of patronage. Simony was one of the abuses criticized at the time of the *Reformation.

Sind, now a province of Pakistan which occupies the predominantly desert region of the lower Indus valley. Several sites, notably *Mohenjo-daro, show that the region was a cradle of early civilization. It was incorporated in the *Mauryan empire, but for long periods was left to its own devices. In AD 711 Arab invasions initiated an era of Islamicization. *Akbar extended Mogul power to Sind in the 16th century, yet isolation and aridity favoured local independence which only ended with British expansion in the 19th century.

Sindhia, a leading *Maratha family, based at Gwalior, in northern India, which dominated the Maratha confederacy in the late 18th century. The founder of the family's fortunes, Ranoji Sindhia (d. 1750), seized control of the Malwa region which he had been appointed to govern. A successor, Mahadaji Sindhia (ruled 1761-94), created a virtually independent kingdom in north India. He became the arbiter of power not only within the Maratha confederacy, but also at the *Mogul court at Delhi, and won victories against the *East India Company as well as the *Rajputs. His successors failed to stem British expansion in north India. Daulat Rao Sindhia was defeated in 1803, and in 1818 Gwalior became a Princely State under the Company's protection.

Sioux *Dakota Indians.

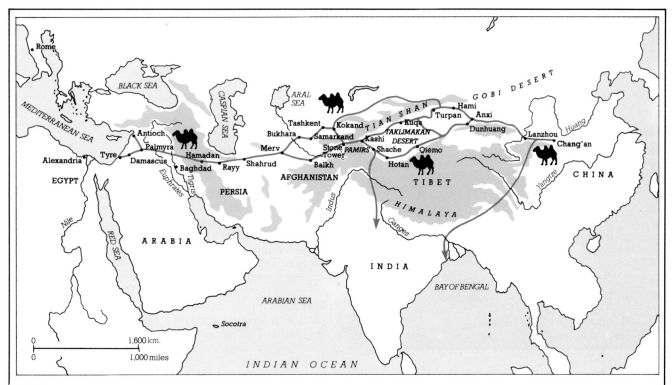

Silk Route

Crossing Asia from eastern China to the Mediterranean, the Silk Route was pioneered by the Chinese in the 2nd century BC. Bactrian camels carried many goods besides silk on the Silk Route, but only a small proportion of the goods, and few merchants, ever travelled its whole length. The route began in the old capital, Chang'an (now Xi'an) and skirted the Taklimakan desert north and south, before climbing into the Pamirs from Kashi (Kashgar) and continuing through Persia to the Mediterranean. From Shache (Yarkand) another route led to India.

Sivaji (c.1630–80), founder of the *Maratha kingdom in India. He was responsible for a revival of Hindu fortunes in western India at a time of Muslim expansion. He was a member of the Bhonsla family who, at an early age planned the downfall of the Muslim sultanates which surrounded his Deccan home. In confrontations with the sultan of Bijapur and the *Mogul emperor, *Aurangzeb, he won unlikely victories, and a reputation for deeds of remarkable daring. When finally defeated by the Mogul army he escaped in disguise to win back and expand his domains. In 1674 he was crowned king and rewarded his subjects' devotion by a fair and firm administration, showing tolerance to all religious communities. His greatest achievements were to withstand Mogul expansion in the west, and to create a nucleus for subsequent Maratha expansion. He is regarded as an Indian hero.

Sixtus IV (Francesco della Rovere, 1414–84), Pope (1471–84). Originally a Franciscan monk and teacher, as pope, he was a patron of the arts, and a powerful secular ruler. He organized two military expeditions against the *Turks between 1472–3, and in 1474–6 attempted re-unification between the Russian *Orthodox and the Catholic churches. In 1478, he authorized the setting up of the *Spanish Inquisition and in 1483 confirmed Tomás de *Torquemada as Grand Inquisitor. Thereafter he concentrated on furthering the ambitions of his family (six of the thirty-four cardinals he created were his nephews), and in strengthening the political and territorial power of the papacy. He was involved in the *Pazzi conspiracy, fought against *Venice in 1483, and spent lavishly on military campaigns. He instigated the building of the Sistine Chapel.

Sixtus V (Felice Peretti, 1521–90), Pope (1585–90). A noted Franciscan preacher, with experience of the Venetian Inquisition (1557–60), he was elected to the papacy as a compromise candidate. He attempted to reform the chaotic *Papal States. The resulting increase in revenues enabled him to engage in an extensive building programme which transformed Rome from a medieval into a baroque city, though at the cost of the destruction of several important antique sites. His reform of central church administration, which included limiting the number of cardinals to seventy, contributed to the success of the *Counter-Reformation.

Slave Kings   *Mameluke.

slavery (in the classical world). The condition of a person being another's property has a history going back to the earliest civilizations. Thus Linear B tablets of the *Mycenaean civilization refer to persons who were held in some form of captivity. Later, invaders from the north reduced complete populations in Greece to slavery. The *helots of Sparta are the best-known example, and the Spartans therefore had no need for the personal ownership of slaves. By contrast at Athens many families owned one or two slaves for domestic purposes, and private slaves were also employed in factories and in the silver mines. There were also public slaves at Athens. Much of the ancient economy relied on slaves and the sacred island

Slavery: a kneeling figure of a young black Roman slave cleaning a boot. This bronze statuette was made between the 1st and 3rd centuries AD. (British Museum, London)

of Delos served as the main slave market of the Aegean. It was very often the practice amongst the peoples of the ancient world to enslave prisoners-of-war, and that was doubtless the major source of slaves, though traders were often able to purchase slaves. Piracy was another means of supply. The expansion of the Roman empire created an enormous number of slaves, as for instance when *Epirus was annexed in 146 BC. On three occasions during the 2nd and 1st centuries BC major slave revolts occurred, and were put down only with great difficulty. The coming of Christianity helped to improve the slave's lot, but slavery proved persistent throughout and beyond antiquity.

slave trade, African, the trade organized by Europeans, mainly to the Americas, where slave labour enabled colonists to establish *plantations. Following their 15th-century discoveries, the Portuguese began taking slaves from Senegambia and Guinea-Bissau: they went to Portugal, the Atlantic islands, and later to the Americas; by the 16th century they came also from Angola, and, occasionally, Mozambique. In the 17th century trading forts, stretching from Arguin (now in Mauritania) to Angola, had been established by slavers from Brandenburg, Denmark, Holland, Courland (on the eastern Baltic), England, France, Genoa, and Sweden. These forts could not have operated without active African participation in supplying slaves for shipment to Brazil, and to British, French, Dutch, and Spanish colonies. It is thought that by the mid-19th century 9.5 million Africans had been transported to the New World, in addition to those who died while being captured or in transit. This figure does not include the Arab slave trade nor the flourishing trade in slaves within Africa. Disgust at this treatment of Africans led to demands for

emancipation of the slaves and the virtual abolition of the slave trade in the 19th century.

slave trade, Arab, a form of commerce that existed in earliest times, as an ancient Semitic institution. The Prophet *Muhammad forbade his followers to enslave Muslims, but did not free slave converts. His legislation insisted on humane treatment and gave slaves rights against oppressive masters. In early Islam slaves were recruited from prisoners-of-war (including women and children), and were acquired by raiding and by purchase in Eastern and Southern Europe, Central Asia, and Central, East, and West Africa. Under the caliphs the trade was brisk. Slaves served a variety of purposes: agricultural, mining, domestic, and clerical, and for military service. Many slave women employed as concubines were given the rights of wives. Men and children often received vocational training after capture. Through international pressure, the trade was virtually abolished during the 19th century.

Slavs, peoples who occupied eastern Europe in ancient times and were known to the Romans as Sarmatians and Scythians. The name is believed to come from *slowo* ('well speaking'), the same root as Slovenes. After the collapse

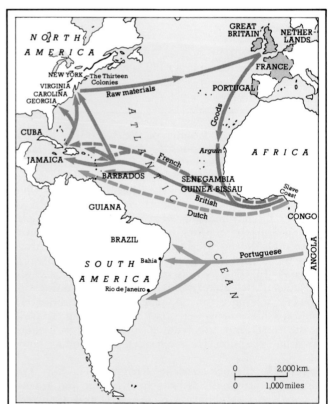

African slave trade

The African slave trade reached its height during the 17th and 18th centuries. Several hundred thousand Africans were transported across the Atlantic each year, to the plantations of the European colonies in the Americas. It was a triangular trade. A typical three-leg voyage set out from Europe to West Africa, carrying cotton goods, hardware, and, increasingly, guns. These goods were exchanged for slaves, who were taken to the West Indies and the southern colonies of North America. The ships returned to Europe with colonial produce, notably sugar.

of the *Huns in the 5th century they migrated westwards to the Elbe, the Baltic, the Danube, the Adriatic, and the Black Sea. In the 9th century the missionaries Cyril and Methodius from Constantinople evangelized the Slovenes or southern Slavs.

Sluys, battle of (24 June 1340), a naval battle in the *Hundred Years War. A force of French, *Genoese, and *Castilian ships intercepted an English force but was defeated by massed archers. Both French commanders were killed and the victory gave the English control of the English Channel.

Smith, Adam (1723-90), Scottish economist and philosopher. He became professor of moral philosophy at Glasgow University in 1752, and later travelled abroad as tutor to the Duke of Buccleuch. He visited Paris and met a number of the *philosophes*, including *Quesnay, whose ideas confirmed his own hostility to *mercantilism. His *Wealth of Nations* (1776) attacked mercantilism and his ideas had great influence, forming the basis for the development of modern economic theory.

Smith, John (1580-1631), founder of *Virginia and promoter of colonization in America. He was born in Lincolnshire and fought against Spain and the Turks (1596-1604). As one of the governing council of Virginia in 1607, his military discipline and firmness with the Indians saved *Jamestown from collapse. After being burnt in an explosion, he returned to England in 1609. Five years later he explored the New England coast. His optimistic *Description of New England* (1616) was a powerful encouragement to emigration there in the 1620s and 1630s.

smuggler, one who moves goods illegally from one country to another, to evade payment of customs duties or in defiance of laws prohibiting the importation of merchandise. Smugglers have existed since civilization began: those smuggling cats out of ancient Egypt were subject to the death penalty. The rapid development of organized smuggling came in the late 17th century with higher *customs duties and the restrictions of the *Navigation Acts. Smugglers were popular in Britain because they made brandy, tea, silks, and perfumes available at reasonable prices. In colonial America they were regarded as patriotic heroes, who were defying the hated Navigation Acts. The reduction of duties in the late 18th century, the spread of free trade, and the development of the coastguard system, caused a decline in smuggling.

Sobieski, John (1624-96), King John III of Poland (1674-96). He was a member of a great Polish noble family and, after defeating the Turks at Khotin in 1673, was elected king and expelled the Turks from southern Poland, including Lvov. In 1683 he was largely responsible for driving them also from Vienna. He regarded his struggle against the *Ottoman empire as a crusade to be pursued at all costs, and can be criticized for not using his great abilities to strengthen the Polish monarchy. He was succeeded by *Augustus II.

socage, the term applied in Anglo-Saxon and Norman England to a free tenure of land which did not require the tenant to perform military service. He might pay a

An etching c.1800 of a group of **smugglers** landing their illicit cargo. (National Maritime Museum, London)

rent in cash or in kind, and perform some ploughing on his lord's estates. He was liable to pay the three feudal dues—twenty shillings when the lord's son came of age and when the lord's daughter married, and one year's rent to redeem his lord from captivity. In contrast to military tenure, no restrictions attached to the inheritance of the tenure nor to the marriage of the heir.

social contract, a term used by *Hobbes, *Locke, and *Rousseau in examining the nature of the state's authority over the individual. They examined the condition of man in a state of nature, that is without law and government, and the transition by means of a 'contract' to ordered society. Hobbes considered man without government to be in such a state of fear that he will accept absolute authority in order to gain security. Locke believed that man is guided by reason and conscience even in a state of nature and that in accepting government he still retained natural rights. This influenced the makers of the American Constitution: *Jefferson held that the preservation of natural rights was an essential part of the social contract. Rousseau disagreed with Locke's description of the state of nature and thought that man, a 'noble savage' in his natural state, acquired a moral and civic sense only when part of a larger democratic community.

Society of Friends   *Quakers.

Socinus, Laelius (or Sozzini, Lelio) (1525-62), Italian theologian. After studying law at Bologna, he travelled throughout Europe before he settled in Zürich (1548) to study Greek and Hebrew. He corresponded with leading *Protestant reformers and conducted his own theological enquiries, but his *Confession of Faith* (1555) showed that he had reached few clear conclusions of his own. His nephew, Faustus Socinus (1539–1604), developed Laelius's views into a well-defined system, which led to the founding of the Socinian sect and contributed to the growth of the Unitarian movement.

Socrates (469–399 BC), Athenian philosopher, one of the most remarkable and influential figures of ancient Greece. He was endowed with great courage, both physical and moral. In his early life he seems to have been preoccupied with scientific philosophy, but later he turned to the question of ethical philosophy, or how best to live one's life. He attracted a wide circle of young men, including *Alcibiades. It was his association with them that may have led to the charges of impiety and corrupting the youth levelled against him in 399 BC. He

A statue of **Socrates** displays the features attributed to him by surviving descriptions: stout and short, with a snub nose, prominent eyes, broad nostrils, and wide mouth. (British Museum, London)

defended himself at his trial, but was condemned to die by drinking hemlock, which he did, rejecting his friends' offers to help him escape.

The dialogues of his disciple *Plato and the *Memorabilia* of *Xenophon give a vivid picture of his personality, ideas, and methods. He usually conducted his enquiries through ruthless but wryly humorous cross-questioning, the 'Socratic method', and was often critical of the *Sophists. Though he professed no fixed set of doctrines, his influence on later thinkers was profound. He believed that virtue was something which could be taught, and applied himself in a systematic philosophical manner to addressing the question of day-to-day conduct and beliefs.

Solemn League and Covenant (1643), the agreement between the English Parliament and Scottish *Covenanters during the *English Civil War. It undertook that the Presbyterian Church of Scotland was to be preserved, and the Anglican Church was to be reformed. The Scots soon realized that Presbyterianism would not be imposed on England by the specially established Westminster Assembly of Divines. This put considerable strain upon the other aspect of the agreement: Scottish military aid for Parliament in return for £30,000 per month.

Solomon (d. *c.*922 BC), King of *Israel (*c.*961–*c.*922 BC). The second son of *David and Bathsheba, he succeeded his father and was the last king of a united *Israel. His riches and wisdom became legendary, and under his rule the nation grew wealthier and alliances were forged with Egypt and Phoenicia. These alliances provoked discontent since they led to the official establishment of foreign religious cults in Jerusalem. He organized the land into administrative districts to facilitate government and introduced a system of forced labour to sustain his extensive building works, including palaces and the Temple at Jerusalem, which became the central sanctuary of the Jewish religion. It was partly because of the high taxes imposed to support court luxury that the northern tribes seceded under *Jeroboam after his death.

Solon (*c.*640–*c.*560 BC), Athenian statesman and poet. He was the author of a written code of laws which introduced major reforms at *Athens in the first quarter of the 6th century BC. These included the cancellation of many debts, the restoration to freedom of many who had been made slaves, economic changes intended to stimulate trade and industry, and the division of the citizens into four classes, based on wealth, each with particular political responsibilities and prerogatives. This undermined aristocratic power which was based solely on birth. He may have created a new council (*boule*) to prepare business for the citizen-assembly, thereby undercutting some of the power of the *Areopagus. He replaced the *Draconian law with a less harsh code, which remained the basis of later classical laws.

Solutrian *Upper Palaeolithic.

Somers, John Somers, Baron (1651–1716), English lawyer and Whig politician. He played a leading part in attacking *James II's illegal acts (especially in defending the seven bishops who had objected to the second *Declaration of Indulgence), and in drafting the *Bill of Rights. In *William III's reign he presided over the

near Hong Kong while trying to avoid capture by a *Mongol fleet, all China for the first time came under non-Chinese rule.

The Song's mastery of technology, superior to that of western Europe, included the making of firearms and bombs, shipbuilding, the use of the compass, and clock making. A form of vaccination (variolation) was practised. It was also a time of great sophistication in literature, painting, and ceramics, and Hangzhou became the artistic centre of China.

Songhay (also Songhai and Songhoi), originally a state on the Niger River, also the name of the people and their language, which is spoken by the majority in Dendi, Gao, Jenné, Jerma, and Timbuktu. Tradition claims that a Berber Christian, al-Yaman, founded it in the 7th century AD on Kukiya Island, below Gao. The rulers became Muslim c.1200, transferring the capital to Gao. In 1325 the *Mali empire annexed Songhay, but in 1335 Sonni Ali Kolon made himself king. In c.1464 *Sonni Ali made Songhay independent and enlarged it greatly. His son, Bakari, was a weakling, and with him the line of al-Yaman failed. In 1493 the new dynasty founded by *Askia Muhammad I replaced him; he made Songhay the most important empire in western Africa, eclipsing Mali. In 1528 or 1529 Askia Muhammad was deposed by his son. He, and the seven other Askias who followed, were weak, cruel, and debauched, and the empire foundered. In 1591 it fell easy prey to a well disciplined and well armed force of *Moroccans who defeated the Songhay army at Tondibi, near Gao. However, they could not control such a large area, and in the 17th century it broke up into a number of smaller states.

Sonni Ali (Ali Ber, Ali the Great) (d. 1492), ruler of *Songhay from 1464 or 1465, and the real founder of the Songhay empire. He made Songhay independent and then enlarged its boundaries. In 1469 he took *Timbuktu, and Jenné in 1473. Timbuktu chroniclers describe him as cruel, irreligious, and of immoral habits, and report that he persecuted men of religion, although pretending to be a Muslim. He died by drowning, and was succeeded by his son Bakari, the last of his line.

Sons of Liberty, American Revolutionary groups, which sprang up in Massachusetts and New York in 1765 to organize colonial opposition to the *Stamp Act. In both colonies, serious riots, propaganda, and boycotts effectively nullified the measure. The organization spread to other colonies and reactions to English 'tyranny' were synchronized by committees of correspondence. Sons of Liberty were later responsible for the *Boston Tea Party (1773), the radicalization of the *Continental Congress (1774), and the tarring and feathering of pro-British loyalists.

Sophia (1630-1714), Electress of Hanover (1658-1714). She was the twelfth child of the Elector Frederick of the Palatinate and Elizabeth, daughter of *James I. She married Ernest Augustus of Brunswick-Lüneburg, who became Elector of Hanover. By the Act of *Settlement (1701), as the only surviving Protestant descendant of the *Stuarts, she was named as *Anne's successor on the British throne. She died a few weeks before Anne and her son became King *George I. She was a cultured woman with a great interest in English affairs.

Painting was among the arts that flourished in China during the **Song** dynasty. This late 12th-century album leaf illustration in ink and colours on silk shows the Han Palace. (National Palace Museum, Taipei)

inner cabinet. Although he was the most trusted of William's advisers he was frequently attacked by jealous political rivals, and was dismissed in 1700. He held office only briefly during Queen Anne's reign.

Somerset, Edward Seymour, 1st Earl of Hertford and Duke of (c.1506-52), Protector of England and effective ruler of England on behalf of *Edward VI (1547-9). On the death of *Henry VIII in 1547 Edward Seymour (brother of Jane *Seymour), took the titles of Duke of Somerset and Lord Protector and won an immediate military success against Scotland at the battle of *Pinkie. His attempts to enforce the use of a Protestant English Prayer Book by Act of Uniformity (1549) sparked off the *Western Rising. *Kett's Rebellion, coinciding with rising discontent among magnates grouped around his rival, the Earl of Warwick, led to his downfall. He was overthrown in 1549 and executed on the orders of the Duke of *Northumberland in 1551.

Song (Sung) dynasty (Northern Song, 960-1126; Southern Song, 1127-1279), a dynasty that reunited much of China after a period of fragmentation after the *Tang dynasty. It never ruled all China, however—the north-west was a Tibetan kingdom, and in the north-east were the *Liao, to whom the Song paid tribute of silk. In 1125 the *Jin ousted the Liao. In 1125-6 Jin horsemen descended on Kaifeng, Northern Song's capital, and took the emperor and 3,000 retainers captive to Manchuria. A Song prince then set up a court in Hangzhou. Hoping to regain the north he called it Xingzai (Temporary Capital). Southern China became the heart of the empire. In 1276 *Kublai Khan captured Hangzhou. Two little princes escaped, but when the last survivor was drowned

Hogarth's satirical comment on the bursting of the **South Sea Bubble**, 1721. While subscribers to the South Sea Company are being taken for a ride, honour and honesty are publicly punished by self-interest and villainy.

sophist (Greek *sophistes*, 'wise man'), an itinerant professional teacher in Greece in the 5th century BC. Sophists offered instruction in a wide range of subjects, rhetoric in particular, in return for fees. Gorgias of Leontini (*c*.483–376 BC) specialized in teaching rhetoric, and his visit to Athens in 427 BC acted as a considerable encouragement to the development of oratory there. Young Athenian democrats needed rhetoric to persuade the democratic assemblies. By questioning the nature of gods, conventions, and morals, and by their alleged ability to train men 'to make the weaker argument the stronger' through rhetoric, they aroused some opposition. During the Roman empire sophists were essentially teachers of rhetoric. The word sophistry, meaning a quibbler or fallacious reasoner, reflects the popular distrust of sophists.

Sousa, Martim Afonso de (1500–64), Portuguese colonist, leader of an exploratory expedition to southern Brazil (1531–3). In 1532 he established the first permanent Portuguese colony in Brazil at São Vicente (near present-day Santos), where he introduced sugar-cane. In his efforts to expel French intruders and also in search of precious metals he explored the coast south from Rio de Janeiro to the Rio de la Plata. In 1534 he was granted the hereditary captaincy of São Vicente but he never returned to Brazil, and later served as governor of India and as a member of the Council of State in Lisbon.

South America, the southern continent of the New World extending from the Isthmus of Panama in the north to Cape Horn in the south. The continent was originally settled by *Amerindians and a number of sophisticated cultures developed, most notably that of the *Incas. Most of South America was colonized by Spain and Portugal. Spain's claim was based on the discoveries of Christopher *Columbus; on his third voyage (1498) he sailed along the Venezuelan coast and landed on Trinidad. In 1500 the Portuguese explorer, Pedro Alvares Cabral, landed near present-day Bahia and took possession of this territory for Portugal. Portuguese *Brazil and the Spanish viceroyalty of *Peru constituted the two principal administrative jurisdictions in South America during the 16th and 17th centuries. In the 18th century, Spain subdivided Peru, adding two new viceroyalties, New Granada (1717) and Rio de la Plata (1776). England, France, and the Netherlands also established small colonies on the north-eastern coast of the continent in the 17th century. Colonial settlement was round the coast and to this day urban centres are concentrated near the coastline and not in the interior.

South Carolina, a colony and state of the USA on the southern Atlantic coast. *Charles II granted a charter to colonize the Carolinas to a syndicate of eight proprietors. The northern and southern parts developed separately and the colony was divided into North and South

Carolina in 1713. Its first settlement was at *Charleston in 1670. Its colonial development was influenced by immigrants from the West Indies and in the 18th century its slave population outnumbered whites. Its coastlands proved ideal for rice and indigo *plantations and it was one of the richest and most valued English colonies. Friction between planters and proprietors and the crisis of Indian attacks in the Yamasee War (1715) resulted in its reversion to crown rule in 1729. The colony was divided over independence and while the British occupied Charleston (1780-2) they enjoyed considerable loyalist support.

Southern cult, a common religious cult and artistic symbolism in the prehistoric *Mississippi and adjacent cultures. It was fully formed by *c.* AD 1000 and appears archaeologically as an interrelated assemblage of ritual objects and design motifs painted or incised on pottery, carved or sculpted on shell, wood, and stone, embossed on thin sheets of native copper, and painted on cloth. These artefacts and motifs are thought to have been used in connection with ceremonies on large earthen temple mounds. Much of the symbolism can ultimately be traced to Mexican religions, but other elements were derived from the indigenous and preceding *Hopewell religion.

South Sea Bubble, an English financial crisis of 1720. A scheme whereby the South Sea Company took over most of the National Debt (*Bank of England) and needed a rise in the value of the shares of the Company. This was achieved not by favourable past trading profits, but by rumours of future ones. Some politicians and even members of the court were bribed with cheap or free South Sea Company shares to promote the company's interests. The shares increased ten-fold in value, and expectations of high dividends rose accordingly. Many bogus companies secured substantial investment by exploiting the speculative fever. When confidence in these bogus companies collapsed, the South Sea Company's shares fell to less than 10 per cent of their peak value, and thousands of investors were ruined. Sir Robert *Walpole began his long period of office by saving the company and restoring financial stability, and managed to limit the political damage to two of George I's ministers who had been implicated in the scandal.

spa town, named after Spa, in Belgium, celebrated since medieval times for the restorative quality of its water. In 18th century England spas were fashionable resorts offering cures and amusements and the upper and middle classes flocked to take the waters and attend the Assembly Rooms in Bath (built around the original Roman thermal baths), Epsom, Tunbridge Wells, Buxton, and Cheltenham. The prosperity generated by these visitors is witnessed by the many fine Georgian buildings which grace these towns, notably the crescents in Bath and the Pantiles in Tunbridge Wells.

Sea bathing also became fashionable in the 18th century, encouraged by doctors who recommended the benefits of salt water and sea air. Margate, Ramsgate, Scarborough, and Weymouth were all flourishing resorts by the 1780s and the patronage of the Prince Regent (later George IV) ensured the social success of Brighton in the 1790s. Spas remained fashionable in Europe in the 19th century, particularly the German resort of Baden-Baden.

George III of England and his family take the waters at the fashionable **spa town** of Cheltenham in 1789, a painting by Peter la Cave. (Cheltenham Art Gallery and Museums)

Spain, a European country occupying most of the Iberian peninsula. It began to come under Roman control after 206 BC, after a period of Carthaginian domination. Roman rule was followed, after AD 415, by that of the *Visigoths, who were themselves toppled by Muslim invaders from Morocco (711-18). *Moorish Spain reached its zenith under the *Umayyad dynasty of al-Andalus (736-1031). During the subsequent political fragmentation, Christian kingdoms became consolidated where Muslim power was weakest, in the north: *Aragon and *Castile were the most significant of these. By 1248 Christian reconquest had been so successful that only *Granada remained in Muslim hands. *Ferdinand II of Aragon and *Isabella of Castile united their respective kingdoms in 1479, reconquered Granada in 1492, and went on to establish unified Spain as a power of European and world significance. Under their rule the vast *Spanish empire overseas began to take shape, and under their 16th-century successors, *Charles V and *Philip II, Spain enjoyed its 'golden age'. Decline set in during the 17th century, culminating in the War of the *Spanish Succession (1701-14), which marked the transition from Habsburg to Bourbon rule.

Spanish Armada, a large naval and military force which *Philip II of Spain sent to invade England at the end of May 1588. It consisted of 130 ships, carrying about 8,000 sailors and 19,000 infantrymen, under the command of the inexperienced Duke of *Medina Sidonia. The Spanish fleet was delayed by a storm off Corunna, and was first sighted by the English naval commanders on 19 July, then harassed by them with long-range guns, until it anchored off Calais. Unable to liaise with an additional force from the Low Countries led by *Farnese, its formation was wrecked by English fireships during the night and as it tried to escape it suffered a further pounding from the English fleet before a strong wind drove the remaining vessels into the North Sea and they

An unknown Dutch master's interpretation of the **Spanish Armada** showing the heavy Spanish galleons being defeated by the more manoeuvrable English ships fitted with port and starboard cannon. (National Maritime Museum, London)

were forced to make their way back to Spain round the north of Scotland and the west of Ireland. Barely half the original Armada returned to port.

Spanish empire, the overseas territories which came under Spanish control from the late 15th century onwards. They included the Canaries, most of the West Indian islands, the whole of central America, large stretches of South America, and the Philippines. Christopher *Columbus laid the foundations of the empire with his four voyages (1492–1504) in search of a western route to the Orient. Then the *conquistadores followed, colonizing by force in *Mexico, *Peru, and elsewhere in the New World. As the wealth of these lands became apparent, private enterprise gradually gave way to direct conciliar rule by the mother power. The Council of the Indies (chartered 1524) stood at the head of the imperial administration until almost the end of the colonial period. Control over all colonial trade was vested in the House of Trade (established in 1503 at Seville). The gold and silver from the New World made 16th-century Spain the richest country in Europe, under Emperor Charles V. The colonies themselves were eventually divided into

viceroyalties: *New Spain (1535), *Peru (1569), *New Granada (1717), and Rio de la Plata (1776). Despite government regulations like the New Laws (1542) it was difficult to prevent exploitation of the native Indians. English, French, and Dutch depredations of the empire were damaging to Spain.

Spanish Inquisition, a council authorized by Pope *Sixtus IV in 1478 and organized under the Catholic monarchs *Ferdinand II and *Isabella of Spain to combat heresy. Its main targets were converted Jews and Muslims, but it was also used against *witchcraft and against political enemies. The first Grand Inquisitor was *Torquemada. Its methods included the use of torture, confiscation, and burning at autos-da-fé. It ordered the expulsion of the Jews in 1492, the attack on the Moriscos (Muslims living in Spain who were baptized Christians but retained Islamic practices) in 1502, and, after the *Reformation attacked all forms of Protestantism. In the 16th century there were fourteen Spanish branches and its jurisdiction was extended to the colonies of the New World, including Mexico and Peru, and to the Netherlands and Sicily. Its activities were enlarged in the reign of *Philip II, who favoured it as a *Counter-Reformation weapon. It was suppressed and finally abolished in the 19th century.

Spanish Main, originally the mainlands of the Americas adjacent to the Caribbean Sea, particularly the northern coastal waters of South America from the Orinoco River on the east to the Isthmus of Panama on the west, settled by the Spanish in the first half of the 16th century. Later it referred to the Caribbean in general, as travelled by Spanish merchantmen. The 'Main' is intimately linked with early struggles to control Caribbean trade, and especially with the English 'sea dogs' John *Hawkins and Sir Francis *Drake. These and others were involved in early slave trading and raided Spanish shipping, even temporarily capturing islands and ports in the late 16th century.

Spanish Netherlands, the southern provinces of the Netherlands ceded to *Philip II of Spain in the Union of Arras (1579), during the *Dutch Revolts. These lands originally included modern Belgium, Luxemburg, part of northern France, and what later became part of the *United Provinces. Although Philip II still intended to re-subjugate the rebellious northern provinces, he granted the sovereignty of the Spanish Netherlands to his daughter Isabella and her husband the Archduke Albert (1598). During the Twelve-Year Truce (1609–21) and the unsuccessful war against the United Provinces (1621–48), the region enjoyed only nominal independence from Spain. A great deal of territory was lost to *Louis XIV

of France during the wars of the 17th century, including Artois and part of Flanders. On the expiry of the Spanish Habsburg dynasty in 1700, the region came under French rule until 1706, when it was occupied by the British and Dutch. By the Peace of *Utrecht (1713) it passed under the sovereignty of the Austrian Habsburg Holy Roman Emperors.

Spanish Succession, War of the (1701–13), a conflict which arose from the problem facing Europe on the death of the childless Charles II of Spain in 1700. One of his sisters had married *Louis XIV, the other Emperor Leopold, so both the French *Bourbons and the Austrian *Habsburgs claimed the right to rule the Spanish empire, which included the southern Netherlands, Milan, Naples, and most of Central and South America. Before Charles II's death *William III took a leading part in negotiations to pre-empt the crisis, and a partition treaty was signed (1698) between *Louis XIV and William, that Spain and its possessions would be shared out between France, Austria, and Joseph Ferdinand, the 7-year old Elector of Bavaria, grandson of Leopold. Charles II meanwhile left all of Spain's empire to Joseph Ferdinand. When he died, Louis and William signed a second partition (1699). However, Charles II left a will bequeathing his whole empire to Louis XIV's second grandson, the future *Philip V. Louis accepted this will and, instead of allaying European fears of French domination, intervened in Spanish affairs, seized the Dutch barrier fortresses, recognized *James II's son as King of England, and refused

An elaborate *auto-da-fé* ('act of faith') of the **Spanish Inquisition**. As the engraving shows, such acts of cruelty against Jews, Muslims, and Protestant heretics were turned into grand public spectacles.

SPAANSCHE INQUISITIE.

War of the Spanish Succession

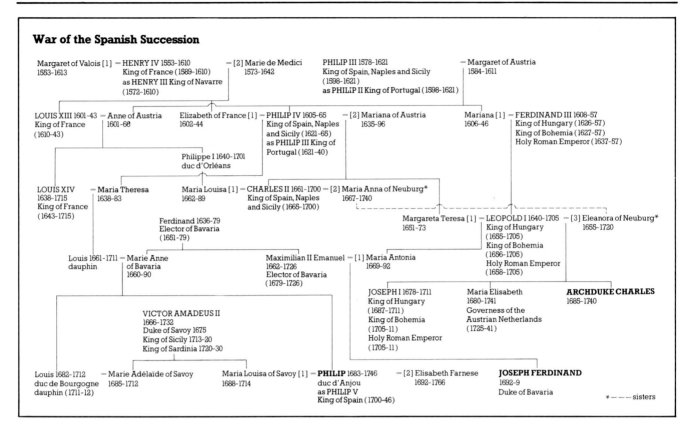

| | | | |

Margaret of Valois [1] 1553-1613 = HENRY IV 1553-1610 King of France (1589-1610) as HENRY III King of Navarre (1572-1610) = [2] Marie de Medici 1573-1642

PHILIP III 1578-1621 King of Spain, Naples and Sicily (1598-1621) as PHILIP II King of Portugal (1598-1621) = Margaret of Austria 1584-1611

LOUIS XIII 1601-43 King of France (1610-43) = Anne of Austria 1601-66

Elizabeth of France [1] 1602-44 = PHILIP IV 1605-65 King of Spain, Naples and Sicily (1621-65) as PHILIP III King of Portugal (1621-40) = [2] Mariana of Austria 1635-96

Mariana [1] 1606-46 = FERDINAND III 1608-57 King of Hungary (1626-57) King of Bohemia (1627-57) Holy Roman Emperor (1637-57)

Philippe I 1640-1701 duc d'Orléans

LOUIS XIV 1638-1715 King of France (1643-1715) = Maria Theresa 1638-83

Maria Louisa [1] 1662-89 = CHARLES II 1661-1700 King of Spain, Naples and Sicily (1665-1700) = [2] Maria Anna of Neuburg* 1667-1740

Ferdinand 1636-79 Elector of Bavaria (1651-79)

Margareta Teresa [1] 1651-73 = LEOPOLD I 1640-1705 King of Hungary (1655-1705) King of Bohemia (1656-1705) Holy Roman Emperor (1658-1705) = [3] Eleanora of Neuburg* 1655-1720

Louis 1661-1711 dauphin = Marie Anne of Bavaria 1660-90

Maximilian II Emanuel 1662-1726 Elector of Bavaria (1679-1726) = [1] Maria Antonia 1669-92

JOSEPH I 1678-1711 King of Hungary (1687-1711) King of Bohemia (1705-11) Holy Roman Emperor (1705-11)

Maria Elisabeth 1680-1741 Governess of the Austrian Netherlands (1725-41)

ARCHDUKE CHARLES 1685-1740

VICTOR AMADEUS II 1666-1732 Duke of Savoy 1675 King of Sicily 1713-20 King of Sardinia 1720-30

Louis 1682-1712 duc de Bourgogne dauphin (1711-12) = Marie Adélaïde of Savoy 1685-1712

Maria Louisa of Savoy [1] 1688-1714 = PHILIP 1683-1746 duc d'Anjou as PHILIP V King of Spain (1700-46) = [2] Elisabeth Farnese 1692-1766

JOSEPH FERDINAND 1692-9 Duke of Bavaria

*———sisters

to make it impossible for Philip also to inherit the French throne.

In 1701 William III formed a grand alliance of the English and Dutch with the Austrian emperor and most of the German princes to put the rival Austrian candidate, the Archduke Charles, on the throne; Savoy and Portugal later joined the alliance. William died in 1702 and the war therefore became Queen Anne's War. Fighting took place in the Netherlands, Italy, Germany, and Spain. France's only allies were Bavaria and the people of Castile, who supported Philip V while Catalonia declared for the Archduke Charles. *Marlborough and *Eugène of Savoy won a series of brilliant victories, including *Blenheim. France was invaded in 1709 and the allies were stronger at sea, taking Gibraltar, in 1704. The war at last came to an end because Castile would not abandon Philip V, there was general war weariness, and when Marlborough fell from power the new Tory government in England began the negotiations which led to the Peace of *Utrecht (1713).

Sparta, the usual name for the state of Laconia in ancient Greece, of which the town of Sparta was the capital. Invading Dorian Greeks occupied Laconia *c*.950 BC, and by about 700 BC the Spartans had emerged as the dominant element among them, with a large slave class of helots working on the land. Sparta had also, in the late 8th century, defeated and annexed the territory of Messenia, its western neighbour, reducing its population to helotry and dividing its land amongst the full Spartiate citizens. The Spartan state of the classical period was headed by two 'kings', who were the hereditary commanders of the army. An assembly of all adult male Spartiates had ultimate sovereignty but it generally followed the lead of the Senate which comprised the kings and twenty-eight 'elders'. The Spartiates, relieved from their daily tasks by the helots, were free to cultivate military skills, and from the age of 7 underwent a rigorous communal physical and military training that produced the finest soldiers in Greece. The stark austerity, militarism, and discipline of Spartan society were traditionally ascribed to a single great legislator, Lycurgus, variously dated *c*.900 and *c*.700 BC; it is likeliest that the fully developed Spartan system took shape somewhere between 700 and 600 BC.

From the 6th century, Sparta became the hub of an alliance which comprised most of Peloponnesian and Isthmian states except its traditional rival, Argos; but many of these allies in the 'Peloponnesian League' were little more than puppets of Sparta. Sparta led the successful Greek resistance in the *Greek–Persian wars, but later came into protracted conflict with *Athens in the *Peloponnesian War. Its final victory in 404 BC left it dominant in Greece and the Aegean; but after crushing defeats by Thebes at Leuctra (371) and Mantinea (362) and the loss of Messina it declined in importance.

The authoritarian and stratified 'closed society' of Sparta was much admired by some Greeks, most notably by *Plato. But the stifling of individual initiative and hostility to new ideas, the ever-present threat of helot revolts, the instability of its unrepresentative governments, and (from the mid-5th century) a steep decline in the numbers of the exclusive caste of full Spartiates undermined Sparta's bid for lasting domination of the Greek world.

Spartacus (d. 71 BC), a gladiator, leader of a slave revolt against Rome. A shepherd, then a Roman military auxiliary, he deserted and on recapture trained as a *gladiator in Capua. In 73 BC he led an uprising that occupied the dormant crater of the volcano Vesuvius, and went on to defeat two Roman armies. His following swelled to 90,000, and they devastated southern Italy. *Crassus and *Pompey defeated him in 71 and Pompey claimed credit for intercepting some of the fugitives. He was killed and many of his followers were crucified. He has since then been a hero to revolutionaries; early German Marxists called themselves 'Spartakists'.

Spence, Thomas (1750–1814), English social reformer who advocated the nationalization of land. In 1775 he published a pamphlet, *The Real Rights of Man*, in which he proposed that all land should be placed in the hands of local corporations, who would charge a fair rent for its use and distribute the money earned among the community. He moved to London in 1792 and set up as a printer of radical tracts. He was imprisoned twice, once for publishing *Paine's *Rights of Man*. After his death his followers continued to meet as members of the socity of Spencean philanthropists.

Spice Islands   *Moluccas.

Spinoza, Baruch (1632–77), Dutch Jewish philosopher and theologian. He was expelled from the Amsterdam synagogue in 1656 for his unorthodox views and lived in some obscurity, making a living from grinding and polishing lenses. He believed in the absolute unity of the universe and God (pantheism) and not in personal immortality. His critical questioning of the Bible was regarded as blasphemy and most of his writings, of which the most important were the *Ethics*, were published posthumously. Politically he argued for a *social contract in which man surrendered part of his natural rights to the state in order to guarantee his security.

Spithead mutiny (April 1797), a mutiny by sailors of the British navy based at Spithead, off the southern coast of Britain. In April 1797 the fleet refused to put to sea, calling for better pay and conditions, including the provision of edible food, improved medical services, and opportunities for shore leave. The Admiralty, acknowledging the justice of the sailors' grievances and fearing that the mutiny would spread further (by May it had already affected the fleet stationed at the *Nore), agreed to their demands and issued a royal pardon.

Spurs, battle of the   *Courtrai, battle of.

squire (or esquire), originally an apprentice *knight in medieval Europe. Usually young men, they served as the personal attendants of fully fledged knights. The title was then one of function rather than birth. It derives from the Latin 'scutarius', referring to the shield-bearing role of the squire. In later medieval England the term came to be applied to all gentlemen entitled to bear arms. By the 17th century, 'squire' had become synonymous with a district's leading landowner, perhaps even the lord of the manor. The considerable local influence, both political and ecclesiastical, of the 'squirearchy' has since diminished.

Sri Lanka (formerly Ceylon), an island in the Indian Ocean off the south-east tip of India. Its early history was shaped by Indian influences and its modern identity by three phases of European colonialism. The origins of the dominant Sinhalese racial group go back to Indo-Aryan invaders from north India, whose successors

dominated the north central plain from the 5th century BC until about AD 1200. During the 2nd century BC *Buddhism spread, following the conversion of the reigning king. An outstanding ruler was Parakramabahu I (1153–86), who exercised strong military and administrative leadership and also reformed the quarrelling Buddhist sects. However, intermittent invasions from south India gradually created an enclave of Tamil Hindu power on the northern Jaffna peninsula and the north-eastern coast. The centre of Sinhalese and Buddhist civilization gradually shifted south-westwards, and political power was divided between a number of kingdoms.

European contacts began in the early 16th century when Portuguese merchants, profiting from the internal disunity, gained trading privileges on the west coast. Dutch traders gradually supplanted Portuguese influence in the 17th century, but were replaced by British forces in 1796. When the embattled interior kingdom of Kandy fell in 1815, the entire island came under British control.

Srivijaya, a river port near Palembang in southern *Sumatra. It was described by a Chinese Buddhist pilgrim of the 7th century as 'a great fortified city'. Its position some 80 km. (50 miles) upstream from the mouth of the Musi River enabled it to trade easily with the mountain peoples of the interior and also protected it from seaborne attack. There were probably two separate states known as Srivijaya, the first flourishing from the 7th to the 9th century, the second from the 10th to the 13th century. The latter grew into a commercial empire extending from Java to Kedah, on the Malay peninsula and had trading contacts with China. Later, raids by Tamils from *Chola in India weakened its power. In the 14th century it succumbed to *Majapahit. Its port, once the greatest in south-east Asia, declined and became a Chinese pirates' base.

Stamford Bridge, battle of (25 September 1066). At a village on the River Derwent in Yorkshire, north-east England, *Harold II of England all but annihilated a large invading army under his exiled brother Tostig and the King of Norway, Harald Hardrada, both of whom were killed. Earlier (20 September) they had inflicted a heavy defeat on the Saxon forces of Earl Edwin of Mercia and Earl Morcar of Northumbria. Harold's army marched south from Stamford to face the Norman invasion and fight the battle of *Hastings.

Stamp Act (1765), a British taxation measure, introduced by *Grenville to cover part of the cost of defending the North American colonies. It required that all colonial legal documents, newspapers, and other items should bear a revenue stamp, as in England. Seen by *Sons of Liberty and many Americans as a first attempt at 'taxation without representation', it was met with widespread resistance. In October 1765, nine colonial delegations met at the Stamp Act Congress in New York and petitioned for repeal. American boycotts of British goods and civil disobedience induced *Rockingham to accede in 1766, though the Declaratory Act reasserted parliamentary power over the colonies. It helped initiate the campaign for American independence.

Stanhope, James, 1st Earl (1673–1721), English soldier and statesman. He served in Spain during the War of the *Spanish Succession and was appointed commander of the British forces there in 1708. He won some successes, but was captured in 1710. On his return to Britain he entered politics, and played a major part in securing the succession of *George I. As a leading minister he organized the government's swift response to the *Fifteen Rebellion. His genius lay in foreign affairs: he ended Britain's isolation, securing a treaty of alliance with its recent enemy France; he put forward a feasible solution to Austro-Spanish rivalry over Italy, and worked for a settlement of the *Northern War. Accused, probably unjustly, of involvement in ministerial corruption arising from the *South Sea Bubble, he died of a stroke while defending himself against his accusers in the House of Lords.

Stanislaus II (formerly Count Stanislaus-Augustus Poniatowski) (1732–98), the last King of Poland (1764–95). He was a lover of *Catherine II of Russia and her candidate for the Polish throne, and as the country was under Russian control he gained the crown. In 1772 at the first partition of Poland, Russia, Austria, and Prussia all took slices of Polish territory. From 1773 to 1792 there was a period of national revival encouraged by Stanislaus. In 1793 he was forced to agree to the second partition

The court of **Star Chamber** in session, with the king presiding over the meeting of judges and privy councillors. From a book published in 1555 recounting the annals of the reign of Henry VII, under whom the court of Star Chamber was founded.

which left him with a truncated kingdom and made him almost a vassal of Russia. A rising led by General *Kosciuszko was crushed, and the third partition completed the destruction of Poland. In November 1795 he was forced to abdicate.

Star Chamber, an English court of civil and criminal jurisdiction primarily concerned with offences affecting crown interests, noted for its summary and arbitrary procedure. It was long thought to have had its origin in a statute of 1487; in fact, however, since the reign of *Edward IV the court of Star Chamber had been developing from the king's council acting in its judicial capacity into a regular court of law. It owed its name to the fact that it commonly sat in a room in the Palace of Westminster that had a ceiling covered with stars. Its judges specialized in cases involving public order, and particularly allegations of riot. Its association with the royal prerogative, and *Charles I's manipulation of legislative powers in the making of decrees during the period of his personal rule, made it unpopular in the 17th century and caused its abolition by the *Long Parliament in 1641.

States-General (or Estates-General), usually a gathering of representatives of the three estates of a realm: the church; the nobility; and the commons (representatives of the corporations of towns). They met to advise a sovereign on matters of policy. The name was applied to the representative body of the *United Provinces of the Netherlands in their struggle for independence from Spain in the 16th century. As an assembly of the various provinces of the Dutch Republic it wielded considerable power, although delegating authority in emergency to the House of *Orange. It was replaced in 1795 by a national assembly, but was restored as a legislative body for the kingdom of the Netherlands in 1814.

In France, it began as an occasional advisory body, usually summoned to register specific support for controversial royal policy. It was developed by Philip IV who held a meeting in 1302 to enlist support during a quarrel with the pope, but throughout the 14th century it was rarely convoked and the first proper States-General in France was in 1484 in the reign of Louis XI. Thereafter it was used by the *Guises during the French Wars of Religion as a political device against the Huguenots, and another was called by the nobles in 1614 to attack Marie de Medici, who, however, turned it to her own advantage. The rise of absolutism saw its neglect in the 17th century, but it was urgently summoned in 1789 in an attempt to push through much needed revenue and administrative reforms. The *Parlements and the nobles had resisted these reforms and *Louis XVI and his minister, *Necker, tried to break the deadlock. Its summoning and composition were based on the precedent of 1614 and its members were encouraged to draw up *cahiers*, representative lists of grievances. Voting was carried out eventually by head rather than by order (as in 1614), giving radicals a majority. The nobles lost control of the States-General, which then formed itself into a *National Assembly, helping to precipitate the *French Revolution.

statholder, a provincial leader in the Netherlands, first appointed by the ruling dukes of *Burgundy in the 15th century. Their duties included presiding over the provincial state assemblies and commanding provincial armies. During the *Dutch Revolts (1568–1684), they were elected by the central States-General and subsequently by the provincial state assemblies. In the *United Provinces the House of *Orange-Nassau came to dominate the statholderates. Within the province of Holland there was protracted dispute between the Orange statholders and the states for overall leadership. In 1795 the office of statholder ceased to exist.

steam power, the use of steam to power machinery, a major factor in the *Industrial Revolution. A steam engine, developed by Thomas Newcomen (1663–1729) by 1712 was used to pump water from Cornish tin mines. Major improvements made by James Watt (1736–1819) greatly increased its efficiency and in 1781 he adapted a steam engine to drive factory machinery, thus providing a reliable source of industrial power. Before this many factories depended on water power and were sited in the countryside near swiftly flowing streams, where transport was difficult, and production always dependent upon the weather. However, steam engines were expensive and only large businesses could afford to install them. Factories were sited near coal mines and large towns grew up to house the factory workers. The use of steam engines in the textile industry and in other manufacturing processes led to a growth in the size of factories while their application in the 19th century to railways and steamships led to both faster and cheaper travel and transport of goods. The steam hammer (1808) enabled much larger pieces of metal to be worked while such developments as the steam-driven threshing machine reduced farmers' reliance on wind- and water-mills.

Stephan Dushan (1308–55), King of *Serbia (1331–55). The greatest ruler of medieval Serbia, he deposed his father in 1331 and took the title of Emperor of the Serbs and Greeks in 1345. He also controlled Bulgaria as a result of a marriage alliance. He fought the Byzantine empire, and seized Macedonia, Albania, and much of Greece, and introduced a new code of laws. His achievements were shortlived as his son could not maintain the Serbo-Greek empire against *Ottoman invasion and regional challenges.

Stephen (c.1096–1154), King of England (1135–54). He seized the crown after his uncle, *Henry I, had persuaded the English barons including Stephen, to recognize his daughter *Matilda as his heir. Stephen was supported by his brother, the Bishop of Winchester, and others who disliked Matilda's husband Geoffrey, Count of Anjou.

Stephen's reign was marked by rebellion and intermittent civil war. In 1138 Matilda's half-brother Robert rebelled, and the Scots invaded northern England. Stephen was captured at Lincoln in 1141 and temporarily deposed, but defeats at Winchester (1141) and Faringdon (1145) forced Matilda to withdraw from England in 1148. However, the year before he died Stephen recognized Matilda's son as his successor, Henry II.

Stephen I, St (975–1036), King of Hungary (997–1038). He was crowned by the authority of Pope Sylvester II in 1000. He was chiefly concerned with the thorough Christianization of the country and with the establishment of a durable code of laws. The reign was troubled by persistent warfare, particularly with the Bulgars, but later with the German emperor, Conrad II, in which the

A print dated 1747 showing the early use of **steam power**. The 'engine to raise water by fire' depicted here is the Newcomen steam pump. Its simple beam action was effective for lifting water out of a mineshaft but not until Watt's improvements could steam engines drive machinery.

king was successful. King Stephen's crown remains the outstanding symbol of Hungarian royalty.

Sterkfontein, a complex of collapsed limestone caves near Johannesburg, South Africa, one of the most important sites for studies of early human evolution in Africa. In 1936 an *australopithecine fossil was found there and successive discoveries have made the site the richest source of fossils of the species *Australopithecus africanus*. It has not yet been possible to date the remains exactly but they seem to be from 3 to 2.5 million years old. With the australopithecines are many fossilized animal bones, which suggest that they were preyed upon by leopards and other large cats and by hyenas.

Stewart *Stuart

stocks *pillory.

Stoic, a member of a philosophical school founded by Zeno of Citium c.300 BC. The name derives from the fact that he taught in the *Stoa poikile* ('Painted Colonnade') at *Athens. Their philosophy had at its core the beliefs that virtue is based on knowledge; reason is the governing principle of nature; individuals should live in harmony with nature. The vicissitudes of life were viewed with equanimity: pleasure, pain, and even death were irrelevant to true happiness.

In time, the idea that only the consummately wise man (the philosopher) could attain virtue was challenged, and Stoicism became more relevant to the reality of politics and statesmen. This, and the Stoic belief in the brotherhood of man, helped the philosophy to make a real impact in later republican Rome, upon such men as the young *Cato (whose suicide brought him a martyr's fame), *Brutus, and *Cicero. Later it underlay much aristocratic opposition to the emperors, but even so its disciples included Seneca, tutor and adviser to *Nero, and the emperor *Marcus Aurelius.

Stone Ages, those periods of the past when metals were unknown and stone was used as the main material for missiles, as hammers, for making tools for such tasks as cutting and scraping and, later, as spear heads. Hard, fine-grained stone was the material most suitable for flaking. Although the best locally available would have been the material of first choice, stone needed for special purposes was occasionally brought from long distances away, even by early toolmakers of up to 2 million years ago as at *Olduvai Gorge and *Koobi Fora in eastern Africa. Flint is popularly associated with flaked stone tools, especially in Europe, but in Africa, where flint is rare, quartz, chert, and volcanic rocks, such as basalt and obsidian (natural glass), were the materials worked long before early Europeans used flint.

In Europe, three Stone Ages are recognized—the Old Stone Age (*Palaeolithic), the Middle Stone Age (*Mesolithic), and the New Stone Age (*Neolithic). In other parts of the world, different subdivisions are used. The Stone Ages are followed by the *Bronze and *Iron Ages. This division of prehistory into three chronological stages, defined by the main material used for tools (stone,

A rare example of a human figure in the rock paintings at Lascaux Cave, south-west France. Animal images predominate, suggesting that the art reflected the great importance of successful hunting for this **Stone Age** society.

bronze, and iron)—the Three Ages System—was first put to practical use for classifying archaeological material in Denmark in 1819. As it spread to other countries, it became necessary to subdivide the three ages.

Stonehenge, a prehistoric monument on Salisbury Plain in England. It consists of a circular bank and external ditch surrounding a ring of narrow pits, the Aubrey holes. This was a henge monument (the name derives from this site) of the late *Neolithic period, c.2500 BC. A short-lived double stone circle was replaced after 2000 BC by the well known stone structure, a horseshoe of seven trilithons, each of a lintel supported on two uprights, within thirty uprights with their lintels forming a continuous circle. Individual blocks of sarsen, a sandstone found on the Plain, are up to 54 tonnes in weight. The whole is duplicated in smaller blocks, up to 4 tonnes, transported in antiquity from Preseli Mountain in Wales, 220 km. (140 miles) away.

The function of this monument remains controversial— whether it was a temple, a secular ceremonial centre, or an astronomical observatory. Certainly it called for a vast outlay of technical effort, and its axis, over the outlying Heel Stone, does indeed align on the midsummer sunrise. Much of the speculation over Druids, Bronze Age kings, and the forecasting of lunar eclipses cannot, however, be proved.

Strafford, Thomas Wentworth, 1st Earl of (1593-1641), English statesman. He entered the service of *Charles I in 1628. He had been a Member of Parliament since 1614, and had opposed royal policies but he was a believer in firm government and accepted preferment in order to uphold the king's power. Thenceforth, as Lord President of the Council of the North (1628) and Lord Deputy of Ireland (1633), he was a principal exponent of the policy of 'Thorough'—putting the royal will into effect with the utmost authority. His autocratic style made him extremely unpopular, and most of his achievements in the north and Ireland turned out to be temporary. On the outbreak of the *Bishops' wars he was recalled by Charles to England. Now at the centre of affairs for the first time, he could not avert the approaching *English Civil War. He was created Earl of Strafford in January 1640, but impeached for treason in the same year. Opposition members of the *Long Parliament believed that he was about to impose a Catholic dictatorship on England. He was executed under Act of *Attainder.

Strathclyde and Cumbria, the Romano-British kingdom of north-west England and south-west Scotland which was formed in the 2nd century AD and lasted until the 11th century, the last kingdom of the Britons to disappear. It survived invasions by Eadbert of Northumbria (750 and 756), plundering by the Danes (870), and the temporary loss of its independence (920) to Edward the Elder of England. Duncan I acquired Strathclyde in 1018 and united it with three other regions to found the kingdom of Scotland (1034).

strips, scattered units of unfenced land, the basis of agriculture in much of England between the 10th and 18th centuries. Tenants received a number of strips (also called selions, lands, loons, ridges, and shots) based on their social and economic status: an average holding was 12 ha. (30 acres) or one virgate. Each strip was about

0.4 ha. (1 acre) in size, usually one furlong (one ploughed furrow long) or 210 m. (220 yards) in length and 20.1 m. (22 yards) in width. Each strip's shape was determined by the amount of land ploughed in a day and the nature of the terrain. Repeated ploughing in one direction produced a 'corrugated' effect ('ridge and furrow') which often survives in pasture-land once under strip cultivation. By the 13th century much of England (except East Anglia and Kent and hill areas mainly in the north and west) had evolved the fully developed common field system based on a three-course *crop rotation designed to prevent the soil from becoming exhausted. One field's rotation might be spring grain (first year), winter grain (second year), untilled or fallow (third year). Scattered strips would benefit from this rotation and would also give each tenant a mixture of good and poor soils. *Enclosures of consolidated land and farming improvements especially during the *Agricultural Revolution destroyed the strip system.

Stuart (or Stewart), the family name of the Scottish monarchs from 1371 to 1714 and of the English monarchs from 1603 to 1714. The founder of the Stuart house was Walter Fitzalan (d. 1177) who was steward (from which the name Stewart derives) to the King of Scotland. His descendant became the first Stewart king of Scotland as *Robert II (ruled 1371–90). By the 16th century the name was more generally spelt Stuart. The marriage of Margaret Tudor, daughter of *Henry VII, to *James IV linked the royal houses of Scotland and England, and on the death of *Elizabeth I without heirs in 1603, James VI of Scotland succeeded to the English throne as *James I. The Stuarts lost the throne temporarily with the execution of Charles I in 1649, regaining it with the *Restoration of Charles II in 1660. The *Glorious Revolution (1688) sent *James II into exile and the crown passed to his daughter Mary and her husband William, then to his second daughter *Anne. Her death without heirs in 1714 resulted in the replacement of the Stuart house by the *Hanoverian family headed by George I. Supporters of the exiled house of Stuart were known as *Jacobites. After the failure of the *Fifteen and the *Forty-Five (1715 and 1745) rebellions the Stuart cause faded, and George III felt able to grant a pension to the last direct Stuart claimant, Henry, Cardinal York, who died in 1807.

Stuarts

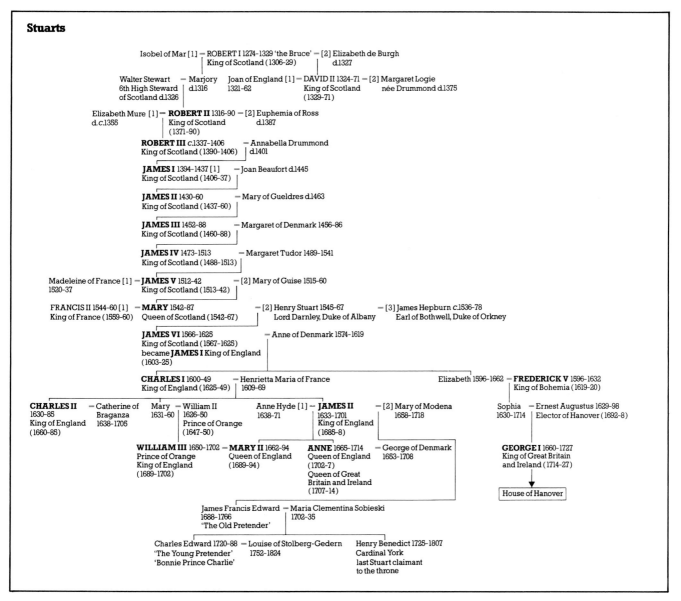

sturdy beggars, those classed in the English *Poor Law of 1531 as able-bodied persons who chose not to work. This presumed, wrongly, that there was enough work for everyone to do. Those who took to the roads, seeking jobs or charity, were severely punished. This was because Tudor governments regarded them as threats to public order, especially returned soldiers who organized themselves into bands and robbed travellers. By the end of the century new poor laws made parishes provide work for the genuinely unemployed, while 'incorrigible rogues' were to be whipped, returned to the parishes whence they had come, or even banished overseas for persistent offences.

Stuyvesant, Peter (1592–1672), Dutch colonial governor of the North American colony of New Netherland (1647–64). Though his authoritarian temperament led to friction with the colonists, he gained control of the rival colony of New Sweden in 1655. His own colony had been so infiltrated by English settlers, however, that he was unable to resist the Duke of York's expedition in 1664. New Netherland became the colony of New York.

Suetonius, Gaius Suetonius Tranquillus (c.AD 69–140), Roman historian of the early emperors. He was a lawyer who became the emperor Hadrian's private secretary. He wrote a biography of each of the first twelve Caesars. Part grounded in fact, part drawn from informed anecdote and court gossip, his work has preserved much valuable information. His main flaw as a historian was his failure to judge the accuracy of his raw material. He was one of the earliest non-Christian writers to record the early following of *Jesus Christ.

The Dutch colonial governor Peter **Stuyvesant**, artist unknown, c.1660, one of the earliest portraits in American history. (Library of Congress)

Sufi, the name of the Muslim mystics, from the word for the simple woollen garment (*suf*) worn by early ascetics. Seeking personal union with God, they have since the early days of *Islam eschewed the formal rituals and orthodox learning of the *ulema* (the doctors of sacred law and theology) in favour of esoteric practices and teachings from sheikhs, whose personal powers equip them to set initiates on the path to communion with God. Sufis have included outstanding poets such as the Persian Rumi, scholars like al-Ghazzali, and massive brotherhoods, among them the Qadiriyya and Tijaniyya. Frequently at odds with the orthodox establishment, they have been influential in spreading the faith in Africa and south and south-east Asia.

Sui (581–618), a dynasty that reunited China after over three centuries of territorial fragmentation, founded by Yang Qien. The first two emperors undertook campaigns against *Taiwan, *Annam, *Champa, and *Srivijaya. The *Great Wall was rebuilt following a different alignment, and canals, some later to form part of the Grand Canal, were dug. Defeats in Korea and never-ending demands for labour bred rebellion. The second emperor fled; betrayed by one of his 3,000 concubines he was strangled. When an ambitious official Li Yuan, founder of the *Tang dynasty, seized Chang'an, the last emperor abdicated. Although the dynasty only ruled for a short period, its establishment of a strong central government greatly assisted the development of the Tang dynasty.

Suleiman I (the Magnificent) (1495–1566), *Ottoman sultan (1520–66). He was known to Europeans as 'the Magnificent' from the brilliance of his court, but his subjects knew him as 'Qanuni'—the Lawgiver—from the many regulations produced by his administration. In three campaigns against Safavid Persia he confirmed Ottoman control of eastern Asia Minor and annexed *Mesopotamia. He drove the Portuguese from the Red Sea, taking Aden in 1538, and harried them in the Gulf. He created *corsair states in north Africa, frustrating Christian ambitions to rule there. He took Rhodes in 1522 and maintained a naval supremacy in the eastern Mediterranean marred only by a costly failure to take Malta in 1565. His most brilliant victory was at *Mohács in 1526, when he crushed the Hungarians and advanced on Vienna, though he could not take it. Nevertheless he brought the middle Danube under Ottoman rule and made Transylvania a dependency. He died on campaign. He was a poet and a patron of the arts, and did much to beautify Constantinople.

Sulla, Lucius Cornelius (c.138–78 BC), Roman soldier and statesman. He assumed the name 'Felix', or 'Fortunate'. Dictator of Rome in 82–79 BC, he served as quaestor (magistrate and paymaster) under *Marius in Africa and was instrumental in securing the betrayal of *Jugurtha. He had been accepted by the dominant aristocratic faction in the Senate by 88 when he was consul. Deprived of a command at Marius' instigation in 88 BC, he marched on Rome to regain it. After successful campaigning in the East he returned to take Rome a second time in 82 BC. He was appointed dictator to reconstitute the state and he restored the Senate's power, by imposing strict controls on the tribunes and other magistrates. Shortly before the end of an apparently dissolute life he retired abruptly.

Suleiman I (the Magnificent), seated, conferring with his army commanders at the siege of Buda during his Hungarian campaigns of 1541–3. Turks and Hungarians confront one another across the River Danube, with siege guns in evidence. From a contemporary miniature. (Topkapi Museum, Istanbul)

Sully, Maximilien de Béthune, duc de (1560–1641), French statesman. He was educated as a Huguenot and narrowly escaped death during the *St Bartholomew's Day Massacre (1572). In 1576 he joined the army of Henry of Navarre; he distinguished himself as a soldier, and contributed significantly to Henry's successful bid for the throne. Under the new king, *Henry IV, he helped to pacify the realm, reorganized the national finances by stringent economies and the reform of abuses, and encouraged agriculture. He was created duc de Sully in 1606, but was forced out of office soon after Henry IV's assassination in 1610.

sultanate, territory subject to sovereign independent *Muslim rule. The word 'sultan' is used in the *Koran and the traditions of the Prophet *Muhammad to mean authority. *Mahmud of Ghazna was the first Muslim ruler to be addressed as sultan by his contemporaries. It thereafter became a general title for the effective holders of power, such as the *Seljuk or *Mameluke dynasties, though it was also used as a mark of respect under the *Ottomans for princes and princesses of the imperial house. The term 'sultanate' is also used of a number of virtually independent centres of Muslim power, such as the sultanate of *Delhi (1206–1526), the predecessor of the *Mogul empire in India, and the Sulu sultanate, a trading empire in the southern Philippines, which flourished between the 16th and 19th centuries.

Sumatra, the westernmost island of *Indonesia. Indian traders early introduced Hinduism and Buddhism. In the first centuries AD small states emerged. When *Srivijaya's long dominance ended (c.1300), *Majapahit and *Bantam held sway over southern Sumatra. By 1300 Islam had a footing in the north, thence spreading over Sumatra, through Malaya, and the Indies. In 1613 the sultanate of *Acheh allowed the English to establish a trading station. The Dutch later secured a base at Padang on the west coast and began to incorporate Sumatra into the *Dutch East Indies.

Sumerians, a people living in southern Mesopotamia in the 4th and 3rd millennia BC. By 3000 BC a number of city states had developed in Sumer, such as *Uruk, *Eridu, and *Ur. The Sumerians are credited with inventing the *cuneiform system of writing, which was originally pictographic but gradually became stylized. Many simple inscriptions survive as evidence of this, and they also directly attest the increase in administration which accompanied urban growth.

The first great empire of Sumer was established by the people of *Akkad, who conquered the area c.2350 under the leadership of Sargon. The dynasty founded by him was destroyed c.2200, and after 2150 the kings of Ur not only re-established Sumerian sovereignty in Sumer but also conquered Akkad. This new empire lasted until c.2000 when pressure from the Elamites and Amorites reached its culmination with the capture and devastation of Ur. The Sumerians at this point disappear from history, but the influence of their culture on the subsequent civilizations of Mesopotamia was far-reaching.

summoner (or apparitor), a minor official, not a cleric, who summoned people before the ecclesiastical courts. Summoners acquired inquisitorial powers in cases leading to excommunication, such as non-payment of tithes, heresy (including incidents of *Lollardy in England and Scotland), usury, slander, and witchcraft. Over 10,000 excommunication writs survive from the 13th century. Summoners were condemned for extortion in the Council of London (1342), Parliament (1378), and by writers including Chaucer in *The Canterbury Tales*.

Sunderland, Robert Spencer, 2nd Earl of (1641–1702), English statesman. He was renowned for his intrigues and double-dealing. He became Secretary of State in 1679 but was dismissed for disloyalty. He won the favour of *James II by adopting Roman Catholicism, but meanwhile he was in secret correspondence with *William of Orange, for whom he reverted to Protestantism. William employed him until 1697, and learned from him to govern with a small inner circle of ministers; but Sunderland was despised by both Whigs and Tories, and he resigned to exert his political influence less openly.

Sundjata Keita, founder and ruler of the *Mali empire (c.1235–55) in west Africa. He was sickly as a child, but became a vigorous warrior. In about 1235 he was called

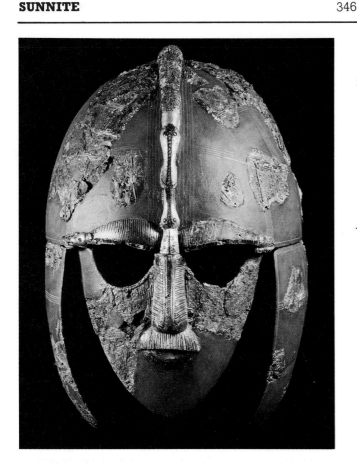

A reconstruction of the 7th-century visored helmet buried at **Sutton Hoo**. Made of iron and covered with tinned bronze, the decorative plates created a silvered effect, while the face-mask features were gilded. Garnets line the under edge of the eyebrows, and the brows and crest terminate in animal ornamentation typical of Anglo-Saxon art. (British Museum, London)

on to fight Soumangourou, King of Sosso, and defeated him at Krina. After further victories, he conquered *Ghana and Walata, and then all the neighbouring gold-bearing regions. After c.1240 he devoted himself to administration. He died accidentally, either from an arrow shot at random during a festival or by drowning in the Sankarani River.

Sunnite, a Muslim follower of the Sunna, the exemplary practice of *Muhammad as recorded in traditions (Hadith) relating to him. Sunnites constitute the majority of Muslims, as against the *Shiite minority. The Sunna, variously translated as 'custom', 'code', or 'usage', means whatever Muhammad, by positive example or implicit approval, demonstrated as the ideal behaviour for a Muslim to follow. It therefore complements the *Koran as a source of legal and ethical guidance. Sunnites also accept the first four *caliphs as legitimate successors to the Prophet's authority, whereas Shiites accept only Ali, the fourth, as such.

Supremacy, Acts of (1534 and 1559), enactments of the English Parliament, confirming respectively the supremacy of *Henry VIII and *Elizabeth I over the *Anglican Church. Henry was styled 'Supreme Head', but Elizabeth, in an attempt to reduce opposition was called 'Supreme Governor'. Under the terms of both Acts

the 'Oath of Supremacy' was demanded of suspected malcontents to ensure their loyalty.

Sutton Hoo, the site, overlooking the estuary of the River Deben in Suffolk, eastern England, where in 1939 a magnificent Anglo-Saxon ship burial was uncovered. It may have been the grave of Raedwald (d. c.624), King of East Anglia. No body was found, but it is believed that this may have been destroyed by the acidity of the soil. The tomb comprises a boat, 24.3m. (80 ft.) long and 4.2m. (14 ft.) across at its widest beam, which was propelled by thirty-eight oarsmen. Byzantine spoons and bowls, Swedish weaponry, Egyptian bowls, East-Anglian jewellery, and a hoard of forty gold coins from Merovingian Gaul make it the greatest single archaeological find in Western Europe.

Suvarov, Alexander Vasilevich, Count Rimniksky, Prince Italysky (1729-1800), Russian field marshal. His brilliant campaigns against the Poles (1769) and the Turks (1773-74) laid the foundations of his reputation. Further successes against the Turks fourteen years later led *Catherine II to appoint him a count in 1788. In 1790 he was placed at the head of the army that subdued the Poles. In 1796 he was dismissed by the new emperor *Paul I, but was recalled to face the French in Italy three years later. After some early successes he was forced to retreat and returned to St Petersburg in disgrace and, worn out and ill, died on 18 May 1800.

Swabia, an area in south-west Germany. Taking its name from the Suevi, a Germanic tribe resident in the area in Roman times, Swabia emerged as a medieval duchy in the 10th century and was ruled by the *Hohenstaufen dynasty from 1079 to 1268, before being divided among other rulers. In 1488 the **Swabian League** was formed by a group of cities and magnates to counteract the growing strength of the Swiss Confederation and the Bavarian Wittelsbach dynasty. It functioned briefly as the mainstay of imperial power in south-west Germany, drawing most of the small states of the area into its ranks, but became divided by religious issues and broke up in the 1530s.

Swanscombe Man, the earliest known human inhabitant of Britain, named after a village in Kent, England, where three skull bones were discovered in river gravels in 1935, 1936, and 1955. They belong to an adult individual, probably a young woman. Around 250,000 years ago, she and her kind camped by the river, while they hunted deer and other animals. The skull seems closely related to a similar skull from Steinheim in West Germany; both probably represent a transitional stage between *Homo erectus and *Homo sapiens. They seem to have more *Neanderthal features than the Arago skull has. A possible interpretation is that they are at the root of a European sidebranch of human evolution.

Sweden, a northern European country comprising the eastern part of the Scandinavian peninsula, along with the Baltic islands of Gotland and Öland. The country's earliest history is shrouded in legend, but Suiones tribesmen are mentioned by Tacitus and were probably the founders of the first unified Swedish state. Swedish *Vikings were active in the Baltic area, and also ventured into Russia and the Arab caliphate of Baghdad. Christianity was introduced in the 9th century but the whole population was not converted until much later. In the 13th century parts of Finland and Karelia were occupied, but from 1397 to 1523 Sweden belonged to the Danish-dominated Union of *Kalmar. *Gustavus Vasa led the successful revolt which ended in independence for Sweden, a national crown for himself and his dynasty, and the introduction of *Lutheranism as the state religion. During the 16th and 17th centuries Sweden expanded territorially and achieved considerable political status, thanks largely to the efforts of *Gustavus Adolphus and Axel *Oxenstierna. The high point was reached after the Treaty of *Westphalia, during the reign of Charles X (1654–60), but the strains of maintaining a scattered empire began to tell during the reign of *Charles XII (1682–1718). After the *Northern War (1700–21) the empire was dismembered, and a form of parliamentary government then prevailed until the coup of Gustavus III in 1772.

Sweyn Forkbeard (d. 1014), King of Denmark (c.986–1014). He revolted against his father Harald Bluetooth to establish himself as ruler of Denmark. He spent much of his time on *Viking raids, attacking England several times, once in alliance with *Olaf Tryggvason; he was paid well by its king *Ethelred to desist. In alliance with Sweden and Norwegian rebels at the battle of Svolde in 1000 he defeated his former ally, King Olaf, who was drowned. In 1013 he invaded England, capturing London, but died suddenly before the conquest could be consolidated. He was the father of *Canute.

Switzerland, a land-locked, largely mountainous country in central Europe occupied by the Celtic *Helvetii in the 2nd century BC. Its position astride vital Alpine passes caused the area to be invaded by the Romans, Alemanni, Burgundians, and Franks before it came under the control of the Holy Roman Empire in the 11th century. In 1291 the cantons (Swiss confederacies) of Uri, Schwyz, and Unterwalden declared their independence of their Habsburg overlords, and the alliance for mutual defence was later joined by Lucerne, Zürich, and Bern. During the 15th century this Swiss Confederation continued to expand, and it fought successfully against Burgundy, France, and the Holy Roman Empire, creating a great demand for its soldiers as mercenaries. During the *Reformation and *Counter-Reformation, its political stability was undermined by civil warfare, but in 1648 the Habsburgs acknowledged its independence in the Treaty of *Westphalia.

Syria, a country in the Middle East, which was settled successively by the Arameans, Akkadians, and Canaanites, and formed a valuable province of successive empires, from the Phoenicians to the Byzantines. After the Arab conquest of the 630s, Damascus became the brilliant capital of the Arab caliphate under the *Umayyads from 661 to 750, but subsequently Syria became a province of other rulers, such as the *Fatimids and the *Mamelukes of Egypt, before passing under *Ottoman control.

Tacitus, Publius Cornelius (c. AD 55–120), Roman historian and consul. He served as governor of Asia in 112. He compiled the earliest known account of the German tribes and a biography of his father-in-law, *Agricola, governor of Britain. His *Annals*, a history of Rome from Augustus to Nero survives in part but most of his *History* covering the period from Galba to Domitian is lost. He is one of the earliest non-Christian writers to record the crucifixion of *Jesus Christ, which he mentions in connection with the persecution of Christians in AD 64. As a historian he has always been admired for his scrupulous accuracy and his epigrammatic style.

taille *tallage.

Taiwan, an island about 160 km. (100 miles) off the coast of China. Portuguese explorers called it Formosa ('the Beautiful Island'). Sparsely populated by a non-Chinese people, it was long a Chinese and Japanese pirate base. In the 17th century the Dutch (1624) and the Spaniards (1626) established trading posts, the Dutch driving out the Spaniards in 1642. With the fall of the *Ming dynasty in 1644 opponents of the *Qing started to settle on the island and in 1661 'Koxinga' (Zheng Chenggong), a Ming patriot, expelled the Dutch. It was conquered by the Qing in 1683 and for the first time became part of China. Fighting continued between its original inhabitants and the Chinese settlers into the 19th century.

Taizong (T'ai-tsung) (596–649), second *Tang Emperor of China (627–49). He was renowned for his military prowess, scholarship, and concern for people. Strong central government was re-established, and he extended Chinese influence in Central Asia, subjugating in 630 the Eastern Turks, who accepted him as their Heavenly Khan. In his capital, Chang'an (now Xi'an), a city of two million people, envoys with tribute from Samarkand, India, and Sumatra mingled with merchants and scholars of many lands and many faiths—Nestorians, Buddhists, Zoroastrians, Manichees, Jews, Muslims, and others. He was a patron of Xuanzang, the Buddhist pilgrim who, in 645, brought back from India Buddhist scriptures which he translated into Chinese.

tallage, a tax in medieval Europe. It was generally imposed by an estate owner upon his unfree tenantry and its amount and frequency varied. In England it was a royal tax from the 12th century onward, which was levied on boroughs and royal lands. It was condemned by the barons in *Magna Carta in 1215 and became less important with the rise of parliamentary taxation, finally being abolished in 1340.

In France the 'taille' was greatly extended in the 14th century to meet the expenses of the *Hundred Years War, though, because it was the monetary equivalent of feudal service, the nobility and clergy were exempted from payment. The main burden of the taille, by now the most important direct tax, lay upon the peasants until it was abolished in the *French Revolution.

A portrait of the Tang emperor **Taizong**, wearing a dragon-embroidered silk robe. (National Palace Museum, Taipei)

Talleyrand-Périgord, Charles Maurice de (1754–1838), French diplomat and statesman. He entered the priesthood in 1778, and two years later he became agent-general of the clergy of France and in 1789 was installed as Bishop of Autun. He was one of a minority of French clergy who tried to reform the Church to serve the nation. In 1791 he resigned his bishopric, began a diplomatic career in London, and six years later was appointed Foreign Minister by the *Directory. He became *Napoleon's trusted adviser and Foreign Minister (1799–1807), and took part in most of the peace negotiations during the Napoleonic wars. He quarrelled with Napoleon in 1814, and, when the allies entered Paris, persuaded them to depose him. He was instrumental in securing the restoration of the Bourbons, whom he represented at the Congress of Vienna. After his resignation in September 1815 he spent fifteen years in semi-retirement until recalled by Louis-Philippe. He crowned his diplomatic career with the signing of the Quadruple Alliance of 1834.

Talmud (literally 'study'), the name given to the codification of Jewish oral tradition and learned commentary upon it. The destruction of the Temple in Jerusalem (AD 70) and the growth of *diaspora prompted a vigorous effort to preserve traditional teachings. *Judah Ha-Nasi *c*. AD 200 compiled the Mishnah, an organized summary of the oral tradition. This, together with subsequent commentary, the Gemara, constitutes the Talmud, of which there are two versions, the incomplete Jerusalem or Palestinian version of *c*.450 and the Babylonian version of *c*.500, which is taken as the authoritative text. Summaries of Talmudic teachings were subsequently prepared by scholars like *Maimonides in the 12th century and Joseph *Caro in the 16th century; the latter's 'Shulhan Arukh' ('Prepared Table') found wide favour as a work of ready reference. Study of the Talmud has been central to Jewish intellectual and religious life ever since its compilation.

Tamerlane (*Timur Leng*, 'Timur the Lame') (1336–1405), Mongol leader, a descendant of *Genghis Khan. He seized Turkistan and from this base conducted destructive operations against Persia (1380–8, 1392–4),

The Mongol leader **Tamerlane** besieging Urganj, from the *Zafar-nama* of Shaval ad-Din Yazdi, copied by Murshid al-Attar of Shiraz, 1523. (British Library, London)

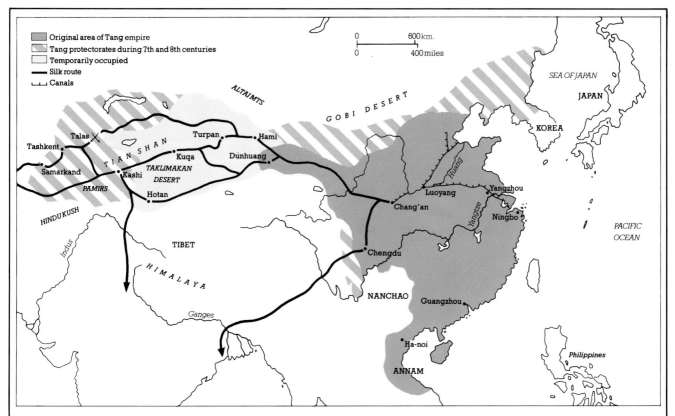

The Tang: 618–906

Under the Tang, China was initially a highly centralized state. An efficient system of canals augmented river transport, and a good road system linked provincial centres with the capital, Chang'an (Xi'an), which was probably the largest and most cosmopolitan city in the world in the 7th century. Standing at the end of the Silk Route, it attracted merchants from Persia, Arabia, and other parts of Asia. The Tang were especially noted for their sculpture and painting.

the *Golden Horde (1388–91, 1395), the sultanate of *Delhi (1398–9), the *Mameluke rulers of Egypt and Syria (1399–1401), and the *Ottomans, whose ruler *Bayezid I he captured at the battle of Ankara (1402). He died marching on China.

His genius was military rather than administrative, his conquests being followed merely by the installation of vassal rulers rather than integration into an imperial system; his descendants, the Timurids, ruled only the central Asian heartland. Fighting in the name of Islam, he usually plundered fellow Muslims. He was a keen patron of the arts, and imported foreign craftsmen to beautify his capital, Samarkand.

Tang (618–907), a Chinese dynasty founded by the *Sui official Li Yuan, which first established its power over China with help from nomad troops commanded by his son, later to become the second Tang emperor *Taizong. The unification of China started by the Sui was extended. Chinese armies penetrated central Asia, *Korea, and *Annam. Until defeated in 751 by the Arabs near the Talas River, West Turkistan, it ruled the largest empire in the world and its vessels voyaged as far as Aden. Printing was invented and gunpowder manufactured for fireworks. The dynasty is famed for its art, literature, and poetry. Neighbouring states, particularly *Korea and *Japan, sought to make their homelands replicas of China. After *Minghuang's reign and the abortive rebellion of An Lushan in 755 decay set in. Nomad invasions and revolts brought to power generals who con-trolled regional armies. When the last emperor abdicated fragmentation under ephemeral 'dynasties' followed.

Taoism, one of the two major Chinese religious and philosophical systems (the other is *Confucianism), tra-ditionally founded by *Laozi in the 6th century BC. The central concept and goal is the *tao*, an elusive term denoting the force inherent in nature and, by extension, the code of behaviour that is in harmony with the natural order. Its most sacred scripture is the *Daodejing* (*Tao Te Ching*; also called *Laozi*), ascribed to Laozi.

A popular cult appeared later. Numerous gods, of whom the Jade Emperor became chief, were attached to it. It acquired monasteries and priests. It associated itself with magic practices and inspired rebels—for instance the Yellow Turbans under the *Han. Taoist alchemists tried to turn cinnabar into gold and discovered magnetism and how to make gunpowder. In time Taoism and certain Buddhist sects became almost indistinguishable. Popular Taoism borrowed the concept of reincarnation from the Buddhists, but the final goal was not nirvana but becoming an 'immortal'. The cult was at times favoured, at times frowned upon, by China's rulers.

Tara, the ancient coronation and assembly place of the High Kings of *Ireland, in county Meath. In the 4th century there were five tribal kingdoms: Ulster, Meath, Leinster, Munster, and Connaught, which nominally acknowledged the overlordship of the High King (the ruler of Tara). Conn was reputedly the first High King

The hill of **Tara**, County Meath, seat of the High Kings of Ireland until the 6th century. Although the walls are long gone, the banks and ditches that once marked the royal buildings can still be discerned, especially the immense royal enclosure 290 by 245 m. (950 by 800 ft.).

('Ard Ri'). Niall of the Nine Hostages, possibly the son of a British prince, ruled there c.400 and his son Leary received St Patrick there in 432.

Tariq ibn Zaid (fl. 700-12), a freedman of Musa ibn Nusayr, *Umayyad governor of North Africa. In 711 he was sent to conquer Spain with 7,000 men, landing near the famous rock which has immortalized his name, Jabal al-Tariq (Mount of Tariq), that is, Gibraltar. On 19 July 711 he defeated the Visigothic king Roderick and went on to conquer half of Spain. In 712 Musa crossed to Spain, and, out of jealousy at Tariq's success, put him in chains. His subsequent fate is unknown.

Tarquin, traditionally the name of the fifth and seventh Etruscan kings of Rome, Priscus (616-579 BC) and Superbus (534-510), both subjects of legend and tradition. The stories were largely symbolic, contrasting the decadence of the monarchy with the idealism of the new *Roman republic. After this time the word 'king' was used by the Romans as a term of political abuse.

Tartars, a name used with some imprecision to cover a number of Central Asian peoples who, over the centuries, were a threat to civilized peoples in Asia and Europe. More specific names, for example Mongol, Turk, Kipchak, emerge for some of these peoples who were constantly moving, often over great distances, and who spoke a variety of related Turki and Mongol languages. The name Tartars (Tatars) is applied specifically to tribesmen living south of the Amur who were defeated by the Ming emperor *Yongle in the early 15th century. Papal envoys (c.1250) to the Mongols consistently called them Tartars, probably by association with Tartarus, the place of punishment in the underworld of Greek mythology. The name was also applied to the *Golden Horde. Some of the Cossacks (originally Kahsaks, 'free men') on the River Dnieper were Tartars. Later any

people of Turkish stock in Russia were called Tartars. In the 16th century the khanates of the Volga Tartars came under Russian rule. In the 15th century the Crimean Tartars formed an independent khanate, tributary to the *Ottoman Turks until annexed by Russia in 1783.

Tasman, Abel Janszoon (c.1603-59), Dutch navigator and European discoverer of New Zealand. In 1642 he sailed round western Australia and then turned due east until he came to land which he named Van Diemen's Land (after the governor of the Dutch East Indies, Anthony van Diemen, sponsor of the expedition), the island later renamed Tasmania. He continued eastwards until he sighted high land, the western coast of New Zealand, which he called Staten Landt. He sailed northwards up the New Zealand coast and continued into the Pacific and discovered several islands in the Tonga and Fiji groups. His next voyage resulted in a survey of the Gulf of Carpentaria and the northern Australian coast. By 1653 he had amassed a considerable fortune in the service of the *Dutch East India Company and settled down to retirement in Batavia.

Taxila, a ruined city north-west of Rawalpindi, Pakistan, which was an important centre of trade, learning, and pilgrimage for some thousand years. In the 5th century BC it was the capital of the Gandhara kingdom. It fell to Alexander the Great in 326 BC, and soon afterwards was incorporated in the *Mauryan empire, becoming a centre of *Buddhism. The Bactrians, Scythians, Parthians, Kushans, and Sassanians in turn invaded and annexed the city. It had already entered a period of decline when destroyed by Hun invasions in the 5th century, and it was never rebuilt.

Taylor, Jeremy (1613-67), Church of England divine. He was chaplain to *Laud and *Charles I and was taken prisoner during the *Engish Civil War. After the *Restoration he was appointed Bishop of Down and Connor and subsequently of Dromore, in Ireland. His writings, especially *Holy Living* (1650) and *Holy Dying* (1651), had great influence on contemporary religious attitudes. He was one of the earliest Anglican leaders to plead the case for religious toleration.

tell, a mound representing an ancient settlement site. Though the name, coming from the Arabic, is commonly

A drawing by the navigator and explorer Abel **Tasman**, showing Dutch ships, native boats, and Fiji islanders.

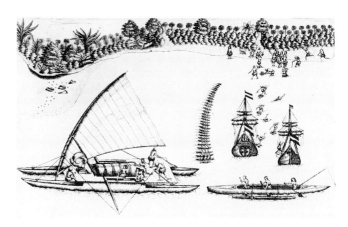

applied only to examples in the Near and Middle East, the phenomenon is known in many other areas where long-lived towns were built in mud brick. This decays quite rapidly and is not worth salvaging, so leads to a rapid accumulation of deposit.

Temple, Sir William (1628–99), English diplomat and author. As English ambassador to The Hague, he negotiated the Triple Alliance of 1668 between England, Holland, and Sweden, and brought about the marriage between William of Orange and Princess Mary, daughter of *James II, in 1677. His attempt to achieve a compromise solution to the political crisis of 1679 by reviving and strengthening the powers of the Privy Council ended in failure. He refused political honours from William III, and devoted the remainder of his life to writing.

Tennis Court Oath, a dramatic incident which took place at Versailles in the first stage of the *French Revolution. On 17 June 1789 the Third Estate of the *States-General under the presidency of Jean Bailly, a representative of Paris, declared themselves the *National Assembly, claiming that they were the only Estate properly accredited and that the First and Second Estates must join them. On 20 June they found their official meeting-place closed and moved to the Tennis Court, a large open hall nearby. The Oath bound them not to separate until they had given France a constitution.

Tenochtitlán, literally 'place of the Tenochca', the island capital of the *Aztecs in Lake Texcoco, now modern Mexico City. Traditionally founded in c.1345, it grew to a population of c.300,000 by the early 16th century. It was laid out in a grid of streets and canals round a huge ceremonial precinct of pyramids, temples, and palaces, and surrounded by artificial islands of gardens called *chinampas*. Three wide causeways stretched out across the lake to the mainland. As the hub of the Aztec empire, its capture by *Cortés on 13 August 1521 was rapidly followed by complete capitulation.

Teotihuacán, the earliest true city in Mesoamerica (Mexico and northern Central America). Its development

A Dutch engraving of 1794 of the large hall at Versailles where the **Tennis Court Oath** was sworn by the National Assembly in 1789.

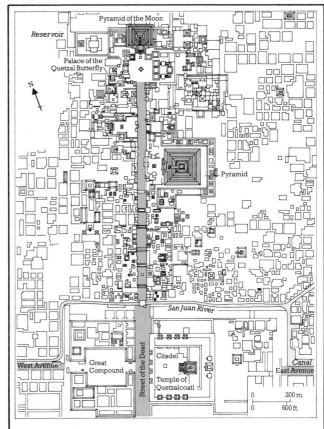

Teotihuacán

Teotihuacán, a genuine city rather than a ceremonial complex, was centred on a north-south axis, on either side of which were a series of plazas and pyramids. The population lived in large, cramped apartment compounds, in groups based probably on male blood relationships as well as social status.

began c.100 BC and for some 600 years it dominated Central Mexico economically and perhaps politically. At its height it covered c.22.5 square km. (c.8.7 square miles) with a regular grid of streets, alleys, and apartment-like residential complexes, and a population of c.200,000.

Teresa of Ávila, St (1515–82), Spanish nun and mystic, the originator of the *Carmelite reform. She entered a Carmelite convent at Ávila in 1535, but it was not until twenty years later that she underwent a religious experience. Thereafter, she opened the first reformed convent where the primitive (original) Carmelite rule was observed, in 1562, and from 1567 worked with St *John of the Cross in nurturing the work of reform. Considerable friction within the order led in 1579 to the granting of independent jurisdiction to Teresa's austere Carmelites, known as Discalced Carmelites. She was canonized in 1622. Her spiritual writings include *The Way of Perfection* and *The Interior Castle*.

Terror, Reign of, a period of the *French Revolution which began in March 1793 when the Revolutionary government, known as the Convention, having executed the king, set about attacking opponents and anyone else considered a threat to the regime. A Revolutionary Tribunal was set up to bring 'enemies of the state' to trial, and the following month the *Committee of Public

This devotional painting of **Teresa of Ávila** shows the saint, aged 61. The dove represents the visitation of the Holy Spirit. (Convent of St Teresa, Seville)

Safety was created. It began slowly but during the ruthless dictatorship that followed the defeat of the *Girondins at least 12,000 political prisoners, priests, and aristocrats were executed, including *Marie Antoinette and Madame *Roland. The Terror was intensified in June 1794 after the execution of *Hébert and *Danton had left *Robespierre supreme. It ended the following month, after the arrest and execution of Robespierre.

Tertullian, Quintus Septimius Florens (*c.* AD 160–240), a citizen of Carthage, the first Latin-speaking Christian writer, who had had a considerable influence on the development of Christian doctrine. He attacked Gnosticism, contributed to the formulation of a doctrine of the Trinity, urged a rigorous asceticism, and believed that only *martyrs were assured of salvation.

Test Acts, laws that made the holding of public office in Britain conditional upon subscribing to the established religion. Although Scotland imposed such a law in 1567, the harsh laws against *recusants in England were sufficient in themselves to deter Roman Catholics and dissenters from putting themselves forward for office. But in 1661 membership of town corporations, and in 1673 all offices under the crown, were denied to those who refused to take communion in an Anglican church. In 1678 all Catholics except the Duke of York (the future *James II) were excluded from Parliament. In the 18th century religious tests in Scotland were not always enforced, except for university posts, and in England the test could be met by occasional communion, but this was not possible for Roman Catholics. The Test Acts were finally repealed in 1829, and university religious tests were abolished in the 1870s and 1880s.

Tetzel, Johann (1465–1519), a German *Domincan friar. An agent for the sale of papal *indulgences (to raise money for the rebuilding of St Peter's, Rome) he

was responsible for inspiring Martin *Luther's protest of 1517, which sparked the *Reformation. He replied to Luther's *Ninety-Five Theses*, but died shortly afterwards, discredited.

Teutonic Knight, a member of a military and religious order whose full title was the Order of the Knights of the Hospital of St Mary of the Teutons in Jerusalem. Founded in 1190 at Acre, it was made up of knights, priests, and lay brothers and was active in Palestine and Syria, though its members retreated to Venice when the *Crusaders failed to contain the *Muslims. The Holy Roman Emperor Frederick II employed the order as missionaries to overcome and convert the pagans beyond the north-eastern border of the empire and in this they were very successful, gaining Prussia in 1229.

In 1234, though in practice independent, they declared that they held the lands they had conquered as a fief from the pope. Joining with the *Livonian Order, they continued to advance around the Baltic coast, amassing huge territories, but their progress was checked decisively when they were defeated at Tannenberg in 1410 by King Ladislas of Poland. In 1525 the Grand Master, Albert of Brandenburg, became a Lutheran, resigned his office, and the order was declared secular, and it remained an order under the control of the Electors of Brandenburg.

Teutons, a Germanic tribe, believed to be from Holstein or Jutland, who migrated to southern Gaul and northern Italy with the Cimbri in the late 2nd century BC. In 102 they were defeated by *Marius at Aquae Sextiae (Aix-en-Provence). They disappeared from history but the name Teuton survived as a synonym for German.

Tewkesbury, battle of (4 May 1471). A battle in the Wars of the *Roses fought between *Edward IV, fresh from his victory at *Barnet and the Lancastrian forces of Margaret of Anjou. Margaret's forces were defeated and her son, Prince Edward, was among those killed.

Thailand *Siam.

thane (also *thegn*), a nobleman of Anglo-Saxon England. Their status, as determined by their *wergild, was usually 1,200 shillings. In return for their services to the crown they received gifts of land which became hereditary. The king's thanes, members of the royal household, were required to do military service, attend the *witan, and assist in government.

Thebes (Egypt), the capital of ancient Egypt in the New Kingdom (*c.*1550–1050 BC), on the site of modern Luxor. On the eastern side of the Nile its remains are divided between *Karnak and Luxor. A sphinx-lined road linked Karnak to Luxor, where Amenhotep III built a magnificent temple to Amun. On the western side lie the two royal necropolises—the *Valley of the Kings and the Valley of the Queens—and various mortuary temples of which the finest may have been that of Amenhotep III (ruled *c.*1411–1372 BC). Of the actual temple building only the foundations remain, but also extant are the two 'colossi of Memnon' which represent the pharaoh.

Thebes (Greece), historically the most important city of *Boeotia in ancient Greece. From *c.*519 BC onwards it

An Egyptian wallpainting from **Thebes**. Discovered in the tomb of Neb-amun, it dates from c.1380 BC, shortly before the pharaoh Akhenaten left Thebes for his new capital of el-Amarna. The painting shows an offering of water being made to the mummies of the dead. (British Museum, London)

was a great rival of Athens, which in that year came to the aid of *Plataea. Thebans fought alongside the Persians in the *Greek–Persian wars, temporarily lost influence as a consequence, and fought in alliance with the Spartans throughout the *Peloponnesian War. The peace of 404 did not satisfy Thebes, however, and in the Corinthian War (395–386) it was allied with Athens, Corinth, and Argos against Sparta. From 382 to 378 the Theban citadel was garrisoned by the Spartans, but in 371 Theban troops routed the apparently invincible army of Sparta at Leuctra, thanks to the generalship of Epaminondas.

He it was who, together with Pelopidas, guided Thebes to its brief period of supremacy in Greece. However, a second great victory over Sparta, at Mantinea in 362, resulted in his death. Thebans fought with Athenians at *Chaeronea, but to no avail. When the city revolted against *Alexander the Great, it was destroyed and was never again a power in Greece.

Themistocles (c.528–462 BC), Athenian statesman and general. He survived several attempts at *ostracism in the 480s and 470s. He made Athens' future naval greatness possible when in 483 he persuaded his fellow citizens to spend the city's wealth on the construction of a fleet of *triremes. As one of Athens' generals in 480 BC, he was responsible for inducing the Persians to fight in the confined waters near *Salamis, and it was his skill that ensured the rebuilding of Athens' walls against Spartan wishes in 479–8 BC. He championed anti-Spartan policies, and was ostracized c.471, going to Argos and later fleeing to Asia Minor, where he was appointed governor of Magnesia-ad-Maeandrum by the Persian king, Artaxerxes I, the son of *Xerxes.

Theodora (c.500–548), Byzantine empress. She was probably the most influential woman in the history of the Byzantine empire. Of humble birth, she married *Justinian I, whom she dominated. A woman of courage, she made him stand against rioters anxious to overthrow him. Henceforth her influence in affairs of state was

A 6th-century mosaic at the church of St Vitale, Ravenna, depicting the Byzantine empress **Theodora** and her attendants processing towards a representation of Jesus Christ, in the apse. She carries a jewelled chalice and is surrounded by state regalia.

decisive, and she was merciless in her use of torture and secret police.

Theodoric I (AD 418–51), King of the *Visigoths. He was defeated by the Romans at Toulouse in 439 after fifteen years warfare. He joined Aetius as an ally against *Attila but was killed on the *Catalaunian Fields. His son Theodoric II ruled the Visigothic kingdom of Spain and parts of Gaul until 466.

Theodoric the Great (c.455–526), King of the Ostrogoths (475–526), and ruler of Italy from 493. He invaded Italy and established his capital at Ravenna. At its greatest extent his empire included not only the Italian mainland but Sicily, Dalmatia, and parts of Germany. An *Arian Christian, his reign brought a degree of authority and stability to the country, though the scholar *Boethius was executed on treason charges. His reign saw the beginning of a synthesis of Roman and Germanic cultures. He is a hero of German literature, figuring in the epic *Nibelungenlied*.

Theodosius, Flavius (the Great) (AD 349–95), Roman emperor in the East (379–94) and emperor (394–95). The son of a famous general, Count Theodosius, Gratian appointed him co-emperor in 378. After failing to defeat the *Goths he formed a treaty with them in 382. He was a champion of strict political and religious orthodoxy. His two sons, Arcadius and Honorius, succeeded him.

Thermopylae, a narrow pass on the Malian Gulf which once controlled entry to central Greece from the north-east. In 480 BC it was the scene of a famous defensive action by the Greeks against the invading Persians. The greatly outnumbered Greeks, led by the Spartan king *Leonidas, resisted a number of frontal assaults, but a traitor, Ephialtes, led the Persians via a mountain track round to attack their rear. Most of the Greeks were sent away, but Leonidas remained with a rearguard. He was killed along with 300 Spartans and 700 Thespians; he has been revered since as a Greek hero. The accompanying Thebans surrendered.

The pass was again outflanked by the Gauls in 279 BC and, in 191 BC, by the Romans under the Elder *Cato, who defeated the forces of the *Seleucid empire there.

Thessaly, a region of central Greece. It was famous in the ancient Greek world for its horsemen, but was isolated from *Boeotia and the rest of Greece. The Thessali conquered the area in the 12th century BC, and by the 6th century were a power of some importance. They fought alongside the Persians in the *Greek–Persian wars, but played little part in the *Peloponnesian War despite being allied to Athens. Their disunity prevented them from putting up effective resistance to *Philip II of Macedonia. Rome later allowed the formation of a Thessalian League, which was incorporated into the province of Macedonia in 148 BC. Diocletian created the Roman province of Thessaly c. AD 300, and it was later part of the Byzantine empire. After suffering the depredations of successive invaders, it entered Turkish rule in 1393. It finally became part of Greece in 1881.

Thing, a regular meeting or parliament in Norse communities, of nobles, priests, and heads of families, to establish or interpret law and to administer justice. The

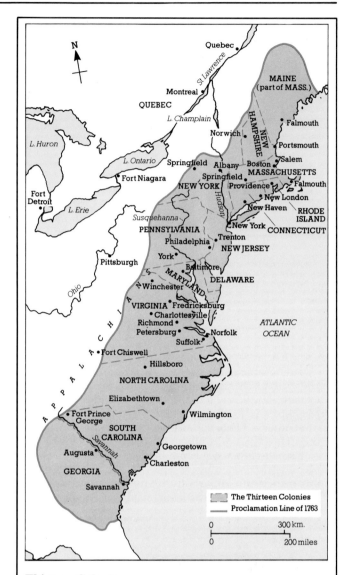

Thirteen Colonies

The thirteen British colonies in North America, strung out along the eastern seaboard from Maine to Georgia, had even fewer ties to one another than to the mother country. However, they did share certain underlying interests, one of which was the potential and desire for westward expansion. The Royal Proclamation of 1763, which forbade settlement beyond the Appalachians, was therefore widely resented, particularly since it came at a time when economic recession made the land beyond the frontier especially attractive. The Proclamation was issued as an emergency measure designed to protect Indian lands, but it was widely ignored by the settlers.

most famous example is the Icelandic Althing, founded in 930; it survived throughout the long period during which the island was under the Norwegian crown.

Thirteen Colonies, the British colonies in North America which ratified the *Declaration of Independence (1776) and thereby became founding states of the USA. They were, with dates of foundation or English colonial status: *Virginia (1607), *Massachusetts (1629), *Maryland (1632), *Connecticut (1635), *Rhode Island (1636), *North Carolina (1663), *South Carolina (1663), *New York (1664), *New Jersey (1664), *Delaware (1664),

*New Hampshire (1679), *Pennsylvania (1681), and *Georgia (1732). By 1776, all were ruled by royal governors except Maryland, Pennsylvania, Delaware, Connecticut, and Rhode Island, and all had representative assemblies. Though there were major differences over slavery or religion, for instance, and often quarrels between neighbouring colonies, they managed to sustain a fragile unity between 1776 and 1783. This improbable cohesion could be described as the greatest achievement of *George III and his ministers, who, in *Franklin's words, 'made thirteen clocks strike as one'.

Thirty-Nine Articles, the set of doctrinal formulae first issued in 1563 and finally adopted by the *Anglican Church in 1571 as a statement of its position. Many of the articles allow a wide variety of interpretation. They had their origin in several previous definitions, required by the shifts and turns of the English Reformation. The Ten Articles (1536) and Six Articles (1539) upheld religious conservatism, but the Forty-Two Articles (1553), prepared by *Cranmer and Ridley, were of markedly Protestant character, and they provided the basis of the Thirty-Nine Articles.

Thirty Years War (1618–48), the name for a series of conflicts, fought mainly in Germany, in which Protestant–Catholic rivalries and German constitutional issues were gradually subsumed in a European struggle. It began in 1618 with the Protestant Bohemian revolt against the future emperor *Ferdinand II; it embraced the last phase of the *Dutch Revolts after 1621; and was concentrated in a Franco-Habsburg confrontation in the years after 1635.

By 1623 Ferdinand had emerged victorious in the Bohemian revolt, and with Spanish and Bavarian help had conquered the *Palatinate of *Frederick V. But his German ambitions and his Spanish alliance aroused the apprehensions of Europe's Protestant nations and also of France. In 1625 Christian IV of Denmark renewed the war against the Catholic imperialists, as the leader of an anti-Habsburg coalition organized by the

Dutch. After suffering a series of defeats at the hands of *Tilly and *Wallenstein, Denmark withdrew from the struggle at the Treaty of Lübeck (1629), and the emperor reached the summit of his power.

Sweden's entry into the war under *Gustavus Adolphus led to imperial reversals. After Gustavus was killed at *Lützen (1632), the Swedish Chancellor *Oxenstierna financed the Heilbronn League of German Protestants (1633), which broke up after a heavy military defeat at Nördlingen in 1634. In 1635 the Treaty of Prague ended the civil war within Germany, but in the same year France, in alliance with Sweden and the United Provinces, went to war with the Habsburgs. Most of the issues were settled after five years of negotiation at the Treaty of *Westphalia in 1648, but the Franco-Spanish war continued until the Treaty of the Pyrenees in 1659.

Thomas of Woodstock (1355–97), Earl of Buckingham and Duke of Gloucester, the youngest son of *Edward III of England. From 1384 he was generally at odds with his nephew *Richard II, and in 1387–8 was one of the five *Lords Appellant who attacked Richard's leading counsellors; for some months afterwards Thomas was the most powerful man in England. In 1397 Richard was able to have him arrested, tried for treason, and executed.

Thrace, an area in the Balkans lying between the Black Sea and ancient Macedonia. It was inhabited by various Indo-European tribes. The Thracian tribes were conquered by the Persians c.516 BC, on whose side some of them fought in the *Greek–Persian wars. Later in the 5th century Teres, King of the Odrysae, extended his rule over a number of the tribes, and his son Sitalces allied himself with Athens against Macedonia. In the

The battle of Lützen in Saxony (16 November 1632) was a typical encounter of the **Thirty Years War**, with massed formations of infantry facing artillery bombardment. The engagement brought Wallenstein face to face with Gustavus Adolphus of Sweden. Although the Swedes won the battle, Gustavus was killed.

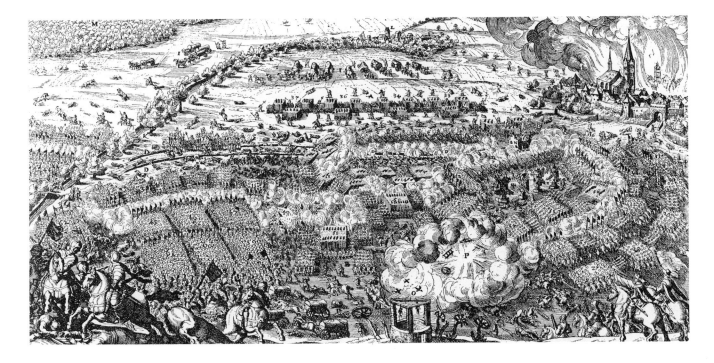

following century *Philip II annexed the area to Macedonia. Following Macedonia's final defeat by Rome in 168 BC, western Thrace was incorporated into the province of Macedonia. The next two centuries saw regular outbreaks of trouble between the Thracians and Rome. It suffered greatly from invasion by Visigoths and Slavs. It is now partitioned between Turkey, Greece, and Bulgaria.

Three Henrys, War of the (March 1585–August 1589), the eighth of the *French Wars of Religion. It was precipitated by the efforts of Duke Henry of *Guise to exclude the Huguenot Henry of Navarre from the succession to the French throne. Guise, backed by Spain and the Catholic League (*Holy League), forced Henry III to capitulate in the Treaty of Nemours (1585), and to accept the Catholic Cardinal de Bourbon as his heir instead of Henry of Navarre. In the ensuing conflict, Henry III lost control of events and Guise acted as if he were king himself. In late 1588, Henry III had both Guise and the Cardinal de Bourbon murdered. When Henry III was murdered in turn (1589), most of France was in League or Huguenot hands, and Henry of Navarre became king as *Henry IV.

Three Kingdoms (220–80), the period in China immediately following the end of the *Han dynasty. Three kingdoms, the Wei in the north, the Wu in the south-east, and the Shu Han in the west, rose and constantly fought each other for supremacy. The period ended when the Wei general Sima Yen seized power and unified China under the *Western Jin dynasty. Many events and legends of this period appear in one of the classics of Chinese literature, *The Romance of the Three Kingdoms*.

Throckmorton Plot (1583), an international Catholic conspiracy, in the manner of the *Ridolfi and *Babington plots, to place *Mary, Queen of Scots on the English throne. Francis Throckmorton (1554–84), a member of a leading English *recusant family, helped to contrive this unlikely scheme: Henry of *Guise would invade England with a French Catholic force, financed by Spain and the papacy; then the English Catholics would depose *Elizabeth I in favour of Mary. In late 1583 *Walsingham's agents uncovered the plans. Reprisals were moderate, but Throckmorton was tried and executed.

Thucydides (c.460–400 BC), Greek historian. He wrote an unfinished history of the *Peloponnesian War (431–404) which breaks off in the year 411. In 424 he failed, as naval commander in the northern Aegean, to save Amphipolis from capture by the Spartan Brasidas, for which he was exiled. Only at the end of the war did he return home. He called his work a 'possession for all time', and he is recognized as one of the greatest exponents of *history writing. He clearly took great care to discover the precise truth of the events about which he was writing. His accounts of the plague which afflicted Athens, the civil strife at Coreyra, the conquest of Melos, and the ill-fated Athenian expedition to Sicily are outstanding examples of his skill.

Thutmose I (or Tuthmosis), Pharaoh of Egypt (c. 1525–1512 BC). He extended his domains deep into Nubia and later penetrated with his army as far as the River Euphrates. He made extensive improvements to the temple of Amun at *Karnak and was the first pharaoh to be buried in the *Valley of the Kings.

Thutmose III (or Tuthmosis) Pharaoh of Egypt (c. 1504–1450 BC). During the first twenty-two years of his reign he was overshadowed by his aunt *Hatshepsut, wife of Thutmose II, who had herself declared regent in 1503. When she died in 1482, he promptly mobilized the army and defeated a coalition of Syrian and Palestinian enemies near Megiddo, gaining nearly all of Syria for his empire. Further successes followed, culminating in the defeat of the powerful Mitanni beyond the Euphrates. He also further extended Egyptian rule in Nubia, but generally he concentrated on the administration of his lands. The thriving prosperity of his reign was reflected in much new building at *Karnak.

Tiahuanaco, a vast ceremonial site and city near Lake Titicaca in Bolivia, and centre of the culture and art style of the same name, between c.AD 600 and 1000. As a religious centre it fostered the cult of the 'staff god', carved on a massive, monolithic lava-stone gateway. He has a halo of puma heads, and holds two staves tipped with eagles' heads. Similar birds and feline figures were painted on pottery, which has been found as far south as the Atacama Desert in Chile, and to the north were connected with the military expansion of the Huari empire in the 7th and 8th centuries.

Tiberius, Claudius Nero (42 BC–AD 37), Roman emperor (AD 17–37). He pursued a brilliant military career in Germany and Pannonia. He was Augustus' stepson, son-in-law, and adoptive son. From 6 BC–AD 2 he lived in virtual exile on Rhodes, while Augustus' own grandsons were promoted. After their deaths, Augustus was obliged to acknowledge Tiberius as the only possible successor. Succeeding as emperor in AD 14 he applied stringent economies and had the makings of a good emperor but his reign was marred by the increased number of treason trials. *Jesus Christ was crucified during his reign. He became a recluse, paranoic about conspiracy, ordering numerous executions and disliking Rome intensely. Finally he was persuaded by Sejanus, the Prefect of the *Praetorian Guard, to leave the city in 26 to live on Capri. His death during a rare excursion to the mainland was rumoured to have been murder.

Tibet (or Xizang), a region in Central Asia, largely independent from China until the *Qing dynasty. In 607 Tibet was first unified and by the 8th century a large empire had been established stretching from Lanzhou in China to Kashgar in Central Asia and south to the site of modern Calcutta in India. For a time it was a serious rival to the *Tang empire. *Kublai Khan conquered eastern Tibet and gave *Lamaism (which controlled every aspect of life) his approval. In the 16th and 17th centuries Mongolia, converted to Lamaism, and China intrigued over the Dalai Lama's succession. The Qing conducted several Tibetan campaigns and in 1720 established a protectorate, incorporating Tibet into the Chinese empire. The Dalai Lama retained a position of temporal and spiritual authority within Tibet.

Ticonderoga, New York, USA, a frontier fortress commanding the Champlain-Hudson valley between

A 17th-century engraving of the Potala, or Bietala, the monastery and palace of the Dalai Lama in **Tibet** from *China monumentis illustrata* (1667) by Athanasius Kircher. The fortress-like building stands on a high ridge overlooking Lhasa, the capital.

Lake Champlain and Lake George. Built by the French as Fort Carillon in 1755, it held out against British attack in 1758, but the next year fell to *Amherst, who renamed it. In the War of Independence it was surprised by Benedict *Arnold and Ethan *Allen in 1775, recaptured by *Burgoyne in 1777, but recovered by the Americans after Saratoga (1777).

Tilly, Johannes Tserklaes, Count of (1559–1632), Flemish soldier. He served under *Farnese in the Netherlands, and then in the army of Emperor Rudolf II against the *Ottoman Turks (1594). In 1610 Duke Maximilian of Bavaria appointed him to create an army which became the spearhead of the Catholic League during the *Thirty Years War. He was victorious at the battle of the White Mountain (1620) and went on to dominate north-western Germany. He crushed the Danes at Lutter (1626), and on *Wallenstein's dismissal, took command of imperial as well as League troops. His brutal destruction of the Protestant city of Magdeburg (1631) blackened his reputation. He was routed by *Gustavus Adolphus at Breitenfeld (1631), and killed.

Timbuktu, a city on the middle River Niger, founded in 1106, according to a 17th-century Arabic source, as a staging post for the then flourishing trans-Saharan trade in gold, ivory, and slaves (which were exchanged for salt, cloth, and pottery from the north). It became one of the main cities of the empire of *Mali and was later a cultural centre of the *Songhay empire, with a university, mosques, and over a hundred schools. Thereafter it declined, being sacked by Moroccan invaders in 1591 and by Tuaregs and others repeatedly afterwards. In the 18th century it was absorbed into the Bambara empire of Segu; but it remained famous, and several European explorers tried to reach it. Alexander Laing (1793–1826), the first of them to do so, was murdered there.

Timur *Tamerlane.

Tipu Sultan (*c*.1753–99), sultan of *Mysore (1782–99). He inherited the kingdom recently created by his father, *Hyder Ali and was a formidable enemy to both the British and neighbouring Indian states. Failure to secure active French support left him without allies in resisting

the British. He was finally besieged in his own capital, Seringapatam, when unfounded rumours that he had secured an alliance with Revolutionary France gave the British the necessary pretext for a final assault. He was killed in the attack.

tithe (Old English, 'tenth'), a payment made by parishioners for the maintenance of the church and the support of its clergy. Levied by the early Hebrews and common in Europe after the synods of Tours (567) and Mâcon (585), tithes were enforced by law in England from the 10th century. They were divided into three categories—praedial (one-tenth of the produce of the soil), personal (one-tenth of the profits of labour and industry), and mixed (a combination of the produce of animals and labour). Attempts to abolish them in England were made in 1653 by the *Barebones Parliament, but they were abolished finally in England in 1936.

Toba Wei *Northern Wei.

Tokugawa, the last Japanese *shogunate (1603–1867). *Tokugawa Ieyasu, its founder, ensured supremacy by imposing severe restrictions on the *daimyo. To avoid the effects of European intrusion, Christianity was proscribed in 1641 after the suppression of the Christian *Shimabara rebellion and all foreigners except a few Dutch and Chinese traders at Nagasaki were excluded. Japanese were forbidden to go overseas. Interest in

A contemporary portrait of **Tipu Sultan** by G. F. Cherry, 1792. Cherry was Persian Secretary to Lord Cornwallis during the latter's mission to Seringapatam in 1792. Lord Cornwallis later presented the portrait to Tipu's mother. (India Office Library, London)

Tokugawa Ieyasu, a portrait of the future master of Japan as a young and aspiring warlord in 1572. (Tokugawa Reimeikai Foundation, Tokyo)

European science and medicine increased during the rule of *Tokugawa Yoshimune.

Two hundred and fifty years of almost unbroken peace and economic growth followed. An economy based largely on barter became a money economy. An influential merchant class emerged whilst some daimyo and their *samurai were impoverished; some married into commercial families. The shogunate was faced with growing financial difficulties but under its rule educational standards improved dramatically.

Tokugawa Ieyasu (1542–1616), the founder of the *Tokugawa shogunate. His base was Edo (now Tokyo). In 1600, at Sekigahara, he defeated *daimyo loyal to *Hideyoshi's son Hideyori. Appointed *shogun in 1603, he abdicated two years later, but still controlled affairs. In 1615 Hideyori and his retainers, after a hard siege, committed suicide in their moated castle in Osaka. Ieyasu then executed Hideyoshi's grandson, Kunimatsu. Hideyoshi's line was extinct, Ieyasu's power complete.

Tokugawa Yoshimune (1684–1751), Japanese *shogun, the eighth *Tokugawa to hold that office (1716–45). He was extremely capable and, though conservative, was interested in science. In 1720 he allowed European books, hitherto excluded, to be imported by Dutch traders at Nagasaki. Religious books were still banned. He also had a Dutch–Japanese dictionary compiled. The introduction of *rangaku* (Dutch learning) had a profound effect on what had been a closed world. There was

particular interest in medicine, cartography, and military science. Yoshimune worked to increase the shogun's authority and improve government finances.

Toleration Act (1689), the granting by the English Parliament of freedom of worship to dissenting Protestants, that is, those who could not accept the authority or teaching of the *Anglican Church. Dissenters were allowed their own ministers, teachers, and places of worship subject to their taking oaths of allegiance and to their acceptance of most of the *Thirty-Nine Articles. The *Test Acts, which deprived dissenters of public office remained, but from 1727 annual indemnity acts allowed them to hold local offices. Roman Catholics were excluded from the scope of the Act, and had to rely on failure to enforce the penal laws.

Toltecs, a northern Mexican tribe who established a military state between the 10th and 12th centuries at Tula, *c*.80 km. (*c*.50 miles) north of modern Mexico City. They played an important part in the downfall of *Teotihuacán and were themselves overrun in the mid-12th century by nomadic Chichimec tribes from the north. One of their kings was Topiltzín-*Quetzalcóatl, a religious leader who in their legendary history was driven from Tula by a military faction and sailed east into the Gulf of Mexico, vowing to return one day.

Tone, Theobald Wolfe (1763–98), Irish nationalist, who played a leading part in the insurrection of 1798. He was born in Dublin of Protestant parents, studied law, and was called to the Bar in 1789. Inspired by the French Revolution, in 1791 he helped to found the Society of *United Irishmen, whose initial aim was to establish a democratic Ireland through parliamentary reform. In 1796 Tone sought military support from France to overthrow British rule and accompanied an abortive expedition to Ireland later that year. During the Irish rebellion of 1798 he enlisted further French aid but was captured by the British and committed suicide while under sentence of death.

Tooke, John Horne (1736–1812), British radical politician and philologist. In 1769 he founded the Society of Supporters of the *Bill of Rights, which was largely designed to pay *Wilkes's debts and get him into Parliament. In 1771 he founded the Constitutional Society to agitate for British parliamentary reform and self-government for the American colonists. After the battle of *Lexington he associated himself with a denunciation of the British forces there as murderers, for which he was imprisoned. He supported *Pitt against *Fox from 1783 until 1790, but the French Revolution led to public hostility to reformers, and as a leading member of the *London Corresponding Society he was tried for treason but was acquitted in 1794.

Topa Inca (d. 1493), Inca emperor (1471–93), son of Pachacuti Inca (ruled 1438–71). While still heir-apparent to the throne he led his father's armies north, to conquer the powerful *Chimú state on the north coast of Peru. After his accession to the throne he extended the empire to the south, conquering the northern half of Chile and part of Argentina. He also built the great fortress of Sacsahuaman, overlooking the imperial city of *Cuzco. His successor, Huayna Capac (1493–1525) extended the northern boundaries even further, conquering Ecuador and founding *Quito as the second Inca capital.

Tordesillas, Treaty of (7 June 1494), an alliance between Spain and Portugal. It settled disputes about the ownership of lands discovered by *Columbus and others. Pope Alexander VI had (1493) approved a line of demarcation stretching between the poles 100 leagues (about 500 km.) west of Cape Verde islands. All to the west was Spanish, to the east Portuguese—an award disregarded by other nations. Portuguese dissatisfaction led to a meeting at Tordesillas in north-west Spain where it was agreed to move the papal line to 370 leagues (about 1,850 km.) west of Cape Verde islands. The pope sanctioned this (1506). It was modified by the Treaty of Zaragossa (1529) which gave the *Moluccas (Spice Islands) to the Portuguese.

Torquemada, Tomás de (1420–98), Spanish Dominican friar. He acted as the notorious first Grand Inquisitor of the *Spanish Inquisition. As such he was responsible for directing its early activities against Jews and Muslims in Spain, and in fashioning its methods, including the use of torture and burnings.

Tory, the name of a British political party traditionally opposed to the *Whigs. In the political crisis of 1679 royalist supporters, who opposed the recall of Parliament and supported the Stuart succession, were labelled Tories (Irish Catholic brigands) by their opponents. In the reign of *James II many Tories preferred passive obedience to open defiance; they supported the royal prerogative, close links between church and state, and an isolationist foreign policy. The Tories had a brief revival under *Harley late in Queen Anne's reign, but were defeated in the 1715 general election and reduced to a 'country' party with about 120 Members of Parliament and no effective leaders. The Hanoverian succession dealt a severe blow to the Tories, as George I and George II preferred to trust the Whigs. The political power struggle in the 1760s was between rival Whig factions, despite pejorative accusations of Toryism levelled at *Bute, *Grafton, and *North. William *Pitt the Younger, the independent Whig, fought the Foxite Whigs, and it was from the independent Whigs that the new Tory party of the 19th century emerged.

In colonial America loyalists were called Tories, the term being used of anyone loyal to the crown.

Toungoo (1539–1752), a *Burman dynasty that brought unified rule after an interregnum following *Pagan's collapse. Its founder, Tabinshweti, and his successor, Bayinnaung, subdued the *Shans and *Mons, conquered Tenasserim, and overran Thai states in *Siam. However, the Thais broke free and in 1600 sacked Pegu, which except for a brief period was the Toungoo capital until 1634. From 1635 when Ava, two months' journey by river from the delta, became the capital, there was growing estrangement between the Toungoo and the Mons. Weakened by raids from Manipur, Toungoo fell when the Mons captured Ava.

tourney (or tournament), an armed combat, usually under royal licence, between knights, designed to show their skills and valour. Tournaments were introduced into England from France in the 11th century. Early

Knights in mounted combat with blunted swords and lances: part of the elaborate ritual of the **tourney** (or tournament), from *Sir Thomas Holme's Book*, a 15th-century manuscript. (British Library, London)

versions tended to be confused occasions of mock battles between groups of knights, but they were formalized in the 15th century. The elaborate ritual, in which *heraldry played an important part, included the issuing and accepting of challenges, conditions of engagement, and points scoring (according to the number of broken lances or blows sustained). Fighting could be on horseback with swords (tourney) or on foot. Simple mounted combat (jousting or tilting) with the two knights charging on either side of an anti-collision barrier (the tilt) was the most dramatic, but there were also mock sieges and assaults on defended places. Blunt weapons and padded armour reduced accidents, but deaths did occur, including that of *Henry II of France in 1559.

Townshend Acts (1767), a British revenue measure, which was introduced after the failure to raise direct taxes in the American colonies by the *Stamp Act. Faced with a parliamentary revolt against his budget, the Chancellor of the Exchequer, Charles Townshend, grandson of Charles *Townshend, the agricultural reformer, sought revenue through taxes on American imports of paint, paper, glass, and tea. To enforce this trade tax, a new American Board of Customs Commissioners and Vice-Admiralty Courts without juries were established. The revenue was to be used to pay salaries of colonial officials, thus making them independent of the colonial assemblies. American resistance led to the repeal of the duties, except for that on tea, in 1770.

Townshend, Charles, 2nd Viscount (1674–1738), English politician and agriculturist. He became Secretary of State in 1714, but quarrelled with George I over foreign policy, and was dismissed from the government in 1716. His brother-in-law *Walpole resigned in sympathy, and Townshend was restored as Secretary of State when Walpole came to power in 1721. Walpole at first gave him a free hand in foreign affairs, but Townshend allowed a quarrel with Austria to get out of hand, and frequent interference from Walpole led to his resignation in 1730. He retired from politics completely and became famous as a pioneer of the *Agricultural Revolution, popularizing four-course rotation of crops and the widespread cultivation of the turnip, which earned him the nickname 'Turnip' Townshend.

Towton, battle of (29 March 1461). A site near Tadcaster in Yorkshire was the scene of a desperate encounter in a snowstorm between the army of the newly crowned *Edward IV and the retreating Lancastrian forces of Queen *Margaret of Anjou and *Henry VI. The armies were large—the Lancastrians numbered over 22,000—and the losses on both sides were heavy, but in a late reverse the Lancastrians were routed and some of their ablest leaders killed.

trained band (or trainband), a unit of armed men based in one of the counties of England. The formation of these companies in 1573 was the result of a decision of *Elizabeth I's government that a selection of the most able *militia men in each county should be properly trained in the use of pikes and firearms. Freemen in each city were also selected. The bands multiplied in numbers and were extremely active in the early years of the *English Civil War. In the 18th century they were replaced by a standing army supported by militia.

Trajan (Marcus Ulpius Nerva Traianus) (c. AD 53–117), Roman emperor (AD 98–117). He was born in Spain and served as a soldier, before becoming consul in 91. Emperor Nerva adopted him as his successor and the *Praetorians backed the choice on Nerva's death in 98. He put down the Parthians and the Armenians and fought two *Dacian wars which are commemorated on the scenes of reliefs on Trajan's column standing in Rome. Many public works were undertaken including a new section of the *Appian Way. He was an excellent administrator and commanded the loyalty of his subjects. He died campaigning against the Parthians.

Transylvania, a historic region of modern Romania. It formed the nucleus of Roman *Dacia after AD 106, and after the Roman evacuation c.270, its history is unrecorded until the Hungarian *Magyars conquered the area c.900. In 1003 it was incorporated into the Hungarian state, but it retained its autonomy, safeguarded by the 'brotherly union' (1437) of its 'three nations' (the noble descendants of Szekler, Saxon, and Magyar colonists). Effective independence was secured in 1566, when Transylvania became an autonomous principality subject to *Ottoman suzerainty. The 17th century was its 'golden age' as an international power, with a reputation for religious tolerance. In 1699 the Ottomans recognized Austrian Habsburg rule in Transylvania. During the 18th century the Habsburgs deprived the three nations of many of their rights and liberties, and the number of Romanians in the region grew until they were more than half of the total population. They were unable to win political recognition as a 'fourth nation', however, and suffered from religious discrimination as Greek Orthodox Christians.

Travancore *Kerala.

Trent, Council of (1545–63). The city of Trento in northern Italy was the scene of an ecumenical council of the Roman Catholic Church, which met in three sessions. It defined the doctrines of the Church in opposition to those of the *Reformation, reformed discipline, and strengthened the authority of the papacy. Its first session (1545–7), held at Trento in northern Italy, produced a ruling against *Luther's doctrine of justification by faith alone. The brief second session (1551–2) included a rejection of the Lutheran and *Zwinglian positions on the Eucharist. By the third session (1562–3), any lingering hopes of reconciliation with the Protestants had disappeared. Various works recommended or initiated by the Council were handed over to the pope for completion; these included the revision of the Vulgate version of the Bible (finally completed in 1592). The Council thus provided the foundation for a revitalized Roman Catholic Church in the *Counter-Reformation.

trial by ordeal *ordeal.

tribune (of the people), a word derived from 'tribe'. In ancient Rome, ten tribunes were elected to protect *plebs from *patricians and could veto decisions of magistrates and, later, the Senate's decrees. They could also propose legislation of their own. Roman emperors also took the title of tribune, which gave them the constitutional rights of tribunes and a popular image. Military tribunes were senior officers of legions, also elected.

Trinidad, an island off north-west Venezuela, discovered by Columbus in 1498 during his third voyage. Trinidad was claimed by Spain but left to its indigenous *Caribs

The Council of **Trent** in session, from a contemporary engraving. Within the enclosed conference area, entered by means of a guarded doorway, the assembled church delegates listen to a succession of speakers. The apostolic legates are seated on the dais, and a secretary (seated, centre) records the council's proceedings. (Biblioteca Nacional, Madrid)

until 1532, when settlement was begun. As it lacked precious metals it remained largely ignored until 1595, when Sir Walter *Raleigh landed there for ship repairs and sacked the newly founded town of San José; the Dutch raided it in 1640, and the French in 1677 and 1690. Sugar and tobacco *plantations were established in the 17th century, worked by imported African slaves. In 1797, during war between England and Spain, a British squadron entered the Gulf of Paria but met little resistance before the island surrendered. In 1802 it was offically ceded to Britain under the Treaty of Amiens.

Tripolitania, the western province of Libya, the richest and most populous. It was settled by Carthaginians and Greeks, and was taken by *Numidia after the fall of Carthage in 146 BC. It was conquered by the Romans in 46 BC and became an important agricultural area. The Vandals occupied it in 429, but Byzantium recovered it in 533. The Arabs took it in 643, and a succession of dynasties administered it from without. In the 16th century the Knights of Malta held it briefly, followed by the Ottoman Turks. The Karamanli Beys (1711–35) made it a *corsair base, provoking the Tripolitan War with the United States (1801–15).

trireme, the principal warship of antiquity from the 6th to the late 4th century BC. Lightly built for speed and manoeuvrability and unable to venture very far from land, each trireme carried a crew of some 200 men, the majority being rowers. They were probably seated three to a bench, the bench being angled so that each rower pulled a separate oar. A beak of metal and wood was set at the front of the galley, ramming being the principal aim of the steersman. Athens' fleet of triremes played a major part in the Greek victory at *Salamis and was instrumental in controlling the *Athenian empire.

Tromp, Maarten Harpertszoon (1597–1653), Dutch admiral. A distinguished seaman, he was largely responsible for Dutch naval greatness in the 17th century. In 1639 he completely defeated the Spanish fleet at the battle of the Downs. In the First *Anglo-Dutch War he defeated *Blake off Dungeness in December 1652, fought magnificently but unsuccessfully to protect Dutch convoys in the English Channel, and was killed trying to break *Monck's blockade of the Dutch coast. His son **Cornelius** (1629–91), *Ruyter's rival, won victories in the Mediterranean, in the Second and Third *Anglo-Dutch wars, and against Sweden at Gotland.

troubadour, a lyric poet or minstrel in the 11th century in southern France, particularly in Provence. Though to be a troubadour was primarily an aristocratic calling, members of troubadour bands were also drawn from lower social levels. Their poetry dealt with courtly love, *chivalry, religion, and politics, but usually the matter was heavily disguised in formal, decorative language. Much of it was heretical, and in the 13th century many of its devotees were persecuted. William IX, Count of Anjou, is said to have been the first troubadour. Bertrand de Born was received at the court of *Eleanor of Aquitaine, and through her influence this Provençal poetry reached northern France.

Troy (or Ilium), the city which, according to Greek legend as told by *Homer in the *Iliad*, was captured by

A manuscript illustration of 1109 showing two male dancers, one with a stringed instrument, performing the lyric poetry of the French **troubadours**. (British Library, London)

the Greeks under Agamemnon after a ten-year siege. Historical Troy was discovered by Heinrich Schliemann at Hissarlik in north-western Asia Minor, a few miles inland from the Aegean sea. The excavations conducted by him from 1870 to 1890, and by others since, have revealed ten periods of occupation.

The first five belong to the Early Bronze Age, ending soon after 2000 BC. Troy II in particular was a flourishing community, with impressive fortifications and domestic buildings, but was destroyed by a major fire. Troy VI saw an influx of new settlers who introduced horses, but an earthquake shattered their city c.1300. It was followed by Troy VIIA, but this phase did not last long before being destroyed by fire. The indications are that this was not an accidental disaster, but accompanied the capture of the city by enemies. The date of destruction, c.1250, coinciding with a flourishing Mycenaean civilization in mainland Greece, indicates that it was this event which lies behind the *Iliad*, and that the conquerors of Troy VIIA were Greeks. Troy remained unoccupied for perhaps 400 years before Troy VIII was established there. Troy IX lasted into the Roman period.

Ts'ao Ts'ao   *Cao Cao.

Tudor, the English royal house which began as a family of Welsh gentry. Its fortunes started to rise when *Henry V's widow, Katherine of Valois, 'fair Kate', married Owen Tudor (c.1400–61), her clerk of the wardrobe. He was executed after the *Yorkists' victory of *Mortimer's Cross (1461) during the Wars of the Roses but his son Edmund (c.1430–56), Earl of Richmond, married Margaret *Beaufort, and their son Henry was thus

Tudors

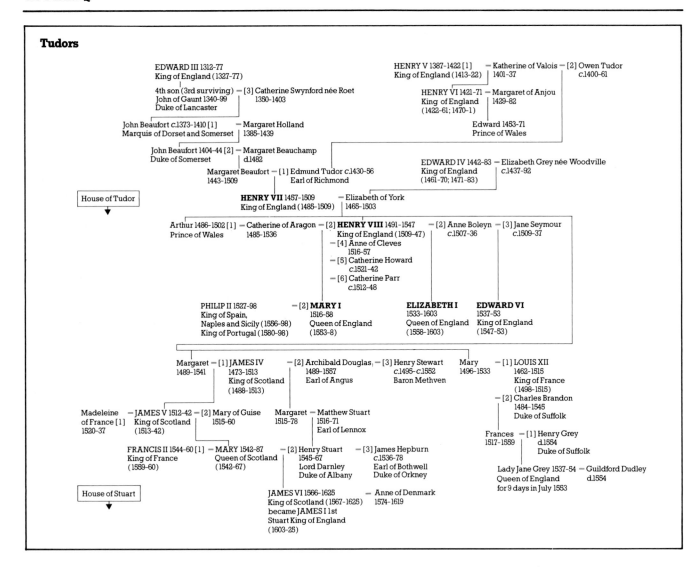

EDWARD III 1312-77
King of England (1327-77)

4th son (3rd surviving) — [3] Catherine Swynford née Roet
John of Gaunt 1340-99   1350-1403
Duke of Lancaster

John Beaufort c.1373-1410 [1] — Margaret Holland
Marquis of Dorset and Somerset   1385-1439

John Beaufort 1404-44 [2] — Margaret Beauchamp
Duke of Somerset   d.1482

Margaret Beaufort — [1] Edmund Tudor c.1430-56
1443-1509   Earl of Richmond

HENRY V 1387-1422 [1] — Katherine of Valois — [2] Owen Tudor
King of England (1413-22)   1401-37   c.1400-61

HENRY VI 1421-71 — Margaret of Anjou
King of England   1429-82
(1422-61; 1470-1)

Edward 1453-71
Prince of Wales

EDWARD IV 1442-83 — Elizabeth Grey née Woodville
King of England   c.1437-92
(1461-70; 1471-83)

House of Tudor

HENRY VII 1457-1509 — Elizabeth of York
King of England (1485-1509)   1465-1503

Arthur 1486-1502 [1] — Catherine of Aragon — [2] **HENRY VIII** 1491-1547 — [2] Anne Boleyn — [3] Jane Seymour
Prince of Wales   1485-1536   King of England (1509-47)   c.1507-36   c.1509-37
— [4] Anne of Cleves
1516-57
— [5] Catherine Howard
c.1521-42
— [6] Catherine Parr
c.1512-48

PHILIP II 1527-98 — [2] **MARY I**   **ELIZABETH I**   **EDWARD VI**
King of Spain,   1516-58   1533-1603   1537-53
Naples and Sicily (1556-98)   Queen of England   Queen of England   King of England
King of Portugal (1580-98)   (1553-8)   (1558-1603)   (1547-53)

Margaret — [1] JAMES IV — [2] Archibald Douglas — [3] Henry Stewart   Mary — [1] LOUIS XII
1489-1541   1473-1513   1489-1557   c.1495-c.1552   1496-1533   1462-1515
King of Scotland   Earl of Angus   Baron Methven   King of France
(1488-1513)   (1498-1515)
— [2] Charles Brandon
1484-1545
Duke of Suffolk

Madeleine — JAMES V 1512-42 — [2] Mary of Guise   Margaret — Matthew Stuart
of France [1]   King of Scotland   1515-60   1515-78   1516-71
1520-37   (1513-42)   Earl of Lennox
    Frances — [1] Henry Grey
    1517-1559   d.1554
    Duke of Suffolk

FRANCIS II 1544-60 [1] — MARY 1542-87 — [2] Henry Stuart — [3] James Hepburn
King of France   Queen of Scotland   1545-67   c.1536-78
(1559-60)   (1542-67)   Lord Darnley   Earl of Bothwell
    Duke of Albany   Duke of Orkney

Lady Jane Grey 1537-54 — Guildford Dudley
Queen of England   d.1554
for 9 days in July 1553

House of Stuart

JAMES VI 1566-1625 — Anne of Denmark
King of Scotland (1567-1625)   1574-1619
became JAMES I 1st
Stuart King of England
(1603-25)

a descendant, though illegitimately, of the House of Lancaster. His claim to the throne became more acceptable after the death of *Henry VI's son Edward in 1471, and *Richard III's loss of the nobility's support paved the way for Henry's invasion of England and taking of the throne as *Henry VII in 1485.

Henry safeguarded his claim to the throne by marrying Elizabeth of York, the Yorkist heiress: she bore him eight children, although four died in infancy. Arthur died soon after marrying Catherine of Aragon, and it was his younger brother who succeeded to the throne, as *Henry VIII. Of his children, his only son, *Edward VI, died in his youth. His elder daughter *Mary died in 1558 after a childless marriage to Philip II of Spain, and *Elizabeth never married. With Elizabeth's death (1603) the House of Tudor ended, and the throne passed to James VI of Scotland, of the House of *Stuart.

Tughluq, a Muslim dynasty which ruled in India for almost a century (1320-1413). Seizure of power was followed by military campaigns which brought the sultanate of Delhi to its greatest territorial extent, including the extreme south of India. The chief architect of this success was the second sultan, Muhammad ibn Tughluq (1325-51). But some of his controversial actions are deemed to have undermined rather than strengthened the empire. Gifted and well intentioned, he nevertheless gained a reputation for extreme cruelty which led to rebellions throughout his territories. The invasion of *Timur in 1398, in which Delhi was devastated, increased the chaos, and in 1413 the *Sayyids seized power from the Tughluqs.

Tuileries, a French royal residence in Paris. In June 1792 during the *French Revolution crowds forced their way into the palace, and on 10 August it was attacked, the Swiss Guard was massacred, and the royal family took refuge with the Assembly. In 1793 the *Committee of Public Safety installed themselves there. The palace was burned down in the 19th century.

Tumulus culture, a civilization, centred on southern Germany, which occupied most of central and eastern Europe in the Middle *Bronze Age, 1800-1500 BC. The name is taken from the burial rite of individual inhumation, later cremation, beneath a round *barrow or tumulus. The graves were often very richly furnished, serving an élite class. These people had a rich and varied range of bronzework, based on German and Bohemian copper and tin, and traded amber from the Baltic. They were succeeded by the *Urnfield cultures.

Tunisia, a North African country, the strategic centre of the Mediterranean. *Phoenicians came first c.1000 BC;

and, traditionally, *Carthage, seat of a sea-borne empire, was founded here in 814 BC. *Berber caravans came north to exchange produce for imports. Carthage fell in 146 BC, and, despite Berber resistance, Rome made the province of Africa Proconsularis rich in corn, olives, and vines. *Vandals from Spain took it in 429, but *Byzantium recovered it in 533. The Berbers, nevertheless, held the interior, giving way only when the Arabs built Kairouan as an inland base to control Africa. The caliphate was replaced by an independent local dynasty, the Aghlabids, in 800, until 909, when the *Fatimids took Kairouan. Another local dynasty, the Zirids, replaced them when they moved to Cairo in 969. In revenge, the Fatimids sent thousands of Arab tribesmen to lay waste the country. In the 12th century the Normans from Sicily held some towns, until the *Almohads expelled them. Then another local dynasty, the Hafsids (1228–1574) emerged, taking Algiers (1235) and Tlemcen (1242). In 1270 they repulsed the Crusaders under St *Louis. From 1574 until 1881 the Regency of Tunis owed nominal allegiance to the *Ottomans, but after 1612 a dynasty of Beys established itself. A period of great prosperity ended when the *corsairs and the *slave trade were suppressed (1819).

tunnage and poundage, duties levied in England on each tun (a large cask) of imported wine and on every pound of most imported or exported merchandise. From the 15th century onward Parliaments granted the revenues from these duties to English kings for life. By 1625 they ranked as the largest items in the crown's ordinary revenue. Then *Charles I's first Parliament granted him the duties for one year only as Members of Parliament, fearing arbitrary customs dues, sought to assert that customs revenue was under parliamentary control. When he continued to collect them, the issue was raised in the *Petition of Right (1628). They were eventually abolished in 1787.

Tupac Amarú (José Gabriel Condocanqui) (c.1742–81), leader of a widespread Indian revolt in the Peruvian highlands (1780–1). As the Indian chief of Tinta, south of Cuzco, Tupac Amarú used his links to the *Inca royal dynasty to develop an Indian base of support and his Spanish connections to attract creole and mestizos to his reformist political movement which espoused Inca nationalism, fairer taxes, better courts, and a more open interregional economy. In 1780, reacting to economic abuses, Tupac Amarú plotted the execution of the local Spanish corregidor and then recruited a large indigenous army, led by non-Indian, middle-level, provincial leaders, which occupied much of the highland area, even threatening Cuzco. Although Tupac Amarú was defeated and executed in May 1781, the revolt spread into Upper Peru, becoming more hostile to non-Indians, and finally provoking a severe repression which retarded the independence movement in Peru.

Turenne, Henri de la Tour d'Auvergne, vicomte de (1611–75), Marshal of France. A soldier of outstanding ability, he made his reputation on the battlefields of the *Thirty Years War. During the *Fronde he was briefly persuaded by *Condé's sister, Madame de Longueville, to join the antiroyalist faction, but finally supported *Mazarin in establishing order. He captured Dunkirk in 1658 and led the brilliant invasions of Flanders in 1667

and the United Provinces in 1672. In 1674, when *Louis XIV was threatened from all sides, he showed his supreme ability in deploying troops against superior forces in defending France's eastern frontier. He died on the battlefield.

Turin Shroud, a cloth, preserved since 1578 in Turin Cathedral in Italy, which is venerated as the shroud in which *Jesus Christ was buried. Although its history can be traced with certainly only to 1354, there are various known statements from the earlier period, going back as far as the 5th or 6th century, that Jesus' burial shroud had been preserved. The cloth bears the imprint of a man, about 2 m. (6 ft). tall, who had been scourged, and with wounds from nails in his wrists and feet. The arguments for and against its authenticity are complex and technical.

Turkistan, an area of central Asia, stretching from the Caspian Sea to China, the bridge between east and west. It includes both deserts and highlands and the fertile Ferghana valley, and among its historic centres are Bukhara, Samarkand, Tashkent, all on the *Silk Route. The western part of the region was under Persian rule from the 6th century BC, Muslim control from the 7th century AD, and Russian overlordship from the 18th century. The eastern portion (*Xinjiang) was contested between the Chinese empire and various nomadic groups.

Turks, originally nomads from *Turkistan. During the 6th century AD they controlled an empire stretching from Mongolia to the Caspian Sea. With the conquest of western Turkistan in the 7th century by the *Abbasids, many were converted to Islam and moved westwards, retaining their distinctive language and culture. In the 11th century, under the *Seljuks they replaced the Arabs as rulers of the Levant and Mesopotamia, then expanded north-west at the expense of *Byzantium. The rival house of Osman continued this trend, founding the *Ottoman empire, which endured for 600 years, and embraced most of the Middle East, North Africa, and the Balkans.

Tuscany, a region in north-central Italy, whose chief city is Florence. Under Roman rule from the 4th century BC, as the Romans declined it suffered invasion from Goths, Byzantines, and Lombards, and stabilized under Frankish rule from 774 to 1115, with Lucca as its capital. It was bequeathed to the papacy and became a centre for *Guelph–Ghibelline rivalry, and for struggles between its successful commercial cities. The *Medici acquired the title of duke after 1569. Medici rule ended in 1737 and Tuscany passed to the Habsburgs.

Tutankhamun, Pharaoh of Egypt (ruled c.1361–1352 BC). Little is known about his reign and premature death and his importance derives largely from the fact that his tomb alone in the *Valley of the Kings escaped looting in antiquity. His tomb was hidden by rubble from the construction of a later tomb and was not discovered until 1922 when it was found almost intact by the British archaeologist Howard Carter and his patron, Lord Carnarvon. His mummified body was inside three coffins, the inner one of solid gold; over his face was a magnificent gold funerary mask, and the burial chamber and other rooms housed a unique collection of jewellery, weapons, and other items.

The southern half of the antechamber 8 by 3·7 m. (26 by 12 ft.) of **Tutankhamun's** tomb discovered on 26 November 1922. Howard Carter described some of the treasures found there, among them, on the right, 'gilded couches, their sides carved in the form of monstrous animals, curiously attenuated in body...but with heads of startling realism...and on the left a confused pile of overturned chariots, glistening with gold and inlay'.

Tuthmosis I   *Thutmose I.

Tuthmosis III   *Thutmose III.

Tyler, Wat (d. 1381), English rebel, leader of the *Peasants' Revolt. He was a former soldier who was accepted as their captain by the Kentish peasants in June 1381. Following him and John *Ball, they marched from Canterbury to Blackheath, just outside London, in only two days, 11–13 June. On 15 June *Richard II met them and ordered Tyler to go home; in an angry exchange of words with the mayor of London, William Walworth, Tyler was pulled from his horse and stabbed to death. Richard proclaimed himself the rebels' leader.

Tyndale, William (1494–1536), English scholar and translator of the Bible. An early Lutheran, he was forced to work abroad by authorities unsympathetic to his Protestantism. His translations of the scriptures into English found a ready illicit English market after 1526. He suffered a heretic's death by strangling at Louvain in the Netherlands. He was immortalized in John *Foxe's 'Book of Martyrs', and in the Authorized Version of the Bible (1611), much of which derives from his translations.

Tyre, one of the main cities of the *Phoenicians, and famous in early times for its skilled bronze-workers. In the 10th century BC it made treaties with *David and *Solomon of Israel. Conflict with Assyria was a feature of the following centuries, Tilgath-Pileser III capturing the city and exacting tribute in 734–732. In 668 the attacks of the Assyrian king *Ashurbanipal were beaten off but the city fell to *Nebuchadnezzar II in 587. It entered the empire of the Achaemenids later that century. Alone of the Phoenician cities it resisted *Alexander the Great, but fell to him after an epic siege in 332. Despite destruction it revived to become a major commercial centre. The Roman emperor Severus made it the capital of Syrian Phoenicia.

Tyrone, Hugh O'Neill, 2nd Earl of (1540–1616), Ulster chieftain. He received his earldom from *Elizabeth I of England in 1585. He established himself as the most powerful chief in Ulster and with the support of other Catholic chiefs rebelled against Elizabeth and her religious policies in 1594. After a famous victory at the Yellow Ford (1598), the initiative slipped to the English, under Charles Blount, Lord Mountjoy. Spanish support for the rebels at Kinsale (1601) proved inadequate, and in 1603 Tyrone surrendered. In 1607, after another abortive insurrection the earls of Tyrone and Tyrconnel, with their households, fled to Flanders, the celebrated 'flight of the earls', and he died in Rome.

Ukraine, a region in south-western Russia. Originally inhabited by Neolithic settlers in the Dnieper and Dniester valleys, the area was overrun by numerous invaders before Varangian adventurers founded a powerful Slav kingdom based on *Kiev in the 9th century. Mongol conquest in the 13th century was followed in the 14th century by Lithuanian overlordship until 1569, when Polish rule brought serfdom and religious persecution, which produced an exile community of Cossacks who resisted both Polish and Russian domination. By 1795 the region, including the *Crimea, under *Ottoman control from 1478, was under Russian control.

Ulster, a province of *Ireland, lying in north-eastern Ireland. As a kingdom it reached its zenith in the 5th century AD, at the beginning of the Christian era. During the Anglo-Norman conquest the de Lacey and de Burgh families held the earldom of Ulster from 1205 to 1333. By the 16th century, the O'Neill clan had reasserted its commanding position in the area, until the failure of *Tyrone's rebellion against *Elizabeth I and her unwelcome religious policy (1594–1601) marked the end of O'Neill supremacy. James I promoted the plantation in Ulster of thousands of Presbyterian Scots and Protestant English and many Catholics were forced off their land. These Protestants supported *William III in his campaign against *James II, which culminated in William's victory at the battle of the *Boyne (1690).

Umar ibn al-Khattab (AD c.581–644), second *caliph of Islam (634–44). He presided over the first major wave of *Arab conquests, which were the work of great captains such as Khalid ibn al-Walid. Hostile at first to *Muhammad, he had become an ardent convert. The bond between Prophet and disciple was strengthened by Muhammad's marriage to his daughter Hafsa. His genius was administrative rather than military, and his achievements included systematizing the rule of his vast territories, establishing the Islamic calendar, organizing state pensions, and upholding justice.

Umayyad, a Muslim dynasty founded by Mu'awiya and centred on Damascus, which wrested control of the Arab empire from Ali, son-in-law of the Prophet *Muhammad, in the civil wars following the death of the caliph *Uthman in 656. They annexed North Africa and Spain, Transoxania and Sind, and twice attempted to take Constantinople. Their military achievements were complemented by success in incorporating diverse peoples and territories into a new social and legal order. Their failure to accommodate tribal conflicts and newly arising sectarian jealousies led to their fall at the hands of the *Abbasids, though one of their number, Abd al-Rahman, escaped to establish an independent Umayyad dynasty in Spain (756–1031).

Umbria, a mountainous region in central Italy. The Umbri were an ancient Italic people with a distinctive culture and territory stretching from Ravenna and the Adriatic to the River Tiber. Defeated first by their *Etruscan neighbours and then by *Rome (c.290 BC), their land became one of eleven regions of Italy under *Augustus, the first emperor. After the disintegration of the Western Roman empire and the rise of the papacy as a political power in Italy after AD 800, it formed part of the *Papal States.

Uniformity, Acts of, a series of English laws intended to secure the legal and doctrinal basis of the *Anglican Church. The first (1549) made the Book of Common Prayer compulsory in church services, with severe penalties on non-compliant clergymen. The second (1552) imposed a revised Prayer Book which was more Protestant in tone, and laid down punishments for *recusants. *Mary I had both Acts repealed, but the third (1559) introduced a third Book of Common Prayer and weekly fines for non-attendance at church. The fourth (1662) presented a further revised, compulsory Book. Under its terms some 2,000 non-compliant clergymen lost their benefices, creating the Anglican-*Nonconformist breach.

Union, Acts of, laws which cemented the political union of Great *Britain and Ireland. Following the complete subjugation of Wales by 1284, the Statute of Rhuddlan, never submitted to a formal Parliament, sanctioned the English system of administration there. Not until 1536 was an Act passed by *Henry VIII which incorporated Wales with England, and granted for the first time Welsh representation in Parliament. The Stuarts united the thrones but not the governments of England and Scotland in 1603. In 1707 an Act of Union between England and Scotland gave the Scots free trade with England, but in return for representation at Westminster they had to give up their own Parliament. The Protestant Irish Parliament enjoyed independence from 1782 to 1800, when legislation (1 August 1800) was introduced to establish the United Kingdom of Great Britain and Ireland (1 January 1801).

United Empire Loyalist, the title adopted by some 50,000 Americans loyal to George III who emigrated to Canada during the War of *Independence. By 1784 about 35,000 were settled in Nova Scotia and some 10,000 in the Upper St Lawrence valley and round Lake Ontario, an area designated Upper Canada (later Ontario) in 1791. They came mainly from New England and New York, and among them were several distinguished loyalists, or Tories as they were called, as well as thousands of farmers and artisans. In 1789 the governor-general ordained that all who had arrived by 1783 could put 'UE' for United Empire Loyalist after their and their descendants' names—later arrivals were called 'Late Loyalists'. These 'marks of honour' were treasured throughout the 19th century.

United Irishmen, a society established in Belfast in 1791 by Wolfe *Tone and others with the aim of bringing about religious equality and parliamentary reform in Ireland. Inspired by the ideals of the French Revolution, it drew its support from both Catholics and Presbyterians. The British government took steps to remove some grievances, notably with the Catholic Relief Act of 1793. However, after the dismissal of Earl Fitzwilliams, the Lord Lieutenant, who sympathized with the demands for religious equality, the society began to advocate violent revolution in order to overthrow British rule and

French troops land on Irish soil in support of the **United Irishmen**. This painting by William Sadler, entitled, *The French in Killala Bay* (1798), illustrates the limited support that was given to the Irish nationalist rebels by Revolutionary France after the failure of the attempt at military invasion two years earlier. (National Gallery of Ireland, Dublin)

establish an Irish republic. It sought military assistance from France, but a French expedition which set forth in 1796 to invade Ireland was scattered by storms. Repression of its members followed. In May 1798 sporadic risings occurred, especially in County Wexford, but two months later another French force was intercepted and Tone captured. Thereafter the society went into decline.

United Kingdom   *Britain, Great, *Union, Acts of.

United Provinces of the Netherlands, the historic state, alternatively known as the Dutch Republic, which lasted from 1579 to 1795, and comprised most of the area of the present kingdom of the Netherlands. It was recognized as an independent state by Spain at the conclusion of the *Dutch Revolts (1648), and power was subsequently shared between the Holland and Zeeland patricians and the *statholder princes of *Orange. During its 'golden age' before 1700, the United Provinces developed the vast *Dutch empire and Dutch merchants traded throughout the world. The United Provinces gave refuge to religious refugees, especially Portuguese and Spanish Jews and French *Huguenots, who made a notable contribution to the country's prosperity. It was a 'golden age' in art and in philosophy as well as in commercial affairs. A series of wars was fought against England and France in the 18th century. The commercial and military fortunes of the Netherlands declined as those of England and France improved. When in 1794-5 France overran the country during the French Revolutionary wars, there was a Dutch popular movement inspired by the ideas of the *Enlightenment ready to

overthrow the ruler, William V of Orange, and to set up a Batavian Republic (1795-1806) under French protection in place of the United Provinces.

United States of America, a North American country. European colonization of the eastern seaboard of North America began in the early 17th century, gaining momentum as the rival nations, most notably the British and French, struggled for control of the new territory. The Treaty of *Paris (1763) marked the final triumph of Britain, but by that time the British colonies, stretching from *New England in the north to *Georgia in the south, had become accustomed to a considerable measure of independence. British attempts to reassert central authority produced first discontent and then open resistance. The First *Continental Congress met in 1774 to consider action to regain lost rights, and the first armed encounters at *Lexington and Concord in April 1775 led directly to full-scale revolt and to the formal proclamation of the separation of the *thirteen colonies from Britain in the *Declaration of Independence (4 July 1776). In the War of *Independence which lasted until 1783, the American cause was assisted by France and Spain. The Peace of *Paris (1783) recognized US independence, and the new state was given legislative form by the *Constitution of 1787.

university, a centre of higher education with responsibilities for teaching and research. Universities evolved in Europe from *Studia generalia*, schools open to scholars from all countries, established to educate priests and monks beyond standards attainable in cathedral or monastic schools. Originally societies or guilds of foreigners who banded together for protection in a strange land, by the 13th century 'universities of scholars' had developed into corporate bodies with well-defined administrative structures. At Bologna and Paris foreign students were granted special privileges, such as right of trial for misdemeanours in an ecclesiastical court and the right to strike in protest against unsatisfactory conditions.

Among the earliest European universities were Bologna (1088), Paris (*c*.1150), Prague (1348), Vienna (1365) and Heidelberg (1386). In Britain Oxford (*c*.1150) and Cambridge (1209), where college organization developed early, were followed by St Andrews (1411) and Glasgow (1451) in Scotland, and in Ireland by Trinity College, Dublin (1591). American universities evolved from colleges founded in the 17th and 18th centuries, Harvard (1636) being the earliest. For long the humanities and jurisprudence were the principal subjects of study with theology the most important. However, in Salerno, influenced by Arab culture, there was a renowned medical school by the 11th century. During the 17th century the *Scientific Revolution forced the slow widening of the curriculum.

Upanishads (Sanskrit, 'sitting near', i.e. at the feet of a master), a collection of more than one hundred spiritual treatises composed in Sanskrit at an uncertain date (probably after about 400 BC). They contain a distillation of the teaching of the *Vedas and the Brahamanas (commentaries on the Vedas) and are therefore known as the Vedanta ('conclusion of knowledge'), but are more philosophical and mystical in character. Scholars identify in the Upanishadic era the first emergence of a concept within *Hinduism of a single supreme God (Brahman) who is knowable by the human self (atman). Hence the *Bhagavadgita*, although part of the later Mahabharata epic, is often classed with the Upanishads as providing the highest and most essential Hindu teaching.

Upper Palaeolithic, the final division of the *Palaeolithic or Old *Stone Age, associated with the appearance of *Homo sapiens sapiens* about 50,000 years ago. It occupies the second half of the last glaciation, ending 10,000 years ago, and contrasts with the Middle Palaeolithic in terms of increased population and larger communities, and in a faster rate of cultural change in which distinctive regional groups appeared for the first time. The most developed cultures were on the steppe and tundra belts south of the ice-sheets, with their plentiful herds of horses, reindeer, and mammoths. Several styles of toolmaking can be distinguished in Europe: the earliest is Aurignacian, with the first blade tools and a plentiful use of bone

for missile heads. This was followed by the Gravettian, stretching from France to the Ukraine, and more localized traditions, such as the Solutrian and Magdalenian, which were responsible for the cave art of south-western France and northern Spain. These cultures disappeared with the onset of warmer conditions at the end of the last Ice Age and the spread of forests, which displaced the large herds of game animals.

Ur (modern Muqayyar), a city of the *Sumerians. It was occupied from the 5th millennium BC, and was at one point damaged by a severe flood. By 3000 BC it was one of a number of sizeable Sumerian cities. It was subject to the rule of *Akkad, but emerged *c*.2150 as the capital of a new Sumerian empire, under the third dynasty established by Ur-Nammu. The city was captured by the Elamites *c*.2000, but continued to thrive under the Chaldean kings of Babylon. It was finally abandoned in the 4th century BC. Remains of the ancient city include a ziggurat (pyramidal tower), a palace and temple built by Nabonidus of Babylon, and private dwellings. *Cuneiform inscriptions found there shed much light on the economy and administration of the city.

Urban II (Odo of Lagery, *c*.1042–99), Pope (1088–99). He was a Cluniac monk who was made Bishop of Ostia near Rome in 1078 and then a cardinal by Pope Gregory VII. At the Council of Clermont in 1095 he preached for the sending of the First *Crusade to recover *Palestine which had fallen to the Muslims, hoping also to achieve thereby the unity of the Christian West, then torn by strife between rival rulers. He continued Gregory's work of church reform and his councils condemned *simony, lay *investiture, and clerical marriage.

Urban VIII (Maffeo Barberini, 1568–1644), Pope (1623–44). He became a cardinal in 1606, and Bishop of Spoleto in 1608. As pope, he canonized Philip *Neri and Ignatius *Loyola, condemned Galileo and *Jan-

The 'peace' side of the 'standard' of **Ur**, *c*.2600 BC, inlaid with shell, lapis lazuli, and red limestone, showing scenes of Sumerian life. The top strip depicts a royal feast and the lower two show farmers with their animals. (British Museum, London)

Peace of Utrecht: Europe in 1713

The Peace of Utrecht ended the long period of conflict between Louis XIV's France and various European alliances, which had been provoked by the fear of French ambitions in Europe. France remained powerful, despite earlier military setbacks, with Louis's grandson assured of the Spanish throne. The Austrian Emperor Charles VI had been forced to accept the failure of his claim to the Spanish throne, but Austria became the dominant power in Italy and gained the Austrian Netherlands (now Belgium). British gains, including the acquisition of Gibraltar and Minorca, and of territory in North America, laid the foundation for the challenge to France as a colonial and commercial power later in the 18th century.

senism, and approved a number of new religious orders. He was a noted poet, scholar, and patron of the arts, but was given to nepotism, appointing his relatives to high office. In diplomacy, his fears of Habsburg domination in Italy led him to favour France during the *Thirty Years War. He also extensively fortified the *Papal States, and fought the War of Castro (1642-4) against the north Italian Farnese Duke of Parma. The result was a humiliating defeat which crippled the papal finances and made him bitterly unpopular with the Roman people, who had already suffered from his lavish expenditure on the beautification of the city.

Urnfield cultures, people who become recognizable as a distinct group in east central Europe in about the 15th century BC, with their origins in Hungary and Romania. Thereafter these people spread widely in the Late *Bronze Age, replacing the *Tumulus culture. Their characteristic feature was the use of cemeteries of flat graves containing the ashes of the cremated dead in urns. They spread into northern Italy and as far as Sicily. In the 11th century they crossed into France, where the cemeteries and associated bronzework are widely known, and in the 8th century on into Spain. Certainly there and probably elsewhere this activity can be attributed to the *Celts. They were followed by the *Hallstatt Iron Age.

Uruk, one of the leading cities of the *Sumerians. A community occupied the site as early as 5000 BC and in the 3rd millennium BC the city was surrounded by a 9.5 km. (6 mile) wall which was attributed in later tradition to the hero *Gilgamesh. Excavation has revealed much, not least ziggurats (pyramidal towers), dedicated to the two main gods, Anu and Inanna. It continued to be inhabited into Parthian times.

Ussher, James (1581-1656), Irish theologian, Archbishop of Armagh from 1625. In 1640 he escaped to England on the outbreak of the Irish Rebellion and settled there. A Calvinist but also a royalist, he was well treated by Oliver *Cromwell. Among his writings on a wide variety of subjects was an influential chronology of scripture (1650-4), which set the date of the creation as 23 October 4004 BC.

Uthman (Osman) (*c.*574–656), third *caliph of Islam (644–56). He restored representatives of the old *Meccan aristocracy to positions of influence, creating considerable discontent. His personal weakness led to rivalry to his authority from Aisha, the youngest wife of *Muhammad, from Ali, his cousin and son-in-law, and others. He was murdered by mutinous troops from Egypt. His lasting memorial was the authorized version of the *Koran, compiled at his order.

Utrecht, Peace of (1713), the treaty which ended the War of the *Spanish Succession. After negotiations between the English and French, a Congress met at Utrecht without Austria and signed the treaties. The Austrian emperor Charles VI found he could not carry on without allies and accepted the terms at Rastadt and Baden in 1714. *Philip V remained King of Spain but renounced his claim to the French throne and lost Spain's European empire. The southern Netherlands, Milan, Naples, and Sardinia went to Austria. Britain kept Gibraltar and Minorca and obtained the right to supply the Spanish American colonies with Negro slaves, the *asiento. From France it gained Newfoundland, Hudson Bay, St Kitts, and recognition of the Hanoverian succession. France returned recent conquests, but kept everything acquired up to the Peace of *Nijmegen in 1679 and also the city of Strasburg. The Duke of Savoy gained Sicily and improved frontiers in northern Italy. The Dutch secured Austrian recognition for their right to garrison 'barrier' fortresses in the southern Netherlands. French domination had been checked but France was still a great power. Britain made significant naval, commercial, and colonial gains and thereafter assumed a much greater role in world affairs.

Uzbeks, a Turkish-speaking people, Mongol by descent and *Sunni Muslim by religion. They moved through Kazakhstan to Turkistan and Transoxania between the 14th and 16th centuries to trouble the *Shiite Safavid rulers of Persia. Initially ruled by the Shaybanids and then the Janids, they later split into dynasties based on Bukhara, Khiva, and Kokand.

vagabonds *sturdy beggars.

Valley Forge (1777–8), an American Revolutionary winter camp 32 km. (20 miles) north-west of Philadelphia. It was occupied by *Washington's army after the British occupation of Philadelphia and defeats such as *Brandywine and *Germantown. A bitter winter and lack of supplies came close to destroying the Continental Army of 11,000 men. The Valley Forge winter was the low point in the American Revolutionary struggle; it hardened the survivors and became a symbol of endurance.

Valley of the Kings, a narrow gorge in western *Thebes containing the tombs of at least sixty pharaohs of the eighteenth to twentieth dynasties (*c.*1550–1050 BC), beginning with *Thutmose I. Although supposedly secret, it was a rich hunting-ground for robbers in antiquity, and of those tombs discovered only that of *Tutankhamun had not been plundered.

Valmy, battle of (1792), the first important engagement of the French Revolutionary wars. Near a small village on the road between Verdun and Paris the French commander-in-chief, Charles-François Dumouriez, with the belated assistance of Marshal François Christophe Kellermann, defeated the Duke of Brunswick's German troops. The French, whose morale was low after a series of defeats, were at first obliged to withdraw, but rallied the next day, and forced Brunswick to retreat.

Valois, a royal family of France, a branch of the *Capetians. When Charles IV died in 1328, an assembly of nobles decided that his daughter could not inherit the throne (*Salic law). Philip of Valois, grandson of Philip III, became *Philip VI. The direct Valois line ended with the death of *Charles VIII, but the dynasty continued under *Louis XII (Valois-Orléans) and *Francis I (Valois-Angoulême). The dynasty was crippled by succession problems and a series of regencies after 1559. The rule of Catherine de *Medici and her three sons coincided with the *French Wars of Religion. The monarchy was challenged by the *Huguenots, the dukes of *Guise, and the Spanish. When Henry III died in 1589, the throne passed to the *Bourbons headed by Henry of Navarre (later *Henry IV).

Vandals, a Germanic tribe that migrated from the Baltic coast in the 1st century BC. After taking Pannonia in the 4th century they were driven further west by the *Huns. With the Suebi and Alemanni they crossed the Rhine into Gaul and Spain where the name Andalusia ('Vandalitia') commemorates them. They were then ousted by the Goths. Taking ship to North Africa under *Genseric they set up an independent kingdom after the capture of Carthage. In 455 they returned to Italy and sacked Rome. *Belisarius finally subjected them in 534.

Vane, Sir Henry (the Younger) (1613–62), leading Parliamentarian, son of Sir Henry Vane (the Elder). He served briefly as governor of Massachusetts (1636–7) then played a leading role in the English *Long Parliament until 1660. A promoter of the *Solemn League and Covenant and the *Self-Denying Ordinance, from 1643 to 1653 he was the civil leader of the Parliamentary cause, while *Cromwell directed the army. He opposed the trial and execution of *Charles I, and in 1653 disagreed with Cromwell over the expulsion of the Rump Parliament. After a period out of politics, he helped to bring about the recall of the Rump (1659). At the *Restoration he was arrested and executed for treason.

Varanasi (Benares), a city in north India, which has long been renowned as a centre of *pilgrimage for *Hindus from all over India. There they seek ritual purification in the sacred River Ganges and cremate their dead on the ghats (flights of steps) which line the river and on other cremation grounds along the river. The city contains hundreds of temples, few of which are older than the 17th century, Muslim invaders having destroyed many earlier shrines. Its importance has always been religious and cultural rather than political.

Varennes, Flight to (20 June 1791), the unsuccessful attempt by *Louis XVI to escape from France and join the exiled royalists. He had been prevented from leaving Paris in April 1791, and elaborate plans for an escape were made. On the night of 20 June the royal party, disguised and with forged passports, left Paris. They were recognized by a postmaster, pursued, and stopped at Varennes. The fugitives were returned and became virtual prisoners in the Tuileries.

vassal, a holder of land by contract from a lord. This tenurial arrangement was one of the essential components of *feudalism. The land received was known as a *fief

The Mogul army's encampment on the banks of the River Ganges, opposite the city of **Varanasi** (once known as Benares) in 1765, during the struggle for control in India between the European powers and local rulers. Lining the city shore are the temples with steps for Hindu pilgrims to reach the waters of the holy river. (India Office Library, London)

The contribution of **Vauban**, military engineer and Marshal of France, to offensive and defensive engineering strategy was immense. Here a city fortified to his pattern is bombarded by artillery and mortars. Vauban also devised trenchworks to protect attacking troops and gunners.

and the contract was confirmed when the recipient knelt and placed his hands between those of his lord. Counts and dukes received their lands from kings in this way, in some cases acquiring more territory and consequently greater power than their nominal masters. Norman kings of England did homage to kings of France for the duchy of *Normandy. The contract could legally be broken only by a formal act of defiance.

Vauban, Sébastien le Prestre de (1633–1707), Marshal of France. He was a brilliant military engineer whose writings on siege warfare have been studied for centuries. He designed fortications for *Louis XIV and also captured fortresses such as Maastricht, Mons, and Namur. His *Projet d'une dîme royale* (1707), advocating a single tax (the *dîme royale*) as fairer than the existing tax system, greatly annoyed the king and was suppressed.

Vedas (Sanskrit, 'wisdom'), the earliest corpus of Hindu sacred literature, composed in Sanskrit at an uncertain date (c.1500–1200 BC), during and following the *Aryan

invasions of India. They comprise four collections of hymns and chants, the Rigveda, Yajurveda, Samaveda, and Atharvaveda, which were used in the Aryans' worship of natural and cosmic forces. During the fire sacrifices which were central to Vedic rituals, priests chanted hymns and spells which were subsequently incorporated into the Vedas. The Rigveda, the oldest collection, is the best known, the Atharvaveda, the latest, carries less authority than the others, but together the four collections constitute the main source of knowledge about the early beliefs and practices of the Aryans in India.

Vendée, a large area of western France, formed in 1790, which was the centre of a series of counter-revolutionary insurrections from 1793 to 1796. The ideas of the Revolution were slow in penetrating this area and the introduction of conscription acts was a signal for widespread rioting. In March republicans were massacred in the towns and the exiled royalists placed themselves at the head of a Catholic and royal army. This Vendéan army, using its superior knowledge of the area and partly equipped with English arms, scored several victories, though republican armies finally crushed the revolts.

Vendôme, Louis Joseph, duc de (1654–1712), soldier and Marshal of France. In 1702 at the beginning of the War of the *Spanish Succession he was made commander in Italy. He defeated Prince *Eugène at Cassano, but in 1708 was defeated by *Marlborough at *Oudenarde. In 1710 he was sent to Spain, where his victories helped to keep *Philip V on the throne.

Veneti, an Italic tribe with their own distinctive language who inhabited north-east Italy from the 1st millennium BC. Always in need of protection against major invasions of Italy by the Gauls and by Hannibal, they remained loyal to the Roman empire. Under the later Roman empire their territory became the province of Venetia. Their geographical position put their cities (for example, Aquileia and Padua) in the path of *Goth and *Hun invasions of Italy. A gradual migration took place towards

the barely inhabited islands, marshes, and sandbanks of the coastal estuaries. The area grew in prosperity largely through trade. In the 9th century, the Republic of St Mark, better known as *Venice was established.

Venice, a city built on the islands of a lagoon on the Adriatic coast of north-east Italy. The islands were first inhabited by the *Veneti as a refuge from barbarian invasion in the 5th century and by the end of the 6th century were permanently inhabited. They were extended and sea defences were built. From 726 rulers, or *doges, were elected and Venice became a republic independent of the Byzantine empire in the 9th century. Venice became a maritime power, defeating pirates and growing rich on the profits from trade with the East. Venice was prominent in the *Crusades: it benefited from the sack of *Constantinople (1204). On the mainland of Italy it ruled the large adjoining province known as Venetia and also many Greek islands.

The republic at the height of its powers dominated the Mediterranean, gaining Cyprus in the 15th century and ruling the towns of Bergamo, Brescia, Padua, Verona, and Vicenza: 'the Veneto'. Government was in the hands of a few great families. Though successful at the naval battle of *Lepanto against the Turks, it lost Cyprus and its power declined thereafter. Succumbing easily to the French invasion of 1797, it was given to Austria, but in the 19th century joined the kingdom of Italy.

Vercingetorix (d. 46 BC), King of the tribe of the Averni in Gaul. Towards the end of the *Gallic wars in 52 BC he revolted against Roman occupation and was acclaimed King of the united Gauls. He was defeated and captured and was finally paraded through Rome as a trophy in *Caesar's triumph (46 BC) and then executed.

A painting of a procession in the Piazza San Marco in **Venice**, by Gentile Bellini, 1496. St Mark is the patron of Venice and of its famous cathedral, constructed in the 11th century and still being embellished in the 15th. (Venezia Accademia)

Verdun, Treaty of (843), the peace made between the Frankish kings Lothar, Louis, and Charles, the grandsons of *Charlemagne. When their father, Louis the Pious, died in 840 he bequeathed them the united *Carolingian empire, but the brothers could not agree on how to divide the inheritance and civil war followed until 842. Long negotiations then culminated in the meeting in Verdun where the empire was divided into three kingdoms. Charles and Louis received West and East Francia (roughly, present-day France and Germany), while Lothar held the middle kingdom, a long strip of territory stretching from the North Sea over the Alps to Rome and bordered in the west by the rivers Scheldt, Meuse, and Saône and in the east by the Rhine. The treaty was not governed by geographical factors but was an attempt to satisfy the claims of each brother for a share in the Carolingian family estates, many of which were in the fertile lands of the middle kingdom, Lotharingia. Lotharingia soon lost its own identity and became a battleground for the embryonic kingdoms of France and Germany.

Vermont, New England, USA, a state in western *New England. Settlement began in the mid-18th century, but as the region was claimed by both New York and New Hampshire, the pioneers formed the *Green Mountain Boys under Ethan *Allen in 1771 to protect their property rights. After the British defeat at *Bennington (1777), the next fourteen years were spent in securing US statehood, through negotiations, threats of joining the British empire, or declaring an independent republic. In 1791 it became the 14th state and attracted small farmers from southern New England.

Vernon, Edward (1684–1757), English admiral. In 1739 he was sent to fight against the Spanish in the Caribbean. He captured Porto Bello in the opening phase of the War of *Jenkins's Ear in 1739, but failed disastrously at Cartagena in 1741. His boat-cloak of grogram (a mixture of silk, mohair, and wool) gave him the nickname 'Old Grog', and 'grog' became the name of the naval rum ration when diluted with three parts of water, as Vernon was in the habit of issuing it to his own sailors. He was removed from the active list in 1746 because he was thought to be the author of anonymous pamphlets attacking the Admiralty.

Verrazzano, Giovanni da (c.1485–c.1528), Florentine navigator in the service of France. He led three expeditions in search of a westward passage into the Pacific and thus to the East. In 1524 he explored the North American coast from North Carolina to New York Bay and continued north to Newfoundland before returning to Dieppe. In 1527 he took a second expedition across the Atlantic and reached Brazil. He set out once more in 1528 but was met in the Antilles by cannibal *Caribs who killed and ate him.

Versailles, the great palace, 16 km. (10 miles) from Paris, built by *Louis XIV of France. In 1682 it became the seat of government; here Louis moved through the day according to a strict pattern of ceremonial, and attendance at court was essential to any nobleman who wished to retain the royal favour. Further reconstruction and redecoration was executed under *Louis XV and *Louis XVI and *Marie Antoinette. The king was forced

The master-gardener André le Nôtre (1613–1700) created one of the finest man-made landscapes in the world for Louis XIV's palace at **Versailles**, illustrated here by Jean-Baptiste Martin's painting of a promenade. (Musée de Versailles)

to leave the palace in October 1789 during the *French Revolution. It was stripped of much of its contents and today the palace is a museum.

Vespasian (Titus Flavius Sabinus Vespasianus) (AD 9–79), Roman emperor (AD 69–79), the first of the Flavian emperors. Consul in 51, he became a governor in North Africa in 63. On the outbreak of the *Jewish Revolt in 66 *Nero sent him to quell Palestine, but on Nero's death he suspended operations, leaving his son Titus in command. He refused to accept Vitellius in the imperial civil war and was proclaimed emperor by his legions in 69. He restored discipline in the army after the civil war, pacified the frontiers, and restored the exhausted economy, largely by taxation. He kept the administration under tight control so that, after a peaceful rule, he died leaving Rome solvent.

Vespucci, Amerigo (1451–1512), Florentine merchant and adventurer. His letters describe four voyages to the New World (America), but his claims as an explorer of the West Indies and South America have been doubted because of the seemingly impossible distances and positions involved. There is no doubt that he was an experienced navigator and he evolved a system for computing nearly exact longitude. The name of America is said to have been derived from his own first name, but this is a matter of speculation.

Vestal virgin, an attendant of Vesta, goddess of fire, hearth, and home. The virgins numbered six, chosen by lot from a short list of aristocratic girls. Under vows of

chastity they served for thirty years, dressed as brides, cleaning Vesta's shrine and tending its fire. Unchaste Vestals were buried alive. They lived in the House of the Vestals in the *Forum at Rome, and wills were deposited with them for safe-keeping.

Vietnam *Annam.

Vijayanagar, a ruined city site on the River Tungabhadra in south India which from the 14th to the 17th century was the centre of a powerful Hindu empire. It was founded in 1336, its name, 'city of victory' reflecting the rise of a new southern dynasty to fill the vacuum left by *Chola collapse. Three dynasties ruled successively, the Sangama (1336–c.1485), the Saluva (c.1485–1505), and the Tuluva (c.1505–65). The empire reached its peak under Krisna Deva Raya (1509–29), but soon afterwards concerted Muslim pressure from the Deccan kingdoms of Bijapur, Ahmadnagar, and Golconda resulted in the massive defeat at the battle of Talikota (1565) in which Vijayanagar was destroyed. Despite some recovery under the upstart Aravidu dynasty, the empire, one of the greatest in India's history, had collapsed by the early 17th century. It had prevented Muslim penetration into south India and had revitalized Hindu religious and literary traditions.

Viking, a Scandinavian trader or pirate of the 8th to 12th centuries. Vikings in the 8th century began one of the most remarkable periods of expansion in history. Setting sail from Denmark and Norway, they voyaged

Vestal virgins at a banquet, pictured on a relief that is thought to have come from the altar of the Piety of the emperor, at the time of Tiberius, AD 22. The Vestals were accorded many honours and privileges and exercised great influence in the Roman state. At such public celebrations they were treated with equal honour as the ladies of the imperial family. (Museo Archeologico, Palermo)

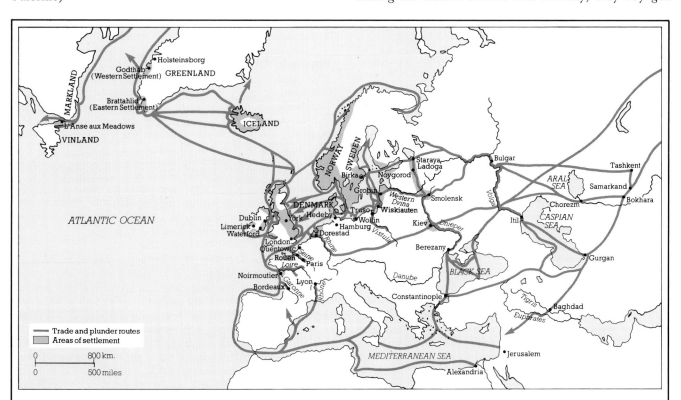

Vikings: European invasions in the 9th and 10th centuries

The Vikings' voyages from Scandinavia were made possible by their well-designed ships, in which they eventually sailed as far as Vinland (Newfoundland), where they settled briefly. The Norwegians crossed the North Sea to Ireland in the early 9th century, founding a settlement at Dublin (841), which they used as a base for further raids in Britain and the continent. The Danes' bases were in the areas around York and Dorestad. The Swedes under Rurik crossed the Baltic Sea and travelled down the Dnieper, founding the first Russian state at Kiev in 882.

westward in their *longships through the Shetlands, Iceland, and Greenland, as far as Vinland. They attacked Britain and Ireland, ravaged the coast of continental Europe as far as Gibraltar, and entered the Mediterranean, where they fought Arabs as well as Europeans. From the Baltic they sailed down the rivers of western Russia to a point from which they threatened Constantinople. In Europe they were able to strike far inland, sailing up the Rhine, Loire, and other rivers. Local rulers often preferred to buy them off, rather than resist.

They were also traders, and showed skills as farmers in the areas they settled, including Normandy, the north of England, and the area around Dublin in Ireland. They were skilled wood- and metalworkers and manufactured superb jewellery. They had a powerful oral poetic tradition, manifest in their sagas. Moreover they were an extremely adaptable people, able to absorb the cultures which they encountered while retaining their own vital qualities. This adaptability was perhaps forced upon them, because they were greatly outnumbered by the native populations; it was easier to modify existing forms than to impose their own. They adopted languages and quickly modified fighting styles to suit land-based operations. From *Rollo's settlement in Normandy descendants of the Vikings (*Normans) were a most powerful element in the *Crusades, and throughout the Mediterranean.

Villanovans, one of the peoples occupying Italy during the *Iron Age. Their territory included Tuscany and Latium as far as Rome, and reached out into Campania. They were great bronzesmiths and were the first in their area to work iron. Evidence for them comes largely from rich cremation cemeteries of the 10th to 8th century BC, when they become recognizable historically as the *Etruscans, whose cities occupy former Villanovan sites.

Villehardouin, Geoffroi de (c.1150–1217), Marshal of Champagne in France and historian. He took part in the disastrous Fourth *Crusade and became Marshal of the Eastern empire. His work *Conquête de Constantinople* described and justified the Fourth Crusade and is one of the earliest examples of French prose.

villein (Latin *villanus*, 'villager'), a medieval peasant entirely subject to a lord or attached to a manor, similar to a *serf. Both groups were part of the *manorial system which dominated Europe between the 4th and 13th centuries. Villeins provided labour services to the lord (in return for tilling their own strips of land). In England these services could vary from region to region. There were few villeins in Kent or East Anglia; labour services in the Midlands were not usually heavy. Ecclesiastical estates, however, tended to make severe demands of their villeins. By the 13th century villeins in England had become unfree tenants. In Europe they had fewer duties and remained essentially free peasants, creating a significant difference in rank to the serfs. By the 15th century, even in England, social and economic changes had blurred the distinctions between free and unfree peasants, leading to a single enlarged class of peasants.

Villiers *Buckingham, George Villiers, 1st Duke of.

Vinland (Vine Land), the Viking name for part of eastern Canada; other parts were Helluland (Stone Land) and Markland (Forest Land). Sagas record their accidental discovery c.985–6 by Bjarni Herjolfsson, blown off course en route to the settlements on *Greenland; and short-lived settlement in the late 10th and early 11th centuries by Leif Ericsson, his brother Thorvald, and Thorfinn Karlsefni. The natives, the Beothuk tribe, were called *Skraellings*. Incontestable traces of a Norse presence are known only from Ellesmere Island and the *L'Anse aux Meadows site in Newfoundland, Canada, and a coin from Maine, USA.

Virginia, a colony and state of the USA, which consists of 'tidewater' land in the mid-Atlantic coastal plain which drains into Chesapeake Bay, and Piedmont, the eastern slopes of the Allegheny Mountains. The first permanent British colony, *Jamestown was settled in 1607 by the Virginia Company of London, who inherited the name from *Raleigh's compliment to the 'virgin queen' Elizabeth I. Survival and prosperity were based on tobacco, first cultivated by white servants (supporters of *Bacon's Rebellion) but after the 1690s mainly by slaves. By then a rural gentry of planters (*plantation) had emerged. In the 18th century the Piedmont was settled by Germans and Scots-Irish, the latter often fervent converts in the *Great Awakening. Virginians were remarkably united in their opposition to British 'tyranny', demonstrated in documents from *Henry's Virginia Resolves (1765) to Jefferson's *Declaration of Independence. The last major action of the War of Independence occurred in Virginia, at *Yorktown (1781). Though deeply divided over the Constitution because of jealousy about states' rights, the 'Old Dominion' provided four out of the first five presidents and became a heartland of the Democratic-Republican party.

Visconti, Gian Galeazzo (1351–1402), Italian statesman and patron of the arts. Succeeding his father Galeazzo II in 1378, Gian Galeazzo ruled Milan jointly with his uncle Bernabò until 1385 when he had the latter put to death and assumed sole control as duke. His expansionist policies united independent cities such as Pisa and Siena under Milanese rule, and made the Visconti dynasty master of northern Italy. He arranged marriage alliances with most of the western European powers. After his death in 1402, the unity of his dominions was temporarily disrupted before being restored by his son Filippo Maria, who ruled as duke from 1412 to 1447. On his death Milan passed to the *Sforzas.

Visigoths, or western *Goths, a people originating in the Baltic area. Migrations in search of farmland took them to the Danube delta and the western Black Sea by the 3rd century AD. They raided Greece and threatened the eastern Mediterranean but were temporarily repulsed by Claudius II 'Gothicus'. *Aurelian conceded *Dacia and the Danube to them. Competition over land with the migrating *Huns drove them south in 376, and they defeated the Roman emperor Valens at Adrianople. A treaty of alliance followed but on the death of *Theodosius I they ravaged the empire and Rome itself under the leadership of *Alaric. They occupied parts of Gaul and Spain (Languedoc and Catalonia), assisting Rome against other barbarians, notably against the *Ostrogoths at the *Catalaunian Fields in 451. Frankish and Muslim invaders defeated and absorbed them in the following two centuries.

vizier (Arabic, *wazir*), a leading court official of a traditional Islamic regime. Viziers were frequently the power behind nominal rulers. At times the office became hereditary, under the early *Abbasids falling into the hands of the Barmakids and under the *Ottomans in the late 17th century held by the *Köprülü family.

Vladimir I, St (956-1015), Grand Duke of *Kiev (978-1015). He was converted to Christianity and married the sister of the Byzantine emperor. By inviting missionaries from Greece into his territories he initiated the Russian branch of the Eastern *Orthodox Church, which was rapidly established. He was canonized and became the patron saint of Russia.

Voltaire, François-Marie Arouet (1694-1778), French writer, one of the dominating figures of the 18th century. He was the son of a lawyer, educated by the Jesuits, and his brilliant wit made him a success in Paris. He was imprisoned in the *Bastille after a quarrel with the chevalier de Rohan and then went into exile in London in 1729. His *Lettres philosophiques sur les Anglais* (1734) express his admiration for English freedom and the English system of government.

He became the chief protagonist of the new philosophy of the *Enlightenment. He was violently opposed to the Roman Catholic Church: he was a deist and regarded belief in the existence of God as necessary for the government of the masses. *Frederick II invited him to his court in 1750, but they quarrelled after two years and he finally settled at Ferney near the Swiss frontier. His satire *Candide* (1759) ridiculed the spirit of optimism found in some 18th-century philosophers and reflected

Voltaire in his study, a contemporary portrait of the French writer and philosopher by an unknown artist. (Musée Carnavalet, Paris)

The burning of the legendary tribal Romano-British king **Vortigern**, an illustration from the *Chronicle of England* by Peter Langtoft, *c.*1307–27. (British Library, London)

his harsh sense of realism. He had a passionate concern for justice, as is shown in his efforts for individuals such as *Lally. His personality and gifts as a writer gave him a pervasive influence over his age, which is often called the 'Age of Voltaire'. He made a triumphant visit to Paris in the year of his death.

Vortigern, a legendary 5th-century Romano-British king said by *Bede to have invited *Hengist and Horsa to Britain as mercenaries in an attempt to withstand the raids by the *Picts and the *Scots. The plan rebounded on Vortigern when Hengist and Horsa turned against him (455) and seized lands in Kent. He was blamed by *Gildas for his misjudgement and also for the loss of Britain.

Wakefield, battle of (30 December 1460). A battle in the Wars of the *Roses fought in Yorkshire. The *Lancastrians defeated and killed the *Yorkist claimant to the throne, Richard, 3rd Duke of York.

Waldenses (or Vaudois), a Christian religious sect founded by Peter Valdes (d.1217) in the 12th century. Assuming a life of poverty and religious devotion, he founded the 'Poor Men of Lyons' in France. They lived simply and taught from vernacular scriptures. They were persecuted as heretics but survived in southern France and in Piedmont in north-eastern Italy. In 1532 they formed an alliance with the Swiss reformed church but were almost destroyed in France, and in 1655 the Piedmontese Waldenses, despite support from Protestant groups, were massacred.

Wales (Welsh, Cymru), the western part of Great Britain and a principality of the United Kingdom. Its population, Celtic in origin, resisted the Romans (who penetrated as far as Anglesey in a campaign against the *Druids), and after the departure of the Romans was increased in size by British refugees from the *Saxon invaders (c.400). By the 7th century Wales was isolated from the other Celtic lands of Cornwall and Scotland. Christianity was gradually spread throughout Wales by such missionaries as St Illtud and St *David, but politically the land remained disunited, having many different tribes, kingdoms, and jurisdictions; Gwynedd, Deheubarth, Powys, and Dyfed emerged as the largest kingdoms, one notable ruler being Hwyel Dda (the Good), traditionally associated with an important code of laws.

From the 11th century the Normans colonized and feudalized much of Wales and Romanized the Church, but the native Welsh retained their own laws and tribal organization. They several times revolted, but as each revolt was crushed the English kings tightened their grip. Although *Llywelyn the Great (ruled 1194–1240) recovered a measure of independence, *Edward I's invasion in 1277 ended hopes of a Welsh state: Llywelyn II was killed in 1282, and in 1301 Edward of Caernavon (*Edward II) was made Prince of Wales. Thereafter Wales was divided between the Principality, royal lands, and virtually independent *marcher lordships. The unsuccessful revolt of Owen *Glendower in the early 15th century revived Welsh aspirations, but *Henry VIII, the son of the Welsh *Henry VII, united Wales with England in 1536, bringing it within the English legal and parliamentary systems. Welsh culture was eroded as the gentry and church became Anglicized, although most of the population spoke only Welsh, given a standard form in the Bible of 1588, until the 19th century.

Wallace, Sir William (c.1270–1305), Scottish soldier who became a national hero for his resistance to English rule. In May 1297 he attacked the English garrison at Lanark and then defeated an English army at Stirling Bridge (1297) after which Wallace was knighted, recaptured Berwick, and raided northern England. He was

then elected governor of Scotland. In 1298 *Edward I invaded Scotland with 88,000 men and defeated Wallace at *Falkirk. He fled to France, but was later captured, tried in London, and beheaded.

Wallenstein, Albrecht Wenzel Eusebius von (1583–1634), Duke of Friedland (1625), Duke of Mecklenburg (1629), Czech magnate and military entrepreneur. At the outbreak of the *Thirty Years War he remained loyal to Emperor Ferdinand II, and in 1621 was made governor of Bohemia, a position from which he profited enormously. He served as commander-in-chief of the emperor's Catholic forces (1625–30), but after driving the Danes from north Germany, he tried to establish his own empire on the Baltic, and the German princes prevailed upon the emperor to dismiss him. He was subsequently recalled as imperial general (1632–4) to deal with the ascendant Swedes and Saxons. Having become embittered by his earlier dismissal, he was determined to build up his personal power, and his secret negotiations with the enemy were discovered. Ferdinand again removed him from command and shortly afterwards connived at his assassination.

Walpole, Sir Robert, 1st Earl of Orford (1676–1745), English statesman. He entered Parliament in 1701 and briefly held the offices of Secretary of War and Treasurer of the Navy, until he was dismissed with his *Whig colleagues in 1710. The *Tories impeached him for corruption in 1711, and expelled him from Parliament, making him a martyr for the Whig cause. On the accession of George I in 1714 he became Paymaster of the Forces, and Chancellor of the Exchequer in 1715,

The assassination of **Wallenstein** on 15 February 1634. His attackers first cut down his officers (above) before murdering Wallenstein in his bedroom (below). (Germanisches National-Museum, Nuremberg)

but resigned in sympathy with his brother-in-law *Townshend in 1717. The *South Sea Bubble crisis brought Walpole to power in 1721, and he remained in office as leading minister until 1742.

During his long period of power he strove for peace abroad and also did his best to avoid political controversy at home, especially on contentious issues like religion; he strengthened the economy, and by his mastery of the House of Commons and the use of patronage he maintained political stability. He regularly presided over *cabinet meetings, and was thus generally regarded as the first effective Prime Minister; he insisted on cabinet loyalty, and moved a long way towards the idea of the collective responsibility of the cabinet. In 1733 his unpopular attempts to impose excise duties on wine and tobacco were defeated in Parliament, but he placed a heavy duty on molasses imported into America, which became a major American grievance. He then faced an increasingly powerful opposition from within his own party in Parliament and lost an important patron when Queen Caroline died in 1737. After reluctantly going to war with Spain in 1739, and seeing Britain increasingly involved in the War of the *Austrian Succession, he accepted a peerage and resigned in 1742.

Walsingham, Sir Francis (c.1530–90), English statesman and diplomat. He entered the service of *Elizabeth I in 1568, becoming her joint Secretary of State in 1573. A zealous Protestant, he constantly advocated aggression toward Catholics at home and abroad. Elizabeth generally preferred the more moderate counsels of *Cecil, yet she valued Walsingham highly—not least because of his efficient network of diplomats and spies which spanned Europe. It ensnared such enemies of the state as *Throckmorton, *Babington, and *Mary, Queen of Scots, and brought him detailed information on the preparation of the *Spanish Armada (1588).

Walter, Hubert (d. 1205), English cleric and statesman. After studying law at the University of Bologna he served first under *Henry II and then went with *Richard I on the Third *Crusade. On his return in 1193 was made Archbishop of Canterbury and Justiciar, or Regent, of England. He was appointed papal legate in 1195. Walter's legal and administrative skills were recognized by both Richard I (whose £100,000 ransom he organized) and *John (whose accession he secured) who made him Chancellor. During Richard's frequent absences abroad he was virtually the ruler of England. Extensive financial and judicial reforms were made, including improvements to the enforcement of law and order (1195). He was Lord Chancellor from 1199.

Wandewash, battle of (22 January 1760). The coastal fortress of Wandewash, in South India, was the site of one of the last great battles between the English and French East India Companies in south India. Although the French commander, the comte de *Lally, had the advantage of numbers, he was hampered by dissensions among his forces. His attack was repulsed by the able British commander, Eyre Coote. The fall of Pondicherry followed.

Wang Anshi (1021–86), Chinese statesman, chief councillor to the *Song (1069–76). He introduced major financial and administrative reforms and reorganized local policing and the militia, known collectively as the 'New Policies'. The prices of commodities were stabilized and farmers benefited from reduced land tax, low-interest state loans, and a reduction in the levy that replaced forced labour. There was much opposition to his reforms, particularly from officials and landowners, and he was dismissed in 1076. Many of his reforms were reversed shortly thereafter.

Warbeck, Perkin (1474–99), *pretender to the throne of England. He was in the service of a merchant who was friendly with *Edward IV of England, and posed from 1491 to 1497 as Richard, Duke of York, the younger of Edward's murdered sons. He was well received in Ireland, France, and Flanders by those hostile to *Henry VII, but soon after he had landed in Cornwall in September 1497 he was captured. His large following dispersed, as he made a full confession and was imprisoned and later executed.

Wardrobe, a department in the household of the English kings, who found the *Exchequer's methods of collecting revenues too cumbersome to meet their financial needs, particularly when travelling. King John had used another household department, the Chamber, but under Henry III the Wardrobe was developed and Edward I treated it virtually as his war treasury to supply military expenditure. By the 15th century, however, the Chamber had become the main financial department.

Warwick, Richard Neville, Earl of (1428–71). He earned the title of 'kingmaker' of England, as a result of the influence he exerted during the Wars of the *Roses. He inherited the earldom of Salisbury from his father Richard (d. 1460) and gained that of Warwick by his marriage to Anne Beauchamp. He had the support of the rest of the Neville family, which was central to the *Yorkist party. He was an ally of Richard Plantagenet, Duke of *York, and was largely responsible for putting his son, *Edward IV, on the throne in 1461. His wealth and power made him easily the most formidable subject of the king until Edward's marriage to the *Lancastrian Elizabeth Woodville, after which he and his family gradually lost influence and had to see Edward move towards alliance with *Burgundy. After Edward declared him a traitor, he fled to France, but invaded England in September 1470 and put *Henry VI back on the throne. He was himself defeated and killed in the battle of *Barnet on 14 April 1471.

Washington, George (1732–99), first President of the United States of America (1789–97). He came from a well-established Virginian planter family. As a young man he saw active service in the last of the *French and Indian wars, leaving the army in 1758 with the rank of colonel. Politically he was a supporter of resistance to British measures, and at the second *Continental Congress he was elected commander-in-chief. Taking command at Cambridge, Massachusetts, he imposed some order on the 16,000 volunteers and in March 1776 drove the British from Boston. In September, after an inept defence of New York, he brilliantly extricated his army; then, for the next five years, he played a waiting game against the British in New York and Philadelphia, with occasional raids and skirmishes such as at Trenton (1776), Princeton, Brandywine, and Germantown (1777), and, after the

George **Washington**'s defeat of the British at Princeton on 3 January 1777, a painting by M. M. Sanford.

*Valley Forge winter, at Monmouth (1778). French troops arrived as allies in 1780, and at last he was able to engage in the *Yorktown campaign, which effectively ended the war.

Washington's main achievement had been to hold his ill-supplied army together while deferring to a divided Congress. Anxious about post-war political anarchy, he encouraged the calling of the *Constitutional Convention in 1787 and supported the resultant Constitution. He was elected unanimously to the presidency, and again for a second term in 1792. Although he tried to keep his office above politics, he became identified with *Federalist policies; and his suppression of the *Whisky Rebellion, his refusal to support Revolutionary France, and his approval of *Jay's Treaty provoked attacks from Jeffersonian Republicans. His farewell address deplored factionalism and called for American neutrality in foreign affairs. The father of his nation, his powers of leadership, his stoicism, and his integrity together earned him the admiration and respect of his countrymen.

watch and ward, the system developed in 13th century England to preserve the peace in local communities. Guards were appointed and the duties of the constables at night (watch) and in daytime (ward) were defined. Town gates were to remain closed from dusk to dawn; strangers had to produce sureties; up to sixteen men maintained the watch in cities, twelve in boroughs, and four in smaller communities. Modifications to the system were made regularly throughout the 13th century and were eventually incorporated in the Statute of Winchester of 1285.

Watling Street, a north-westerly Roman road of Britain which ran from Dubris (Dover), via Londinium (London) and Verulamium (St Albans), to Deva (Chester). Much of it was built c. AD 60–70 for the advance north of the *Fosse Way. Its name derives from the Anglo-Saxon name for Verulamium ('Waeclingacaester') and for a paved road ('street').

Wayne, Anthony (1745–96), American Revolutionary general, who led a Pennsylvanian regiment in the abortive invasion of Canada (1776). He displayed conspicuous initiative in such actions as Brandywine, Germantown, and Monmouth. In 1779 his brilliantly planned and executed night attack on Stony Point, a British-held fort on the Hudson, won him the nickname of 'Mad Anthony'. In 1794 he commanded the western army, which defeated the Indians at the battle of Fallen Timbers and opened up the *Northwest Territory to settlement.

Wenceslas II (1361–1419), King of *Bohemia (1378–1419), and Germany (1378–1400), and Holy Roman Emperor (uncrowned, 1378–1400). A weak king, he was overcome by the ambitions of the imperial princes, the town leagues, and by his brother *Sigismund. A Bohemian revolt, starting in 1394, deposed him in favour of Sigismund in 1402, and he lost the German throne in 1400 in favour of Rupert III, retaining the title but losing power. Much of his reign was disturbed by the *Hussite movement; Wenceslas supported Huss, and tried to prevent his burning, which was ordered by Sigismund.

Wenceslas, St (c.907–29), Duke of *Bohemia. He became the second Christian duke on the death of his father, Uratislas, though part of the dukedom passed to

The earliest known representation of St **Wenceslas**, from *Gumpold's Legend* (c.1006), written and illuminated by Emma, wife of Boleslav II, Duke of Bohemia. (Herzog August Bibliotek, Wolfenbüttel)

Boleslav, his younger non-Christian brother. He was named king by the Holy Roman Emperor, Henry I, over an area which corresponds approximately to the present Czechoslovakia, but soon afterwards was murdered by his brother in the church of St George in Prague. This event precipitated war and the intervention of the emperor, and resulted in the permanent Christianization of the country under Boleslav and his son, Boleslav II.

Wentworth, Thomas *Strafford.

wergild (or 'man-price'), the term used in Anglo-Saxon England to signify the variable monetary value of each freeman which had to be paid in compensation by a murderer to the kinsmen of the victim. The amount of an individual's wergild was fixed in law and varied according to his rank in society. An ordinary freeman (ceorl) was valued at 200 shillings, a nobleman (earl) at 1,200 shillings, and the king at 7,200 shillings. Although the unfree had no wergild, compensation for their murder was paid to their owners.

Wesley, John (1703-91), founder of Methodism. He was an Anglican priest, the leader of an earnest, devout, and scholarly group in Oxford whose members (including his brother Charles) became known as *Methodists. Their methodical prayer and Bible readings were a reaction against the worldliness of student life rather than a revolt against the Church. He went to the American colony of Georgia to act as priest to the settlers, and although his tactlessness led to his early return to England, his

fellowship meetings, the hymn-singing, and the contact with the Moravians (an *Anabaptist sect) brought him invaluable experience. After his profound spiritual conversion in 1738 he began widespread preaching, taking to the open air as early as 1739. At Bristol in the same year he established a chapel and a school for the religious societies. As the *Anglican Church showed indifference to his work, or even became actively opposed to him, he began to found Methodist societies wherever he preached, and he chose lay preachers as full-time helpers. In 1744 he drew up his strict rule-book for the regulation of Methodist societies. The American War of Independence had cut his contact with the colonies and in 1784 it became necessary for him to ordain twenty-seven preachers for America, the Anglican Church having refused to do so.

During his 50-year ministry he travelled 250,000 miles, mostly on horseback, and delivered 50,000 sermons. He preached in industrial areas where the parish system had broken down. The religious revival for which he was responsible, the 'Great Awakening', not only created Methodism but challenged the Anglican Church into reviewing its own evangelism.

His brother **Charles Wesley** (1707-88), was the co-founder of Methodism. He accompanied John to America and on his return carried out exhausting preaching tours, but his genius lay in hymn-writing.

Wessex, the kingdom of the West Saxons (*Gewisse*, 'companions'), whose royal dynasty achieved the unification of England by the early 10th century. According to the *Anglo-Saxon Chronicle*, Wessex was founded in 495 by two chieftains, Cerdic and Cynric, who landed near Southampton Water before advancing inland over Hampshire and the basin of the upper Thames. Under Ceawlin (560-91) whose overlordship was recognized by other Saxon leaders, the policy of expansion continued.

A portrait of the Methodist leader John **Wesley**, by William Hamilton, 1788. (National Portrait Gallery, London)

A victory over the Romano-British at Dyrham (577) extended West Saxon control over the towns of Gloucester, Bath, and Cirencester. However, in the 7th century the growth of Wessex was challenged by the rise of its northern neighbour *Mercia, particularly under its kings Penda (ruled c.626–55) and Offa (ruled 757–96). Wessex continued to expand along the south coast under Ine (688–726), and by the reign of Egbert had established its authority in southern England. *Alfred (871–99) resisted the Danish invasions (871–86), and Athelstan not only ousted the Danes but was the first ruler of all England (926). Despite a temporary Danish recovery the authority of Wessex was not seriously challenged, as the reign of Edgar the Peaceable (959–75) confirmed.

West Indies   *Caribbean

Western Jin (Western Chin) (265–316), a dynasty that briefly unified China after the period of the *Three Kingdoms. It was established by Sima Yan, a general of the Wei kingdom. He attempted to curb the power of the dominant families but after his death in 290, Sima princes feuded with each other and the central government collapsed. In the resulting chaos, a *Xiongnu invasion occurred and Luoyang (311) and Chang'an (316) were sacked. One Sima prince in the south-eastern kingdom of Wu established the Eastern Jin dynasty (317–420), but it, too, lacked strong central government. During this period, the colonization of southern China by *Han Chinese greatly increased.

Western Neolithic, the period after 4000 BC during which the indigenous hunting and gathering peoples living along the western coasts of Europe merged with incoming farmers from central Europe (related to the *Bandkeramik culture) to form new groups of farmers with a basically similar material culture. Their simple round-based pottery has sometimes given them the name 'bowl cultures', but they are better known for the construction of monumental tombs out of large boulders or *megaliths.

Western Rising (July–August 1549), an English rebellion in Cornwall and Devon, at the same time as *Kett's Rebellion in Norfolk. The West Country rebels' main grievance was the imposition, by Act of *Uniformity, of the first Book of Common Prayer in English. Demanding that no such religious changes should be made until *Edward VI came of age, they laid siege to Exeter on 2 July. It was not until mid-August that government troops under Lord Russell scattered them, killing some 4,000. The leaders were executed in London; lesser rebels hanged throughout the West Country.

Westphalia, Treaty of (1648), the treaty signed at Münster and Osnabrück, which brought the *Thirty Years War to a conclusion. By its terms, the Habsburgs acknowledged the independence of Switzerland and the separation of the *United Provinces from the *Spanish Netherlands; France secured undefined rights in Alsace and retained the bishoprics of Metz, Toul, and Verdun; Sweden acquired West Pomerania and the bishoprics of Bremen and Verden: and Bradenburg acquired East Pomerania and the succession to the archbishopric of Magdeburg. The full sovereignty of the German states was recognized, thus marking the failure of the Holy Roman Emperor to turn Germany into a centralized Catholic monarchy.

Whig, a British political party traditionally opposed to the *Tories. The Whigs owed their name, like the Tories to the *exclusion crisis of *Charles II's reign. Those who petitioned for the recall of Parliament in 1679 were named Whigs (Scottish Covenanting brigands) by their Tory opponents. The Whigs suffered defeat in Charles's reign, but joined with the Tories in inviting *William of Orange to England, and they alternated with the Tories in power until 1714. Their principles were to maintain the power and privileges of Parliament, to show sympathy with religious dissent, keeping links between church and state to a minimum, and to play an active role in Europe.

From the accession of *George I the Hanoverian kings placed their trust in the Whigs, and there followed the long period of Whig supremacy. From the mid-1720s there were Whigs in opposition to *Walpole and the development of factions within the party became increasingly acute by the mid-century, bringing political instability in the 1760s. The *Rockingham faction, which formed the core of *Fox's followers, became the basis of the new Whig party in the late 18th century.

Whisky Rebellion (1794), a rising of farmers in western Pennsylvania, USA, in protest at Secretary of the Treasury Alexander *Hamilton's excise tax of 1791. The frontiersmen who made whisky considered the tax discriminatory. President *Washington called out 15,000 troops to quell the rioting, proving the federal government's power to enforce the country's laws, and earning the frontiersmen's hatred of the *Federalists' policies.

Whitby, Synod of (664), a church council that resolved the differences between the Celtic and Roman forms of Christian worship in England, particularly the method used for calculating the date of Easter. The *Celtic Church had its own method of fixing the date of Easter; this was a matter of dispute after the arrival of St *Augustine's mission. The Celtic case was presented by St Colman, Bishop of Lindisfarne. The Roman case was put forward by St Wilfrid of Ripon, whose arguments were finally accepted by King *Oswy of Northumbria. This decision was crucial, severing the connection with the Irish church and allowing for the organization of the English church under Roman discipline. Theodore, Archbishop of Canterbury, summoned an assembly of the whole English church at Hertford in 672.

Whitefield, George (1714–70), British evangelist. He was one of the early *Methodists, whom he joined at Oxford in 1732. He followed the *Wesleys to Georgia in 1738, and on his return began open-air preaching, encouraging the Wesleys to follow his example. His Calvinist views caused a temporary breach with John Wesley but the two men were reconciled. He spent a lifetime of preaching in Britain and America.

White Lotus Society (or Incense Smelling Society), a Chinese secret society. It had religious affiliations, tracing its origins to a Buddhist monk of the 4th century AD. (The lotus, springing unsullied from the mud, is a Buddhist symbol.) In times of trouble, its leaders preached of the coming of the Buddha and of the establishment of

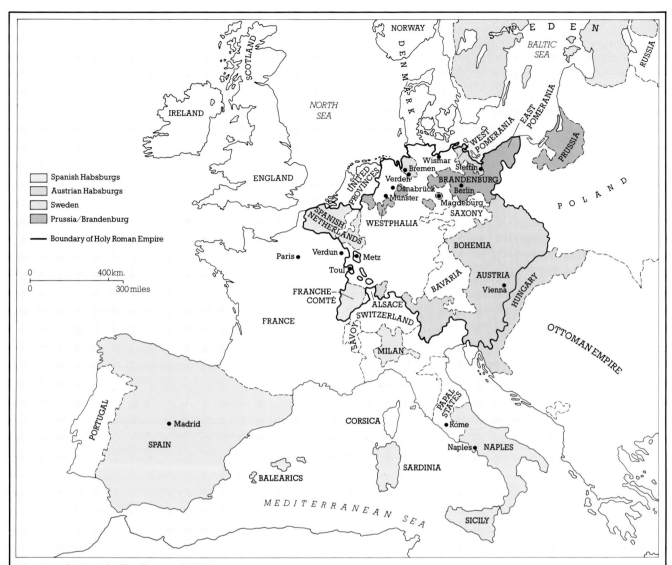

NORWAY • DENMARK • SWEDEN • RUSSIA • BALTIC SEA • NORTH SEA • IRELAND • SCOTLAND • ENGLAND • POLAND • WEST POMERANIA • EAST POMERANIA • PRUSSIA

Spanish Habsburgs
Austrian Habsburgs
Sweden
Prussia/Brandenburg
—— Boundary of Holy Roman Empire

0 ——— 400 km.
0 ——— 300 miles

UNITED PROVINCES • Wismar • Bremen • Stettin • Verden • BRANDENBURG • Osnabrück • Berlin • Münster • Magdeburg • SAXONY • WESTPHALIA • SPANISH NETHERLANDS • BOHEMIA • Paris • Verdun • Metz • Toul • BAVARIA • AUSTRIA • HUNGARY • Vienna • FRANCHE-COMTÉ • ALSACE • SWITZERLAND • SAVOY • FRANCE • MILAN • OTTOMAN EMPIRE • CORSICA • PAPAL STATES • Rome • Madrid • SPAIN • PORTUGAL • Naples • NAPLES • SARDINIA • BALEARICS • MEDITERRANEAN SEA • SICILY

Treaty of Westphalia: Europe in 1648

The Treaty of Westphalia ended the Thirty Years War and laid down the political framework of Europe until well into the 18th century. It marked the beginning of French ascendancy and the defeat of the imperial ambitions of the Habsburgs. Germany, economically devastated by the long war, was split into more than 300 sovereign states, and the Holy Roman Empire existed in little more than name. France now emerged as the dominant power in Europe, and although French territorial acquisitions were small, they were strategically important. Sweden, whose intervention against the Habsburgs had played a decisive role in the war, gained territories in central Europe to add to its growing empire.

---

a new dynasty. Its supporters, bound by blood ceremonies and claiming magic powers, came from an impoverished peasantry. Major risings occurred in the mid-14th century during the decline of the *Yuan dynasty, and from 1796 to 1804, when they successfully opposed *Manchu troops in southern Shaanxi province. This set-back weakened Chinese belief in the invincibility of the Manchu troops and the authority of the *Qing dynasty.

Wilderness Road, a frontier trail in North America opened by Daniel *Boone in 1775 across the Allegheny Mountains. It ran nearly 482 km. (300 miles) from western Virginia through the Cumberland Gap to the upper Kentucky River at Boonesborough. It was commissioned by the Transylvania Company, an association of land speculators, to encourage settlement in Kentucky, Tennessee, and the Ohio valley. For fifty years it was a major route into the eastern Mississippi valley.

Wilkes, John (1727–97), British journalist and politician. He was hailed in both Britain and America as a champion of liberty. In 1763 in number 45 of his paper the *North Briton* he attacked George III's ministers, and by implication the king himself, but when arrested for seditious libel he claimed the privileges of a Member of Parliament to contest the legality of his arrest, which had been made under a general warrant, not specifying him by name. The government then managed to expel him from Parliament on grounds of obscenity, particularly for the publication of his *Essay on Woman*, and Wilkes fled to France in 1764. He returned in 1768 to fight the general election and to serve a twenty-two month sentence for his earlier offences. Controversy raged when Parliament refused to let him take his seat, even though he was elected Member of Parliament for Middlesex on four consecutive occasions. He was at last allowed back into Parliament in 1774, since this was a new Parliament, and sat for Middlesex until 1790. He supported the parliamentary reform movement, and declared an interest in and sympathy for the American cause. All this time he enjoyed the support of the populace and a mob could always be called out to rally to his cause. In 1780, after some hesitation, he supported the action being taken to suppress the *Gordon riots, and his conversion to respectability was completed when he opposed the French Revolution.

William I (the Conqueror) (1028–87), the first Norman King of England (1066–87). The illegitimate son of Robert the Devil, Duke of Normandy, he succeeded to the dukedom in 1035. His early life was fraught with danger—three of his closest advisers were murdered and an attempt was made on his own life. Twice he faced major rebellions. In 1047 he was saved only by the intervention of the French king, Henry I, who helped him in battle. In 1053–4 Henry failed to seize Normandy for himself. William's claim to the English throne was based on the promise allegedly given to him in 1051 by *Edward the Confessor. With papal backing he landed in England and defeated *Harold II at the battle of *Hastings (1066). While the English leaders considered their next move William laid waste parts of Sussex, Surrey, and Hertfordshire. He was crowned on Christmas Day at Westminster Abbey.

The period 1067–71 was characterized by a number of rebellions against his rule in Northumbria, Wessex, Mercia, and the Isle of Ely. His reaction was ruthlessly effective. Much of his later life he spent in Normandy fighting against the French king *Philip I but before his

death he initiated the *Domesday survey (1086). He died on the battlefield, fighting Philip I of France.

William I (the Lion) (1143–1214), King of Scotland (1165–1214), succeeding his brother Malcolm IV. He helped to establish the independence of the Church of Scotland (1188), formulated the first major alliance between his country and France (1168), and stimulated Scotland's urban development. Initially loyal to *Henry II of England, William's determination to recover the earldom of Northumberland led to his support of Henry's three sons in a disastrous civil war which resulted in William's capture at Alnwick (1175), the acknowledgement of Henry as his feudal superior by the Treaty of Falaise, and, in 1189, the payment of 10,000 marks to *Richard I to recover his independence.

William I (the Silent) (1533–84), Prince of *Orange, and Count of Nassau-Dillenburg. He is regarded as the founding father of the *United Provinces of the Netherlands. He was trusted by Emperor *Charles V, and initially by *Philip II of Spain, who made him *statholder of Holland, Zeeland, and Utrecht (1559) and then of Franche-Comté (1561). Nevertheless he emerged in the 1560s as the leader of the aristocratic opposition to Philip's centralizing absolutism. On *Alba's arrival in the Netherlands (1567) he became the key figure in the first phase of the *Dutch Revolts. He was never a great general in the field, but his strengths lay in negotiating financial and military aid from abroad and in providing leadership in a country often torn by rivalries. He was recognized as statholder by the Estates of Holland (1572) and joined the Calvinist church (1573). His dream of a united Netherlands under a national government seemed close to realization with the signature of the Pacification of *Ghent (1576); but he was powerless to prevent the permanent north–south division of 1579. In 1580 he was outlawed by Philip II, and four years later he was assassinated by a Catholic fanatic.

William I (the Conqueror): this section of the Bayeux Tapestry shows Harold Godwinson in Normandy in 1064 swearing to uphold Duke William's claim to the throne of England. The tapestry was embroidered after William's defeat of Harold's army at Hastings had made him King of England. (Musée de Bayeux)

William II (Rufus, 'red-faced') (*c*.1056-1100), King of England (1087-1100). He was the second son of *William the Conqueror and Matilda. His succession was challenged by some Norman barons led by his uncle, Bishop Odo of Bayeux, who preferred his elder brother *Robert, Duke of Normandy. This rebellion (1088) was crushed as was a second revolt in 1095. Robert's departure on the First *Crusade (1096) gave William the opportunity to secure Normandy for himself. His successes there against the French king and in neighbouring Maine did much to secure the boundaries of Normandy. William's resistance to *Anselm's appointment as Archbishop of Canterbury contributed to his unpopular image with the chroniclers. His death from an arrow when hunting in the New Forest may have been arranged by his younger brother, who succeeded him as *Henry I.

William III (1650-1702), King of England, Ireland, and Scotland (1689-1702). He was *stadtholder of Holland, and took over effective rule of the *United Provinces (1672-1702) after the crisis of the French invasion in 1672. In 1677 he married his cousin, *Mary of England, and was invited in 1688 by seven leading English politicians to save England from his Roman Catholic father-in-law, *James II. In what became known as the *Glorious Revolution, he landed at Torbay, met with virtually no resistance, and in 1689 jointly with Mary accepted from Parliament the crown of England. He defeated James II's efforts to establish a base in Ireland by the victory of the *Boyne, and suppressed the highlanders of Scotland. He commanded the Dutch army in the Netherlands and although he scored only one victory, at Namur in 1695, he was able to win a favourable peace at *Ryswick two years later. He was never popular in England and relied heavily on Dutch favourites, such as *Keppel. Although he preferred the *Whigs to the *Tories, he tried to avoid one-party government. His reputation was affected by the broken promises after the Treaty of Limerick (a treaty (1691) in which William guaranteed political and religious freedom to Irish Catholics, but which he did not honour), and the massacre of *Glencoe (1692).

William of Malmesbury (*c*.1095-*c*.1143), historian of 12th-century England. A librarian at the Benedictine abbey of St Aldhelm in Malmesbury, Wiltshire, he was the author of a history of the English church to 1125 (*Gesta Pontificum Anglorum*). He is best known for his other historical work the *Gesta Regum Anglorum* ('Acts of the English Kings') dealing with the period from *Bede to 1120, and the *Historia Novella* ('Modern History') which continues the account to 1142. His work is notable for its attempt to understand and interpret events rather than record them in an uncritical fashion.

William of Occam (or Ockham) (*c*.1290-*c*.1347), English theologian and scholastic philosopher. He was a Franciscan friar who developed an anti-papal theory of the state, denying the pope secular authority and was *excommunicated in 1328, living thereafter in Munich under Emperor Louis IV's protection. His form of nominalist philosophy saw God as beyond human powers of reasoning, and things as provable only by experience or by (unprovable) scriptural authority. Hence his famous maxim, 'Occam's razor', that the fewest possible assumptions should be made in explaining a thing.

William of Wykeham (1324-1404), English administrator and Bishop of Winchester (1367-1404). He is thought to have been a *serf's son; in the 1350s, as a royal clerk, he gained *Edward III's favour, and in the next decade he became the king's right-hand man. As Bishop of Winchester and Chancellor of England from 1367, he aroused parliamentary opposition which forced him out of the chancellorship in 1371; *Richard II brought him back as Chancellor (1389-91). He was responsible for two educational foundations, New College, Oxford (1379) and Winchester College (1382).

Williams, Roger (*c*.1603-83), English founder of the North American colony of *Rhode Island. He emigrated to Massachusetts in 1631, and, as a minister at Salem, he soon quarrelled with John Cotton and Governor John *Winthrop. In 1635 he was banished. He founded Providence, Rhode Island, where he developed friendly relations with local Indians and established the Baptist Church. He obtained a parliamentary charter for Rhode Island in 1644. Though free of religious persecution, the colony was beset by feuding for most of Williams's life.

window tax, an English tax on any window or window-like opening, which was in force from 1695 to 1851. It was originally imposed to pay for the losses of the great recoinage of 1695, and was retained—and increased six times in the 18th century, particularly by *Pitt the Younger—until the 19th century. The tax was eventually applied to all windows in excess of six, and windows bricked up to avoid the tax can still be seen in older houses throughout Britain.

Winstanley, Gerrard (b. *c*.1609, d. after 1660), English radical Puritan. He was the leader of the *Diggers, who cultivated common land in Surrey in 1649-50, when food prices had risen sharply, and were then forced off their land by the authorities. Winstanley later became prominent as a pamphleteer with communistic ideas. In dedicating his most famous pamphlet, *The Law of Freedom in a Platform*, to *Cromwell in 1652, he showed surprising naïvety in thinking that Cromwell would approve the thesis that the *English Civil War had been fought against all who were enemies of the poor, including landlords and priests.

Winthrop, John (1588-1649), first governor of the North American colony of *Massachusetts. Born in Suffolk, he became a member of the Puritan Massachusetts Bay Company and left England in 1630 to establish the new colony, based on Boston. Under him a representative government of church members was established, the law was codified, Congregationalism defined, and *Arminianism bitterly attacked. He left a vivid account of the period from his departure from England until his death in his *Journal*, sometimes called *The History of New England*.

Winthrop, John (1606-76), son of John *Winthrop, and governor of *Connecticut from 1657 until his death. He emigrated from England to Massachusetts in 1631. In 1635 he established the colony of Saybrook, Connecticut, and became the leader of Connecticut colonies. He encouraged industrialization, erected an ironworks at Saugus, Massachusetts (1644), established an efficient administration, and in 1662 obtained the charter for the unified colony of Connecticut and New Haven. A notable physician and scientist, he became in 1663 the first Fellow of the *Royal Society to be resident in America.

Wishart, George (*c.*1513–46), Scottish Protestant preacher and martyr. He fled to England in 1538, then spent some time in Europe before returning to Scotland in 1543. There he opposed the anti-English policies of Cardinal *Beaton, and was possibly involved in plots to assassinate him. In 1546 John *Knox became his disciple. Wishart was burned to death, after being charged with heresy.

witan (from Old English *witenagemot*, 'moot', or meeting, of the king's councillors), the council summoned by the Anglo-Saxon kings. The meetings of the witan in the 10th and 11th centuries were a formalization of the primitive councils which existed in the early Saxon kingdoms of the 7th century. These formal gatherings of *aldermen, *thanes, and bishops discussed royal grants of land, church benefices, charters, aspects of taxation, defence and foreign policy, customary law, and the prosecution of traitors. The succession of a king had usually to be acknowledged by the witan.

witchcraft, the believed use of supernatural powers for harmful or evil ends. Most witch trials in early modern Europe arose from harm which a villager claimed had been inflicted on him or her by a neighbour (usually an old woman). In the face of disasters like the ruin of crops by abnormal weather or outbreaks of plague like the

*Black Death, popular ignorance and fear blamed witches, who were frequently tortured or burned to death. At first the Church tried to prevent this but by the 15th century belief in Satan's power led to heresy being associated with witchcraft. Pope Innocent VIII instructed the Inquisition (1484) to try witches and a book *Malleus Maleficarum* ('Hammer of the Witches') (1487) listed their habits, supposed to include flying on broomsticks. In England roughly a thousand people were hanged for witchcraft, mostly under Elizabeth I and James I; the last execution was in 1685.

In America the belief in witchcraft was rife but the *Salem witch trials (1692) in the end caused a general revulsion. The growth of scientific knowledge in the 17th century led educated opinion to reject belief in witchcraft though popular belief survived much longer.

Witt, Jan de (1625–72), Dutch statesman, an opponent of William II of Orange. In 1653 he was made grand pensionary of Holland and was the effective leader of the United Provinces during the minority of *William III, dominating the other provinces by his political skill and his knowledge of foreign affairs. The republican party sought to limit the powers of the Orange family and in 1654 members were excluded from state offices. In 1668 he signed the defensive Triple Alliance with England and Sweden to thwart *Louis XIV's designs on the Netherlands. When Louis invaded in 1672 de Witt was caught unprepared and William proved himself an able commander of the Dutch forces. De Witt's power was undermined and he and his brother Cornelius, a naval officer, were attacked and killed by a mob in 1672.

Wittelsbach, a German family which formed a ruling dynasty in Bavaria between 1180 and 1918. By a marriage of 1214, the Rhenish Palatinate was added to the family holdings. Duke Louis II (*c.*1283–1347) was elected Holy Roman Emperor, and divided the Wittelsbach succession between a younger branch (which received Bavaria), and

This woodcut from a German pamphlet dating from 1555 shows the burning at the stake of three women who have been accused of practising **witchcraft**. They are condemned to fire both on earth and in the afterworld (right), and the presence of a demon beside them confirms their guilt in the view of the witch-hunters whose persecutions were frequent in Europe at this time.

A portrait by an unknown artist of the English prelate and statesman **Wolsey**, wearing the red robes and hat of a cardinal. (National Portrait Gallery, London)

a senior one (which inherited the Rhenish and Upper Palatinate and was given an electoral title in 1356). During the *Reformation, the Bavarian branch remained staunchly Catholic. The Palatinate branch espoused the new Protestant faith, and during the *Thirty Years War forfeited both its electoral vote and the Upper Palatinate to Bavaria. By the Treaty of *Westphalia (1648) a new electoral vote was created for the Rhenish Palatinate. The Elector Charles Albert of Bavaria (1697-1745) became Holy Roman Emperor (1742-5).

Wolfe, James (1727-59), British general. He was appointed second-in-command to *Amherst in 1758 at the age of 32, after seventeen years' outstanding military service in Europe and Scotland. He led the landing on Cape Breton Island in 1758, which forced the surrender of *Louisburg. In the next year his daring surprise attack on Quebec and his defeat of Montcalm on the *Plains of Abraham won Canada for Britain. His training of his troops in disciplined fire-power brought them victory; his determination to lead from the front resulted in his death on the Plains of Abraham.

Wolsey, Thomas (*c*.1474-1530), English prelate and statesman. He rose from humble origins to the favour of *Henry VII and was made royal almoner by *Henry VIII (1509). As privy councillor (1511), then Lord Chancellor (1574), he virtually ruled on Henry's behalf while as cardinal (1515) and papal legate (1518) he was effectively head of the English church. He acquired a

string of rich benefices, culminating in the archbishopric of York (1514), which supported the grandeur of his lifestyle in palaces such as Hampton Court.

Wolsey's ambitions in international affairs were marked by the Treaty of London (1518), which established a temporary peace between France and Spain. Although he generally favoured a peaceful foreign policy Henry committed England to a policy of European warfare, which contributed to Wolsey's unpopularity. His despotism and personal ambition fed the growing anti-clericalism of the times. He supported Henry's wish to divorce Catherine of Aragon, but tried to persuade him to remarry into the French royal house, and his failure to secure the divorce finally discredited him with the king. He was arrested on a trumped-up charge of treason, but died before he could be brought to trial.

wool staple, one of the towns in England, Wales, Ireland, or in Continental Europe, through which wool merchants traded. They were set up by *Edward III of England as a means of controlling the principal English export, wool, so that he could be guaranteed the tax due on it at the customs point. By the Ordinance of the Staple (1353) fifteen British staple towns were established, but in 1363 Calais was made the wool staple through which all wool exports had to pass; a very profitable monopoly in the wool trade was given to the *Merchant Staplers. A continental staple existed until the ban on exports of wool in 1617.

workhouse, a public institution where people unable to support themselves were housed and (if able-bodied) made to work. The 1601 *Poor Law Act made parishes responsible for their own workhouses, but often they were hard to distinguish from the houses of correction, set up to discipline vagrants. The 1723 Workhouse Act denied relief to able-bodied paupers who refused to enter workhouses.

Worms, Diet of (1521). The city of Worms, on the River Rhine, in Germany was the scene of a meeting between *Luther and *Charles V. Luther committed

Martin Luther (right) appearing before Charles V's imperial Diet at **Worms** where he refused to recant and was declared an outlaw, a woodcut printed in Augsburg in 1521. (Stadtarchiv, Worms)

himself to the cause of Protestant reform, and on the last day of the Diet his teaching was formally condemned in the Edict of Worms.

writing systems, the means by which the spoken word may be converted to a durable and transmittable form. Writing is closely associated with the appearance of *civilization, since in simple societies speech and memory were sufficient and there was no need for writing. It was essential, however, for the administration on which civilized states depend. The simplest form was the *quipus* of the *Incas, which were bundles of variously knotted strings; these served the purposes of accounting but lacked the flexibility of all other writing systems, whether carved, painted, scratched, impressed, handwritten, or printed.

The earliest system, used by the *Sumerians *c*.3400 BC, was a series of pictographs, little drawings of the concepts intended. That soon had to be elaborated by the addition of ideograms (symbols for the concepts), followed by arbitary signs for words, and phonetic elements or symbols that represents the sound of a word or part of it. All ancient scripts were of this complex form.

Later development has followed one of two opposite directions. In the Chinese script, the ideograms proliferated and were standardized. In all others, phonetic signs were as far as possible the only ones retained, either syllabic, as in the linear scripts of Crete and modern Japanese, or more efficiently alphabetic. These all derive from a single common ancestor, the *Phoenicians' alphabet in use from 1000 BC.

Wyatt's Rebellion (February 1554), a protest in England against *Mary I's projected marriage to the future *Philip II of Spain. Its leader was a Kentish landowner, Sir Thomas Wyatt. Convinced that the marriage would turn England into 'a cockleboat towed by a Spanish galleon', he led 3,000 Kentishmen in a march on London. The rebellion's ultimate aims are uncertain, as is the involvement of Mary's half-sister, Princess Elizabeth, but it led to the execution of Lady Jane *Grey. Wyatt found most Londoners' loyalty to Mary stronger than their antipathy to Spain. He surrendered, and was executed with a hundred others.

Wyclif, John (or Wycliffe) (*c*.1330-1384), English church reformer. He was for centuries thought to have been the sole translator of the Bible (from Latin into English) in the 14th century; today he is regarded as one of several men who shared in the joint enterprise of producing the 'Wyclif' or *Lollard Bible. An Oxford academic and ecclesiastic, he wrote various logical and philosophical works, but achieved fame through his theological writings. These writings were condemned as heretical (1382) but gained him the interest and support of the royal family and the king's council in the later years of *Edward III and the minority of *Richard II. He died without ever having been imprisoned, and the Lollards kept alive many of his beliefs.

XYZ Affair, the name given to an episode (1797-8) in American-French diplomatic relations. A three-man mission was sent to France to resolve a dispute caused by America's unwillingness to aid France in the French Revolutionary wars in spite of treaty obligations made in 1778. *Talleyrand refused to see the delegation and indirect suggestions of loans and bribes to France came through Mme de Villette, a friend of Talleyrand. Negotiations were carried on through her with X (Jean Conrad Hottinguer), Y (a Mr Bellamy, an American banker in Hamburg), and Z (Lucien Hauteval). A proposal that the Americans should pay Talleyrand $250,000 created outrage in America. President *Adams, however, ignored calls for war and reached agreement with the French at the Convention of Mortefontaine (30 September 1800).

Xenocrates (396-314 BC), Greek philosopher. He was a student under *Plato and subsequently became head of the *Academy (339-314 BC). He was chosen to be an ambassador on Athens' behalf to the Macedonian king Antipater in 322 BC. As a philosopher he was an imitative follower of Plato, with a strong interest in the nature of the gods and in establishing a practical morality.

Xenophon (*c*.430-*c*.355 BC), Greek historian, essayist, and military commander. He left Athens in 401 to join the army of *Cyrus, who was attempting to win the

Head of **Xenophon**, the Greek historian best known for his description of the Anabasis, the retreat from Persia of 10,000 Greek mercenaries to the Black Sea. (Greek Museum, Alexandria)

Persian throne. After defeat at Cunaxa, Xenophon was chosen as a general by the 10,000 Greek mercenaries, who had been hired by Cyrus, and he helped lead them back across Asia Minor to safety. He later recorded his experiences in the *Anabasis*. He was exiled from Athens *c*.399, and in 396–394 he fought under the Spartan king Agesilaus. Soon afterwards the Spartans allocated him an estate near Olympia, from where he later moved to Corinth, and finally in 366–5 back to Athens, where he spent the rest of his life.

Apart from the *Anabasis*, he wrote a number of other works which have survived, most notably the *Hellenica*, a historical work which continues from where *Thucydides left off in 411 down to 362, and the *Memorabilia*, in which he recorded his memories of *Socrates.

Xerxes I (the Great), ruler of the Achaemenid Persian empire from 486 to 465 BC. He personally led the great expedition against Greece, but after watching his fleet being defeated at *Salamis in 480 BC, he withdrew, leaving behind Mardonius under whose command the army was defeated at *Plataea in 479. The subsequent activities of the *Delian League deprived him of many Greek cities in Asia Minor. The latter part of his reign was marked by intrigues, one of which led to his murder.

Xia (Hsia) (*c*.21st century–*c*.16th century BC), by tradition the first dynasty to rule in China. It was reputedly founded by Yu the Great, a model ruler, who, it is said, attempted to control flooding by irrigation schemes. As yet no evidence authenticates the Xia. But it may graduate from myth to history—as did China's second dynasty, the *Shang, whose existence was verified in the 1920s.

Xinjiang (Sinkiang), a region in the west of China, lying between Tibet and Mongolia, sometimes called Chinese Turkistan. It was inhabited by various Turki-speaking peoples of which the Uighur became the most numerous. Through its oasis towns ran the *Silk Route connecting east and west Asia. From *Han times the more powerful Chinese dynasties exercised control there. Part of the *khanate of Turkistan, it was later brought within the *Qing empire during the 18th century.

Xiongnu, nomad horsemen who began harrying northern Chinese states *c*.300 BC. Their homelands were in southern Siberia and Mongolia. It was to fend off their incursions that some Chinese states built walls, later joined together to form the *Great Wall. The most serious attacks came under the early *Han, after the Xiongnu had formed a league under their Shan Yu (Heavenly Ruler). The Han at first attempted to buy off these great chieftains by conferring titles on them and giving them Chinese princesses as wives. Later, the policy of the Han emperor Wudi of isolating them by making alliances with other Asian peoples was in general successful. In the 1st century BC one group moved westward. The others, nominal vassals of China, were often employed by the later Han as frontier troops. To some extent they adopted Chinese ways. In the confusion following the Han's collapse, claiming descent from Chinese princesses, they set up ephemeral dynasties in northern China. Thereafter Chinese records make no reference to them. The Eastern Turks, who submitted to the Tang emperor *Taizong, are thought to be their descendants.

Y

Yamato, the name of the clan from which all the emperors of Japan are descended. Claiming the sun-goddess as ancestress, they had their chief shrine at Ise. Gradually they established control over rival clans and by the 5th century AD much of Japan was subject to them. They were influenced by Chinese culture, initially learning of China through southern Korean. *Buddhism and the study of Chinese language and literature were introduced in the 7th century and Prince *Shotuku produced administrative systems based on *Sui China. The Yamato chief assumed the title emperor and built capitals based on *Tang Chinese designs first at Nara then at Kyoto. By the 9th century the *Fujiwara family controlled the imperial court and during the period of the *shoguns the imperial family had little power.

Yangshao (in northern Henan province, China), a good example of a site of the Chinese *Neolithic period, though Banpo near Xi'an is a more representative settlement. There were square and round houses of timber post construction with thatched roofs, in villages up to 5 ha. (12 acres) in size. The red burnished pottery, often painted in black, was handmade but finished on a slow wheel. Polished stone axes and knives were in general use. The staple crop was millet, grown on terraces, and pigs and dogs were the commonest domestic animals. This culture was distributed over much of the middle Huang He (Yellow River) valley during the 4th and early 3rd millennia BC.

Yemen, a country in the mountainous south-west of the arid Arabian peninsula. From *c*.950 to 115 BC it was a flourishing region called Saba—the site of the kingdom

A neolithic pot of the **Yangshao** culture, decorated with geometric circles and a human figure. Yangshao pottery chiefly consisted of such jars, generally decorated on the top half only and intended for funerary deposits. (Art and History Museum, Shanghai)

of the biblical queen of Sheba. Because of its summer rains it was known to Rome as Arabia Felix ('Happy Arabia'), but it declined with its irrigation system around the 6th century AD. It was converted to *Islam in the 7th century and came under the rule of the Muslim *caliphate. In the 16th century it became part of the *Ottoman empire.

yeoman, a person qualified by possessing free land of an annual value of forty shillings to serve on juries, vote for knights of the shire, and exercise other rights. In the 13th and 14th centuries yeomen in England were freehold peasants, but by 1400, as many peasants became richer, the term was coming to be applied to all prosperous peasants, whether freeholders or not, as well as to *franklins (freehold farmers); in the 15th century some yeoman farmers, leasehold as well as freehold, entered the ranks of the gentry.

Yongle (Yung-lo) (1359-1424), *Ming Emperor of China (1403-24). He was a usurper who seized the throne when the second Ming emperor, his young nephew, disappeared in a mysterious palace fire. A man of great enterprise, he obliged Japan to pay tribute and extended the empire by campaigns in the steppes and in *Annam. He and his successor sent *Zheng He on prestigious voyages as far as the east coast of Africa. In 1421 he transferred the main capital from Nanjing to Beijing, his power base, and assured its food supplies by restoring the Grand Canal, linking it to the Huang and Yangtze rivers. He built the great halls and palaces of the Forbidden City in Beijing and arranged for the preparation of definitive editions of the Confucian classics.

York, Richard Plantagenet, 3rd Duke of (1411-60). He was the son of Richard, Earl of Cambridge, and Anne Mortimer, and until 1453 was heir to the throne of England. He led the opposition to *Henry VI, especially after the death of *Humphrey, Duke of Glou-

cester in 1447, and in 1455 he captured Henry at the first battle of *St Albans and became Protector of the kingdom. In October 1460 he claimed the English throne, but two months later he was defeated and killed by *Lancastrian forces in the battle of *Wakefield. The *Yorkist party survived him, and triumphed at the second battle of *St Albans.

Yorkist, a descendant or adherent of Richard, 3rd Duke of *York (their badge was a white rose). Despite Richard's death at the battle of *Wakefield (1460), his party was soon afterwards successful against the *Lancastrians and his son Edward became king as *Edward IV. The House of York continued on the throne with *Edward V, and then with Edward IV's younger brother, *Richard II, until *Henry VII began the *Tudor dynasty after his victory at *Bosworth Field in 1485. Henry prudently married the Yorkist heiress, Edward IV's eldest daughter, Elizabeth of York.

Yorktown, Virginia, USA, an American Revolutionary battlefield on the peninsula between the York and James rivers. In August 1781 *Cornwallis, his southern army exhausted after the vain pursuit of Nathanael *Greene, seized and fortified the area for winter quarters. Thanks to the promise of French naval support from the Caribbean, *Washington was persuaded by the French to march from the Hudson and concentrate all his forces on a siege. With relief by sea cut off by thirty-six French warships, Cornwallis was forced to surrender on 19 October. American independence was assured.

Yuan (1279-1368), the *Mongol dynasty that ruled China following *Kublai Khan's defeat of the *Song

The British troops marching in good order to surrender at **Yorktown**, effectively conceding defeat in the American War of Independence, a painting of the battle by Van Blarenberghe, 1781. (Musée de Versailles)

dynasty. He established a strong central government that in many ways reflected Chinese rather than Mongol practice. His reign and that of his grandson Temur (1294–1307) saw the growth of internal and foreign trade and the re-establishment of direct links with the West along the *Silk Route (for a while *Marco Polo worked for the Yuan dynasty). Many religions were tolerated and vernacular literature, particularly drama and novels, flourished. However, after the death of Temur, the empire suffered from neglect, disorder, and rebellion. From 1348 there was continuous conflict, with rebel Chinese armies fighting each other as well as the Mongols. In 1368 the last Yuan emperor fled to Mongolia when the rebel leader Zhu Yuanzhang captured Khanbaligh (now Beijing) and founded the *Ming dynasty.

Yucatán, the peninsula of eastern Mexico, and the state forming the northern part. In prehistory it was the northern area of *Maya civilization, including several long-occupied, powerful cities, some being occupied as early as *c.* 750 BC. From *c.* AD 800 many Maya migrated from the Southern Lowlands (Guatemala) into the Northern Lowlands of the peninsula and founded new states at Chichén Itzá, Izamal, *Mayapán, and Uxmal, linked by political alliances and trade. At first Chichén Itzá dominated, then Mayapán, and seventeen small principalities were formed in the late 15th century. Fernández de Córdoba explored the coast in 1516–17 sighting several cities, but resistance to conquest was strong and they were not fully subdued until the 1540s.

Yunnan, a province in south-west China. It was the area from which, over several millennia, Malay and Thai peoples moved into south-east Asia. In the 8th century Thai people set up a kingdom, Nanchao, there. They defied Chinese armies but were overcome by *Kublai Khan, accelerating Thai expansion to the south. At this time a Mongol general introduced Islam to Yunnan. Inhabited by a variety of racial groups, it was only integrated into China by the *Qing dynasty in the 17th century, becoming the site of a major copper industry.

The Castillo, the tallest pyramid at the Toltec-Maya city of Chichén Itzá in **Yucatán**. Flights of ninety-one steps lead up the four sides of the pyramid to the temple.

Z

Zagwe, an *Ethiopian dynasty founded in 1137, which derives its name from its founder. Zagwe and his successors are reckoned as usurpers because they were of Agau origin from Lasta, and not descended from the Solomonic kings of Ethiopia. The first five are said to have been Jewish, the later kings Christian. Theirs was a turbulent period in which other dynasties arose, and the issue of currency from *Axum came to an end.

Zanzibar, an island off the East African coast. Little of it is known in early times: *c.*1100 it was importing pottery from the Persian Gulf and became a base for Arab traders. In 1506, when the Portuguese demanded tribute, it was poor and thinly populated. The Portuguese established a trading post and a Catholic mission, but were displaced by the sultanate of *Oman who took it in 1698. It began to prosper *c.*1770 as an entrepôt for Arab and French slave traders.

Zarathustra (or Zoroaster) (7th–6th century BC), Persian religious reformer, the founder of the *Zoroastrian religion. Born in an aristocratic family, and probably a priest, he is said to have received a vision from Ahura Mazda ('the Wise Lord'), one of many gods then worshipped, urging belief in one god. After King Vishtaspa's conversion (*c.*588 BC) the religion spread.

It is difficult to distinguish between his actual teachings and later legends, for the details of his life cannot be reconstructed with accuracy. Speculations about his influence on Greek, Judaic, and Christian thought have been exaggerated. He certainly brought Ahura Mazda, the creator, to the centre of worship as the principle of 'good', but retained the ancient fire cult while abolishing orgiastic sacrificial practices. The *Gathas* (hymns) contain his teachings, but they too are unauthenticated.

Zaria (or Zazzau), one of the original *Hausa states of West Africa. Tradition claims that it had sixty rulers before the *Fulani conquest of the early 19th century, and that it controlled a large territory. It was converted to Islam in the 14th and 15th centuries, when the city walls were nearly 16 km. (10 miles) round, containing the Great Mosque, a palace, and private estates and houses. The emirate was chosen in turn from three families and had several vassals.

Zealots, named for their fanaticism or 'zeal', the party of revolt among the Jews of Roman Palestine. Also known as Canaans after the early inhabitants of Palestine, they have been identified with the 'Daggermen' ('Sicarii') of the *Jewish Revolt of AD 66–70 and the defenders of *Masada. Simon, one of the *disciples of *Jesus Christ was also known as 'the Zealot', meaning either that he was a member of the party, or equally likely, that he was of a 'zealous' disposition.

Zen, a *Buddhist sect of major importance in Japan. Strongly influenced by *Taoism, it originated in China and spread to Japan during the *Kamakura period. In sharp contrast to popular sects such as Pure Land Buddhism, it seeks salvation through enlightenment—revelation of the Buddha-nature which, it says, is innate to all people. Enlightenment comes suddenly, perhaps after some trivial physical or mental shock. Meditation under a master, intellectual exercises, and physical endurance are stressed. With its strict discipline it appealed to the *samurai. It flowered under the *Ashikaga, when its masters, emphasizing harmony with nature, had much influence on aesthetics. It was associated with such refinements as the tea ceremony, which emerged under the Ashikaga. Some masters were active in affairs of state and had extensive contacts, often through trading missions, with China.

Zeno of Citium (c.335–c.262 BC), Greek philosopher. He was the founder of the *Stoics. He attended the *Academy in *Athens, devoting himself first to the Cynic philosophy and then to the Socratic method of enquiry. The school of thought which he originated included a theory of knowledge, ethics, and physics, and a new system of logic. He taught that virtue, the one true good, is the only important thing, the virtue of a wise man cannot be destroyed, and that the vicissitudes of life are irrelevant to a man's happiness.

Zeno of Elea (born c.490 BC), Greek philosopher. He was a pupil of Parmenides and a keen advocate of his theory of monism (the theory that there is only a single ultimate principle or kind of being). He wrote a famous work which drew pairs of contradictory conclusions from the presuppositions of his rivals. This led *Aristotle to credit him with the invention of dialectic. His most renowned paradox is that of Achilles and the tortoise, by which it is shown that once Achilles has given the tortoise a start he can never overtake it, since by the time he arrives where it was it has already moved on.

Zheng He (Cheng Ho) (d. c.1433), Chinese admiral and explorer, a Muslim court *eunuch from Yunnan province. He commanded seven remarkable voyages (1405–33) undertaken by order of the Ming emperor *Yongle and his successor. His first voyage in 1405–7, consisting of sixty-two ships, called in at *Malacca and reached India. Subsequent voyages went to the Persian Gulf and his last voyage in 1431–3 reached the coast of East Africa. His voyages were made possible by the use of the compass and by Chinese advances in navigation and shipbuilding. The purpose of the voyages is unclear, however, as they were used neither to develop trade nor political influence with the countries visited, although Zheng He did return with tribute in the form of gifts to the emperor, including giraffes, ostriches, and zebras from *Mogadishu in East Africa.

Zhou (Chou) (c.11th century BC–256 BC), the second historically authenticated dynasty in China. It was founded by Wu the Martial, who justified his overthrow of the *Shang by claiming that its oppressive rule had led Heaven to transfer its mandate to a new 'Son of Heaven' as the emperor of China is styled. (The same claim was made by all subsequent founders of dynasties.) Western Zhou ruled over feudal vassal states in the Wei valley and far beyond from their capital at Hao near Xi'an until 771 BC. Eastern or Later Zhou, based in Luoyang, never exercised real power. While the emperor

A bronze ritual wine bucket of the **Zhou** dynasty, c.11th–early 10th century BC. The dragons at the neck of the vessel along with the intertwining of parts of the design and elaborate birds are typical of the period. The inscription reads: 'Kung has made this precious ritual vessel to be used in perpetuity by sons and grandsons.'

performed the ritual sacrifices, his vassals strove for mastery of the kingdom. In 403 BC the era of the Warring States began. In 256 BC the last Zhou sovereign was overwhelmed by the armies of his most powerful vassal, Prince Zheng of the state of *Qin.

Out of the turmoil of the Eastern Zhou came much that became identified with Chinese civilization. There was a flowering of philosophical thought whose prime exponents were *Confucius and *Mencius. Agriculture developed through the building of irrigation systems and trade grew, encouraged by demand for silk, the use of coins, and transport by canals. The use of bronze spread to southern China and iron was introduced, first for weapons and then for ploughs.

Zhoukoudian (Choukoutien), a cave complex southwest of Beijing, made famous by fossils of *Peking Man. The fossils are now believed to represent a Chinese variant of *Homo erectus (Homo erectus pekinensis), but were once classified in a separate genus, Sinanthropus. More than forty male and female Homo erectus individuals are now known from the caves. Added to this hominid collection are many tens of thousands of simple flaked stone tools as well as fossilized bones of more than 100 animal species and so-called ash layers that were once interpreted as cooking hearths. The caves were occupied by Peking Man from about 500,000 to 250,000 years ago.

Zhuangzi (Chuang-tzu) (c.369–c.286 BC), Chinese philosopher. A native of the state of Meng on the border of present-day Shandong and Henan, he is believed to have written at least part of the Zhuangzi (Chuang-tzu), the 3rd-century BC classic of *Taoism. The *Zhuangzi* asserts that all ideas and conventions used to judge truth are relative and that individual freedom comes from the identification with the *tao*. His ideas greatly influenced *Zen Buddhism.

Zimbabwe, a country in south-east Africa, named after the ancient palace city of Great Zimbabwe, a 24 ha. (64-acre) site, that dates from the 11th to the 15th centuries. Gold and copper were exported from more than a thousand mines by the 10th century AD, the trade passing through Sofala, in Mozambique, to Arab hands. In the early 15th century the region's riches enabled the rise of the Shona (Karanga) empire, with the stone-built city as its capital. The sovereign had an elaborate court and constitution, and trade links with both sides of Africa; but after Portuguese incursions in the 16th century, Zimbabwe's fortunes steadily declined. In 1629 an attempt to expel the Portuguese resulted in the installation of a puppet ruler. After 1693 the territory was absorbed by the *Rozvi empire.

'Zinjanthropus', the name originally given to an East African *australopithecine fossil from *Olduvai Gorge, Tanzania from about 1.8 million years ago, now usually called *Australopithecus boisei*. An almost complete skull of this species has been popularly known as 'Nutcracker Man', because of the very rugged build of its teeth and jaw-bone.

Zoroastrian, follower of the religious doctrines originally disseminated in Persia by *Zarathustra in the 6th century BC. Zoroastrianism was the state religion in Persia under the *Sassanian dynasty (3rd–7th century AD), but Islamic invasion resulted in the persecution and emigration of believers. Isolated groups have survived there, but the focus of emigration was the west coast of India (Gujarat) where Hindus tolerated the new cult whose followers readily adapted themselves to the environment. There

The Swiss Protestant reformer Ulrich **Zwingli**, a woodcut from the painting by Hans Asper, 1550.

they became known as Parsees (from the Persian *parsi*, 'Persian'). Characteristic religious practices include preservation of the sacred fire, and disposal of the dead by exposure on 'towers of silence'.

Zwingli, Ulrich (1484–1531), Swiss *Protestant reformer. He studied at Vienna and Basle before becoming pastor at Glarus in Switzerland (1506). He taught himself Greek and was considerably influenced by the humanist precepts of *Erasmus. He also served as chaplain to Swiss mercenaries involved in the Italian wars. As pastor at Einsiedeln (1516–18), he began to reach Lutheran conclusions, possibly independently of *Luther himself. After becoming Common Preacher at the Great Minster in Zürich (1518), he played a significant part in converting the city's inhabitants to the reformed religion. At a public disputation with a papal representatives in 1523 Zwingli presented his doctrines in sixty-seven theses, and these were adopted by the general council of Zürich as official doctrine. The Protestant Reformation thereafter proceeded to make great headway in Switzerland.

The last period of his life was marked by dispute and conflict. He was unable to reach agreement with Luther at Marburg (1529) about the nature of the Eucharist. The division on this matter was so deep that any union of the Protestant branches was impossible. He was killed at the battle of Kappel in which Zürich was defending itself against the Catholic cantons (provinces) of Switzerland. John *Calvin then became the principal champion of the Reformation in Switzerland.

The solid stone-built conical tower at Great **Zimbabwe** stands in the ruins of the great elliptical enclosure, which was once the religious and political heart of the settlement. The enclosure walls reach 10 m. (33 ft.) high and, with the tower, exhibit the most impressive of the several architectural styles employed by the builders of Great Zimbabwe.

Acknowledgements

Photographs

Abbreviations: t = top; b = bottom; c = centre; l = left; r = right.

The illustrations on the following pages are reproduced by Gracious Permission of Her Majesty the Queen, 16, 93, 126, 140, 241.

The illustration on page 5t is reproduced by permission of His Grace the Archbishop of Canterbury and the Trustees of Lambeth Palace Library.

The illustration on page 237 is reproduced by kind permission of the Master and Fellows of Sidney Sussex College, Cambridge.

The following have also kindly given permission to reproduce paintings: Marquess of Anglesey, 43; Lady Cholmondeley, 302; Viscount de L'Isle VC, KG, 114; J. More-Molyneux, 184; Earl Rosebery, 69; Duke of Roxburghe, 99t; Miss M. L. A. Strickland, 85, 239.

Aerofilms, 63, 350t.

Alinari, 129.

The Ancient Art and Architecture Collection, 99b, 101, 107, 174.

Anderson, 142.

Archiv für Kunst und Geschichte, Berlin, 264.

Armémuseum, Sweden, 324.

Artothek, 45t.

Ashmolean Museum, 10, 146t, 253.

Bayerische Staatsbibliothek, Munich, 265.

BBC Hulton Picture Library, 5b, 225.

Bibliothèque Nationale, 24t, 135, 139, 189t, 245, 272, 281, 313, 337.

Bildarchiv Preussischer Kulturbesitz, 191r, 249.

Bodleian Library, 221.

Bord Fáilte, 228.

Boston Museum of Fine Arts, 84.

Bridgeman Art Library, 172, 335.

British Library, 25, 38, 39, 44, 59, 75, 80, 90, 96, 109, 118, 128, 131, 138, 150, 189b, 193, 220, 234, 242, 269t and b, 273b, 289, 290, 303b, 305l, 319, 322, 340, 348t, 359, 361, 370b, 375r.

British Museum, 37, 40, 180t, 232, 250, 260, 278, 295, 297, 311t, 332, 353t, 367.

Cambridge University Library, 111b.

Collection Ferrand, 12.

Colorphoto Hinz, 342.

Comune di Firenze, 215.

Eastern National Park and Monument Association; 191l.

E. T. Archive, 36, 201, 282, 357b.

B. Fleming, 315b.

Werner Forman Archive, 67, 71, 79, 97, 197, 203, 233b, 240, 255, 261, 268, 276, 298t, 315t, 387; Werner Forman/Robert Aberman, 391l.

Forschungsbibliothek Gotha, 218.

Fotomas Index, 56, 190, 208, 270, 357t.

Alison Frantz, 2.

Christina Gascoigne, 325.

Germanisches Nationalmuseum, Nuremberg, 133, 376.

Graphische Sammlung Albertina, 70.

Sonia Halliday Photographs, 35, 185, 263, 345; Sonia Halliday and Laura Lushington, 18, 224; Sonia Halliday Photographs/Jane Taylor, 1.

Robert Harding Picture Library, 94, 390.

Herzog August Bibliothek, Wolfenbüttel, 379t.

Historical Society of Pennsylvania, 275.

Michael Holford, 9, 11, 24b, 26, 49, 50, 61, 62, 65, 81, 155t and b, 158t, 166, 183, 194, 288, 330, 346, 382, 389.

Hutchison Library/Michael MacIntyre, 13.

International Society for Educational Information, Tokyo, 165.

Larousse, 284.

Library of Congress, 344.

Mansell Collection, 21, 28, 34, 47, 58, 77, 95, 119, 124, 143, 148, 158b, 162l, 195t, 210, 216, 258, 277, 291t, 304, 320, 334, 341, 350b, 351, 384, 391r

Mas, 7, 105, 212, 279, 352, 360.

Leonard von Matt, 57.

Metropolitan Museum of Art, 45b, 223.

Ministry of Public Building and Works, 85, 154, 239.

J. Montet, 78.

Museo Archeologico Palermo, 373.

Museum and Art Gallery, Leipzig, 213.

Museum of London, 123.

National Galleries of Scotland, 69, 287, 291b.

National Gallery of Ireland, 48, 366.

National Maritime Museum, 14, 195b, 206, 298b, 331, 336.

National Monuments Record, 111t.

Nationalmuseum, Copenhagen, 182l.

National Museum of Ireland, 66.

National Palace Museum, Taipei, 200, 333, 348l.

National Portrait Gallery, 51, 161, 162r, 163, 199b, 204, 209, 246, 262, 283, 379b, 385t.

National Portrait Gallery, Smithsonian Institution, 286.

New York Public Library, 88, 273t.

Novosti Press Agency, 182l.

Order of St John, 198, 303t.

Photographie Giraudon, 68, 103, 205, 271, 305r, 311b, 314, 375l.

Photoresources, 199t, 309.

Photo Meyer, 222.

Pontifical Commission for Sacred Archaeology, 64.

Public Archives of Canada, 180b, 252.

Rainbird Publishing Group Ltd, 43, 83, 99t, 146b, 151, 159, 184, 187, 233t, 243, 244, 292, 302, 355, 370t, 386, 390.

Reale Fotografia Giacomelli, 60.

Réunion des Musées Nationaux, 30, 117, 211, 230, 280, 372, 388.

Roger-Viollet, 113.

Scala, 86, 121, 227, 285, 317, 353b, 371.

Science Museum, 178.

Society for Cultural Relations with the USSR, 89, 259.

Society of Antiquaries of London, 127.

South American Pictures/Tony Morrison, 175.

Staatliche Museen zu Berlin, 33.

Stadtarchiv, Worms, 385b.

Walter Steinkopf, 132.

The Times, 364.

Tokugawa Remeikai Foundation, 358.

Torino Museum, 299.

Uppsala Universitetsbibliothek, 76.

USIS, 53, 378.

Victoria and Albert Museum, 6, 32.

Weidenfeld and Nicolson, 173.

Yale Center for British Art, Paul Mellon Collection, 145.

Yale University Art Gallery, 100.

The publishers have made every attempt to contact the owners of the photographs appearing in this book. In the few instances where they have been unsuccessful they invite the copyright holders to contact them direct.

Picture researcher: Sheila Corr

Maps and Illustrations

Creative Cartography Ltd/Terry Allen and Nick Skelton: 4, 15, 72, 75, 107, 120, 151, 257, 309.

Eugene Fleury: 42, 46, 63, 116, 125, 153, 156, 172, 235, 237, 255, 259, 310, 338, 343, 351, 362, 373, 381.

Vanna Haggerty: 20, 22, 92, 219, 248.

Creative Cartography Ltd and Eugene Fleury: 9, 19, 52, 55, 98, 122, 137, 147, 168, 175, 176, 177, 238, 266, 274, 301, 307, 323, 329, 330, 349, 354, 368.

Creative Cartography Ltd, Eugene Fleury, and Vanna Haggerty: 31, 226.

From 1001 to 1550

ASIA	EUROPE	REST OF THE WORLD
	1017 Canute king of Denmark and England	
1030 Height of Ghaznavid power in north India		
	1054 East-West Schism between Greek and Latin churches	**1054** End of kingdom of Ghana in West Africa
	1066 Norman conquest of England	**1061** Almoravids conquer North Africa
1071 Battle of Manzikert	**1073** Pope Gregory VII	
1096-9 First Crusade	**1098** Cistercian Order founded	
	1100 First European universities	
1126 Chinese Jin dynasty		
1127 Chinese Southern Song dynasty		
1147-9 Second Crusade	**1154** Henry II rules Angevin empire	**1150** Destruction of Toltec capital Tula
	1155 Frederick Barbarossa crowned emperor	
1189-92 Third Crusade	**1170** Murder of Thomas à Becket	**1171** Saladin conquers Egypt
1192 Japanese Kamakura shogunate	**1198** Pope Innocent III	
1200 Khmer empire at height of power		**1200** Inca dynasty founded in Peru
1204 Crusaders sack Constantinople		
1206 Genghis Khan begins conquest of Asia, sultanate of Delhi founded	**1209** Albigensian Crusade	
	1215 Magna Carta	
	1226 Death of St Francis	**1240** Foundation of Mali empire in West Africa
1234 Mongols destroy Jin empire	**1241** Mongols invade Europe	**1250** Mameluke dynasty in Egypt
1258 Mongols sack Baghdad		
1279 Chinese Yuan dynasty	**1274** Death of St Thomas Aquinas	
	1282 Sicilian Vespers	
1291 Fall of Acre		
	1314 Battle of Bannockburn	
1336 Foundation of Vijayanagar in south India		
1339 Japanese Ashikaga shogunate	**1338** Start of Hundred Years War	
c.1339 Death of Nanak, founder of Sikh religion		
	1348 Black Death reaches Europe	**1345** Foundation of Tenochitlán in Mexico
	1360 Treaty of Brétigny	
1368 Chinese Ming dynasty		
	1378 Great Schism between Rome and Avignon	
	1381 Peasants' Revolt	
1396 Battle of Nicopolis	**1397** Union of Kalmar	
1402 Timur defeats Ottomans, foundation of Malacca		
1405-33 Zheng He's voyages of exploration	**1410** Battle of Tannenberg	
	1415 Battle of Agincourt, death of Huss	
	1431 Death of Joan of Arc	**1438** Start of Inca ascendancy
1453 Ottomans take Constantinople	**1453** End of Hundred Years War	**1441** Fall of Mayapan
	1455-85 Wars of the Roses	
	1469 Marriage of Ferdinand and Isabella	
1480 Ivan III overthrows Mongols	**1477** Death of Charles the Bold	**1476** Incas conquer Chimú
		1492 Columbus reaches the West Indies
	1494 Treaty of Tordesillas	**1493** Askia emperor of Songhay in West Africa
1498 Vasco da Gama reaches India		**1497** Cabot reaches Newfoundland
1501 Safavid dynasty in Persia		
1511 Portuguese take Malacca	**1517** Start of Protestant Reformation	**1517** Ottomans conquer Egypt
		1519-21 Cortés conquers Aztec empire
	1521 Diet of Worms	**1520** Magellan crosses Pacific
1526 Babur founds Mogul dynasty	**1526** Battle of Mohács	
	1534 Henry VIII breaks with Rome	**1532-3** Pizarro conquers Inca empire
	1545-63 Council of Trent	
	1547 Ivan the Terrible crowned Tsar	